Hydroscience and Engineering

Series Editors

This Springer book series Hydroscience and Engineering addresses multidisciplinary advancements in water-related science and engineering practice. Main scope of the book series includes fundamental research and cutting-edge engineering technology in water resources and river/reservoir management in relation with hydraulic structures, erosion control, sediment transport, river basin management and planning, flood control, geological/hydrological risk assessment, offshore and costal engineering. The book series focuses on developing innovative, sustainable and environment-friendly materials and structures to improve infrastructural functionality and sustainability, and to enhance adaptability to changing environmental conditions. Recent advancements and successful insights obtained from engineering practice in China and worldwide will also be thoroughly introduced.

The overarching goal of the book series is to provide combined knowledge on foremost research and solid engineering expertise on hydroscience and engineering from an interdisciplinary standpoint. The individual book volumes in the series are thematic, each specializing a focused topic, such as a new dam type and innovative dam construction materials. As a collection, the book series provides valuable resources targeting on a wide spectrum of audience, including researchers in academia, practicing engineers in engineering community and students who work on expanding their knowledge in the related areas.

Qiwei Han · Mingmin He

Non-equilibrium Transport of Non-uniform Suspended Load

Springer

Qiwei Han
Department of Sediment Research
China Institute of Water Resources
and Hydropower Research
Beijing, China

Mingmin He
Department of Sediment Research
China Institute of Water Resources
and Hydropower Research
Beijing, China

ISSN 2730-9002 ISSN 2730-9010 (electronic)
Hydroscience and Engineering
ISBN 978-981-97-1122-2 ISBN 978-981-97-1123-9 (eBook)
https://doi.org/10.1007/978-981-97-1123-9

This Springer imprint is published by the registered company Springer Nature Singapore Pte Ltd.
The registered company address is: 152 Beach Road, #21-01/04 Gateway East, Singapore 189721, Singapore

Preface

Non-equilibrium transport of non-uniform suspended load is an important topic of the sediment transport study and a major theoretical basis for sedimentation engineering and mathematical modeling of sediment transport. This book presents a summary of the results and findings of the authors' scientific research in this field. The main contents of this book are translated from the authors' book named *Theory of Non-equilibrium Transport of Non-uniform Suspended Load* that was published in Chinese. Example of advances of the authors' research includes the following: developing a theory for non-equilibrium transport of non-uniform suspended load; establishing general boundary conditions for the diffusion equation of suspended load based on sediment exchange intensity; introducing the gradation of the carrying-sediment capacity, the effective bed material gradation (water volume percentage), and the saturation recovery coefficients for non-equilibrium transport of non-uniform suspended load; and deriving a complete equation system for one-dimensional non-equilibrium transport of non-uniform suspended load.

The study on the non-equilibrium sediment transport was conducted based on the authors' statistical theory of sediment movement. Due to the complexities of sediment transport, individual and semi-empirical studies were typically conducted for specific problems without a systematic approach. By combining mechanics and probability theory, the authors derived the theoretical expressions for the parameters to describe sediment movement under various motion states, analyzed the random process of sediment movement based on the Markov stochastic theory, and developed the theory and formulas for sediment exchange intensity. The concept of sediment exchange intensity is important because it can be used to formulate the sediment transport rate under various motion states. The exchange of coarse and fine sediment is a common phenomenon during sediment transport, and thus the theory of sediment exchange intensity reflects the fundamental law of sediment movement. Because most authors' literatures of statistical studies of sediment movement are written in Chinese, an appendix is included in this book for the main elements of the statistical theory that are translated in English.

Based on the theories of sediment exchange intensity and non-equilibrium sediment transport, the authors presented answers to a number of long-standing and debatable questions. These questions include: the similarities and differences between the wash load and the bed material load, the reasonable criterion for determining the threshold particle size for scour and deposition during sediment movement, the mechanism of the long-distance channel erosion downstream of a dam, and another mode of bed material coarsening that is caused by sediment exchange.

The authors' study is based on extensive field data that was collected in many rivers in China. They conducted long-term field investigations of natural streams, reservoirs with variable backwater zones, and channel scour downstream of dams and set up an experiment station at the variable backwater zone of the Danjiangkou Reservoir to collect a large amount of hydrologic and sediment data. Based on their field investigations and analysis of the field data, the authors conducted an in-depth study on the phenomena and mechanisms of the exchange between coarse and fine sediment and non-equilibrium sediment transport.

These research results and findings have been applied to sedimentation problems in hydraulic projects. Based on the equations for non-equilibrium sediment transport, the authors developed a mathematical model for reservoir sedimentation and downstream channel evolution. This model has been used since the late 1970s to predict sedimentation in the Three Gorges Reservoir. The formulas and parameters in this model remain mostly unchanged, and the model results show good agreement with the field data collected after the Three Gorges Project was completed. This model has also been applied to the Danjiangkou Reservoir, four large reservoirs in the lower reach of the Jinsha River, the middle and lower reaches of the Yellow River, and many other hydraulic projects in China.

The authors research has been funded by the Yellow River Key Natural Science Foundation, programs of the Ministry of Water Resources of China, and the foundation from State Key Laboratory of Simulation and Regulation of Water Cycle in River Basin (SKL-WAG), China Institute of Water Resources and Hydra-Electric Research Institute (IWHR). The authors also received support and help from numerous scientific experts including RichangTang, Zhian Xiang, Jianheng Xie, Ning Qian, Zezhen Zhang, Bingnan Lin, Shangfu Kuang, and Chunhomg Hu et al. This book was translated by Dr. Wenkai Qin. We would like to record here our gratitude to those who made the writing of this book possible.

Beijing, China

Qiwei Han
Mingmin He

About This Book

This book introduces the author's achievements in the study of non-equilibrium transport of non-uniform suspended load. This research offers a breakthrough in the field of sediment transport with considerable potential for application in sediment engineering.

In nature, sediment is always non-uniform and shows non-equilibrium transport. In addition, a significant fraction of the sediment is suspended load. Therefore, the non-equilibrium transport of the non-uniform suspended load is an important topic for investigation in the field of sediment transport. A majority of previous studies are limited to equilibrium sediment transport and non-equilibrium transport of uniform sediment. In addition, results based on these research activities have been applied to the non-equilibrium transport of non-uniform sediment. So that the research results do not fit in well with reality.

The authors have been conducting a series of theoretical studies based on statistical theory of sediment movement, particularly the theory of transition intensities, on non-equilibrium transport of non-uniform suspended load, which includes establishing the boundary conditions for the suspended load diffusion equation using transition intensities; deriving the one-dimensional (1-D) transport equation based on the two-dimensional (2-D) diffusion equation and its boundary conditions; introducing the carrying-sediment capacity, its size distribution, water percentage (gradation of effective bed material), and the theoretical expressions of saturation recovery coefficients; and deriving a complete sediment transport equation system for non-equilibrium transport of non-uniform suspended load. Thus, a nearly complete theory of non-equilibrium transport of non-uniform sediment has been established.

The proposed theory could resolve certain controversial sediment transport problems that have long remained unsolved, such as the general mathematical model of reservoir sedimentation and river channel revolution, long-distance scour in the reach downstream of a reservoir, similarities and differences between wash load and bed material load, and the multi-value of carrying-sediment capacity, and the mechanism of bed material coarsening.

In addition, a significant amount of field data was collected to verify the theoretical results.

Contents

Chapter 1
Introduction

Non-equilibrium transport of non-uniform suspended load is an important topic of the study for sediment transport. Study of this topic will advance the science for sediment transport and river dynamics, provide and improve the theoretical basis for channel evolution and reservoir sedimentation, and provide crucial supports to the channel improvement and solving of the sedimentation engineering problems.

Based on the chronological order, the studies for suspended load transport can be categorized into four topics. They include sediment concentration and carrying-sediment capacity, vertical distribution of sediment concentration, one-dimensional (in the longitudinal direction) and vertically two-dimensional non-equilibrium sediment transport, and non-equilibrium transport of non-uniform sediment. Prior to 1960s, the studies for the first three topics were conducted mainly for uniform sediment. By doing that, the problems were simplified, the results were relatively easy to obtain, and some results were also applied to some extents. However, sediment in rivers is not uniform, and the variation range of the particle size is typically more than three orders of magnitude. Hence, the results for uniform sediment do not agree with reality, and the error can be significant. Of course, the bottom boundary condition that cannot be correctly described is also one reason for the solution of the non-equilibrium sediment transport, particularly for the solution of the vertically two-dimensional non-equilibrium transport, to have a great discrepancy from the reality. The problems of sediment transport are very complicated for the rivers and reservoirs, and the study of the non-equilibrium transport of non-uniform sediment started in China as early as in 1960s. In addition to the vertical distribution of sediment concentration and carrying-sediment capacity, the study for non-uniform sediment transport involves more subjects. These include the size distribution of suspended load, the bed material gradation, the carrying-sediment capacity and the corresponding size distribution under the non-equilibrium condition, the bottom boundary condition, and the exchange of sediment, with various states, on the bed.

Q. Han and M. He, *Non-equilibrium Transport of Non-uniform Suspended Load*,
Hydroscience and Engineering, https://doi.org/10.1007/978-981-97-1123-9_1

The study for non-equilibrium transport of non-uniform sediment covers many subjects, and the advance in each subject is different. Five main subjects are reviewed in this Chapter. These subjects include: vertically two-dimensional non-equilibrium sediment transport with the bottom boundary condition, one-dimensional non-equilibrium sediment transport, the size distribution for carrying-sediment capacity under the equilibrium condition, the size distribution for carrying-sediment capacity and the effective bed material gradation for non-equilibrium sediment transport, and the sediment saturation recovery coefficient. It is worthy to point out that these five subjects are selected not only because of the availability of the study results for these subjects, but also because of their importance in the study for non-equilibrium transport of non-uniform sediment.

1.1 Vertically Two-Dimensional Non-equilibrium Sediment Transport and Bottom Boundary Condition

Two approaches are typically used to investigate the non-equilibrium transport of suspended load from the aspect of the variation of sediment concentration. One approach is to solve the two-dimensional (2-D) convection-diffusion equation for suspended load with certain simplified conditions. The other is to solve the one-dimensional equation for non-equilibrium sediment transport that is directly derived based on the sediment mass balance under certain assumptions. The 2-D equations include the horizontally 2-D equations and the vertical 2-D equations. The studies for the vertically 2-D diffusion equation started in the early and middle of twentieth century [5, 36, 55], with more analyses emerging in 1960s to 1980s. Most of these studies were limited to the uniform sediment. The vertical distributions of the convection and diffusion coefficients were significantly simplified in order to obtain analytical solutions. Also, the bottom boundary condition was not well defined. As a result, while the results of these studies have some physical meaning, they are not consistent with reality that limits its application to the engineering problems. In the recent twenty years, numerical solutions were developed for the horizontally and vertically 2-D diffusion equations in some 2-D mathematical models for sediment transport. However, due to the special characteristics of the relevant parameters, initial conditions and boundary conditions, and to the complexity of the channel cross-section, the numerical solutions were only developed for certain specific conditions, without being used to analyze the general law for sediment transport. In the horizontally 2-D mathematic models, the term in the governing equation that is related to the bed change still depends on the bottom boundary condition as used in the vertically 2-D diffusion equation. Hence, the following review of the 2-D non-equilibrium sediment transport focuses on the vertically 2-D case. It is noted that three-dimensional (3-D) mathematical models have been recently developed. However, the numerical solutions to the 3-D diffusion equation have not been well established, which limits these

3-D models to certain special conditions without being used to analyze sediment transport under general conditions.

Prior to 1970s, most studies on the vertically 2-D non-equilibrium sediment transport focused on the analytical solutions to the vertical 2-D diffusion equation with assumptions, and later switched to the numerical solutions. The representative studies on the vertically 2-D non-equilibrium sediment transport, with the bottom boundary condition being defined, include Kalinske [36], Dobbins [5], Takaharu [55], Rosinske and Kuzmin [52], Mei [47], Jobson and Sarye [35], Hjelmfelt and Lenau [30], Sarikaya [53], Partheniades [50, 51], Sumer [54], Kerssens et al. [39], van Rjin [57, 58], Zhang [62], Hou [31], Han [11], Han and He [24, 26], Bechteler and Schrimpf [1], Jiang and Fan [34], Ismail and Rodi [33], Celik and Rodi [3], Cao [2], and Cui [4].

The studies on the vertically 2-D non-equilibrium sediment transport have the following three main problems:

(1) The accurate bottom boundary condition is difficult to be theoretically determined. Various bottom boundary conditions have been proposed. Most of them lack of generality and some were developed without strict theoretical derivations. Since the bottom boundary condition largely determine the bed erosion or deposition, the unreliability of the bottom boundary condition devalues the efforts to solve for the solutions to the equations and the applications of the results. This is the reason why the results of solving the vertically diffusion equation have not been directly used in the studies on non-equilibrium sediment transport.
(2) Strong assumptions were generally used in the earlier studies in order to obtain analytical solutions. These assumptions include the uniform flow assumption, assuming no variation of the longitudinal flow velocity over the water depth [5, 30, 31, 36, 47, 62], and the constant diffusion coefficient assumption [5, 31, 34, 36, 47, 62]. Then, as the numerical solutions became popular, the vertical distribution of the flow velocity could be considered, and the non-constant diffusion coefficient could be used, which are more consistent with the reality [3, 4, 33, 35]. However, numerical solutions usually do not clearly show some characteristics of sediment transport, and some parameters vary from case to case. As a result, the application of these numerical solutions to the investigation of non-equilibrium sediment transport is limited.
(3) Most studies on the 2-D non-equilibrium sediment transport in the past were for uniform sediment, which is significantly different from the actual rivers and reservoirs where non-uniform sediment exists. As a result, the results are hard to be used. Then, the transport of non-uniform sediment was analyzed by summing up the fractional sediment transport rates for various size classes. Einstein determined the total bed load transport capacity by summing up the fractional transport capacities with the bed material gradation. However, it is only true for the strong equilibrium condition. It is much more complicated for non-equilibrium sediment transport. A more in-depth method for suspended load transport was developed by Han [11] in his study for the one-dimensional

(1-D) non-equilibrium sediment transport, which was also used by Han and He [24] when integrating the vertically 2-D diffusion equation. Introducing the concept of the water percentages, Han and He [26] developed the diffusion equation for non-uniform sediment, and provided the precise definition for the size distribution of carrying-sediment capacities. The summation of the fractional sediment transport rate theoretically involves the expression of the diffusion equation for non-uniform sediment, which was approved by Han and He [26] with the water percentages. The uniform sediment and non-uniform sediment are essentially different. The summation of the fractional sediment concentrations needs the size distribution of suspended sediment and the summation of the fractional carrying-sediment capacities will be discussed later in this book.

The bottom boundary condition is very important for non-equilibrium transport of suspended load. Different types of the bottom boundary condition are reviewed and analyzed as follows. The mechanism of non-uniform sediment transport and the size distribution for carrying-sediment capacity, which need to be investigated for both the 1-D and 2-D non-equilibrium sediment transport, will be reviewed later in this chapter.

The bottom boundary conditions proposed in various studies can be he grouped into three approaches with six types. The three approaches include the gradient approach, flux approach, and composite approach. The first type of bottom boundary condition is to set the bottom sediment concentration S_b to be a constant, which equals the carrying-sediment capacity at the bottom S_b^* [1, 30, 50, 51, 52, 55, 62], or,

$$S|_{y=a} = S_b = S_b^* \tag{1.1}$$

where a is the reference distance above the bed where suspended load starts, S is the sediment concentration, S_b is the bottom sediment concentration, and S_b^* is the bottom carrying-sediment capacity. The physical meaning of this equation is that the sediment concentration at the bottom always instantaneously recover to the equilibrium condition, or instantaneously reaches the carrying-sediment capacity at the bottom.

The second type of bottom boundary condition is to assume no sediment pick-up at the bottom [1, 34, 50, 53, 54, 62], or

$$\left.\frac{\partial S}{\partial y}\right|_{y=a} = 0 \tag{1.2}$$

The physical meaning of this equation is that the flow is so weak that no sediment is picked up by the turbulent flow, or the bed material is not erodible.

The third type of bottom boundary condition is to assume the sediment pick-up term at the bottom caused by the turbulence equals the diffusion term under the equilibrium condition ωS_b^* [1, 31, 34, 51, 62], or

$$\varepsilon_y \frac{\partial S}{\partial y}\bigg|_{y=a} + \omega S_b^* = 0 \tag{1.3}$$

This can be rewritten as

$$\varepsilon_y \frac{\partial S}{\partial y}\bigg|_{y=a} = -\omega S_b^* \tag{1.4}$$

or

$$\varepsilon_y \frac{\partial S}{\partial y}\bigg|_{y=a} + \omega S_b = \omega\left(S_b - S_b^*\right) \tag{1.5}$$

The bottom boundary condition proposed by Celik and Rodi [3] and by Cao [2] also belongs to this type. Considering the difference between the non-erodible bed and the erodible bed, Celik and Rodi [3] proposed the following relation:

$$-\varepsilon_y \frac{\partial S}{\partial y}\bigg|_{y=a} = \begin{cases} \omega S_b^* & \text{Erodible bed} \\ \min\left(\omega S_b, \omega S_b^*\right) & \text{Non-erodible bed} \end{cases} \tag{1.6}$$

The lower part of the right side of Eq. (1.6) indicates: (1) For $S_b > S_b^*$, the diffusion term in the upward direction equals ωS_b^*, and thus the net sedimentation quantity $\omega(S_b - S_b^*) > 0$, resulting channel deposition; (2) For $S_b < S_b^*$, the diffusion term equals ωS_b, and thus the net sedimentation quantity $\omega\left(S_b - S_b^*\right) = 0$, resulting in an equilibrium sediment transport. This is true for the non-erodible bed. It is noted that Eq. (1.6) is consistent with the equation developed by Han [14] for the bottom sediment transport capacity in streams with pebble bed or rock bed, which is

$$S_b^* = \min\left(S_b, S_b^*\right) \tag{1.7}$$

The assumption associated with Eq. (1.4) is that the sediment diffusion depends only on the flow condition and sediment properties, and is independent of the equilibrium conditions. In other words, no matter erosion or deposition occurs (or no matter the sediment concentration is high or low), Eq. (1.4) is always satisfied for given flow discharge and sediment fall velocity. This has been questioned by some researchers. From the theoretical view point, this should be further investigated. However, based on the results, if $-\varepsilon_y \frac{\partial S}{\partial y}\Big|_{y=a}$ is independent of the sediment concentration, then Eq. (1.4) is practically adopted.

While questioning Eq. (1.4), some researchers directly developed the sediment pick-up function by combining theoretical derivation with empirical relationship. For example, van Rijn [57, 58] proposed the following pick-up function based on experimental data:

$$E_a = -\varepsilon_y \frac{\partial S}{\partial y}\bigg|_{y=a} = 0.030 \frac{D_{50}}{4} \frac{T^{1.5}}{D_*^{0.5}} = 0.015 \frac{D_{50} T^{1.5}}{a D_*^{0.5}} \tag{1.8}$$

in which

$$T = \frac{\left(u'_*\right)^2 - \left(u_{*.c}\right)^2}{\left(u'_*\right)^2} \tag{1.9}$$

$$D_* = D_{50} \frac{\left(\frac{\rho_s - \rho}{\rho} g\right)^{\frac{1}{3}}}{\nu^2} \tag{1.10}$$

$$\left(u'_*\right)^2 = \mu u_*^2 \tag{1.11}$$

where E_a is the upward sediment flux, the reference height a can be set to an half of the sand ripple height Δ, or the roughness coefficient K_s, D_{50} is the median particle size of bed material, ρ_s and ρ are the density of sediment and water, respectively, g is the gravity acceleration, ν is the kinematic viscosity coefficient, u_* is the shear velocity, $u'_{*.c}$ is the shear velocity corresponding to the threshold bed shear stress according to Shields, u'_* is the shear velocity after considering the impact of the sand ripples, and

$$\mu = \left(\frac{C}{C'}\right)^2 = \frac{\lg\left(12 \frac{h}{K_s}\right)}{\lg\left(12 \frac{h}{D_{50}}\right)} \tag{1.12}$$

Based on van Rijn's concept and considering the characteristics of turbulent bursting, Cao [2] proposed the following formulation for the upward diffusion term:

$$-\varepsilon_y \frac{\partial S_v}{\partial y}\bigg|_{y=a} = \begin{cases} E_a & \text{Erodible bed} \\ \min(E_c, \omega S_b) & \text{Non-erodible bed} \end{cases} \tag{1.13}$$

Based on qualitative analysis, Cao further obtained

$$E_a = P\left[\left(\frac{\rho_s}{\rho} - 1\right) g\right]^{0.5} d^2 \left(\frac{F}{F_c} - 1\right) F \tag{1.14}$$

$$P = \frac{\lambda C_0 \left(\frac{\rho_s}{\rho} - 1\right) g^{0.5}}{\nu T_B^*} \tag{1.15}$$

$$F = \frac{u_*^2}{\frac{\rho_s - \rho}{\rho} g d} \tag{1.16}$$

$$F_c = \frac{u_{*.c}^2}{\frac{\rho_s - \rho}{\rho} gd} \tag{1.17}$$

where T_B^* is the dimensionless period of turbulent bursting, with an empirical value of 100, C_0 is the volumetric sediment concentration on the bed, $\lambda = 0.02$, S and E_a are measured in volume, u_* is the shear velocity, and P = (370–480) $m^{-1.5}$ under general conditions. The advantage of this type of boundary condition is that the upward sediment flux E_a is no longer controlled by the relationship

$$\varepsilon_y \frac{\partial S_v}{\partial y}\bigg|_{y=a} = -\omega S_b^* \tag{1.18}$$

which indicates the upward flux always equals the downward flux regardless of the equilibrium condition. Different from previous studies, this type of boundary condition involves some new ideas. However, the formulation of the pick-up function is independent of the equilibrium condition, and thus a concept similar to Eq. (1.18) is still implied in this type of boundary condition. For example, the following relationship was developed by Cao [2] for the equilibrium sediment transport

$$\begin{aligned} S_b^* &= P\left(\frac{\rho_s}{\rho} - 1\right)^{0.5} d^2\left(\frac{F}{F_c} - 1\right)\frac{F}{\omega_0(1 - S_a^*)^m} \\ &= P\left(\frac{\rho_s}{\rho} - 1\right)^{0.5} d^2\left(\frac{F}{F_c} - 1\right)\frac{F}{\omega} = \frac{E_a}{\omega} \end{aligned} \tag{1.19}$$

where ω_0 is the sediment fall velocity in clear water, ω is the fall velocity in water with sediment. For erodible bed, based on Eqs. (1.13) and (1.14)

$$-\varepsilon_y \frac{\partial S}{\partial y}\bigg|_{y=a} = P\left(\frac{\rho_s}{\rho} - 1\right)^{0.5} d^2\left(\frac{F}{F_c} - 1\right)F = E_a \tag{1.20}$$

Comparing Eqs. (1.19) and (1.20), the diffusion term under a general condition $-\varepsilon_y \frac{\partial S}{\partial y}\Big|_{y=a} = E_a$ equals the diffusion term under the equilibrium condition E_c, and it is always true

$$E_c = \omega S_{v.a}^*$$

or

$$-\varepsilon_y \frac{\partial S}{\partial y}\bigg|_{y=a} = E_a = E_c = \omega S_{v.a}^* = \omega S_{v.b}^* \tag{1.21}$$

It is seen that while the pick-up function can be described in Eq. (1.20), this type of boundary condition still implies that the sediment diffusion term is independent of the equilibrium condition. Some characteristics of the turbulent bursting were considered in Cao's [2] derivation of the formula for the bottom carrying-sediment capacity S_b^*. While the derivation seems comprehensive, the advantage of this formula is hard to be shown when it is used as a quantitative relation. In general, the results of carrying-sediment capacity computed with the formulas that have been intensively investigated show good agreement with the measured data for sediment concentration and sediment transport rate for the case of equilibrium transport. Based on Cao's formula, $\omega \propto d^2$ for the finer sediment, and $\omega \propto \sqrt{d}$ for the coarser sediment. As shown in Eq. (1.19), the bottom carrying-sediment capacity S_b^* is inversely proportional to $\omega^{-1/2}$ in the laminar flow region, and is proportional to ω in the turbulent regime. This is significantly different from most formulas for carrying-sediment capacity, and contradicts the general concepts.

Similarly, van Rijn developed a formula for the bottom sediment concentration (or carrying-sediment capacity) that reads

$$S_{v.a}^* = 0.15\frac{D_{50}T^{1.5}}{aD_*^{0.5}\omega} = 0.15\frac{T^{1.5}D_{50}^{0.5}\nu}{a\left(\frac{\rho_s-\rho}{\rho_s}g\right)^{\frac{1}{3}}\omega} \tag{1.22}$$

For the finer sediment, $S_{v.a}^* \propto \omega^{-3/4}$. For the coarser sediment, $S_{v.a}^* \propto \omega^0$, or the carrying-sediment capacity is independent of the sediment fall velocity ω, which significantly contradicts with the general concept for carrying-sediment capacity.

The fourth type of bottom boundary condition is to assume the net sedimentation quantity is proportional to ωS_b [1, 34, 35, 50, 51, 57], or

$$-\varepsilon_y \frac{\partial S}{\partial y}\bigg|_{y=a} = B\omega S_b \tag{1.23}$$

This can be rewritten as

$$\varepsilon_y \frac{\partial S}{\partial y}\bigg|_{y=a} + \omega S_b = (1-B)\omega S_b = A\omega S_b \tag{1.24}$$

where B is a constant, representing a probability. In Jobson and Sarye [35], $B = 1-A$ where A is the probability for a sediment particle to settle down and deposit on the bed. In Partheniades [50], $B = P_r$, where P_r is the probability for sediment particles or flocks to stick to the bed. In Bechteler and Schrimpf [1], $B = 1-\alpha$, where α is a re-suspension coefficient. In Jiang and Fan [34], $B = \alpha(1-\varepsilon_0)$, where α is the settling probability, and $(1-\varepsilon_0)$ is the probability for sediment to cease movement. In Eq. (1.24), $A\,\omega S_b$ is the difference between the sediment settling term due to the gravity and the sediment diffusion term caused by the turbulence, or the net sedimentation quantity. Since A and B were considered as probabilities in these studies, A, $B < 1$ and this boundary condition only applies to the case of deposition. It can be used for the case of scour only if the constrain of $B < 1$ is removed.

The fifth type of bottom boundary condition [51] is

$$\varepsilon_y \frac{\partial S}{\partial y}\bigg|_{y=a} + \omega S_b = P_r \omega S_b - E_a \tag{1.25}$$

where E_a is the scour rate per unit area of the bed surface, which is also the rate of sediment pick-up or the upward sediment flux, and ωS_b is the sediment settling term. ωS_b times the probability P_r is the sediment flux that deposits on the bed. This equation can be applied to both the case of deposition and the case of scour. Unfortunately, Partheniades did not provide a general expression for E_a.

The sixth type of bottom boundary condition [11, 24, 26] is

$$\left(\varepsilon_y \frac{\partial S}{\partial y} + \omega S\right)\bigg|_{y=a} = \alpha_1 \omega S_b - \alpha_2 \omega S_b^* = \omega\left(\tilde{\alpha}\overline{S} - \alpha^* \overline{S}^*\right) \tag{1.26}$$

where $\tilde{\alpha}$ and α^* are the saturation recovery coefficients, and α_1 and α_2 are composite factors. This relation was first proposed by Han [11], and was then verified by Han and He [24, 26] using their statistical theory for sediment movement with the concept of sediment exchange intensity (transition intensity) on the bed. Under certain conditions, Eq. (1.26) can be simplified to

$$\left(\varepsilon_y \frac{\partial S}{\partial y} + \omega S\right)\bigg|_{y=a} = \alpha_1 \omega\left(S_b - S_b^*\right) \tag{1.27}$$

In summary, the following conclusions can be made for these six types of the bottom boundary condition for the vertically 2-D non-equilibrium sediment transport problem:

(1) Except for the sixth type of bottom boundary condition, the other five types are lack of strict derivation and cannot comprehensively represent the reality while they hold certain physical meanings and can reflect some aspects of the real world problems. The bottom boundary condition determines the sediment scour or deposition. The sediment movement on the bed and the resulting channel scour or deposition are complicated, which not only depend on the flow and sediment conditions, but also are impacted by the interactions between sediment particles with various states. As a result, it is difficult to comprehensively represent the laws for sediment exchange on the bed and the resulting channel scour or deposition with simple relations or a certain probability.
(2) The first, second, and fourth types of the bottom boundary condition are limited to certain conditions, and cannot be used for a general condition. The first type assumes the bottom sediment concentration can be recovered instantaneously so that $S_b = S_b^*$ at any time. This is not true in reality because channel scour or deposition needs to be developed during a period, which is particularly true

for finer sediment particles. The second type does not consider the sediment pick-up, and thus can only be applied to the case of deposition, such as in the desilting basin with extremely weak flow. In the fourth type of bottom boundary condition, B is a probability and is less than 1, resulting in $A > 0$. Hence, this boundary condition can only be applied to the case of deposition.

(3) The fifth type of bottom boundary condition theoretically can be applied to both the case of scour and the case of deposition. However, the formulation for the scour rate of the bed material is not provided, and this will significantly limits its application.

(4) The third type of bottom boundary condition can represent various conditions of channel scour or deposition as S_b and S_b^* varies. In fact, it represents a special case of Eqs. (1.26) and (1.27) for $\alpha_1 = \alpha_2 = \alpha = 1$. From the physical phenomenon, $\alpha_1 = 1$ means all sediment particles depositing on the bed will change to the bed material, and none of them will change to the bed load or be re-suspended. This apparently contradicts with the reality. Also, α_1, α_2, and α are actually less than 1 under most conditions, and α is rarely greater than 1.

(5) The first five types of bottom boundary condition can all be considered as special cases for the sixth type of bottom boundary condition. The first type is equivalent to Eq. (1.26) with $\alpha_1 = \alpha_2$, and

$$\varepsilon_y \frac{\partial S}{\partial y}\bigg|_{y=a} = \omega S_b^*$$

This is the case of equilibrium sediment transport, with $S_b = S_b^*$. The second type is equivalent to Eq. (1.26) with $\alpha_1 = \alpha_2 = 1$ and $S_b^* = 0$. The third type is equivalent to Eq. (1.26) with $\alpha_1 = \alpha_2 = 1$. The fourth type is equivalent to Eq. (1.26) with $S_b^* = 0$, and $\alpha_1 = B$. The fifth type is equivalent to Eq. (1.26) with $\alpha_1 = P_r$, and $\alpha_2 S_b^* \omega = E_a$.

1.2 1-D Non-equilibrium Sediment Transport

The earlier studies on 1-D non-equilibrium sediment transport were conducted by Russian scientists. Prior to 1960s, the governing equations 1-D non-equilibrium sediment transport were intuitively derived based on the mass balance of sediment [7, 36, 37, 48]. Without linking to the sediment diffusion equation, these earlier studies were considered empirical. Then, the governing equations were developed by integrating the diffusion equation over the water depth [11, 12, 13]. For uniform sediment, if the time period is short and $\partial S/\partial t$ can be neglected, these governing equations for 1-D non-equilibrium sediment transport can be written in the following form

$$\frac{dS}{dx} = -\frac{\alpha\omega}{q}(S - S^*) \tag{1.28}$$

For uniform flow and uniform sediment, integrating Eq. (1.28) yields

$$S = S^* + (S - S^*)\exp\left(-\frac{\alpha \omega x}{q}\right) \tag{1.29}$$

where S is the depth-averaged (or cross-section averaged) sediment concentration, S^* the carrying-sediment capacity of the flow, α the saturation recovery coefficient. The representative studies for 1-D non-equilibrium sediment transport, including Mikhayev and Yunevich [48], Karauschev [38], Dou [7], Karasev [37], Han [11, 12, 13], are reviewed as follows.

1. **Mikhayev and Yunevich** [48]

Mikhayev and Yunevich [48] assumed that the upward sediment pick-up due to turbulence was proportional to $u\uparrow(S^* - S)$, and the sediment settling flux was proportional to $n\omega(S - S^*) = -n\omega(S^* - S)$. The upward component of the turbulent flow velocity is

$$u\uparrow = 1.25\left(\frac{R}{u}\right)^{\frac{1}{4}}(RJ)^{\frac{4}{7}} \tag{1.30}$$

where R is the hydraulic radius, U the mean flow velocity, J the water surface slope, n the compaction factor of bed material, and $n = 3$ based on experimental data. Hence, the following governing equation, in the form of Eq. (1.28), can be obtained:

$$\frac{dS}{dx} = \frac{1}{q}(u\uparrow - n\omega)(S^* - S) = -\frac{\alpha\omega}{q}(S - S^*) \tag{1.31}$$

For steady uniform flow with carrying-sediment capacity remaining constant along the flow path, integrating Eq. (1.31) yields

$$\begin{aligned} S &= S^* + (S_0 - S^*)\exp[-(u\uparrow - \omega n)x] \\ &= S^* + (S_0 - S^*)\exp\left(-\frac{\alpha\omega x}{q}\right) \end{aligned} \tag{1.32}$$

This is the same as Eq. (1.29) with

$$\alpha = \frac{u\uparrow}{\omega} - n \tag{1.33}$$

2. **Karauschev** [38]

Setting

$$Q\frac{dS}{dx} = (q_{s.u} - q_{s.d})B \tag{1.34}$$

where $q_{s.u} = K(S_B - S)$ is the sediment pick-up rate, caused by the turbulence, per unit area of the bed surface, $q_{s.d} = \alpha S$ the sediment settling rate per unit area, and S_B the bottom sediment concentration, Karauschev [38] obtained

$$\frac{dS}{dx} = -\frac{\omega S - K(S_B - S)}{q} \tag{1.35}$$

which leads to

$$S = \frac{KS_B}{\omega + K} + \left(S_0 - \frac{KS_B}{\omega + K}\right)\exp\left(-\frac{\omega + K}{q}x\right) \tag{1.36}$$

For $dS/dx = 0$, $S = S^*$, and it can be derived from Eq. (1.35) that

$$\frac{K}{\omega + K}S_B = S^* \tag{1.37}$$

from which, Eqs. (1.35) and (1.36) can be rewritten in the forms of Eqs. (1.28) and (1.29), respectively, with

$$\alpha = \frac{\omega + K}{\omega} \tag{1.38}$$

3. **Dou** [7]

Using Eq. (1.34) as Karauschev [38] did, and assuming

$$q_u = \alpha\omega S \tag{1.39}$$

$$q_d = \alpha\omega S^* \tag{1.40}$$

Dou [7] obtained Eqs. (1.28) and (1.29), where α is the sediment settling probability that is

$$\alpha = P\left[V_y' < \omega\right] = \frac{1}{\sqrt{2\pi}}\int_{-\infty}^{\frac{\omega}{\sigma_y}} e^{-\frac{t^2}{2}}dt \tag{1.41}$$

4. **Karasev** [37]

Karasev [37] proposed the following equation for non-equilibrium sediment transport:

$$\frac{QdS}{dx} = \left(S_{B,\eta=1} \frac{0.755V\varepsilon\eta\sqrt{g}}{C} - S\frac{0.755V\sqrt{g} + 0.5\omega C}{6C} \right) B$$
$$= \left(AS_{B,\eta=1} - \Pi S \right) B \tag{1.42}$$

where S is the average sediment concentration,

$$A = \frac{0.755\eta\varepsilon V\sqrt{g}}{C} = 0.755\eta\varepsilon u_* \tag{1.43}$$

η is the entrainment probability, ε is the suspension probability, V is the mean flow velocity, C is the Chezy coefficient, and

$$\Pi = \frac{0.755V\sqrt{g} + 0.5\omega C}{6C} = \frac{0.755u_*}{6} + \frac{\omega}{12} \tag{1.44}$$

For the equilibrium condition, $dS/dx = 0$, and it can be derived from Eq. (1.42) that

$$S_{B,\eta=1} \frac{0.755V\varepsilon\eta\sqrt{g}}{C} = S^* \frac{0.755V\sqrt{g} + 0.5\omega C}{6C} \tag{1.45}$$

where S^* is the average carrying-sediment capacity, or the sediment concentration under the equilibrium condition. Plugging Eq. (1.45) into Eq. (1.42) yields

$$\frac{dS}{dx} = -\frac{\Pi}{q}\left(S - S^*\right) = -\frac{\alpha\omega}{q}\left(S - S^*\right) \tag{1.46}$$

Integrating Eq. (1.42) and considering Eq. (1.45) yields

$$S = \frac{AS_{B,\eta=1}}{\Pi} + \left(S_0 - \frac{AS_{B,\eta=1}}{\Pi} \right) \exp\left(-\frac{\Pi x}{q} \right) = S^* + \left(S_0 - S^* \right) \exp\left(-\frac{\Pi x}{q} \right) \tag{1.47}$$

where

$$\alpha = \frac{\Pi}{\omega} = \frac{0.755\frac{V}{\omega}\sqrt{g} + 0.5C}{6C} = \frac{0.755\frac{V}{\omega} + 0.5\frac{C}{\sqrt{g}}}{6\frac{C}{\sqrt{g}}} = 0.126\frac{u_*}{\omega} + 0.083 \tag{1.48}$$

It is noted that Eqs. (1.46) and (1.47) are consistent with Eqs. (1.28) and (1.29), respectively.

5. Han's Studies for Non-equilibrium Sediment Transport

The governing equations, which were proposed by Mikhayev and Yunevich [48], Karauschev [38], Dou [7], and Karasev [37] for the 1-D non-equilibrium sediment transport, were developed directly based on the sediment mass balance, and were not related to the convection-diffusion equation for suspended load transport. The upward sediment flux picked up from the bed and the sediment settling flux were arbitrarily determined in these studies. As a result, these equations are lack of solid theoretical basis. Alternatively, by integrating the 2-D sediment diffusion equation over the water depth, Han [11, 13] developed the following governing equation for 1-D non-equilibrium sediment transport in steady uniform flow:

$$\frac{dq_{s.l}}{dx} = q\frac{dS_l}{dx} = -\varepsilon_y \frac{\partial S_l}{\partial y}\bigg|_{y=0} - S_{b.l}\omega_l \tag{1.49}$$

or

$$\frac{dS_l}{dx} = \frac{\omega}{q}\left(S_{b.l} - S_{b.l}^*\right) = -\frac{\omega_l}{q}\left(\tilde{\alpha}_l S_l - \alpha_l^* S_l^*\right) = -\frac{\omega_l}{q}\left(\tilde{\alpha}_l P_{4.l} S - \alpha_l^* P_{4.l}^* S^*\right) \tag{1.50}$$

where $S_{b.l}$ is the fractional sediment concentration at the bottom, $S_{b.l}^*$ is the bottom carrying-sediment capacity (or the bottom sediment concentration under the equilibrium condition), S is the depth-averaged (or cross-section averaged) sediment concentration, S^* is the depth-averaged carrying-sediment capacity, $P_{4.l}$ is the size distribution of suspended load (i.e., the weight percent of the size class of D_l relative to the total suspended load), $P_{4.l}^*$ is the size distribution for carrying-sediment capacity, $\tilde{\alpha}_l = S_{b.l}/S_l$, and $\alpha_l^* = S_{b.l}^*/S^*$. Neglecting the difference between $\tilde{\alpha}_l$ and α_l^*, Eq. (1.50) becomes

$$\frac{d(P_{4.l}S)}{dx} = -\frac{\alpha_l\omega_l}{q}\left(P_{4.l}S - P_{4.l}^*S^*\right) \tag{1.51}$$

It is noted that $S_b = S_b^*$ for the equilibrium condition is used in the derivation of Eq. (1.51), or

$$-\varepsilon_y \frac{\partial S}{\partial y}\bigg|_{y=0} = S_b^*\omega \tag{1.52}$$

Let neglect the difference in $\tilde{\alpha}_l$ and α_l^* for various size classes, and simply denote them as $\tilde{\alpha}$ and α^*. It is noted that $\tilde{\alpha}$ and α^* have specific physical meaning and are both greater than 1. Han [11] assumed the vertical distribution of sediment concentration only varied slightly for non-equilibrium sediment transport, and thus $\tilde{\alpha} = \alpha^* = \alpha$. Based on calibration with the field data, Han estimated $\alpha \approx 0.25$ for the case of deposition, and $\alpha \approx 1$ for the case of scour [12, 13]. It is noted that α varies in natural streams, and also includes the impacts of other factors. As a result, the calibrated value of α can be small based on the field data. For example, $\alpha = 0.01$ was used for the lower Yellow River [60]. Further analysis found that the difference between the theoretical value of α and the calibrated result is caused by the assumption that the sediment particles settling on the bed (ωS_b) all change to the bed material. In reality, it is only one of the multiple possibilities for the sediment particles settling on the bed to change to the bed material. These particles can also be re-suspended, or changes to the rolling or hopping states. When these particles settling on the bed are assumed to completely change to the bed material, α is great than 1.

Based on their statistical theory for sediment movement, and particularly on the theory for sediment exchange (transition) intensity on the bed, Han and He [24] derived a general bottom boundary condition for the diffusion equation, and then obtained the theoretical formulation for the sediment saturation recovery coefficient under the equilibrium condition [26]. The governing equation that was developed by Han [13] for the longitudinal variation of sediment concentration for 1-D non-equilibrium sediment transport is

$$\begin{aligned} S = S^* &+ \left[S_0 \sum_{l=1}^{n} P_{4.l.0} \exp\left(-\frac{\alpha_l \omega_l x}{q}\right) - S^* \sum_{l=1}^{n} P^*_{4.l.0} S^* \exp\left(-\frac{\alpha_l \omega_l x}{q}\right) \right] \\ &+ \left\{ \sum_{l=1}^{n} P^*_{4.l.0} S^*_0 \left[1 - \exp\left(-\frac{\alpha_l \omega_l x}{q}\right)\right] - S^* \sum_{l=1}^{n} P^*_{4.l} S^* \left[1 - \exp\left(-\frac{\alpha_l \omega_l x}{q}\right)\right] \right\} \end{aligned} \tag{1.53}$$

This is the integral of Eq. (1.51). For massive deposition or massive scour, $P^*_{4.l} = P_{4.l}$ or $P^*_{4.l} \approx P_{4.l}$. Further assuming $\alpha_l = \alpha$, then Eq. (1.53) can be simplifies to Han [11]

$$\begin{aligned} S = S^* &+ \left[S_0 \sum_{l=1}^{n} P_{4.l.0} \exp\left(-\frac{\alpha \omega_l x}{q}\right) - S^*_0 \sum_{l=1}^{n} P^*_{4.l.0} S^* \exp\left(-\frac{\alpha \omega_l x}{q}\right) \right] \\ &+ \left\{ \sum P_{4.l.0} S^*_0 \frac{q}{\alpha \omega_l x} \left[1 - \exp\left(-\frac{\alpha \omega_l x}{q}\right)\right] \right. \\ &\left. - \sum P_{4.l} S^* \frac{q}{\alpha \omega_l x} \left[1 - \exp\left(-\frac{\alpha \omega_l x}{q}\right)\right] \right\} \end{aligned} \tag{1.54}$$

where $P_{4.l}$ and $P^*_{4.l}$ are the size distributions for suspended load and carrying-sediment capacity, respectively, the subscript "0" denotes the values at the entrance section, the parameters without the subscript "0" denote the values at the exit section, and x is the distance between the entrance and the exit sections. α is set to 0.25 for deposition, and 1 for scour. The theoretical analysis [26] shows these assigned values for α is acceptable. Equations (1.53) and (1.54) have been widely used in the 1-D mathematical models for sediment transport. The equations for the size distribution of suspended load at the exit section $P_{4.l}$ and for the variation of the bed material gradation within a reach were also developed in Han [11]. For uniform sediment, based on Eq. (1.53), Han [11] obtained

$$S = S^* + \left(S_0 - S_0^*\right)\exp\left(-\frac{\alpha\omega x}{q}\right) + \frac{q}{\alpha\omega x}\left[1 - \exp\left(-\frac{\alpha\omega x}{q}\right)\right]\left(S_0^* - S^*\right) \quad (1.55)$$

If the carrying-sediment capacity remains constant along the flow path,

$$S = S^* + \left(S_0 - S_0^*\right)\exp\left(-\frac{\alpha\omega x}{q}\right) \quad (1.56)$$

It is noted that Eq. (1.55) covers the results of [7, 37, 38, 48], or Eqs. (1.32), (1.36) and (1.47).

6. **Dou et al.** [8]**, Dou and Zhao** [9]

In order to consider the non-equilibrium transport of non-uniform sediment and the size distribution for carrying-sediment capacity, Dou et al. [8], Dou and Zhao [9] updated Dou's [7] equations for 1-D non-equilibrium sediment transport, and developed new equations for fractional sediment transport. The equation developed by Dou et al. [8] is

$$\frac{dS_l}{dx} = \frac{-\alpha_l\omega_l}{q}P_{1.l.1}^{\beta}\left(S_l - S_l^*\right) = -\frac{\alpha_l\omega_l}{q}P_{1.l.1}^{\beta}\left(P_{4.l}S - P_{4.l}^*S^*\right) \quad (1.57)$$

where α is the saturation recovery coefficient, which is set to the settling probability. The formula for carrying-sediment capacity is

$$S_l^* = P_{4.l}^*\frac{K_1}{C_0^2}\frac{\gamma_s\gamma}{\gamma_s - \gamma}\frac{V^3}{gh\omega} \quad (1.58)$$

The formula for the size distribution for carrying-sediment capacity is

$$P_{4.l}^* = \frac{\left(\frac{P_{4.l}}{\omega_l}\right)^{\gamma}}{\sum\left(\frac{P_{4.l}}{\omega_l}\right)^{\gamma}} \quad (1.59)$$

$$\beta = \begin{cases} 0 & S_l \geq S_l^* \\ 1 & S_l < S_l^* \end{cases} \quad (1.60)$$

where $\gamma = 1/6$. The average sediment fall velocity is determined with

$$\omega = \sum P_{4.l}\omega_l \quad (1.61)$$

Considering the bed scour or deposition is determined by the difference between the upward sediment pick-up flux $q_l \uparrow$ and the settling flux $q_l \downarrow$, Dou and Zhao [9] concluded that the sediment pick-up flux is proportional to $S_l^* - S_l$, the surplus of sediment carrying-capacity for a given size class, and are also impacted by the quantity of this size class on the bed, or

$$S_{b.l}^* = P_{1.l.1}\left(S_l^* - S_l\right) \tag{1.62}$$

For $S_l^* < S_l$, the pick-up flux is

$$q_l \uparrow = \alpha_l \omega_l \left(S_l - S_{b.l}^*\right) = \alpha_l \omega_l S_l - \alpha_l \omega_l P_{1.l.1}\left(S_l^* - S_l\right) \tag{1.63}$$

For $S_l^* > S_l$, the pick-up quantity is

$$q_l \uparrow = \alpha_l \omega_l S_l^* \tag{1.64}$$

The settling flux is

$$q_l \downarrow = \alpha_l \omega_l S_l \tag{1.65}$$

Thus,

$$q_l \downarrow - q_l \uparrow = \begin{cases} \alpha_l \omega_l \left(S_l - S_l^*\right) & \left(S_l > S_l^*\right) \\ -\alpha_l \omega_l P_{1.l.1}\left(S_l^* - S_l\right) & \left(S_l < S_l^*\right) \end{cases} \tag{1.66}$$

For steady flow, the governing equation for non-equilibrium sediment transport is

$$\frac{dS_l q}{dx} = -(q_l \downarrow - q_l \uparrow)$$

or

$$\frac{dS_l}{dx} = -\frac{\alpha_l \omega_l}{q} \beta_l \left(P_{4.l} S - P_{4.l}^* S^*\right) = -\frac{\alpha_l \omega_l}{q} \beta_l \left(S_l - S_l^*\right) \tag{1.67}$$

where

$$P_{4.l}^* = \frac{\left(\frac{P_{1.l.1}}{\omega_l}\right)^\gamma}{\sum \left(\frac{P_{1.l.1}}{\omega_l}\right)^\gamma} \tag{1.68}$$

$$\beta_l = \begin{cases} 1 & (S_l \geq S_l^*) \\ P_{1.l.1} & (S_l < S_l^*) \end{cases} \tag{1.69}$$

and the carrying-sediment capacity

$$S^* = K_s \frac{V^3}{h\omega} \tag{1.70}$$

In fact, Eq. (1.67) is the same as Eq. (1.57), while Eq. (1.67) was developed with slightly more rigorous derivation, and with slightly different formulations for related parameters.

One problem occurs when applying Dou's (1997) formula for the case of scour. In Dou's derivation, Eq. (1.62) was assumed to emphasize the role of the bed material gradation in scour. Because the formulation is not reasonable, however, this purpose is not achieved. For the case of scour, $S_l < S_l^*$, and

$$\frac{dS_l}{dx} = -\frac{\alpha_l \omega_l}{q} \mathrm{P}_{1.l.1}\left(P_{4.l} S - P_{4.l}^* S^*\right) \tag{1.71}$$

This formula apparently contradicts with the general equations for non-equilibrium sediment transport, such as Eqs. (1.50) and (1.51). In Eq. (1.71), two size distributions multiply each other, with $\sum P_{1.l.1} P_{4.l} < 1$, and $\sum P_{1.l.1} P_{4.l}^* < 1$, which is obviously unreasonable. In particular, assuming there are n size classes in total, and the percent for each size class is the same with $P_{1.l} = 1/n$, then based on Eq. (1.71),

$$\frac{dS_l}{dx} = -\frac{\alpha_l \omega_l}{q} \frac{1}{n} (P_{4.l} S_l - P_{4.l}^* S^*)$$

Summing up for the n size classes yields

$$\frac{dS}{dx} = -\frac{1}{nq}\left(\overline{\alpha\omega} S - \overline{\alpha\omega^*} S^*\right) \tag{1.72}$$

where

$$\overline{\alpha\omega} = \sum P_{4.l} \alpha_l \omega_l \tag{1.73}$$

$$\overline{\alpha\omega^*} = \sum P_{4.l}^* \alpha_l \omega_l \tag{1.74}$$

It is noted that the results of dS_l/dx and dS/dx depend on n, the number of size classes. For a large n, $\overline{\alpha\omega}$ and $\overline{\alpha\omega^*}$ are approximately constant, dS/dx is entirely determined by n, and dS/dx decreases as n increases. This extremely unreasonable result indicates that Eqs. (1.57) is incorrect.

The reason for this problematic result can be traced back to Eq. (1.62), which is proved to be an incorrect assumption. In fact, $S_l^* - S_l$ itself represents the sediment saturation degree for the l-th size class. It is unreasonable to multiply $S_l^* - S_l$ with the bed material gradation to represent the surplus of sediment concentration for this size class.

7. **Zhou and Lin** [64]**, and Zhou** [63]

Based on Eq. (1.71) that was developed by Dou and Zhao [9], Zhou and Lin [64], and Zhou [63] proposed the following equation for 1-D non-equilibrium sediment transport:

$$\frac{dS_l}{dx} = -\frac{\alpha_l \omega_l}{q}(S_l - \Phi_l) \tag{1.75}$$

in which

$$\Phi_l = P_{1.l.1}\varphi_l + (1 - P_{1.l.1})\mathrm{Min}(S_l, \varphi_l) \tag{1.76}$$

$$\phi_l = P_l \phi = P^*_{4.l}\phi \tag{1.77}$$

$$\phi = K_s \left(\frac{V^3}{gh\omega}\right)^m = S^*(\omega) \tag{1.78}$$

$$\omega = \left(\sum_{l=1}^{n} P^*_{4.l}\omega_l^m\right)^{\frac{1}{m}}$$

The size distribution for carrying-sediment capacity is determined with the formula developed by Li [41] that is

$$P^*_{4.l} = \frac{\beta_l P_{1.l.1}}{\sum \beta_l P_{1.l.1}} \tag{1.79}$$

where

$$\beta_l = \frac{(1 - A_l)\left[1 - \exp\left(-\frac{6\omega_l}{\kappa u_*}\right)\right]}{\omega_l} \tag{1.80}$$

A_l and β_l were determined in Li [41] for the strong equilibrium sediment transport. For deposition, $S_l > \varphi_l$, and Eq. (1.75) is the same as the general form for non-equilibrium sediment transport. For scour, $S_l < \varphi_l$, and $\Phi_l = P_{1.l.1}\varphi_l + (1 - P_{1.l.1})S_l$ according to Eq. (1.76), and thus (1.75) becomes

$$\frac{dS_l}{dx} = -\frac{\alpha_l \omega_l}{q}[S_l - P_{1.l.1}\phi_l - (1 - P_{1.l.1})\phi_l] = -\frac{\alpha_l \omega_l}{q} P_{1.l.1}\left[P_{4.l}S - P^*_{4.l}S^*(\omega^*)\right] \tag{1.81}$$

which is the same as Eq. (1.71). Also, for deposition, Eq. (1.75) is the same as Eq. (1.67). Hence, Eq. (1.74) developed by Zhou and Lin [64] and Zhou [63] is actually the same as Eq. (1.67) developed by Dou and Zhao [9]. The only difference was that Zhou and Lin [64] and Zhou [63] determined the carrying-sediment capacity based on Li's [41] formula for the size distribution for carrying-sediment capacity under the strong equilibrium condition. The following comments are made:

(1) The equation that was developed by Zhou and Lin [64] and Zhou [63] for non-equilibrium sediment transport is basically the same as Dou and Zhao's [9] formula except for the coefficient α and some differences in the derivation and expression of the equation. As a result, for scour, two size distributions multiply each other, with $\sum P_{1.l.1}P_{4.l} < 1$, and $\sum P_{1.l.1}P^*_{4.l} < 1$. These unreasonable results are the same as those for Dou and Zhao [9].
(2) Assuming there are n size classes in the non-uniform sediment mixture, and the percent for each size class is the same with $P_{1.l} = 1/n$, then the longitudinal gradient of the sediment concentration is inversely proportional to n based on the formula developed by Zhou and Lin [64] and Zhou [63]. This incorrect result is also the same as the mistake associated with Dou and Zhao's [9] formula.

1.3 Size Distribution for Carrying-Sediment Capacity under Equilibrium Condition

1. **Han** [11, 13]

For equilibrium transport of non-uniform suspended sediment, the characteristics of the size distributions can be represented by the size distribution of suspended load $P_{4.l}$ and the bed material gradation $P_{1.l.1}$. Han [11] proved that the following equation holds for the strong equilibrium sediment transport:

$$S^*_l = P^*_{4.l}S^* = P_{1.l}S^*(l) = P_{1.l.1}S^*(l) \tag{1.82}$$

or

$$P_{4.l} = P^*_{4.l} = \frac{S^*(l)}{S^*(\omega)}P_{1.l} = \frac{S^*(l)}{S^*(\omega)}P_{1.l.1} = P_{1.l.1}\left(\frac{\omega^*}{\omega_l}\right)^m \tag{1.83}$$

$$P_{1.l.1} = \frac{S^*(\omega^*)}{S^*(l)}P_{4.l} = P_{4.l}\left(\frac{\omega_l}{\omega^*}\right)^m \tag{1.84}$$

where $S^*(l) = S^*(\omega_l)$ is the capacity for carrying the uniform sediment with the grain size of D_l, S^* is the total carrying-sediment capacity for the non-uniform sediment mixture, ω^* is the average sediment fall velocity for carrying-sediment capacity, $P_{1.l}$ is the effective bed material gradation, and $P_{1.l}$ does not equal the actual bed material gradation $P_{1.l.1}$ under a general (non-equilibrium) condition.

As shown in Eqs. (1.83) and (1.84), the size distribution for carrying-sediment capacity $P^*_{4.l}$ is not needed in the summation of the carrying-sediment capacity for uniform sediment $S^*(l)$, or in the determination of the fractional carrying-sediment capacity S^*_l and the total carrying-sediment capacity S^* for the non-uniform sediment mixture. In fact, for the strong equilibrium condition, the size distribution for carrying-sediment capacity $P_{4.l}^*$ equals the size distribution of suspended load $P_{4.l}$, and the effective bed material gradation $P_{1.l}$ equals the actual bed material gradation $P_{1.l.1}$. If the carrying-sediment capacity $S^*(l)$ for the uniform sediment with the grain size of D_l is determined with

$$S^*(l) = \kappa\left(\frac{V^3}{gh\omega_l}\right)^m = f(V,h)\frac{1}{\omega_l^m} \tag{1.85}$$

then the following relations can be derived from Eq. (1.82)

$$S_1^* = \sum_{l=1}^{n} P_{1.l}S^*(l) = f(V,h)\sum\frac{P_{1.l}}{\omega_l^m} = f(V,h)\frac{1}{\omega_1^m} \tag{1.86}$$

$$\left(S_2^*\right)^{-1} = \sum_{l=1}^{n} P_{4.l}/S^*(l) = f^{-1}(V,h)\sum_{l=1}^{n} P_{4.l}\omega_l^m = f^{-1}(V,h)\omega_2^m \tag{1.87}$$

$$S_2^* = \frac{1}{\sum_{l=1}^{n} P_{4.l}/S^*(l)} = \frac{f(V,h)}{\sum P_{4.l}\omega_l^m} = \frac{f(V,h)}{\omega_2^m} \tag{1.88}$$

It can be derived from Eqs. (1.86) and (1.84) that

$$S_1^* = \sum_{l=1}^{n} P_{1.l}S^*(l) = \sum_{l=1}^{n}\frac{S^*}{S^*(l)}P_{4.l}S^*(l) = S^*$$

Considering Eqs. (1.83), (1.87) can be rewritten as

$$\left(S_2^*\right)^{-1} = \sum_{l=1}^{n}\frac{P_{4.l}}{S^*(l)} = \sum_{l=1}^{n}\frac{S^*(l)P_{1.l}}{S^*S^*(l)} = \frac{1}{S^*} \tag{1.89}$$

or

$$S_1^* = S_2^* \tag{1.90}$$

Hence, $S_1^* = S_2^*$ with the corresponding $\omega_1 = \omega_2$, and the total carrying-sediment capacity summed up with Eq. (1.86) based on the bed material gradation is the same as that summed up with Eq. (1.88) based on the size distribution of suspended load.

The approach of summing up the fractional suspended load transport rate (or concentration) that was adopted in the U.S. Army Corps' [56] HEC-6 model, which was further discussed in Feldman [10], is also consistent with the results of Eq. (1.82), and can be considered as a special case of Eq. (1.82). Based on different sediment exchange model, Li [41] developed a relationship between the size distribution of suspended load and the bed material gradation for equilibrium sediment transport, which is also consistent with Eq. (1.82). It is seen that these studies on the capacity for carrying the non-uniform sediment mostly focused on the strong equilibrium condition, under which the size distribution for carrying-sediment capacity equals the size distribution of suspended load. These studies actually did not considered the size distribution for carrying-sediment capacity for a general non-equilibrium condition, while the size distribution of suspended load was named with different terminologies, or was even directly referred to as the size distribution for carrying-sediment capacity.

2. **Li** [41]

Li [41] conducted an in-depth analysis of the relationship between the size distribution of suspended load and the bed material gradation for equilibrium sediment transport. Li assumed that, under the actions of the vertical component of the turbulent velocity and of the gravity, the sediment pick-up quantity within a time interval of ΔT on a unit area is

$$\begin{aligned} g_{s.l.u} &= \Delta T \int_{\omega_l}^{\infty} \frac{\overline{V_l}}{\sqrt{2\pi}\sigma_y}(V_y' - \omega_l)\exp\left(-\frac{V_y'^2}{2\sigma_y^2}\right)dV_y' \\ &= \overline{V}_l \Delta T \left\{ \frac{\sigma_y}{\sqrt{2\pi}} \exp\left(-\frac{\omega_l^2}{2\sigma_y^2}\right) - \omega_l\left[1 - \Phi\left(\frac{\omega_l}{\sigma_y}\right)\right]\right\} \overline{V}_l \end{aligned} \tag{1.91}$$

and the settling quantity is

$$\begin{aligned} g_{s.l.d} &= -\Delta T \int_{-\infty}^{\omega_l} \frac{S_{b.l}}{\sqrt{2\pi}\sigma_y}(V_y' - \omega_l)\exp\left(-\frac{V_y'^2}{2\sigma_y^2}\right)dV_y' \\ &= \Delta T \left[\frac{\sigma_y}{\sqrt{2\pi}} \exp\left(-\frac{\omega_l^2}{2\sigma_y^2}\right) + \omega_l \Phi\left(\frac{\omega_l}{\sigma_y}\right)\right] S_{b.l} \end{aligned} \tag{1.92}$$

where $\Phi(\omega_l/\sigma_y)$ is the Gaussian integral, V_y' the vertical velocity fluctuation at the bottom, σ_y the variance of V_y', $S_{b.l}$ the concentration of suspended load at the bottom, and $\overline{V}_l$ the quantity of the bed material of the l-th size class that participates in the sediment exchange, and

$$\overline{V_l} = P_{1.l.1}\overline{V} \tag{1.93}$$

in which $\overline{V}$ is the total quantity of the bed material that participates in the sediment exchange, and $P_{1.l.1}$ the bed material gradation. Setting Eq. (1.91) be equal to Eq. (1.92) obtains

$$\begin{aligned} S_{b.l} &= \frac{\overline{V_i}\left\{\frac{\sigma_y}{\sqrt{2\pi}} \exp\left(-\frac{\omega_l^2}{2\sigma_y^2}\right) - \omega_l\left[1 - \Phi\left(\frac{\omega_l}{\sigma_y}\right)\right]\right\}}{\frac{\sigma_y}{\sqrt{2\pi}} \exp\left(-\frac{\omega_l^2}{2\sigma_y^2}\right) + \omega_l \Phi(\frac{\omega_l}{\sigma_y})} \\ &= \overline{V}_l(1 - A_l) = P_{1.l.1}\overline{V}(1 - A_l) \end{aligned} \tag{1.94}$$

where

$$A_l = \frac{\omega_l}{\frac{\sigma_y}{\sqrt{2\pi}}\exp\left(-\frac{\omega_l^2}{2\sigma_y^2}\right) + \omega_l \Phi\left(\frac{\omega_l}{\sigma_y}\right)} \tag{1.95}$$

Summing up Eq. (1.94) for l yields

$$S_b = \sum_{l=1}^{n} S_{b.l} = \sum_{l=1}^{n} P_{1.l.1}(1 - A_l)\overline{V} = \overline{V}A \tag{1.96}$$

in which

$$A = \sum_{l=1}^{n} P_{1.l.1}(1 - A_l) \tag{1.97}$$

Thus,

$$\overline{V} = \frac{S_b}{A} \tag{1.98}$$

Comparing Eq. (1.94) with Eq. (1.96), the size distribution of suspended load at the bottom is

$$P_{b.l} = \frac{S_{b.l}}{S_b} = P_{1.l.1}\frac{1 - A_l}{A} = \frac{P_{1.l.1}(1 - A_l)}{\sum P_{1.l.1}(1 - A_l)} \tag{1.99}$$

This equation illustrates the relationship between the size distribution of suspended load at the bottom and the bed material gradation.

Also, assuming the vertical distribution of sediment concentration under the equilibrium condition satisfies

$$S_l = S_{b.l}\exp\left(-\frac{6\omega_l}{\kappa u^*}\frac{y}{h}\right) \tag{1.100}$$

The depth-averaged sediment concentration is

$$\overline{S_l} = S_{b.l}\frac{\kappa u^*}{6\omega_l}\left[1 - \exp\left(-\frac{6\omega_l}{\kappa u^*}\right)\right] \tag{1.101}$$

Summing up Eq. (1.101) for l, and plugging in Eq. (1.99) for $S_{b.l}$, one obtains

$$\overline{S} = \sum_{l=1}^{n} S_{b.l}\frac{\kappa u^*}{6\omega_l}\left[1 - \exp\left(-\frac{6\omega_l}{\kappa u^*}\right)\right] = S_b\sum_{l=1}^{n} P_{b.l}\frac{\kappa u^*}{6\omega_l}\left[1 - \exp\left(-\frac{6\omega_l}{\kappa u^*}\right)\right] \tag{1.102}$$

Comparing Eq. (1.102) with Eq. (1.101), it can be derived that

$$P_{4.l} = \frac{\overline{S}_l}{\overline{\overline{S}}} = \frac{S_{b.l}}{S_b} \frac{\frac{1}{\omega_l}[1 - \exp(-\frac{6\omega_l}{\kappa u^*})]}{\sum_{l=1}^{n} \frac{P_{4.l.b}}{\omega_l}[1 - \exp(-\frac{6\omega_l}{\kappa u^*})]} \tag{1.103}$$

Plugging Eq. (1.99) into Eq. (1.103) yields

$$P_{4.l} = \frac{P_{1.l.1}\frac{1-A_l}{\omega_l}[1 - \exp(-\frac{6\omega_l}{\kappa u^*})]}{\sum_{l=1}^{n} P_{1.l.1}\frac{1-A_l}{\omega_l}[1 - \exp(-\frac{6\omega_l}{\kappa u^*})]} \tag{1.104}$$

This equation illustrate the relationship between the size distribution of suspended load $P_{4.l}$ and the bed material gradation $P_{1.l}$.

It is worthy to point out that Eq. (1.104) is consistent with Eq. (1.82) that was developed by Han [11]. In fact, under the strong equilibrium condition, $\overline{S} = S^*$, $\overline{S}_l = S_l^*$, and $P_{4.l} = P_{4.l}^*$. For this condition, the following relation can be derived based on Eqs. (1.101) and (1.99):

$$S_l^* = S_b P_{1.l.1} \frac{1 - A_l}{A} \frac{\kappa u^*}{6\omega_l}\left[1 - \exp\left(-\frac{6\omega_l}{\kappa u^*}\right)\right] \tag{1.105}$$

For the uniform sediment with the grain size of D_l,

$$S^*(l) = S_b \frac{1 - A_l}{A} \frac{\kappa u^*}{6\omega_l}\left[1 - \exp\left(-\frac{6\omega_l}{\kappa u^*}\right)\right] \tag{1.106}$$

Hence,

$$S_l^* = P_{1.l.1} S^*(l) \tag{1.107}$$

Considering Eqs. (1.105) and (1.107), Eq. (1.103) can be rewritten as

$$\begin{aligned} P_{4.l}^* = P_{4.l} &= \frac{S_l^*}{S^*} = \frac{\overline{S}_l}{\overline{S}} = \frac{\overline{S}_l}{\sum \overline{S}_l} = \frac{P_{1.l.1}\frac{1-A_l}{\omega_l}[1 - \exp(-\frac{6\omega_l}{\kappa u^*})]}{\sum P_{1.l.1}\frac{1-A_l}{\omega_l}[1 - \exp(-\frac{6\omega_l}{\kappa u^*})]} \\ &= \frac{P_{1.l.1} S_b \frac{1-A_l}{\omega_l}\frac{\kappa u^*}{6\omega_l}[1 - \exp(-\frac{6\omega_l}{\kappa u^*})]}{\sum P_{1.l.1} S_b \frac{1-A_l}{\omega_l}\frac{\kappa u^*}{6\omega_l}[1 - \exp(-\frac{6\omega_l}{\kappa u^*})]} = \frac{P_{1.l.1} S^*(l)}{\sum P_{1.l.1} S^*(l)} = \frac{P_{1.l.1} S^*(l)}{S^*} \end{aligned} \tag{1.108}$$

Therefore,

$$P_{4.l}^* S^* = P_{1.l.1} S^*(l)$$

which is the same as Eq. (1.82).

1.4 Size Distribution for Carrying-Sediment Capacity and Effective Bed Material Gradation for Non-equilibrium Sediment Transport

The size distribution for carrying-sediment capacity and the effective bed material gradation are two important and pioneering topics for the studies on non-equilibrium transport of non-uniform sediment. The first study on these two topics were conducted by Han, and the groundbreaking studies can be found in Han [14], Han [13], Han and He [25], and He and Han [28, 29]. These two topics have become popular in sediment transport study in China since 1987. Some scientists acknowledged the importance of these two topics in the study on the non-equilibrium transport of non-uniform suspended load, and also proposed their formulations for the size distribution for carrying-sediment capacity. Example studies include Dou et al. [8], Dou and Zhao [9], Ouyang et al. [49], Huang [32], Yang [59], Yu et al. [61], Liu et al. [42], Lu and Zhang [46], Dong and Lu [6] and Le [40]. Lu and Zhang [46] and Dong and Lu [6] acknowledged the concept of the size distribution for carrying-sediment capacity. Lu and Zhang [46] and Dong and Lu [6] directly adopted the full set of formulas, which were developed by Han and He for the size distribution for carrying-sediment capacity and for the effective bed material gradation, in their mathematical model. The model verification showed these formulas agreed well with the reality. They further extended this approach to bed load, and obtained the size distribution for the bed load transport capacity that was different from the size distribution of bed load. Liu et al. [44] also applied Han and He's concept and formulas to the middle and lower reaches of the Yellow River, and obtained satisfactory results. Their results clarified how the bed material impacts the carrying-sediment capacity, demonstrated the interaction of carrying-sediment capacities among the finer sediment, the coarser sediment and the bed material, and thus dynamically showed the variation of the size distribution for carrying-sediment capacity. Recently, more analyses have been conducted for the size distribution for carrying-sediment capacity, with more applications in the mathematical models. The size distribution for carrying-sediment capacity has become an important topic for non-equilibrium transport of non-uniform sediment and a popular subject for the theoretical analysis of sediment transport.

As a new subject, the non-equilibrium transport of non-uniform suspended load is essentially different from the equilibrium transport of uniform sediment or the equilibrium transport of non-uniform sediment. Some concepts for equilibrium sediment transport do not apply to the non-equilibrium transport of non-uniform sediment. Limited by conventional concepts, the complete meaning of the non-equilibrium transport of non-uniform suspended load is hard to be profoundly understood. As a result, many results are specious and questionable, and some of them are unreasonable or incorrect. While the studies abroad generally focuses on the transport capacity for uniform sediment, the studies on the non-equilibrium transport of non-uniform sediment in China are advanced.

The concepts of the size distribution for carrying-sediment capacity and of the effective bed material gradation were developed through the in-depth analysis of the non-equilibrium transport of non-uniform suspended load. The most critical one is the size distribution for carrying-sediment capacity, which has a one-to-one relationship with the effective bed material gradation, and which has a clear physical meaning. The effective bed material gradation is a basis for the concept of the size distribution for carrying-sediment capacity.

The following review focuses on the representative studies on the size distribution for carrying-sediment capacity and corresponding effective bed material gradation. The purpose is for the reader to have a correct, comprehensive, and in-depth understanding of this topic so that more effective and profound studies can be performed in the future.

1. **Han** [13, 14]**, Han and He** [25]**, and He and Han** [28, 29]

The concept of the size distribution for carrying-sediment capacity was firstly developed by Han [13]. The general expressions for the size distribution for carrying-sediment capacity and for the effective bed material gradation for non-equilibrium sediment transport were formulated by Han [14], which have been widely used in the mathematical models and sedimentation engineering applications [25, 44]. For non-equilibrium transport of non-uniform sediment, the size distribution for carrying-sediment capacity does not equal the size distribution of suspended load or the size distribution of suspended load determined by bed material gradation with Eq. (1.83). In fact, for a flow on the rock or pebble bed, with an extremely low sediment concentration that is much lower than the carrying-sediment capacity corresponding to the average fall velocity determined from the size distribution of suspended load, no channel scour or deposition would occur. Apparently, the flow still has surplus carrying-sediment capacity. However, the carrying-sediment capacity would be zero if it was computed with the bed material gradation. Hence, for this case, the carrying-sediment capacity should be computed using the size distribution of suspended load instead of the bed material gradation. For the initial scour phase as the clear water enter an alluvial stream, the carrying-sediment capacity would be zero if it was computed based on the size distribution of suspended load, which could not explain the reason for scour. For this case, the carrying-sediment capacity should be determined using the bed material gradation. For a general condition, the size distribution for carrying-sediment capacity $P^*_{4.l}$ cannot be represented solely by the size distribution $P^*_{4.l.1}$ corresponding to the bed material gradation, or solely by the size distribution of suspended load $P_{4.l}$. Instead, it should be a weighted sum of these two size distributions, which is

$$P^*_{4.l} = \beta P_{4.l} + (1 - \beta) P^*_{4.l.1} \tag{1.109}$$

where $\beta < 1$, and $P^*_{4.l.1}$ is the size distribution for the part of carrying-sediment capacity that corresponds to the bed material gradation, which, similar to Eq. (1.83), is

$$P^*_{4.l.1} = \frac{S^*(l)}{S^*} P_{1.l.1} \tag{1.110}$$

Equations (1.9) and (1.110) were firstly developed by Han [14]. Based on the concept of water percentage, and starting from the fractional carrying-sediment capacity, Han [14] developed a formula for the size distribution for carrying-sediment capacity. It consists of three parts. Two parts represent the contributions of the size distributions for the coarser particles and for the finer particles, respectively, of the suspended load. The third part represents the contribution of the sediment eroded from the bed material. Equation (1.109) is a simplified expression with two terms. Assuming part of the bed material, which has $D_l \leq D_n$, or $l \leq n$, can be partially suspended; the finer particles, with $l \leq k$ or similar to wash load, of the suspended load influx can directly change to the carrying-sediment capacity; and the coarser particles, with $k + 1 \leq l \leq n$ or similar to bed material load, first become the bed material and then are eroded, the fractional carrying-sediment capacity can be expressed as

$$P_{4.l}^{*} S^{*}\left(\omega^{*}\right) = S P_{4.l} P_{4.l.1} + S P_{4.2} P_{4.l.2} \frac{S^{*}(l)}{S^{*}(\omega_{1.1}^{*})} + \left[1 - \frac{P_{4.1} S}{S^{*}(\omega_{1})} - \frac{S P_{4.2}}{S^{*}(\omega_{1.1}^{*})}\right] P_{1} P_{4.l.1}^{*} S^{*}(\omega_{1.1}^{*}) \tag{1.111}$$

Thus, the carrying-sediment capacity and the corresponding size distribution are

$$S^{*}(\omega^{*}) = S P_{4.1} + S P_{4.2} \frac{S^{*}(\omega_{2}^{*})}{S^{*}(\omega_{1.1}^{*})} + \left[1 - \frac{P_{4.1} S}{S^{*}(\omega_{1})} - \frac{P_{4.2} S}{S^{*}(\omega_{1.1}^{*})}\right] P_{1} S^{*}(\omega_{1.1}^{*}) \tag{1.112}$$

$$P_{4.l}^{*} = P_{4.1} P_{4.l.1} \frac{S}{S^{*}(\omega^{*})} + P_{4.2} P_{4.l.2} \frac{S}{S^{*}\left(\omega_{1.1}^{*}\right)} \frac{S^{*}(l)}{S^{*}(\omega^{*})} + \left[1 - \frac{P_{4.1} S}{S^{*}(\omega_{1})} - \frac{S P_{4.2}}{S^{*}\left(\omega_{1.1}^{*}\right)}\right] \frac{P_{1} P_{4.l.1.1}^{*} S^{*}(\omega_{1.1}^{*})}{S^{*}(\omega^{*})} \tag{1.113}$$

and the effective bed material gradation is

$$P_{1.l} = P_{4.l}^{*} \frac{S^{*}(\omega^{*})}{S^{*}(l)} = P_{4.1} P_{4.l.1} \frac{S}{S^{*}(l)} + P_{4.2} P_{4.l.2} \frac{S}{S^{*}\left(\omega_{1.1}^{*}\right)} + \left[1 - \frac{P_{4.1} S}{S^{*}(\omega_{1})} - \frac{S P_{4.2}}{S^{*} \omega_{1.1}^{*}}\right] P_{1} P_{1.l.1.1} \tag{1.114}$$

$$P_{4.l.1} = \begin{cases} \frac{P_{4.l}}{P_{4.1}} = \frac{P_{4.l}}{\sum_{l=1}^{k} P_{4.l}} & (l \leq k) \\ 0 & (l \geq k) \end{cases} \tag{1.115}$$

$$P_{4.l.2} = \begin{cases} \frac{P_{4.l}}{P_{4.2}} = \frac{P_{4.l}}{\sum_{l=k+1}^{n} P_{4.l}} & (n \geq l \geq k+1) \\ 0 & (l \leq k) \end{cases} \tag{1.116}$$

$$P_{1.l.1.1} = \begin{cases} \frac{P_{1.l.1}}{\sum_{l=1}^{n} P_{1.l.1}} = \frac{P_{1.l.1}}{P_l} & (l \le n) \\ 0 & (l > n) \end{cases} \tag{1.117}$$

$$\frac{1}{S^*(\omega_1^*)} = \sum_{l=1}^{n} \frac{P_{4.l.1}}{S^*(l)} \tag{1.118}$$

$$S^*(\omega_{1.1}^*) = \sum_{l=1}^{n} P_{1.l.1.1} S^*(l) \tag{1.119}$$

$$S^*(\omega_2^*) = \sum_{l=k+1}^{n} P_{4.l.2} S^*(l) \tag{1.120}$$

$$P_{4.l.1.1}^* = \frac{S^*(l)}{S^*(l)} P_{1.l.1.1} \tag{1.121}$$

where $P_{4.l.1.1}^*$ is the size distribution for the part of carrying-sediment capacity that corresponds to the bed material gradation $P_{1.l.1.1}$, and P_l is the proportion of the bed material that can be suspended.

As shown in Eqs. (1.112) through (1.114), the carrying-sediment capacity, its size distribution, and the effective bed material gradation all consists of three parts, which represent the different transport and sedimentation characteristics of the coarser and the finer particles of the sediment influx, and represent the exchange between suspended load and the bed material. The first term on the right side of Eq. (1.112) represents the contribution from the finer particles of sediment influx, which are similar to wash load and which do not exchange with the bed material. The second term represents the contribution from the coarser particles of the sediment influx, which deposit on the bed first and then are re-suspended, and thus partially transferred to the carrying-sediment capacity with $S^*(\omega_2^*) < S^*(\omega_{1.1}^*)$. The third term represents the contribution from the bed material, in which $S^*(\omega_{1.1}^*)$ is the carrying-sediment capacity computed with the bed material gradation.

The constraint condition for Eqs. (1.111) through (1.114) is

$$\frac{P_{4.1}S}{S^*(l)} + \frac{P_{4.2}S}{S^*(\omega_{1.1}^*)} \le 1 \tag{1.122}$$

Otherwise, the scour of the bed material is not needed. There are two cases. One case is that the left side of Eq. (1.122) is one, which is the strong equilibrium condition. The other is that the left side of Eq. (1.122) is greater than 1, which is the case of deposition. For the strong equilibrium sediment transport,

$$S^*(\omega^*) = S \tag{1.123}$$

$$P_{4.l}^* = P_{4.l} \tag{1.124}$$

The formula for the effective bed material gradation $P_{1.l}$ corresponding to $P_{4.l}^*$ was derived by Han [14] and by He and Han [29]. They also proved that the water percentage K_l equals the effective bed material gradation for the non-equilibrium sediment transport, or

$$P_{1.l} = K_l \tag{1.125}$$

and developed an important relationship for non-equilibrium transport of non-uniform sediment, which is

$$P_{4.l}^* S^*(\omega^*) = P_{1.l} S^*(l) = K_l S^*(l) \tag{1.126}$$

For the strong equilibrium sediment transport, this equation becomes Eq. (1.82). For the case of deposition, the water percentage needs to be adjusted, while the other parameters can still be formulated with Eqs. (1.111) through (1.114). This will be discussed in detail later in the review of Le's [40] study and Chap. 7.

The results of Han [14] illustrate some mechanisms of the non-equilibrium transport of non-uniform sediment, and particularly the characteristics for the exchange between the coarser particles and the finer particles of the suspended load and the bed material. These results have been applied to the mathematical models with satisfactory results [25], and were further thoroughly investigated by He and Han [28, 29]. The two shortcomings of these results are as follows: (1) The results are mainly for the case of scour and for the case of mild scour or mild deposition. For massive deposition or massive scour, the formulations for the size distribution for carrying-sediment capacity were provided by Han [14], He and Han [28, 29] without a unified derivation and elaboration. (2) The results are for a given cross-section, not for a reach. These two problems of size distribution for carrying-sediment capacity will be specifically studied in Chap. 7.

2. **Dou et al.** [8]**, and Dou et al.** [9]

The following comments are made for the results of the size distribution for carrying-sediment capacity by Dou et al. [8] and Dou and Zhao [9]:

(1) It is incorrect to determine ω of Dou's formula Eq. (1.58) utilizing Eq. (1.61) with $P_{4.l}$. Based on Eq. (1.82).

$$P_{1.l.1} = \frac{P_{4.l}^* S^*(\omega^*)}{S^*(l)}$$

For the uniform sediment with $P_{4.l}^* = 1$, $S_l^* = S^*(l)$, and $\omega = \omega_l$, Eq. (1.58) becomes

$$S^*(l) = \frac{K_1}{C_0^2} \frac{\gamma_s \gamma}{\gamma_s - \gamma} \frac{V^3}{gh\omega_l}$$

Combining these two equations and summing up for l, it can be derived that

$$S^* = \frac{1}{\sum \frac{P_{4.l}^*}{S^*(l)}} = \frac{1}{\sum \frac{P_{4.l}^*}{\frac{K_1}{C_0^2}\frac{\gamma_s\gamma}{\gamma_s-\gamma}\frac{V^3}{h\omega_l}}}$$
$$= \frac{K_1}{C_0^2}\frac{\gamma_s\gamma}{\gamma_s-\gamma}\frac{V^3}{gh\sum P_{4.l}^*\omega_l} = \frac{K_1}{C_0^2}\frac{\gamma_s\gamma}{\gamma_s-\gamma}\frac{V^3}{gh\omega^*} = S^*\left(\omega^*\right) \quad (1.127)$$

where K_1 is the coefficient of carrying-sediment capacity, C_0 the dimensionless Chezy coefficient, and γ_s and γ are the specific gravities of sediment particles and of water, respectively. It is noted that, after introducing the size distribution for carrying-sediment capacity, the average fall velocity for carrying-sediment capacity is

$$\omega^* = \sum P_{4.l}^*\omega_l \quad (1.128)$$

which is different from the average fall velocity ω that is determined using Eq. (1.61) with the size distribution of suspended load. If Eq. (1.61) is used, the results are correct only for the strong equilibrium condition or for massive deposition, but not for a general non-equilibrium sediment transport.

(2) Equation (1.68) is also questionable. Based on Dou's Eq. (1.58), it is true that

$$S_l^* = P_{4.l}^* S^*(\omega^*) \quad (1.129)$$

with ω^* being replaced with ω. On the other hand, according to Eq. (1.82)

$$S_l^* = P_{1.l.1} S^*(l) = K_{1.l} S^*(l) \quad (1.130)$$

which shows the scoured sediment is proportional to the bed material gradation, the same as that presented by Dou (Eq. 1.63). Based on Eqs. (1.129) and (1.130), Eq. (1.82) can be written as

$$P_{4.l}^{*\gamma} S^{*\gamma}\left(\omega^*\right) = P_{1.l.1}^{\gamma} S^{*\gamma}\left(\omega_l\right) \quad (1.131)$$

Thus,

$$S^{*\gamma}\left(\omega^*\right)\sum_{l=1}^{n} P_{4.l}^{*\gamma} = \sum_{l=1}^{n} P_{1.l.1}^{\gamma} S^{*\gamma}\left(\omega_l\right)$$

or

$$\frac{P_{4.l}^{*\gamma}}{\sum_{l=1}^{n} P_{4.l}^{*\gamma}} = \frac{\frac{P_{1.l.1}^{\gamma}}{\omega_l^{\gamma}}}{\sum_{l=1}^{n} \frac{P_{1.l.1}^{\gamma}}{\omega_l^{\gamma}}} \quad (1.132)$$

It is noted that Eq. (1.132) has a similar form as Eq. (1.68), and the difference lies in the exponential component γ for $P_{4.l}^*$. Hence, Eq. (1.68) is only correct for $\gamma = 1$ and for this case Eq. (1.132) is the same as the results that were derived by Han [11] for the strong equilibrium condition with $m = 1$. However, based on Dou et al.'s [8], Dou and Zhao [9] analysis, $\gamma = 1/6$, the difference between $\left(P_{4.l}^*\right)^{1/6}$ and $P_{4.l}^*$ is significant, and the bed material gradation only has insignificant impact on the size distribution for carrying-sediment capacity.

(3) For the strong equilibrium condition, $P_{4.l} = P_{4.l}^*$, and based on Eq. (1.59),

$$P_{4.l} = P_{4.l}^* = \frac{\left(\frac{P_{4.l}}{\omega_l}\right)^\gamma}{\sum\left(\frac{P_{4.l}}{\omega_l}\right)^\gamma} = \left(\frac{\omega'}{\omega_l}\right)^\gamma P_{4.l}^\gamma$$

This result is paradoxical, which can be written as

$$P_{4.l}^{1-\gamma} = \left(\frac{\omega'}{\omega_l}\right)^\gamma \tag{1.133}$$

where

$$\frac{1}{\omega'^\gamma} = \sum\left(\frac{P_{4.l}}{\omega_l}\right)^\gamma \tag{1.134}$$

This condition is difficult to be satisfied. Based on Dou's analysis, $\gamma = 1/6$, then $1 - \gamma > 0$, and $P_{4.l}^{1-\gamma} < 1$. For the finer sediment particles with $\omega_l < \omega'$, $(\omega'/\omega_l)^\gamma > 1$, and at least Eq. (1.133) cannot be satisfied. The left side of this equation is less than 1, while the right side is greater than 1. Hence, Eq. (1.59), from which Eq. (1.133) is derived, is also questionable.

3. **Ouyang et al.** [49]

Inspired by Dou et al.'s [8, Dou and Zhao 9] studies, Ouyang et al. [49] proposed the following formula for the size distribution for carrying-sediment capacity:

$$P_{4.l}^* = \frac{\left(\frac{P_{4.l}}{\omega_l}\right)^\gamma}{\sum\left(\frac{P_{4.l}}{\omega_l}\right)^\gamma} \tag{1.135}$$

and the following formula for the fractional carrying-sediment capacity:

$$S_l^* = P_{4.l}^* S^* = P_{4.l}^* k\left(\frac{V^3}{gh\omega}\right)^m \tag{1.136}$$

$$\omega^m = \sum P_{4.l}\omega_l^m \tag{1.137}$$

However, the differentiation equation for the non-equilibrium sediment transport adopted by Ouyang et al. was Eq. (1.51) that was developed by Han [11, 13], not Eq. (1.57) that was developed by Dou et al. [8]. The problems associated with Eq. (1.135) are summarized as follows:

(1) Equation (1.135) was directly proposed without derivation. Although Dou's formula for the size distribution for carrying-sediment capacity was adopted, Ouyang et al. used a different governing equation for non-equilibrium sediment transport, and thus the results for scour are different between Dou et al.'s [8], Dou and Zhao [9] and Ouyang et al. [49]. For the case of scour, Dou considered the impact of the bed material gradation $P_{1.l.1}$ on the non-equilibrium sediment transport, while Ouyang et al. did not. Although Ouyang et al. somewhat considered the impact of $P_{1.l.1}$ in the computation of $P_{4.l}$, this is different from Dou's analysis.
(2) For the equilibrium condition, $P_{4.l}^* = P_{4.l}$, and Eq. (1.135) yields the same paradoxical results as Eq. (1.59) does. Based on Eq. (1.135),

$$P_{4.l} \neq \frac{\left(\frac{P_{4.l}}{\omega_l}\right)^{\gamma}}{\sum\left(\frac{P_{4.l}}{\omega_l}\right)^{\gamma}} = P_{4.l}^* \tag{1.138}$$

which contradicts with the strong equilibrium condition. If Eq. (1.135) is forced to be true for the equilibrium condition, then in the neighborhood of the equilibrium point

$$\lim_{S_l^*-0\to S_l} P_{4.l}^* = \frac{\left(\frac{P_{4.l}}{\omega_l}\right)^{\gamma}}{\sum\left(\frac{P_{4.l}}{\omega_l}\right)^{\gamma}} \tag{1.139}$$

$$\lim_{S_l^*+0\to S_l} P_{4.l}^* = \frac{\left(\frac{P_{4.l}}{\omega_l}\right)^{\gamma}}{\sum\left(\frac{P_{4.l}}{\omega_l}\right)^{\gamma}} \tag{1.140}$$

It is noted that while the limits of $P_{4.l}^*$ on the left and on the right of the equilibrium point are the same, they are different from the value $P_{4.l}$ at equilibrium point. In the neighborhood of the equilibrium point, the value of $P_{4.l}^*$ jumps from $\left(\frac{P_{4.l}}{\omega_l}\right)^{\gamma} / \sum\left(\frac{P_{4.l}}{\omega_l}\right)^{\gamma}$ for deposition, to $P_{4.l}$ for the equilibrium condition, and then jump back to $(\frac{P_{4.l}}{\omega_l})^{\gamma} / \sum(\frac{P_{4.l}}{\omega_l})^{\gamma}$ for scour. As a result, dS_l/dx in the governing equation for non-equilibrium sediment transport is discontinuous at the equilibrium point.

(3) As pointed out before, the carrying-sediment capacity should be computed with the average fall velocity ω^* determined based on the size distribution for carrying-sediment capacity, not with the average fall velocity ω based on the size distribution of suspended load.

(4) It is still not satisfactory to consider Eq. (1.135) as an empirical formula. In addition to the theoretical problems, the results from this formula with $\gamma = 1/6$ for the approximate equilibrium condition show a great and unidirectional discrepancy from the data. Table 1.1 list the comparison of measured $P_{4.l}$ and $P^*_{4.l}$ computed with Eq. (1.135) for the Yichang, Hankou, and Datong stations along the Yangtze River. The following can be observed from this table:

(i) $P_{4.l}$ listed in this table is the long-term averaged size distribution of suspended load. Since sediment transport is approximately equilibrium at these three stations, the computed $P^*_{4.l}$ should be close to $P_{4.l}$. However, as listed in this table, $P^*_{4.l}$ is greater than $P_{4.l}$ for the finer sediment particles with $D < 0.01$ mm and for the coarser particles with $D > 0.25$ mm, while $P^*_{4.l}$ is less than $P_{4.l}$ for the intermediate size class.

(ii) ω^* is generally greater than ω for these three stations, indicating a unidirectional discrepancy.

(iii) The carrying-sediment capacity is incorrectly computed with ω instead of ω^*. However, this theoretically incorrect method ironically reduces the error in the computed carrying-sediment capacity. As the Ouyang et al.'s formula for $P^*_{4.l}$ is essential the same as the formula developed by Dou et al. [8] and Dou and Zhao [9], the problems associated with Dou et al.'s formula also apply to Ouyang et al.'s formula for $\gamma = 1/6$.

For comparison, Table 1.2 lists the results of $P^*_{4.l}$ computed with Eq. (1.82) that was developed by Han [17]. As listed in this table, the computed $P^*_{4.l}$ shows good agreement with $P_{4.l}$ except for the finest size class.

4. **Huang** [32]

Huang [32] proposed a formula for the fractional carrying-sediment capacity. For a given reach with known water and sediment influxes and size distribution of

Table 1.1 Comparison between $P_{4.l}$ and $P^*_{4.l}$ computed with Ouyang's formula

Diameter (mm)	ω_l (m/s)	Yichang		Hankou		Datong	
		$P_{4.l}$	$P^*_{4.l}$	$P_{4.l}$	$P^*_{4.l}$	$P_{4.l}$	$P^*_{4.l}$
< 0.005	9.50×10^{-6}	0.084	0.2519	0.144	0.2794	0.117	0.2720
0.005–0.01	5.36×10^{-5}	0.111	0.1978	0.137	0.2077	0.147	0.2118
0.01–0.025	0.000299	0.189	0.1623	0.226	0.1695	0.215	0.1694
0.025–0.05	0.00134	0.271	0.1342	0.235	0.1329	0.248	0.1351
0.05–0.10	0.00536	0.233	0.1039	0.184	0.1013	0.222	0.1053
0.10–0.25	0.0172	0.087	0.0726	0.07	0.0710	0.045	0.0664
0.25–0.5	0.0404	0.024	0.0508	0.004	0.0382	0.005	0.0400
0.5–1.0	0.0828	0.001	0.0265	0	0	0	0
ω or ω^* (m/s)		0.00389	0.0550	0.00251	0.00311	0.00237	0.00313

Table 1.2 Comparison between $P_{4.l}$ and $P^*_{4.l}$ formula computed with Han's formula

Diameter (mm)	Prior to reservoir operation		Scour in year 1–10		Scour in year 11–20	
	$P_{4.l}$	$P^*_{4.l}$	$P_{4.l}$	$P^*_{4.l}$	$P_{4.l}$	$P^*_{4.l}$
< 0.005	0.144	0.130	0.171	0.215	0.187	0.300
0.005–0.01	0.137	0.123	0.149	0.151	0.162	0.145
0.01–0.025	0.226	0.203	0.156	0.141	0.169	0.140
0.025–0.05	0.235	0.212	0.087	0.075	0.089	0.074
0.05–0.10	0.184	0.184	0.104	0.087	0.060	0.048
0.10–0.25	0.070	0.141	0.326	0.323	0.323	0.282
0.25–0.5	0.004	0.007	0.007	0.009	0.011	0.011
0.5–1.0	0	0	0	0	0	0

suspended load at the entrance section, the fractional carrying-sediment capacity for the l-th size class can be computed with

$$S_l^* = P_{0.l} S^*(l) \tag{1.141}$$

where S_l^* is the fractional carrying-sediment capacity for the l-th size class of the non-uniform sediment mixture, $P_{0.l}$ is the size distribution of suspended load at the entrance section, $S^*(\omega_l) = S^*(l)$ is the capacity for carrying the uniform sediment with the grain size equal to the l-th size class, which can be determined with

$$S^*(\omega_l) = \kappa\left(\frac{V^3}{gh\omega_l}\right)$$

It is noted that Eq. (1.141) is inconsistent with Eq. (1.86) or Eq. (1.107). In other words, it is different from

$$S_l^* = P_{1.l} S^*(\omega_l) \tag{1.142}$$

As shown in Eq. (1.141), the sum of the fractional carrying-sediment capacity is based on the weighting factor of the size distribution of suspended load. According to Huang [32], $P_{0.l}$ in Eq. (1.41) is actually the size distribution of suspended load that is denoted as $P_{4.l}$ in this section. Thus,

$$P_{4.l} S^*(\omega) = S_l^* = P_{0.l} S^*(\omega_l) = P_{4.l} S^*(\omega_l) \tag{1.143}$$

which leads to

$$S^*(\omega) = S^*(\omega_l) \tag{1.144}$$

Apparently, the average fall velocity ω cannot equal to the fall velocity ω_l for each size class. It is thus concluded that Eq. (1.141) is incorrect.

5. **Liu et al.** [42]

Liu et al. [42] proposed a method for summing up the fractional carrying-sediment capacity for non-uniform sediment after considering the case of high sediment concentration. The proposed formula for the total carrying-sediment capacity is

$$S_V^* = \kappa\left[\frac{\gamma_m}{\gamma_s - \gamma_m}\frac{V^3}{gh\omega^*}\left(1 + \frac{\gamma_s - \gamma_m}{\gamma_m}S_{V_u}\right)\mu_r\right]^{m_1} \tag{1.145}$$

where γ_m is specific weight of the muddy water, μ_r the correction factor for sediment fall velocity, S_V^* the volumetric carrying-sediment capacity, and S_{V_u} the volumetric sediment concentration from upstream. For strong equilibrium sediment transport, $S_{V_u} = S_V^*$, and

$$S_V^* = \kappa\left[\frac{\gamma_m}{\gamma_s - \gamma_m}\frac{V^3}{gh\omega^*}\left(1 + \frac{\gamma_s - \gamma_m}{\gamma_m}S_V^*\right)\mu_r\right]^{m} \tag{1.146}$$

The formula proposed by Liu et al. for the fractional carrying-sediment capacity is

$$S_{V.l}^* = \kappa P_{n.l}^m\left[\frac{\gamma_m}{\gamma_s - \gamma_m}\frac{V^3}{gh\omega_l}\left(1 + \frac{\gamma_s - \gamma_m}{\gamma_m}\frac{S_{V.l}^*}{P_{n.l}}\right)\mu_r\right]^{m} \tag{1.147}$$

where $P_{n.l}$ is the power percentage, and Liu et al. assumed $P_{n.l} = P_{1.l.1}$. Hence,

$$S_{V.l}^* = \kappa P_{1.l.1}^m\left[\left(\frac{\gamma_m}{\gamma_s - \gamma_m}\frac{V^3}{gh\omega_l}\right)\left(1 + \frac{\gamma_s - \gamma_m}{\gamma_m}\frac{S_{v.l}^*}{P_{1.l}}\right)\mu_r\right]^{m} \approx P_{1.l.1}^m S_V^*(l) \tag{1.148}$$

where $S_V^*(l)$ is the "potential" carrying-sediment capacity for the l-th size class, which is in fact the capacity for carrying the uniform sediment with the grain size equal to the l-th size class. The method proposed by Liu et al. for summing up the carrying-sediment t capacity are discussed as follows:

(1) Liu et al. assumed the power percentage equals the bed material gradation, or

$$P_{n.l} = P_{1.l.1} \tag{1.149}$$

This is true for the strong equilibrium condition, as proved by Han and He [26]. However, this is not true for non-equilibrium sediment transport. In fact, summing up the fractional carrying-sediment capacity that is expressed with Eq. (1.148), or

$$S_{V.l}^* \approx P_{1.l.1}^m S_V^*(l)$$

for all size classes yields

$$S_V^* = \sum_{l=1}^{n} S_{V.l}^* = \sum_{l=1}^{n} P_{1.l.1}^m S_V^*(l) \tag{1.150}$$

It is noted from Eq. (1.150) that the carrying-sediment capacity is independent of the size distribution of suspended load for the case of deposition and is only a function of the bed material gradation. This is lack of theoretical basis.

(2) In Eq. (1.148), $m \neq 1$, and thus the sum of the weighting factor $\sum_{l=1}^{n} P_{1.l.1}^m \neq 1$. This is questionable, as discussed before. Although differentiating $P_{4.l}$ from $P_{4.l}^*$ and using the average fall velocity ω^*, Liu et al. determined ω^* with

$$\omega^* = \sum_{l=1}^{n} P_{1.l.1}^m \omega_l = \sum_{l=1}^{n} P_{4.l}^{*m} \omega_l \tag{1.151}$$

in which m is an exponent. Since $\sum_{l=1}^{n} P_{4.l}^{*m} \neq 1$, Eq. (1.151) is also lack of theoretical basis.

(3) Liu et al. considered their results as expressed in Eq. (1.148), or

$$S_{V.l}^* = P_{1.l.1}^m S_V^*(l) \tag{1.152}$$

is inconsistent with

$$S_{V.l}^* = P_{1.l.1} S_V^*(l) \tag{1.153}$$

Hence, Liu et al. concluded Eq. (1.153) has no theoretical or experimental basis without thorough derivation. In fact, Eq. (1.153) is the same as Eq. (1.82) that was developed by Han [11, 13]. Equation (1.153) can actually be proved with the concepts developed by Liu et al. [42]. Based on Eq. (1.148), the fractional carrying-sediment capacity for the l-th size class can be written as

$$S_V^*(l) = \kappa\left[\left(\frac{\gamma_m}{\gamma_s - \gamma_m}\frac{V^3}{gR\omega_l}\right)\mu_r\left(1 + \frac{\gamma_s - \gamma_m}{\gamma_m}S_{V.l}\right)\right]^m \approx k\left(\frac{\gamma_m}{\gamma_s - \gamma_m}\frac{V^3}{gR\omega_l}\right)^m \tag{1.154}$$

The approximately equal sign is used here to denote the case of low sediment concentration, with $1 + \frac{\gamma_s - \gamma_m}{\gamma_m} S_{V.l}^* \approx 1$. Summing up Eq. (1.154) for l with the weighting factor of the power percentage $P_{1.l.1}$ yields

$$\begin{aligned} S_V^* &= \sum_{l=1}^{n} S_{V.l}^* = \sum_{l=1}^{n} P_{1.l.1} S_V^*(l) = k\left[\left(\frac{\gamma_m}{\gamma_s - \gamma_m}\right)^m \sum \frac{P_{1.l.1}}{\omega_l^m}\right] \\ &= k\left(\frac{\gamma_m}{\gamma_s - \gamma_m}\right)^m \left(\frac{V^3}{gR\omega^*}\right)^m \end{aligned} \tag{1.155}$$

in which

$$\frac{1}{\omega^{*m}} = \sum_{l=1}^{n} \frac{P_{1.l.1}}{\omega_l^m} \tag{1.156}$$

Equation (1.155) indicates that Eq. (1.153) is correct based on the derivation with Liu et al.'s results, at least for the case of low sediment concentration, and that Eq. (1.152) is questionable. Also, considering the definition of the fractional carrying-sediment capacity and Eq. (1.154), one obtains

$$S_{V.l}^* = P_{4.l}^* S_V^* = P_{1.l.1} S_V^*(l) \tag{1.157}$$

Then,

$$\frac{1}{S_V^*} = \sum \frac{P_{4.l}^*}{S_V^*(l)} = \frac{1}{k\left(\frac{\gamma_m}{\gamma_s - \gamma_m}\right)^m \left(\frac{V^3}{gR}\right)^m} \sum_{l=1}^{n} P_{4.l}^* \omega_l^m$$

or

$$S_V^* = k\left(\frac{\gamma_m}{\gamma_s - \gamma_m}\right)^m \left(\frac{V^3}{gR\omega^*}\right)^m \tag{1.158}$$

$$\omega^{*m} = \sum_{l=1}^{n} P_{4.l}^* \omega_l^m \tag{1.159}$$

This is the generally accepted result. Hence, it lacks of generality and basis to replace $P_{1.l}$ with $P_{1.l}^m$. Also, Eqs. (1.156) or (1.159) is different from Liu et al.'s formula for the average sediment fall velocity for sediment transport, which is

$$\omega^* = \sum_{l=1}^{n} P_{1.l.1}^m \omega_l = \sum_{l=1}^{n} P_{4.l}^{*m} \omega_l \tag{1.160}$$

6. **Yang** [59]**, and Yu et al.** [61]

Yang [59] and Yu et al. [61] proposed the following formula for the fractional carrying-sediment capacity

$$S_{V.l}^* = k_0 \mu_r^\alpha \left[\frac{\omega_l}{\omega}(P_{4.l} + P_{1.l.1})\right]^\beta \left(\frac{V^3}{\frac{\gamma_s - \gamma}{\gamma} gR\omega_l}\right)^m \tag{1.161}$$

where γ is the specific weight of the muddy water, $S_{V.l}^*$ the fractional carrying-sediment capacity in terms of the volumetric concentration, ω the average fall velocity of the non-uniform sediment mixture, μ_r the relative viscosity coefficient, and the exponents β and m are approximately 1 if the sediment grain size is very small.

According to Yang [59] and Yu et al. [61], $P_{4.l} = 0$ for the case with no sediment influx from upstream, and $P_{1.l.1} = 0$ for the case with no local sediment supplement from the bed material. Discussions on Eq. (1.161) are as follows:

(1) For the uniform sediment with the grain size of D_l, Eq. (1.161) becomes

$$S_V^*(l) = 2^\beta k_0 \mu_r^\alpha \left(\frac{V^3}{\frac{\gamma_s - \gamma}{\gamma} g R \omega_l} \right)^m \tag{1.162}$$

Comparing Eq. (1.161) with Eq. (1.162), one obtains

$$S_{V.l}^* = \left(\frac{\omega_l}{\omega} \frac{P_{4.l} + P_{1.l.1}}{2} \right)^\beta S_V^*(l) \tag{1.163}$$

It is noted that

$$\sum_{l=1}^{n} \left(\frac{\omega_l}{\omega} \frac{P_{4.l} + P_{1.l.1}}{2} \right)^\beta \neq 1 \tag{1.164}$$

which contradicts with the general criterion for weighted summation.

(2) For the case with no sediment influx from upstream, the fractional carrying-sediment capacity is

$$S_{V.l}^* = \left(\frac{\omega_l}{\omega} \frac{P_{1.l.1}}{2} \right)^\beta S_V^*(l) \tag{1.165}$$

This is unreasonable even for $\beta = 1$, because only an half of the bed material gradation is used as the weighting factors. Similarly, for the case with no sediment supplement from the bed material,

$$S_{V.l}^* = \left(\frac{\omega_l}{\omega} \frac{P_{4.l}}{2} \right)^\beta S_V^*(l) \tag{1.166}$$

in which only an half of the size distribution of suspended load is used as the weighting factor.

(3) For uniform sediment, $S_{V.l}^* = S_V^* = S_V^*(l)$, and Eqs. (1.165) and (1.166) both give

$$S_V^* = \left(\frac{1}{2} \right)^\beta S_V^*(l) \neq S_V^*(l) \tag{1.167}$$

which is self-contradictory.

(4) Even neglecting the discrepancy in these coefficients, it is incorrect to sum up the fractional carrying-sediment capacity using the bed material gradation as the weighting factor. In fact, for the case with no sediment influx from upstream, $\beta = 1$, and $m = 1$, it can be derived from Eq. (1.161) that

$$S_{V.l}^{*} = k_0 \mu_r P_{1.l.1} \left(\frac{V^3}{\frac{\gamma_s - \gamma}{\gamma} g R \omega} \right) = P_{1.l.1} S^{*}(\omega) \tag{1.168}$$

On the other hand, according to the definition of the fractional carrying-sediment capacity,

$$S_{V.l}^{*} = P_{4.l}^{*} S^{*}\left(\omega^{*}\right) \approx P_{4.l} S^{*}(\omega)$$

Comparing these two equations, one obtains

$$P_{1.l.1} = P_{4.l}^{*} \tag{1.169}$$

This relation is incorrect, indicating a mistake exists in Eq. (1.161). This mistake lies in the bed material being considered as a part of the carrying-sediment capacity. However, the correct approach is to only consider part of the bed material that actually contributes to the carrying-sediment capacity, as discussed before.

7. **Le's** [40] **Study**

Le [40] considered the carrying-sediment capacity is needed to carry both the wash load and bed material load of suspended load, while their roles in sediment transport and sedimentation process and the corresponding formulations are different. Le introduced the following saturation coefficients for the total load, and for the fractional sediment of the l-th size class

$$N = \frac{S}{\rho} \tag{1.170}$$

$$N_l = \frac{S_l}{\rho_l} \tag{1.171}$$

where S is the total sediment concentration, S_l is the fractional sediment concentration for the l-th size class, ρ the total carrying-sediment capacity of the flow, $\rho(l)$ the capacity for carrying the uniform sediment with the grain size of D_l.

Under the saturated condition,

$$N = \sum_{l=1}^{n} N_l = \sum \frac{S_l}{\rho(l)} = 1 \tag{1.172}$$

Also, the over-saturated condition was defined by Le as $\sum N_l > 1$, and the sub-saturated condition as $\sum N_l < 1$. Defining the size distribution of suspended load

$$P_l = \frac{S_l}{S} \tag{1.173}$$

then the total carrying-sediment capacity under the saturated condition can be derived from Eqs. (1.172) and (1.173), which is

$$S = \rho = \left[\sum_{l=1}^{n} \frac{P_l}{\rho(l)}\right]^{-1} = \left[\sum_{l=1}^{n} \frac{P_l}{\kappa\left(\frac{V^3}{gh\omega_l}\right)^m}\right]^{-1} = \kappa\left(\frac{V^3}{gh\omega}\right)^m = f(\omega) = \rho(\omega) \tag{1.174}$$

with

$$\omega^m = \sum_{l=1}^{n} P_l \omega_l^m \tag{1.175}$$

The carrying-sediment capacities for bed material load and for wash load are $\rho_b(\omega_b)$ and $\rho_w(\omega_w)$, respectively, with

$$\omega_w^m = \sum_{l=1}^{k} \frac{P_l \omega_l^m}{P_w} = \sum_{l=1}^{k} \frac{P_l \omega_l^m}{\sum_{l=1}^{k} P_l} \tag{1.176}$$

$$\omega_b^m = \sum_{l=k+1}^{n} \frac{P_l \omega_l^m}{P_b} = \sum_{l=k+1}^{n} \frac{P_l \omega_l^m}{\sum_{l=k+1}^{n} P_l} \tag{1.177}$$

where P_w and P_b are the percentages by weight for wash load and for bed material load, respectively. The sediment in the size classes of $l \leq k$ is wash load, and the sediment in the size classes of $l \geq k + 1$ is bed material load. For the saturated condition, it can be derived from Eq. (1.172) that

$$\frac{S_w}{\rho_w} + \frac{S_b}{\rho_b} = 1 \tag{1.178}$$

The total sediment concentration is the sum of the concentration of bed material load and the concentration of wash load. Considering Eq. (1.178), the total sediment concentration can be written as

$$S = \rho = S_w + S_b = S_w + \left(1 - \frac{S_w}{\rho_w}\right)\rho_b \tag{1.179}$$

Equation (1.179) is the formula for the total carrying-sediment capacity under the saturated condition, which is equivalent to Eq. (1.174). Le considered the sediment concentration of wash load S_w did not directly participate in channel forming, and had no impact on channel evolution. The carrying-sediment capacity for bed material load is ρ_b, and $\left(1-\frac{S_w}{\rho_w}\right)\rho_b$ is the contribution of the carrying-sediment capacity for bed material load to the total sediment carrying-sediment capacity. Hence, Le called $\left(1-\frac{S_w}{\rho_w}\right)\rho_b$ "the effective carrying-sediment capacity for bed material load", which directly participate in channel forming.

Le adopted van Rjin's [58] relation for the carrying-sediment capacity for bed material load ρ_b and the corresponding size distribution $P_{*.l}$, which reads

$$P_{*.l}\rho_b = i_l\rho_l \tag{1.180}$$

where i_l is the bed material gradation. Thus,

$$\rho_b = \sum_{l=k+1}^{n} i_l\rho_l \tag{1.181}$$

van Rjin [58] seemed to assume all the bed material can be suspended. Otherwise, i_l should be normalized. Hence, Eq. (1.180) lacks of generality. The carrying-sediment capacity associated with the bed material, which was determined by van Rjin [58], was actually the capacity for sediment pick-up from the bed, or

$$\rho_* = \kappa\left(\frac{V^3}{gh\omega_*}\right)^m \tag{1.182}$$

$$\omega_*^m = \sum_{l=r+1}^{n} P_{*.l}\omega_l^m \tag{1.183}$$

For the equilibrium condition,

$$\rho_* = \rho_b \tag{1.184}$$

Thus, the total carrying-sediment capacity is

$$\rho = S_w + \left(1-\frac{S_w}{\rho_w}\right)\rho_* = S_w + \left(1-\frac{S_w}{\rho_w}\right)\rho_b \tag{1.185}$$

It is noted that Le did not presented the formula for the size distribution for carrying-sediment capacity $P_{*.l}$. This is not necessary for the strong equilibrium condition, in which $P_{*.l}$ can be determined with the bed material gradation or the size distribution of suspended load. However, this is not true for non-equilibrium sediment transport.

Le's [40] formulations for the total and fractional carrying-sediment capacities under various sedimentation conditions are discussed as follows, with further consideration of the case with only part of the bed material suspendable. Let denote the proportion of the bed width, within which the bed material can be suspended, is P_s. For sub-saturated sediment transport, or for the case of channel scour, the proportion of the bed width, within which the bed material cannot be suspended, is $(1-P_s)$. The carrying-sediment capacity within this width of $(1-P_s)$ is the sediment concentration of the influx, S_0, and thus its contribution to the sediment concentration of the entire width is $S_0(1-P_s)$. For the width (P_s) with the suspendable bed material, the sediment concentration consists of S_0 and the sediment supplement from the bed, $\left(1-\frac{S_w}{\rho_w}-\frac{S_b}{\rho_b}\right)\rho_*$, or, the total carrying-sediment capacity within this width of P_s is $S_0+\left(1-\frac{S_w}{\rho_w}-\frac{S_b}{\rho_b}\right)\rho_*$. Hence, the total carrying-sediment capacity for the entire width or cross-section is

$$\rho = S_0(1-P_s)+\left[S_0+\left(1-\frac{S_w}{\rho_w}-\frac{S_b}{\rho_b}\right)\rho_*\right]P_s = S_0+\left(1-\frac{S_0}{\rho_0}\right)\rho_* P_s \quad (1.186)$$

where ρ_0 is the carrying-sediment capacity corresponding to S_0, which can be determined with

$$\frac{S_0}{\rho_0}=\frac{S_w}{\rho_w}+\frac{S_b}{\rho_b} \quad (1.187)$$

Without directly deriving Eq. (1.187), Le [40] only qualitatively illustrated this relation for the equilibrium condition. The exact formulation of ρ_0 is derived as follows. In fact, Eq. (1.187) can be rewritten as

$$S_0=\rho_0\frac{P_w S_0\sum_{l=1}^{k}\frac{P_l\omega_l^m}{\rho_w}}{\kappa\left(\frac{V^3}{gh}\right)^m}+\rho_0\frac{P_b S_0\sum_{l=k+1}^{n}\frac{P_l\omega_l^m}{\rho_b}}{\kappa\left(\frac{V^3}{gh}\right)^m}=\rho_0 S_0\frac{\sum_{l=1}^{n}P_l\omega_l^m}{\kappa\left(\frac{V^3}{gh}\right)^m}$$

which leads to

$$\rho_0=\kappa\left(\frac{V^3}{gh\omega}\right)^m \quad (1.188)$$

with

$$\omega=\sum_{l=1}^{n}P_l\omega_l^m \quad (1.189)$$

It is noted that ω, which is used to determine ρ_0, is computed with the size distribution of suspended load. In Eq. (1.186), S_0 is equivalent to wash load, and the supplement to the sediment concentration only comes from the sediment eroded from the bed material. For the case of deposition, Le [40] recommended using Eq. (1.185).

Le [40] developed different formulas for the fractional carrying-sediment capacity for three sedimentation conditions, while no formula was developed for the corresponding size distribution. These formulas for the fractional carrying-sediment capacity were reviewed in the following. In addition, the size distribution for carrying-sediment capacity is derived based on the formulas for the fractional carrying-sediment capacity, and the results are compared with Han et al.'s [13, 14] analysis.

(1) For the case of pure deposition, $S_b > \left(1 - \frac{S_w}{\rho_w}\right)\rho_b$, and $\rho_* < \rho_b$. The fractional carrying-sediment capacity is

$$\rho_l = P_{w.l} P_w S + \left(1 - \frac{S_w}{\rho_w}\right) P_{b.l} P_b \rho_b \tag{1.190}$$

where $P_{w.l}$ and $P_{b.l}$ are the normalized size distributions of wash load and of bed material load, respectively. Summing up Eq. (1.190) over l yields

$$\rho = P_w S + \left(1 - \frac{S_w}{P_w}\right) P_b \rho_b \tag{1.191}$$

The size distribution of carrying-sediment capacity for total load is

$$P^*_{\rho.l} = \frac{\rho_l}{\rho} = P_{w.l} P_w \frac{S}{\rho} + \left(1 - \frac{S_w}{\rho_w}\right) P_{b.l} P_b \frac{\rho_b}{\rho} \tag{1.192}$$

This is also the size distribution of suspended load. It is noted that $P_{w.l}$ is the size distribution of wash load after normalizing P_l among the size classes of $1 \le l \le k$, and $P_{b.l}$ the size distribution of bed material load after normalizing P_l among the size classes of $k + 1 \le l \le n$.

(2) For the case of pure scour, $S_b < \left(1 - \frac{S_w}{\rho_w}\right)\rho_b$, and $\rho_* < \rho_b$. The fractional carrying-sediment capacity is

$$\rho_l = P_l S_0 + \left(1 - \frac{S_0}{\rho_0}\right) P_s P_{*.l} \rho_* \tag{1.193}$$

where P_s is the proportion of the bed material that is erodible. Summing up over l yields

$$\rho = S_0 + \left(1 - \frac{S_0}{\rho_0}\right) P_s \rho_* \tag{1.194}$$

Thus, the size distribution for carrying the total load capacity is

$$P_{4.l}^{*} = P_l \frac{S_0}{\rho} + \left(1 - \frac{S_0}{\rho_0}\right) P_s \frac{\rho_*}{\rho} P_{*.l} \tag{1.195}$$

Summing up Eq. (1.195) over l yields

$$\left(1 - \frac{S_0}{\rho_0}\right) P_s \frac{\rho_*}{\rho} = 1 - \frac{S_0}{\rho} \tag{1.196}$$

Then, Eq. (1.195) can be rewritten as

$$P_{4.l}^{*} = P_l \frac{S_0}{\rho} + \left(1 - \frac{S_0}{\rho}\right) P_{*.l} \tag{1.197}$$

(3) For the case with coarser sediment particles depositing and finer particles being eroded, $\rho_* > \rho_b$. The fractional carrying-sediment capacity is

$$\rho_l = P_{w.l} P_w S + \left(1 - \frac{S_w}{\rho_w}\right) P_{*.l} \rho_* \tag{1.198}$$

Summation over l is

$$\rho = P_w S + \left(1 - \frac{S_w}{\rho_w}\right) \rho_* \tag{1.199}$$

The size distribution for carrying total load capacity is

$$P_{4.l}^{*} = \frac{\rho_l}{\rho} = P_{w.l} P_w \frac{S}{\rho} + \left(1 - \frac{S_w}{\rho_w}\right) \frac{\rho_*}{\rho} P_{*.l} \tag{1.200}$$

It is noted that the fractional capacity for carrying the total load is denoted as ρ_l in Eqs. (1.190), (1.193), and (1.198), while it was denoted as $\rho_{*.i}$ in Le's [40]. This is done in order to differentiate ρ_l from the fractional capacity for carrying-bed material load that was also denoted as $\rho_{*.i}$. The denotation of the capacity for carrying-the uniform sediment with the grain size equivalent to the l-th size class is changed from ρ_l to $\rho(l)$.

Discussion on Le's [40] **Study**

Le's [40] analysis has some new ideas, such as introducing the sediment saturation, and computing the carrying-sediment capacity separately for wash load and bed material load. By analyzing his ideas and results, it is concluded that Le's [40] analysis is similar to part of the analyses by Han [11], Han et al. [14], and He and Han [28, 29], or is a special case of these studies. Part of Le's [40] analysis is still questionable, and overall his analysis for the size distribution for carrying-sediment capacity is incomplete. The similarities and differences between Le's [40] analysis and the analyses by Han et al., and the incompleteness of Le's [40] analysis are discussed as follows.

(1) The saturation coefficient introduced by Le [40] can be changed to the following form used by Han et al.

$$\begin{aligned} N_l &= \frac{S_l}{\rho_l} = \frac{P_{4.l} S}{S^*(l)} = \frac{S}{S^*(\omega)} \frac{P_{4.l} S^*(\omega)}{S^*(l)} \\ &= \frac{S}{S^*(\omega)} \frac{P_{4.l}\omega_l^m}{\omega^m} = \frac{S}{S^*(\omega)} P_{1.l.1} = \frac{S}{S^*(\omega)} K_l \end{aligned} \tag{1.201}$$

where K_l is the percentage of the water to carry the l-th size class. Equations (1.84) and (1.125) are used in the derivation of Eq. (1.201). For the equilibrium condition, $S = S^*(\omega)$, and thus

$$N_l = K_l \tag{1.202}$$

Hence, for the equilibrium condition, Eq. (1.172), or $\sum N_1 = 1$ can also be derived from $\sum K_l = 1$. However, $\sum N_i = 1$ is only true for the equilibrium condition, while $\sum K_l = 1$ is true for both the equilibrium and the non-equilibrium conditions. Hence, Eq. (1.172) is only a special case for $\sum K_l = 1$.

(2) For the strong equilibrium condition, $K_l = P_{1.l.1}$, and it can be derived from Eq. (1.84) that

$$\begin{aligned} 1 &= \sum_{l=1}^{n} K_l = \sum_{l=1}^{n} \frac{P_{4.l}\omega_l^m}{\omega^m} = \sum_{l=1}^{k} \frac{P_{4.l}\omega_l^m}{\omega^m} + \sum_{l=k+1}^{n} \frac{P_{4.l}\omega_l^m}{\omega^m} \\ &= \frac{P_{4.w}}{\omega^m} \sum_{l=1}^{k} \frac{P_{4.l}\omega_l^m}{P_{4.W}} + \frac{P_{4.b}}{\omega^m} \sum_{l=k+1}^{n} \frac{P_{4.l}\omega_l^m}{P_{4.b}} \\ &= P_{4.w}\left(\frac{\omega_w}{\omega}\right)^m + P_{4.b}\left(\frac{\omega_b}{\omega}\right)^m = P_{4.w}\frac{S^*(\omega)}{S^*(\omega_w)} + P_{4.b}\frac{S^*(\omega)}{S^*(\omega_b)} \\ &= P_{4.w}\frac{S}{S^*(\omega_w)} + P_{4b}\frac{S}{S^*(\omega_b)} = \frac{S_w}{S^*(\omega_w)} + \frac{S_b}{S^*(\omega_b)} \end{aligned} \tag{1.203}$$

This is the same as Eqs. (1.178) and (1.179). Thus, it is concluded that these two equations are only correct for the strong equilibrium condition.

(3) For pure scour, as $k = n, P_{4.w} = 1$, $P_{4.l.1} = P_{4.l}$, and $S^*(\omega_l) = S(\omega)$; $P_{4.2} = 0$, $\frac{1}{S^*(\omega_1)} = \sum_{l=1}^{n} \frac{P_{4.l}}{S^*(l)}$, $S^*(\omega^*) = \rho$, $S^*(\omega_{1.1}^*) = \rho_*$. Then, based on Eq. (1.112),

$$S^*(\omega^*) = S + \left(1 - \frac{S}{S^*(\omega)}\right) P_1 S^*(\omega_{1.1}^*) = S_0 + \left(1 - \frac{S_0}{\rho_0}\right) P_s \rho_* = \rho \tag{1.204}$$

With the corresponding relations of $S^*(\omega^*_{1.1}) = \rho_*$, $S = S_0$, $P_{4.l} = P_l$, and $P_1 = P_s$, Eq. (1.204) is equivalent to Eq. (1.194). Hence, Eq. (1.194) is a special case of Eq. (1.112). Based on Eq. (1.113), the size distribution for carrying-sediment capacity is

$$\begin{aligned} P^*_{4.l} &= P_{4.l}\frac{S}{S^*(\omega^*)} + \left[1 - \frac{S}{S^*(\omega)}\right] P_1 P_{1.l.1}\frac{S^*(\omega^*_{1.1})}{S^*(\omega^*)} \\ &= P_{\rho.l}\frac{S_0}{\rho} + \left(1 - \frac{S_0}{\rho_0}\right) P_s P_{*.l}\frac{\rho_*}{\rho} \end{aligned} \tag{1.205}$$

which is the same as Eq. (1.195). Hence, Eq. (1.195) is a special case of Eq. (1.114) with only two terms included.

(4) For pure deposition, the third term of Eq. (1.112) vanishes, and the condition $K_1 + K_2 > 1$ is satisfied. Thus,

$$S^*\left(\omega^*\right) = SP_{4.1} + P_{4.2}S\frac{S^*\left(\omega^*_2\right)}{S^*\left(\omega^*_{1.1}\right)} \tag{1.206}$$

Based on Eq. (1.114), the sum of the water percentages is not equal to 1, or

$$\begin{aligned} \sum_{l=1}^{n} P_{1.l} &= P_{4.l}S\sum_{l=1}^{k}\frac{P_{4.l.1}}{S^*(l)} + P_{4.2}S\sum_{l=k+1}^{n}\frac{P_{4.l.2}}{S^*\left(\omega^*_{1.1}\right)} \\ &= \frac{P_{4.1}S}{S^*(\omega_1)} + \frac{P_{4.2}S}{S^*\left(\omega^*_{1.1}\right)} = K_1 + K_2 > 1 \end{aligned}$$

Hence, an adjustment is needed to satisfy $K_1 + K_2 = 1$, and Eq. (1.206) should be changed to

$$S^*\left(\omega^*\right) = P_{4.l}S_1 + \left[1 - \frac{P_{4.1}S}{S^*(\omega_1)}\right] S^*\left(\omega^*_2\right) \tag{1.207}$$

It is noted that $S^*(\omega^*_2)$ is determined with Eq. (1.120), and is the capacity for carrying the coarser suspended sediment particles that settle down on the bed and then are eroded. It is different from the carrying-sediment capacity that directly corresponds to the concentration of bed material load, which is

$$S^*(\omega_2) = \sum_{l=k+1}^{n}\frac{P_{4.l.2}}{S^*(\omega_l)} \tag{1.208}$$

If the settling of the coarser suspended particles during the sediment exchange process is neglected, and these particles are assumed to directly change to the carrying-sediment capacity, then $S^*(\omega_2) = S^*(\omega^*_2)$. Considering the corresponding relationships of the parameters between Le [40] and Han et al.'s [13, 14], such as

$S_w = P_1 S$, $P_w = S * (\omega_1)$, and $P_b = S^*$, Eq. (1.207) or Eq. (1.112) becomes

$$S^*(\omega^*) = P_w S + \left(1 - \frac{S_w}{\rho_w}\right)\rho_b \tag{1.209}$$

This is the same as Eq. (1.191). Considering $P_{4.l.1} = \frac{P_{4.l}}{P_{4.1}}\frac{P_{4.l}}{P_w} = P_{w.l}$, and $P_{4.l.2} = \frac{P_{4.l}}{P_{4.2}}\frac{P_{4.l}}{P_b} = P_{b.l}$, the size distribution for carrying-sediment capacity can be derived from Eq. (1.113), and is

$$\begin{aligned} P^*_{4.l} &= P_{4.l} P_{4.l.1} \frac{S}{S^*(\omega^*)} + \left[1 - \frac{P_{4.1} S}{S^*(\omega_1)}\right] \frac{S^*(\omega_2)}{S^*(\omega^*)} P_{4.l.2} \\ &= P_{w.l} P_w \frac{S}{\rho} + \left(1 - \frac{S_w}{\rho_w}\right) P_{b.l} P_b \frac{\rho_b}{\rho} \end{aligned} \tag{1.210}$$

This is the same as Eq. (1.192). Hence, Eq. (1.192), which was developed by Le [40] for pure deposition, is still a special case for Han's results for the case with no sediment exchange between the suspended load and the bed material.

(5) For the case with coarser sediment particles depositing and finer particles being eroded, the second term of Eq. (1.112) vanishes, or $P_{4.2} = 0$. Assuming all bed material are suspendable, and considering $S^*(\omega^*_{1.1}) = S^*(\omega^*_1) = \rho_*$, and $P^*_{4.l.1.1} = P^*_{4.l.1} \rho_{*.l}$ for this case, then

$$S^*(\omega^*) = S P_{4.1} + \left[1 - \frac{P_{4.1} S}{S^*(\omega)}\right] S^*(\omega^*_{1.1}) = P_w S + \left(1 - \frac{S_w}{\rho_w}\right)\rho_* \tag{1.211}$$

This is the same as Eq. (1.199). Correspondingly, the size distribution for carrying-sediment capacity can be derived from Eq. (1.113), and is

$$\begin{aligned} P^*_{4.l} &= P_{4.l} P_{4.l.1} \frac{S}{S^*(\omega^*)} + \left[1 - \frac{P_{4.1} S}{S^*(\omega_1)}\right] P^*_{4.l.1} \frac{S^*(\omega^*_1)}{S^*(\omega^*)} \\ &= P_{w.l} P_w \frac{S}{\rho} + \left(1 - \frac{S_w}{\rho_w}\right) P_{*.l} \frac{\rho_*}{\rho} \end{aligned} \tag{1.212}$$

This is the same as Eq. (1.200). Hence, Le's [40] analysis is also a special case of Han's results for the case with coarser sediment particles depositing and finer particles being eroded.

The following conclusions can be made based on the comparisons above:

(1) Assimilating Han et al.'s [11, 14], He and Han [28, 29] results, but with somewhat different derivation, Le [40] obtained a full set of formulations for the total and fractional carrying-sediment capacities for non-uniform sediment, and part of his results are consistent with Han et al.'s analyses. The cross-check of these studies increases the reliability of the results. While Le's analysis has some new ideas, it is not comprehensive and profound.

(2) Compared to Le's results, the analyses by Han et al. apply to more general cases.
(3) Le's analysis lacks of completeness and integrity. Although his analysis involves many aspects, these aspects are not integrated and do not cover the complete characteristics of the capacity for carrying non-uniform sediment. The main problems of Le's analysis are discussed as follows:

(i) The three sediment transport modes proposed by Le [40], including pure deposition, pure scour, and the mode with coarser sediment particles depositing and finer particles being eroded, do not cover all patterns for sediment transport. Also, these three modes are discrete, with unknown gaps in between. For example, a gap exists between his pure deposition mode and pure scour mode. Based on Eq. (1.194) for the pure scour mode, as $\rho_* \to \rho_b$

$$\rho \to S_0 + \left(1 - \frac{S_0}{\rho_0}\right) P_s \rho_b$$

which does not approach Eq. (1.191) that is for the pure deposition mode.

(ii) The coarser and finer sediment particles exchange in natural streams, which was barely considered in Le's [40] analysis. For the case of deposition, Le considered the finer sediment particles ($l < k$) do not settles on the bed, which is consistent with the traditional concept for wash load. However, it is difficult to understand the coarser sediment particles ($l > k$) also do not settles on the bed for the case of scour, as assumed by Le. Even the bed material gradation, the size distribution of suspended load, and the hydraulic parameters are the same, the sediment concentration still should be different between the case of deposition and the case of scour. For the case of scour, the sediment pick-up capacity ρ_* as defined by Le does not include the finer sediment particles, while they exist in the bed material. Le used "pure" to set a constraint for the modes included in his analysis. While his analysis already reached a certain in-depth extent, it is unfortunate for Le to still limit his analysis to pure deposition and pure scour, without extending to a general case.
(iii) Le [40] considered that wash load did not directly participate in bed forming, and had no impact on channel evolution. This is consistent with the traditional concept. On the other hand, in the calculation of capacity for carrying the total load, Le used Eq. (1.175) to determine the average fall velocity, which included the finer sediment particles. Le also defined the capacity ρ_w for carrying wash load. It is conceptually contradictory to consider capacity exists for carrying wash load while wash load does not participate in bed forming. For example, for $S < \rho$, scour occurs. If S_w is higher, then S_b and ρ_b are smaller, resulting in less scour; if S_w is lower, then S_b and ρ_b become larger, resulting in more scour. This indicates that S_w indeed impacts the channel evolution, and it is unreasonable to assume it does not participate in bed forming. Even this impact is indirectly, it should not be ignored. The traditional concept of differentiating wash load from bed load did not involve the capacity for carrying wash load,

and thus the above contradiction did not exist. When Le corrected part of this traditional concept, the contradiction emerged. Also, for the case of scour, the finer particles of bed material are eroded, and the scour can be significant if these finer particles are exposed. Hence, the wash load does participate in bed forming in reality. Le only introduced the carrying capacity, or pick-up capacity, for the sediment eroded from the bed material, and ignored the possibility for the wash load to be eroded. In order to obtain more reasonable formulations for sediment transport and resulting deposition or scour, a new concept should be developed. This new concept should acknowledge the obvious differences in phenomena between wash load and bed material load, and, on the other hand, should consider the wash load follows the same law of transport as the bed material load does, has transport capacity, and participate in bed forming. This can be seen from the first term $SP_{4.l}$ and the third term $S^*\left(\omega^*_{1.1}\right)$ on the right side of Eq. (1.112). The in-depth discussions on this new concept will be presented in Chaps. 7 and 10.

(iv) Le [40] divided sediment transport into three modes: pure deposition, pure scour, and the mode with coarser sediment particles depositing and finer particles being eroded. As discussed before, these three modes cannot cover all patterns for sediment transport, and cannot be integrated into one unity with a common law. As these three modes were considered separately by Le, some conditions that lack of basis were added, resulting in some misunderstandings and problems. In fact, ρ is the carrying-sediment capacity under the strong equilibrium condition, ρ_b is the carrying-sediment capacity for bed material, and ρ_* and $\rho = S_0 + (1 - S_0/\rho_0)\rho_*$ are the carrying-sediment capacities under the non-equilibrium condition. Based on Le's study, formulations for general carrying-sediment capacity and corresponding size distribution can be further developed for the non-equilibrium condition. With various parameters, the sediment transport can be divided into the three modes as discussed above and even more modes, with corresponding carrying-sediment capacities and the size distributions. This can be seen in Eqs. (1.112) and (1.113), or in Chap. 7 for details. The major irrationality and disputation for the formulas developed by Le [40] for the carrying-sediment capacity are summarized as follows:

(a) For the case of pure scour,

$$\rho = S_0 + \left(1 - \frac{S_0}{\rho_0}\right) P_s \rho_* > \rho_c = S_w + \left(1 - \frac{S_w}{\rho_w}\right) \rho_b$$

should be true. This is because $\rho = \rho_c$ for the strong equilibrium condition, and thus ρ should be greater than ρ_c for the case of scour. Assuming $S_w = S_0$, and considering $\rho > \rho_c$, then $P_s \rho_* > \rho_b$. For $P_s = 1$, $\rho_* > \rho_b$. This is different from the relation Le proposed, which was $\rho_* < \rho_b$.

(b) For the case of deposition, the carrying-sediment capacity is $\rho = S_w + (1 - S_w/\rho_w)\rho_b$. Based on this formula, S_w/ρ_w is less than 1 for the strong equilibrium condition, which is reasonable. However, for some special conditions of non-equilibrium sediment transport, S_w/ρ_w may exceed 1, and thus the formula for carrying-sediment capacity needs to be corrected.

(c) The requirement of $\rho_* < \rho_b$ for pure deposition is not necessary. In fact, for $S_w = S_0$, it is true for $\rho < \rho_c$. Then, for $P_s = 1$, $\rho_* < \rho_b$ is always satisfied.

1.5 Sediment Saturation Recovery Coefficient

The sediment saturation recovery coefficient, α, is included in the governing equations for non-equilibrium suspended sediment transport, no matter these equations are one-dimensional (1-D), two-dimensional (2-D), or three-dimensional (3-D). For the 1-D equation, α reflects the rate of variation in dS/dx, or how fast the average sediment concentration adjusts from the non-equilibrium condition to the equilibrium condition. For the vertically 2-D equation, α represents how fast the bottom carrying-sediment capacity S_b^* can be recovered from the bottom sediment concentration. For the horizontal 2-D equation, α is included in the term for sediment deposition or scour. If it is integrated over the water depth, then α changes to the saturation recovery coefficient for the 1-D equation. For the 3-D equation, α is included in the bottom boundary condition. Of course, the meanings and the formulations for α vary for different conditions. The following review focuses on α_l that is used in the terms for sediment deposition or scour, such as $\alpha_l\omega_l(S_l - S_l^*)$, $\alpha_l\omega_l(S_{b.l} - S_{b.l}^*)$, where $S_{b.l}$ and $S_{b.l}^*$ are the fractional bottom sediment concentration and carrying-sediment capacity, respectively, and $\omega_l(\alpha_l S_l - \alpha_l^* S_l^*)$ for the more general cases.

1. **Mikhayev and Yunevich** [48]

As discussed before, the formula (Eq. 1.33) developed by Mikhayev and Yunevich [48] for the saturation recovery coefficient is

$$\alpha = \left(\frac{u\uparrow}{\omega} - n\right) = \left(\frac{u\uparrow}{\omega} - 3\right)$$

As shown in this equation, α can be greater or lower than 1. α is greater for the finer sediment particles, and is smaller for the coarser particles. The constraint condition for this equation is $u\uparrow/\omega > 3$.

2. **Karauschev** [38]

The formula developed by Karauschev [38] for α is Eq. (1.38), or

$$\alpha = \frac{\omega + K}{\omega}$$

Based on this equation, α is always greater than 1, and α increases as ω decreases, or the saturation recovery coefficient is greater for the finer sediment particles.

3. **Dou** [7]

Dou [7] considered α to be the sediment settling probability as shown in Eq. (1.41),

$$\alpha = P\left[V_y' < \omega\right] = \frac{1}{\sqrt{2\pi}} \int_{-\infty}^{\frac{\omega}{\sigma_y}} e^{-\frac{t^2}{2}} dt$$

Based on this equation, α is less than 1/2, and α increases as the particle size decreases.

4. **Karasev** [37]

The formula developed by Karasev [37] for α is Eq. (1.48), or

$$\alpha = \frac{\Pi}{\omega} = 0.083 + 0.126\frac{u_*}{\omega}$$

Based on this formula, α can be greater or less than 1, and α increases as the particle size decreases.

5. **Zhang** [62]

Zhang [62] developed the following vertically 2-D sediment diffusion equation for steady flow

$$u\frac{\partial S}{\partial x} = \varepsilon_y \frac{\partial^2 S}{\partial y} + \frac{\partial(\omega S)}{\partial y} \tag{1.213}$$

The boundary condition at the entrance specified by Zhang is

$$x = 0,\ S(0, y) = 0 \tag{1.214}$$

or

$$x = 0,\ S(0, y) = S_{b.0} \exp\left(-\frac{\omega}{\varepsilon} y\right) \tag{1.215}$$

The boundary condition at the bottom is

$$y = 0,\ \frac{\partial S}{\partial y} = 0 \tag{1.216}$$

or

$$y = 0,\ S(x, 0) = S_b^* \tag{1.217}$$

or

$$y = 0, \varepsilon \frac{\partial S}{\partial y} = -\omega S_b^* \tag{1.218}$$

The boundary condition at the water surface is

$$y = h, \varepsilon_y \frac{\partial S}{\partial y} + \omega S = 0 \tag{1.219}$$

Zhang sought solutions of Eq. (1.214) using Dobbins's [5] analysis for reference. Solving Eq. (1.213) with the boundary conditions, including Eqs. (1.214), (1.215), and (1.219), Zhang obtained the distribution of sediment concentration $S(x, y)$ for the case of clear-water scour. Furthermore, Zhang derived the series solutions for the depth-averaged sediment concentration $\overline{S}(x)$ and its longitudinal gradient $d\overline{S}/dx$. For the case with very fine sediment, only keeping the first order term, Zhang obtained

$$\frac{d\overline{S}}{dx} = \left(\frac{\beta_n^2}{K_1} + \frac{K_1}{4}\right)\frac{\omega}{q}(S^* - \overline{S}) \approx \left(\frac{\pi^2}{K_1} + \frac{K_1}{4}\right)\frac{\omega}{q}(S^* - \overline{S}) = -\frac{\alpha\omega}{q}(\overline{S} - S^*) \tag{1.220}$$

where S^* is the depth-averaged carrying-sediment capacity, $\overline{S}$ is the depth-averaged sediment concentration, and the saturation recovery coefficient is

$$\alpha = \frac{\pi^2}{K_1} + \frac{K_1}{4} \tag{1.221}$$

in which

$$K_1 = \frac{\omega h}{\varepsilon_y} = \frac{6\omega}{\kappa u_*} \tag{1.222}$$

Similarly, for the case of pure deposition that is described with Eqs. (1.213), (1.215), (1.216), and (1.219), Zhang obtained

$$\alpha = 1 + \frac{K_1}{2} \tag{1.223}$$

Table 1.3 lists the relationship between α and K_1 for the cases of clear-water scour and of pure deposition, which are described with Eqs. (1.221) and (1.223), respectively. While Zhang derived the series solution of the vertically 2-D sediment diffusion equation using the method of separation of variables, the determination of the saturation recovery coefficient is arbitrary with strong assumptions, as listed in Table 1.3. Furthermore, the bottom boundary condition used by Zhang is too simple and unreasonable. As a result, the saturation recovery coefficient is always greater than 1, which is inconsistent with reality. This is particularly true for the case of pure scour, in which the minimum of α is π at $K_1 = 2\pi$.

Table 1.3 Relationship between α and K_1 for the cases of clear-water scour and of pure deposition

K_1	0.01	0.1	0.2	0.5	1	2	5	6.28	10	30
α (clear-water scour)	990	99	60	20.6	10.15	7.92	3.97	3.14	3.49	7.83
α (pure deposition)	1.005	1.05	1.10	1.25	1.5	2.0	3.50	4.14	6.0	16.0

6. **Hou** [31]

Hou also conducted an in-depth study on the vertically 2-D sediment diffusion equation. For the case of clear-water scour, Hou [31] obtained the same results as Zhang [62]. However, Hou pointed that Eq. (1.221) is approximately true only for the case with strong turbulence and fine sediment, or with a relatively large ε/ω. For a general case of deposition or scour, which is described with Eqs. (1.213), (1.215), (1.218) and (1.219), Hou derived the formula for the saturation recovery coefficient, which is the same as Eq. (1.223), by only keeping the first order term of the series solution and assuming the turbulence is weak.

Based on Hou's results of α for the general condition of scour or deposition, α for scour is greater than that for deposition, and α is greater than 1 for both cases, which is significantly different from the reality. Hence, these results are barely utilized in mathematical models for sediment transport. The main reason of the significant discrepancy in α lies in the unreasonable boundary conditions being used by Hou.

7. **Han** [11] **and Han et al.** [17]

For the vertically 2-D case, Han [11] integrated Eq. (1.213) over the water depth, and obtained

$$\frac{d(Sq)}{dx} = -\left[\omega S + \varepsilon_y \frac{\partial S}{\partial Z}\right]_{q=0} = \omega\left(S_b - S_b^*\right) \tag{1.224}$$

where S is the average sediment concentration that is defined as $S = \frac{1}{q}\int_0^h uSdy$, S_b is the bottom sediment concentration, and S_b^* is the bottom carrying-sediment capacity. Defining

$$\alpha = \frac{S_b}{S} \tag{1.225}$$

$$\alpha^* = \frac{S_b^*}{S^*} \tag{1.226}$$

then Eq. (1.224) can be written as

$$\frac{dS}{dx} = -\frac{\omega}{S}\left(\alpha S - \alpha^* S^*\right) \tag{1.227}$$

Compared to that under the equilibrium condition, the vertical distribution of sediment concentration is less uniform for the case of scour, and more uniform for the case of deposition. Thus, $\alpha > \alpha^*$ for scour, $\alpha < \alpha^*$ for deposition, and α for scour is greater than that for deposition. Also, α is greater than 1 for both cases of scour and of deposition. This is qualitatively consistent with Zhang's [62] results. Neglecting the difference between α and α^*, it was assumed $\alpha = \alpha^*$ in some analyses. However, massive field data indicates α^* is much less than 1 no matter for scour or for deposition. Based on field data analysis, Han [11–13] suggested $\alpha = 1$ for the case of scour, and $\alpha = 0.25$ for the case of deposition. These two values have been widely used in the mathematical models. It is noted that the empirical value of α is much less than the theoretical value, which has caused a long term controversy in the academic community of sediment study in China.

Using the exchange intensities (the number of exchanged particles per unit area per unit time) for sediment particles to change from the rest state to suspension and vice versa, which were derived from the statistical theory of sediment exchange on the bed, as the boundary conditions, Han and He [26] integrated the vertically 2-D sediment diffusion equation, and obtained

$$\frac{dS_l}{dx} = -\frac{\omega_l}{q}\left(\tilde{\alpha}_l S_l - \alpha_l^* S_l\right) = -\frac{\omega_l}{q}\left(\tilde{\alpha}_l P_{4.l} S - \alpha_l^* P_{4.l}^* S^*\right) \approx -\alpha_l \frac{\omega_l}{q}\left(S_l - S_l^*\right) \tag{1.228}$$

where $P_{4.l}$ and $P_{4.l}^*$ are the size distributions for suspended load and for carrying-sediment capacity, respectively, S and S^* are the depth-averaged sediment concentration and carrying-sediment capacity, respectively, and $\tilde{\alpha}_l$ and α_l^* are the saturation recovery coefficients for the non-equilibrium and equilibrium conditions, respectively. Also, Han and He obtained

$$\alpha_l^* = \frac{(1 - \varepsilon_{0.l})(1 - \varepsilon_{4.l})}{L_{4.l}^*} \frac{q}{\omega_l} \tag{1.229}$$

where $\varepsilon_{0.l}$ is the probability for sediment to cease movement, $\varepsilon_{4.l}$ the suspension probability, $L_{4.l}^*$ the single-step movement distance of suspended load under the equilibrium condition, which is a function of the vertical distributions of sediment concentration and flow velocity with a complicated formulation. It is noted that α_l^* depends on the sediment grain size D_l and the relative flow intensity that is defined as u_*/ω_l. Based on the computed results for the cases with various combinations between eight grain sizes, which are finer than 1 mm, and six ω_l/u_*, α_l^* has the maximum value of 5.88, the minimum value of 0.0214, and is less than 1 for most cases. The mean value of α_l^*, which is averaged over the fifteen cases with the hydraulic conditions that typically occur, is approximately 0.5. Considering $\tilde{\alpha}_l > \alpha_l^*$ for scour, and $\tilde{\alpha}_l < \alpha_l^*$ for deposition, and based on their analysis, Han and He [26] suggested $\alpha_l = 2\alpha_l^* = 1.0$ for the case of scour, and $\alpha_l = 0.5\alpha_l^* = 0.25$ for the case of deposition. These results are consistent with the empirical values that were suggested by Han [11–13] based on the field data, and also cover the cases with extremely small values of α_l, as low as of 0.01, which are used in the mathematical models for the lower Yellow River [44].

8. **Zhou and Lin** [64]**, and Zhou** [63]

Keeping the first order term of Zhang's [62] analytic solution of the 2-D sediment diffusion equation, Zhou and Lin [64], and Zhou [63] obtained an approximate solution for the saturation recovery coefficient, which is

$$\alpha = \frac{K_1}{4} + \frac{\beta_1^2}{K_1} \tag{1.230}$$

Based on this equation, for scour, α is always greater than 1, and increases rapidly as K_1 decreases. For deposition, α is slightly greater than 1, and increases slowly as K_1 increases. These characteristics of α is for Zhang's [62] solution, which are particularly true for the approximation with the first order term. Considering the sediment concentration and carrying-sediment capacity also related to the cross-sectional variations of the water depth and the flow velocity, due to the flow velocity being a function of the water depth as shown in the Manning's equation, and integrating Zhang's equation over the channel width, Zhou et al. [63, 64] obtained

$$\beta^S \frac{\partial S}{\partial t} + U\frac{\partial S}{\partial x} = -\alpha^s \frac{\omega}{h}(s - \Phi) \tag{1.231}$$

where Φ is determined with Eq. (1.76), h is a local water depth. Also,

$$\alpha^S = \frac{\int_0^1 \alpha\eta^{(3\upsilon-1)m} d\xi}{K_u \int_0^1 \eta^{(3\upsilon-1)m+\upsilon+1} d\xi} \tag{1.232}$$

$$\beta^S = \frac{\int_0^1 \eta^{(3\upsilon-1)m+1} d\xi}{K_u \int_0^1 \eta^{(3\upsilon-1)m+\upsilon+1} d\xi} \tag{1.233}$$

where m is the exponent in the formula for sediment transport capacity, ξ denotes the relative coordinate in the cross-sectional direction, which is

$$\xi = \frac{z}{H} \tag{1.234}$$

in which H is the average water depth. The depth-averaged flow velocity was assumed to have the following relationship with the water depth

$$u(h) = mh^\upsilon = m(H\eta)^\upsilon = \frac{U\eta^\upsilon}{\int_0^1 \eta^{1+\upsilon} d\xi} = K_u U \eta^\upsilon \tag{1.235}$$

where U is the cross-sectional average flow velocity, and

$$K_u = \left(\int_0^1 \eta^{\upsilon+1} d\xi \right)^{-1} \quad (1.236)$$

α^s was considered by Zhou et al. as the saturation recovery coefficient. Considering the impact of the cross-sectional variations of sediment concentration and carrying-sediment capacity on the saturation recovery coefficient, Zhou et al.'s analysis has certain originality.

The following discussions are made for Zhou et al.'s analysis:

(1) Equation (1.231) is different from the general form of the equation for 1-D unsteady non-equilibrium sediment transport, which is

$$\frac{\partial S}{\partial t} + U\frac{\partial S}{\partial x} = -\alpha\frac{\omega}{H}\left(S - S^*\right) \quad (1.237)$$

The main difference is the coefficient β^s being added for $\partial S/\partial t$.

(2) The saturation recovery coefficient α should be used to represent the recovery of sediment saturation caused by the diffusion process for non-equilibrium transport of suspended load, for instance, as used in Eq. (1.222) developed by Zhang [62]. It should not be used to emphasize the impact of the cross-sectional variations of sediment concentration and carrying-sediment capacity.

(3) Only the first order term of Eq. (1.220) is adopted, which can cause significant errors for some cases.

(4) Both α and α^s are greater than 1 for the case of scour. This does not agree with the field data. For deposition, α is also greater than 1. It is hard to only use the cross-sectional variations of sediment concentration and carrying-sediment capacity to explain that fact of $\alpha^s < 1$, including $\alpha \approx 0.01$ based on the field data for the lower Yellow River. Also, for a rectangular cross-section, it is difficult to reduce α to below 1% by only considering the impact of the cross-sectional variations of sediment concentration and carrying-sediment capacity.

(5) Due to sediment diffusion in the cross-sectional direction, the cross-sectional distribution of sediment concentration should be much more uniform than that of carrying-sediment capacity. Zhou et al. assumed these two distributions have the same uniformity, which lacks of basis.

9. **Liu et al.** [42, 43]

Liu et al. [42, 43] argued that the bed material coarsening process would develop rapidly during scour, resulting in a more significant impact on the recovery of sediment concentration and a smaller saturation coefficient. Liu et al. proposed a concept of bed material exchange rate, analyzed the bed material coarsening process in the surface layer of the bed and the sediment concentration recovery process for various bed material exchange rates, and concluded the scour process would result in a smaller saturation recovery coefficient. Obviously, bed material coarsening impacts the recovery rate of sediment concentration during scour. This is mainly because the carrying-sediment capacity becomes smaller after bed material coarsening. In fact, the recovery of sediment concentration should be caused by the sediment diffusion process instead of bed material coarsening. The study on non-equilibrium sediment transport not only includes solving the sediment diffusion equation, but also includes the analysis of the bed material coarsening or refining, derivation of corresponding equations and finding solutions. The sediment diffusion equation is solved with the known bed material gradation, and the recovery of sediment concentration caused by diffusion also corresponds to the given bed material gradation. In other words, as a representative of the sediment diffusion process and the resulting recovery of sediment concentration, the saturation recovery coefficient should be caused by the factors other than the bed material coarsening, and the impact of bed material coarsening is exerted as the bottom boundary condition. Also, bed material coarsening is not only caused by scour. It can also be caused by the exchange between the coarser sediment particles from upstream and the finer particles on the local bed. In fact, this second kind of bed material coarsening causes the suspended load to become finer along the flow path, resulting in a long scour distance downstream of a reservoir, as indicated by the field data [29].

10. **Summary**

The above reviews of sediment saturation recovery coefficient can be summarized as follows:

(1) The saturation recovery coefficient is an important parameter for non-equilibrium suspended load transport. In addition to the theoretical meaning, it is also needed in the 1-D and 2-D mathematical models for sediment transport.
(2) The studies for saturation recovery coefficients can be categorized into four approaches. The first approach is to summarize some physical concepts, and give the expression for the saturation recovery coefficient for the 1-D diffusion equation that is derived based on the mass balance of sediment. The representative studies included Mikhayev and Yunevich [48], Karauschev [38], Dou [7], and Karauschev [38]. The second approach is to solve for the analytic solution of the vertically 2-D diffusion equation with certain simplified conditions, and then determine the saturation recovery coefficient for the corresponding 1-D diffusion equation. The derivation in this approach is mathematically strict. However, because of the unreasonable boundary condition being used, the derived α is always greater than 1, which shows a great discrepancy from the reality, and is thus hard to be applied. The representative studies included Zhang [62] and Hou [31]. The third approach is to utilize the solution of the diffusion equation that is derived with the second approach, and then lower the value of α, by considering other factors, in order to solve the inconsistence between the "theoretical" value and the reality. The representative studies included Zhou and Lin [64], Zhou [63], and Liu et al. [42]. These studies only considered the impacts of some factors, did not involve the bottom boundary condition, and thus could not essentially solve the problems. The fourth approach is to conduct more in-depth theoretical analysis based on the field data, and start from the

boundary condition to obtain more reasonable results for α. For instance, while the empirical results of α that were proposed by Han [13] lacked theoretical basis, they showed good agreement with the reality as a result of massive field data being used in the calibration for these values. Furthermore, with more theoretical analysis and explanations, these empirical values have been used in numerous mathematical models for sediment transport.

(3) Designating the sediment settling term is ωS, and assuming the sediment pick-up caused by turbulence is $\varepsilon \frac{\partial S}{\partial y}\Big|_{y=0} = -\omega S_b^*$, then the bottom boundary condition, or the net sedimentation quantity per unit area on bed surface per unit time, is

$$\varepsilon \frac{\partial S}{\partial y}\Big|_{y=0} + \omega S_b = \omega S_b - \omega S_b^*$$

Based on this boundary condition, no matter α is determined after solving 2-D diffusion equation for $S(x, y)$, or is determined by directly integrating the 2-D diffusion equation [58], the derived α is always greater than 1. Hence, in order to further solve this problem of α always greater than 1, additional analysis should be conducted for the bottom boundary condition.

(4) The bottom boundary condition $\varepsilon \frac{\partial S}{\partial y}\Big|_{y=0} + \omega S_b$ is developed based on the sediment diffusion equation, by approaching the bed from the water. On the other hand, the deposition or scour on the bed should also depends on the exchange between the bed material and suspended load, and even the exchange between the bed material and bed load. Hence, in order to obtain a reasonable bottom boundary condition, the analysis should starts from the sediment exchange on the bed, which is pioneered by Han and He [26].

1.6 Main Topics for Further Studies on Non-equilibrium Transport of Non-uniform Suspended Load

Great progress has been made in the studies for non-equilibrium transport of non-uniform suspended load, with pioneering work being done by Chinese scientists, both in-depth and in-breadth, for various topics. These topics include the carrying-sediment capacity for non-uniform sediment, sediment exchange on the bed, sediment saturation recovery coefficient, size distributions for carrying-sediment capacity and for suspended load, and the bed material gradation, the effective bed material gradation, and their relation. While the major topics have been covered in the existing studies for non-equilibrium sediment transport, future in-depth studies should focus on the following subjects:

(1) Further investigation of the mechanism for non-uniform suspended load transport, including the convection-diffusion equation for non-uniform sediment, the criteria and basis for the summation for various size classes, and the size distribution of suspended load.
(2) The form of the carrying-sediment capacity for non-uniform sediment, the multi-valued nature of carrying-sediment capacity for the case of scour or deposition, conditions for carrying-sediment capacity to be single-valued and the corresponding formulation, the relationship of the carrying-sediment capacities between non-uniform and uniform sediment.
(3) The physical meanings, the theoretical bases, and the proper and general expressions for the size distribution for carrying-sediment capacity and for the effective bed material gradation; and investigation, clarification and further development of the existing study results of these subjects.
(4) The physical meaning of the bed surface layer, the state transitions among sediment particles with various states, the expressions for the transition probabilities and transition intensities, the general bottom boundary condition that can be derived from the transition intensities and its impact on the solution of the sediment diffusion equation, and the characteristics and theoretical expression for the saturation recovery coefficient.
(5) Pioneering studies have been conducted by scientists in China in recent years for various subjects of non-equilibrium sediment transport. These subjects include the carrying-sediment capacity under the strong or weak equilibrium condition, the carrying-sediment capacity, the vertical distribution of sediment concentration, and the saturation recovery coefficient under the non-equilibrium condition, bed material coarsening, the exchange between the coarser and finer sediment particles, and the re-investigation of the concepts of bed material load and wash load. Further in-depth analysis can be continued based on the original results of these pioneering studies [15, 16, 19, 20, 21, 22]. This will not only complete the theoretical system for non-equilibrium sediment transport, but also further illustrate the complicated mechanisms and discover the laws for sediment transport.
(6) The subjects mentioned above are important for non-equilibrium transport of non-uniform suspended load. Original results have been obtained for some subjects, while the others are still in exploratory stage. On one hand, active studies, with wide views, are being conducted for these subjects. On the other hand, some studies do not incorporate the new ideas and concepts, resulting in unclear, unreasonable, or self-contradictory results. This is because the past studies were mainly limited to the equilibrium transport of uniform sediment, while the present studies focus on the non-equilibrium transport of non-uniform sediment, in which the new concepts and ideas are essentially different from the past studies. Hence, in future studies, it is necessary to re-analyze and have thorough discussions on the existing results, identify the problems, and make corresponding clarification or correction.

1.7 Summary of Contents

Chapter 1

The studies on the non-equilibrium transport of non-uniform suspended load and on the capacity for carrying non-uniform sediment are reviewed. The review focuses on five topics with more study results available, while the results for other topics are discussed in corresponding chapters. It is noted that our review may not be complete, and comments are welcome.

Chapter 2

The statistical theory for sediment exchange on the bed is presented. Since Einstein (1950), the concept for exchange between bed load and bed material has been proposed, and formulas have been derived for bed load transport rate. While this concept has been used in many studies, an in-depth illustration of this "simple" exchange between bed load and bed material and well accepted formulations for the exchange quantity are still not available. Based on their statistical theory for sediment movement, Han and He [23, 24] investigated the transitions among the four movement states of sediment particles in the bed layer, and the results are summarized in Chap. 2. The main subjects are as follows:

(1) The non-uniform sediment movement on the bed can be categorized as four states: at rest, rolling, hopping, and suspension; and the threshold conditions for the transitions among these four states and corresponding sixteen transition probabilities are proposed.
(2) For the case with independent longitudinal and vertical fluctuating velocities at the bottom, the six basic probabilities are analyzed for incipient sediment motion (entrainment), for particles to cease movement, for particles to transit from resting to hopping, for particles to transit from a moving state to hopping, for incipient suspension, and for keeping suspension. The expressions and numerical results are derived for these transition probabilities. The numerical results are proved to be reasonable and are consistent with empirical values.
(3) A 4×4 transition probability matrix is further developed for the case with independent longitudinal and vertical bottom fluctuating velocities. Furthermore, the state probabilities are derived for the sediment particles in the bed layer to be in resting, rolling, hopping and suspension states. The results are proved to be reasonable.
(4) The life distributions for various sediment states, and the distributions and mathematical expectations for the number of particles with state transitions are introduced. The sixteen sediment exchange intensities are derived for the transitions among the four states of sediment particles in the bed layer, for both uniform and non-uniform sediment.
(5) The derived sediment exchange intensities are physically more underlying parameters than sediment transport rates. They can be used to derive the capacities for carrying suspended load and bed load, and to develop the bottom

boundary condition for the equation of non-equilibrium sediment transport. They are useful parameters for sediment transport studies, as shown in the derivations of equations that are presented later in this book.

(6) For the case with dependent longitudinal and vertical bottom fluctuating velocities, the adjustment and order of the threshold flow velocities are analyzed. The 2×2 and 4×4 transition probability matrixes and the corresponding state probabilities are developed for various conditions.

Chapter 3

The convection-diffusion equation for non-uniform suspended load and the characteristics of this equation are studied. As a critical factor to determine the sediment deposition or scour, the bottom boundary condition is investigated. Various kinds of the bottom boundary conditions have been proposed. However, most of them are empirical and lack of theoretical basis, and thus the results are usually hard to be applied. Focusing on the governing equation for non-uniform suspended load transport and the bottom boundary condition, the contents of Chap. 3 are summarized as follows:

(1) The convection-diffusion equation for uniform suspended load is extended to non-uniform sediment, the water volume percentage is introduced, and the size distribution of suspended load is defined.

(2) The three-dimensional (3-D) convection-diffusion equation for non-uniform sediment is integrated over the channel width and over the water depth, respectively, under certain conditions. The horizontally and vertically two-dimensional (2-D) equations are derived, respectively, with new terms such as $\overline{V_x S_l}$ and $\overline{V_y S_l}$ being introduced. These new terms represents 3-D impacts.

(3) Based on the statistical theory for sediment movement, particularly for the sediment exchange on the bed, the bottom boundary condition for the diffusion equation is developed. This bottom boundary condition can be applied to the 3-D and vertically 2-D diffusion equations, provides the sedimentation term for the horizontally 2-D equation, and be used to derive the governing equation for 1-D non-equilibrium sediment transport.

Chapter 4

This chapter presents the in-depth theoretical analysis for carrying-sediment capacity, which is one of the most important topics for sediment transport that lacks thorough illustrations of the mechanisms and strict theoretical expressions. The in-depth theoretical study on this topic has been carried out in this chapter. The main contents include:

(1) Based on Han and He's [23, 24] theory for the sediment exchange intensities in the bed surface layer, and their study on the single-step process for suspended load movement (picked up from the bed, suspended up to the maximum suspension height, and then settles down on the bed), a theoretical system is developed for the carrying-sediment capacity for non-uniform sediment. A set of more than

twenty equations are derived for the carrying-sediment capacity and related parameters for non-uniform sediment under equilibrium transport condition. The mechanisms, parameters, and expressions of these equations are completely different from the results in the past.

(2) Considering sediment pick-up by the turbulent bursting, and based on existing studies, the impact of the turbulent bursting on the incipient suspension period is analyzed.

(3) The newly derived formula for carrying-sediment capacity is compared to the data collected in flume experiments and in natural streams. The formulas show good agreement with the data, without any parameters that need to be calibrated. It is also proved that the total carrying-sediment capacity determined with the summation based on the bed material gradation is the same as that determined with the summation based on the size distribution of suspended load. This solves a past dispute about which size distribution should be used in the summation in order to determine the total carrying-sediment capacity for non-uniform sediment mixture.

(4) This newly derived formula shows the carrying-sediment capacity is not only a function of $\frac{V^3}{gh\omega}$. It is also impacted by the sediment grain size, the shear velocity u_*, and the water depth. This new formula can be applied to more general conditions than the existing formulas.

(5) This newly derived formula can be used to encapsulate, within certain ranges, some formulas that are generally used, and to illustrate the limitations of these formulas.

(6) As indicated by this newly derived formula, the upper limit of sediment concentration is approximately 400–500 kg/m^3, which is consistent with the empirical upper limit for the carrying-sediment capacity of the turbulent flow. If the sediment concentration is higher than this limit, sediment transport is no longer solely caused by the turbulence.

Chapter 5

Considering the characteristics of non-uniform sediment, an equation system is formed for 1-D non-equilibrium transport of non-uniform sediment. It consists of the equations for 1-D non-equilibrium sediment transport, for the size distribution of suspended load, and for the bed material gradation. These equations are solved in a coupled manner, based on which the non-equilibrium transport of non-uniform sediment is further analyzed. The main contents are summarized as follows:

(1) A 1-D equation for non-equilibrium sediment transport is derived by integrating the vertically 2-D sediment diffusion equation. The longitudinal variation of sediment concentration is determined for non-uniform sediment under the constant flow condition. As shown in the solution, the sediment concentration at a given section is comprised of three components: the local carrying-sediment capacity at this section, the residual part of the surplus sediment concentration at the upstream section (or the value of the sediment concentration at the upstream section minus the local carrying-sediment capacity at this section)

after decaying along the flow path, and the sediment adjustment due to the longitudinal variation of the carrying-sediment capacity.
(2) Equations are developed for the variations of the size distribution for suspended load during the scour or deposition process. These equations can represent the suspended load sorting during deposition and the bed material supplement during scour.
(3) Equations for the bed material gradation are developed. These equations can represent the bed material sorting and coarsening during a scour process.
(4) The theoretical results are compared to, and show good agreement with, the field data collected in the reservoirs, channels, and desilting canals and basins along the Yangtze River and the Yellow River, China. The corresponding mathematical models have been widely applied to the sedimentation engineering for the Yangtze River, the Yellow River and many other rivers in China.

Chapter 6

The sediment saturation recovery coefficient is important for the study on non-equilibrium sediment transport and for the deposition or scour computation in the mathematical models. A wide range of the saturation recovery coefficient, ranging from 0.01 to 1000, has been proposed in various studies. Chapter 6 presents the in-depth analysis of the saturation recovery coefficient that has been conducted by the author of this book. The main contents of this chapter include:

(1) Based on the theoretical analysis, the saturation recovery coefficients α_l^* and $\tilde{\alpha}_l$ for equilibrium and non-equilibrium sediment transport, respectively, and the composite saturation recovery coefficient α_l are proposed.
(2) Four theoretical characteristics of the saturation recovery coefficient are illustrated.
(3) Based on the analyses of the upward and settling velocities of suspended load, the single-step distance, the suspension height, and the flow discharge involved in sediment suspension, a formula is derived for the saturation recovery coefficient for equilibrium sediment transport. The numerical results are computed based on this formula, which are consistent with the empirical values.
(4) The calculation of the saturation recovery coefficient for non-equilibrium sediment transport is complicated. This is because the suspension height depends on the vertical distribution of sediment concentration, which is different between the non-equilibrium sediment transport and the equilibrium sediment transport. Based on the vertical distribution of sediment concentration as discussed in Chap. 8, the saturation recovery coefficients is computed for non-equilibrium sediment transport. It can be greater or less than 1. For the general hydraulic and sediment conditions, the saturation recovery coefficient is less than 1 for most cases, and can even be 0.01 or less. This theoretically solves the past dispute for the reasonable values of the saturation recovery coefficient.
(5) The composite saturation recovery coefficient is relatively stable. It shows less variation with the scour or deposition process, while it varies significantly with ω/u_*. Hence, it is more convenient to utilize the composite saturation

recovery coefficient. The theoretical values of the composite saturation recovery coefficient show good agreement with the field data of the lower Yellow River.

Chapter 7

This chapter focuses on the size distribution for carrying-sediment capacity and the effective bed material gradation. For strong equilibrium sediment transport, it is sufficient to only use the bed material gradation and the size distribution of suspended load to describe the non-uniform suspended load transport. However, for non-equilibrium condition, it is necessary to introduce the size distribution for carrying-sediment capacity and the effective bed material gradation, which are important and critical for non-equilibrium transport of non-uniform suspended load. Until now there is no study on this topic presented except two size distributions under equilibrium transport condition. This chapter presents the pioneering and systematic study conducted by the author of this book for these subjects. The main contents of this chapter are as follows:

(1) Based on the analysis of some typical cases for non-equilibrium transport of non-uniform sediment, the size distribution for carrying-sediment capacity and the effective bed material gradation are introduced, and the necessity of introducing these size distributions is discussed.
(2) Ten properties of the size distribution for carrying-sediment capacity are illustrated and verified theoretically. They represent the theoretical meaning of this size distribution.
(3) General expressions are developed for the carrying-sediment capacity, its size distribution, and the effective bed material gradation at a given section. These expressions cover the cases with sediment inflow all or partially changing to carrying-sediment capacity and with the bed material all or partially suspendable during a scour or deposition of non-uniform sediment.
(4) The size distribution for carrying-sediment capacity and the effective bed material gradation are thoroughly investigated using new approaches with the water volume percentage and the sediment exchange on the bed.
(5) The mechanisms and simplified expressions for the size distribution for the carrying-sediment capacity, the effective bed material gradation, and the size distribution of suspended load are investigated for the cases of massive scour and of massive deposition. These expressions will simplify their applications.
(6) The size distribution for carrying-sediment capacity and the effective bed material gradation are determined for the exit section of a relatively long reach with significant scour or deposition. As $x \to 0$ and scouring percentage $\lambda^* \to 0$, the results for a reach approaches the results for a section.
(7) Some special cases of sediment transport are thoroughly analyzed and explained with the application the size distribution for carrying-sediment capacity. This indicates that the concept and the expression for the size distribution for carrying-sediment capacity are consistent with the reality.

Chapter 8

This chapter presents the numerical results for the vertically 2-D sediment diffusion equation, and the approximate vertical distribution of sediment concentration for non-equilibrium sediment transport. As discussed in Chapter 1, massive studies have been conducted for the analytical and numerical solutions to the vertically 2-D diffusion equation. However, strong assumptions were generally used in order to derive the analytical solutions, and the numerical solutions mainly aimed to solve the sedimentation engineering problems. Particularly, the bottom boundary conditions used in these studies were not consistent with the reality. The main contents of Chap. 8 consist of:

(1) The bottom boundary condition, as derived in Chap. 3 based on the sediment exchange intensity on the bed, is adopted for the vertically 2-D diffusion equation. This boundary condition has solid theoretical basis, with parameters being widely used and verified in the 1-D equation for non-equilibrium sediment transport.
(2) The parameters in the bottom boundary condition, such as the suspension height and the saturation recovery coefficient, depend on the vertical distribution of sediment concentration and the saturation condition. In other words, this bottom boundary condition is coupled with the diffusion equation. Hence, the numerical solution of the diffusion equation is solved in a coupled manner.
(3) A series of numerical solutions are computed. Analyses show that these numerical solutions are reasonable, and can represent the vertical distribution of sediment concentration and its longitudinal variation along the flow path for the case of scour or deposition.
(4) Massive studies have been conducted for the vertical distribution of sediment concentration under the equilibrium condition. For non-equilibrium sediment transport, the distribution of sediment concentration was typically obtained by solving the vertically 2-D diffusion equation. Introducing the un-saturation factor C, and considering the variation of the bottom sediment concentration, a simplified formulation is derived in this chapter to approximate the vertical distribution of sediment concentration under the non-equilibrium condition. This simplified distribution shows good agreement with the numerical solutions for most cases except for the saturation coefficient $S/S^* > 5$.

Chapter 9

The bed material coarsening is discussed in this chapter. Most of the former studies on bed material coarsening are empirical, without clear illustration of the mechanisms. Also, these studies attribute bed material coarsening solely to scour, without thorough investigation of other factors that may impact bed material coarsening. The computations of bed material coarsening in many mathematical models are usually limited by the sedimentation results of suspended load, without including the change to the coarser sediment particles that cannot be suspended. Also, most studies do not provide specific equations for bed material coarsening. Focusing on the study on the bed material coarsening, the main contents of Chap. 9 include:

(1) Massive field data for bed material coarsening are collected and analyzed. The data can be grouped into two categories (coarsening due to scour, and coarsening due to sediment exchange) with six types. The latter category includes the bed material coarsening on coarse sand beds, sand mixed with pebble or gravel beds, and pebble or gravel beds.
(2) Based on the theory of sediment exchange intensities, the mechanisms for bed material coarsening and the factors impacting bed material coarsening are investigated. In addition to the impacts of the entrainment velocity and the corresponding entrainment probability, which are already included in most of the existing studies, bed material coarsening is also impacted by the threshold velocity for incipient suspension, the velocity for keeping suspension, the probability for sediment to cease movement and the corresponding transition probability, the bed load transport rate, the suspended load transport rate, the sediment entrainment period, the single step distance and the single step velocity for sediment movement.
(3) Utilizing the equation for bed load transport rate and the general equation for bed material coarsening, a formula is theoretically derived for bed material coarsening for the case of bed load scour. This formula is verified with the field data of bed material coarsening on pebble beds.
(4) For bed material coarsening on a sand bed, the derived equation for bed material coarsening for the case of bed load scour is compared to the equation for the case of massive suspended load scour, which is derived in Chap. 5. It is found that bed material coarsening on a sand bed is mainly caused by the scour of suspended load, not by the scour of bed load. This is because for a sand bed, the bed material gradation show little difference from the size distribution of the eroded coarser sediment particles associated with the bed load scour, while it is significantly different from the size distribution of the eroded particles associated with the suspended load scour.
(5) The thickness of bed armor layer, and the proportion of finer particles in the armor layer are investigated.
(6) The bed material coarsening caused by sediment exchange is studied. A typical model is developed for this type of bed material coarsening. The study indicates that bed material coarsening occurs when the deposition volume of the coarser sediment particles equal the scour volume of the finer particles. From a general concept, the sediment transport volume is equilibrium. However, bed material coarsening still occurs. Further investigation shows this equilibrium is a weak equilibrium, not a strong equilibrium, because the carrying-sediment capacity still exceeds the sediment concentration for this case.

Chapter 10

The exchange between the coarser and finer sediment particles and the similarity and differences between the bed material load and the wash load are discussed in this chapter. It was considered in the past that the laws for sediment transport were considered essentially different between the coarser and finer particles. As a result, the suspended load is divided into the bed material load and the wash load. On the other hand, the total load is generally used in the mathematical models. This actually refutes the theory that the bed material load and the wash load follow different laws of transport. In this chapter, the one-sidedness is theoretically illustrated for the view that no carrying-sediment capacity exists for the wash load. The differences in the transport phenomena between the bed material load and the wash load, and the similarities in the basic laws of transport are discussed. The main contents of this chapter is summarized as follows:

(1) Analysis is conducted on the massive field data collected in alluvial streams, downstream channels and variable backwater zones of reservoirs. The results indicate that the exchange between the coarser sediment particles and the finer particles constantly occur during the channel scour or deposition. During the sediment exchange, the finer particles are eroded, and the coarser particles deposit. Within a certain length of a reach, the threshold particle size for scour or deposition is relatively stable.
(2) Based on the data analysis, and utilizing the formulas derived for the size distribution for carrying-sediment capacity and for the non-equilibrium sediment transport, the inevitability of the exchange between the coarser and finer particles is proved.
(3) The importance of the coarser and finer particle exchange in sediment transport and channel evolution is theoretically illustrated. This exchange increases the carrying-sediment capacity along the flow path, causes the exchange and coarsening of the bed material, and is responsible for the long distance scour downstream of a reservoir, which can reach the mouth of the stream in some cases.
(4) The relative water volumes to carry the bed material load and the wash load are determined based on the water volume percentage. As a result, the capacities for carrying the total load, the bed material load, and the wash load are separated. These carrying-sediment capacities are proved to have the same form of expressions and to follow the united laws. The results are verified with the field data of the Yangtze River and the Yellow River of China.
(5) The theoretical values of the threshold particle size to differentiate the coarser particles and the finer particles agree with the empirical results that are practically used.
(6) It is illustrated that the total sediment concentration can increase along the flow path, resulting in a long distance channel scour downstream of a reservoir, for the case with approximately constant hydraulic parameters and with fining size distribution of bed material along the flow path. This is verified with the field data of the lower Yellow River downstream of the Sanmenxia Reservoir and of lower Han River downstream of the Danjiangkou Reservoir. The scour of the lower Yellow River reaches the mouth of this river, and the scour of the lower Han River reaches Hankou at the Yantze River.

(7) Based on the united form of the capacities for carrying the bed material load, the wash load, and the total load, and on the sorting relationship during non-equilibrium sediment transport and resulting scour or deposition, the differences in the transport phenomena between the coarser particles and the finer particles are identified. The differences include: (a) transport of wash load only expends a small amount of flow energy, (b) sediment carried by the density flow in a reservoir is typically wash load, and (c) the threshold particle size for the coarser and finer sediment exchange, the threshold particle size to differentiate the bed material load and the wash load, and the threshold particle size for plunging of density flow are basically equivalent. In other words, while the bed material load and the wash load are proved to follow the same laws of transport, the differences in the transport phenomena can still be illustrated based on the theories for non-equilibrium transport and corresponding carrying-sediment capacity for non-uniform sediment. Hence, it is unnecessary to differentiate the bed material load and the wash load except to divide the coarser and finer particles in the computation of the carrying-sediment capacity for non-uniform sediment.

Chapter 11

This chapter discusses the multi-valued nature of carrying-sediment capacity. There have been different opinions on whether the carrying-sediment capacity is single-valued or is multi-valued. Based on the concept of carrying-sediment capacity, it equals the sediment concentration under the equilibrium condition, and thus is considered to be single-valued. Due to the errors in the measured sediment data, whether the carrying-sediment capacity is single-valued or multi-valued cannot be simply determined based on the data. While most scientists accept the opinion that the carrying-sediment capacity is single-valued, some scientist consider the carrying-sediment capacity to be multi-valued. Focusing on the multi-valued nature of carrying-sediment capacity, the main contents of Chap. 11 include:

(1) The multi-valued nature of carrying-sediment capacity reflects the complicated mechanisms for carrying-sediment t capacity. Study on this multi-valued nature can further understand and illustrate the characteristics of the carrying-sediment capacity, improve the computation accuracy in the sedimentation engineering and in mathematical models for sediment transport. A useful approach for studying the carrying-sediment capacity is to first acknowledge the multi-valued nature, then analyze it, and finally introduce new variables to eliminate this multi-valued nature. This approach will gradually enhance the understanding of carrying-sediment capacity based on the existing knowledge.

(2) Based on the analysis of the long-term field data of the Yangtze River and the Yellow River, and removing the non-equilibrium impact, the multi-valued nature of carrying-sediment capacity is proved. In addition, it is theoretically determined that the carrying-sediment capacities for the case of scour and for the case of deposition are approximately 0.65–0.75 times and 1.25–1.34 times, respectively, of that for the strong equilibrium condition for the lower reach of the Yellow River.

(3) The mechanisms for the multi-valued nature of carrying-sediment capacity are illustrated. This multi-valued nature is caused by three factors: the multi-valued nature of the threshold flow velocities including the incipient suspension velocity and the suspension velocity, the multi-valued nature of the saturation recovery coefficients, and the multi-valued nature of the size distribution for carrying-sediment capacity. The last two natures were proposed by the author first. The multi-valued nature of saturation recovery coefficients not only causes the carrying-sediment capacity to be multi-valued, but also changes the boundary limit between scour and deposition.
(4) The impacts of the threshold flow velocities and of the size distribution for carrying-sediment capacity are combined in a theoretical formula. As these impacts can be quantified, the multi-valued nature of carrying-sediment capacity that are caused by these factors can be eliminated. As a result, the carrying-sediment capacity can be single-valued, which will provide a new theoretical basis for the study on the carrying-sediment capacity.

Appendix A Includes Statistical Theory of Sediment Movement

The statistical theory developed by the authors is the foundation of the study on the non-equilibrium transport of non-uniform suspended load. Since most of these study results were published in Chinese, for the convenience of English readers we have introduced the main contents in the Appendix A, include:

(1) According to the characteristics of sediment movement, the particle movement can be divided into four categories: rest, rolling, hopping, and suspension. Moreover, each of these motion states are divided into single steps and single events. The mechanical and statistical parameters of single step motion are derived by means of mechanics and probability theory.
(2) Based on Markov stochastic process theory, the stochastic process of sediment movement is established and the theoretical expression of sediment transition intensities and transport rates are introduced.
(3) The theoretical study results are verified by a lot of flume experiment data and measured field data.

Appendix B Includes Mechanics and Statistical Law of Incipient Sediment Motion

Sediment entrainment is one of the most important topics in statistical theory of sediment movement and it is widely used in engineering technology. But up to now how to define the incipient flow velocity and how to determine the entrainment threshold criterion has not been a unified understanding. Based on the study of statistical theory of sediment motion, the authors give the exact concept of incipient flow velocity and the general threshold criterion of sediment entrainment in Appendix B. Besides, Appendix B includes:

(1) The incipient flow velocity of fine particles caused by adhesive force and Additional Film Water Pressure.
(2) Initial dry density and compact dry density and dry density for non-uniform sediment.
(3) The influence of pebble shape on incipient motion.

References

1. Bechteler W, Schrimpf W (1983) Mathematical model of non-uniform suspended sediment transport. In: Proceedings of the second international symposium on river sedimentation, Nanjing, China. Water Resources and Electric Power Press, pp 390–401
2. Cao ZX (1997) Boundary condition near bottom layer of mathematical model for sediment, II: Transport of non-uniform sediment. J Hydraul Eng (1):20–24 (in Chinese)
3. Celik I, Rodi W (1991) Suspended sediment capacity for open channel flow. J Hydraul Eng 117(2):191–204
4. Cui X (1987) Calculations of the equilibrium and the non-equilibrium transportation. J Sediment Res (3):81–89 (in Chinese)
5. Dobbins WE (1944) Effect of turbulence on sedimentation. Trans ASCE 19–109
6. Dong Z, Lu YJ (2003) Primary study on 2-D sediment mathematical model and animated drawing of the velocity field. J Sediment Res (2):32–38 (in Chinese)
7. Dou GR (1963) Suspended sediment movement by tidal currents and bed deformation calculations. J Hydraul Eng 4:31–38 (in Chinese)
8. Dou GR, Zhao SQ, Huang YF (1987) Preliminary results for the 2-D mathematical model for suspended sediment transport in the Chongqing reach for the 170-meter and the 180-meter plans of the three Georges project. Technical report, Nanjing Hydraulic Research Institute (NHRI), Nanjing, China (in Chinese)
9. Dou GR, Zhao SQ (1993) National key scientific research project for the seventh five-year plan, No. 75-16-01, study and preliminary application of 2-D mathematical model for total load transport in the backwater zone of the three Georges reservoir. Research papers for key sediment and navigational technologies of the three Georges project, vol 2. Wuhan University of Technology Press, Wuhan, China, pp 147–166 (in Chinese)
10. Feldman A (1981) HEC models for water resources system simulation, theory and experience. Academic Press, New York
11. Han QW (1973) Preliminary studies for the non-equilibrium transport of sediment in reservoirs. Report on sedimentation in reservoir. Task Group on Yellow River Sedimentation Research, pp 145–168 (in Chinese)

12. Han QW (1979a) Reservoir sedimentation, vol 1. The Yangtze Valley Planning Office, Hydrology Department (in Chinese)
13. Han QW (1979b) A study of the non-equilibrium transportation of non-uniform suspended load. Chin Sci Bull 24(17):804–808 (in Chinese)
14. Han QW (1982) Reservoir sedimentation vol 5. The Yangtze Valley Planning Office, Hydrology Department (in Chinese)
15. Han QW (2005) Size distribution of carrying sediment capacity. In: Proceedings of 6th national symposium on basic theory of sedimentation. Water Conservancy of Yellow River Press, pp 3–20 (in Chinese)
16. Han QW (2006a) Theoretical study for non-equilibrium transport of non-uniform suspended load. Technical report. China Institute of Water Resources and Hydropower Research (IWHR) (in Chinese)
17. Han QW (2006b) Only deposition in river reach from Chenglingji to Wuhan after impounding of TGR. J Hydroelectr Eng (6):79–90 (in Chinese)
18. Han QW (2007) Percentage of flow discharge and its application in non-equilibrium transport of non-uniform suspended load. Adv Water Sci 18(5):633–640 (in Chinese)
19. Han QW (2010) Theoretical study for non-equilibrium transport of non-uniform suspended load. China Institute of Water Resources and Hydropower Research (IWHR), Key Laboratories of Hydraulics, Sediment Sciences and River Harness, China Ministry of Water Resources (in Chinese)
20. Han QW, Chen XJ (2008) Theoretical method for calculating the saturation recovery coefficient. J Sediment Res 6:8–16 (in Chinese)
21. Han QW, Chen XJ, Xue XC (2010) Vertical distribution of sediment concentration in non-equilibrium transportation. Adv Water Sci 21(4):512–523 (in Chinese)
22. Han QW, Guan JC (2011) Multiple value of carrying-sediment capacity. In: Proceedings of 8th national symposium on basic theory of sedimentation. Hohai University Press, Nanjing, China, pp 3–16 (in Chinese)
23. Han QW, He MM (1981) Statistical characters of sediment exchange. J Hydraul Eng (1):9–22 (in Chinese)
24. Han QW, He MM (1984) Stochastic theory of sediment motion. Science Publishing House, Beijing (in Chinese)
25. Han QW, He MM (1987) Mathematical modeling of reservoir sedimentation and fluvial process. J Sediment Res 3:14–29 (in Chinese)
26. Han QW, He MM (1997a) Study on the two-dimensional equations for non-uniform and non-equilibrium suspended sediment transport and the boundary conditions. J Hydraul Eng 1:1–10 (in Chinese)
27. Han QW, He MM (1997b) Preliminary study on saturation recovery coefficients. J Sediment Res 3:32–40 (in Chinese)
28. He MM, Han QW (1989) Concepts about grain size distribution of carrying capacity and effective bed material. J Hydraul Eng 3:7–16 (in Chinese)
29. He MM, Han QW (1990) The determination of the size distributions of carrying capacity and effective bed material. J Hydraul Eng (3):l–12 (in Chinese)
30. Hjelmfelt AT, Lenau CW (1970) Non-equilibrium transport of suspended sediment. J Hydraul Div ASCE (7):1567–1586
31. Hou HC (1982) Basic problems of river dynamic. Hydraulic and Electric Power Press, Beijing, China (in Chinese)
32. Huang JC (1992) The study on some problems for mathematical model of non-equilibrium sediment transport. In: Proceedings of national symposium on basic theory of sedimentation, vol I. Chinese Hydraulic Engineering Society (CHES), Sediment Professional Committee, Beijing, pp 586–593 (in Chinese)
33. Ismail C, Rodi W (1988) Modeling suspended sediment transport in non-equilibrium situation. J Hydraul Eng 114:1157–1191
34. Jiang RQ, Fan JH (1983) Deposition of muddy water with salt and 2-D numerical calculation. In: Proceedings of the 2nd international symposium on river sedimentation. Water Resources and Electric Power Press, Beijing, China, pp 118–127 (in Chinese)

35. Jobson HH, Sarye WW (1970) Prediction concentration profiles in open channels. J Hydraul Div ASCE 96(10):1983–1996
36. Kalinske AA (1940) Suspended-material transportation under non-equilibrium condition. Trans AGU 21(19):613–617
37. Karashev IF (1965) Translational capacity of turbulent flows and deformation of river channels in cohesive soils. In: the Proceedings of the Russian academy of sciences, 124
38. Karauschev AV (1960) Problems of the dynamics of natural flows. Hydrometeorological Publishing House, Russia
39. Kerssens PJM, Prins A, van Rjin LC (1979) Model for suspended sediment transport. J Hydraul Div ASCE (HY5) 461–476
40. Le PJ (2001) The riverbed evolution of analog anthology. Tianjin Science and Technology Press, Tianjin, China (in Chinese)
41. Li YT (1987) Bed material load in equilibrium between deposition and scouring. J Sediment Res (1):82–87 (in Chinese)
42. Liu JM, Wang SQ Wang GQ (2000) Effect of coarsening surface bed material on non-equilibrium sediment transport process during river degradation. Adv Water Sci (3):229–234 (in Chinese)
43. Liu JM, Wang GQ, Wang SQ (2003) Study on non-equilibrium sediment transport in the sandy river erosion. Adv Water Sci 14(5):563–568 (in Chinese)
44. Liu YL, Yu X, Jiang EH (2010) Calculation for non-equilibrium transport of non-uniform suspended load in yellow river. In: Proceedings of symposium on morphological. Chinese Hydraulic Engineering Society (CHES), Sediment Professional Committee, Beijing, and Yellow River Institute of Hydraulic Research, Zhengzhou, China (in Chinese)
45. Liu XN, Cao SY, Huang E, Peng QE (2000) Carrying-sediment capacity of coarse and fine sediment. J Sediment Res 4:35–39
46. Lu YJ, Zhang HQ (1992) Numerical simulation of planar two-dimensional riverbed deformation. In: Proceedings of national symposium on basic theory of sedimentation, vol I. Chinese Hydraulic Engineering Society (CHES), Sediment Professional Committee, Beijing, pp 669–680 (in Chinese)
47. Mei CC (1969) Non-uniform diffusion of suspended sediment. Proc ASCE 95:581–584
48. Mikhayev NV, Yunevich, DL (1959) River channel regulation for land reclamation purpose. Agricultural Publishing house, Russia
49. Ouyang LT, Huang YL, Luo MS, Lu JY, Wu ZY (1993) Impact of the channel scour downstream of the three Georges project on flood protection and navigation. Research papers for key sediment and navigational technologies of the three Georges project, vol 2. Wuhan University of Technology Press, Wuhan, China
50. Partheniades E (1971) Results of recent investigations on erosion and deposition of cohesive sediments. Sedimentation 20-1–20-2
51. Partheniades E (1986) The present state of knowledge and needs for research on cohesive sediment dynamics. In: Proceedings of the 3rd international symposium on river sedimentation, pp 3–25
52. Rosinski KI, Kuzmin IA (1965) Solution of diffusion equation applied to suspended load. In: River morphology. Science Press, Beijing, China (in Chinese)
53. Sarikaya HZ (1973) Numerical calculation of removed ratio of suspended matter in setting basins. In: Proceedings of the 15th congress of IAHR, Istanbul, pp 207–214
54. Sumer BM (1977) Sediment of solid particles in open channel flow. Proc ASCE 103(HY11):1323–1337
55. Takaharu N (1943) River science. Jirendo Press, Tokyo, Japan
56. U.S. Army Corps of Engineers (1976) HEC-6, scour and deposition in river and reservoir. Computer program 723-G2-L2470. Users manual, exhibits theoretical basis
57. van Rijn LC (1984) Sediment pick-up function. J Hydraul Eng 110(10):1490–1494
58. van Rijn LC (1986) Mathematical modeling of suspended sediment in non-uniform flows. J Hydraul Eng 112(6):433–455

59. Yang GL (1999) Sediment carrying capacity of flow. Sediment research on the three Georges project. Water Resource and Electric Power Press, Beijing, China, pp 127–134 (in Chinese)
60. Yellow River Conservancy Commission (YRCC), Survey and Design Institute (1998) Numerical model for sedimentation of lower yellow river (in Chinese)
61. Yu MH, Yang GL, Liu GF, Liu JJ (2001) A preliminary study on the formula of non-uniform sediment carrying capacity. J Sediment Res (3):65–68 (in Chinese)
62. Zhang QS (1980) Diffusion process of sediment in open channel and its application. J Sediment Res 1:37–52 (in Chinese)
63. Zhou J (2000) Study on the optimal operation plan to alleviate sedimentation in the three Georges reservoir and increase flood storage capacity. Technical report, Tsinghua University, Beijing, China
64. Zhou J, Lin B (1998) One-dimensional mathematical model for suspended sediment by lateral integration. J Hydraul Eng ASCE 124(7):712–717

Chapter 2
Statistical Characteristics of Sediment Exchange on Bed Surface

As discussed in the previous chapter, the bottom boundary condition is important for the study on the suspended sediment transport, which includes solving the sediment diffusion equation. The bottom boundary condition has a close relation with the motion state change for sediment on the bottom, or the exchange among sediment with various motion states. As shown in Fig. 2.1, a thin layer with a thickness of δ between the bed surface *A–A* and surface *B–B* is called the bottom layer (bed layer). Sediment within this layer has four motion states: (1) at rest, (2) rolling, (3) hopping and (4) suspension. The particles at rest are called bed material. While the rolling particles are in constant contact with the bed, the hopping particles only contact with the bed intermittently. The suspended particles are known as suspended load. The rolling and hopping particles are called bed load. The layer above the bottom layer is solely occupied by the suspended load. The thickness of the bottom layer δ depends on the maximum hopping height of the particles. The study domain of the sediment diffusion equation focus on the layer above surface *B–B*. Otherwise, the exchange between the bed load and the suspended load will make the study too complicated. Therefore, the bottom boundary for the diffusion equation should be on surface *B–B*. This is similar to what was done in previous studies for the vertical distribution of sediment concentration, which typically started from a small distance of δ above the bed. Since δ is much less than the water depth H, it can be neglected when integrating over the water depth. The suspension height is h and $h \leq H$. The movement of a particle can be divided into events and steps. A single event consists of several single steps. Particle at the end point of a single step transform or maintain their original state, while particle at the end point of a single event change their motion state. The distance, height, and the number of step of an event for each particle movement are all random.

Sediment in the bottom layer has four motion states. These states constantly change, or the sediment with various states constantly exchange among each other. Changing from the other states to suspended load or vice versa apparently determines one aspect of sediment movement on the bottom and the bottom boundary condition for the

Q. Han and M. He, *Non-equilibrium Transport of Non-uniform Suspended Load*, Hydroscience and Engineering, https://doi.org/10.1007/978-981-97-1123-9_2

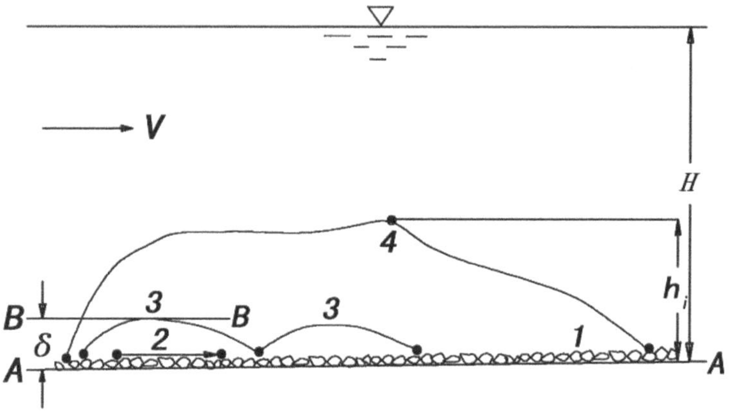

Fig. 2.1 Schematic diagram for sediment rolling, hopping and suspension

diffusion equation. The bottom boundary for the diffusion equation is at $y \to \delta + 0$, while the boundary for sediment exchange in the bed layer is at $y \to \delta - 0$.

The concepts for transitions between the four sediment motion states and the exchange intensity were first introduced in 1978 by 3, 4].

2.1 Basic Conditions and Probabilities for State Transitions of Non-uniform Sediment in Bottom Layer

2.1.1 State Transitions and Basic Transition Conditions and Probabilities for Sediment in the Bottom Layer

Sediment particles in rivers have four states: at rest, rolling, hopping and suspension. The particles at rest are called the bed material. The rolling and hopping particles are known as the bed load, and the suspended particles are called the suspended load. River sediment is generally divided into the bed material, bed load and suspended load in sediment mechanics and river morphology studies. But from the mechanics aspect, rolling (coarse particles, and especially pebbles always move in this state) and hopping (median to fine sands and small gravels mostly move in this state) have different mechanisms for movement. Thus, it is necessary and more appropriate to divide sediment into four states [3, 4, 7]. As did conventionally, Han and He [6, 7] and Han [1] divided the sediment into three motion states, and investigated the corresponding motion states, transition conditions, and transition and state probabilities. Among these three states, the parameters for the bed load can be determined as the summation of the parameters for the rolling state and the hopping state, and thus the parameters for the conventionally used three states can also be derived from those for the four states. The parameters for the four states were investigated by Han [2] and Han and He [3, 4], and are discussed as follows.

Let's define the bed layer (bottom layer) as a layer between the bed surface and the maximum hopping height of the bed load. Sediment below the bed surface is the bed material, and sediment above the bed layer is only the suspended load. The transitions among the four states only occur within the bed layer. The motion states transition chains for sediment in the bed layer is shown in Fig. 2.2. Each state can transit to the other three states and itself. The transition probability matrix is

$$\begin{bmatrix} \beta_{1.1} & \beta_{1.2} & \beta_{1.3} & \beta_{1.4} \\ \beta_{2.1} & \beta_{2.2} & \beta_{2.3} & \beta_{2.4} \\ \beta_{3.1} & \beta_{3.2} & \beta_{3.3} & \beta_{3.4} \\ \beta_{4.1} & \beta_{4.2} & \beta_{4.3} & \beta_{4.4} \end{bmatrix} \tag{2.1}$$

where $\beta_{i.j}(i, j = 1,2,3,4)$ is the transition probability; subscript 1 through 4 represent the rest, rolling, hopping and suspension states, respectively; the first subscript i represents the state before transition, and the second subscript j represents the state after transition. For example, $\beta_{1.2}$ represents the probability for the transition from rest (state 1) to rolling (state 2). For simplicity, six independent transition conditions [2, 3, 6, 7] are introduced in this section. The derivation of independent transition conditions is described in Appendix A.

The condition for incipient sediment motion (entrainment) is

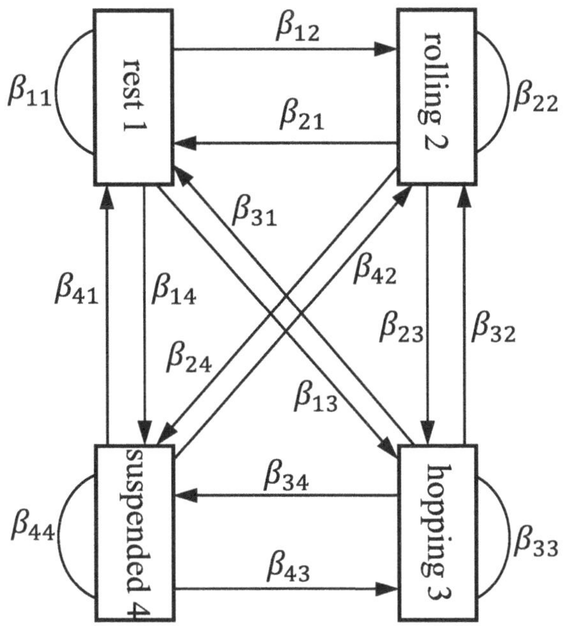

Fig. 2.2 Transition chains among four states of sediment in the bed layer

$$B_{1.l} = [\xi_{V_b} > V_{b.c.1.l}] \tag{2.2}$$

where ξ is a general notation of random variable, the random variable ξ_{V_b} corresponds to the longitudinal bottom flow velocity V_b, $V_{b.c.1.l}$ is the threshold bottom flow velocity for sediment entrainment, event $B_{1.l}$ is the condition for incipient sediment motion, which can be rolling, hopping or suspension, and l is the number of particle size class.

The condition for sediment to cease motion is

$$B_{0.l} = [\xi_{V_b} < V_{b.c.0.l}] \tag{2.3}$$

where $V_{b.c.0.l}$ is the threshold bottom flow velocity for sediment to cease movement. When the moving particles transit to the rest state, no film water interaction and thus no cohesive force or additional pressure of film water exists among the particles. Therefore, the threshold flow velocity for ceasing sediment movement is different from the velocity for sediment entrainment, and $V_{b.c.0.l} \leq V_{b.c.1.l}$.

The sediment hopping may start from the rest state or from a moving state. The condition for incipient hopping from the rest state is

$$B_{3.l} = \left[\xi_{V_b} > V_{b.c.2.l}\right] \tag{2.4}$$

where $V_{b.c.2.l}$ is the threshold bottom flow velocity for incipient sediment hopping from the rest state, which is different than that from a moving state (with an initial velocity), and the difference will be discussed later.

The condition for sediment suspension is

$$B_{4.l} = [\xi_{V_{b.y}} > \omega_l] \tag{2.5}$$

where $\xi_{V_{b.y}}$ is the random variable corresponding to $V_{b.y}$, the vertical flow velocity at the bottom, ω_l is the sediment fall velocity, and $B_{4.l}$ is the event that the loosen sediment particles can be suspended due to the vertical turbulent flow velocity $V_{b.y}$ at the bottom.

For the fine sediment particles resting at the bed to be suspended, additional pressure of water film and cohesive force needed to be overcome, and thus the loosing condition should also be satisfied. The condition for the incipient sediment suspension is

$$B_{5.l} = \left[\xi_{b.y} > V_{b.y.c.3.l}\right] \tag{2.6}$$

where $V_{b.y.c.3.l}$ is the threshold vertical flow velocity at the bottom for incipient sediment suspension.

The threshold flow velocities $V_{b.c.1.l}$, $V_{b.c.0.l}$, $V_{b.c.2.l}$, and $V_{b.y.c.3.l}$ are discussed as follows. The instant bottom flow velocity for sediment entrainment $V_{b.c.1.l}$, is

$$V_{b.c.1.l} = \varphi(\Delta')\omega_{1.l} \tag{2.7}$$

where $\Delta' = \frac{\Delta}{R}$ is the relative value for the sediment particle's location Δ on the bed surface and is a random variable, and $\omega_{1.l}$ is the characteristic velocity for sediment entrainment [4, 7]

$$\begin{aligned}\omega_{1.l} &= \sqrt{\frac{4}{3C_x}\left(\frac{\gamma_s - \gamma}{\gamma}\right)gD_l + \frac{3\times 10^{-7}}{D_l}(1+0.85H)} \\ &= \sqrt{53.9D_l + \frac{3\times 10^{-7}}{D_l}(1+0.85H)}\end{aligned} \tag{2.8}$$

in which H is the water depth, and D_l is the sediment particle size. For $C_x = 0.4$ and $\gamma_s = 2.65\ t/m^3$, $\frac{4}{3C_x}\left(\frac{\gamma_s-\gamma}{\gamma}\right)g = 53.9$. Also,

$$\varphi(\Delta') = \sqrt{\frac{\sqrt{2\Delta' - \Delta'^2}}{\left(\frac{4}{3} - \Delta'\right) + \frac{1}{4}\left(\frac{1}{3} + \sqrt{2\Delta' - \Delta'^2}\right)}} \tag{2.9}$$

is a parameter to represent the particle's location on the bed,, and its mathematic expectation $\overline{\varphi}(\Delta') = 0.916$.

The threshold bottom flow velocity for not ceasing sediment movement is [7]

$$V_{b.c.0.l} = \varphi(\Delta')\omega_{0.l} \approx 0.916\omega_{0.l} \tag{2.10}$$

$$\omega_{0.l} = \sqrt{\frac{4}{3C_x}\frac{\gamma_s - \gamma}{\gamma}gD_l} = \sqrt{53.9D_l} \tag{2.11}$$

Both Eqs. (2.8) and (2.11) are in metric units m, s. By comparing these two equations, it is noted that the difference between the threshold flow velocity for ceasing sediment movement and that for sediment entrainment lies the terms for the additional pressure of water film and for the cohesive force.

The condition and threshold flow velocity for incipient sediment hopping from a moving state are [1, 2, 6, 8].

$$V_b > \sqrt{V_{1.l}^2 - \mu^2 u_{3.y.0.l}^2} \tag{2.12}$$

$$V_{b.c.2.l} = \sqrt{V_{1.l}^2 - \mu^2 u_{3.y.0.l}^2}$$

where $u_{3.y.0.l}$ is the vertical velocity of the sediment particle after collision with bed material

$$u_{3.y.0.1} = u_{x.0.1}(1+\kappa_0)(1-\Delta')\sqrt{2\Delta' - \Delta'^2} + u_{y.0.1}[1-(1+\kappa_0)(1-\Delta')] \tag{2.13}$$

$$\mu^2 = \frac{C}{C_y}\exp\left[-\frac{3C}{2}\left(\frac{\gamma}{\gamma_s+\gamma/2}\right)\right]\cdot\left\{1-\exp\left[-\frac{3C}{2}\left(\frac{\gamma}{\gamma_s+\gamma/2}\right)\right]\right\}^{-1} \tag{2.14}$$

where γ_s and γ are the specific weights of the sediment and water, respectively, $u_{3.y.0.l}$ is the incipient hopping vertical velocity of the sediment particle, $u_{x.0.l}$ and $u_{y.0.l}$ are the longitudinal and vertical components, respectively, of the step end velocity of a particle before it collides with the bed material, κ_0 is the recovery coefficient after collision, $C = 1.2$ is the resistance coefficient for sediment settling, $C_y = 0.1$ is the lift coefficient and $V_1^2 = C_x\omega_{0.l}^2/C_y = 4\omega_{0.l}^2$ where C_x is the drag coefficient; $\omega_{0.l}$ is the velocity parameter for sediment entrainment when cohesive force and additional pressure of water film do not exist, or the entrainment velocity when $\varphi(\Delta') = 1$. Obviously, the initial velocities are different for incipient hopping from the rest, rolling, hopping or suspension state. For an incipient hopping from the rest state, the velocity before the collision $u_{x.0.l} = u_{y.0.l} = 0$, and it can be derived from Eq. (2.12) that

$$V_b > V_{1.l} = 2\omega_{0.l} = V_{b.c.2.0.l} \tag{2.15}$$

where $V_{b.c.2.0.l}$ is the instant hopping velocity when the particle hops from the rest state. The incipient velocities for hopping from the other three moving states are complicate. For simplicity, it is assumed that the hopping from the initial rolling, hopping and suspension states can all be simplified as the hopping from the rolling state [2, 8]. This is because the sediment particles with initial suspension or hopping state typically roll for a short distance before hopping. The initial velocity before collision is approximated by

$$u_{x.0.l} = V_b - V_{b.c.0.l}$$

Or, because the longitudinal velocity is much larger than the vertical velocity, the velocity can also be approximated by the rolling velocity when the sediment falls on the bed. For this case, $u_{y.0.l} = 0$, the mathematical expectation of $\xi_{\Delta'}$ is $\overline{\Delta'} = M(\xi_{\Delta'}) = 0.433$, the bouncing coefficient after the collision $\kappa_0 = 0.5$. Also, $\gamma_s = 2.65\ \text{g/cm}^3$, and $\gamma = 1.00\ \text{g/cm}^3$. Plugging these parameters into Eqs. (2.14) and (2.13), we have $\mu^2 = 1.297 \times 12$, and

$$u_{3.y.0.l} = 0.701(V_b - V_{b.c.0.l})$$

So the condition for incipient sediment hopping, as expressed in Eq. (2.12), can be written as

$$V_b^2 > 4\omega_{0.l}^2 - 7.65(V_b - V_{b.c.0.l})^2$$

For the threshold condition, $V_b = V_{b.c.2.l}$, then

$$V_{b.c.2.l}^2 + 7.65(V_{b.c.2.l} - V_{b.c.0.l})^2 - 4\omega_{0.l}^2 = 0$$

or

$$1.131V_{b.c.2.l}^2 - 2V_{b.c.0.l}V_{b.c.2.l} + (V_{b.c.0.l}^2 - 0.523\omega_{0.l}^2) = 0$$

Solving this quadratic equation, the threshold flow velocity for incipient hopping from an initial moving state is [8]

$$\begin{aligned} V_{b.c.2.l} &= 0.884V_{b.c.0.l} \\ &\quad + \sqrt{0.782V_{b.c.0.l}^2 - 0.884\left[V_{b.c.0.l}^2 + 0.523\left(\frac{V_{b.c.0.l}}{0.916}\right)^2\right]} \\ &= 0.884 + 0.670V_{b.c.0.l} = 1.55V_{b.c.0.l} \end{aligned} \quad (2.16)$$

It is noted that Han [2] conducted further investigation for the sediment hopping from an initial moving state, and determined $K(\Delta') = \frac{V_{b.c.2.l}}{V_{b.c.0.l}}$ for various Δ'. The average value $\overline{K}(\Delta') = 1.620$, which is similar to the value of 1.55 in Eq. (2.16).

It should be emphasized that Eq. (2.5) is correct only for the bed material that is not consolidated. Consolidation typically occurs for the fine sediment, resulting in the cohesive force and additional pressure of water film. For this case, the threshold conditions for sediment suspension is not only limited to $\xi_{V_{b.y}} > \omega_l$. It also includes the condition of $B_{5.l}$, or the sediment loosening requirement.

Han and He [4] investigated the threshold condition for suspension of material initially at rest on the bed. Considering the balance among the lift force, the vertical hydrodynamic force caused by $\xi_{V_{b.y}}$, the net gravity force, the cohesive force and the additional pressure of water film, and assuming a negatively linear correlation between the longitudinal velocity V_b and the vertical velocity $V_{b.y}$, they derived the following condition for loosening sediment

$$B_{5.l} = \left[\frac{4}{3}V_{b.y}^2 - \frac{1}{3}\overline{V_b}V_{b.y} + \left(\frac{1}{12}\overline{V_b^2} - \frac{1}{3}\omega_{1.l}^2\right)\right] > 0 \quad (2.17)$$

where $\overline{V}_b$ is the average (over time) longitudinal bottom flow velocity, $V_{b.y}$ is the vertical component of the instantaneous bottom flow velocity. Setting Eq. (2.17) to zero and solving this quadratic equation, the threshold condition for loosening sediment, or the threshold vertical flow velocity for loosening sediment can be expressed as

$$V^{(1)}_{b.y.c.3.l} = \frac{\overline{V_b}}{8} + \sqrt{\frac{\omega^2_{1.l}}{4} - \frac{3}{64}\overline{V}^2_b} \tag{2.18}$$

$$V^{(2)}_{b.y.c.3.l} = \frac{\overline{V_b}}{8} - \sqrt{\frac{\omega^2_{1.l}}{4} - \frac{3}{64}\overline{V}^2_b} \tag{2.19}$$

So Eq. (2.17) can be written as

$$B_{5.l} = \left[(V_{b.y} - V^{(1)}_{b.y.c.3.l})(V_{b.y} - V^{(2)}_{b.y.c.3.l})\right] > 0 \tag{2.20}$$

The terms in the two parenthesis of Eq. (2.20) must be both positive or both negative in order to satisfy this equation. Let's represent the vertical flow velocity with $\xi_{V_{b.y}}$. When both terms are positive, noting $V^{(1)}_{b.y.c.3.l} > V^{(2)}_{b.y.c.3.l}$, Eq. (2.20) becomes

$$B_{5.1.l} = \left[\xi_{V_{b.y}} > V^{(1)}_{b.y.c.3.l}\right] \cap \left[\xi_{V_{b.y}} > V^{(2)}_{b.y.c.3.l}\right] = \left[\xi_{V_{b.y}} > V^{(1)}_{b.y.c.3.l}\right] \tag{2.21}$$

where $\cap$ denotes the intersection of two sets. When both terms are negative, Eq. (2.20) becomes

$$B_{5.2.l} = \left[\xi_{V_{b.y}} < V^{(1)}_{b.y.c.3.l}\right] \cap \left[\xi_{V_{b.y}} < V^{(2)}_{b.y.c.3.l}\right] = \left[\xi_{V_{b.y}} < V^{(2)}_{b.y.c.3.l}\right] \tag{2.22}$$

Therefore

$$B_{5.l} = B_{5.1.l} \cup B_{5.2.l} = \left[\xi_{V_{b.y}} > V^{(1)}_{b.y.c.3.l}\right] \cup \left[\xi_{V_{b.y}} < V^{(2)}_{b.y.c.3.l}\right] \tag{2.23}$$

where $\cup$ means the sum of two sets. It means that in the derivation of Eqs. (2.21) and (2.22) from the sediment loosening condition, or Eq. (2.17) the square root terms of Eqs. (2.18) and (2.19) should be real numbers, or

$$\frac{\omega^2_{1.l}}{4} - \frac{3}{64}\overline{V}^2_b \geq 0 \text{ or } \overline{V}_b \leq \frac{4}{\sqrt{3}}\omega_{1.l} \tag{2.24}$$

On the other hands, if the square root terms are imaginary numbers, Eq. (2.17), or the threshold sediment loosening condition, is always satisfied, and thus sediment can be suspended due to Eq. (2.5) being satisfied. From the physical viewpoint, the bottom flow velocity $\overline{V}_b$ is so large that the sediment loosening condition is always satisfied. Therefore, the condition for sediment initially at rest on the bed to get loosen and suspended can be summarized as [4]

$$B_{4.l} \cap B_{5.l} = \begin{cases} [\xi_{V_{b.y}} > \omega_l] = B_{4.l} & \left(\overline{V}_b > \frac{4}{\sqrt{3}}\omega_{1.l}\right) \\ [\xi_{V_{b.y}} > \omega_l] \cap [\xi_{V_{b.y}} > V^{(1)}_{b.y.c.3.l}] & \\ \cup[\xi_{V_{b.y}} > \omega_l] \cap [\xi_{V_{b.y}} < V^{(2)}_{b.y.c.3.l}] & \left(\overline{V}_b \le \frac{4}{\sqrt{3}}\omega_1\right) \end{cases} \tag{2.25}$$

2.1.2 Basic Transition Probabilities

Assuming normal distributions for the vertical component of the bottom flow velocity $V_{b.y}$ and for the longitudinal component V_b, the transition probabilities can be derived from the six basic transition conditions as discussed above. The probability of sediment entrainment [1, 2, 4] is

$$\begin{aligned} \varepsilon_{1.l} = P[B_{1.l}] = P[\xi_{V_b} > V_{b.c.1.l}] &= \frac{1}{\sqrt{2\pi}} \int\limits_{\frac{V_{b.c.1.l} - \overline{V}_b}{\sigma_x}}^{\infty} e^{-\frac{t^2}{2}} dt \\ &= \frac{1}{\sqrt{2\pi}} \int\limits_{\frac{V_{b.c.1.l}}{2u_*} - 2.7}^{\infty} e^{-\frac{t^2}{2}} dt \end{aligned} \tag{2.26}$$

According to Nikitin's experiments [10], $\overline{V}_b = 5.6u_*$ at a height of D (particle diameter) above the bed, and $\overline{V}_b = 3.73u_*$ at a height of $2/3D$. Here, we use $\overline{V}_b = 3.73u_*, \sigma_x = 0.37\overline{V}_b$, and thus $\overline{V}_b/\sigma_x = 2.7$. Also, $\sigma_y = 1.05u_* \approx u_*$ based on Nikitin's experiments. The probability for sediment to cease movement is

$$P[B_{0.l}] = 1 - \varepsilon_{0.l}$$

where

$$\varepsilon_{0.l} = P[\xi_{V_b} > V_{b.c.0.l}] = \frac{1}{\sqrt{2\pi}} \int\limits_{\frac{V_{b.c.0.l} - \overline{V}_b}{\sigma_x}}^{\infty} e^{-\frac{t^2}{2}} dt = \frac{1}{\sqrt{2\pi}} \int\limits_{\frac{V_{b.c.0.l}}{2u_*} - 2.7}^{\infty} e^{-\frac{t^2}{2}} dt \tag{2.27}$$

is the probability for sediment not to cease movement. The probability for sediment hopping from the rest state is

$$\varepsilon_{2.0.l} = P[\xi_{V_b} > V_{b.c.2.0.l}] = \frac{1}{\sqrt{2\pi}} \int_{\frac{V_{b.c.2.0.l} - \overline{V}_b}{\sigma_x}}^{\infty} e^{-\frac{t^2}{2}} dt = \frac{1}{\sqrt{2\pi}} \int_{\frac{V_{b.c.2.0.l}}{2u_*} - 2.7}^{\infty} e^{-\frac{t^2}{2}} dt \tag{2.28}$$

The probability for sediment hopping from a motion state is

$$\varepsilon_{2.l} = P[\xi_{V_b} > V_{b.c.2.l}] = \frac{1}{\sqrt{2\pi}} \int_{\frac{V_{b.c.2.l} - \overline{V}_b}{\sigma_x}}^{\infty} e^{-\frac{t^2}{2}} dt = \frac{1}{\sqrt{2\pi}} \int_{\frac{V_{b.c.2.l}}{2u_*} - 2.7}^{\infty} e^{-\frac{t^2}{2}} dt \tag{2.29}$$

The probability for sediment suspension is

$$\varepsilon_{4.l} = P[B_{4.l}] = \frac{1}{\sqrt{2\pi}} \int_{\frac{\omega_l}{\sigma_y}}^{\infty} e^{-\frac{t^2}{2}} dt = \frac{1}{\sqrt{2\pi}} \int_{\frac{\omega_l}{u_*}}^{\infty} e^{-\frac{t^2}{2}} dt \tag{2.30}$$

The probability for incipient sediment suspension can be divided into several cases according to Eq. (2.25)

$$\beta_l = \begin{cases} \frac{1}{\sqrt{2\pi}} \int_{\frac{\omega_l}{u_*}}^{\infty} e^{-\frac{t^2}{2}} dt = \varepsilon_{4.l} & \left(\overline{V}_b > \frac{4\omega_{1.l}}{\sqrt{3}}, or \overline{V}_b \le \frac{4\omega_{1.l}}{\sqrt{3}} \; and \; V^{(1)}_{b.y.c.3.l} \le \omega_l\right) \\ \frac{1}{\sqrt{2\pi}} \int_{\frac{V^{(1)}_{b.y.c.3.l}}{u_*}}^{\infty} e^{-\frac{t^2}{2}} dt = \varepsilon_{5.1.l} & (\overline{V}_b \le \frac{4\omega_{1.l}}{\sqrt{3}}, \; V^{(2)}_{b.y.c.3.l} \le \omega_l < V^{(1)}_{b.y.c.3.l}) \\ \frac{1}{\sqrt{2\pi}} \int_{\frac{V^{(1)}_{b.y.c.3.l}}{u_*}}^{\infty} e^{-\frac{t^2}{2}} dt & \\ + \frac{1}{\sqrt{2\pi}} \int_{\frac{\omega_l}{u_*}}^{\frac{V^{(2)}_{b.y.c.3.l}}{u_*}} e^{-\frac{t^2}{2}} dt & (\overline{V}_b \le \frac{4\omega_{1.l}}{\sqrt{3}}, \; \omega_l < V^{(2)}_{b.y.c.3.l}) \\ = \varepsilon_{5.1.l} + \varepsilon_{4.l} - \varepsilon_{5.2.l} & \end{cases} \tag{2.31}$$

where

$$\varepsilon_{5.1.l} = \frac{1}{\sqrt{2\pi}} \int\limits_{\frac{V^{(1)}_{b.y.c.3.l}}{u_*}}^{\infty} e^{-\frac{t^2}{2}} dt \tag{2.32}$$

$$\varepsilon_{5.2.l} = \frac{1}{\sqrt{2\pi}} \int\limits_{\frac{V^{(2)}_{b.y.c.3.l}}{u_*}}^{\infty} e^{-\frac{t^2}{2}} dt \tag{2.33}$$

2.2 Characteristics of Basic Transition Probabilities

The numerical values for the four basic transition probabilities ($\varepsilon_{0.l}$, $\varepsilon_{1.l}$, $\varepsilon_{4.l}$, β_l) are listed in Tables 2.1, 2.2, 2.3, 2.4, 2.5 and 2.6 [6]. As shown in formulas and in these tables, $\varepsilon_{0.l}$ and $\varepsilon_{4.l}$ are independent of the water depth, and only depend on D_l and u_*. But $\varepsilon_{1.l}$ and β_l depend not only on D_l and u_*, but also on the water depth. In addition, they are impacted by the sediment dry density when the dry density of the sediment on the bed is different from that of the bed material, or it is not under the normal dry density condition [7]. The results in Tables 2.1, 2.2, 2.3, 2.4, 2.5 and 2.6 are only for the normal dry density condition. The characteristics of the four basic transition probabilities are illustrated in these table, and some characteristic are summarized as follows:

Table 2.1 Probability for sediment not to cease movement $\varepsilon_{0.l}$

D (mm)	u_* (cm/s)								
	1.00	2.00	3.00	5.00	8.00	10.00	12.00	15.00	20.00
0.001	0.9910	0.9940	0.9950	0.9960	0.9960	0.9960	0.9960	0.9960	0.9960
0.005	0.9740	0.9900	0.9930	0.9950	0.9950	0.9960	0.9960	0.9960	0.9960
0.010	0.9490	0.9850	0.9900	0.9940	0.9950	0.9950	0.9950	0.9960	0.9960
0.025	0.8460	0.9680	0.9840	0.9910	0.9940	0.9940	0.9950	0.9950	0.9960
0.050	0.6260	0.9350	0.9720	0.9870	0.9920	0.9930	0.9940	0.9940	0.9950
0.100	0.2540	0.8460	0.9430	0.9790	0.9890	0.9910	0.9920	0.9930	0.9940
0.250	0.0045	0.5170	0.8230	0.9490	0.9790	0.9850	0.9880	0.9900	0.9920
0.500	0.0000	0.1450	0.5770	0.8840	0.9610	0.9740	0.9810	0.9860	0.9900
1.000	0.0000	0.0045	0.2000	0.7170	0.9150	0.9490	0.9650	0.9770	0.9850

Table 2.2 Probability for sediment suspension $\varepsilon_{4.l}$

D (mm)	u_* (cm/s)								
	1.00	2.00	3.00	5.00	8.00	10.00	12.00	15.00	20.00
0.001	0.5000	0.5000	0.5000	0.5000	0.5000	0.5000	0.5000	0.5000	0.5000
0.005	0.4990	0.5000	0.5000	0.5000	0.5000	0.5000	0.5000	0.5000	0.5000
0.010	0.4980	0.4990	0.4990	0.5000	0.5000	0.5000	0.5000	0.5000	0.5000
0.025	0.4840	0.4920	0.4950	0.4970	0.4980	0.4980	0.4990	0.4990	0.4990
0.050	0.4380	0.4690	0.4790	0.4880	0.4920	0.4940	0.4950	0.4960	0.4970
0.100	0.2680	0.3780	0.4180	0.4510	0.4690	0.4750	0.4790	0.4840	0.4880
0.250	0.0012	0.0644	0.1560	0.2720	0.3520	0.3810	0.4000	0.4200	0.4400
0.500	0.0000	0.0003	0.0100	0.0811	0.1910	0.2420	0.2800	0.3210	0.3640
1.000	0.0000	0.0000	0.0000	0.0087	0.0686	0.1170	0.1610	0.2140	0.2760

Table 2.3 Probability for sediment entrainment $\varepsilon_{1.l}$ ($H = 10$ m)

D (mm)	u_* (cm/s)								
	1.00	2.00	3.00	5.00	8.00	10.00	12.00	15.00	20.00
0.001	0.0000	0.0000	0.0000	0.0000	0.0000	0.0000	0.0001	0.0071	0.1220
0.005	0.0000	0.0000	0.0000	0.0000	0.0524	0.2240	0.4280	0.6530	0.8340
0.010	0.0000	0.0000	0.0000	0.0141	0.3600	0.6000	0.7450	0.8570	0.9300
0.025	0.0000	0.0000	0.0065	0.3410	0.7750	0.8730	0.9200	0.9520	0.9730
0.050	0.0000	0.0019	0.1520	0.6780	0.9030	0.9430	0.9610	0.9750	0.9840
0.100	0.0000	0.0649	0.4560	0.8440	0.9500	0.9680	0.9770	0.9840	0.9890
0.250	0.0000	0.1810	0.6150	0.8950	0.9640	0.9760	0.9820	0.9870	0.9900
0.500	0.0000	0.0755	0.4770	0.8520	0.9520	0.9690	0.9780	0.9840	0.9890
1.000	0.0000	0.0030	0.1750	0.6980	0.9090	0.9460	0.9630	0.9760	0.9840

Table 2.4 Probability of sediment entrainment $\varepsilon_{1,l}$ ($H = 1$ m)

D (mm)	u_* (cm/s)								
	1.00	2.00	3.00	5.00	8.00	10.00	12.00	15.00	20.00
0.001	0.0000	0.0000	0.0000	0.0000	0.0590	0.2380	0.4430	0.6650	0.8400
0.005	0.0000	0.0000	0.0084	0.3610	0.7850	0.8790	0.9230	0.9540	0.9730
0.010	0.0000	0.0033	0.1810	0.7020	0.9100	0.9470	0.9640	0.9760	0.9840
0.025	0.0000	0.2080	0.6390	0.9020	0.9660	0.9770	0.9830	0.9870	0.9910
0.050	0.0037	0.5040	0.8180	0.9480	0.9790	0.9850	0.9880	0.9900	0.9920
0.100	0.0183	0.6200	0.8650	0.9590	0.9820	0.9870	0.9890	0.9910	0.9930
0.250	0.0012	0.4330	0.7840	0.9390	0.9760	0.9830	0.9870	0.9900	0.9920
0.500	0.0000	0.1280	0.5570	0.8780	0.9590	0.9730	0.9800	0.9860	0.9900
1.000	0.0000	0.0041	0.1940	0.7130	0.9130	0.9480	0.9650	0.9770	0.9850

Table 2.5 Incipient suspension probability β_l ($H = 1$ m)

D (mm)	u_* (cm/s)								
	1.00	2.00	3.00	5.00	8.00	10.00	12.00	15.00	20.00
0.001	0.0000	0.0000	0.0000	0.0003	0.0000	0.0001	0.0006	0.0044	0.0277
0.005	0.0000	0.0000	0.0000	0.0062	0.0145	0.0488	0.1870	0.5000	0.5000
0.010	0.0000	0.0000	0.0034	0.0730	0.0871	0.5000	0.5000	0.5000	0.5000
0.025	0.0000	0.0010	0.0211	0.4880	0.4980	0.4980	0.4990	0.4990	0.4990
0.050	0.0000	0.0029	0.0392	0.4510	0.4920	0.4940	0.4950	0.4960	0.4970
0.100	0.0000	0.0005	0.0143	0.2720	0.4690	0.4750	0.4790	0.4840	0.4880
0.250	0.0000	0.0000	0.0016	0.0478	0.3520	0.3810	0.4000	0.4200	0.4400
0.500	0.0000	0.0000	0.0000	0.0069	0.1910	0.2420	0.2800	0.3210	0.3640
1.000	0.0000	0.0000	0.0000	0.0000	0.0686	0.1170	0.1610	0.2140	0.2760

Table 2.6 Incipient Suspension probability β_l ($H = 10$ m)

D (mm)	u_*(cm/s) 1.00	2.00	3.00	5.00	8.00	10.00	12.00	15.00	20.00
0.001	0.0000	0.0000	0.0000	0.0000	0.0000	0.0000	0.0000	0.0000	0.0000
0.005	0.0000	0.0000	0.0000	0.0000	0.0000	0.0000	0.0005	0.0039	0.0258
0.010	0.0000	0.0000	0.0000	0.0000	0.0002	0.0024	0.0096	0.0350	0.2730
0.025	0.0000	0.0000	0.0000	0.0002	0.0130	0.0445	0.1530	0.4990	0.4990
0.050	0.0000	0.0000	0.0000	0.0049	0.0748	0.4940	0.4950	0.4960	0.4970
0.100	0.0000	0.0000	0.0006	0.0294	0.4690	0.4750	0.4790	0.4840	0.4880
0.250	0.0000	0.0000	0.0027	0.0640	0.3520	0.3810	0.4000	0.4200	0.4400
0.500	0.0000	0.0000	0.0008	0.0326	0.1910	0.2420	0.2800	0.3210	0.3640
1.000	0.0000	0.0000	0.0000	0.0059	0.0686	0.1170	0.1610	0.2140	0.2760

(1) For $u_* = 1.0$ cm/s and $D_l \geq 0.25$ mm, or $u_* = 2.0$ cm/s and $D_l \geq 1$ mm, $\varepsilon_{0.l} \cong 0$ (as shown below the bold line in Table 2.1), or the probability for sediment to cease movement is close to 1. This indicates that there is no sediment movement under these conditions, or specifically, the moving sediment will continuously cease movement until no bed load exists.

(2) For $u_* = 1.0$ cm/s and $D_l \geq 0.25$ mm, $u_* = 2.0$ cm/s and $D_l \geq 0.5$ mm, or $u_* = 3 - 5$ cm/s and $D_l \geq 1.0$ mm, $\varepsilon_{4.l} < 0.01$ (as shown below the bold line in Table 2.2), indicating the sediment suspension can be ignored. This range of condition corresponds to $Z = \frac{\omega}{ku_*} \geq 5$, or $\frac{\omega}{u_*} \geq 2$, which is also the range for no sediment suspension based on the sediment diffusion theory. This indicates that $\varepsilon_{4.l}$ is indeed an important parameter to describe the suspended sediment transport.

(3) $\varepsilon_{1.l}$ is smaller for both coarse and fine sediment particles, and reaches the maximum value for an intermediate particle size. For a water depth of $H = 10$ m, $\varepsilon_{1.l}$ reaches the maximum value for $D = 0.25$ mm (as shown in the cells enclosed within the dashed lines in Table 2.3). For $H = 1$ m, $\varepsilon_{1.l}$ reaches the maximum value for $D = 0.10$ mm (as shown in the cells enclosed within the dashed lines in Table 2.4). For the sediment finer than the particle size corresponding to the maximum value of $\varepsilon_{1.l}$, $\varepsilon_{1.l}$ decreases as the particle size decreases because the increased cohesive force and the additional pressure of water film result in a larger threshold velocity of incipient motion. On the other hand, $\varepsilon_{1.l}$ decreases as the particle size increases for the coarser particles because the greater gravity results in a larger threshold velocity of incipient motion.

(4) The cohesive force and the additional pressure of water film decrease as the particle size increases. Therefore, $\varepsilon_{0.l}$ and $\varepsilon_{1.l}$ are approximately the same if the particle is coarse ($D \geq 0.5$ mm) and u_* is not too small. As listed in Table 2.4, for $H = 1$ m, and $D = 0.5$–1 mm, $\varepsilon_{0.l}$ and $\varepsilon_{1.l}$ are almost the same for $u_* \geq 5.0$ cm/s. As listed in Table 2.3, for $H = 10$ m, and $D = 0.5$–1 mm, $\varepsilon_{0.l}$ and $\varepsilon_{1.l}$ are almost the same for $u_* \geq 10.0$ cm/s.

(5) Assuming the sediment entrainment can be neglected for $\varepsilon_{1.l} < 0.01$, the range for sediment entrainment is shown in the bottom and right portion of the bold lines in Tables 2.3 and 2.4.

(6) Assuming the incipient sediment suspension can be neglected for $\beta_l < 0.01$, all sediment particles cannot be suspended for $u_* = 1$–2 cm/s, as shown in Table 2.5. This is the reason why only sediment deposition (without erosion) occurs under the reservoir backwater condition when u_* is very small. For $u_* = 5$ cm/s, sediment suspension will only start for particle size of $D = 0.01$–0.25 mm. This illustrates the fact that the very fine or viscous particles are hard to be suspended after being consolidated. The ranges of possible incipient sediment suspension ($\beta_l > 0.01$) are shown in the bottom and right portion of the bold lines in Tables 2.5 and 2.6.

(7) Similar to $\varepsilon_{1.l}$, β_l also reaches the maximum value for an intermediate particle size. But the maximum point varies significantly with u_*. For example, it varies between $u_* = 5.0$–20.0 cm/s for $D = 0.01$–0.025 mm, as listed in Table 2.5. It is noted that β_l reaching the maximum value is only for the sediment at the bed layer. It does not mean this size class accounts for the largest proportion of the suspended load. This is because the sediment transport rate of a certain size class is dependent on many factors other than just on the incipient suspension probability β_l.

(8) If $u_* \geq 8.0$ cm/s, for the particles coarser than a certain particle size there is $\beta_l = \varepsilon_{4.l}$, as shown in the bottom portion below the dashed lines in Tables 2.5 and 2.6.

2.3 State Transition and State Probabilities for the Case with Independent Longitudinal and Vertical Fluctuating Velocities

As shown in Fig. 2.2, sediment has four states with 16 transition chains and 16 corresponding transition probabilities. If the rolling and hopping particles are grouped as bed load, sediment has three states with nine transition probabilities. It is obvious that the six basic transition probabilities discussed in the previous sections are not sufficient, and are not even the common transition probabilities for the transition chains. The sixteen transition probabilities will be derived from the six basic probabilities in this section. It is noted that the state transition probabilities and state probabilities with subscript "l" denote the probabilities for the particle size of D_l.

2.3.1 State Transition Probabilities

The longitudinal and vertical fluctuating velocities at the bottom are closely correlated with varying correlation coefficients. Two extreme cases are included in the following analysis. The case with independent fluctuating velocities is analyzed in this section. The case with a correlation coefficient of –1 will be discussed in Sect. 2.5. To compensate the drawback of assuming independence of fluctuating velocities, the longitudinal and vertical hydrodynamics were considered in the analysis of the sediment loosing condition, and thus the correlation of the fluctuating velocities is included.

1. For Four Sediment States with Independent Longitudinal and Vertical Bottom Fluctuating Velocities

The transition probability for sediment to remain at rest on the bed is

$$\beta_{1.1.l} = P[\xi_{V_b} \leq V_{b.c.1.l} \cap \xi_{V_b} \notin (B_{4.l} \cap B_{5.l})] = (1 - \varepsilon_{1.l})(1 - \beta_l) \quad (2.34)$$

which indicates that sediment particles cannot be entrained or be suspended in order to maintain the previous rest state. The probability for particles at rest to transit to the rolling state is

$$\beta_{1.2.l} = P[(V_{bc.1.l} \leq \xi_{Vb} \leq V_{b.c.2.0.l}) \cap \xi_{Vb,y} \notin (B_{4.l} \cap B_{5.l})] = (\varepsilon_{1.l} - \varepsilon_{2.0.l})(1 - \beta_l) \quad (2.35)$$

which indicates that sediment particles shall be entrained, but without hopping or being suspended, in order for the particles to transit from the rest state to the rolling state. The probability for particles to transit from the rest state to the hopping state is

$$\beta_{1.3.l} = P[(\xi_{V_b} > V_{b.c.1.l}) \cap \xi_{V_b} \notin (B_{4.l} \cap B_{5.l})] = \varepsilon_{2.0.l}(1 - \beta_l) \quad (2.36)$$

which indicates that particles shall be able to start hopping but not be suspended in order for the particles to transit from the rest state to the hopping state. The probability for particles at rest to be suspended is

$$\beta_{1.4.l} = P(B_{4.l} \bigcap B_{5.l})] = \beta_l \quad (2.37)$$

which indicates that sediment will become suspended load as long as the particles are loosen and suspended. The probability for sediment with an initial moving state $i\ (i \neq 1)$ to transit to the rest state is

$$\beta_{i.1.l} = P[(\xi_{Vb} \leq V_{b.c.0.l} \cap \xi_{Vy} \leq \omega_l] = (1 - \varepsilon_{0.l})(1 - \varepsilon_{4.l}) \quad (i = 2, 3, 4) \quad (2.38)$$

which indicates that the sediment particles with an initial moving state i ($i = 2$, 3, 4) will become the bed material when they are no longer suspended and cease movement. Similarly,

$$\begin{aligned}\beta_{i.2.l} &= P\left[\left(V_{b..c.0.l} < \xi_{V_b} \le V_{b.c.2.l}\right) \cap \xi_{V_y} \le \omega\right] \\ &= (\varepsilon_{1.l} - \varepsilon_{2.l})(1 - \varepsilon_{4.l}) \quad (i = 2,\ 3,\ 4)\end{aligned} \tag{2.39}$$

which indicates that the particles with an initial moving state i ($i = 2,3,4$) will change to the rolling state when they are in motion, but do not hop or are not suspended. Also

$$\beta_{i.3.l} = P\left[\left(\xi_{V_b} > V_{b.c.2.l}\right) \cap \xi_{V_{b.y}} \le \omega_l\right] = \varepsilon_{2.l}(1 - \varepsilon_{4.l}) \quad (i = 2, 3, 4) \tag{2.40}$$

which indicates that the particles with an initial state i will be in hopping state when they start hopping but are not suspended. It is obvious that the probability for sediment in moving state to transit to the suspension state is

$$\beta_{i.4} = P[\xi_{V_b} > \omega_l] = \varepsilon_{4.l} \quad (i = 2, 3, 4) \tag{2.41}$$

So far the 16 transition probabilities have been derived, and the transition probability matrix for the sediment on the bed can be written as

$$(\beta_{i.j.l}) = \begin{pmatrix} (1-\varepsilon_{1.l})(1-\beta_l) & (\varepsilon_{1.l}-\varepsilon_{2.0.l})(1-\beta_l) & \varepsilon_{2.0.l}(1-\beta_l) & \beta_l \\ (1-\varepsilon_{0.l})(1-\varepsilon_{4.l}) & (\varepsilon_{0.l}-\varepsilon_{2.l})(1-\varepsilon_{4.l}) & \varepsilon_{2.l}(1-\varepsilon_{4.l}) & \varepsilon_{4.l} \\ (1-\varepsilon_{0.l})(1-\varepsilon_{4.l}) & (\varepsilon_{0.l}-\varepsilon_{2.l})(1-\varepsilon_{4.l}) & \varepsilon_{2.l}(1-\varepsilon_{4.l}) & \varepsilon_{4.l} \\ (1-\varepsilon_{0.l})(1-\varepsilon_{4.l}) & (\varepsilon_{0.l}-\varepsilon_{2.l})(1-\varepsilon_{4.l}) & \varepsilon_{2.l}(1-\varepsilon_{4.l}) & \varepsilon_{4.l} \end{pmatrix} \tag{2.42}$$

2. For Three Sediment States with Independent Longitudinal and Vertical Bottom Fluctuating Velocities

For this case, the second state is bed load movement, including rolling and hopping, and the third state is suspension. The transition probability $\beta_{i.1.l}(i = 2, 3)$ remains the same as for the four states, $\beta_{1.3.l} = \beta_l$, and $\beta_{i.3.l}(i = 2, 3) = \varepsilon_{4.l}$. Also,

$$\beta_{1.2.l} = P[(\xi_{V_b} > V_{b.c.1.l} \cap \xi_{V_{b.y}} \notin (B_{4.l} \cap B_{5.l})] = \varepsilon_{1.l}(1 - \beta_l) \tag{2.43}$$

is the probability for sediment not to cease movement but is not suspended, and

$$\beta_{i.2.l} = P\left[\left(\xi_{V_b} > V_{b.c.0.l}\right) \cap \xi_{V_{b.y}} \notin (B_{4.l} \cap B_{5.l})\right] = \varepsilon_{0.l}(1 - \varepsilon_{4.l}), \quad (i = 2, 3) \tag{2.44}$$

Therefore, the transition probability matrix can be written as

$$(\beta_{i.j.l}) = \begin{pmatrix} (1-\varepsilon_{1.l})(1-\beta_l) & \varepsilon_{1.l}(1-\beta_l) & \beta_l \\ (1-\varepsilon_{0.l})(1-\varepsilon_{4.l}) & \varepsilon_{0.l}(1-\varepsilon_{4.l}) & \varepsilon_{4.l} \\ (1-\varepsilon_{0.l})(1-\varepsilon_{4.l}) & \varepsilon_{0.l}(1-\varepsilon_{4.l}) & \varepsilon_{4.l} \end{pmatrix} \tag{2.45}$$

2.3.2 State Probabilities

1. For Four Sediment States With Independent Longitudinal and Vertical Bottom Fluctuating Velocities

As indicated in the transition probability matrix, the sediment state transition is an aperiodic and irreducible Markov's chain, and thus the limit probabilities exist. The limit probability (or the state probability) derived from the Kolmogorov equation is

$$R_{j.l} = \sum_{i=1}^{4} R_{i.l}\beta_{i.j.l} \quad (j = 1, 2, 3, 4) \tag{2.46}$$

Then from the transition probability matrix, or Eq. (2.42), Eq. (2.46) turns to be

$$\begin{cases} R_{1.l} = (1-\varepsilon_{1.l})(1-\beta_l)R_{1.l} + (1-\varepsilon_{0.l})(1-\varepsilon_{4.l})R_{2.l} \\ \qquad + (1-\varepsilon_{0.l})(1-\varepsilon_{4.l})R_{3.l} + (1-\varepsilon_{0.l})(1-\varepsilon_{4.l})R_{4.l} \\ R_{2.l} = (\varepsilon_{1.l}-\varepsilon_{2.0.l})(1-\beta_l)R_{1.l} + (\varepsilon_{0.l}-\varepsilon_{2.l})(1-\varepsilon_{4.l})R_{2.l} \\ \qquad + (\varepsilon_{0.l}-\varepsilon_{2.l})(1-\varepsilon_{4.l})R_{3.l} + (\varepsilon_{0.l}-\varepsilon_{2.l})(1-\varepsilon_{4.l})R_{4.l} \\ R_{3.l} = \varepsilon_{2.0.l}(1-\beta_l)R_{1.l} + \varepsilon_{2.l}(1-\varepsilon_{4.l})R_{2.l} \\ \qquad + \varepsilon_{2.l}(1-\varepsilon_{4.l})R_{3.l} + \varepsilon_{2.l}(1-\varepsilon_{4\cdot l})R_{4.l} \\ R_{4.l} = \beta_l R_{1.l} + \varepsilon_{4.l}R_{2.l} + \varepsilon_{4.l}R_{3.l} + \varepsilon_{4.l}R_{4.l} \end{cases} \tag{2.47}$$

with the constraint condition

$$R_{1.l} + R_{2.l} + R_{3.l} + R_{4.l} = 1 \tag{2.48}$$

The state probabilities can be determined by solving Eqs. (2.47) and (2.48). Plugging Eq. (2.48) into the first equation of (2.47) yields

$$R_{1.l} = (1-\varepsilon_{1.l})(1-\beta_l)R_{1.l} + (1-\varepsilon_{0.l})(1-\varepsilon_{4.l})(1-R_{1.l})$$

Thus

$$\begin{aligned} R_{1.l} &= \frac{(1-\varepsilon_{0.l})(1-\varepsilon_{4.l})}{1+(1-\varepsilon_{0.l})(1-\varepsilon_{4.l})-(1-\varepsilon_{1.l})(1-\beta_l)} \\ &= \frac{(1-\varepsilon_{0.l})(1-\varepsilon_{4.l})}{A_l} = \frac{\beta_{2.1.l}}{A_l} = \frac{A_{1.l}}{A_l} \end{aligned} \tag{2.49}$$

where

$$A_l = 1 + (1-\varepsilon_{0.l})(1-\varepsilon_{4.l}) - (1-\varepsilon_{1.l})(1-\beta_l) \tag{2.50}$$

$$A_{1.l} = (1 - \varepsilon_{0.l})(1 - \varepsilon_{4.l}) \tag{2.51}$$

Plugging Eq. (2.48) into the fourth equation of (2.47) yields

$$R_{4.l} = \beta_l R_{1.l} + \varepsilon_{4.l}(1 - R_{1.l})$$

Substituting $R_{1.l}$ with Eq. (2.49) yields

$$\begin{aligned} R_{4.l} &= \frac{1}{A_l}\{\varepsilon_{4.l}[1 - (1 - \varepsilon_{1.l})(1 - \beta_l)] + \beta_l(1 - \varepsilon_{0.l})(1 - \varepsilon_{4.l})\} \\ &= \frac{A_{2.l}}{A_l}\varepsilon_{4.l} + \frac{A_{1.l}}{A_l}\beta_l \\ &= \frac{A_{2.l}}{A_l}\beta_{4.4.l} + \frac{A_{1.l}}{A_l}\beta_{1.4.l} \end{aligned} \tag{2.52}$$

where

$$\frac{A_{1.l}}{A_l} = \frac{1}{A_l}(1 - \varepsilon_{0.l})(1 - \varepsilon_{4.l})$$

$$\frac{A_{2.l}}{A_l} = \frac{1}{A_l}[1 - (1 - \varepsilon_{1.l})(1 - \beta_l)] = 1 - \frac{A_{1.l}}{A_l}$$

This indicates that the state probability $R_{4.l}$ is the sum of the state probabilities for the transitions from rest to suspension and from the moving states (rolling, hopping or suspension) to suspension.

Plugging Eq. (2.48) into the second equation of (2.47) yields

$$R_{2.l} = (\varepsilon_{1.l} - \varepsilon_{2.0.l})(1 - \beta_l)R_{1.l} + (\varepsilon_{0.l} - \varepsilon_{2.l})(1 - \varepsilon_{4.l})(1 - R_{1.l}) \tag{2.53}$$

Substituting $R_{1.l}$ with Eq. (2.49) yields

$$\begin{aligned} R_{2.l} &= (\varepsilon_{0.l} - \varepsilon_{2.l})(1 - \varepsilon_{4.l}) + \frac{1}{A_l}[(\varepsilon_{1.l} - \varepsilon_{2.0.l})(1 - \beta_l) \\ &\quad - (\varepsilon_{0.l} - \varepsilon_{2.l})(1 - \varepsilon_{4.l})](1 - \varepsilon_{0.l})(1 - \varepsilon_{4.l}) \\ &= \frac{1}{A_l}\{(\varepsilon_{0.l} - \varepsilon_{2.l})(1 - \varepsilon_{4.l})[1 - (1 - \varepsilon_{1.l})(1 - \beta_l)] \\ &\quad + (\varepsilon_{1.l} - \varepsilon_{2.0.l})(1 - \beta_l)(1 - \varepsilon_{0.l})(1 - \varepsilon_{4.l})\} \\ &= \frac{A_{1.l}}{A_l}\beta_{1.2.l} + \frac{A_{2.l}}{A_l}\beta_{2.2.l} \end{aligned} \tag{2.54}$$

Plugging Eq. (2.48) into the third equation of (2.47) yields

$$R_{3.l} = \varepsilon_{2.0.l}(1-\beta_l)R_{1.l} + \varepsilon_{2.l}(1-\varepsilon_{4.l})(1-R_{1.l})$$

Substituting $R_{1.l}$ with Eq. (2.49) yields

$$\begin{aligned} R_{3.l} &= \varepsilon_{2.l}(1-\varepsilon_{4.l}) + \frac{1}{A_l}[\varepsilon_{2.0.l}(1-\beta_l) - \varepsilon_{2.l}(1-\varepsilon_{4.l})] \\ &= \frac{1}{A_l}\{\varepsilon_{2.l}(1-\varepsilon_{4.l})[1-(1-\varepsilon_{1.l})(1-\beta_l)] \\ &\quad + \varepsilon_{2.0.l}(1-\beta_l)(1-\varepsilon_{0.l})(1-\varepsilon_{4.l})\} \\ &= \frac{A_{1.l}}{A_l}\varepsilon_{2.0.l}(1-\beta_l) + \frac{A_{2.l}}{A_l}\varepsilon_{2.l}(1-\varepsilon_{4.l}) \\ &= \frac{A_{1.l}}{A_l}\beta_{1.3.l} + \frac{A_{2.l}}{A_l}\beta_{3.3.l} \end{aligned} \tag{2.55}$$

The Eqs. (2.46) for state probabilities have definite physical meaning. Take the rolling state as an example, there is

$$R_2 = R_1\beta_{1.2} + R_2\beta_{2.2} + R_3\beta_{3.2} + R_4\beta_{4.2}$$

where R_{1l} is the probability for sediment particles at rest state, $\beta_{1.2}$ is the transition probability from rest to rolling, $R_1\beta_{1.2}$ represents the rolling state probability transited from the rest state probability. Similarly, $R_i\beta_{i.2}$ represents the part of R_i by transition from i state to rolling. In other words, the probability in rolling state equals the weighted summation of state probabilities in states of rest, rolling, hopping and suspension respectively with the weighted coefficients $\beta_{i.2}$, the transition probabilities from those states respectively to rolling.

Equations (2.49), (2.52), (2.54) and (2.55) represent the probability for sediment on the bed to be in rest, rolling, hopping, and suspension states, respectively. Summation of these equations yields

$$
\begin{aligned}
& R_{1.l} + R_{2.l} + R_{3.l} + R_{4.l} \\
& = \frac{1}{A_l}\{(1-\varepsilon_{0.l})(1-\varepsilon_{4.l}) + (\varepsilon_{0.l} - \varepsilon_{2.l})(1-\varepsilon_{4.l})[1-(1-\varepsilon_{1.l})(1-\beta_l)] \\
& + (\varepsilon_{1.l} - \varepsilon_{2.l})(1-\beta_l)(1-\varepsilon_{0.l})(1-\varepsilon_{4.l}) \\
& + \varepsilon_{2.l}(1-\varepsilon_{4.l})[1-(1-\varepsilon_{1.l})(1-\beta_l)] \\
& + \varepsilon_{2.0.l}(1-\beta_l)(1-\varepsilon_{0.l})(1-\varepsilon_{4.l}) + (1-\varepsilon_{0.l})(1-\varepsilon_{4.l})\beta_l \\
& + [1-(1-\varepsilon_{1.l})(1-\beta_l)]\varepsilon_{4.l}\} \\
& = \frac{1}{A_l}\{(1-\varepsilon_{0.l})(1-\varepsilon_{4.l}) + \varepsilon_{0.l}(1-\varepsilon_{4.l})[1-(1-\varepsilon_{1.l})(1-\beta_l)] \\
& + \varepsilon_{1.l}(1-\beta_l)(1-\varepsilon_{0.l})(1-\varepsilon_{4.l}) \\
& + (1-\varepsilon_{0.l})(1-\varepsilon_{4.l})\beta_l + [1-(1-\varepsilon_{1.l})(1-\beta_l)]\varepsilon_{4.l}\} \\
& = \frac{1}{A_l}\{(1-\varepsilon_{0.l})(1-\varepsilon_{4.l})[(1+\beta_l) + (1-\beta_l)\varepsilon_{1.l}] \\
& + [1-(1-\varepsilon_{1.l})(1-\beta_l)][\varepsilon_{0.l}(1-\varepsilon_{4.l}) + \varepsilon_{4.l}]\} \\
& = \frac{1}{A_l}\{(1-\varepsilon_{0.l})(1-\varepsilon_{4.l}) + \varepsilon_{0.l}(1-\varepsilon_{4.l})[1-(1-\varepsilon_{1.l})(1-\beta_l)] \\
& + \varepsilon_{1.l}(1-\beta_l)(1-\varepsilon_{0.l})(1-\varepsilon_{4.l}) \\
& + (1-\varepsilon_{0.l})(1-\varepsilon_{4.l})\beta_l + [1-(1-\varepsilon_{1.l})(1-\beta_l)]\varepsilon_{4.l}\} \\
& = \frac{1}{A_l}\{(1-\varepsilon_{0.l})(1-\varepsilon_{4.l})[(1+\beta_l) + (1-\beta_l)\varepsilon_{1.l}] \\
& + [1-(1-\varepsilon_{1.l})(1-\beta_l)][\varepsilon_{0.l}(1-\varepsilon_{4.l}) + \varepsilon_{4.l}]\} \\
& = \frac{1}{A_l}\{(1-\varepsilon_{0.l})(1-\varepsilon_{4.l})[1+\varepsilon_{1.l} + \beta_l(1-\varepsilon_{1.l})] \\
& + [1-(1-\varepsilon_{1.l})(1-\beta_l)][1-(1-\varepsilon_{0.l})(1-\varepsilon_{4.l})]\} \\
& = \frac{1}{A_l}\{(1+\varepsilon_{0.l})(1-\varepsilon_{4.l}) + [1-(1-\varepsilon_{1.l})(1-\beta_l)] \\
& [(1-\varepsilon_{0.l})(1-\varepsilon_{4.l}) + 1 - (1-\varepsilon_{0.l})(1-\varepsilon_{4.l})]\} \\
& = \frac{A_{1.l}}{A_l} + \frac{A_{2.l}}{A_l} = 1
\end{aligned}
$$

This is expected, because a particle on the bed must be in one of these four states.

2. For Three Sediment States with Independent Longitudinal and Vertical Bottom Fluctuating Velocities

For three states with nine transition probabilities, the state probabilities satisfy the following relations based on the Kolmogorov equation

$$R_{j.l} = \sum_{i=1}^{3} R_{i.l}\beta_{i.j.l} \quad (i, j = 1, 2, 3) \tag{2.56}$$

with the constraint condition

$$\sum_{i=1}^{3} R_{i.l} = 1 \tag{2.57}$$

Plugging the transition probabilities, or Eq. (2.45), into Eq. (2.56)

$$\begin{cases} R_{1.l} = (1-\varepsilon_{1.l})(1-\beta_l)R_{1.l} + (1-\varepsilon_{0.l})(1-\varepsilon_{4.l})R_{2.l} \\ \quad\quad +(1-\varepsilon_{0.l})(1-\varepsilon_{4.l})R_{3.l} \\ R_{2.l} = \varepsilon_{1.l}(1-\beta_l)R_{1.l} + \varepsilon_0(1-\varepsilon_{4.l})R_{2.l} + \varepsilon_{0.l}(1-\varepsilon_{4.l})R_{3.l} \\ R_{3.l} = \beta_l R_{1.l} + \varepsilon_{4.l}R_{2.l} + \varepsilon_{4.l}R_{3.l} \end{cases} \tag{2.58}$$

Solving Eqs. (2.57) and (2.58) for the state probabilities

$$R_{1.l} = \frac{(1-\varepsilon_{0.l})(1-\varepsilon_{4.l})}{1+(1-\varepsilon_{0.l})(1-\varepsilon_{4.l})-(1-\varepsilon_{1.l})(1-\beta_l)} = \frac{A_{1.l}}{A_l} \tag{2.59}$$

$$\begin{aligned} R_{2.l} &= \frac{(1-\varepsilon_{0.l})(1-\varepsilon_{4.l})\varepsilon_{1.l}(1-\beta_l) + [1-(1-\varepsilon_{1.l})(1-\beta_l)]\varepsilon_{0.l}(1-\varepsilon_{4.l})}{1+(1-\varepsilon_{0.l})(1-\varepsilon_{4.l})-(1-\varepsilon_{1.l})(1-\beta_l)} \\ &= \frac{A_{1.l}}{A_l}\varepsilon_1(1-\beta_l) + \frac{A_{2.l}}{A_l}\varepsilon_{0.l}(1-\varepsilon_{4.l}) \\ &= \frac{A_{1.l}}{A_l}\beta_{1.2} + \frac{A_{2.l}}{A_l}\beta_{2.2} \end{aligned} \tag{2.60}$$

$$\begin{aligned} R_{3.l} &= \frac{(1-\varepsilon_{0.l})(1-\varepsilon_{4.l})\beta_l + [1-(1-\varepsilon_{1.l})(1-\beta_l)]\varepsilon_{4.l}}{1+(1-\varepsilon_{0.l})(1-\varepsilon_{4.l})-(1-\varepsilon_{1.l})(1-\beta_l)} \\ &= \frac{A_{1.l}}{A_l}\beta_l + \frac{A_{2.l}}{A_l}\varepsilon_{4.l} = \frac{A_{1.l}}{A_l}\beta_{1.3.l} + \frac{A_{1.l}}{A_l}\beta_{3.3.l} \end{aligned} \tag{2.61}$$

These are the probabilities for the sediment on the bed to be in rest, bed load movement, and suspension states, respectively. Obviously, summation of these three state probabilities satisfies Eq. (2.57).

The following can be observed from Eqs. (2.59) through (2.61): (1) If $\varepsilon_{0.l} = 0$ and $\varepsilon_{4.l} = 0$, then $\varepsilon_{1.l} = 0$, $\beta_l = 0$, and thus $R_{1.l} = 1$, $R_{2.l} = R_{3.l} = 0$; (2) If $\varepsilon_{0.l} = 1$, or the ceasing probability is 0, then $R_{1.l} = 0$; (3) If the cohesive force and the additional pressure of water film do not exist, or $\varepsilon_{4.l} = \beta_l$ and $\varepsilon_{0.l} = \varepsilon_{1.l}$, then $R_{1.l} = (1-\varepsilon_{0.l})(1-\varepsilon_{4.l})$, $R_{2.l} = \varepsilon_{0.l}(1-\varepsilon_{4.l})$ and $R_{3.l} = \varepsilon_{4.l}$; (4) If $\beta_l = \varepsilon_{4.l}$, then $R_{3.l} = \varepsilon_{4.l}$. It noted that some studies simply used the transition probabilities to represent the state probabilities, such as $R_{3.l} = \varepsilon_{4.l}$, which shall be limited to a few special cases.

2.4 Transition Intensities for the Case with Independent Longitudinal and Vertical Fluctuating Velocities [4]

The transition probability matrix discussed in Sect. 2.3 can be used to describe the sediment state transitions if the sediment movement is considered as a Markovian chain, which physically implies that the state transitions do not need time and can occur at any time. But this is not true. For example, the suspended load and the bed load cannot change to the other states, such as rolling and rest, right after being entrained from the bed. Thus, sediment movement should be considered as a continuous Markovian process with discrete states and continuous time period.

Compared to the transition probabilities, the life distributions and the derived transition intensities can describe the state transition for sediment movement in more depth. The transition intensities were investigated by Han and He [3, 4]. Some of their results are introduced in this section. To simplify the subscript denotations, the equations were first developed for the uniform sediment, and then were revised for the non-uniform sediment.

2.4.1 *Life Distributions*

Sediment on the bed has four states, each with a certain life and life distribution. For a single moving particle ($i = 2, 3, 4$), or a particle in rolling, hopping or suspension state, its life distribution can be approximated by an exponential function

$$F_{t.i.1}(t) = P[\xi_{t.i.1} \le t] = 1 - \mathrm{e}^{-(1-\beta_{i.i})\mu_i U_i t} = 1 - \mathrm{e}^{-(1-\beta_{i.i})t/T_i} \quad (i = 2, 3, 4) \tag{2.62}$$

where F is the life distribution function; subscripts i denotes the initial state, 1 indicates a single particle, $\beta_{i.i}$ is the probability for a transition from state i to i, i.e. the probability for a particle to remain its initial state, t the time, $\xi_{t.i.1}$ the time interval for a single unit movement of a particle with state i, $T_i = \frac{1}{\mu_i U_i}$ the averaged single-step movement interval, $\mu_i = \frac{1}{l_i}$ the reciprocal of the averaged single-step movement distance, and U_i the averaged movement speed. The single-event in state i consists of several continuous steps in state i, and after the end event the particle change its state to state j, $j \ne i$. The distribution of the time of the single-event movement reflects the life (time interval) for a particle to be in state i, and thus is called the life distribution for this state, representing the probability for state transitions. For example, when $t = 0$, the probability for a particle to change its state is 0; and the probability for the state transitions increases with time t. Assuming the transition probability $\beta_{i.i} = 0.5$, the probability for this particle to change its state is 0.393 after a single-step movement, and is 0.993 after ten single-steps. For K_i particles moving simultaneously, the probability for no particle to change its state is

$$\left[e^{-(1-\beta_{i.i})\mu_i U_i t}\right]^{K_i} = e^{-(1-\beta_{i.i})K_i \mu_i U_i t}$$

The probability for at least one particle to change its state is

$$F_{t.i.K_i}(t) = P\left[\xi_{t.i.K_i} \le t\right] = 1 - e^{-(1-\beta_{i.i})\mu_i U_i K_i t} = 1 - \exp\left(-\frac{(1-\beta_{i.i})K_i t}{T_i}\right) \tag{2.63}$$

A particle at rest also has life distribution, which is

$$F_{t.1.1}(t) = P[\xi_{t.1.1} \le t] = 1 - e^{-\frac{(1-\beta_{1.1})t}{T_1}} \tag{2.64}$$

with

$$\frac{1}{T_1} = \sum_{j=2}^{4} \frac{\beta_{1.j}}{1-\beta_{1.1}} \frac{1}{t_{0.j}} \tag{2.65}$$

where $\xi_{t.1.1}$ is the interval for a particle at rest to change its state, T_1 the averaged time for a particle to be detached from the bed layer after being entrained, and $t_{0.j}$ the time interval for a particle to be detached from the bed layer after being transited from the rest state to state j. Similarly, the life distribution for K_1 particles is

$$F_{t.1.K_1}(t) = P\left[\xi_{t.1.K_1} \le t\right] = 1 - \exp\left(\frac{-(1-\beta_{1.1})K_1 t}{T_1}\right) \tag{2.66}$$

where $F_{t.1.K_1}$ is the probability for at least one particle at rest to be entrained within time t, $\xi_{t.1.K_1}$ the time interval for one particle to be entrained among the K_1 particles resting on a unit area of the bed surface, and $K_1 = R_1 n_0$ the number of particles resting on a unit area of the bed surface, or

$$K_1 = R_1 n_0 = \frac{R_1 m_0}{\frac{\pi}{4} D^2} \tag{2.67}$$

where $m_0 = 0.4$ is the static packing coefficient, D the particle diameter, R_1 the state probability for sediment resting on the bed (Eq. 2.49).

From Eqs. (2.63) and (2.66) the densities of the life distributions for multiple particles with state i can be derived, which are $(1-\beta_{i.i})K_i\mu_i U_i e^{-(1-\beta_{i.i})K_i\mu_i U_i t}$ and $(1-\beta_{1.1})\frac{K_1}{T_1}e^{-(1-\beta_{1.1})\frac{K_1}{T_1}}$, respectively. Their mathematical expectations are

$$\begin{aligned} M\left[\xi_{t.i.K_i}\right] &= \int_0^\infty t\left[(1-\beta_{i.i})\mu_i U_i K_i e^{-(1-\beta_{i.i})\mu_i U_i K_i t}dt\right] \\ &= (1-\beta_{i.i})\mu_i U_i K_i\left\{\frac{e^{-(1-\beta_{i.i})\mu_i U_i K_i t}}{[(1-P_{i.i})\mu_i U_i K_i]^2}[-(1-\beta_{i.i})\mu_i U_i K_i t-1]_0^\infty\right\} \\ &= \frac{1}{(1-\beta_{i.i})\mu_i U_i K_i} = \frac{T_i}{(1-\beta_{i.i})K_i} \quad (i=2,3,4) \end{aligned} \tag{2.68}$$

$$M\left[\xi_{t.1.K_1}\right] = \int_0^\infty t\left[1-e^{-(1-\beta_{1.1})\frac{t}{T_1}}\right]dt = \frac{T_1}{(1-\beta_{1.1})K_1} \quad (i=1) \tag{2.69}$$

2.4.2 *Distributions and Expectations for the Number of Particles with State Transitions*

Consider a water column with a height of h and a unit area on the bed surface, within which the number of particles with moving state i is K_i. K_i is a random number. Among the K_i particles, m particles change states during a time interval from 0 to t, and m is also a random number. The conditional probability for m particles to change their states under the condition of $\xi_{K_i} = K_i$ is

$$W_{m.K_i}(t) = P\left[\xi_{m.i}(t) = m \middle| \xi_{K_i} = K_i\right] \tag{2.70}$$

If $m = 0$, $W_{0.K_i}(t)$ becomes the probability for none of the K_i particles with state i to change its state within time t. The event of no particle to change its state is the inverse of the event with at least one particle changing its state. Therefore, it can be derived from Eqs. (2.63) and (2.66) that

$$\begin{aligned} W_{0.K_i}(t) &= P\left[\xi_{t.i.K_i} > t\right] = 1 - P\left[\xi_{t.i.K_i} \le t\right] \\ &= e^{-\frac{(1-\beta_{i.i})K_i t}{T_i}} = e^{-\lambda_i t}(i=1,2,3,4) \end{aligned} \tag{2.71}$$

with

$$\lambda_i = \frac{(1-\beta_{i.i})K_i}{T_i} \quad (i=1,2,3,4) \tag{2.72}$$

where T_1 can be determined from Eq. (2.65), and $T_i = \frac{1}{\mu_i U_i}, (i = 2, 3, 4)$. In addition to the assumptions in deriving Eq. (2.71), it is also assumed that the number of particles with state transition shall not exceed one as $\Delta t \to 0$, or the probability for two or more particles to change their states is a higher order term of Δt. It can be proved that the probability for one particle to change its state within Δt is

$$\begin{aligned} W_{1.K_i}(\Delta t) &= 1 - W_{0.K_i}(\Delta t) + 0(\Delta t) \\ &= 1 - [1 - \lambda_i \Delta t + \cdots\cdots] + 0(\Delta t) = \lambda_i \Delta t + 0(\Delta t) \end{aligned}$$

Therefore,

$$\begin{aligned} W_{m.K_i}(t + \Delta t) &= W_{m-1.K_i}(t) W_{1.K_i}(\Delta t) + W_{m.K_i}(t) W_{0.K_i}(\Delta t) + 0(\Delta t) \\ &= W_{m-1.K_i}(t) \lambda_i \Delta t + W_{m.K_i}(t)(1 - \lambda_i \Delta t) + 0(\Delta t) \quad (m \geq 1) \end{aligned}$$

As $\Delta t \to 0$,

$$\frac{dW_{m.K_i}(t)}{dt} = \lim_{\Delta t \to 0} \frac{W_{m.K_i}(t + \Delta t) - W_{m.K_i}(t)}{\Delta t} = \lambda_i \left[W_{m-1.K_i}(t) - W_{m.K_i}(t) \right] \tag{2.73}$$

which is a linear equation and with the initial condition, i.e., $t = 0$, $W_{m.K_i}(0) = 0 (m > 1)$. It can be solved as

$$W_{m.K_i}(t) = e^{-\lambda_i t} \int_0^t \lambda_i e^{\lambda_i \tau} W_{m-1.K_i}(\tau) d\tau \tag{2.74}$$

Setting

$$W_{m.K_i}(t) e^{\lambda_i t} = R_{m.K_i}(t) \tag{2.75}$$

then Eq. (2.74) becomes

$$R_{m.K_i}(t) = \int_0^t \lambda_i e^{\lambda_i \tau} W_{m-1.K_i}(\tau) d\tau = \lambda_i \int_0^t R_{m-1.K_i}(\tau) d\tau \tag{2.76}$$

This is the recurrence formula for $R_{m.K_i}(t)$. For $m = 1$, plugging Eq. (2.71) into Eq. (2.76) yields

$$R_{1.K_i}(t) = \lambda_i \int_0^t e^{\lambda_i \tau} e^{-\lambda_i \tau} d\tau = \lambda_i t$$

For $m = 2$, plugging the above equation into Eq. (2.76) yields

$$R_{2.K_i}(t) = \lambda_i \int_0^t \lambda_i \tau d\tau = \frac{(\lambda_i t)^2}{2!} \tag{2.77}$$

More generally,

$$R_{m.K_i}(t) = \frac{(\lambda_i t)^m}{m!}$$

Also, it can be derived from Eqs. (2.71) and (2.75) that

$$R_{0.K_i}(t) = W_{0.K_i}(t) e^{\lambda_i t} = e^{-\lambda_i t} e^{\lambda_i t} = 1$$

Plugging $R_{m.K_i}$ into Eq. (2.75) yields

$$W_{m.K_i} = e^{-\lambda_i t} \frac{(\lambda_i t)^m}{m!} \quad (m = 0, 1, 2 \cdots) \tag{2.78}$$

It is noted that the condition distribution of m is a Poisson distribution when the number of particles with state transitions K_i, is sufficiently large (or t is sufficiently long). So the conditional expectation of m is

$$\begin{aligned} \overline{m}(K_i) &= \sum m W_{m.K_i}(t) = \sum_{m=0}^{\infty} m e^{-\lambda_i t} \frac{(\lambda_i t)^m}{m!} = \lambda_i t \sum_{m=1}^{\infty} e^{-\lambda_i t} \frac{(\lambda_i t)^{m-1}}{(m-1)!} \\ &= \lambda_i t \sum_{m=0}^{\infty} e^{-\lambda_i t} \frac{(\lambda_i t)^m}{m!} = \lambda_i t = \frac{(1-\beta_{i.i}) K_i}{T_i} t \quad (i = 1, 2, 3, 4) \end{aligned} \tag{2.79}$$

The unconditional expectation is

$$\begin{aligned}\overline{m}_i &= \sum_{K_i=0}^{\infty} \overline{m}(K_i)P(K_i) = \sum \lambda_i t P(K_i) \\ &= \sum_{K_i=0}^{\infty} \frac{(1-\beta_{i.i})K_i}{T_i} t P(K_i) \\ &= \frac{(1-\beta_{i.i})}{T_i}\overline{K}_i t \quad (i = 1, 2, 3, 4)\end{aligned} \tag{2.80}$$

The number of particles to change states per unit time is

$$\frac{\overline{m}_i}{t} = \frac{(1-\beta_{i.i})\overline{K}_i}{T_i} = \overline{\lambda}_i \quad (i = 1, 2, 3, 4) \tag{2.81}$$

2.4.3 Transition Intensities for Uniform Sediment

Noting that

$$T_i = \frac{1}{\mu_i U_i} \quad (i = 2, 3, 4) \tag{2.82}$$

$$1 - \beta_{i.i} = \sum_{j \neq i} \beta_{i.j}$$

the intensity for ending state i (the number of particles with initial state i changing to the other states) can be derived from Eq. (2.81) written as

$$\overline{\lambda}_i = \frac{(1-\beta_{i.i})K_i}{T_i} = \begin{cases} \sum\limits_{j \neq i} \beta_{i.j}\overline{K}_i \mu_i U_i = \sum\limits_{j=i} \overline{\lambda}_{i.j} \ (i = 2, 3, 4) \\ \sum\limits_{j \neq 1} \beta_{1.j}\frac{\overline{K}_1}{t_{0.j}} = \sum\limits_{j \neq 1} \overline{\lambda}_{1.j} \qquad (i = 1) \end{cases} \tag{2.83}$$

where

$$\overline{\lambda}_{i.j} = \begin{cases} \beta_{i.j}\overline{K}_i \mu_i U_i \ (i = 2, 3, 4;\ j = 1, 2, 3, 4;\ i \neq j) \\ \beta_{1.j}\frac{\overline{K}_1}{t_{0.j}} \qquad (i = 1;\ j \neq 1) \end{cases} \tag{2.84}$$

is called the intensity for transition (or exchange intensity) from state i to state j. Plugging Eq. (2.42) for the transition probabilities $\beta_{i.j}$ into Eq. (2.84), and applying Eq. (2.67) yields

$$\begin{cases} \overline{\lambda}_{1.2} = (\varepsilon_1 - \varepsilon_{2.0})(1-\beta)\dfrac{R_1 n_0}{t_{0.2\ \ i}} \\ \overline{\lambda}_{1.3} = \varepsilon_{2.0}(1-\beta)\dfrac{R_1 n_0}{t_{0.3}} \\ \overline{\lambda}_{1.4} = \beta\dfrac{R_1 n_0}{t_{0.4}} \\ \overline{\lambda}_{2.1} = (1-\varepsilon_0)(1-\varepsilon_4)\overline{K}_2\mu_2 U_2 \\ \overline{\lambda}_{2.3} = \varepsilon_2(1-\varepsilon_4)\overline{K}_2\mu_2 U_2 \\ \overline{\lambda}_{2.4} = \varepsilon_4\overline{K}_2\mu_2 U_2 \\ \overline{\lambda}_{3.1} = (1-\varepsilon_0)(1-\varepsilon_4)\overline{K}_3\mu_3 U_3 \\ \overline{\lambda}_{3.2} = (\varepsilon_0-\varepsilon_2)(1-\varepsilon_4)\overline{K}_3\mu_3 U_3 \\ \overline{\lambda}_{3.4} = \varepsilon_4\overline{K}_3\mu_3 U_3 \\ \overline{\lambda}_{4.1} = (1-\varepsilon_0)(1-\varepsilon_4)\overline{K}_4\mu_4 U_4 \\ \overline{\lambda}_{4.2} = (\varepsilon_0-\varepsilon_2)(1-\varepsilon_4)\overline{K}_4\mu_4 U_4 \\ \overline{\lambda}_{4.3} = \varepsilon_2(1-\varepsilon_4)\overline{K}_4\mu_4 U_4 \end{cases} \tag{2.85}$$

These 12 transition intensities are derived from Eq. (2.84) for $j \neq i$. For $j = i$, the transition intensities can be defined as

$$\overline{\lambda}_{i.i} = \beta_{i.i}\overline{K}_i\mu_i U_i \quad (i = 2, 3, 4) \tag{2.86}$$

or

$$\begin{aligned} \overline{\lambda}_{2.2} &= (\varepsilon_0-\varepsilon_2)(1-\varepsilon_4)\overline{K}_2\mu_2 U_2 \\ \overline{\lambda}_{3.3} &= \varepsilon_2(1-\varepsilon_4)\overline{K}_3\mu_3 U_3 \\ \overline{\lambda}_{4.4} &= \varepsilon_4\overline{K}_4\mu_4 U_4 \end{aligned}$$

For $\lambda_{1.1}$

$$\overline{\lambda}_{1.1} = \beta_{1.1}\frac{\overline{K}_1}{T} = \beta_{1.1}\overline{K}_1\frac{1}{1-\beta_{1.1}}\left(\frac{\beta_{1.2}}{t_{0.2}} + \frac{\beta_{1.3}}{t_{0.3}} + \frac{\beta_{1.4}}{t_{0.4}}\right) \tag{2.87}$$

Summation of the first equation in (2.84) and Eq. (2.86) yields

$$\begin{aligned}\sum_{j\neq i}\bar{\lambda}_{i.j}+\bar{\lambda}_{i.i}&=\sum_{i\neq j}\beta_{i.j}\overline{K}_i\mu_i U_i+\beta_{i.j}\overline{K}_i\mu_i U_i\\&=\sum_{j=1}^{4}\beta_{i.j}\overline{K}_i\mu_i U_i\\&=\overline{K}_i\mu_i U_i=\frac{\overline{K}_i}{T_i}\quad(i=2,3,4)\end{aligned}\tag{2.88}$$

which is the sum of the transition intensity from state i to state $j(i\neq j)$ and the intensity for remaining in state i per unit time and per unit area. Summation of the second equation in (2.84) and Eq. (2.87) yields

$$\begin{aligned}&\sum_{j\neq 1}\bar{\lambda}_{1.j}+\bar{\lambda}_{1.1}\\&=\sum_{j\neq 1}\beta_{1.j}\frac{\bar{K}_i}{t_{0.j}}+\beta_{1.1}\bar{K}_1\frac{1}{1-\beta_{1.1}}\left(\frac{\beta_{1.2}}{t_{0.2}}+\frac{\beta_{1.3}}{t_{0.3}}+\frac{\beta_{1.4}}{t_{0.4}}\right)\\&=\frac{1-\beta_{1.1}}{1-\beta_{1.1}}\bar{K}_1\sum_{j=2}^{4}\frac{\beta_{1.j}}{t_{0.j}}+\frac{\beta_{1.1}}{1-\beta_{1.1}}\bar{K}_1\sum_{j=2}^{4}\frac{\beta_{1.j}}{t_{0.j}}\\&=\frac{\bar{K}_1}{1-\beta_{1.1}}\sum_{j=2}^{4}\frac{\beta_{1.j}}{t_{0.j}}=\frac{\bar{K}_1}{T_1}\end{aligned}\tag{2.89}$$

This is the sum of the transition intensity from state 1 to the other states and the intensity for remaining in its initial state, and this result is similar to Eq. (2.88) for $i = 2, 3, 4$.

The transition intensity discussed above has definite physical meaning. It is part of the results of Han and He's [3, 4] statistical theory for sediment transport. Starting from the transition probability, state transition intensity and life distribution, Han and He derived the conditional distribution and expectation for the number of particles with state transitions, and then the transition intensity. Their derivation was in thorough detail and mathematically strict, but was complicated and not intuitive. For easier understanding of the derivation, the random process of sediment state transitions is considered as a deterministic process, and the transition intensity from state i to state j is derived in a more intuitive approach in the following. Assuming the number of particles with state i is $\overline{K}_i$ within a unit water column, and the averaged single-event movement distance is $1\big/[(1-\beta_{i.i})\mu_i]$ with the averaged time interval of $1\big/[(1-\beta_{i.i})\mu_i U_i]$, then all of the $\overline{K}_i$ particles will change their states over the interval of $1\big/[(1-\beta_{i.i})\mu_i U_i]$. Therefore, the number of particles that change their state i per unit time is

$$\overline{\lambda}_i = \frac{\overline{K}_i}{\frac{1}{(1-\beta_{i.i})\mu_i U_i}} = (1-\beta_{i.i})\overline{K}_i\mu_i U_i = \sum_{j\neq i}\beta_{i.j}\overline{K}_i\mu_i U_i = \sum_{j\neq i}\overline{\lambda}_{i.j} \quad (i=2,3,4) \tag{2.90}$$

This is the same as the first equation in (2.83). These particles will transit from state i to j according to the ratio of the number of particles $\frac{\beta_{i.j}}{1-\beta_{i.i}}$, or

$$\overline{\lambda}_{i.j} = \frac{\beta_{i.j}}{1-\beta_{i.i}}\overline{\lambda}_i = \beta_{i.j}\overline{K}_i\mu_i U_i \quad (i=2,3,4;\ j\neq i) \tag{2.91}$$

This is the same as the first equation in Eq. (2.84). For sediment at rest, assuming the period for entrainment is T_1, a particle can be entrained for t/T_1 times over time t. Assuming the number of particles at rest on a unit area of the bed surface is $\overline{K}_1$, and the ratio for particle number to end the rest state is $1-\beta_{1.1}$, then the number of particles that can be entrained at single-event time is $(1-\beta_{1.1})\overline{K}_1$. As a particle can be entrained t/T_1 times over time t, the number of particles that are entrained over time t is

$$\overline{m}_1 = (1-\beta_{1.1})\overline{K}_1\frac{t}{T_1}$$

The number of particles that are entrained per unit time is

$$\overline{\lambda}_1 = \overline{m}_1/t = (1-\beta_{1.1})\overline{K}_1/T_1 \tag{2.92}$$

Let's assume the period for a particle at rest to change to state j ($j=2,3,4$) is $t_{0.j}$, and the ratio of the number of particles to transit from rest to state j is $\beta_{1.j}$. Then among the $\overline{K}_1$ particles at rest on a unit area of the bed surface, the number of particles that change to state j over time t is

$$\overline{m}_{1.j} = \beta_{1.j}\frac{K_1}{t_{0.j}}t$$

The intensity for particles to transit from rest to state j per unit time is

$$\overline{\lambda}_{1.j} = \frac{\overline{m}_{1.j}}{t} = \frac{\beta_{1.j}\overline{K}_1}{t_{0.j}} \quad (j\neq 1) \tag{2.93}$$

The following relation can be developed by combining Eqs. (2.92) and (2.93)

$$\frac{(1-\beta_{1.i})\overline{K}_1}{T_1} = \overline{\lambda}_1 = \sum_{j=2}^{4} \overline{\lambda}_{1.j} = \sum_{j=2}^{4} \frac{\beta_{1.j}\overline{K}_1}{t_{0.j}}$$

which is the same as the second equation in (2.83). The expression of T_1 can also be written as

$$\frac{1}{T_1} = \frac{1}{(1-\beta_{1.1})} \sum_{j=2}^{4} \frac{\beta_{1.j}}{t_{0.j}} \tag{2.94}$$

2.4.4 Transition Intensities for Non-uniform Sediment

The exchange intensities for non-uniform sediment were derived by Han [2] and Han and He [3, 4]. Dividing the non-uniform sediment into n size classes, the basic transition probabilities are $\varepsilon_{1.l}$, $\varepsilon_{0.l}$, $\varepsilon_{2.l}$, $\varepsilon_{4.l}$, and β_l for each size class l with a particle size of D_l. The corresponding transition probability is $\beta_{i.j.l}$, and the number of particles at rest on a unit area of the bed surface is

$$\overline{K}_{1.l} = R_{1.l} n_{0.l} = \frac{P_{1.l} R_{1.l} m_0}{\frac{\pi}{4} D_l^2} \tag{2.95}$$

where $R_{1.l}$ is the probability for the particle with the size of D_l that is at rest on the bed, $P_{1.l}$ the gradation of the bed material, and $n_{0.l} = \frac{P_{1.l} m_0}{\frac{\pi}{4} D_l^2}$ the number of particles with the size of D_l on a unit area of the bed surface. Also the number of particles in the l-th size class that are in state i is

$$\overline{K}_{i.l} = P_{i.l} \overline{K}_i \quad (i = 2, 3, 4) \tag{2.96}$$

where $P_{i.l}$ is the gradation for the rolling, hopping and suspended sediment, respectively, or

$$P_{i.l} = \frac{K_{i.l}}{\sum\limits_{l=1}^{n} K_{i.l}} = \frac{K_{i.l}}{K_i} \quad (i = 2, 3, 4) \tag{2.97}$$

The transition intensity for the l-th size class is

$$\overline{\lambda}_{i.j.l} = \begin{cases} \beta_{i.j.l} \overline{K}_{i.l} \mu_{i.l} U_{i.l} & (i = 2, 3, 4;\ j \neq i;\ l = 1, 2, \cdots n) \\ \beta_{1.j.l} \frac{\overline{K}_{1.l}}{t_{0.j.l}} = \beta_{1.j.l} \frac{R_{1.l} P_{1.l} m_0}{t_{0.j.l} \frac{\pi}{4} D_l^2} & (i = 1;\ j \neq 1;\ l = 1, 2, \cdots n) \end{cases} \tag{2.98}$$

Plugging the transition probability $\beta_{i.j.l}$ in which the transition intensity for non-uniform sediment can be written as

$$\begin{cases} \overline{\lambda}_{1.2.l} = (\varepsilon_{1.l} - \varepsilon_{2.0.l})(1-\beta_l)\dfrac{m_0 R_{1.l} P_{1.l}}{\frac{\pi}{4} D_l^2 t_{0.2.l}} \\ \overline{\lambda}_{1.3.l} = \varepsilon_{2.0.l}(1-\beta_l)\dfrac{m_0 R_{1.l} P_{1.l}}{\frac{\pi}{4} D_l^2 t_{0.3.l}} \\ \overline{\lambda}_{1.4.l} = \beta_l \dfrac{m_0 R_{1.l} P_{1.l}}{\frac{\pi}{4} D_l^2 t_{0.4.l}} \\ \overline{\lambda}_{2.1.l} = (1-\varepsilon_{0.l})(1-\varepsilon_{4.l})\overline{K}_{2.l}\mu_{2.l}U_{2.l} \\ \overline{\lambda}_{2.3.l} = \varepsilon_{2.l}(1-\varepsilon_{4.l}) = \overline{K}_{2.l}\mu_{2.l}U_{2.l} \\ \overline{\lambda}_{2.4.l} = \varepsilon_{4.l}\overline{K}_{2.l}\mu_{2.l}U_{2.l} \\ \overline{\lambda}_{3.1.l} = (1-\varepsilon_{0.l})(1-\varepsilon_{4.l})\overline{K}_{3.l}\mu_{3.l}U_{3.l} \\ \overline{\lambda}_{3.2.l} = (\varepsilon_{0.l}-\varepsilon_{2.l})(1-\varepsilon_{4.l})\overline{K}_{3.l}\mu_{3.l}U_{3.l} \\ \overline{\lambda}_{3.4.l} = \varepsilon_4 \overline{K}_{3.l}\mu_{3.l}U_{3.l} \\ \overline{\lambda}_{4.1.l} = (1-\varepsilon_{0.l})(1-\varepsilon_{4.l})\overline{K}_{4.l}\mu_{4.l}U_{4.l} \\ \overline{\lambda}_{4.2.l} = (\varepsilon_{0.l}-\varepsilon_{2.l})(1-\varepsilon_{4.l})\overline{K}_{4.l}\mu_{4.l}U_{4.l} \\ \overline{\lambda}_{4.3.l} = \varepsilon_{2.l}(1-\varepsilon_{4.l})\overline{K}_{4.l}\mu_{4.l}U_{4.l} \end{cases} \tag{2.99}$$

2.5 Transition and State Probabilities for the Case with Linearly Correlated Longitudinal and Vertical Fluctuating Velocities [9]

It is more realistic to consider the linear correlation between the longitudinal and the vertical fluctuating velocities in the calculation of the transition probabilities among various sediment states. The correlation coefficient is close to -1 on the bed. For this case we have

$$\xi_{Vb} - \overline{V}_b = -\frac{\sigma_x}{\sigma_y}\xi_{Vb,y} = -2\xi_{Vb,y} \tag{2.100}$$

where $\overline{V}_b$ is the time average longitudinal flow velocity at the bottom, ξ_{V_b} and $\xi_{V_{b_y}}$ are the random longitudinal and vertical fluctuating velocities, respectively, $\sigma_x \approx 2u_*$, and $\sigma_y \approx u_*$. Equation (2.100) can be rewritten as

$$\xi_{Vb,y} = \frac{\overline{V}_b - \xi_{Vb}}{2} \tag{2.101}$$

The parameters of this section are mainly expressed in terms of the vertical flow velocity $V_{b.y}$ and the corresponding threshold velocities, with the subscript y. The following analysis is conducted for uniform sediment only in order to reduce the number of subscripts for the parameters.

2.5.1 Adjustment of Threshold Flow Velocities

The transition probabilities for the case with correlated longitudinal and vertical fluctuating velocities are related to the ranking of magnitudes of the threshold flow velocities, which is complicated because of the threshold velocities $V_{b.c.1}$ depend on water depth. Therefore, simplified formulations are adopted for the threshold velocities in this section. Taking the average value of $\varphi(\Delta')$ in Eq. (2.11), the threshold flow velocities for incipient motion are

$$V_{b.c.1} = \overline{\varphi}(\Delta')\omega_1 = 0.916\omega_1 = 0.916\sqrt{53.9D + \frac{3\times 10^{-7}}{D}(1+0.85H)} \quad (2.102)$$

where

$$\omega_1 = \sqrt{53.9D + \frac{3\times 10^{-7}}{D}(1+0.85H)} = K\sqrt{53.9D} = K\omega_0 \quad (2.103)$$

The sediment fall velocity is

$$\omega = \sqrt{\frac{4}{3C}\left(\frac{\gamma_s - \gamma}{\gamma}\right)gD} \quad (2.104)$$

where

$$\frac{\omega_0}{\omega} = \sqrt{\frac{C}{C_x}} = \sqrt{\frac{1.2}{0.4}} = \sqrt{3} \quad (2.105)$$

So

$$\omega_1 = K\omega_0 = \sqrt{3}K\omega \quad (2.106)$$

where

$$K = \frac{\sqrt{53.9D + \frac{3\times 10^{-7}}{D}(1+0.85H)}}{\sqrt{53.9D}} \quad (2.107)$$

The value of K is related with water depth and can be determined from Eq. (2.107). More generally, let's assume K is approximately 1.2, and then

$$\omega_1 = 1.2\sqrt{3}\omega = 2.08\omega \tag{2.108}$$

The threshold velocity for sediment entrainment can be approximated as

$$V_{b.c.1} = 0.916\omega_1 = 1.90\omega \tag{2.109}$$

which is independent of water depth.

The relation between $V_{b.y.c.3}^{(2)}$ and ω also changes under condition of Eq. (2.108). From Eq. (2.19), we have

$$V_{b.y.c.3}^{(2)} = \frac{\overline{V}_b}{8} - \sqrt{\frac{\omega_1^2}{4} - \frac{3}{64}\overline{V}_b^2}$$

Plugging Eq. (2.24), or $\overline{V}_b < \frac{4\omega_1}{\sqrt{3}}$ into which, and then plugging Eq. (2.108) for ω_1, we obtain

$$\begin{aligned} V_{b.y.c.3}^{(2)} &= \frac{\overline{V}_b}{8} - \sqrt{\frac{\omega_1^2}{4} - \frac{3}{64}\overline{V}_b^2} < \frac{1}{8}\frac{4}{\sqrt{3}}\omega_1 - \sqrt{\frac{\omega_1^2}{4} - \frac{3}{64}\left(\frac{4}{\sqrt{3}}\omega_1\right)^2} \\ &= \frac{\omega_1}{2\sqrt{3}} = \frac{2.08\omega}{2\sqrt{3}} < \omega \end{aligned} \tag{2.110}$$

It is seen that $\left[\xi_{V_{b.y}} > \omega \bigcap \xi_{V_{b.y}} < V_{b.y.c.3}^{(2)}\right]$ in Eq. (2.25) is an empty set. Also, for $V_b = \frac{4\omega_1}{\sqrt{3}}$ and $\omega_1 = 1.2\sqrt{3}\omega$, we have

$$\begin{aligned} V_{b.y.c.3}^{(1)} &= \frac{\bar{V}_b}{8} + \sqrt{\frac{\omega_1^2}{4} - \frac{3}{64}\bar{V}_b^2} = \frac{1}{8}\frac{4\omega_1}{\sqrt{3}} + \sqrt{\frac{\omega_1^2}{4} - \frac{3}{64}\frac{16\omega_1^2}{3}} \\ &= \frac{1}{8}\frac{4}{\sqrt{3}}1.2\sqrt{3}\omega = 0.6\omega \end{aligned}$$

For $0.6\omega \le V_{b.y.c.3}^{(1)} \le \omega$ and $\overline{V}_b < \frac{4\omega_1}{\sqrt{3}} = 4.8\omega$, the transition probability from rest to suspension is still $P\left[\xi_{V_{b.y}} > \omega\right]$ based on Eq. (2.31). Therefore,

$$B_4 \cap B_5 = \begin{cases} [\xi_{V_{b.y}} > \omega] & (V_{b.y.c.3}^{(1)} < \omega) \\ [\xi_{V_{b.y}} > V_{b,y.c.3}^{(1)}] & (V_{b.y.c.3}^{(1)} \ge \omega) \end{cases} \tag{2.111}$$

and the transition probability from rest to suspension can be written as

$$\beta = \beta_{1.4} = \begin{cases} P[\xi_{V_{b.y}} > \omega] = \varepsilon_4 & (V^{(1)}_{b.y.c.3} < \omega) \\ P[\xi_{V_{b.y}} > V^{(1)}_{b,y.c.3}] = \varepsilon_{5.1} & (V^{(1)}_{b.y.c.3} \geq \omega) \end{cases} \tag{2.112}$$

If $V_b > \frac{4\omega_1}{\sqrt{3}}$, from Eq. (2.31) there is $\beta = \beta_{1.4} = \varepsilon_4$. It is noted that the incipient suspension probability $\beta = \varepsilon_{51}$ is only used for a transition from rest to suspension, while the suspension probability $\varepsilon_4 = P[\xi_{V_{b.y}} > \omega]$ was used for a transition from a moving state to suspension.

2.5.2 Order of Threshold Flow Velocities

1. Order of the four threshold flow velocities

As shown in Eqs. (2.5) through (2.10), (2.15) and (2.16), the ranking of magnitudes of the four threshold flow velocities, $V_{b.c.0}$, $V_{b.c.1}$, $V_{b.c.2}$, and $V_{b.c.2.0}$, is fixed, which is

$$V_{b.c.0} < V_{b.c.1} < V_{b.c.2} < V_{b.c.2.0} \tag{2.113}$$

By noting Eq. (2.101) we have

$$V_{b.c.0} = \overline{V}_b - 2V_{b.y.c.0} \tag{2.114}$$

$$V_{b.c.1} = \overline{V}_b - 2V_{b.y.c.1} \tag{2.115}$$

When $V_{b.c.0} < V_{b.c.1}$, we have

$$\overline{V}_b - 2V_{b.y.c.0} < \overline{V_b} - 2V_{b.y.c.1} \tag{2.116}$$

or, $V_{b.y.c.0} > V_{b.y.c.1}$. Therefore the order of the threshold flow velocities in the vertical direction is different than that in the longitudinal direction, and is

$$V_{b.y.c.0} > V_{b.y.c.1} > V_{b.y.c.2} > V_{b.y.c.2.0} \tag{2.117}$$

According to Nikitin's analysis, V_b=5.6u_* for $y = D$. Assuming the bottom flow velocity acts on the particles at $y = 2D/3$, then $\overline{V}_b = 3.73u_*$. Plugging this $\overline{V}_b$ and $\omega_1^2 = 1.44 \times 3 \times \omega^2 = 4.32\omega^2$ into Eq. (2.18) yields

$$\frac{V^{(1)}_{b.y.c.3}}{u_*} = 0.466 + \sqrt{1.08\left(\frac{\omega}{u_*}\right)^2 - 0.652} \tag{2.118}$$

From Eqs. (2.114), (2.7), and (2.15), and $V_b = 3.73u_*$, we obtain

$$\frac{V_{b.y.c.0}}{u_*} = \frac{\overline{V}_b - V_{b.c.0}}{2u_*} = \frac{\overline{V}_b - 0.916\omega_0}{2u_*} = \frac{\overline{V}_b - 0.916\sqrt{3}\omega}{2u_*} = 1.865 - 0.793\frac{\omega}{u_*} \tag{2.119}$$

From Eqs. (2.108), (2.109), and (2.115), we have

$$\begin{aligned}\frac{V_{b.y.c.1}}{u_*} &= \frac{\overline{V}_b - V_{b.c.1}}{2u_*} = \frac{\overline{V}_b - 0.916\omega_1}{2u_*} \\ &= \frac{3.73u_*}{2u_*} - \frac{0.916 \times 1.2 \times \sqrt{3}\omega}{2u_*} \\ &= 1.865 - 0.952\frac{\omega}{u_*}\end{aligned} \tag{2.120}$$

Similarly,

$$\begin{aligned}\frac{V_{b.y.c.2}}{u_*} &= \frac{\overline{V}_b - V_{b.c.2}}{2u_*} = \frac{\overline{V}_b - 1.55V_{b.c.0}}{2u_*} \\ &= \frac{\overline{V}_b - 1.55 \times 0.916 \times \sqrt{3}\omega}{2u_*} \\ &= 1.865 - 1.229\frac{\omega}{u_*}\end{aligned} \tag{2.121}$$

$$\frac{V_{b.y.c.2.0}}{u_*} = \frac{\overline{V}_b - V_{b.c.2.0}}{2u_*} = \frac{\overline{V}_b - 2\omega_0}{2u_*} = \frac{\overline{V}_b - 2\sqrt{3}\omega}{2u_*} = 1.865 - 1.732\frac{\omega}{u_*} \tag{2.122}$$

2. Relation between sediment fall velocity ω, $V^{(1)}_{b.y.c.3}$, and the four threshold flow velocities

Due to different values of ω and $V^{(1)}_{b.y.c.3}$, the relation between ω/u_* and the four threshold flow velocities varies, which impacts calculation of the transition probabilities. For example, for $\omega/u_* < 0.6826$, it can be derived from Eq. (2.122) that

$$\frac{V_{b,y,c,2.0}}{\omega} = 1.865\frac{u_*}{\omega} - 1.732 > 1.0002 > 1$$

Therefore, $\omega < V_{b,y,c,2.0}$, and

$$\omega < V_{b,y,c,2.0} < V_{b,y,c,2} < V_{b,y,c,1} < V_{b,y,c.0} \tag{2.123}$$

Table 2.7 Ranking of magnitudes of the threshold velocities, ω and $V_{b.y.c.3}^{(1)}$

$\frac{\omega}{u_*}$	Order of the basic threshold velocities	Eq. Nos.
< 0.6826	$\omega < V_{b.y.c.2.0} < V_{b.y.c.2} < V_{b.y.c.1} < V_{b.y.c.0}$	(1)
0.6827–0.7535	$V_{b.y.c.2.0} < \omega < V_{b.y.c.2} < V_{b.y.c.1} < V_{b.y.c.0}$	(2)
0.7536–0.7770	$V_{b.y.c.2.0} < V_{b.y.c.2} < \omega < V_{b.y.c.1} < V_{b.y.c.0}$	(3)
0.7771–0.7785	$V_{b.y.c.3}^{(1)} < V_{b.y.c.2.0} < V_{b.y.c.2} < \omega < V_{b.y.c.1} < V_{b.y.c.0}$	(4)
0.7786–0.849	$V_{b.y.c.2.0} < V_{b.y.c.3}^{(1)} < V_{b.y.c.2} < \omega < V_{b.y.c.1} < V_{b.y.c.0}$	(5)
0.850–0.8679	$V_{b.y.c.2.0} < V_{b.y.c.2} < V_{b.y.c.3}^{(1)} < \omega < V_{b.y.c.1} < V_{b.y.c.0}$	(6)
0.8680–0.9238	$V_{b.y.c.2.0} < V_{b.y.c.2} < \omega < V_{b.y.c.3}^{(1)} < V_{b.y.c.1} < V_{b.y.c.0}$	(7)
0.9239–0.9554	$V_{b.y.c.2.0} < V_{b.y.c.2} < \omega < V_{b.y.c.1} < V_{b.y.c.3}^{(1)} < V_{b.y.c.0}$	(8)
0.9555–0.9804	$V_{b.y.c.2.0} < V_{b.y.c.2} < V_{b.y.c.1} < \omega < V_{b.y.c.3}^{(1)} < V_{b.y.c.0}$	(9)
0.9805–1.0407	$V_{b.y.c.2.0} < V_{b.y.c.2} < V_{b.y.c.1} < \omega < V_{b.y.c.0} < V_{b.y.c.3}^{(1)}$	(10)
> 1.0408	$V_{b.y.c.2.0} < V_{b.y.c.2} < V_{b.y.c.1} < V_{b.y.c.0} < \omega < V_{b.y.c.3}^{(1)}$	(11)

This order of velocities is important for calculation of transition probabilities. It is noted that suspension from the rest state is also related to the values of ω and $V_{b.y.c.3}^{(1)}$. With the similar analysis, the orders of ω and $V_{b.y.c.3}^{(1)}$ were determined for various ω/u_*, as listed in Table 2.7. It is worthy to point out that Eqs. (2.111) and (2.112) only apply for

$$3.73u_* = \overline{V}_b > \frac{4\omega_1}{\sqrt{3}} = \frac{4 \times 1.2 \times \sqrt{3}\omega}{\sqrt{3}} = 4.8\omega$$

or, $\frac{\omega}{u_*} < \frac{3.73}{4.8} = 0.77708 < 0.7771$. Therefore, the incipient suspension probability is the first formula of Eq. (2.112) for $\frac{\omega}{u_*} < 0.7770$, and is the second formula of Eq. (2.112) for $\frac{\omega}{u_*} > 0.7771$. According to Eq. (2.118), for $0.7771 < \frac{\omega}{u_*} < 1.0408$, $\frac{V_{b,y,c,3}^{(2)}}{\omega} < 1,\ or\ \omega > V_{b,y,c,3}^{(2)}$. Combining with Eq. (2.106), it can be found that $\left[\xi_{Vb,y} > \omega \cap \xi_{Vb,y} < V_{v,y,c,3}^{(2)}\right]$ is an empty set, as discussed before.

It is noted that $\frac{\omega}{u_*} < 0.7771$ is the condition for the root of Eq. (2.118) to be an imaginary number, and thus is independent of $V_{b,y,c,3}^{(1)}$.

2.5.3 Calculation of Transition Probabilities

1. $\frac{\omega}{u_*} < 0.6826$

For this case in Table 2.7 the probability for the transition from the rest state to the four states (at rest, rolling, jumping, and suspension) is

$$\beta_{1.1} = P[\xi_{V_b} < V_{b.c.1} \cap \xi_{V_{b.y}} \leq \omega]$$

Noting that $\xi_{V_b} < V_{b.c.1}$ we have

$$\frac{\xi_{V_b} - \overline{V}_b}{2} < \frac{V_{b.c.1} - \overline{V_b}}{2} \tag{2.124}$$

Combining Eq. (2.101) it can be derived that

$$\xi_{V_{b.y}} = \frac{\overline{V}_b - \xi_{V_b}}{2} > \frac{\overline{V_b} - V_{b.c.1}}{2} = V_{b.y.c.1} \tag{2.125}$$

So the set of $\xi_{V_b} < V_{b.c.1}$ is the same as the set of $\xi_{V_{b.y}} > V_{b.y.c.1}$, and

$$\beta_{1.1} = P[\xi_{V_{b.y}} > V_{b.y.c.1} \cap \xi_{V_{b.y}} \leq \omega]$$

It is noticed from Eq. (2.123) that $\omega < V_{b.y.c.1}$, and thus the set above is an empty set, or

$$\beta_{1.1} = P[\xi_{V_{b.y}} > V_{b.y.c.1} \cap \xi_{V_{b.y}} \leq \omega] = 0 \tag{2.126}$$

Similarly,

$$\begin{aligned}\beta_{1.2} &= P[V_{b.c.1} < \xi_{V_b} \leq V_{b.c.2.0} \cap \xi_{V_{b.y}} < \omega] \\ &= P[V_{b.y.c.2.0} \leq \xi_{V_{b.y}} < V_{b.y.c.1} \cap \xi_{V_{b.y}} < \omega] = 0\end{aligned} \tag{2.127}$$

This is demonstrated in Fig. 2.3. The zones hatched with slashes do not intersect, indicating the set is empty. For $\beta_{1.3}$ and $\beta_{1.4}$ there are

$$\begin{aligned}\beta_{1.3} &= P[\xi_{V_b} > V_{b.c.2.0} \cap \xi_{V_{b.y}} < \omega] \\ &= P[\xi_{V_{b.y}} < V_{b.y.c.2.0} \cap \xi_{V_{b.y}} < \omega] \\ &= P[\xi_{V_{b.y}} < \omega] = 1 - P[V_{b.y.c.2.0} > \omega] = 1 - \varepsilon_4\end{aligned} \tag{2.128}$$

and

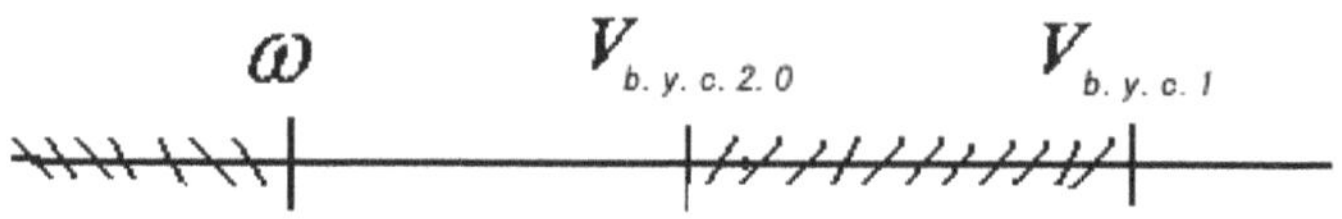

Fig. 2.3 Ranking of magnitudes for ω, $V_{b.y.c.2.0}$ and $V_{b.y.c.1}$

Fig. 2.4 Ranking of magnitudes of ω and $V_{b.y.c.2}$

$$\beta_{1.4} = P[\xi_{V_{b.y}} > \omega] = \varepsilon_4 \tag{2.129}$$

Similarly, for i $\neq$ 1

$$\begin{aligned}\beta_{i.1} &= P\left[\xi_{V_b} \leq V_{b.c.0} \cap V_{b.y} \leq \omega\right] \\ &= P\left[\xi_{V_{b.y}} > V_{b.y.c.0} \cap V_{b.y} \leq \omega\right] = 0\end{aligned} \tag{2.130}$$

$$\begin{aligned}\beta_{i.2} &= P\left[V_{b.c.0} < \xi_{V_b} \leq V_{b.c.2} \cap \xi_{V_{b.y}} \leq \omega\right] \\ &= P\left[V_{b.y.c.0} > \xi_{V_{b.y}} \geq V_{b.y.c.2} \cap \xi_{V_{b.y}} \leq \omega\right] = 0\end{aligned} \tag{2.131}$$

$$\begin{aligned}\beta_{i.3} &= P\left[\xi_{V_b} > V_{b.c.2} \cap \xi_{V_{b.y}} \leq \omega\right] \\ &= P\left[\xi_{V_{b.y}} < V_{b.y.c.2} \cap \xi_{V_{b.y}} \leq \omega\right] = 1 - \varepsilon_4\end{aligned} \tag{2.132}$$

$$\beta_{i.4} = P\left[\xi_{V_{b.y}} > \omega\right] = \varepsilon_4 \tag{2.133}$$

As shown in Fig. 2.4, the intersection between the two zones hatched with slashes and back slashes, respectively, is the intersection between the sets of $\xi_{Vb,y} < V_{b,y,c.2}$ and $\xi_{Vb,y} < \omega$ in Eq. (2.132).

In summary, the transition probability matrix for $\frac{\omega}{u_*} < 0.6826$ is

$$\beta_{i.j} = \begin{bmatrix} 0 & 0 & 1-\varepsilon_4 & \varepsilon_4 \\ 0 & 0 & 1-\varepsilon_4 & \varepsilon_4 \\ 0 & 0 & 1-\varepsilon_4 & \varepsilon_4 \\ 0 & 0 & 1-\varepsilon_4 & \varepsilon_4 \end{bmatrix} \tag{2.134}$$

The transition chain for this case is shown in Fig. 2.5. The transition chains for states 1 and 2 are irreversible, which means the other states cannot transit to these two states. The physical concept is that the hopping and suspended sediment under a strong flow condition cannot change to states 1 and 2, or cannot change to the rest or rolling states. Even though some particles are initially at rest or in rolling state, they will soon change to the states 3 and 4 without reverse transitions. In other words, only suspended and hopping particles exist at bed layer, and the sediment transport is characterized as the movement of the bed load layer. Of course, the normal exchange between suspended load and hopping particles still occurs. Thus, sediment practically only has two states, and the transition probability matrix is

$$\left[\beta_{i.j}\right] = \begin{bmatrix} 1-\varepsilon_4 & \varepsilon_4 \\ 1-\varepsilon_4 & \varepsilon_4 \end{bmatrix} \tag{2.135}$$

2. $0.6827 < \frac{\omega}{u_*} \leq 0.7535$

For this range of $\frac{\omega}{u_*}$, the ranking of magnitudes for the velocities, ω_i, $V_{b,y,c.2.0}$, *and* $V_{b,y,c.2}$ has been changed from $\omega < V_{b.y.c.2.0} < V_{b.y.c.2}$ to $V_{b.y.c.2.0} < \omega < V_{b.y.c.2}$, but the transition probability matrix is similar to Eq. (2.134), except the expression of

$$\begin{aligned}
\beta_{1.2} &= P[V_{b.c.1} < \xi_{V_b} \leq V_{b.c.2.0} \cap \xi_{V_{b.y}} < \omega] \\
&= P[V_{b.y.c.2.0} \leq \xi_{V_{b.y}} < V_{b.y.c.2} \cap \xi_{V_{b.y}} < \omega] \\
&= P[V_{b.y.c.2.0} \leq \xi_{V_{b.y}} < \omega] = 1 - \varepsilon_{20} - \varepsilon_4 \\
\beta_{1.3} &= P[\xi_{V_b} > V_{b.c.2.0} \cap \xi_{V_{b.y}} < \omega] \\
&= P[\xi_{V_{b.y}} < V_{b.y.c.2.0} \cap \xi_{V_{b.y}} < \omega] = \varepsilon_{2.0}
\end{aligned}$$

After neglecting the transitions to states 1 and 2, the actual transition matrix is also the same as Eq. (2.135).

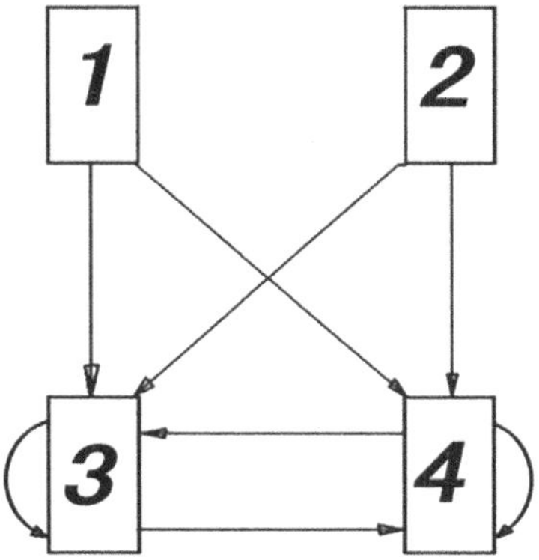

Fig. 2.5 Irreversible transitions for states 1 and 2

3. $0.7536 < \frac{\omega}{u_*} < 0.8679$

According to Table 2.7, $V_{b.y.c.3}^{(1)}$ varies significantly for this range of $\frac{\omega}{u_*}$, and four variation ranges of $\frac{\omega}{u_*}$ should be used for the ranking of velocities. It can derived from Eq. (2.118) that $\frac{V_{b.y.c.3}^{(1)}}{\omega} > 1$ for $\frac{\omega}{u_*} < 0.8679$. Based on the second formula of Eq. (2.112), $P[\xi_{V_{b.y}} > \omega \cap \xi_{V_{b.y}} > V_{b,y.c.2}^{(1)}] = P\left[\xi_{V_{b.y}} > \omega\right] = \varepsilon_4$. Thus, $V_{b.y.c.3}^{(1)}$ is needed for determining the probability for incipient suspension only for $\frac{\omega}{u_*} > 0.8680$. For $0.7770 < \frac{\omega}{u_*} < 0.8679$, $V_{b.y.c.3}^{(1)}$ is not needed and there is no need to calculate the incipient suspension probability individually for four ranges of $\frac{\omega}{u_*}$. The transition probabilities are

$$\begin{aligned} P_{1.1} &= P[\xi_{V_b} < V_{b.c.1} \cap \xi_{V_{b.y}} < V_{b.y.c.3}^{(1)}] \\ &= P[\xi_{V_{b.y}} > V_{b.y.c.1} \cap \xi_{V_{b.y}} < \omega] = 0 \end{aligned} \tag{2.136}$$

$$\begin{aligned} P_{1.2} &= P[V_{b.c.1} \le \xi_{V_b} < V_{b.c.2.0} \cap \xi_{V_{b.y}} < \omega] \\ &= P[V_{b.y.c.1} \ge \xi_{V_{b.y}} > V_{b.y.c.2.0} \cap \xi_{V_b} < \omega] \\ &= P[V_{b.y.c.2.0} < \xi_{V_{b.y}} \omega] = 1 - \varepsilon_4 - \varepsilon_{2.0} \end{aligned} \tag{2.137}$$

$$\begin{aligned} P_{1.3} &= P[\xi_{V_b} > V_{b.c.2.0} \cap \xi_{V_{b.y}} < \omega] \\ &= P\left[\xi_{V_{b.y}} < V_{b.y.c.2.0} \cap \xi_{V_{b.y}} < \omega\right] = \varepsilon_{2.0} \end{aligned} \tag{2.138}$$

$$P_{1.4} = P[\xi_{V_{b.y}} > \omega] = \varepsilon_4 \tag{2.139}$$

For $i \neq 1$,

$$\begin{aligned} P_{i.1} &= P[\xi_{V_{b.y}} > V_{b.y.c.1.0} \cap \xi_{b.y} < \omega] = 0 \\ P_{i.2} &= P[V_{b.y.c.2} \le \xi_{V_{b.y}} < V_{b.y.c.1.0} \cap \xi_{V_{b.y}} < \omega] \\ &= P[V_{b.y.c.2} \le \xi_{V_{b.y}} < \omega] \\ &= 1 - \varepsilon_4 - \varepsilon_2 \\ P_{i.3} &= P[\xi_{V_{b.y}} < V_{b.y.c.2} \cap \xi_{V_{b.y}} < \omega] = P[\xi_{V_{b.y}} < V_{b.y.c.2}] = \varepsilon_2 \\ P_{i.4} &= P[\xi_{V_{b.y}} > \omega] = \varepsilon_4 \end{aligned}$$

Therefore, the transition probability matrix is

$$[P_{i.j}] = \begin{bmatrix} 0 & 1 - \varepsilon_4 - \varepsilon_{2.0} & \varepsilon_{2.0} & \varepsilon_4 \\ 0 & 1 - \varepsilon_4 - \varepsilon_2 & \varepsilon_2 & \varepsilon_4 \\ 0 & 1 - \varepsilon_4 - \varepsilon_2 & \varepsilon_2 & \varepsilon_4 \\ 0 & 1 - \varepsilon_4 - \varepsilon_2 & \varepsilon_2 & \varepsilon_4 \end{bmatrix} \tag{2.140}$$

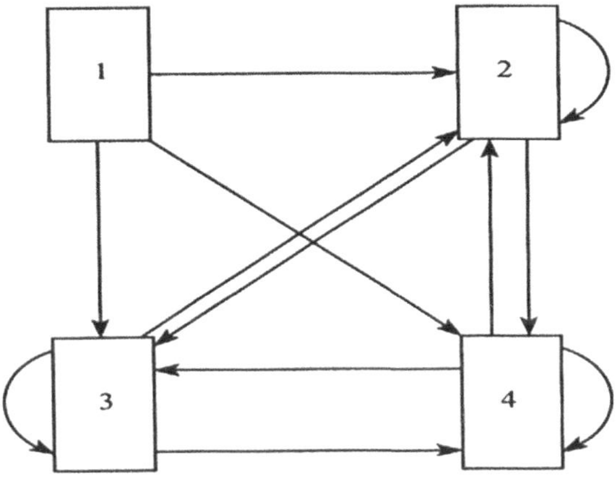

Fig. 2.6 Irreversible transitions for state 1

The transition chain is shown in Fig. 2.6. The transition for state i is irreversible. It physically means that no sediment particles are at rest on the bed (even if some particles initially are at rest, they will be transited to the other states immediately) under a strong flow condition. The transition probability matrix is actually a 3×3 matrix without particles at rest on the bed, which reads

$$P_{i,j} = \begin{bmatrix} 1-\varepsilon_2-\varepsilon_4 & \varepsilon_2 & \varepsilon_4 \\ 1-\varepsilon_2-\varepsilon_4 & \varepsilon_2 & \varepsilon_4 \\ 1-\varepsilon_2-\varepsilon_4 & \varepsilon_2 & \varepsilon_4 \end{bmatrix} \tag{2.141}$$

4. $0.8680 < \frac{\omega}{u_*} < 0.9554$.

Actually the transition probabilities are determined by the order between ω and $V^{(1)}_{b.y.c.3}$, which is changed from $\omega > V^{(1)}_{b.y.c.3}$ for case 3 to $\omega < V^{(1)}_{b.y.c.3}$ for cases 4–6. Then the transition probabilities are

$$\begin{aligned}\beta_{1.1} &= P[\xi_{V_b} < V_{b.c.1} \cap \xi_{V_{b.y}} < V^{(1)}_{b.y.c.3}] \\ &= P[\xi_{V_{b.y}} > V_{b.y.c.1} \cap \xi_{V_{b.y}} < V^{(1)}_{b.y.c.3}] = 0\end{aligned}$$

$$\begin{aligned}\beta_{1.2} &= P[V_{b.c.1} \le \xi_{V_b} < V_{b.c.2.0} \cap \xi_{V_{b.y}} < V^{(1)}_{b.y.c.3}] \\ &= P[V_{b.y.c.1} \ge \xi_{V_{b.y}} > V_{b.y.c.2.0} \cap \xi_{V_b} < V^{(1)}_{b.y.c.3}] \\ &= P[V_{b.y.c.2.0} < \xi_{V_{b.y}} < V^{(1)}_{b.y.c.3}] = 1 - \varepsilon_{5.1} - \varepsilon_{2.0}\end{aligned}$$

$$\begin{aligned}\beta_{1.3} &= P[\xi_{V_b} > V_{b.c.2.0} \cap \xi_{V_{b.y}} < V^{(1)}_{b.y.c.3}] \\ &= P\Big[\xi_{V_{b.y}} < V_{b.y.c.2.0} \cap \xi_{V_{b.y}} < V^{(1)}_{b.y.c.3}\Big] = \varepsilon_{2.0}\end{aligned}$$

$$\beta_{1.4} = P[\xi_{V_{b.y}} > V^{(1)}_{b.y.c.3}] = \varepsilon_{5.1}$$

For $i \neq 1$,

$$\begin{aligned}
\beta_{i,1} &= P\left[\xi_{Vb,y} > V_{b,y,c.0} \cap \xi_{b,y} < \omega\right] \\
\beta_{i,2} &= P\left[V_{b,y,c.2} \leq \xi_{Vb,y} < V_{b,y,c.0} \cap \xi_{Vb,y} < \omega\right] \\
&= P\left[V_{b,y,c.2} \leq \xi_{Vb,y} < \omega\right] \\
&= 1 - \varepsilon_4 - \varepsilon_2 \\
\beta_{i,3} &= P\left[\xi_{Vb,y} < V_{b,y,c.2} \cap \xi_{Vb,y} < \omega\right] \\
&= \beta_{i,1} = P\left[\xi_{Vb,y} < V_{b,y,c.2}\right] = \varepsilon_2 \\
\beta_{i,4} &= P\left[\xi_{V,b} > \omega\right] = \varepsilon_4
\end{aligned}$$

Therefore, the transition probability matrix is

$$[\beta_{i.j}] = \begin{bmatrix} 0 & 1 - \varepsilon_{5.1} - \varepsilon_{2.0} & \varepsilon_{2.0} & \varepsilon_{5.1} \\ 0 & 1 - \varepsilon_4 - \varepsilon_2 & \varepsilon_2 & \varepsilon_4 \\ 0 & 1 - \varepsilon_4 - \varepsilon_2 & \varepsilon_2 & \varepsilon_4 \\ 0 & 1 - \varepsilon_4 - \varepsilon_2 & \varepsilon_2 & \varepsilon_4 \end{bmatrix}$$

The actual transition probability matrix is the same as Eq. (2.141).

5. $0.9555 < \frac{\omega}{u_*} < 1.0407$

The transition probabilities are

$$\begin{aligned}
\beta_{1.1} &= P[\xi_{V_b} < V_{b.c.1} \cap \xi_{V_{b.y}} < V^{(1)}_{b.y.c.3}] \\
&= P[\xi_{V_{b.y}} > V_{b.y.c.1} \cap \xi_{V_{b.y}} < V^{(1)}_{b.y.c.3}] \\
&= P[V_{b.y.c.1} < \xi_{V_{b.y}} < V^{(1)}_{b.y.c.3}] = 1 - \varepsilon_{5.1} - \varepsilon_1
\end{aligned}$$

$$\begin{aligned}
\beta_{1.2} &= P[V_{b.c.1} < \xi_{V_b} \leq V_{b.c.2.0} \cap \xi_{V_{b.y}} < V^{(1)}_{b.y.c.3}] \\
&= P[V_{b.y.c.1} > \xi_{V_{b.y}} \geq V_{b.y.c.2.0} \cap \xi_{V_{b.y}} < V^{(1)}_{b.y.c.3}] \\
&= P[V_{b.y.c.2.0} \leq \xi_{V_{b.y}} < V_{b.y.c.1}] \\
&= P[V_{b.c.2.0} > \xi_{V_b} > V_{b.c.1}] = \varepsilon_1 - \varepsilon_{2.0}
\end{aligned}$$

$$\begin{aligned}
\beta_{1.3} &= P[\xi_{V_b} > V_{b.y.c.2.0} \cap \xi_{V_b} < V^{(1)}_{b.y.c.3}] \\
&= P[\xi_{V_{b.y}} < V_{b.y.c.2.0} \cap \xi_{V_b} < V^{(1)}_{b.y.c.3}] \\
&= P[\xi_{V_{b.y}} < V_{b.y.c.2.0}] = \varepsilon_{2.0}
\end{aligned}$$

$$\beta_{1.4} = P[\xi_{V_b} > V^{(1)}_{b.y.c.3}] = \varepsilon_{5.1}$$

For $i \neq 1$,

$$\begin{aligned}\beta_{i.1} &= P[\xi_{V_b} < V_{b.c.0} \cap \xi_{V_{b.y}} < \omega] \\ &= P[\xi_{V_{b.y}} > V_{b.y.c.0} \cap \xi_{V_{b.y}} < \omega] = 0\end{aligned}$$

$$\begin{aligned}\beta_{i.2} &= P[V_{b.y.c.2} < \xi_{V_{b.y}} < V_{b.y.c.0} \cap \xi_{V_{b.y}} \leq \omega] \\ &= P[V_{b.y.c.2} < \xi_{V_{b.y}} \leq \omega] = 1 - \varepsilon_4 - \varepsilon_2\end{aligned}$$

$$\begin{aligned}\beta_{i.3} &= P[\xi_{V_b} > V_{b.c.2} \cap \xi_{V_{b.y}} < \omega] \\ &= P[\xi_{V_{b.y}} < V_{b.y.c.2} \cap V_{b.y} < \omega] = \varepsilon_2\end{aligned}$$

$$\beta_{i.4} = P[\xi_{V_b} > \omega] = \varepsilon_4$$

Therefore, the transition probability matrix is

$$\beta_{i.j} = \begin{bmatrix} 1 - \varepsilon_{5.1} - \varepsilon_1 & \varepsilon_1 - \varepsilon_{2.0} & \varepsilon_{2.0} & \varepsilon_{5.1} \\ 0 & 1 - \varepsilon_4 - \varepsilon_2 & \varepsilon_2 & \varepsilon_4 \\ 0 & 1 - \varepsilon_4 - \varepsilon_2 & \varepsilon_2 & \varepsilon_4 \\ 0 & 1 - \varepsilon_4 - \varepsilon_2 & \varepsilon_2 & \varepsilon_4 \end{bmatrix} \tag{2.142}$$

which is the transition probability matrix for first step with $\beta_{1.1}(1) = (1-\varepsilon_{5.1}-\varepsilon_1) \neq 0$. For the second step, $\beta_{1.1}(2) = (1 - \varepsilon_{5.1} - \varepsilon_1)^2$. For the n-th step, $\beta_{1.1}(n) = (1 - \varepsilon_{5.1} - \varepsilon_1)^n$, which approaches to zero, indicating that there is no probability for sediment to change back to state 1. Therefore, the actual transition probability matrix is that for the other three states (states 2, 3, and 4) as Eq. (2.141).

6. $\frac{\omega}{u_*} > 1.0408$

In the case if $i \neq 1$, the transition probabilities are

$$\begin{aligned}\beta_{i.1} &= P[\xi_{V_{b.y}} > V_{b.y.c.0} \cap \xi_{V_{b.y}} \leq \omega] \\ &= P[V_{b.y.c.0} < \xi_{V_{b.y}} < \omega] \\ &= 1 - \varepsilon_0 - \varepsilon_4\end{aligned}$$

$$\begin{aligned}\beta_{i.3} &= P[\xi_{V_{b.y}} < V_{b.y.c.2} \cap \xi_{V_{b.y}} \leq \omega] = P[\xi_{V_{b.y}} < V_{b.y.c.2}] = \varepsilon_2 \\ \beta_{i.4} &= P[\xi_{V_{b.y}} > \omega] = \varepsilon_4\end{aligned}$$

If $i = 1$ the transition probabilities, not related with the order between ω and $V^{(1)}_{b.y.c.3}$, are the same as last section. Then the transition probability matrix is

$$\beta_{i.j} = \begin{bmatrix} 1-\varepsilon_1-\varepsilon_{5.1} & \varepsilon_1-\varepsilon_{2.0} & \varepsilon_{2.0} & \varepsilon_{5.1} \\ 1-\varepsilon_0-\varepsilon_4 & \varepsilon_0-\varepsilon_2 & \varepsilon_2 & \varepsilon_4 \\ 1-\varepsilon_0-\varepsilon_4 & \varepsilon_0-\varepsilon_2 & \varepsilon_2 & \varepsilon_4 \\ 1-\varepsilon_1-\varepsilon_{5.1} & \varepsilon_0-\varepsilon_2 & \varepsilon_2 & \varepsilon_4 \end{bmatrix} \quad (2.143)$$

The six cases discussed above can actually be categorized into three cases: (1) For $\frac{\omega}{u_*} \le 0.7535$, the transition probability matrix is Eq. (2.134); (2) For $0.7536 \le \frac{\omega}{u_*} \le 1.0407$, the transition probability matrix is (2.141); and (3) For $\frac{\omega}{u_*} > 1.0408$, the transition probability matrix is Eq. (2.143). A weight summation can be derived from the 4×4 matrixes for the six cases using the occurrence probability of $\frac{\omega}{u_*}$ as the weighting factor, which represents an averaged transition probability matrix. Given u_*, the distribution function $P(\omega)$ can be derived for the fall velocity ω based on the sediment size distribution $P(D)$. Further the probability functions of $\frac{\omega}{u_*}$ can be determined, and they are

$$\begin{cases} P\left[\frac{\omega}{u_*} < 0.6826\right] = P_1 \\ P\left[0.6827 < \frac{\omega}{u_*} \le 0.7535\right] = P_2 \\ \vdots \\ P\left[\frac{\omega}{u_*} > 1.0408\right] = P_9 \end{cases} \quad (2.144)$$

The elements of the weight averaged transition probability matrix are

$$\overline{\beta}_{i.j} = \sum_{k=1}^{8} P_k \beta_{i.j}^{(k)} \quad (2.145)$$

2.5.4 Analysis of State Probabilities

1. State Probabilities

Using Eq. (2.143) as an example, the transition probability matrix is an aperiodic and irreducible Markov chain. According to the Markov theorem, the limiting probabilities (or the state probabilities) R_i (i=1,2,3,4) exist, and they can be solved from equation group (2.46). Plugging the probability elements of Eq. (2.143) into Eq. (2.46) and applying the constraint condition (Eq. 2.48), we get

$$\begin{cases} R_1 = (1 - \varepsilon_1 - \varepsilon_{5.1}) + R_2(1 - \varepsilon_0 - \varepsilon_4) \\ \quad + R_3(1 - \varepsilon_0 - \varepsilon_4) + R_4(1 - \varepsilon_0 - \varepsilon_4) \\ R_2 = (\varepsilon_1 - \varepsilon_{2.0}) + R_2(\varepsilon_0 - \varepsilon_2) + R_3(\varepsilon_0 - \varepsilon_2) + R_4(\varepsilon_0 - \varepsilon_2) \\ R_3 = R_1\varepsilon_{2.0} + R_2\varepsilon_2 + R_3\varepsilon_2 + R_4\varepsilon_2 \\ R_4 = R_1\varepsilon_{5.4} + R_2\varepsilon_4 + R_3\varepsilon_4 + R_4\varepsilon_4 \end{cases} \tag{2.146}$$

The state probabilities can be derived by solving this equation group, and they are

$$R_1 = \frac{1 - \varepsilon_0 - \varepsilon_4}{1 - \varepsilon_0 - \varepsilon_4 + \varepsilon_1 + \varepsilon_{5.1}} = \frac{A_1}{A} \tag{2.147}$$

$$R_2 = \frac{(\varepsilon_1 - \varepsilon_0)(1 - \varepsilon_0 - \varepsilon_4) + (\varepsilon_0 - \varepsilon_2)(\varepsilon_1 - \varepsilon_{5.1})}{1 - \varepsilon_0 - \varepsilon_4 + \varepsilon_1 + \varepsilon_{5.1}} = \frac{A_1 P_{12} + A_2 P_{22}}{A} \tag{2.148}$$

$$R_3 = \frac{\varepsilon_{2.0}(1 - \varepsilon_0 - \varepsilon_4)\varepsilon_2(\varepsilon_2 + \varepsilon_{5.1})}{1 - \varepsilon_0 - \varepsilon_4 + \varepsilon_1 + \varepsilon_{5.1}} = \frac{A_1 P_{13} + A_2 P_{33}}{A} \tag{2.149}$$

$$R_4 = \frac{\varepsilon_{5.1}(1 - \varepsilon_0 - \varepsilon_4) + \varepsilon_4(\varepsilon_1 + \varepsilon_{5.1})}{1 - \varepsilon_0 - \varepsilon_4 + \varepsilon_1 + \varepsilon_{5.1}} = \frac{A_1 P_{14} + A_2 P_{44}}{A} \tag{2.150}$$

Where

$$A = 1 - \varepsilon_0 - \varepsilon_4 + \varepsilon_1 + \varepsilon_{5.1} \tag{2.151}$$

$$A_1 = 1 - \varepsilon_0 - \varepsilon_4 \tag{2.152}$$

$$A_2 = \varepsilon_1 + \varepsilon_{5.1} = A - A_1 \tag{2.153}$$

Obviously the sum of these four state probabilities is one. In fact

$$\begin{aligned} R_1 + R_2 + R_3 + R_4 &= \frac{A_1}{A}(1 + P_{1.2} + P_{1.3} + P_{1.4}) + \frac{A_2}{A}(P_{2.2} + P_{3.3} + P_{4.4}) \\ &= \frac{A_1}{A}\left[(1 + (1 - P_{1.1} + \frac{A_2}{A}(\varepsilon_0 - \varepsilon_2 + \varepsilon_2 + \varepsilon_4)\right] \\ &= \tfrac{A_1}{A}(1 + \varepsilon_0 + \varepsilon_{5.1}) + \frac{A - A_1}{A}(\varepsilon_0 + \varepsilon_4) \\ &= \tfrac{A_1}{A}(1 - \varepsilon_0 - \varepsilon_4 + \varepsilon_1 + \varepsilon_{5.1}) + (\varepsilon_0 + \varepsilon_4) \\ &= A_1(1 - A_1) = 1 \end{aligned} \tag{2.154}$$

Similarly, the state probabilities for the transition probability matrix for two states, i.e., Eq. (2.135), are

$$R_3 = R_3(1 - \varepsilon_4) + R_4(1 - \varepsilon_4) = 1 - \varepsilon_4 \tag{2.155}$$

$$R_4 = R_r\varepsilon_4 + R_3\varepsilon_4 = \varepsilon_4 \tag{2.156}$$

The state probabilities for the transition probability matrix for three states, i.e., Eq. (2.140), are

$$\begin{aligned} R_2 &= R_2(1 - \varepsilon_2 - \varepsilon_4) + R_3(1 - \varepsilon_2 - \varepsilon_4) + R_4(1 - \varepsilon_2 - \varepsilon_4) \\ &= 1 - \varepsilon_2 - \varepsilon_4 \end{aligned} \tag{2.157}$$

$$R_3 = \varepsilon_2 R_2 + \varepsilon_2 R_3 + \varepsilon_2 R_4 = \varepsilon_2 \tag{2.158}$$

$$R_4 = \varepsilon_4 R_4 + \varepsilon_4 R_3 + \varepsilon_4 R_2 = \varepsilon_4 \tag{2.159}$$

2. Analysis of State Probabilities

Equations (2.156) and (2.159) both have a state probability $R_4 = \varepsilon_4$. According to Table 2.7, $\omega < V_{b.y.c.3}^{(1)}$ is always true for $\frac{\omega}{u_*} > 1.0408$, and thus $\varepsilon_4 > \varepsilon_{5.1}$. Based on Eq. (2.150), we have

$$R_4' = \frac{\varepsilon_{5.1}(1 - \varepsilon_0 - \varepsilon_4) + \varepsilon_4(\varepsilon_1 + \varepsilon_{5.1})}{1 - \varepsilon_0 - \varepsilon_4 + \varepsilon_1 + \varepsilon_{5.1}} \leq \frac{\varepsilon_4[(1 - \varepsilon_0 - \varepsilon_4) + (\varepsilon_1 + \varepsilon_{5.1})]}{1 - \varepsilon_0 - \varepsilon_4 + \varepsilon_1 + \varepsilon_{5.1}} = \varepsilon_4$$

Accordingly

$$R_4' \geq \varepsilon_{5.1}$$

Thus

$$\varepsilon_4 \geq R_4' \geq \varepsilon_{5.1} \tag{2.160}$$

where R' represents the state probabilities for the case with correlated vertical and longitudinal velocities. It can be seen that $R_4' \leq \varepsilon_4$ when the longitudinal and vertical components of the bottom velocity are correlated with a correlation coefficient of –1. For the two-state and three-state transitions, or Eqs. (2.156) and (2.159), respectively, $R_4' = \varepsilon_4$. Table 2.8 shows the comparison between $R_4' = \varepsilon_4$ and R_4 for the case with independent vertical and longitudinal velocities (at $H = 1$ m and $H = 20$ m). For coarse sediment particles and great u_*, R_4' is close to R_4. For weak u_* and fine particles, the difference between R_4' and R_4 is significant, and $R_4' = \varepsilon_4$ for the case with correlated vertical and longitudinal velocities is greater than that for the case with independent velocities. This indicates that this condition has significant impact on sediment suspension.

Table 2.8 Comparisons between R_4' and R_4

Parameters	Grain diameter D(mm)	u_*(cm/s)					
		$u_* = 1$	$u_* = 3$	$u_* = 5$	$u_* = 8$	$u_* = 12$	$u_* = 20$
$R_4(H = 1\ \mathrm{m})$	0.01	0.0004	0.4856	0.4979	0.4989	0.5000	0.5000
	0.025	0.0002	0.4889	0.4949	0.4980	0.4990	0.4990
	0.05	0.0076	0.4710	0.4880	0.4920	0.4950	0.4970
	0.10	0.0087	0.4041	0.4510	0.4690	0.4790	0.4880
	0.25	0.0000	0.1334	0.2720	0.3520	0.4000	0.4400
	0.5	0.0000	0.0064	0.0775	0.1910	0.2800	0.3640
$R_4' = \varepsilon_4$	0.01	0.4980	0.4990	0.5000	0.5000	0.5000	0.5000
	0.025	0.4840	0.4950	0.4970	0.4980	0.4990	0.4990
	0.05	0.4380	0.4790	0.4880	0.4920	0.4950	0.4970
	0.10	0.2680	0.4180	0.4510	0.4690	0.4790	0.4880
	0.25	0.0012	0.1560	0.2720	0.3520	0.4000	0.4400
	0.5	0.0000	0.0100	0.0811	0.1910	0.2800	0.3640
$R_4(H = 20\ \mathrm{m})$	0.01	0.0020	0.0039	0.0049	0.0049	0.0050	0.4989
	0.025	0.0001	0.0014	0.4612	0.4951	0.4976	0.4990
	0.05	0.0001	0.1793	0.4792	0.4896	0.4940	0.4970
	0.10	0.0000	0.3439	0.4436	0.4665	0.4790	0.4880
	0.25	0.0000	0.1164	0.2616	0.3520	0.4000	0.4400
	0.5	0.0000	0.0055	0.0743	0.1910	0.2808	0.3640

The 4 × 4 transition probability matrices are compared in the following for the two conditions. Noting $\varepsilon_4 > \beta$, it can be derived from Eq. (2.52) that

$$R_4 = \frac{\varepsilon_4[1 - (1 - \varepsilon_1)(1 - \beta)] + \beta(1 - \varepsilon_0)(1 - \varepsilon_4)}{A} \leq \varepsilon_4$$

$$R_4 \geq \beta$$

Therefore,

$$\varepsilon_4 \geq R_4 \geq \beta \tag{2.161}$$

Noting $\varepsilon_{5.1} \leq \varepsilon_4$, it can be derived from Eq. (2.150) that.

$$\varepsilon_4 \geq R_4' \geq \varepsilon_{5.1}$$

It can be observed that R_4 and R_4' have similar variation ranges, with a lightly wider range for R_4'. Noting $\varepsilon_2 > \varepsilon_{2.0}$ and $\varepsilon_4 \geq \beta$, it can be derived from Eq. (2.55) that

$$R_3 = \frac{\varepsilon_2(1-\varepsilon_4)[1-(1-\varepsilon_1)(1-\beta)] + \varepsilon_{2.0}(1-\beta)(1-\varepsilon_4)}{A} \leq \varepsilon_2(1-\beta)$$

Similarly,

$$R_3 \geq \varepsilon_{2.0}(1-\varepsilon_4)$$

or

$$\varepsilon_2(1-\beta) \geq R_3 \geq \varepsilon_{2.0}(1-\varepsilon_4) \tag{2.162}$$

Noting $\varepsilon_2 < \varepsilon_{2.0}$, it can be derived from Eq. (2.149) that

$$\varepsilon_2 \geq R_3' \geq \varepsilon_{2.0} \tag{2.163}$$

It is noted that $R_3 \leq R_3'$ for a wider range, indicating the state probability R_3 (with independent longitudinal and vertical velocities) is equal to or less than R_3' (with correlated velocities) for most cases. By comparing the state probabilities for states 3 and 4, it is found that the probabilities for hopping and suspension under the condition of correlated velocities are generally higher than those under the condition of independent velocities. This is reasonable, and is consistent with previous findings (no sediment at rest or rolling on the bed for $\frac{\omega}{u_*} < 0.7535$, and no sediment at rest for $\frac{\omega}{u_*} < 1.0407$).

2.6 Numerical Results of State Probabilities for the Case with Independent Longitudinal and Vertical Fluctuating Velocities

In order to illustrate the concepts for the state probabilities of sediment near the bottom, the numerical results of the three probabilities, which are derived from Eqs. (2.59), (2.60) and (2.61), are listed in Table 2.9. The bed load is typically used to represent the rolling and hopping particles. This is consistent with the three states that are discussed here. Similar results of state probabilities can also be derived from Eqs. (2.49), (2.52), (2.54) and (2.55) for the four states. The state probabilities for particles to be at rest or in suspension state can be derived from either four states or three states, and the results are the same. The characteristics of R_1, R_2 and R_3 can be observed from the results listed in Table 2.9. Some characteristics are summarized in the following:

(1) For $u_* \leq 1$ cm/s, the probability for sediment to be at rest on the bed exceeds 0.996, indicating almost all sediment particles are at rest. R_1 decreases as u_* increases. For $u_* = 5 - 12$ cm/s, R_1 is close to zero for certain particle sizes.

Table 2.9 State probabilities for sediment on the bed

H	D	$u_* = 1$cm/s			$u_* = 3$ cm/s			$u_* = 5$ cm/s			$u_* = 8$ cm/s			$u_* = 12$ cm/s			$u_* = 20$ cm/s		
(m)	(mm)	R_1	R_2	R_3	R_1	R_2	R_3	R_1	R_2	R_3	R_1	R_2	R_3	R_1	R_2	R_3	R_1	R_2	R_3
1.0	0.010	0.9992	0.0004	0.0004	0.0269	0.4875	0.4856	0.0042	0.4979	0.4979	0.0027	0.4984	0.4989	0.0025	0.4975	0.5000	0.0020	0.4980	0.5000
	0.025	0.9997	0.0002	0.0002	0.0125	0.4987	0.4889	0.0050	0.5001	0.4949	0.0031	0.4989	0.4980	0.0025	0.4985	0.4990	0.0020	0.4990	0.4990
	0.050	0.9826	0.0098	0.0076	0.0174	0.5115	0.4710	0.0068	0.5052	0.4880	0.0041	0.5039	0.4920	0.0030	0.5020	0.4950	0.0025	0.5005	0.4970
	0.100	0.9676	0.0237	0.0087	0.0367	0.5592	0.4041	0.0117	0.5373	0.4510	0.0059	0.5251	0.4690	0.0042	0.5168	0.4790	0.0031	0.5089	0.4880
	0.250	0.9988	0.0012	0.0000	0.1595	0.7071	0.1334	0.0374	0.6906	0.2720	0.0136	0.6344	0.3520	0.0072	0.5928	0.4000	0.0045	0.5555	0.4400
	0.500	1.0000	0.0000	0.0000	0.4289	0.5648	0.0064	0.1076	0.8149	0.0775	0.0316	0.7774	0.1910	0.0137	0.7063	0.2800	0.0064	0.6296	0.3640
	1.000	1.0000	0.0000	0.0000	0.8048	0.1952	0.0000	0.2818	0.7100	0.0082	0.0793	0.8521	0.0686	0.0294	0.8096	0.1610	0.0109	0.7131	0.2760
5.0	0.010	0.9992	0.0004	0.0004	0.9254	0.0374	0.0372	0.0169	0.4915	0.4915	0.0037	0.4981	0.4981	0.0028	0.4985	0.4987	0.0020	0.4980	0.5000
	0.025	0.9997	0.0001	0.0001	0.0626	0.4734	0.4640	0.0069	0.4995	0.4936	0.0033	0.5001	0.4965	0.0026	0.4984	0.4990	0.0020	0.4990	0.4990
	0.050	0.9999	0.0000	0.0001	0.0313	0.5047	0.4640	0.0078	0.5078	0.4844	0.0042	0.5038	0.4920	0.0031	0.5019	0.4950	0.0025	0.5005	0.4970
	0.100	0.9999	0.0001	0.0000	0.0461	0.5549	0.3990	0.0123	0.5411	0.4466	0.0059	0.5251	0.4690	0.0042	0.5168	0.4790	0.0031	0.5089	0.4880
	0.250	0.9999	0.0001	0.0000	0.1734	0.6965	0.1301	0.0384	0.6957	0.2659	0.0137	0.6343	0.3520	0.0072	0.5928	0.4000	0.0045	0.5555	0.4400
	0.500	1.0000	0.0000	0.0000	0.4458	0.5481	0.0060	0.1089	0.8145	0.0766	0.0317	0.7773	0.1910	0.0137	0.7063	0.2800	0.0064	0.6296	0.3640
	1.000	1.0000	0.0000	0.0000	0.8122	0.1878	0.0000	0.2838	0.7081	0.0081	0.0795	0.8519	0.0686	0.0294	0.8096	0.1610	0.0109	0.7131	0.2760
20.0	0.01	0.9960	0.0020	0.0020	0.9923	0.0038	0.0039	0.9902	0.0049	0.0049	0.9902	0.0049	0.0049	0.9901	0.0049	0.0050	0.0023	0.4988	0.4989
	0.025	0.9998	0.0001	0.0001	0.9971	0.0014	0.0014	0.0720	0.4668	0.4612	0.0058	0.4991	0.4951	0.0030	0.4994	0.4976	0.0021	0.4989	0.4990
	0.050	0.9999	0.0000	0.0001	0.6256	0.1950	0.1793	0.0179	0.5028	0.4792	0.0051	0.5053	0.4896	0.0032	0.5028	0.4940	0.0025	0.5005	0.4970
	0.100	1.0000	0.0000	0.0000	0.1772	0.4789	0.3439	0.0166	0.5398	0.4436	0.0064	0.5271	0.4665	0.0042	0.5168	0.4790	0.0031	0.5089	0.4880
	0.250	1.0000	0.0000	0.0000	0.2546	0.6290	0.1164	0.0423	0.6961	0.2616	0.0139	0.6341	0.3520	0.0073	0.5927	0.4000	0.0045	0.5555	0.4400
	0.500	1.0000	0.0000	0.0000	0.5145	0.4805	0.0050	0.1144	0.8112	0.0743	0.0320	0.7770	0.1910	0.0137	0.7063	0.2800	0.0064	0.6296	0.3640
	1.000	1.0000	0.0000	0.0000	0.8377	0.1623	0.0000	0.2916	0.7008	0.0076	0.0800	0.8514	0.0686	0.0294	0.8096	0.1610	0.0109	0.7131	0.2760

For $u_* = 20\,\text{cm/s}$, R_1 is almost zero for all particle sizes, which indicates that all the sediment on the bed are moving, resulting in a massive layer movement.

(2) For certain particle sizes, the probability for bed load movement R_2, first increases and then decreases as u_* increases. The former increase is caused by the decrease of R_1, and the latter decrease is due to the increase of R_3. For $0.05\,\text{mm} \leq D \leq 0.5\,\text{mm}$, and $u_* = 3\,\text{cm/s}$ R_2 shows little variation, with a value of approximately 0.5 or greater. For coarser sediment with $0.5\,\text{mm} \leq D \leq 1.0\,\text{mm}$, R_2 exceeds 60% for $u_* \geq 5\,\text{cm/s}$, and sediment on the bed is mostly bed load.

(3) For $u_* = 3 - 5\,\text{cm/s}$, the suspension probability R_3 reaches the maximum for a finer particle size, and then decreases as the particle size increases. For $u_* \geq 8\,\text{cm/s}$, R_3 always decreases as the particle size increases, indicating the gravity dominates the other factors. If using $R_3 < 0.01$ as the upper limit of the particle size for suspended load, the particles with $D \geq 0.5\,\text{mm}$ cannot be suspended for $u_* = 3\,\text{cm/s}$, the particles with $D \geq 1\,\text{mm}$ cannot be suspended for $u_* = 5\,\text{cm/s}$, and the particles with $0.05\,\text{mm} \leq D \leq 0.25\,\text{mm}$ (resulting $R_3 > 0.116$) can all be suspended for $u_* > 3\,\text{cm/s}$. The last case is common for alluvial rivers. As listed in Table 2.9, R_3 is small for fine particles with $0.01\,\text{mm} \leq D \leq 0.025\,\text{mm}$, and $u_* = 1\,\text{cm/s}$, after considering the consolidation of fine sediment. If the sediment inflow from the upstream reach is large, and deposition occurs on the bed with no consolidation of fine sediment, $R_3 = \varepsilon_4$ and exceeds 0.49 for particle sizes with $D \leq 0.025\,\text{mm}$ for $u_* > 3\,\text{cm/s}$, or all the sediment with $D \leq 0.025\,\text{mm}$ can be suspended. This is consistent with the major portion of the particle sizes for the suspended load in alluvial rivers.

(4) To further investigate the relation between the state probabilities and the basic transition probabilities, Fig. 2.7 and 2.8 show the state probabilities versus u_* for a representative fine sediment particle size ($D = 0.01$ mm) and for a coarse particle size ($D = 1.0\,\text{mm}$), respectively. The following can be observed from Fig. 2.8: (a) For $u_* = 2 - 3$ cm/s, the probability of incipient motion ε_1 and the probability for sediment to cease movement $1 - \varepsilon_0$ are both small. But it is hard for the moving particles to stop and thus the probability for bed load movement is greater than the probability of incipient motion. (b) For $u_* = 2 - 8$ cm/s, the probability for incipient suspension β is small and can even be neglected. But the suspension probability ε_4 is close to 0.5. This is because the suspension probability for the particles that have been entrained but have not ceased movement or for the particles that have stopped suspension but have been loosen plays a significant role. (c) For $u_* \geq 3\,\text{cm/s}$ R_2 and R_3 are both close to 0.5. (d) For $u_* \geq 4\,\text{cm/s}$, the sediment entrainment probability is greater than the probability for bed load movement ε_1 and the probability for sediment not to cease movement ε_0, which indicates part of the entrained particles are transited to suspension state.

(d) For $u_* \geq 4cm/s$, the sediment entrainment probability ε_1 is greater than the probability for bed load movement ε_1 and the probability for sediment not to cease movement ε_0, which indicates part of the entrained particles are transited to suspension state.

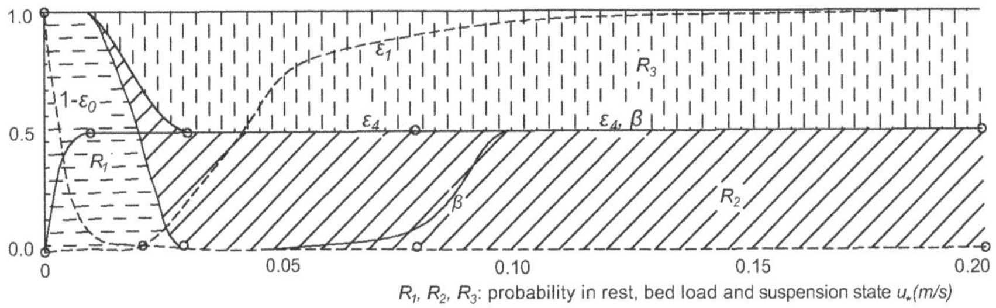

Fig. 2.7 Probabilities versus u_* for fine sediment ($H = 1$ m, $D = 0.01$ mm)

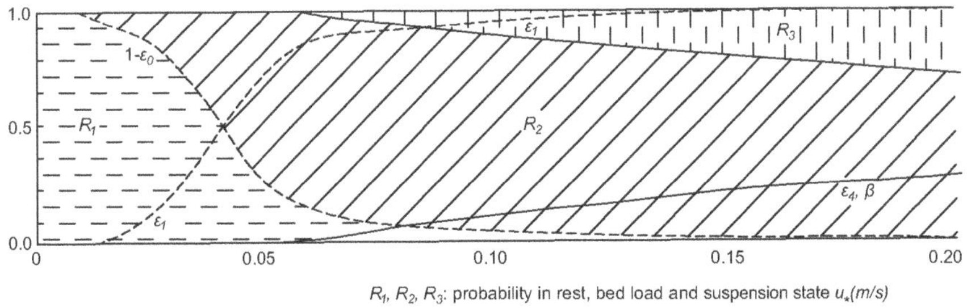

Fig. 2.8 Probabilities versus u_* for coarse sediment ($H = 1$ m, $D = 1.0$ mm)

As shown in Fig. 2.7, the relationship between the probabilities is relatively simple for $D = 1.0$ mm ($H = 1$ m). For this case, $\varepsilon_4 = \beta = R_3$, and sediment entrainment probability ε_1 and the movement ceasing probability ε_0 are close to each other and are approximately symmetric about $R = 0.5$. For $u_* \geq 5$ cm/s, $R_2 > 0.7$.

References

1. Han QW (2003) Reservoir sedimentation. Science Press, Beijing (in Chinese)
2. Han QW (2021) Statistical theory of bed load. Science Press, Beijing (in Chinese)
3. Han QW, He MM (1981) Statistical characters of sediment exchange. J Hydraul Eng 1:9–22 (in Chinese)
4. Han QW, He MM (1984) Stochastic theory of sediment motion. Science Publishing House, Beijing (in Chinese)
5. Han QW, He MM (1987) Exchange and transport rate of bed load. In: Encyclopedia of Fluid Mechanics, vol 6, pp 761–792. Gulf Publishing Company, Huston, TX
6. Han QW, He MM (1999) Study on state probabilities and the ratio of bed load to suspended load. J Hydraul Eng 30(10):7–16 (in Chinese)
7. Han QW, He MM (1999) Characteristics of incipient sediment motion and incipient velocity. Science Press, Beijing (in Chinese)
8. Han QW, He MM (2000) Stochastic theory of sediment motion, Keynote Lectures. In: Proceedings of the 8th International Symposium on Stochastic Hydraulics, Beijing, China, pp 62–84

9. He MM, Han QW (2007) A study on transition probabilities of correlated components of fluctuate velocity. In: Proceedings of 10th International Symposium on River Sedimentation, Moscow, Russia, pp 144–151
10. Nikitin IK (1963) Turbulence process near river bed layer. Press of Falkland Academy of Sciences, Kiev

Chapter 3
Diffusion Equation and Boundary Conditions for Non-uniform Suspended Load

3.1 General Diffusion Equation for Non-uniform Sediment

Diffusion equation was originally developed for uniform sediment, with assumptions that the presence of suspended load does not affect the flow, and that the interaction between sediment particles is negligible. These assumptions are approximately satisfied when the sediment concentration is low and the sediment particle size is not very coarse. Let's divide the non-uniform sediment into n particle size classes. The particle size variation in the l-th size class is very limited, and all the particles in this size class can be approximated as uniform sediment with a single, representative particle size. Based on the assumptions for uniform sediment, the diffusion equation can also be applied to each size class. The equation is derived as follows [3].

Assuming the instantaneous flow velocity vector is $\boldsymbol{V}$, the instantaneous velocity vector for the suspended load $\boldsymbol{u}$ is

$$\boldsymbol{u} = V_x \boldsymbol{i} + (V_y - \omega_l)\boldsymbol{j} + V_z \boldsymbol{k} \tag{3.1}$$

where x, y, z are the Cartesian coordinate directions, with x being the flow direction and y being the upward direction; $\boldsymbol{i}, \boldsymbol{j}, \boldsymbol{k}$ are the Cartesian unit vectors along the x, y, z directions; ω_l is the sediment fall velocity for the l-th size class.

Let σ be a closed surface in flow. The net sediment outflow for the l-th size class (outflow minus inflow) that pass through surface σ within a time interval of dt is

$$dt \iint S_l \boldsymbol{u} \cdot \boldsymbol{n} d\sigma$$

where $\boldsymbol{n}$ is the outward unit normal of σ, and S_l is the instantaneous sediment concentration for the l-th size class. The change to the sediment volume within surface σ during this time interval is $-dt \iiint \frac{\partial S_l}{\partial t} d\tau$. Based on the law of mass conservation

Q. Han and M. He, *Non-equilibrium Transport of Non-uniform Suspended Load*, Hydroscience and Engineering, https://doi.org/10.1007/978-981-97-1123-9_3

$$dt \iint S_l \boldsymbol{u} \cdot \boldsymbol{n} d\sigma = -dt \iiint \frac{\partial S_l}{\partial t} d\tau \tag{3.2}$$

where τ is the water volume within surface σ. After changing the surface integral into volume integral, Eq. (3.2) becomes

$$\iiint \left[\frac{\partial (S_l V_x)}{\partial x} + \frac{\partial [S_l (V_x - \omega_1)]}{\partial y} + \frac{\partial (S_l V_z)}{\partial z} \right] dxdydzdt$$
$$= - \iiint \left[\frac{\partial S_l}{\partial t} \right] dxdydzdt \tag{3.3}$$

Since surface σ is arbitrarily chosen, the integrands of both sides of Eq. (3.3) should be equal, or

$$\frac{\partial S_l}{\partial t} + \frac{\partial (S_l V_x)}{\partial x} + \frac{\partial \left[S_l (S_l V_z) \right]}{\partial z} = 0 \tag{3.4}$$

The instantaneous variables can be decomposed into their mean parts ($\overline{V}_x$, $\overline{V}_z$, $\overline{V}_y$, $\overline{S}_l$) and fluctuating parts (V_x', V_y', V_z', S_l'). Assuming that the diffusion equation for uniform sediment also applies to each size class of non-uniform sediment, we have

$$\left. \begin{aligned} \overline{S_l' V_x'} &= -\varepsilon_x \frac{\partial \overline{S_l}}{\partial x} \\ \overline{S_l' V_y'} &= -\varepsilon_y \frac{\partial \overline{S_l}}{\partial y} \\ \overline{S_l' V_z'} &= -\varepsilon_x \frac{\partial \overline{S_l}}{\partial z} \end{aligned} \right\} \tag{3.5}$$

So averaging Eq. (3.4) over time yields

$$\frac{\partial S_l}{\partial t} + \frac{\partial (\overline{S_l V_x})}{\partial x} + \frac{\partial (\overline{S_l V_y})}{\partial y} + \frac{\partial (\overline{S_l V_z})}{\partial z}$$
$$= \frac{\partial}{\partial x} \left(\varepsilon_x \frac{\partial \overline{S_l}}{\partial x} \right) + \frac{\partial}{\partial y} \left(\varepsilon_y \frac{\partial \overline{S_l}}{\partial y} \right) + \frac{\partial}{\partial z} \left(\varepsilon_z \frac{\partial \overline{S_l}}{\partial z} \right) + \frac{\partial (\overline{S_l} \omega_l)}{\partial y}$$

After removing the symbol "–" that denotes the mean value, this equation turns to

$$\frac{\partial S_l}{\partial t} + \frac{\partial (V_x S_l)}{\partial x} + \frac{\partial (V_y S_l)}{\partial y} + \frac{\partial (V_z S_l)}{\partial z}$$
$$= \frac{\partial}{\partial x} \left(\varepsilon_x \frac{\partial S_l}{\partial x} \right) + \frac{\partial}{\partial y} \left(\varepsilon_y \frac{\partial S_l}{\partial y} \right) + \frac{\partial}{\partial z} \left(\varepsilon_z \frac{\partial S_l}{\partial z} \right) + \frac{\partial S_l \omega_l}{\partial y} \quad (l = 1, 2, \ldots n_l) \tag{3.6}$$

It is noted that this equation has the same form as the equation for the uniform sediment. However, S_l is the sediment concentration for the l-th size class of the non-uniform sediment, and thus its value is much less than the sediment concentration for uniform sediment.

Applying summation for Eq. (3.6) for all the size classes and noticing that

$$S(x, y, z, t) = \sum_{l=1}^{n} S_l(x, y, z, t) \tag{3.7}$$

we have

$$\begin{aligned} &\frac{\partial S}{\partial t} + \frac{\partial (V_x S)}{\partial x} + \frac{\partial (V_y S)}{\partial y} + \frac{\partial (V_z S)}{\partial z} \\ &= \frac{\partial}{\partial x}\left(\varepsilon_x \frac{\partial S}{\partial x}\right) + \frac{\partial}{\partial y}\left(\varepsilon_y \frac{\partial S}{\partial y}\right) + \frac{\partial}{\partial z}\left(\varepsilon_z \frac{\partial S}{\partial z}\right) + \frac{\partial (S\omega)}{\partial y} \end{aligned} \tag{3.8}$$

It is noted that the following relations are used in deriving Eq. (3.8):

$$\sum_{l=1}^{n} \frac{\partial}{\partial y}(S_l \omega_l) = \frac{\partial}{\partial y}\left[S \sum_{l=1}^{n} \frac{S_l}{S}\omega_l\right] = \frac{\partial}{\partial y}(S\omega) \tag{3.9}$$

$$\omega(x, y, z, t) = \sum_{l=1}^{n} \frac{S_l}{S}\omega_l = \sum_{l=1}^{n} P_{4.l}(x, y, z, t)\omega_l \tag{3.10}$$

$$P_{4.l}(x, y, z, t) = \frac{S_l(x, y, z, t)}{S(x, y, z, t)} \tag{3.11}$$

where $P_{4.l}(x, y, z, t)$ is the suspended load gradation, and $\omega(x, y, z, t)$ the averaged fall velocity of suspended load. Comparing Eq. (3.8) with the equation for uniform sediment, it is found that the equation for the total sediment concentration of non-uniform sediment has the same form as the equation for the uniform sediment after defining the averaged suspended load fall velocity with Eq. (3.10). The difference in the equations for uniform and non-uniform sediment lies in the sediment fall velocity ω. While ω is constant for uniform sediment, it is a variable for non-uniform sediment and is unknown before the sediment concentration is determined for each size class. Therefore, the sediment concentration cannot be directly solved from Eq. (3.8) for non-uniform sediment. This is the major challenge for studies of non-uniform sediment transport.

Let's introduce the water volume percentage K_l, and demonstrate an important characteristic for steady uniform flow with strong-equilibrium sediment transport. It is assumed that the sediment concentration (or the carrying-sediment capacity) can be divided into n groups. Each group has a uniform sediment grain size of D_l with a corresponding carrying-sediment capacity of $\overline{S}^*(l)$. The mass of sediment within each group is assumed to be G_l^* for a given study reach with a length of L, width of B

and depth of h. Based on the basic assumptions for the convection-diffusion equation, the impact of sediment on flow and the interactions between sediment particles are negligible when the sediment concentration is not high. It is assumed that the water body can be divided into n small water volumes. Each water volume only carries uniform sediment with a particle size of D_l with a sediment mass of G_l^*. The water volume that carries sediment of the l-th size class is

$$Q_l = \frac{G_l^*}{\overline{S}_l^*} \tag{3.12}$$

The total water volume is

$$Q = \sum_{l=1}^{n} Q_l \tag{3.13}$$

So the water volume percentage is defined as

$$K_l = \frac{Q_l}{Q} = \frac{Q_l}{\sum_{l=1}^{n} Q_l} \tag{3.14}$$

Now let's divide the water body into n small volumes using the water volume percentage. Each small water volume has a water volume of $Q_l = K_l LBh$ and a sediment mass of G_l^*. So the fractional carrying-sediment capacity for the l-th size class is

$$\overline{S_l^*} = \frac{G_l^*}{Q} = \frac{G_l^*}{Q}\frac{Q_l}{Q} = K_l\overline{S}^*(l) \tag{3.15}$$

It can be derived from Eq. (3.14) that

$$\sum K_l = \sum \frac{Q_l}{Q} = 1 \tag{3.16}$$

It is noted that $\overline{S}_l^* = P_{4,l}^*\overline{S}^*$. So

$$P_{4,l}^*\overline{S}^* = \overline{S}_l^* = K_l\overline{S}^*(l) = K_l\overline{S}(l) \tag{3.17}$$

Now we have developed a virtual distribution for sediment concentration by assuming that a small water volume of Q_l only carries sediment of the uniform sediment with particle size D_l with a mass of G_l^*. This distribution is the same as the actual distribution of the l-th size class of non-uniform sediment.

Summation of Eq. (3.17) indicates that the total capacity for carrying non-uniform sediment is the summation of the capacity for carrying uniform sediment $\overline{S}(l)$ multiplied by the corresponding water volume percentage. This has more explicit physical meanings. The concept of water volume percentage was first introduced by Han [1], and was further demonstrated and explained by Han and He [3] and Han [2]. The water volume percentage is an important tool for the study of non-uniform suspended load transport. But Eq. (3.17) is not deterministic for non-equilibrium sediment transport. This is because the relationship between $S(l)$ and the flow condition and sediment characteristics is not deterministic and thus $S(l)$ cannot be derived using existing formulas. As a result, Q_l and K_l can not be determined from Eq. (3.12) even if S, $P_{4,l}$ or S_l are known.

For strong-equilibrium sediment transport, the fractional carrying-sediment capacity $S(l)$, the sediment gradation $P_{4,l}$ and the total sediment concentration satisfy the following relationship

$$S_l(x, y, z, t) = S_l^*(x, y, z, t) \tag{3.18}$$

or

$$P_{4.l}(x, y, z, t) = P_{4.l}^*(x, y, z, t) \tag{3.19}$$

$$S(x, y, z, t) = S^*(x, y, z, t) \tag{3.20}$$

where S_l^* and $P_{4,l}^*$ is the fractional carrying-sediment capacity and gradation, respectively, for gain class l, and S^* is the total carrying-sediment capacity for equilibrium sediment transport. For this case, Eq. (3.17), which is for the carrying-sediment capacity, can also be used for sediment concentration.

$$P_{4,l}^*\overline{S} = \overline{S}_l^* = K_l\overline{S}^*(l) = K_l\overline{S}(l) = P_{4,l}\overline{S} = \overline{S}_l \tag{3.21}$$

Equation (3.17) is of important significance. It illustrates the relationship between the fractional capacity for carrying non-uniform sediment and the capacity for carrying uniform sediment. This allows us to make full use of study results for uniform sediment transport.

3.2 Integration of 3-D Diffusion Equation Over River Width and Over Water Depth

3.2.1 Integration Over Channel Width

For a rectangular channel with a uniform width B, assuming diffusion coefficients ε_x, ε_y and ε_z are independent of z (along the channel width), integrating Eq. (3.6) over the channel width yields

$$\int_0^B \frac{\partial S_l}{\partial t}dz + \int_0^B \left[\frac{\partial(V_x S_l)}{\partial x} + \frac{\partial(V_y S_l)}{\partial y} + \frac{\partial(V_z S_l)}{\partial z}\right]dz$$
$$= \int_0^B \left[\frac{\partial}{\partial x}\varepsilon_x \frac{\partial S_l}{\partial x} + \frac{\partial}{\partial y}\varepsilon_y \frac{\partial S_l}{\partial y} + \frac{\partial}{\partial z}\varepsilon_z \frac{\partial S_l}{\partial z} + \frac{\partial}{\partial y}(\omega_l S_l)\right]dz \tag{3.22}$$

It is noted that

$$\overline{S_l} = \frac{1}{B}\int_0^B S_l dz \tag{3.23}$$

$$\overline{V_y S_l} = \frac{1}{B}\int_0^B V_y S_l dz \tag{3.24}$$

$$\overline{V_y S_l} = \frac{1}{B}\int_0^B V_y S_l dz \tag{3.25}$$

$$\int_0^B \frac{\partial(V_z S_l)}{\partial Z}dz = V_z S_l|_{z=B} - V_z S_l|_{z=0} \tag{3.26}$$

$$\int_0^B \frac{\partial}{\partial y}\omega_l S_l dz = \frac{\partial}{\partial y}B\omega_l \overline{S_l} \tag{3.27}$$

$$\int_0^B \frac{\partial}{\partial x}\varepsilon_x \frac{\partial S_l}{\partial x}dz = \frac{\partial}{\partial x}\varepsilon_x \frac{\partial}{\partial x}\left(B\overline{S_l}\right) \tag{3.28}$$

$$\int_0^B \frac{\partial}{\partial y}\varepsilon_y \frac{\partial S_l}{\partial y}dz = \frac{\partial}{\partial y}\varepsilon_y \frac{\partial}{\partial y}(B\overline{S_l}) \tag{3.29}$$

$$\int_0^B \frac{\partial}{\partial y}\omega_l S_l dz = \frac{\partial}{\partial y}B\omega_l \overline{S_l} \tag{3.30}$$

Then Eq. (3.22) can be rewritten as

$$\frac{\partial(\overline{S}_l B)}{\partial t}+\frac{\partial(\overline{V_x S_l} B)}{\partial x}+\frac{\partial(\overline{V_y S_l} B)}{\partial y}+V_z S_l|_{z=B}-V_z S_l|_{z=0}$$
$$=\frac{\partial}{\partial x}\varepsilon_x\frac{\partial(\overline{S}_l B)}{\partial x}+\frac{\partial}{\partial y}\varepsilon_y\frac{\partial(\overline{S}_l B)}{\partial y}+\varepsilon_z\frac{\partial S_l}{\partial z}\bigg|_{z=B}-\varepsilon_z\frac{\partial S_l}{\partial z}\bigg|_{z=0}+\varepsilon_z\frac{\partial}{\partial y}(\overline{S}_l\omega_l B) \tag{3.31}$$

It is noted that the diffusion equation integrated over the channel width includes not only S_l on lateral channel boundaries, but also terms $\overline{S_l}$, $\overline{V_y S_l}$ and $\overline{V_x S_l}$. For a uniform rectangular channel, sediment can not pass through both sides of the channel. So the sediment convection and diffusion terms are zero on these lateral boundaries, or

$$V_z S_l|_{z=B}-V_z S_l|_{z=0}=0 \tag{3.32}$$

$$\varepsilon_z\frac{\partial S_l}{\partial z}\bigg|_{z=B}-\varepsilon_z\frac{\partial S_l}{\partial z}\bigg|_{z=0}=0 \tag{3.33}$$

Plugging Eqs. (3.32) and (3.33) into Eq. (3.31) and eliminating B, we have

$$\frac{\partial\overline{S}_l}{\partial t}+\frac{\partial\left(\overline{V_x S_l}\right)}{\partial x}+\frac{\partial\left(\overline{V_y S_l}\right)}{\partial y}=\frac{\partial}{\partial x}\left(\varepsilon_x\frac{\partial\overline{S}_l}{\partial x}\right)+\frac{\partial}{\partial y}\left(\varepsilon_y\frac{\partial\overline{S}_l}{\partial y}\right)+\frac{\partial}{\partial_y}\left(\overline{S}_l\omega_l\right) \tag{3.34}$$

Assuming the variations of V_x and V_y along the channel width is negligible for a rectangular channel with a uniform width, we have

$$\overline{V_x S_l}=\frac{1}{B}\int_0^B V_x S_l dz\approx\overline{V_x S_l} \tag{3.35}$$

$$\overline{V_y S_l}=\frac{1}{B}\int_0^B V_y S_l dz\approx\overline{V_y S_l} \tag{3.36}$$

Then Eq. (3.34) can be further simplified as

$$\frac{\partial\overline{S}_l}{\partial t}+\frac{\partial(\overline{V_x S_l})}{\partial x}+\frac{\partial(\overline{V_y S_l})}{\partial y}=\frac{\partial}{\partial x}\left(\varepsilon_x\frac{\partial\overline{S}_l}{\partial x}\right)+\frac{\partial}{\partial y}\left(\varepsilon_y\frac{\partial\overline{S}_l}{\partial y}\right)+\frac{\partial}{\partial y}\left(\overline{S}_l\omega_l\right) \tag{3.37}$$

It is noted that there is only one unknown $\overline{S}_l$ in Eq. (3.37) in addition to the hydraulic parameters $\overline{V}_x$ and $\overline{V}_y$. Since $\overline{S}_l$, $\overline{V}_x$ and $\overline{V}_y$ have the values averaged over the channel width, Eq. (3.37) is consistent with the two-dimensional vertical diffusion equation that is later derived in Sect. 3.3. But Eq. (3.37) includes the three-dimensional effect by replacing $\overline{S_l}$ with a width-averaged value $\overline{S}_l$.

3.2.2 Integration Over Water Depth

Integration of the three-dimensional diffusion equation over water depth yields the two-dimensional horizontal (or depth-averaged) diffusion equation, or the so-called shallow water equation. Integrating Eq. (3.6) over the water depth, and noting

$$\overline{S_l} = \frac{1}{h}\int_0^h S_l dy \tag{3.38}$$

$$\overline{V_x S_l} = \frac{1}{h}\int_0^h V_x S_l dy \tag{3.39}$$

$$\overline{V_z S_l} = \frac{1}{h}\int_0^h V_z S_l dy \tag{3.40}$$

$$\int_0^h \frac{\partial\left(V_y S_l\right)}{\partial y} dy = V_y S_l\Big|_{y=h} - V_y S_l\Big|_{y=0} \approx 0 \tag{3.41}$$

$$\int_0^h \frac{\partial}{\partial x}\left(\varepsilon_x \frac{\partial S_l}{\partial x}\right) dy = \frac{\partial}{\partial x}\left[\varepsilon_x \frac{\partial (h\overline{S_l})}{\partial x}\right] \tag{3.42}$$

$$\int_0^h \frac{\partial}{\partial z}\left(\varepsilon_z \frac{\partial S_l}{\partial z}\right) dy = \frac{\partial}{\partial z}\left[\varepsilon_z \frac{\partial (h\overline{S_l})}{\partial z}\right] \tag{3.43}$$

$$\begin{aligned}\int_0^h \left[\frac{\partial}{\partial y}\left(\varepsilon_y \frac{\partial S}{\partial y} + \omega_l S\right)\right] dy &= \left(\varepsilon_y \frac{\partial S_l}{\partial y} + \omega_l S_l\right)\Big|_{y=h} - \left(\varepsilon_y \frac{\partial S_l}{\partial y} + \omega_l \overline{S_l}\right)\Big|_{y=0} \\ &= -\varepsilon_y \frac{\partial S_l}{\partial y}\Big|_{y=0} - \omega_l S_l\big|_{y=0}\end{aligned} \tag{3.44}$$

we obtain

$$\frac{\partial}{\partial t}(h\overline{S_l}) + \frac{\partial(h\overline{V_x S_l})}{\partial x} + \frac{\partial(h\overline{V_z S_l})}{\partial z} = \frac{\partial}{\partial x}\left[\varepsilon_x \frac{\partial(h\overline{S_l})}{\partial x}\right] + \frac{\partial}{\partial z}\left[\varepsilon_z \frac{\partial(h\overline{S_l})}{\partial z}\right] - \left.\left(\varepsilon_y \frac{\partial S_l}{\partial y} + \omega_l \overline{S_{l.b}}\right)\right|_{y=0} \quad (3.45)$$

It is noted that the diffusion and fall terms in Eq. (3.44) are set to zero at the water surface ($y = h$) as no sediment can pass through this boundary.

Or,

$$\left.\left(\varepsilon_y \frac{\partial S_l}{\partial y} + \omega S_l\right)\right|_{y=h} = 0$$

Using the following approximation

$$\overline{V_x S_l} = \frac{1}{h}\int_0^h V_x S_l dy \approx \overline{V_x} \cdot \overline{S_l} \quad (3.46)$$

$$\overline{V_z S_l} = \frac{1}{h}\int_0^h V_z S_l dy \approx \overline{V_z} \cdot \overline{S_l} \quad (3.47)$$

and introducing the sedimentation rate (sedimentation mass per unit time per unit area)

$$\Delta S_l = \left.\left(\varepsilon_y \frac{\partial S_l}{\partial y} + \omega_l S_l\right)\right|_{y=0} = \gamma'_{s.l} \frac{\partial y_s}{\partial t} \quad (3.48)$$

Equation (3.45) can be written as

$$\frac{\partial}{\partial t}(h\overline{S_l}) + \frac{\partial}{\partial x}(h\overline{V_x S_l}) + \frac{\partial}{\partial z}(h\overline{V_z S_l}) - \frac{\partial}{\partial x}\left[\varepsilon_x \frac{\partial(h\overline{S_l})}{\partial x}\right] - \frac{\partial}{\partial z}\left(\varepsilon_z \frac{\partial(h\overline{S_l})}{\partial z}\right) + \gamma'_{s.l} \frac{\partial y_s}{\partial t} = 0 \quad (3.49)$$

where y_s is the channel bed elevation. This is a shallow water horizontal diffusion equation. It is noted that Eq. (3.49) is derived based on a rectangular channel assumption and on the approximations in Eqs. (3.46) and (3.47).

3.3 Vertical Two-Dimensional Diffusion Equation and Longitudinal Variation of Sediment Transport Rate

For vertical two-dimensional flow and sediment transport, the sediment diffusion equation can be derived from Eq. (3.6) and reads

$$\frac{\partial S_l}{\partial t}+\frac{\partial (V_x S_l)}{\partial x}+\frac{\partial (V_y S_l)}{\partial y}=\frac{\partial}{\partial x}\left(\varepsilon_x\frac{\partial S_l}{\partial x}\right)+\frac{\partial}{\partial y}\left[\varepsilon_y\frac{\partial (S_l)}{\partial y}+\omega_l S_l\right] \tag{3.50}$$

The sediment diffusion along the flow direction $\frac{\partial}{\partial x}\left(\varepsilon_x\frac{\partial S_l}{\partial x}\right)$ usually can be neglected compared to the convection term $\frac{\partial (V_x S_l)}{\partial x}$, and assume the flow is steady with a steady non-equilibrium sediment transport, or $\frac{\partial S_l}{\partial t}=0$, Eq. (3.50) can be written as

$$\frac{\partial (V_x S_l)}{\partial x}+\frac{\partial (V_y S_l)}{\partial y}=\frac{\partial}{\partial y}\left(\varepsilon_y\frac{\partial S_l}{\partial y}+\omega_l S_l\right) \tag{3.51}$$

Equation (3.51) is the commonly used governing equation for the vertical two-dimensional sediment diffusion. The characteristics of this equation are investigated in the following.

3.3.1 Relation Between the Longitudinal Variation of Sediment Transport Rate per Unit Width and the Bottom Boundary Value of the Diffusion Equation in Steady Non-uniform Flow

With reference to Fig. 3.1, the water surface profile $h_2(x)$ and the channel bed profile $h_1(x)$ both change linearly with x-axis for steady non-uniform flow. The x-axis is in the horizontal direction. After integrating Eq. (3.51) over the water depth, the left term of the equation becomes the sediment transport rate per unit width. As shown in Fig. 3.1, the sediment transport rate per unit width, or the sediment flux S_l, can be expressed as

$$
\begin{aligned}
q_{s.l} = q\overline{S_l} = qP_{4.l}\overline{S} &= \int_{h_1(x)}^{h_2(x)} V_n S_l dl \\
&= \int_{h_1(x)}^{h_2(x)} \sqrt{\overline{V}_x^2 + \overline{V}_y^2} S_l dl \\
&= \int_{h_1(x)}^{h_2(x)} \sqrt{\overline{V}_x^2 + \overline{V}_y^2} S_l \frac{dy}{\cos\theta}
\end{aligned} \tag{3.52}
$$

where θ is the angle between the flow velocity and the horizontal direction. As illustrated in Fig. 3.2

$$
\sqrt{V_x^2 + V_y^2} = V_x\sqrt{1 + \left(\frac{V_y}{V_x}\right)^2} = V_x\sqrt{1 + \tan^2\theta} \tag{3.53}
$$

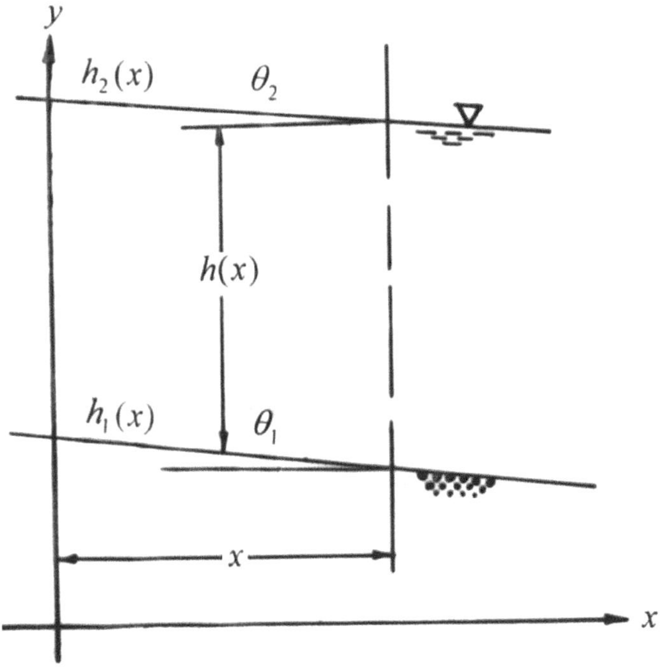

Fig. 3.1 Schematic diagram for steady non-uniform flow

$$\cos\theta = \frac{1}{\sqrt{1 + \left(\frac{dx}{dy}\right)^2}} = \frac{1}{\sqrt{1 + \tan^2\theta}} \tag{3.54}$$

Therefore

$$q_{s.l} = \int_{h_1(x)}^{h_2(x)} S_l V_x (1 + \tan^2\theta) dy \tag{3.55}$$

Because tanθ is much less than 1/100 for most alluvial rivers, Eq. (3.55) can be written as

$$q_{s.l} = \int_{h_1(x)}^{h_2(x)} S_l V_x dy \tag{3.56}$$

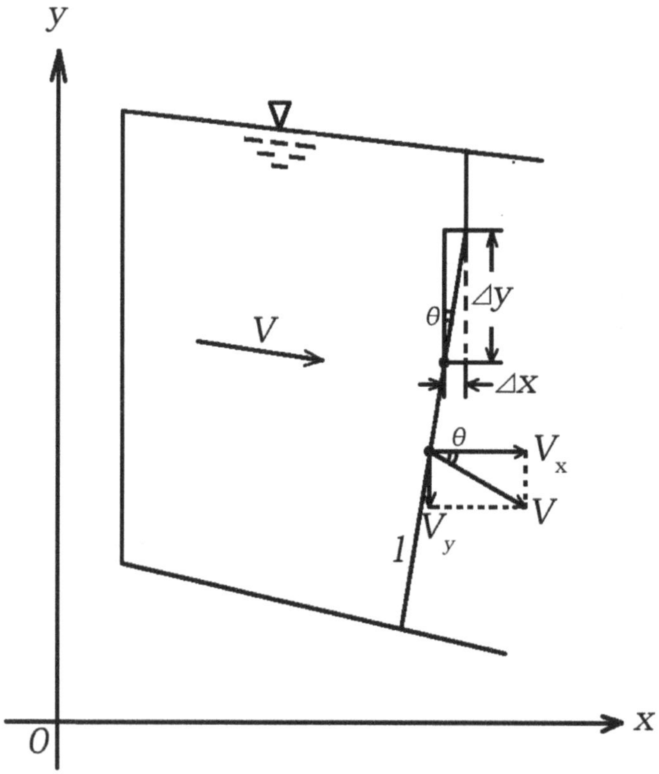

Fig. 3.2 Flow velocity components for steady non-uniform flow

The derivative of Eq. (3.56) shows that the longitudinal variation of $q_{s.l}$ is

$$\begin{aligned}\frac{dq_{s.l}}{dx} &= \int_{h_1(x)}^{h_2(x)} \frac{\partial}{\partial x}(\overline{S_l V_x})dy + (S_l V_x)|_{y=h_2(x)} h_2'(x) - (S_l V_x)|_{y=h_1(x)} h_1'(x) \\ &= \int_{h_1(x)}^{h_2(x)} \frac{\partial}{\partial x}(S_l V_x)dy + (S_l V_y)\big|_{y=h_2(x)} - (S_l V_y)\big|_{y=h_1(x)} \\ &= \int_{h_1(x)}^{h_2(x)} \left[\frac{\partial}{\partial x}(S_l V_x) + \frac{\partial}{\partial y}(S_l V_y)\right]dy \end{aligned} \tag{3.57}$$

The following relation is used to obtain Eq. (3.57)

$$\begin{aligned}&(S_l V_x)|_{y=h_2(x)} h_2'(x) - (S_l V_x)\big|_{y=h_1(x)} h_1'(x) \\ &= (S_l V_x)|_{y=h_2(x)} \frac{dh_2}{dx} - (S_l V_x)|_{y=h_1(x)} \frac{dh_1}{dx} \\ &= (S_l V_y)\big|_{y=h_2(x)} - (S_l V_y)\big|_{y=h_1(x)} = \int_{h_1(x)}^{h_2(x)} \frac{\partial(S_l V_y)}{\partial y} dy\end{aligned}$$

On the other hand, neglecting $\frac{\partial S_l}{\partial t}$ and the longitudinal diffusion $\frac{\partial}{\partial x}\left(\varepsilon_x \frac{\partial S_l}{\partial x}\right)$ in Eq. (3.37) and then integrating over the water depth yield

$$\frac{dq_{s.l}}{dx} = \int_{h_1(x)}^{h_2(x)} \left[\frac{\partial}{\partial x}(S_l V_x) + \frac{\partial}{\partial y}(S_l V_y)\right]dy = -\varepsilon_y \frac{\partial S_l}{\partial y}\bigg|_{y=h_1(x)} - \omega_l S_{b.l} \tag{3.58}$$

This indicates that the longitudinal variation (along the x-axis) of the sediment transport rate per unit width is equal to the bottom boundary value of the diffusion equation.

3.3.2 Relation Between the Longitudinal Variation of Sediment Transport Rate and the Bottom Boundary Value of the Diffusion Equation in Three-Dimensional Steady Uniform Flow with Non-equilibrium Sediment Transport

Integration of the vertical two-dimensional Eq. (3.37), which is derived from the three-dimensional diffusion equation, over water depth yield results similar to the above, but with more strict conditions. As do for Eq. (3.37), we integrate Eq. (3.34) over depth after neglecting $\frac{\partial \overline{S_l}}{\partial t}$ and the longitudinal diffusion $\frac{\partial}{\partial x}\left(\varepsilon_x \frac{\partial \overline{S_l}}{\partial x}\right)$. Assuming the channel is a uniform rectangular channel with uniform flow and a constant $\theta = \theta_0$, we have

$$
\begin{aligned}
&\int_0^h \left[\frac{\partial(\overline{V_x S_l})}{\partial x} + \frac{\partial(\overline{V_y S_l})}{\partial y}\right] dy \\
&= \left[\varepsilon_y \frac{\partial(\overline{S_l})}{\partial y} + \omega_l \overline{S_l}\right]\Bigg|_{y=h} - \left[\varepsilon_y \frac{\partial(\overline{S_l})}{\partial y} + \omega_l S_l\right]\Bigg|_{y=0} \\
&= -\left[\varepsilon_y \frac{\partial(\overline{S_l})}{\partial y}\right]\Bigg|_{y=0} - \omega_l \overline{S}_{b,l}
\end{aligned}
\tag{3.59}
$$

The left term of Eq. (3.59) can be further written as

$$
\begin{aligned}
\int_0^h \left[\frac{\partial(\overline{V_x S_l})}{\partial x} + \frac{\partial(\overline{V_y S_l})}{\partial y}\right] dy &= \int_0^h \frac{\partial(\overline{V_x S_l})}{\partial x} dy + \overline{V_y S_l}\Big|_{y=h} - \overline{V_y S_l}\Big|_{y=0} \\
&\approx \int_0^h \frac{\partial(\overline{V_x S_l})}{\partial x} dy = \frac{d}{dx}\int_0^h \overline{V_x S_l} dy = \frac{dq_{s.l}}{dx}
\end{aligned}
\tag{3.60}
$$

where

$$
q_{s.l} = \int_0^h \overline{V_x S_l} dy = h\overline{\overline{V_x S_l}}
\tag{3.61}
$$

is the suspended sediment transport rate per unit width, and $\overline{\overline{S_l V_x}}$ is the width- and depth-averaged value of $S_l V_x$. Plugging Eq. (3.60) into Eq. (3.59) yields

$$\frac{dq_{s.l}}{dx} = \frac{d(q\overline{\overline{S}}_l)}{dx} = -\varepsilon_y \left.\frac{\partial \overline{S}_l}{\partial y}\right|_{y=0} - \omega_l \overline{S}_{b,l} \tag{3.62}$$

Therefore, for a uniform rectangular channel with steady uniform flow and non-equilibrium sediment transport, integration of the three-dimensional diffusion equation over depth and width also indicates that the longitudinal variation of the sediment transport rate per unit width is equal to the bottom boundary value of the diffusion equation.

It is noted that the above conclusion is also true for steady flow and non-equilibrium sediment transport without any additional assumptions. It can be directly derived from the conservation of mass for sediment. The purpose of the derivation described above is to emphasize and illustrate the relationship between various parameters.

3.3.3 Solution to Two-Dimensional Diffusion Equation for Strong-Equilibrium Sediment Transport and Related Characteristics

For strong-equilibrium transport of non-uniform sediment there is

$$\frac{dq_{s.l}}{dx} = q\frac{dS_l}{dx} = 0, \quad (l = 1, 2, \ldots, n) \tag{3.63}$$

Then from Eq. (3.58) the integrand must equal to zero:

$$\frac{\partial}{\partial x}(S_l V_x) + \frac{\partial}{\partial y}\left(S_l V_y\right) = 0 \quad (0 \le y \le h; l = 1, 2, \ldots n) \tag{3.64}$$

From Eq. (3.51) exists

$$\frac{dq_{s.l}}{dx} = q\frac{dS_l}{dx} = \frac{\partial}{\partial y}\left(\varepsilon_y \frac{\partial S_l}{\partial x} + \omega_l S_l\right) = 0, \quad (0 \le y \le h) \tag{3.65}$$

where q is the uniform flow discharge per unit width. It is noted that the averaged symbol "–" is removed hereafter, i.e., $\overline{S}$ is replaced with S, unless otherwise specified.

Integrating Eq. (3.65) over water depth and applying the boundary condition at the water surface, i.e., $\left(\varepsilon_y \frac{\partial S_l}{\partial x} + \omega_l S_l\right) = 0$ at $y = h$, we have

$$\varepsilon_y \frac{\partial S_l}{\partial y} + \omega_l S_l = 0 \tag{3.66}$$

Integrating Eq. (3.65) over water depth and applying the bottom boundary condition ($S_l = S_{b.l}$ at $y = 0$) yield the solution

$$S_l(y) = S_{b.l} \exp\left(-\int_0^y \frac{\omega_l}{\varepsilon_y} dy\right) = S^*_{b.l} \exp\left(-\int_0^y \frac{\omega_l}{\varepsilon_y} dy\right) = S^*_l(y) \tag{3.67}$$

Equation (3.67) represents the vertical profile of the fractional sediment concentration for the l-th size class for equilibrium sediment transport. The sediment concentration under the equilibrium transport condition is also called the saturation sediment concentration, or the carrying-sediment capacity. So a symbol of * is added to the concentration to represent the carrying-sediment capacity, as shown in Eq. (3.67). Hereafter, the sediment concentration for equilibrium sediment transport is represented with the carrying-sediment capacity S^*. It is noted that the vertical profile of the fractional sediment concentration for each size class of the sediment mixture is similar to that of the uniform sediment, but with different meanings.

It is worthy pointing out that equilibrium sediment transport can also be in weak-equilibrium status in addition to strong-equilibrium condition. The weak-equilibrium sediment transport is defined as

$$\frac{d\sum S_l}{dx} = \frac{dS}{dx} = 0 \tag{3.68}$$

This means that the fractional sediment concentration S_l for each size class may vary along the flow path, but the total concentration S for the sediment mixture is constant.

Let's further investigate and illustrate the meaning of the water percentage coefficient for vertically two-dimensional sediment transport. The vertical profile of the carrying-sediment capacity for uniform sediment with a particle size D_l is

$$S^*(y, l) = S^*_b(l) \exp\left(-\int_0^y \frac{\omega_l}{\varepsilon_y} dy\right) \tag{3.69}$$

So its depth-averaged value is

$$\begin{aligned} S^*(l) &= \frac{1}{hV_m} \int_0^h V S^*(y, l) dy = \int_0^1 \frac{V}{V_m} S^*_b(l) \exp\left(-\int_0^y \frac{\omega_1}{\varepsilon_y} dy\right) d\eta \\ &= S^*_b(l) \int_0^1 \frac{V}{V_m} \exp\left(-\int_0^\eta \frac{\omega_l h}{\varepsilon_y} d\eta\right) d\eta = S^*_b(l) B_l \end{aligned} \tag{3.70}$$

The depth-averaged fractional carrying-sediment capacity is

$$\overline{S_l^*} = S_{b,l}^* B_l \tag{3.71}$$

where $\eta = \frac{y}{h}$, h is the water depth, V is the flow velocity at y, V_m is the depth-averaged flow velocity, and

$$B_l = \int_0^1 \frac{V}{V_m} \exp\left(- \int_0^\eta \frac{\omega_l h}{\varepsilon_y} d\eta \right) d\eta$$

Let's define

$$K_l = \frac{S_{b.l}^*}{S_b^*(l)} \tag{3.72}$$

where $S_b^*(l)$ is the carrying-sediment capacity at the bottom for uniform sediment with particle size of D_l, while $S_{b.l}^*$ is the fractional carrying-sediment capacity at the bottom for the l-th size class of the non-uniform sediment mixture. Plugging Eq. (3.72) into Eq. (3.70) and then applying Eq. (3.71) yields

$$K_l S^*(l) = K_l S_b^*(l) B_l = S_{b.l}^* B_l = \overline{S_l^*} \tag{3.73}$$

or

$$K_l = \frac{\overline{S_l^*}}{\overline{S}^*(l)} \tag{3.74}$$

Thus,

$$\overline{S^*} = \sum_{l=1}^{n} \overline{S_l^*} = \sum_{l=1}^{n} K_l \overline{S}^*(l) \tag{3.75}$$

Also, plugging Eq. (3.72) into Eq. (3.69) yields

$$K_l S^*(y, l) = S_{b.l}^* \exp\left(- \int_0^y \frac{\omega_l}{\varepsilon_y} dy \right) = S_l^*(y) \tag{3.76}$$

or

$$K_l = \frac{S_l^*(y)}{S^*(y, l)} \tag{3.77}$$

Summation of Eq. (3.77) for all particle size classes, turns to

$$\sum K_l S^*(y, l) = S^*(y) \tag{3.78}$$

So far we have obtained the important relationships between the total carrying-sediment capacity S^* for equilibrium transport of non-uniform sediment, the fractional carrying-sediment capacity S_l^*, and the carrying-sediment capacity $S^*(l)$ for uniform sediment. They have the following relations and features: (1) As shown in Eqs. (3.72), (3.74) and (3.77), the fractional carrying-sediment capacity for the l-th size class of the non-uniform sediment mixture, both $S_l^*(y)$ at an arbitrary depth y and the depth-average value $\overline{S_l^*}$, equal the carrying-sediment capacity $S^*(y, l)$ and $\overline{S}^*(l)$ for the uniform sediment with particle size of D_l multiplied the water percentage coefficient K_l; (2) As shown in Eqs. (3.78) and (3.75), respectively, the total carrying-sediment capacity $S^*(y)$ at an arbitrary depth y and the depth-averaged value S^* are the linear weighted summations of the carrying-sediment capacity $S^*(y, l)$ and $S^*(l)$, respectively for the corresponding uniform sediment with the weighting factor of K_l; ; (3) The water percentage coefficient K_l is independent of y, and it is related to the gradation of the total carrying-sediment capacity at the bottom and the gradation of the bed material, as shown in Eq. (3.72); (4) By introducing K_l, the relation between the total carrying-sediment capacity for non-uniform sediment mixture and the carrying-sediment capacity for uniform sediment is derived, and thus the results for uniform sediment transport can be extended to non-uniform sediment.

Introducing the gradation for the depth-averaged carrying-sediment capacity for sediment mixture, Eq. (3.74) can be rewritten as

$$K_l\overline{S}^*(l) = \overline{S_l^*} = \frac{\overline{S_l^*}}{\overline{S^*}}\overline{S^*} = \overline{P_{4.l}^* S} \tag{3.79}$$

where $\overline{P_{4.l}^*} = \frac{\overline{S_l^*}}{\overline{S^*}}$ is defined as the gradation for the depth-averaged carrying-sediment capacity for suspended load. Equation (3.79) illustrates an important relationship between K_l and $\overline{P_{4.l}^*}$, and relates the depth-averaged carrying-sediment capacity of non-uniform sediment to that of the uniform sediment.

It is noted that K_l has another physical meaning. It will be proved later that K_l is the bed material gradation for equilibrium sediment transport and the effective bed material gradation for non-equilibrium transport. This will give another explanation for Eq. (3.79).

3.3.4 Discussion on the New Boundary Conditions Developed Based on Exchange Intensity

For strong-equilibrium sediment transport from Eq. (3.58) the boundary condition turns to

$$-\varepsilon_y \left.\frac{\partial S_l}{\partial y}\right|_{y=0} = \omega_l S_{b.l}^* \tag{3.80}$$

plugging which into Eq. (3.58) to eliminate $-\varepsilon_y \frac{\partial S_l}{\partial y}\Big|_{y=0}$ yields

$$q\frac{d\overline{S}_l}{dx} = \omega_l\left(S_{b.l}^* - S_{b.l}\right) \tag{3.81}$$

Defining

$$\tilde{\beta}_l = \frac{S_{b.l}}{\overline{S}_l} \tag{3.82}$$

$$\beta_l^* = \frac{S_{b.l}^*}{\overline{S}_l^*} \tag{3.83}$$

then Eq. (3.81) can be rewritten as

$$\frac{dS_l}{dx} = -\frac{\omega_l}{q}\left(\tilde{\beta}_l\overline{S}_l - \beta_l^*\overline{S}_l^*\right) = -\frac{\tilde{\beta}_l\omega_l}{q}\left(\overline{S}_l - \frac{\beta_l^*}{\tilde{\beta}_l}\overline{S}_l^*\right) \approx -\frac{\beta_l\omega_l}{q}\left(\overline{S}_l - \overline{S}_l^*\right) \tag{3.84}$$

where β_l is related to $\tilde{\beta}_l$ and β_l^*. The difference among these three parameters is negligible when sediment transport is close to strong-equilibrium condition.

Since the sediment concentration at the bottom is always higher than the depth-averaged concentration, $\tilde{\beta}$ and β^* are both greater than one and considered as the saturation recovery coefficients. By solving the diffusion equation with Eq. (3.81) as the bottom boundary condition, Zhang [8] also concluded that $\tilde{\beta}_l$ is greater than one. Equation (3.84) was first derived by Han [1], but a saturation recovery coefficient much greater than one was not realistic. Based on the analysis of field data, Han [1] suggested $\tilde{\beta}_l = \beta_l^*$, with a value of 0.25 for sediment deposition and 1.0 for erosion. This temporarily solved the conflicts between the theory and the reality. These empirical values for the saturation recovery coefficient have been widely used in the numerical models of sediment transport. YRCC [7] also suggested a saturation recovery coefficient of 0.01 for the Yellow River based on field data.

The conflicts described above indicate that it is problematic to use Eq. (3.81) as the bottom boundary condition. It is correct to use the difference between the turbulent diffusion term and the settling term, or $-\varepsilon_y \frac{\partial S_l}{\partial y} - \omega_l S_l$, to represent the net sediment settling for any location above the bed surface ($y > 0$) and $y \to +0$. But it may not be true for the bed surface, especially for $y \to -0$, because of the discontinuity on the bed surface. In fact, the settling term on the bed $\omega_l S_l$ doesn't necessarily become the bed material. It can be re-suspended, or in particular, can turn into bed load. In addition, the sediment picked up from the bed by turbulent diffusion may also become bed load. Therefore, the bottom boundary condition should depend on the sediment exchange on the bed. Next, a new bottom boundary condition will be derived based on the statistical theory of sediment exchange on the bed.

Based on the statistical theory for sediment movement [3, 5], and in particular, from the 10th formula in Eq. (2.99), the number of the sediment particles in the l-th size class that settle down on a unit area of bed surface per unit time is

$$\begin{aligned}\overline{\lambda}_{4.1.l} &= P_{4.1.l}K_{4.l}\mu_{4.l}U_{4.l} \\ &= (1-\varepsilon_{0.l})(1-\varepsilon_{4.l})\frac{1}{L_{4.l}}U_{4.l}K_{4.l} \\ &= (1-\varepsilon_{0.l})(1-\varepsilon_{4.l})\frac{q}{L_{4.l}}\frac{P_{4.l}S}{\frac{\pi}{6}D_l^3\gamma_s}\end{aligned} \tag{3.85}$$

According to the 3rd formula in Eq. (2.99), the number of the sediment particles picked up from the bed surface is

$$\overline{\lambda}_{1.4.l} = \frac{\beta_l K_{1.l}}{t_{0.4.l}} = \frac{4m_0}{\pi D_l^2}\frac{P_{1.l}\beta_l R_{1.l}}{t_{0.4.l}} \tag{3.86}$$

where $1-\varepsilon_{0.1}$ represents the ceasing probability of sediment on the bed (sediment being stopped from motion on the bed); β_l is the probability for incipient sediment suspension (still particles on the bed being suspended); $\varepsilon_{4.l}$ the probability for sediment suspension; $R_{1.l}$ the probability for particles resting on the bed; $t_{0.4.l}$ the time period for incipient suspension of sediment particle, or the time after which the suspended particle is off the bed; $L_{4.l}$ the average single-step moving distance, within which a sediment particle is suspended from the bed, reaches the suspension height, and then drops to the bed; $U_{4.l}$ the average particle velocity along the flow direction; $K_{4.l}$ the number of the suspended particles per unit water column (per unit area times the water depth); q the flow discharge per unit width; D_l the particle diameter; $m_0 = 0.4$ the compactness factor; $P_{4.l}$ the gradation for suspended load; and $P_{1.l}$ the gradation for bed material; the subscript "l" indicates the particle size class, and the subscripts "4" and "1" indicate suspended load and bed material, respectively; $K_{4.l}U_{4.l} = \frac{6qP_{4.l}S}{\pi\gamma_s D_l}$ represents the sediment transport rate (in terms of number of particles) per unit width for the l-th size class; and $K_{1.l} = \frac{4m_0}{\pi D_l^3}R_{1.l}P_{1.l}$ represents the number of particles in the l-th size class that rest on the per unit area of bed surface.

The physical meaning of the Eq. (3.85) is the following: the number of sediment particles that settle from the suspended load and then rest on a unit area of bed surface per unit time is proportional to the sediment concentration (in terms the number of particles) $K_{4.l}$, the settling probability $1-\varepsilon_{4.l}$, and the ceasing probability $1-\varepsilon_{0.l}$; and is inversely proportional to the single-step suspension period $T_{4.l} = \frac{L_{4.l}}{U_{4.l}}$. The physical meaning of Eq. (3.84) is as follows: the number of sediment particles that are picked up from a unit area of bed surface per unit time is proportional to the number of particles on the bed surface $P_{1.l}\frac{4m_0}{\pi d_l^2}$, the particle resting probability $R_{1.l}$, and the probability of incipient suspension β_l and is inversely proportional to the time period for incipient suspension. Equations (3.85) and (3.84) both have explicit and specific physical meanings. Combining Eq. (3.62) with Eqs. (3.84) and (3.85) yields

$$q\frac{d\overline{S}_l}{dx}=-\varepsilon_y\frac{\partial S_l}{\partial y}\bigg|_{y=0}-\omega_l S_{b.l}=(\overline{\lambda}_{1.4.l}-\overline{\lambda}_{4.1.l})\frac{\pi}{6}\gamma_s D_l^3$$
$$=\frac{2}{3}m_0\gamma_s\frac{P_{1.l}R_{1.l}\beta_l D_l}{t_{0.4.l}}-(1-\varepsilon_{0.l})(1-\varepsilon_{4.l})\frac{qP_{4.l}\overline{S}}{L_{4.l}} \tag{3.87}$$

This equation is the general bottom boundary condition for suspended load transport. The first term on the right side is the mass of sediment that is suspended from a unit area of bed surface per unit time, and the second term is the mass of the bed material that settles from the suspended load. For steady non-equilibrium sediment transport, the difference of these two terms is the longitudinal gradient of transport rate of suspended load for the l-th size class. The diffusion boundary condition $\varepsilon_y\frac{\partial S_l}{\partial y}\Big|_{y=0}+\omega_l S_{b.l}$ of Eq. (3.87) is the value on the bed surface when approaching downward from the water column. The right side of Eq. (3.87), which is the difference in mass between sediment exchanges, is the value on the bed surface when approaching upward from the bed. It is noted that Eq. (3.87), which is derived from the statistical theory for sediment exchange, does not depend on Eq. (3.81). This is the major difference between this boundary condition and the previous results.

For strong equilibrium sediment transport, $\frac{d\overline{S}_l}{dx}=0$, and Eq. (3.87) indicates that

$$\frac{2}{3}m_0\gamma_s\frac{P_{1.l}R_l\beta_l D_l}{t_{0.4.l}}=(1-\varepsilon_{0.l})(1-\varepsilon_{4.l})\frac{qP_{4.l}^*S^*}{L_{4.l}^*} \tag{3.88}$$

Plugging Eq. (3.88) into Eq. (3.87), we have [4]

$$q\frac{d\overline{S}_l}{dx}=-\varepsilon_y\frac{\partial S_l}{\partial y}\bigg|_{y=0}-\omega_l S_{b.l}=(1-\varepsilon_{0.l})(1-\varepsilon_{4.l})q\left[\frac{P_{4.l}^*\overline{S}^*}{L_{4.l}^*}-\frac{P_{4.l}\overline{S}}{L_{4.l}}\right] \tag{3.89}$$

Let's introduce saturation recovery coefficients

$$\tilde{\alpha}_l=(1-\varepsilon_{0.l})(1-\varepsilon_{4.l})\frac{q}{\omega_l L_{4.l}} \tag{3.90}$$

$$\alpha_l^*=(1-\varepsilon_{0.l})(1-\varepsilon_{4.l})\frac{q}{\omega_l L_{4.l}^*} \tag{3.91}$$

where $\tilde{\alpha}_l$ is the saturation recovery coefficient for non-equilibrium sediment transport, or referred to as saturation recovery coefficient for sediment concentration; α_l^* is the saturation recovery coefficient for equilibrium sediment transport, or referred to as the saturation recovery coefficient for carrying-sediment capacity. The single-step moving distance $L_{4.l}$ for the suspended load and $L_{4.l}^*$ for the carrying-sediment capacity will be discussed in the next chapter. Plugging Eqs. (3.90) and (3.91) into Eq. (3.89) yields

$$
\begin{aligned}
q\frac{d\overline{S}_l}{dx} &= -\varepsilon_y \frac{\partial S_l}{\partial y}\bigg|_{y=0} - \omega_l S_{b.l} = \omega_l\left(\alpha_l^* P_{4.l}^* \overline{S}^* - \tilde{\alpha}_l P_{4.l}\overline{S}\right) \\
&= \omega_l\left(\alpha_l^* \overline{S}_l^* - \tilde{\alpha}_l \overline{S}_l\right)
\end{aligned}
\tag{3.92}
$$

The right side of this equation is the boundary condition for the sediment diffusion equation that is derived from the statistical theory for the bed material exchange. This boundary condition has three characteristics: (1) In addition to hydraulic parameters, the boundary condition is mainly controlled by the sediment exchange on the bed, which depends on the bed erosion or sedimentation for steady flow; (2) Compared to using parameters for the bed, it is more stable and reliable to use the depth-averaged parameters, such as $\overline{S}_l^*$, $\overline{S}_l$, to determine the bed erosion or sedimentation; and (3) the trial computation for the boundary condition shall be used to solve the equation for sediment transport.

Plugging Eqs. (3.82) and (3.83) into Eq. (3.92) gives

$$
q\frac{d\overline{S}_l}{dx} = -\varepsilon_y \frac{\partial S_l}{\partial y}\bigg|_{y=0} - \omega_l S_{b.l} = \omega_l\left(\frac{\alpha_l^*}{\beta_l^*} S_{b.l}^* - \frac{\tilde{\alpha}_l}{\tilde{\beta}_l} S_{b.l}\right)
\tag{3.93}
$$

Let's introduce the sediment saturation recovery coefficients for the bottom

$$
\tilde{\gamma}_l = \frac{\tilde{\alpha}_l}{\tilde{\beta}_l} = \frac{\overline{S}_l}{S_{b.c}}\tilde{\alpha}_l
\tag{3.94}
$$

$$
\gamma_l^* = \frac{\alpha_l^*}{\beta_l^*} = \frac{\overline{S}_l^*}{S_{b.l}^*}\alpha_l^*
\tag{3.95}
$$

in which $\tilde{\gamma}_l$ and γ_l^* turn to α_1 and α_2, respectively, in Eq. (1.26) for uniform sediment. Then Eq. (3.93) becomes

$$
q\frac{d\overline{S}_l}{dx} = -\varepsilon_y \frac{\partial S_l}{\partial y}\bigg|_{y=0} - \omega_l S_{b.l} = \omega_l\left(\gamma_l^* S_{b.l}^* - \tilde{\gamma}_l S_{b.l}\right)
\tag{3.96}
$$

Because both $\tilde{\beta}_l$ and β_l^* are greater than 1, the bottom saturation recovery coefficients $\tilde{\gamma}_l$ and γ_l^* are generally less than one. This is particularly true for coarse sediment.

Differences can be observed by comparing the right sides between Eqs. (3.84) and (3.96). The saturation recovery coefficients $\tilde{\beta}_l$ and β_l^* are greater than 1 based on Eq. (3.84) together with Eqs. (3.82) and (3.83). However, these coefficients can be greater or less than 1, and are less than 1 for the vast majority of cases, if Eq. (3.96) is used. More importantly, Eq. (3.96) demonstrates the mechanism that sediment settling onto the bed does not immediately turn into the bed material. It results in decreased saturation recovery coefficients and slower recovery of sediment. A further in-depth study of saturation recovery coefficients is presented in Chap. 6.

For unsteady sediment transport, $\frac{\partial A\overline{S}_l}{\partial t}$ and $\frac{\partial h\overline{S}_l}{\partial t}$ are not zero, and thus the formulas that have $q\frac{d\overline{S}_l}{dx}$ shall be adjusted. After replacing $q = \frac{d\overline{S}_l}{dx}$ with $\frac{\partial(h\overline{S}_l)}{\partial t}+q\frac{d\overline{S}_l}{dx}$, Eq. (3.92) becomes

$$q\frac{\partial \overline{S}_l}{\partial x}+\frac{\partial(h\overline{S}_l)}{\partial t}=-\varepsilon_y\frac{\partial S_l}{\partial y}\bigg|_{y=0}-\omega_l S_{b.l}=\omega_l\left(\alpha_l^* P_{4.l}^*\overline{S}^*-\tilde{\alpha}_l P_{4.l}\overline{S}\right) \tag{3.97}$$

As shown in Fig. 3.3, for a water body with a width of B, a length of Δx, and a height of the water depth h, the amount of sediment that are picked up from the bed within a time interval of Δt is

$$-\left(\varepsilon_y\frac{\partial S_l}{\partial y}\bigg|_{y=0}+\omega_l S_{b.l}\right)B\Delta x\Delta t$$

It should be equal to the increase in the sediment transport rate

$$\left[\left(Q_{S.l}+\frac{\partial Q_{S.l}}{\partial x}\Delta x\right)-Q_{S.l}\right]\Delta t$$

adding the increase in the sediment volume in the water body

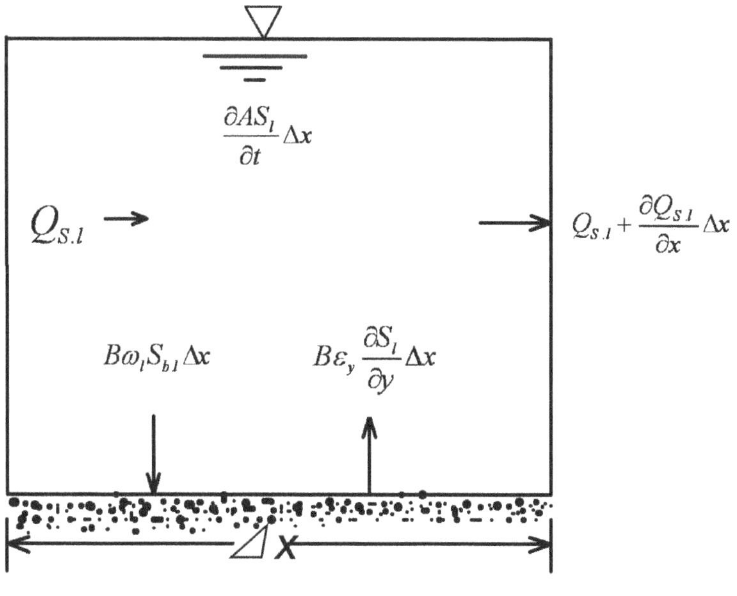

Fig. 3.3 Schematic diagram for river bed deformation

$$\left[\left(hBS_l + B\frac{\partial(hS_l)}{\partial t}\Delta t\right) - hBS_l\right]\Delta x$$

So

$$\frac{\partial Q_{S.l}}{\partial x}\Delta x\Delta t + \frac{\partial(AS_l)}{\partial t}\Delta x\Delta t = -B\left(\varepsilon_y\frac{\partial S_l}{\partial y}\bigg|_{y=0} + \omega_l S_b\right)\Delta x\Delta t$$

$$\frac{\partial(AS_l)}{\partial t} + \frac{\partial Q_{S.l}}{\partial x} + B\left(\varepsilon_y\frac{\partial S_l}{\partial y}\bigg|_{y=0} + \omega_l S_{b.l}\right) = 0 \tag{3.98}$$

or

$$\begin{aligned}&\frac{\partial(hS_l)}{\partial t} + q\frac{\partial S_l}{\partial x} + \left(\varepsilon_y\frac{\partial S_l}{\partial y}\bigg|_{y=0} + \omega_l S_{b.l}\right)\\ &= \frac{\partial(hS_l)}{\partial t} + q\frac{\partial S_l}{\partial x} - \omega_l\left(\alpha_l^* P_{4.l}^* \overline{S}^* - \tilde{\alpha}_l P_{4.l}\overline{S}\right) = 0\end{aligned} \tag{3.99}$$

This is the governing equation for sedimentation. The following equation for bed deformation, which was first given by Han and Huang [6], can also be derived from Eqs. (3.48) and (3.99)

$$\frac{\partial(\gamma_s' y_s)}{\partial t} + \omega_l\left(\alpha_l^* P_{4.l}^* \overline{S}^* - \tilde{\alpha}_l P_{4.l}\overline{S}\right) = 0 \tag{3.100}$$

References

1. Han QW (1973) Preliminary studies for the non-equilibrium transport of sediment in reservoirs. Report on sedimentation in reservoir, task group on yellow river sedimentation research, pp 145–168 (in Chinese)
2. Han QW (2003) Reservoir sedimentation. Science Press, Beijing, China (in Chinese)
3. Han QW, He MM (1997) Study on the two-dimensional equations for non-uniform non-equilibrium suspended sediment transport and the boundary conditions. J Hydrodyn Eng 1:1–10 (in Chinese)
4. Han QW, He MM (1997b) Preliminary study on saturation recovery coefficients. J Sediment Res (3):33–40 (in Chinese)
5. Han QW, He MM (1981) Statistical characters of sediment exchange. J Hydrodyn Eng (1):9–22 (in Chinese)
6. Han QW, Huang YL (1974) Methods of computation of sedimentation and erosion in the reservoir and utilization of digital computer. Selected papers on Yangtze water conservancy and hydroelectric power research, pp 45–85 (in Chinese)
7. Yellow River Conservancy Commission, Survey and Design Institute (1998) Numerical model for sedimentation of lower yellow river (in Chinese)
8. Zhang QS (1980) Diffusion process of sediment in open channel and its application. J Sediment Res (1):37–52 (in Chinese)

Chapter 4
Carrying-Sediment Capacity for Non-uniform Sediment

Carrying-sediment capacity is an important topic for river dynamics. The diffusion theory for suspended load transport only provides the vertical profile of the relative sediment concentration, not the absolute values such as the bottom sediment concentration or the depth-averaged concentration. Therefore, the carrying-sediment capacity cannot be expressed with theoretical integrity solely based on the diffusion theory. If reliable bottom boundary conditions are not available, or these boundary conditions cannot be theoretically expressed, the vertical profile of the absolute sediment concentration cannot be derived from the diffusion theory.

Most investigations on the carrying-sediment capacity are based on the energy approach, with the pioneer studies being conducted by Velikanov [27] and Zhang et al. [29]. Most formulas used in China have similar form as Zhang's formula. Massive and in-depth studies have been conducted for this approach, but with no breakthrough. These studies do not have solid theoretical basis. The formulas cannot be directly derived from the energy theory. Instead, they were developed using qualitative analyses, supplemented by empirical relations and coefficients.

Massive formulas have been developed based on the energy approach with the concepts of the pioneers of this approach. However, the parameters of these formulas cannot be generally applicable due to the limited data used for the parameter calibration. The carrying-sediment capacity has also been investigated using other approaches. Some examples include (1) the analysis conducted by Lane [19] and Kalinske [15] that combined the sediment settling with the turbulent diffusion, (2) the approach used by Karauschev [16] that determined the bottom sediment concentration by considering the balance between the sediment settling and the sediment pick-up by the vertical turbulent fluctuating velocity, (3) the Einstein method (1950) that set the suspended sediment concentration on the bottom equal to the bed load concentration, and (4) the Zhang's approach [29] that derived the bottom sediment concentration using regression or empirical relations. In summary, the carrying-sediment capacity of flow, particularly that the capacity for carrying non-uniform sediment, has not yet been theoretically solved. In spite of the many empirical

Q. Han and M. He, *Non-equilibrium Transport of Non-uniform Suspended Load*, Hydroscience and Engineering, https://doi.org/10.1007/978-981-97-1123-9_4

formulas that have been developed for the carrying-sediment capacity, only few or none has general applications. This has brought difficulties for the sedimentation engineering applications, and extensive investigations and verifications are needed in order to select a proper formula for carrying-sediment capacity.

In their statistical study for sediment movement, Han and He [9, 10] determined the intensities for 16 transitions between various sediment (motion) states, which include the intensities for the exchanges between the suspended load and the bed material. The carrying-sediment capacity for the suspended load and transport capacity for the bed load can be derived based on this approach [7, 9, 11, 12]. This approach has advantage in illustrating the mechanism and characteristics of the carrying-sediment capacity, and the theoretical formula for the carrying-sediment capacity can be obtained [8]. Most parameters of this formula are reasonable and well agree with reality [3, 7, 8] except for the suspension period $t_{0.4}$, which is the time for the suspended load to leave the bed after being suspended. Because the turbulent bursting near the bed was not directly considered, the value of $t_{0.4}$ has discrepancy from the reality. One alternative is to introduce more empirical coefficients to adjust the formula so as to make results have better agreement with reality. But this alternative is not recommended. Progress has been made in a recent investigation of the suspension time [8, 13], and this will help solve one challenging topic in the study of carrying-sediment capacity with the statistical theory.

4.1 Carrying-Sediment Capacity From the Statistical Theory

The sediment exchange intensities derived from the statistical theory were described in Chap. 2. As discussed in Chap. 3, the number of suspended load particles of the l-th size class that settle on a unit area of the bed surface per unit time can be determined from the exchange intensity from the bed material to the suspended load, and is

$$\begin{aligned}\lambda_{4.1.l} &= \beta_{4.1.l} K_{4.l} \mu_{4.l} U_{4.l} \\ &= (1-\varepsilon_{0.l})(1-\varepsilon_{4.l})\frac{1}{L_{4.l}} U_{4.l} K_{4.l} \\ &= (1-\varepsilon_{0.l})(1-\varepsilon_{4.l})\frac{q}{L_{4.l}} \frac{P_{4.l} S}{\frac{\pi}{6} D_l^3 \gamma_s}\end{aligned} \tag{3.85}$$

after using the following relation

$$U_{4.l} K_{4.l} = \frac{q_{s.l}}{\frac{\pi}{6} D_l^3 \gamma_s} = \frac{q P_{4.l} S}{\frac{\pi}{6} D_l^3 \gamma_s}$$

where $K_{4.l}$ is the number of suspended particles of the l-th size class per unit water column, $U_{4.l}\,K_{4.l}$ is the number of particles within a water column that has a bottom area of $U_{4.l}\times 1$ and a water depth of h, or the sediment transport rate in number of particles per unit width, $qSP_{4.l}$ is the transport rate in weight, and S is the depth-averaged sediment concentration, which divided by $\frac{\pi}{6}D_l^3\gamma_s$ becomes the transport rate in number of particles. The number of particles that are picked up from the bed is

$$\lambda_{1.4.l}=\frac{\beta_l K_{1.l}}{t_{0.4.l}}=\frac{4m_0 P_{1.l}\beta_l R_l}{\pi D_l^2 t_{0.4.l}} \tag{3.86}$$

Assuming $\lambda_{1.4.l}$ equals $\lambda_{4.1.l}$ for the strong equilibrium condition, we obtain

$$\frac{4m_0}{\pi D_l^2}\frac{P_{1.l}\beta_l R_L}{t_{0.4.l}}\frac{\pi}{6}D_l^3\gamma_s=\frac{2}{3}m_0\gamma_s D_l\frac{P_{1.l}\beta_l R_l}{t_{0.4.l}}=(1-\varepsilon_{0.l})(1-\varepsilon_{4.l})\frac{q}{L_{4.l}}P_{4.l}S \tag{4.1}$$

It is noted that $S=S*$, $S_l=S_l^*$, and $L_{4.l}=L_{4.l}^*$ for the strong equilibrium condition, thus [7, 10]

$$S_l^*=P_{4.l}S*=\frac{2}{3}m_0\gamma_s\frac{P_{1.l}\beta_l D_l R_l}{(1-\varepsilon_{0.l})(1-\varepsilon_{4.l})}\frac{L_{4.l}^*}{qt_{0.4.l}} \tag{4.2}$$

For uniform sediment, $P_{1.l}=1$, $P_{4.l}=1$, and

$$S*(l)=\frac{2}{3}m_0\gamma_s\frac{\beta_l D_l R_l}{(1-\varepsilon_{0.l})(1-\varepsilon_{4.l})}\frac{L_{4.l}^*}{qt_{0.4.l}} \tag{4.3}$$

The fractional carrying-sediment capacity for non-uniform sediment is

$$S_l^*=P_{4.l}S*=P_{1.l}S*(l) \tag{4.4}$$

Equations (4.3) and (4.2) are the theoretical formulas of the carrying-sediment capacity for uniform and non-uniform sediment, respectively. It is noted that $S*(l)$, and $S(l)$ are the carrying-sediment capacity and concentration, respectively, for uniform sediment with a particle size of D_l; while S_l^* and S_l are the fractional carrying-sediment capacity and concentration, respectively, for the *l-th* size group of the non-uniform sediment mixture. For simplicity, a notation of $X(l)=X_l$ is used for the other parameters.

The parameters in Eqs. (3.85), (3.86), and (4.1) through (4.3) have complicated expressions with definite physical meanings. Some of these parameters are listed in the following. The probability for sediment to cease movement $1-\varepsilon_{0.l}$ can be written as

$$1-\varepsilon_{0.l}=\frac{1}{\sqrt{2\pi}}\int_{-\infty}^{\frac{V_{b.c.0.l}-\overline{V}_b}{\sigma_x}}e^{-\frac{t^2}{2}}dt=1-\frac{1}{\sqrt{2\pi}}\int_{\frac{V_{b.c.0.l}-\overline{V}_b}{\sigma_x}}^{\infty}e^{-\frac{t^2}{2}}dt$$
$$=f_1\left(\frac{\overline{V}_b}{u_*},\frac{V_{b.c.0.l}}{u_*}\right) \tag{4.5}$$

where $V_{b.c.0.l}$ is the threshold bottom flow velocity for sediment to cease movement, $\overline{V}_b$ is the time-averaged longitudinal bottom flow velocity. The resistance for sediment suspension not only includes the particle's gravity force, but also includes the cohesive force and additional pressure of film water for fine particles. Therefore, even if the vertical component of the fluctuating velocity is greater than the particle fall velocity ω_l, sediment may not be suspended. The determination of incipient sediment suspension and the corresponding probability is complicated, as discussed in Chap. 2. The probability for incipient suspension is expressed in Eq. (2.31), which is

$$\beta_l=\begin{cases}\frac{1}{\sqrt{2\pi}}\int_{\frac{\omega_l}{u_*}}^{\infty}e^{-\frac{t^2}{2}}dt=\varepsilon_4 & (\overline{V}_b>\frac{4\omega_1}{\sqrt{3}}; or \overline{V}_b<\frac{4\omega_1}{\sqrt{3}}\text{ and }V_{b.y.c.3.l}^{(1)}<\omega_l)\\ \frac{1}{\sqrt{2\pi}}\int_{\frac{V_{b.y.c.3.l}^{(1)}}{u_*}}^{\infty}e^{-\frac{t^2}{2}}dt=\varepsilon_{5.1} & (\overline{V}_b\le\frac{4\omega_1}{\sqrt{3}}\text{ and }V_{b.y.c.3.l}^{(2)}\le\omega_l<V_{b.y.c.3.l}^{(1)})\\ \frac{1}{\sqrt{2\pi}}\int_{\frac{\omega}{u_*}}^{\frac{V_{b.y.c.3.l}^{(2)}}{u_*}}e^{-\frac{t^2}{2}}dt+\int_{\frac{V_{b.y.c.3.l}^{(1)}}{u_*}}^{\infty}e^{-\frac{t^2}{2}}dt=\varepsilon_{5.1}+\varepsilon_4-\varepsilon_{5.2} & \\ \quad (\overline{V}_b\le\frac{4\omega_1}{\sqrt{3}}\text{ and }\omega_l<V_{b.y.c.3.l}^{(2)}) & \end{cases} \tag{2.31}$$

where $\varepsilon_{5.1.l}$ and $\varepsilon_{5.2.l}$ are determined from Eqs. (2.32) and (2.33), respectively, i.e.,

$$\varepsilon_{5.1}=\frac{1}{\sqrt{2\pi}}\int_{\frac{V_{b.y.c.3}^{(1)}}{u_*}}^{\infty}e^{-\frac{t^2}{2}}dt \tag{2.32}$$

$$\varepsilon_{5.2.l}=\frac{1}{\sqrt{2\pi}}\int_{\frac{V_{b.y.c.3.l}^{(2)}}{u_*}}^{\infty}e^{-\frac{t^2}{2}}dt \tag{2.33}$$

σ_x and σ_y are the standard deviations of the longitudinal and vertical components of the bottom flow velocity, respectively. According to Nikitin's experiments [24], $\sigma_y = 1.05u_*, \sigma_x = 0.37\overline{V}_b$, or $\sigma_x = 2\sigma_y = 2.1u_*$, and the vertical variation of these standard deviations can be neglected. u_* is the shear velocity. For high flow velocity, the loosening condition for coarse sediment may already be satisfied, and thus the incipient sediment suspension probability β_l approaches the suspension probability

$$\varepsilon_{4.l} = \frac{1}{\sqrt{2\pi}} \int_{\frac{\omega_l}{\sigma_y}}^{\infty} e^{-\frac{t^2}{2}} dt = f\left(\frac{\omega_l}{\sigma_y}\right) = f_2\left(\frac{\omega_l}{u_*}\right)$$

The state probability for sediment resting on the bed [10, 13] can be determined with Eq. (2.59),

$$R_{1.l} = \frac{(1-\varepsilon_{0.l})(1-\varepsilon_{4.l})}{1-(1-\beta_l)(1-\varepsilon_{1.l})+(1-\varepsilon_{0.l})(1-\varepsilon_{4.l})} \tag{2.59}$$

where $\varepsilon_{1.l}$ is the incipient motion probability for sediment resting on the bed,

$$\varepsilon_{1.l} = \frac{1}{\sqrt{2\pi}} \int_{\frac{V_{b.c.1.l}-\overline{V}_b}{\sigma_x}}^{\infty} e^{-\frac{z^2}{2}} dz = f_3\left(\frac{\overline{V}_b}{u_*}, \frac{V_{b.c.1.l}}{u_*}\right) \tag{4.6}$$

$V_{b.c.1.l}$ is the instantaneous flow velocity for incipient sediment motion, which can be approximated by [10]

$$V_{b.c.1.l} = \overline{\varphi}(\Delta')\sqrt{53.9D_l + \frac{3\times 10^{-7}}{D_l}(1+0.85h)} \tag{4.7}$$

Δ' is the relative location of the sediment particle at the bed; $\overline{\varphi}(\Delta')$ the average value of the relative location, which is approximately 0.916; and h the water depth. All parameters in Eq. (4.7) are in metric (SI) units. The threshold flow velocity for moving particle to not cease movement is

$$V_{b.c.0.l} = \overline{\varphi}(\Delta')\sqrt{\frac{4}{3c_x}\frac{\gamma_s-\gamma}{\gamma}gD_l} = 0.916\sqrt{53.9D_l} \tag{4.8}$$

and $t_{0.j}$ the time interval for a particle to be detached from the bed layer after being transited from the rest state to moving state j.

The average dimensionless time interval for a particle to be detached from the bed layer after being transited from the rest state to suspension can be determined from the following equation [10].

$$t_{0.4.l} = \frac{1}{\sqrt{2\pi}\beta_l} \int_{\frac{\omega_l}{\sigma_y}}^{\infty} \tau_{0.4.l}\left(z, \frac{\omega_l}{\sigma_y}\right) e^{-\frac{z^2}{2}} dz \tag{4.9}$$

where $z = \frac{V_{b.y}}{\sigma_y}$, and $\tau_{0.4.l}$ is the actual dimensionless time for a suspended particle to be detached from the bed layer.

$$\begin{aligned} \tau_{0.4.l} &= \frac{4}{3C} \frac{\rho_s + \frac{\rho}{2}}{\rho} \left[Arcth\left(\frac{V_{b.y} - u_{4.y.Dl}}{\omega_l} \right) - Arcth\left(\frac{V_{b.y}}{\omega_l} \right) \right] \\ &= \frac{4}{3C} \frac{\rho_s + \frac{\rho}{2}}{\rho} f\left(\frac{V_{b.y}}{\omega_l}, \frac{u_{4.y.Dl}}{\omega_l} \right) = f_4\left(\frac{V_{b.y}}{\sigma_y}, \frac{u_{4.y.Dl}}{\sigma_y}, \frac{\omega_l}{\sigma_y} \right) \end{aligned} \tag{4.10}$$

$V_{b.y}$ is the vertical component of the instantaneous flow velocity at the bottom, *Arcth* is the inverse hyperbolic cotangent function, and $u_{4.y.Dl}$ is the vertical velocity of a particle with particle size D_l at the height D_l from the bed, can be determined from the following equation:

$$\begin{aligned} &\frac{2}{3C} \frac{\rho_s + \frac{\rho}{2}}{\rho} \left\{ \left[\ln \frac{(V_{b.y} - u_{4.y.D_l})^2 - \omega_l^2}{v_{b.y} - \omega_l^2} \right] \right. \\ &\left. + 2\frac{V_{b.y}}{\omega} \left[Arcth \frac{V_{b.y} - u_{4.y.D_l}}{\omega_l} - Arcth \frac{V_{b.y}}{\omega_l} \right] \right\} \\ &= f_5\left(\frac{V_{b.y}}{\sigma_y}, \frac{u_{4.y.D_l}}{\sigma_y}, \frac{\omega_l}{\sigma_y} \right) = 1 \end{aligned} \tag{4.11}$$

ρ and ρ_s are the densities for fluid and for sediment, respectively, and C is the resistance coefficient for sediment that is approximately 1.2. The procedure to compute $t_{0.4.l}$ is given below: determining $\frac{u_{4.y.D_l}}{\sigma_y} = f\left(\frac{V_{b.y}}{\sigma_y}, \frac{\omega_l}{\sigma_y}\right)$ based on Eq. (4.11) with $\frac{\omega_l}{\sigma_y}$ as the parameters; plugging $\frac{u_{4.y.D_l}}{\sigma_y}$, into Eq. (4.10) to solve for $\tau_{0.4.l}$; and then plugging $\tau_{0.4.l}$ into Eq. (4.9) to derive $t_{0.4.l}$.

The average single-step distance for the suspended sediment $L_{4.l}$ is derived as below. The single-step distance is related to the height of suspension. The maximum height of suspension does not necessary equal the whole water depth. Instead, it may equal or less than the water depth. Assuming the single-step suspension height is h_l, the horizontal movement distance, $x_{u.l}$, for a particle to be suspended to the height of h_l can be determined from the following equation: (see Appendix A)

$$\frac{dx_{u.l}}{V(y)} = \frac{dy}{\overline{u}_{y.u.l}} \tag{4.12}$$

where $V(y)$ is the flow velocity in the longitudinal direction, $\overline{u}_{y.u.l}$ is the average velocity for sediment to move upwards. Therefore,

$$x_{u.l} = \frac{1}{\overline{u}_{y.u.l}} \int_0^{h_l} V(y)dy = \frac{q_1(h_l)}{\overline{u}_{y.u.l}} \tag{4.13}$$

Similarly, the horizontal movement distance, $x_{d.l}$, for a particle to settle from the height of h_l to the bed can be determined with the following equation:

$$\frac{dx_{d.l}}{V(y)} = \frac{dy}{\overline{u}_{y.d.l}} \tag{4.14}$$

where $\overline{u}_{y.d.l}$ is the average velocity for sediment to settle down. This leads to

$$x_{d.l} = \frac{1}{\overline{u}_{y.d.l}} \int_0^{h_l} V(y)dy = \frac{q(h_i)}{\overline{u}_{y.d.l}} \tag{4.15}$$

where $q(h_l)$ is the flow discharge per unit width between the bed and the suspension height h_l, or the flow discharge associated with the sediment suspension. The single-step distance for a sediment particle, $L_{4.l}$, is the sum of $x_{u.l}$ and $x_{d.l}$, or

$$L_{4.l} = q_1(h_l)\left[\frac{1}{\overline{u}_{y.u.l}} + \frac{1}{\overline{u}_{y.d.l}}\right] \tag{4.16}$$

Assuming a parabolic velocity distribution, we obtain

$$q(h_l) = qq(\eta_l) = \int_0^{h_l} V(y)dy = q\left[0.647\eta_l + 0.529\eta_l^2 - 0.126\eta_l^3\right] = qq(\eta_l) \tag{4.17}$$

where q is the flow discharge per unit width, h_l is the suspension height with certain vertical distribution of S_l, h is the water depth and $\eta_l = h_l/h$ is the relative suspension height. Because the sediment concentration distribution can be either equilibrium or non-equilibrium, the corresponding suspension height can be expressed as $h_l(S_l)$ or $h_l^*(S_l^*)$. However, $q(\eta_l)$ only depends on the vertical distribution of the flow velocity. Thus, $q(\eta_l)$ and $q(\eta_l^*)$ are the same function with different independent variable. Sediment suspension is mainly supported by large-scale vortices of turbulent flow. After a sediment particle is picked up from the bed, it is affected by the vertical component of the fluctuating velocity and the horizontal component of the time-average flow velocity. The vertical velocity for a sediment particle is $u_{y.l} = V_y - \omega_l$, where V_y is the vertical component of the instantaneous flow velocity, which can be assumed constant along the water depth based on experimental data, or $V_y = V_{b.y}$. Assuming V_y is constant within every single step, and is a random parameter for different time steps, the mathematical expectation for the upward velocity of sediment is [10, 13]

$$
\begin{aligned}
\overline{u}_{y.u.l} &= M[V_y - \omega_l | V_y - \omega \geq 0] \\
&= \frac{1}{\sqrt{2\pi}\sigma_y \varepsilon_{4.l}} \int_{\omega_l}^{\infty} (V_y - \omega_l) \exp\left(-\frac{V_y^2}{2\sigma_y^2}\right) dV_y \\
&= -\frac{\sigma_y}{\sqrt{2\pi}\varepsilon_{4.l}} \int_{\frac{\omega_l}{\sigma_y}}^{\infty} \exp\left(-\frac{t^2}{2}\right) d\left(-\frac{t^2}{2}\right) - \frac{\omega_l}{\sqrt{2\pi}\varepsilon_{4.l}} \int_{\frac{\omega_l}{\sigma_y}}^{\infty} \exp\left(-\frac{t^2}{2}\right) dt \\
&= \frac{\sigma_y}{\sqrt{2\pi}\varepsilon_{4.l}} \exp\left[-\frac{1}{2}\left(\frac{\omega_l}{\sigma_y}\right)^2\right] - \omega_l
\end{aligned}
$$

Replacing σ_y with u_*

$$
\frac{\overline{u}_{y.u.l}}{\omega_l} = \frac{1}{\sqrt{2\pi}\varepsilon_{4.l}} \frac{u_*}{\omega_l} \exp\left[-\frac{1}{2}\left(\frac{\omega_l}{u_*}\right)^2\right] - 1 \tag{4.18}
$$

Similarly, the mathematical expectation for the downward velocity is

$$
\begin{aligned}
\frac{\overline{u}_{y.d.l}}{\omega_l} &= \frac{1}{\sqrt{2\pi}(1 - \varepsilon_{4.l})\omega_l \sigma_y} \int_{-\infty}^{\omega_l} (\omega_l - V_y) \exp\left[-\frac{1}{2}\left(\frac{V_y}{u_*}\right)^2\right] dV_y \\
&= \frac{1}{\sqrt{2\pi}(1 - \varepsilon_{4.l})} \frac{u_*}{\omega_l} \exp\left[-\frac{1}{2}\left(\frac{\omega_l}{u_*}\right)^2\right] + 1
\end{aligned} \tag{4.19}
$$

Plugging Eqs. (4.18) and (4.19) into Eq. (4.16), the single-step distance for a suspended particle can be written as

$$
\begin{aligned}
L_{4.l} = \frac{q(\eta_l)}{\omega_l} \Bigg\{ &\left\{ \frac{1}{\sqrt{2\pi}} \frac{u_*}{\omega_l \varepsilon_{4.l}} \exp\left[-\frac{1}{2}\left(\frac{\omega_l}{u_*}\right)^2\right] - 1 \right\}^{-1} \\
&+ \left\{ \frac{1}{\sqrt{2\pi}} \frac{u_*}{\omega_l(1 - \varepsilon_{4.l})} \exp\left[-\frac{1}{2}\left(\frac{\omega_l}{u_*}\right)^2\right] + 1 \right\}^{-1} \Bigg\}
\end{aligned} \tag{4.20}
$$

It is noted that the mean square deviation σ_y of the vertical fluctuating velocity is assumed to be constant over the water depth in the derivation of Eqs. (4.18) and (4.19). This assumption is reasonable based on previous studies. Both the vertical distribution of the flow velocity $V(y)$ and the average suspension height need to be determined before calculating $q(\eta_l)$. According to the statistical theory for sediment movement [10] (see Appendix A), the average sediment suspension height for the *l.th* size class is

$$h_l = \int_0^h \frac{2y}{h\overline{S_l}} S_l(y)dy \tag{4.21}$$

where $S_l(y)$ is the vertical distribution of the fractional sediment concentration for the *l-th* size class, $\overline{S}_l$ is the depth-averaged concentration, y is the vertical coordinate starting from the bed. Assuming the sediment concentration follows an exponent distribution as shown in Eq. (6.7), the relative suspension height can be derived by plugging Eq. (6.7) into Eq. (4.21), which is

$$\eta_l = \frac{h_l}{h} = 2\left\{\frac{\kappa u_*}{6\omega_l} - \exp\left(-\frac{6\omega_l}{\kappa u_*}\right) \times \left[1 - \exp\left(-\frac{6\omega_l}{\kappa u_*}\right)\right]^{-1}\right\} \tag{4.22}$$

Because the exponent distribution of sediment concentration is for an equilibrium condition, it is actually a distribution for the carrying-sediment capacity. Therefore, a superscript of "*" can also be added to these parameters such as h_l, $L_{4.l}$ and η_l.

So far, all the parameters in Eq. (4.4) and for $\lambda_{1.4.l}$ and $\lambda_{4.1.l}$ have been theoretically determined. It is noted that the single-step distance in the boundary condition depends on the suspension height, and thus is related to the distributions of the flow velocity and of the sediment concentration. Even for the steady condition, in which the distribution of the flow velocity is stationary, the single-step distance still depends on the distribution of the sediment concentration.

Plugging the probability $R_{1.l}$ for sediment resting on the bed, which is represented by Eq. (2.59), into Eqs. (4.3) and (4.4), we obtain

$$\begin{aligned} S_l^* &= \frac{2}{3} m_0 \gamma_s \frac{D_l P_{1.l} \beta_l}{t_{0.4.l} q} \frac{L_{4.l}^*}{1 - (1 - \varepsilon_{1.l})(1 - \beta_l) + (1 - \varepsilon_{0.l})(1 - \varepsilon_{4.l})} \\ &= \frac{2}{3} m_0 \gamma_s P_{1.l} \Omega_l \frac{D_l}{\omega_l t_{0.4.l}} \frac{L_{4.l}^*}{q} = P_{1.l} S*(l) \\ S_l^* &= P_{4.l}^* S^* = S_l \end{aligned} \tag{4.23}$$

where $S*(l)$ is the carrying-sediment capacity for uniform sediment with a diameter of D_l,

$$\Omega_l = \frac{\beta_l}{1 - (1 - \varepsilon_{1.l})(1 - \beta_l) + (1 - \varepsilon_{0.l})(1 - \varepsilon_{4.l})} \tag{4.24}$$

is the composite probability, $\frac{D_l}{\omega_l t_{0.4.l}}$ is the inverse of the relative time for sediment to get suspended and $\frac{L^*_{4.l}\omega_l}{q}$ is the relative single-step distance. The total carrying-sediment capacity can be determined from Eq. (4.23) with summation over l, which is

$$S^* = \sum_{l=1}^{n} P^*_{4.l} S^* = \frac{2}{3} m_0 \gamma_s \sum_{l=1}^{n} P_{1.l} \Omega_l \frac{D_l}{\omega_l t_{0.4.l}} \frac{L^*_{4.l}\omega_l}{q} = \sum_{l=1}^{n} P_{1.l} S^*(l) \tag{4.25}$$

This equation is a summation of the fractional carrying-sediment capacity based on the gradation of the bed material. If summing the fractional carrying-sediment capacity based on the gradation of the carrying-sediment capacity, the total carrying-sediment capacity is

$$S^* = \frac{1}{\sum_{l=1}^{n} \frac{P^*_{4.l}}{S*(l)}} \tag{4.26}$$

It is noted that the above two summations are equivalent for a strong equilibrium condition, yielding the same value for S^*. For a strong equilibrium condition, replacing $P_{1.l}$ with $P_{1.l.1}$, $P^*_{4.l}$ with $P_{4.l}$, Eqs. (4.25) and (4.26) are still valid.

After validating with data and thorough investigation of the theory and concepts [3, 7, 10, 13], it was found that the formulas for carrying-sediment capacity based on the statistical theory, which consists of Eqs. (4.23), (4.25), and (4.26), are reasonable with profound illustration of the mechanisms, rigorous derivations, and no relations or factors that need to be determined arbitrarily. However, the computed time for sediment to get suspended from the bed, $t_{0.4.l}$ is smaller for fine particles. Further investigations are required to determine the impact of the turbulent bursting on $t_{0.4.l}$, and the value of $t_{0.4.l}$. Equations (4.9) through (4.11) that are used to determine $t_{0.4.l}$ in this section will no longer be applied.

4.2 Near-Bed Turbulent Bursting and Incipient Suspension Period

4.2.1 Turbulent Bursting Phenomena

Turbulent bursting is a major part of the development of the turbulence quasi-order structure near the bed [14, 21, 22]. The turbulence typically is fully developed in natural rivers, and the turbulent bursting is a major mechanism for incipient suspension of sediment on bed.

Kline et al. [18] discovered the turbulent bursting during their experimental study for turbulence. Their study indicated that the turbulent energy in the turbulent boundary layer is mostly generated by the bursting in the low-speed zone. The bursting occurs when the flow velocity increases to a certain value. Previous studies [5, 26, 25] showed that bursting also occurs on the rough bed as does on the smooth bed. Although each bursting phenomena may vary, the bursting process is similar. The turbulent bursting process can be divided into four stages, which include alternating streaks of high and low-speed fluid, lift-up of n-shaped vortices, ejections, and sweeps.

During the turbulent bursting process, the sediment particles in the low-speed streaks are lifted up by bursting flow together with the raising and bursting n-shaped vortices and then are ejected into the turbulent region [23]. Figure 4.1 shows the mechanism for the sediment suspension near the bed caused by the turbulent bursting.

The turbulent bursting process is somewhat periodic for certain bed roughness conditions and flow Reynolds numbers. With the particle tracing technology, individual ejections and sweeps can be observed when the Reynolds number is low. As the Reynolds number increases, the turbulent bursting is mostly characterized by mass ejections [21].

Figure 4.2 shows a snapshot of individual ejections of the turbulence bursting that was observed in a flume experiment of turbulent flow [21] with a Reynolds number $Re_R = \frac{RV}{\upsilon} = 6260$, where R is the hydraulic radius, V is the cross-sectional average flow velocity, and ν is the kinematic viscosity. Hydrogen bubbles were used as tracer particles in their experiment. The suspension of the tracer particles caused by the

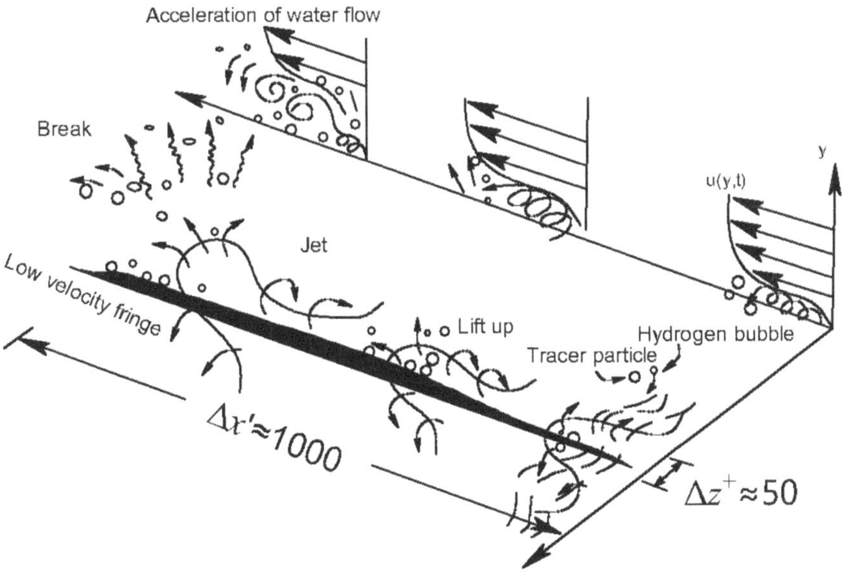

Fig. 4.1 Mechanism for suspension of near-bed sediment

ejections of the turbulent bursting can be clearly observed in Fig. 4.2. Mao et al. [21] found that turbulent bursting would occur more frequently as Re_R increases, and the bursting can be characterized by multiple ejections. If considering multiple ejections as ejections within one bursting period, the bursting period T_B increase with Re_R. Figure 4.3 shows the multiple ejections for turbulent bursting.

The Reynolds number Re_R impacts the ejection patterns and thus the period of the turbulent bursting. Mao et al. [21] found that T_B not only depends on Re_R, but also is impacted by the bed surface roughness that can be represented by the roughness Reynolds number

$$Re_\Delta = \Delta \times \frac{u_*}{\nu} \tag{4.27}$$

where Δ is the equivalent roughness of the bed surface that is approximated by the particle diameter D in this chapter, and u_* is the shear velocity.

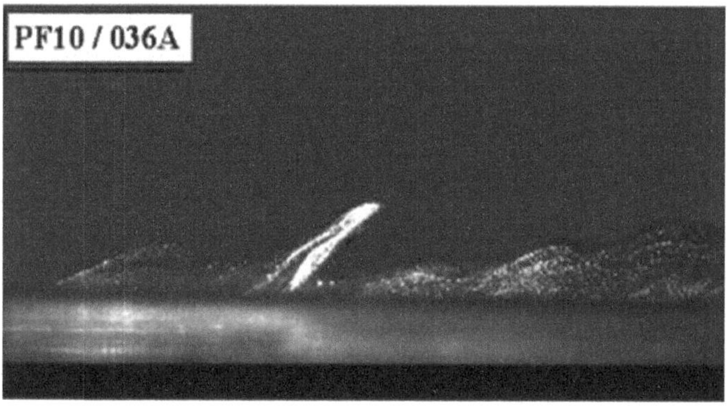

Fig. 4.2 A single turbulent jet

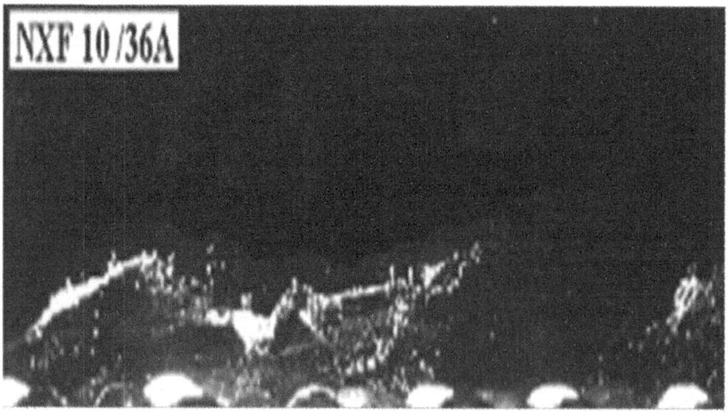

Fig. 4.3 A group of turbulent jet

4.2.2 Turbulent Bursting Period

The turbulent bursting period is defined as the interval between two consecutive ejections during the turbulent bursting process near the bed. Mao et al. [21] expressed the turbulent bursting period as

$$T_B = T_b \frac{h}{u_*} \tag{4.28}$$

where T_b is a dimensionless coefficient of bursting period that is a function of Re_Δ. Based on the experiments [21], T_b is approximately 0.0249–0.042 for $Re_\Delta < 1$. But T_b increases significantly when $Re_\Delta > 5$, and can be as high as 0.0706 [22]. Based on the relation between Re_Δ and T_b that was shown in Mao et al. [22], Guan [6] developed a fitting formula that is

$$T_b = 0.0353 Re_\Delta^{0.1509} \tag{4.29}$$

As shown in Fig. 4.4, Eq. (4.29) reasonably represents the relation between Re_Δ and T_b. For given D, ν, h and u_*, the turbulent bursting period T_B can be determined from Eqs. (4.28) and (4.29).

Another commonly used formula for bursting period was developed by [17], which is

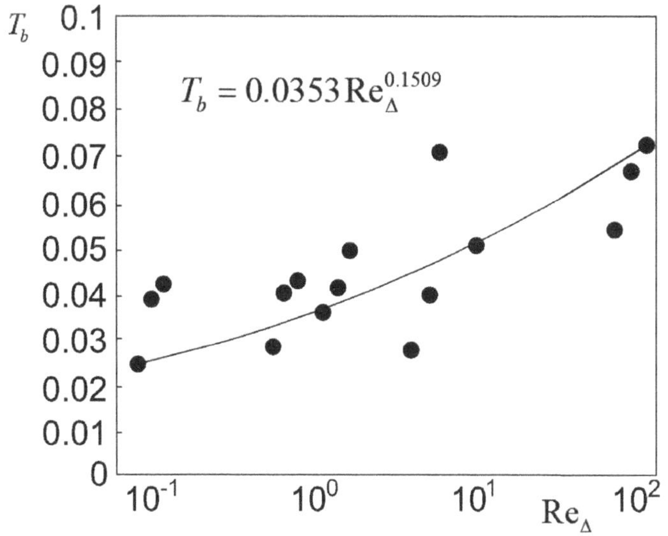

Fig. 4.4 Relation between T_b and Re_Δ

$$T_B = T_B^+ \frac{\nu}{u_*^2} \tag{4.30}$$

where T_B^+ is a dimensionless parameter, generally varies between 90 and 100 and increases with the Reynold number [1]. The average bursting area per unit area, or δ, is approximately 0.016 [2, 20]. Assuming bursting occurs everywhere within a unit area, and the impact of δ is represented by T_B^+, T_B^+ then varies between 5625 and 6250. In his study for the low-speed streaks in the near wall region of open channel turbulent flow, Wang [28] concluded that the dimensionless period of the low-speed streaks, or t^+, should not exceed $T_B^+/2$, and derived the relation between t^+ and shear Reynolds number Re_* based on experiments. For the smooth near-wall region,

$$t^+ = 3.2 \times 10^{-5}\text{Re}_*^2 + 2.0 \tag{4.31}$$

and for the transition region

$$t^+ = 0.021Re_* + 21.4 \tag{4.32}$$

where the shear Reynolds number Re_* is defined as

$$R_{e*} = \frac{u_* h}{\nu} \tag{4.33}$$

Guan [6] compared Eqs. (4.28) and (4.29) with Eqs. (4.30) through (4.32). He computed T_B using Eqs. (4.28) and (4.29), solved T_B^+ with Eq. (4.30), and then compared T_B^+ with t^+ that was determined from Eqs. (4.31) and (4.32). Good correlation was found between T_B^+ and t^+, indicating both types of bursting periods can be used to describe the turbulent bursting. It is noted that the turbulent bursting period determined from Eqs. (4.28) and (4.29) is used in the following analysis.

4.2.3 Incipient Suspension Period for Particles on the Bed

4.2.3.1 Relationship Between Incipient Sediment Suspension Period and Turbulent Bursting Period

$T_{0.4.l}$ determined from Eqs. (4.9) through (4.11) is the time for a particle to be lifted from the bed to a height of a particle size, after which the bed will resume its normal condition, on which the particle can be suspended. $t_{0.4.l}$ is typically shorter than the incipient suspension period $T_{0.4.l}$, and it equals $T_{0.4.l}$ only when the particles are entrained in suspension continuously. For fine sediment and strong flow condition, $T_{0.4.l}$ could be replaced with $t_{0.4.l}$. But more generally, $\varepsilon_4 T_{0.4.l} = t_{0.4.l}$. This is because the time for sediment to be entrained in suspension within $T_{0.4.l}$ is only $\varepsilon_4 T_{0.4.l}$, and thus the incipient suspension period $T_{0.4.l} = t_{0.4.l} / \varepsilon_4$. Since sediment entrainment is

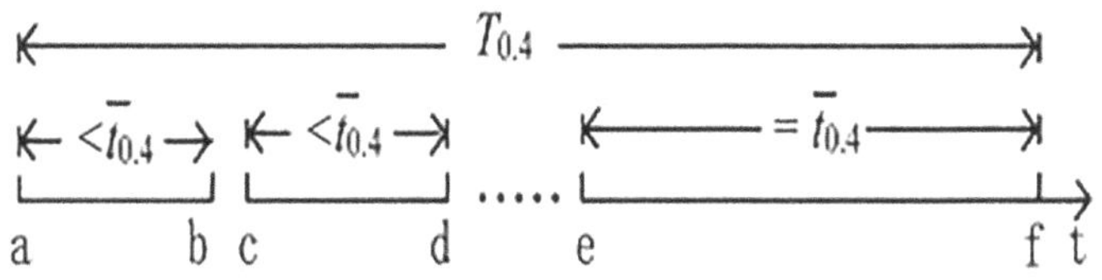

Fig. 4.5 Relation between $T_{0.4.l}$ and $\bar{t}_{0.4.l}$

directly caused by the turbulent bursting, $T_{0.4.l}$ is analyzed in the following from the aspect of the turbulent bursting period.

It takes time for a sediment particle to transit between various states. The state transition occurs only when the flow condition required for a state transition lasts for a sufficient amount of time. If the flow condition always satisfies the critical condition for incipient sediment suspension, the average time required for the incipient suspension shall equal the average value of the theoretical incipient time $\bar{t}_{0.4.l}$. However, the real flow condition is unstable during various stages of the turbulent bursting. The instantaneous flow condition may not always satisfy the critical condition for incipient suspension. Therefore, the actual incipient suspension period $T_{0.4.l}$ is longer than the theoretical time for an incipient suspension $\bar{t}_{0.4.l}$. The relation between $\bar{t}_{0.4}$ $T_{0.4.l}$ and $\bar{t}_{0.4.l}$ is illustrated in Fig. 4.5.

As shown in Fig. 4.5, $T_{0.4.l} \geq \bar{t}_{0.4.l}$. Within the entire period of [a,f], although the flow condition statifies the critical condition for incipient sediment suspension for two period segments [a,b] and [c,d], sediment is not suspended because these two peiords are shorter than $\bar{t}_{0.4.l}$. Only in period [e,f], the flow condition that statifies the critical condition for incipient sediment suspension lasts the time of $\bar{t}_{0.4.l}$, and thus the sediment particle can be suspended. This process of sediment entraiment repeats duing the tubulent busrting process. That's why $T_{0.4.l}$ is called the incipient sediment suspension period.

Figure 4.5 illustrate the average condition for the turbulent busting process. The difference between $T_{0.4.l}$ and $\bar{t}_{0.4.l}$ is caused by the instability of the flow condition in turbulent busting, which was observed in many studies (as e.g., [23]). As shown in Fig. 4.2, incipient suspension of the traced particles is not noticiable at the locations where no bursting ejections occur. In fact, the locations with no ejections are dominated by streaks of low-speed fluid. The streaks of low-speed fluid and ejections are two different stages of turbulent busting, with noticeable difference in the flow condition. According to Wang [28], the duration of low-speed fluid streaks can be one half of the entire period of the turbulent bursting at most, as shown in Fig. 4.6. This implies that noticeable incipient suspension of the bed material may not occur for one half of the time in a turbulent bursting process. This conclusion was made based on a single ejection of turbulent bursting. For flow with a sufficiently large Reynolds number, turbulence is fully developed, and the turbulent bursting is characterized as massive ejections. According to Mao et al. [23], the massive ejections stage lasts longer as the Reynolds number increases. But there is still a certain period with relatively weak intensity of turbulent bursting (as e.g., sweeping) after each massive

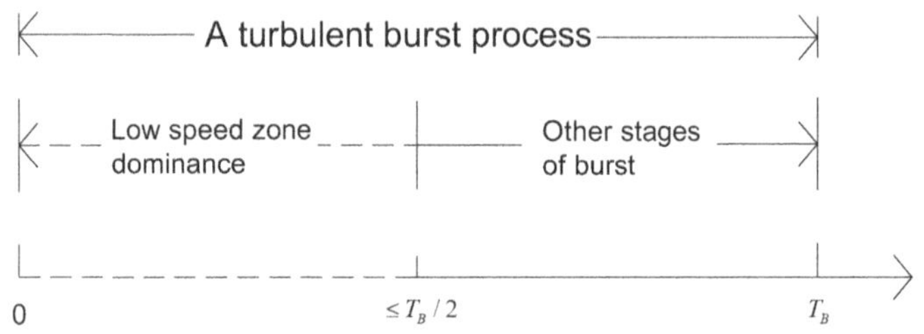

Fig. 4.6 Burst process division

ejections. This indicates that the flow condition in a turbulent bursting process is instable no matter for a single ejection or for massive ejections, and shows periodic behavior.

The turbulent bursting process is periodic, which implies the variation of the flow condition is also periodic. As a result, the actual incipient suspension of the bed material also shows periodic behavior, and the incipient suspension period $T_{0.4.l}$ is closely related to the turbulent bursting period T_B.

4.2.3.2 Derivation of Incipient Sediment Suspension Period

Different size classes of the non-uniform sediment mixture show different responses to a given turbulent bursting process during their incipient suspension. As shown in Eqs. (2.31) and (2.30), for a given flow condition, a sediment particle with a smaller fall velocity has a greater suspension probability $\varepsilon_{4.l}$ (or β_l) and a shorter incipient suspension period than a particle with a larger fall velocity.

As u_* increases, $\varepsilon_{4.l}$(or β_l) and the mean value of the vertical fluctuating velocity acting on a particle increases, and thus the particles are easier to get suspended with a shorter incipient suspension period $T_{0.4.l}$. In fact, u_* in Eq. (2.31) is an approximation for the root-mean-square deviation of the vertical fluctuating velocity of the fluid. Its magnitude represents the intensity of the action of the turbulent diffusion on the particles near the bed. Therefore, $T_{0.4.l}$ is typically inversely proportional to σ_y(or u_*).

In summary, the incipient sediment suspension period $T_{0.4.l}$ is proportional to the turbulent bursting period T_B near the bed, increases with ω, but decreases as u_* increases. According to Chang and Xie (1989), $\omega_{0.l}/u_*$ represents the ratio of the gravity to the turbulent diffusion. As shown in Eq. (2.30), $\omega_{0.l}/u_*$ is the integration limit of the suspension probability. It can represent the intensity of sediment suspension, and the ratio of the gravity to the turbulent diffusion. Therefore, the incipient suspension period for a particle with size of D_l can be written as

$$T_{0.4.l} = KT_B = a\left(\frac{\omega_{0.l}}{u_*}\right)^b T_B \tag{4.34}$$

where T_B is the turbulent bursting period that can be determined with Eqs. (4.28) and (4.29). The coefficient K is

$$K = a\left(\frac{\omega_{0.l}}{u_*}\right)^b \tag{4.35}$$

where $a = 0.05$, $b = 2$ based on the calibration with field data. The shear velocity can be determined with the following equation [11, 12]

$$u_* = V\left[6.5\left(\frac{h}{D}\right)^{\frac{1}{4+\lg(h/D)}}\right]^{-1} \tag{4.36}$$

T_B is the turbulent bursting period expressed by Eqs. (4.28) and (4.29) and $\omega_{0.l}$ represents the ratio of the fluid force to the gravity for given exposure condition, which is dependent on the particle size [11, 12].

$$\omega_{0.l} = \sqrt{\frac{4}{3C_x}\frac{\gamma_s - \gamma}{\gamma} g D_l} \tag{4.37}$$

where C_x is the drag coefficient, and $C_x = 0.4$. $\omega_{0.l}$ is an important parameter for incipient sediment motion. It is used over the sediment fall velocity in order to avoid the complexity caused by the varying resistance coefficients for fine and coarse sediment. Based on the difference in the resistance coefficients,

$$\omega_{0.l} = \sqrt{3}\omega_l \tag{4.38}$$

For fully developed turbulent flow area (region of quadratic resistance law), $\omega_{0.l}$ and ω_l are interchangeable.

4.3 Carrying-Sediment Capacity for Non-uniform Suspended Load Based on Statistical Theory and Its Verification

4.3.1 Carrying-Sediment Capacity from Statistical Theory After Considering Turbulent Bursting

Considering the impact of turbulent bursting on the time for incipient sediment suspension, the formulas for carrying suspended load capacity consist of Eqs. (4.23) through (4.26), (4.5) through (4.8), (2.30) through (2.33), (2.26), (4.13), (4.15) through (4.21), (4.36), (4.37), and Eqs. (4.34), (4.28), (4.29) and (4.39) for the incipient suspension period $T_{4.0.l}$. Among these 27 equations, the most critical equations include Eq. (4.23) and

$$T_{4.0.l} = KT_B = 0.05\left(\frac{\omega_{1.l}}{u_*}\right)^2 T_B = 1.765 \times 10^{-3} Re_{\Delta_l}^{0.1509}\left(\frac{\omega_l}{u_*}\right)^2\left(\frac{h}{u_*}\right) \quad (4.39)$$

It is noted that, $t_{0.4.l}$ in Eqs. (4.23) and (4.25) should be replaced with $T_{0.4.l}$. It is also noted that the carrying-sediment capacity was typically determined in the flume experiments after running the experiments for a sufficiently long time until the bed reaches equilibrium with no noticeable change to the gradations for the influx sediment and the sediment concentration. These flume experiments actually satisfied strong equilibrium condition. The field data used for natural rivers also correspond to the condition with no deposition and erosion that is approximated to the equilibrium condition. The sediment delivered from the upstream reach is carried by the water, or settles down to the bed but is re-suspended before consolidating with the bed material. For this type of sediment, it will be suspended as long as the suspension probability ε_4 is satisfied. Thus it also belongs to or close to the strong equilibrium transport condition, and β in the formulas for Ω_l should be replaced with $\varepsilon_{4.l}$

$$\Omega_l = \frac{\varepsilon_{4.l}}{1-(1-\varepsilon_{4.l})(\varepsilon_{0.l}-\varepsilon_{1.l})} = \Omega_{0.l} \quad (4.40)$$

$$S_{0.l}^* = \frac{2}{3} m_0 \gamma_s P_{1.l} \Omega_{0.l} \frac{D_l}{\omega_l t_{0.4.l}} \frac{L_{4.l}^* \omega_l}{q} \quad (4.41)$$

$$S_0^* = \sum \frac{2}{3} m_0 \gamma_s P_{1.l} \Omega_{0.l} \frac{D_l L_{4.l}^*}{q t_{0.4.l}} = \sum P_{1.l} S_0^*(l) \quad (4.42)$$

where $\Omega_{0.l}$ is the composite probability when no cohesive force or additional pressure of film water exists, and S_0^* is the corresponding carrying-sediment capacity, or the carrying-sediment capacity from the upstream reach. The theoretical formulas of carrying-sediment capacity discussed in the section is the twenty-seven equations with Eqs. (4.23) through (4.25) being replaced with Eqs. (4.40) through (4.42). The formulas for more general (non-equilibrium) condition, which use Eqs. (4.23) through (4.25) for the composite probability, are discussed in Chap. 11 where the multiple values of the carrying-sediment capacity are discussed.

4.3.2 Verification [6]

The information for the data used in the verification is summarized in Table 4.1.

Figure 4.7 shows the computed carrying-sediment capacity compared to the sediment concentration measured in Tsinghua's flume experiments for silt sand.

Table 4.1 Data used in verification

Source	Reach	Sediment concentration (kg/m^3)	Particle size (mm)	Water depth (m)	Mean flow velocity (m/s)	Slope × 10^4	No. of data sets
Yellow River	Downstream	0.100–565	0.005–1	0.76–0.51	0.26–4.41	1.00–4.60	533
	Upper reach of Upstream Riv	0.072–13.8	0.007–1	1.16–3.74	0.382–2.23	0.17–6	80
Yangtze River	Xinchang	0.093–3.76	0.007–0.5	3.56–13.3	0.77–2.09	0.086–0.836	40
Han River	Danjinagkou	0.50–9.49	0.010–2	0.87–19.7	0.54–2.4	0.10–2.77	27
Tsinghua Flume	Fine sediment	25.68–229.32	0.02–0.20	0.116–0.231	0.433–1.143	17.20–129.00	22
	Mixture	13.75–118.18	0.04–0.40	0.094–0.203	0.448–1.546	17.20–70.00	35

Figure 4.8 shows the results for Tsinghua's experiments for sediment mixture. Figures 4.9, 4.10, 4.11, and 4.12 shows the results for the field data collected at Xinchang Station of the Yangtze River, at Danjiangkou Station, at the downstream reach of the Yellow River, and at the lower reach of the upstream Yellow River, respectively.

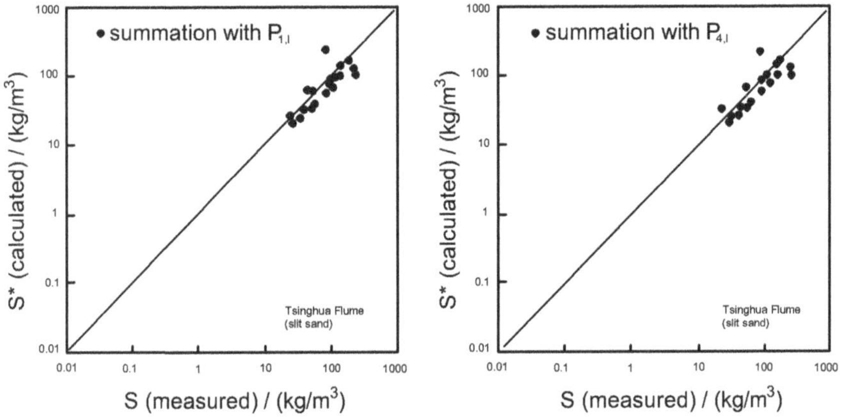

Fig. 4.7 Comparison between computed carrying-sediment capacity and sediment concentration measured in Tsinghua's Flume experiments for silt sand

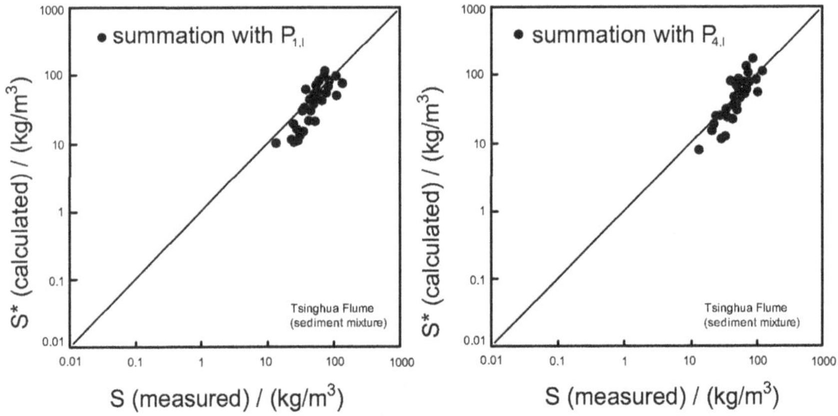

Fig. 4.8 Comparison between computed carrying-sediment capacity and sediment concentration measured in Tsinghua's Flume experiments for sediment mixture

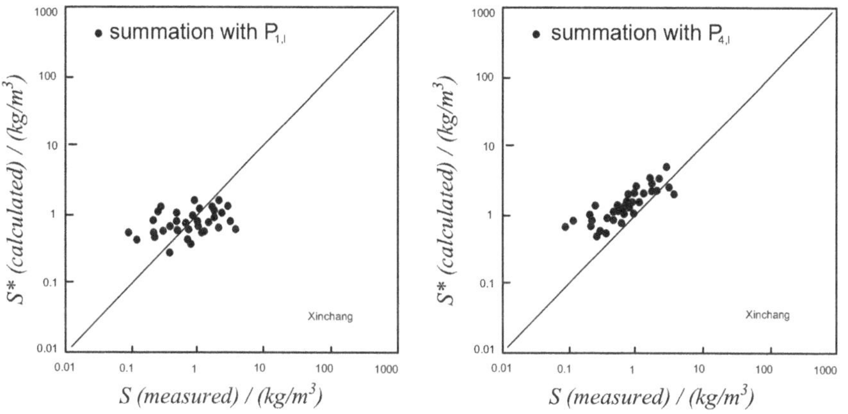

Fig. 4.9 Comparison between computed carrying-sediment capacity and sediment concentration measured at Xinchang Station, Yangtze River

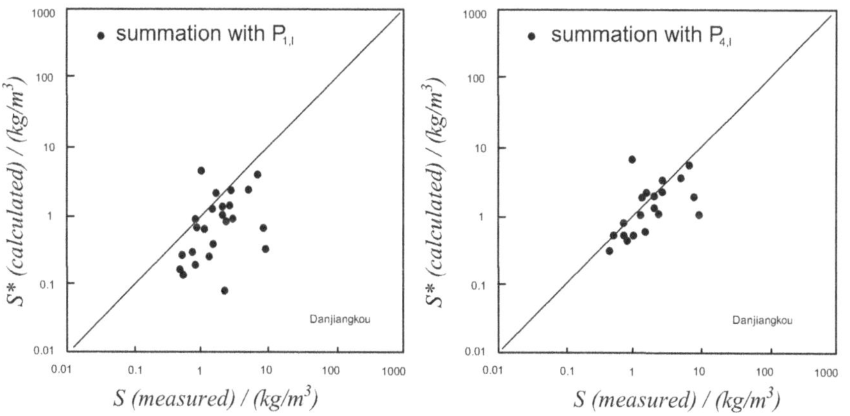

Fig. 4.10 Comparison between computed carrying-sediment capacity and sediment concentration measured at Danjiangkou Station

4.3.3 Analysis of Verification Results

4.3.3.1 Summary of Data Comparison

(1) The horizontal axis of Figs. 4.7, 4.8, 4.9, 4.10, 4.11, 4.12 and 4.13 is the measured sediment concentration, or the carrying-sediment capacity S^* represented by Eq. (4.43). The vertical axis is the carrying-sediment capacity computed with the statistical theory. The subplot a of each figure shows the computed total carrying-sediment capacity by summing up the fractional carrying-sediment capacities based on the gradation of the bed material $P_{1,l}$, and the subplot b shows the

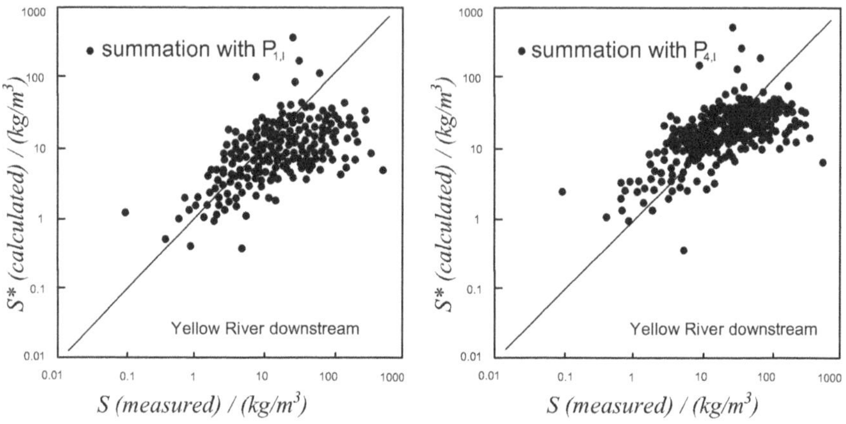

Fig. 4.11 Comparison between computed carrying-sediment capacity sediment concentration measured at the downstream reach of the Yellow River

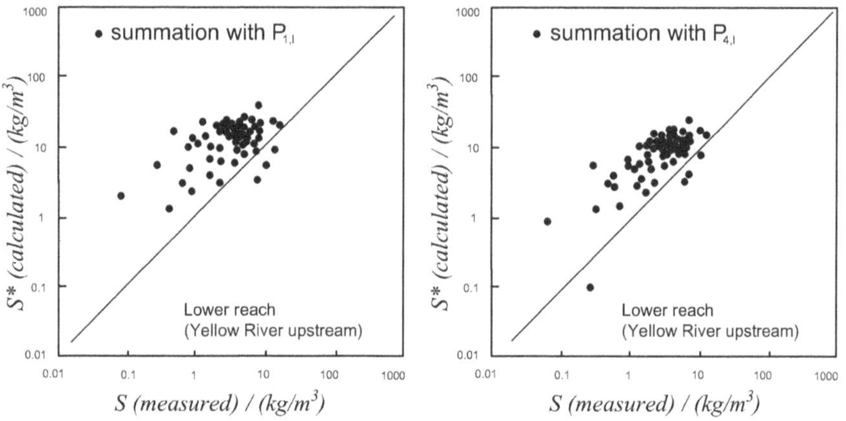

Fig. 4.12 Comparison between computed carrying-sediment capacity and sediment concentration measured at the lower reach of the Upstream Yellow River??

results based on the summation with the gradation of the sediment concentration $P_{4,l}$. As shown in these figures, the statistical theory for the carrying-sediment capacity $S*$, represented by Eqs. (4.26), (4.39), and (4.40) through (4.42), overall shows reasonable agreement with the data measured in flume experiments and collected in the Yangtze River and in the Yellow River. A limited discrepancy can also be observed. For example, the field data collected in the lower reach of the upstream Yellow River is smaller than the computed results. Coarse sediment particles with diameter larger than 1 mm exist in this reach. Thus, not all of the particles can be suspended, resulting in a sediment concentration less than the

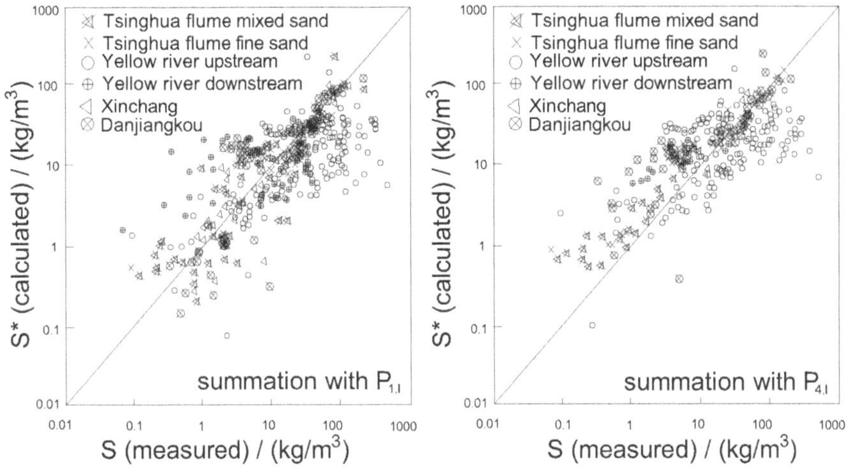

Fig. 4.13 Comparison between computed carrying-sediment capacity and measured sediment concentration (all data sets)

results computed with the formulas that do not take into account the impact of coarser particles.

(2) The computed carrying-sediment capacity compared to the data measured in Tsinghua's flume experiments are listed in Table 4.2. Compared to the field data, the data measured in these laboratory experiments are more reliable. As shown in this table, the computed results, no matter based on the summation with the bed material gradation or based on the sediment concentration gradation (carrying-sediment capacity gradation), agree well with the measured data. This also validates our conclusion that the carrying-sediment capacities based on these two summation approaches are equivalent in the strong equilibrium condition, or Eqs. (4.25) and (4.26) are correct. It is noted that the gradation for the total suspended load was used to represent $P_{4.l}$ in the calculation. This may cause some error in the results, but the error should be insignificant.

(3) In conclusion, the formulas derived from the statistical theory for sediment movement show reasonable agreement with field data, and these formulas can be used to compute the carrying-sediment capacity in engineering applications.

4.3.3.2 Merits of the Theoretical Formulas for Carrying-Sediment Capacity Based on Statistical Theory

(1) The theoretical system for the carrying-sediment capacity based on the statistical theory illustrates the mechanisms of the carrying-sediment capacity for non-uniform sediment mixture. These include: the mechanism for a single sediment particle to be suspended up to the maximum suspension height, settles down to the bed, and then changes to the bed material; re-suspension of settled particles;

Table 4.2 Verification with Tsinghua's Flume experiments

Fine sediment			Sediment mixture		
Measured S (kg/m^3)	Computed S^* (kg/m^3)		Measured S (kg/m^3)	Computed S^* (kg/m^3)	
	Summation of $P_{1.l}$	Summation of $P_{4.l}$		Summation of $P_{1.l}$	Summation of $P_{4.l}{}^*$
26.52	20.50	20.88	55.50	77.42	56.42
96.40	90.28	89.54	99.34	95.21	86.31
50.90	62.05	64.93	78.49	66.77	83.09
50.62	35.23	36.04	45.13	44.90	35.84
54.48	39.55	41.49	64.87	57.08	59.76
34.56	24.83	26.70	37.36	32.72	24.20
39.67	32.13	34.03	29.08	12.28	11.69
25.68	25.56	28.32	49.13	54.50	46.91
52.23	36.25	39.36	25.16	10.23	16.67
141.39	140.94	146.53	78.29	73.49	68.39

the exchange mechanism between massive bed material and suspended load; the relations between the carrying-sediment capacity for uniform sediment, the fractional carrying-sediment capacity for non-uniform sediment mixture, and the total carrying-sediment capacity; impact of turbulent bursting near the bed to sediment suspension; the relation between the period of incipient suspension and the turbulent bursting period; and etc. Because of these comprehensive mechanisms being included in the theory, the formula system for the carrying-sediment capacity consists of twenty-seven formulas. These include Eqs. (2.30) through (2.33), (4.5) through (4.8), (4.13), (4.15) through (4.21), (4.26), (2.26), (4.36), (4.37), (4.28), (4.29), (4.34), (4.39), and (4.40) through (4.42). These formulas were developed based on two advanced studies: 23 formulas based on the Han and He's [9, 10] statistical theory for sediment movement, and four based on the turbulent bursting theory [6].

(2) These formulas were developed based on theoretically rigorous derivation. They were verified, without adding any adjusted coefficients. The representative laboratory and field data used for verification were not pre-screened.

(3) These formulas were developed for the total carrying-sediment capacity of non-uniform sediment mixture. But they were also verified and analyzed for the fractional carrying-sediment capacity, as shown in Table 4.2. The results indicate that both the coarse and fine sediment follows the same laws for carrying-sediment capacity. This provides a theoretical justification for using the total carrying-sediment capacity in the mathematical model for sediment transport.

(4) Comparison with other formulas for carrying-sediment capacity is described in the following. One of the most representative formulas for the carrying-sediment capacity was developed by Zhang et al. [29]. This formula reads

$$S^* = \kappa\left(\frac{V^3}{gh\omega}\right)^m = K\left(\frac{V^3}{h\omega}\right)^m \tag{4.43}$$

where m and κ are functions of $\frac{V^3}{gh\omega}$. As $\frac{V^3}{gh\omega}$ increases from 4 to 4000, m varies from 1.48 to 0.40, and κ increases from 0 to 4.

The formula adopted by Han in his mathematical model, which was derived based on Eq. (5.27) and on the results of Han [7], has also been widely used. This formula is

$$S^* = \kappa\left(\frac{V^3}{gh\omega}\right)^{0.92} = K\left(\frac{V^3}{h\omega}\right)^{0.92} \tag{4.44}$$

Han's formula has similar form as Zhang's formula. After calibration with massive field data, Han determined that $m = 0.92$ for the sediment mixture, $\kappa = 0.245$ (or $K = 0.03$) for the Yellow River, and $\kappa = 0.139$ (or $K = 0.017$) for the Yangtze River. These values of K were later verified with the statistical theory [7]. The comparison between these two formulas and Eqs. (4.23) and (4.25), which were derived based on the statistical theory, are described in the following:

(i) For Eqs. (4.43) and (4.44), no matter κ and m are constants or variables, S^* is a single-valued function of $\kappa\left(\frac{V^3}{gh\omega}\right)$ and is independent of the fall velocity ω, the water depth h, and the shear velocity u_* On the other hand, as shown in Fig. 4.14, S^* computed from the statistic theory not only is a function of $\frac{V^3}{gh\omega}$, but also depends on h, u^* and D. S^* is not a single-valued function of $\frac{V^3}{gh\omega}$. Instead, it varies within a range. This is significantly different from Eqs. (4.43) and (4.44).
(ii) As shown in Fig. 4.14, both Zhang's formula and Han's formula are within the variation range of the results computed with the statistic theory, and in fact pass through the region with dense data points.

The data points shown in Fig. 4.14 cover the results computed for various conditions, which are hard to illustrate some characteristics clearly. Therefore, these data points are characterized into eight groups based on the particle sizes (from $D < 0.005$ mm to 0.50 mm–1.00 mm), and the results are shown in Fig. 4.15, 4.16, 4.17, 4.18, 4.19, 4.20, 4.21 and 4.22, respectively. The data points in each figure can be divided into six slanted columns, which represent the condition for $u_* = 0.005$ m/s, 0.01 m/s, 0.02 m/s, 0.05 m/s, 0.10 m/s, and 0.25 m/s, from left to right, respectively. The five points in each column represent the condition for $h = 1$ m, 2 m, 5 m, 10 m, and 20 m, from bottom to top, respectively. The results from Han's formula include both the condition for the Yellow River ($\kappa = 0.245$) and that for the Yangtze River ($\kappa = 0.139$). The following can be observed from these figures: (a) The results predicted with Zhang's and Han's formulas are higher than the data for fine sediment, as shown in Figs. 4.15 and 4.16. This is because the exponents of these two formulas might be too large for fine sediment. (b) For the other six groups of coarser sediment ($d > 0.01$ mm), the results of these two formulas are close to the results computed with the

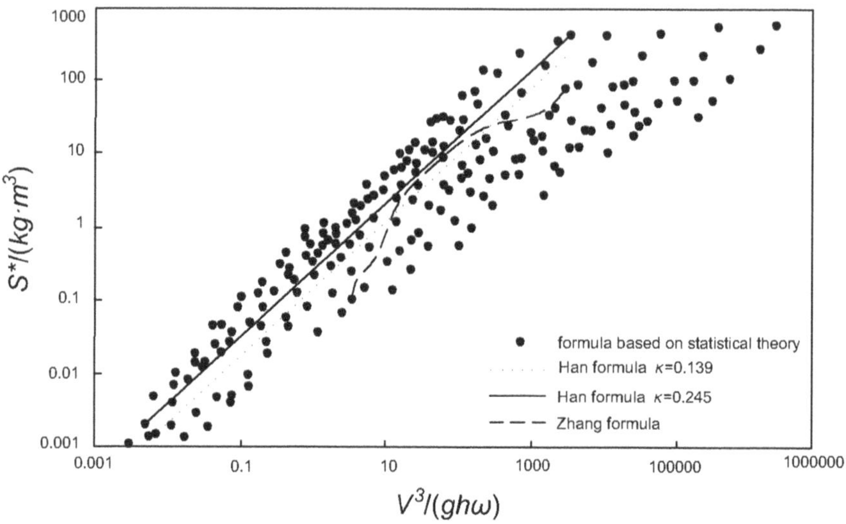

Fig. 4.14 Relation between S^* and $V^3/(gh\omega)$ based on various formulas

statistical theory. It is noted that Han's formula with $\kappa = 0.245$ shows the least discrepancy from the results of the statistical theory.

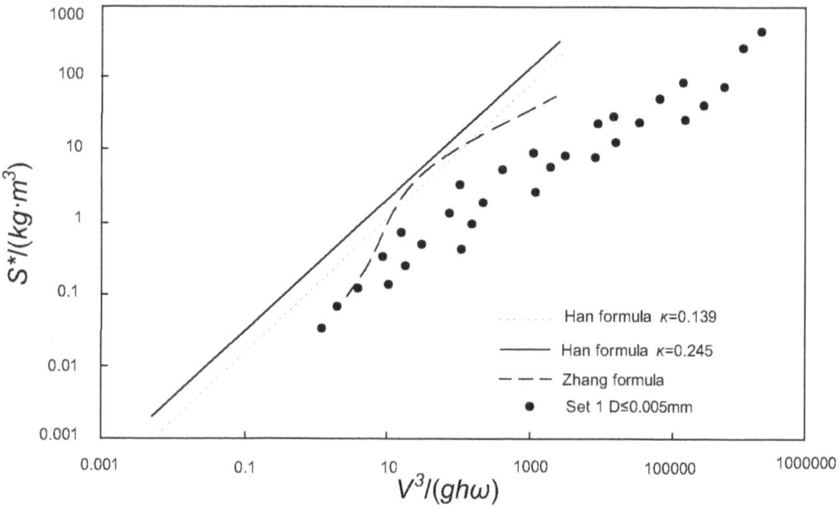

Fig. 4.15 Relation between S^* and $V^3/(gh\omega)$ for $D \leq 0.005$ mm

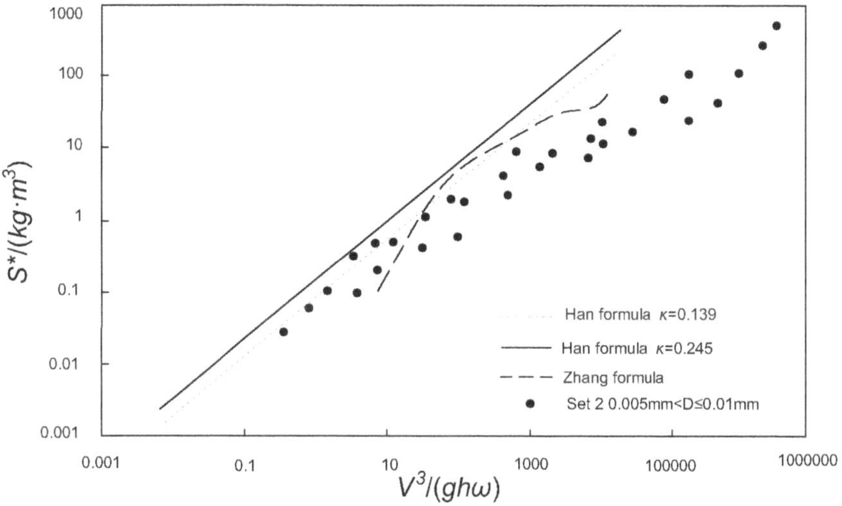

Fig. 4.16 Relation between S^* and $V^3/(gh\omega)$ for 0.005 mm $\leq D \leq$ 0.01 mm

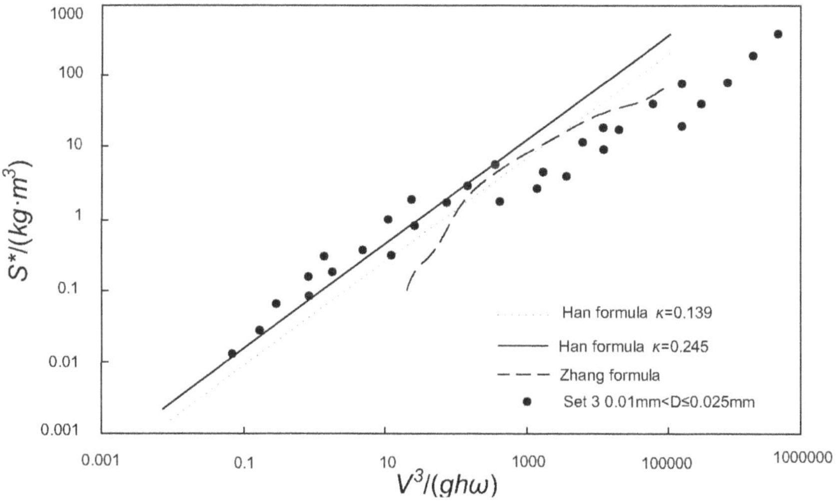

Fig. 4.17 Relation between S^* and $V^3/(gh\omega)$ for 0.01 mm $\leq D \leq$ 0.025 mm

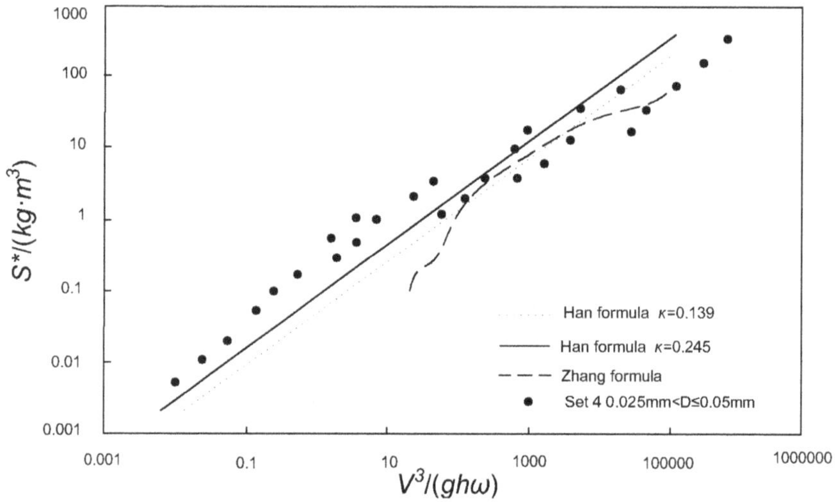

Fig. 4.18 Relation between S^* and $V^3/(gh\omega)$ for 0.025 mm $\leq D \leq$ 0.05 mm

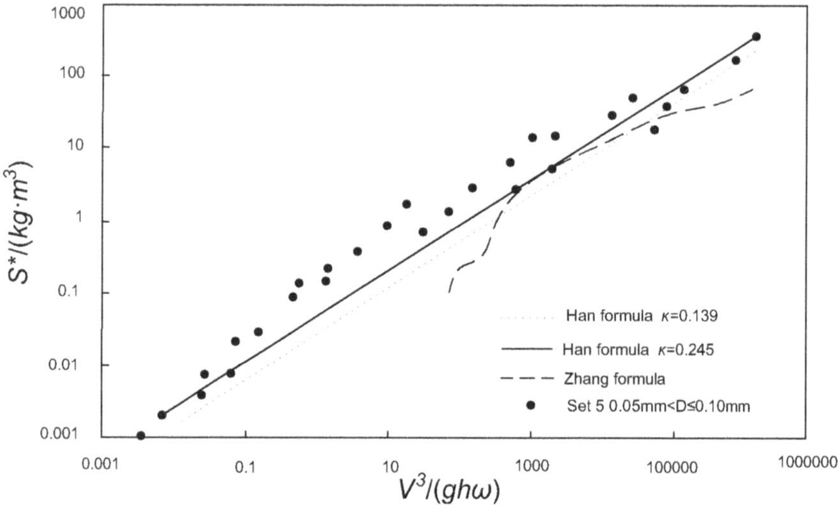

Fig. 4.19 Relation between S^* and $V^3/(gh\omega)$ for 0.05 mm $\leq D \leq$ 0.1 mm

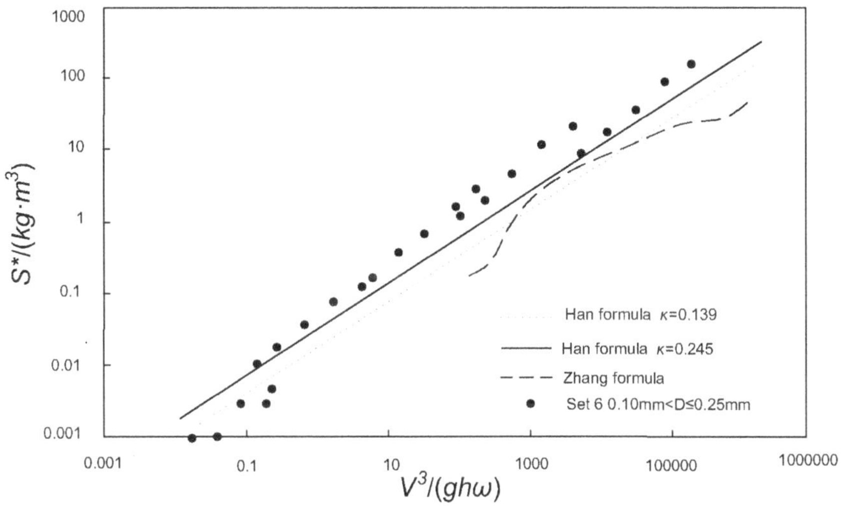

Fig. 4.20 Relation between S^* and $V^3/(gh\omega)$ for 0.1 mm $\leq D \leq 0.25$ mm

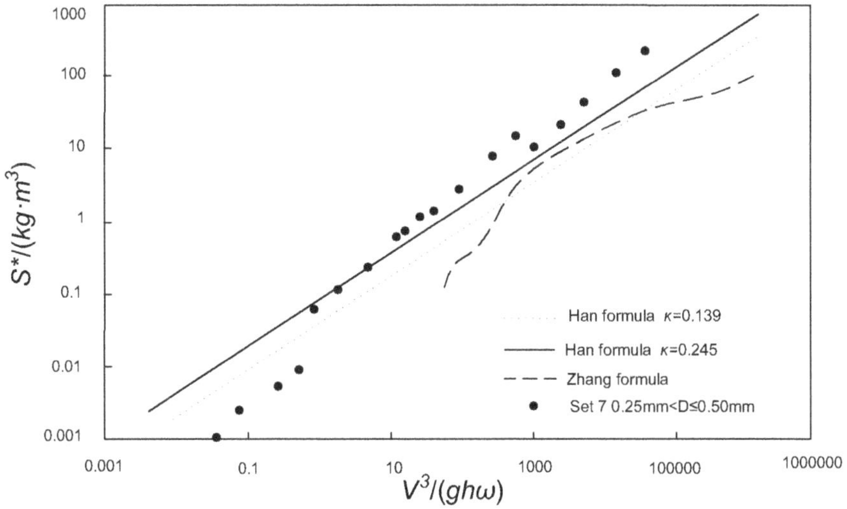

Fig. 4.21 Relation between S^* and $V^3/(gh\omega)$ for 0.25 mm $\leq D \leq 0.5$ mm

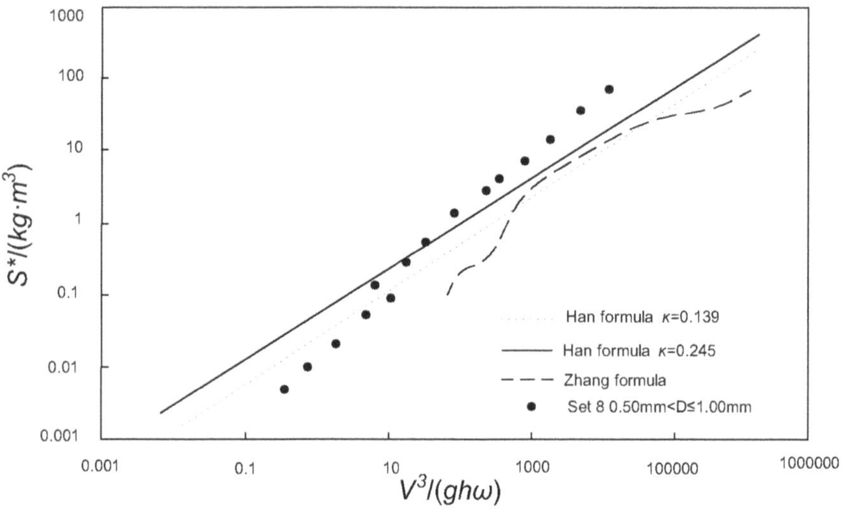

Fig. 4.22 Relation between S^* and $V^3/gh\omega$ for 0.5 mm $\leq$ D $\leq$ 1.0 mm

Zhang's formula has been widely applied for many rivers in China. Han's formula has been used for the Jinsha River, the Chuan River, Three Gorges Reservoir, the middle and lower reaches of the Yangtze River, the Han River, and the middle and lower reaches of the Yellow River, with the coefficients being determined. Both Zhang's and Han's formulas are within the variation range of the results of the statistical theory. This implies that the sediment transport formulas derived from the statistical theory are consistent with the reality.

(iii) For fine sediment and very large $\frac{V^3}{gh\omega}$, S^* computed with Eq. (4.43) may be larger than the dry density of sediment, which is unrealistic and indicates the limitations of these formulas. For example, the values of coefficients m and κ of Zhang's formula are valid for bed material load only, with S^* not exceeding 100 kg/m^3. Han's formula also limits the sediment concentration not exceeding 100 kg/m^3. Otherwise, adjustment is needed for the case with high sediment concentration, which is shown in Eq. (4.45). On the other hand, with the statistical theory, S^* for the strong flow condition only reaches 567 kg/m^3 for $u_* = 0.25$ m/s and h as small as 1 m. For typical condition in natural rivers S^* does not exceed 200–300 kg/m^3. Thus, the statistical theory can be applied to natural rivers with no limitation. This is because various factors are considered and constraint conditions are applied in the statistical theory, which prevents unrealistic results. For example, the state probability R_1 decreases as the flow condition becomes stronger, and thus limits the increase to the carrying-sediment capacity. Of course, the limit of 200–300 kg/m^3 does not include the adjustment for the sediment fall velocity and for the density for high sediment concentration, and it is equivalent to 400 kg/m^3 with adjustments. This might be the upper limit for sediment transport caused by the flow turbulence. If linear regressions are applied to the calculated data points in Fig. 4.15, 4.16, 4.17, 4.18, 4.19, 4.20, 4.21 and 4.22, the slope of these linear lines m increases from 0.625 to 1.25 as the particle size increases, with an average value of 0.942, which is close to 0.92 in Eq. (4.44). It is also observed from these figures that $\frac{V^3}{gh\omega}$ decreases as the particle size increases, and the slope m of Eq. (4.43) increases as $\frac{V^3}{gh\omega}$ decreases. This also qualitatively validates Eq. (4.43), which states m decreases as $\frac{V^3}{gh\omega}$ increases.

(iv) Han's formula, or Eq. (4.44), was used in previous mathematical model. Now updated formulas for carrying-sediment capacity, represented by Eq. (4.23) for the fractional carrying-sediment capacity, have been derived from the statistical theory. Which formulas shall we use now? Should the previous formula be replaced or still be used? As discussed before, the results of Eq. (4.44) are generally consistent with the results from the statistical theory, no matter for the Yangtze River or for the Yellow River. This is particularly true if the fractional carrying-sediment capacity is summated with the gradation of the sediment concentration. Guo et al. [4] used the sediment concentration data collected in the Yellow River to verify [7] formula that can be applied for the cases with high-, medium-, and low-concentrations. As shown in Figs. 4.23 and 4.24, this formula shows reasonable agreement with data. Han [7] formula reads

$$S^* = \kappa\left[1 + (\frac{\gamma_s - \gamma_0}{\gamma_0\gamma_s})\frac{S}{\rho}\right]^m \frac{1}{\left(1 - \frac{S}{\beta\gamma_0}\right)^{(K+1)m}}\left(\frac{V^3}{gh\omega_0}\right)^m \tag{4.45}$$

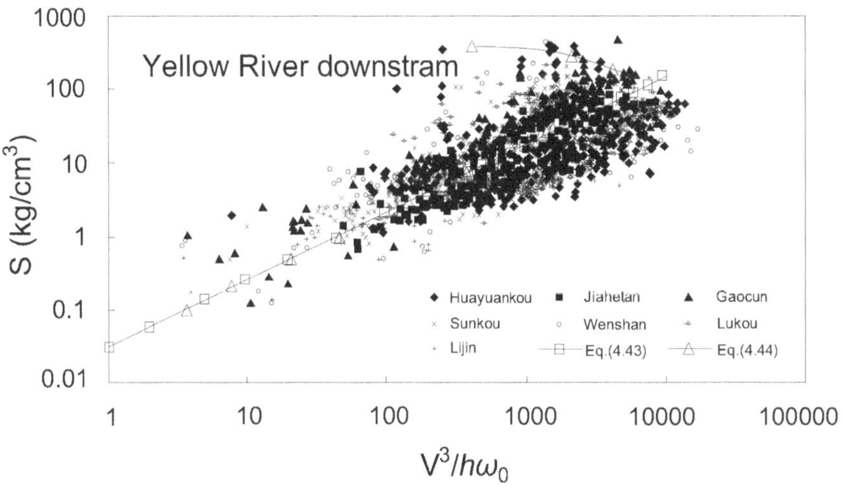

Fig. 4.23 Relation between S and $V^3/(h\omega_0)$ at downstream reach of the Yellow River

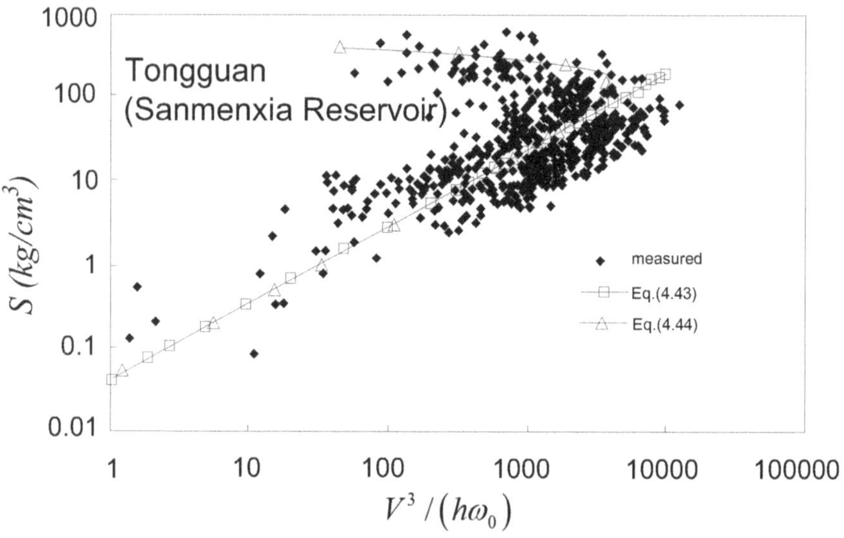

Fig. 4.24 Relation between S and $V^3/(h\omega_0)$ at Tongguan Station, Sanmenxia Reservoir

where γ_s and γ_0 is the density (kg/m^3) of sediment and water, respectively, S the sediment concentration, β the ratio of the net sediment volume to the sediment volume with film water, ω_0 the sediment fall velocity in clear water, and the values of κ and m were previously determined for the Yellow River. Guo et al. [4] also tested Han [7] formula with $\kappa = 0.139$ and $m = 0.92$ using data collected at Hankou Station of Yangtze River and at Xiantao Station of Han River. As shown in Fig. 4.25 and 4.26, the agreement between the formulas and the data are also satisfactory. It is noted that a correction term, which is proposed by Han for the case with high sediment concentration, is included in Eq. (4.45) for Figs. 4.23 and 4.24. As a results, Eq. (4.45) predicts a reverse trend for the relation between the sediment concentration and $\frac{V^3}{gh\omega}$ when the concentration is high. From the theoretical and application aspects, the statistical theory has great advantages compared to Eq. (4.43), and it should be the direction for future studies and applications. When Han [7] first developed Eq. (4.44), he adopted the form of Zhang's formula, and calibrated K and m using field data. The results are in agreement with reality. From the engineering aspect, this formula can still be applied.

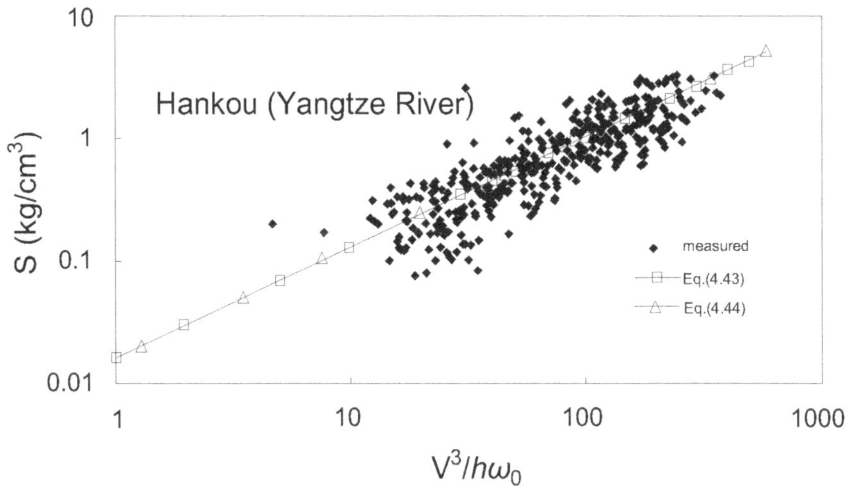

Fig. 4.25 Relation between S and $V^3/(h\omega_0)$ at Hankou Station, Yangtze River

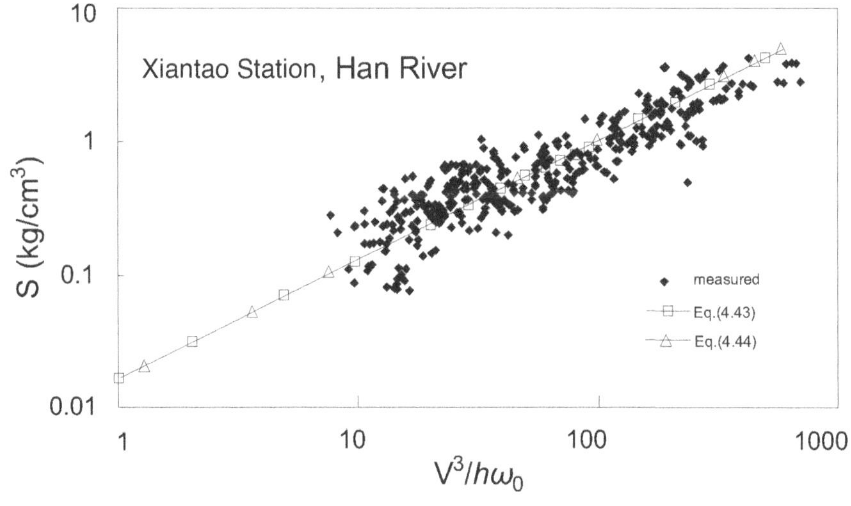

Fig. 4.26 Relation between S and $V^3/(h\omega_0)$ at Xiantao Station, Han River

References

1. Cao ZX (1996) Sediment flux picked-up from loose bed based on turbulent bursting. J Hydraul Eng 1996(5):18–21 (in Chinese)
2. Cao WH, Zhang QS (2000) A study on capacity for carrying suspended sediment under combined actions of waves and tidal current. J Sedim Res 05:16–20 (in Chinese)
3. Fang CM, Han QW, He MM (1998) Calculation of carrying-sediment capacity based on statistical theory and verification. J Hydraul Eng 2:68–71 (in Chinese)
4. Guo QC, Mao JX, Han QW (2009) The adaption of the formulae of carrying-sediment capacity to natural river. In: Proceedings of Symposium on the theory of Non-Equilibrium Transportation of Suspended Load, Sedimentation Department of Chinese Hydraulic Engineering Society, pp 20–30
5. Grass AJ (1971) Structural features of turbulent flow over smooth and rough boundaries. J Fluid Mech 50(2):233–255
6. Guan JC (2009) Study on carrying-sediment capacity for non-uniform suspended load in non-equilibrium transport process. Doctoral dissertation
7. Han QW (2003) Reservoir sedimentation. Science Press, Beijing, China (in Chinese)
8. Han QW (2010) Theoretical study for non-equilibrium transport of non-uniform suspended load. Research Report, Key Laboratories of Hydraulics, Sediment Sciences and River Harness, China Ministry of Water Resources (in Chinese)
9. Han QW, He MM (1981) Statistical characters of sediment exchange. J Hydraul Eng 1:9–22 (in Chinese)
10. Han QW, He MM (1984) Stochastic theory of sediment motion. Science Publishing House, Beijing (in Chinese)
11. Han QW, He MM (1999a) Study on state probabilities and the ratio of bed load to suspended load. J Hydraul Eng 10:7–16 (in Chinese)
12. Han QW, He MM (1999b) Characteristics of incipient sediment motion and incipient velocity. Science Press, Beijing (in Chinese)
13. Han QW, He MM (2000) Stochastic theory of sediment motion. Keynote lectures. In: Proceedings of the 8th International Symposium on Stochastic Hydraulics, Beijing, pp 63–84
14. Jing YF, Dao ZW (1999) Influence of coherent structure on scalar diffusion in turbulent boundary layer. J Hydrodyn (Ser B) 11(4):98–102
15. Kalinske AA (1940) Suspended-material transportation under non-equilibrium condition. Transaction AGU 21(19):613–617
16. Karauschev AV (1960) *Problems of the dynamics of natural flows*. Hydrometeorological Publishing House, Russia
17. Kim HT, Kline SJ, Reynolds WC (1971) The production of turbulence near a smooth wall in a turbulent boundary layer. J Fluid Mech 50(1):133–160
18. Kline SJ, Reynolds WC, Schraub FA (1967) The structure of turbulent boundary layers. J Fluid Mech 50:233–255
19. Lane WE (1937) Stable channels in erodible material. Trans Am Soc Civil Eng 102:123–142
20. Liang ZC (1988) Turbulence mechanics. Henan Science Press, Henan (in Chinese)
21. Mao Y, Yang H, Yuan XM (2002a) Characteristics of turbulent bursting phenomenon in open channel with a flat and smooth bottom. J Hydrodyn 04:413–421 (in Chinese)
22. Mao Y, Yang H, Yuan XM (2002b) Bed surface roughness and bursting phenomena in open channel. J Hydraul Eng 6:53–69 (in Chinese)
23. Mao Y, Zhang ZJ, Xu XR (1999) Review of the study on coherent structure of turbulence in open channel flow and its interaction with sediment movement. J Hohai Univ 4:24–29 (in Chinese)
24. Nikitin IK (1963) Turbulence process near river bed layer. Press of Falkland Academy of Sciences, Kiev
25. Robinson SK (1991) Coherent motions in the turbulent boundary layer. Annu Rev Fluid Mech 23:601–639

26. Sechet P, Le Guennec B (1999) Bursting phenomenon and incipient motion of solid particles in bed load transport. J Hydraul Res 37(5):683–696
27. Velikanov MA (1955) Dynamics of river, Chap. II. Science and Technical Libraries, Moscow
28. Wang DC (2000) On the Spatial Structure of low-speed streaks and particle motion in the near wall region of turbulent flow in open channel. Doctoral dissertation, Tsinghua University, Department of Hydraulic Engineering, Beijing (in Chinese)
29. Zhang RJ, Xie JH, Wang MF, Huang JT (1989) Dynamics of river sedimentation. Water and Power Press, Beijing (in Chinese)

Chapter 5
One-Dimensional Non-equilibrium Sediment Transport

One-dimensional (1-D) non-equilibrium sediment transport is a basic topic of the study for non-equilibrium sediment transport. It is also common in sedimentation engineering applications such as reservoir sedimentation, bed scour and channel evolution. Due to these applications, particularly in light of the investigations for the sedimentation in the Three Gorges Reservoir and channel scour in its downstream reach, Han has studied 1-D non-equilibrium sediment transport since 1960s, and has obtained some pioneering results [3, 4].

Reservoir and channel sedimentation is complex, and studies on this subject are far from satisfactory. Han considered reservoir sedimentation is a result of non-.equilibrium sediment transport. In order to obtain the main characteristics of the complex reservoir sedimentation and channel evolution and conduct an in-depth study on reservoir sedimentation, the general law for non-equilibrium sediment transport, particularly for the relatively simple 1-D transport, should be investigated.

Most studies on the non-equilibrium transport of suspended load have been conducted by investigating the longitudinal variation of sediment concentration. Two approaches are typically used in these studies. One approach is to solve the vertically two-dimensional (2-D) sediment diffusion equation with certain boundary conditions to determine the longitudinal and vertical distributions of sediment concentration. The other approach is to use 1-D concepts to investigate the longitudinal variation of the section-averaged sediment concentration with certain assumptions. The solutions of the first approach are more comprehensive, but the bottom boundary condition of this approach is hard to determine and some study results have not been connected with the application. This was already discussed in Chaps. 1 and 3. The second approach takes into account the principal characteristics of the longitudinal variation of sediment concentration, and the results are simpler. But different researches develop the governing equations from different aspects, without connection to the sediment diffusion equation. The results are only applied for uniform flow. One common problem of these two approaches is that they are not verified with sufficient field data.

Q. Han and M. He, *Non-equilibrium Transport of Non-uniform Suspended Load*, Hydroscience and Engineering, https://doi.org/10.1007/978-981-97-1123-9_5

For non-equilibrium suspended load transport, in addition to the sediment concentration, the size distribution of suspended load and the bed material gradation also show longitudinal variations. The variations of the size distribution of suspended load and the bed material gradation impact the variation of sediment concentration. Therefore, a comprehensive study on non-equilibrium sediment transport should include not only the sediment concentration, but also the size distribution of suspended load and the bed material gradation. Some studies only investigated the non-equilibrium transport of uniform sediment, and thus avoid variations of the suspended load size distribution and the bed material gradation. Some studies investigated the non-equilibrium transport of non-uniform sediment by applying the results for uniform sediment to the fractional sediment transport of non-uniform sediment mixture. Without investigating the variation of the suspended load size distribution, these studies had to assume the section-averaged carrying-sediment capacity, or the carrying-sediment capacity at the bottom, is constant along the flow path, which is inconsistent with the concept of varying gradations. In these studies, the variation of the fractional sediment concentration was determined without considering the gradation variation, and the gradation variation was calculated later based on the variation of the fractional sediment concentration. Tsvezizova [11] has considered the simultaneous variations of size distribution of the concentration and the carrying-sediment capacity while the formulation for the size distribution was not verified.

Starting from investigation and calculation of reservoir sedimentation, Han proposed to comprehensively study all three topics for the 1-D non-equilibrium sediment transport in non-.uniform flow, rather than only analyzing an individual topic (such as the concentration variation for uniform sediment) of 2-D non-equilibrium sediment transport. To better serve engineering applications, particularly the sedimentation engineering for the Three Gorges Project, Han started the investigation of 1-D non-equilibrium sediment transport in 1960s, and has obtained some results with theoretical integrity and reasonable agreement with reality [3–6]. The mathematical model for reservoir sedimentation and scour of the downstream reach [8, 7], which was used in the sedimentation study, design and operation for the Three Gorges Project, was developed based on Han's theory for 1-D non-equilibrium sediment transport. The basic content and major parameters of this theory have been widely used in the past 30–40 years. This theory is discussed in the following four sections. The content of this chapter is only discussed the case of massive erosion and massive deposition, where the gradation of sediment concentration is equal to that of carrying-sediment capacity. The discussion of general carrying-sediment capacity and its gradation and effective bed sediment gradation will be carried out in this chapter and Chap. 6.

5.1 Longitudinal Variation of Average Sediment Concentration

5.1.1 Longitudinal Variation of Sediment Concentration for Uniform-Sediment

Many studies have been conducted to investigate the longitudinal variation of the section-.averaged concentration using the 1-D approach, with relatively simple and consistent differential equations being derived for uniform flow. With different derivations, some studies did not relate their derivations to the sediment diffusion equation. Based on the diffusion equation, the governing equations for the longitudinal variation of the section-averaged sediment concentration will be derived in the following for non-uniform flow. The results will illustrate the theoretical consistence between the 1-D approach and the 2-D approach.

The schematic diagram for 2-D non-equilibrium sediment transport is shown in Fig. 5.1. The 2-D diffusion equation for steady and non-uniform flow is

$$\frac{\partial(V_x S)}{\partial x}+\frac{\partial(V_y S)}{\partial y}=\frac{\partial}{\partial y}\left(\varepsilon_y\frac{\partial S}{\partial y}\right)+\frac{\partial(\omega S)}{\partial y}$$

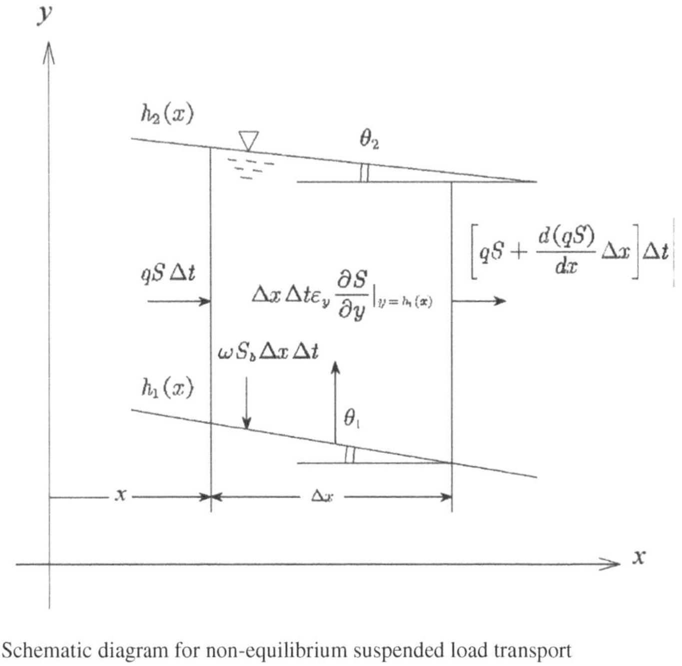

Fig. 5.1 Schematic diagram for non-equilibrium suspended load transport

As discussed in Chap. 3, integration of this equation with the surface and bottom boundary conditions yields Eq. (3.62) for the l-th size class. Equation (3.62) for uniform sediment turns to

$$\frac{dq_s}{dx} = q\frac{dS}{dx} = -\varepsilon_y \frac{\partial S}{\partial y}\Big|_{y=h_1(x)} - \omega S_b \tag{5.1}$$

where q_s is the sediment discharge per unit width, q is the flow discharge per unit width, S_b is the sediment concentration on the bottom, S is the depth-averaged sediment concentration, and $h_2(x) - h_1(x)$ is the water depth h.

It is noted that Eq. (5.1) can also be directly derived based on sediment mass balance, as shown in Fig. 5.1, for uniform sediment. From the sediment mass balance within the control volume there is

$$\left(qS + \frac{d(qS)}{dx}\Delta x\right)\Delta t + \omega S_b \Delta x \Delta t = qS\Delta t - \varepsilon_y \frac{\partial S}{\partial y}\Big|_{y=h_1(x)} \Delta x \, \Delta t$$

Thus

$$\frac{d(qS)}{dx} = -\varepsilon_y \frac{\partial S}{\partial y}\Big|_{y=h_1(x)} - \omega S_b$$

In order to determine $\varepsilon_y(\partial S/\partial y)$, the bottom carrying-sediment capacity S_b^* is introduced. When the bed material can be picked up, it is assumed that the vertical gradient of the sediment concentration on the bottom only depends on the flow condition, and is independent of the relative value between S_b and S_b^*, or the saturation condition of the bottom sediment concentration. Replacing $\varepsilon_y \frac{\partial S}{\partial y}\big|_{y=h_1(x)}$ with the downward sediment influx ωS_b^* in the equilibrium condition, Eq. (5.1) becomes

$$\frac{dS}{dx} = -\frac{\omega}{q}\left(S_b - S_b^*\right)$$

Denoting $\tilde{\alpha} = \frac{S_b}{S}$ and $\alpha^* = \frac{S_b^*}{S^*}$, (see Sect. 3.3.4) then

$$\frac{dS}{dx} = -\frac{\omega}{q}\left(\tilde{\alpha}S - \alpha^* S^*\right) \tag{5.2}$$

This is the governing equation for the longitudinal variation of sediment concentration for 2-D steady non-uniform flow. As discussed above, Eq. (5.2) can be derived either by integrating the 2-D sediment diffusion equation or based on the sediment mass balance as shown in Fig. 5.1.

When [3] first derived Eq. (5.2), the statistical theory of sediment exchange on the bed was not applied. It was assumed that the sediment immediately changes to the bed material after it settles down on the bed, without chances to the bed load or be re-suspended. As a result, $\tilde{\alpha}$ and α^* are the ratios of the sediment concentration and of the carrying-sediment capacity, respectively, between the values on the bottom and the depth-averaged values. Therefore $\tilde{\alpha}$ and α^* have specific physical meaning for the 2-D situation, and they are both greater than 1. It's much complicated for the 3-D situation, particularly due to the approximations and some factors that were not included in previous derivation. For non-equilibrium sediment transport in true flow condition, $\tilde{\alpha}$ and α^* are more like comprehensive coefficients. It is approximated that $\tilde{\alpha} = \alpha^* = \alpha$, with α known as the saturation recovery coefficient, and thus

$$\frac{dS}{dx} = -\alpha\frac{\omega}{q}\left(S - S^*\right) \tag{5.3}$$

Based on calibrations with filed data, the recommended value of α for truly 3-D flow is $\alpha = 1$ for the scour situation and $\alpha = 0.25$ for the deposition situation.

The boundary condition, or Eq. (3.92), which was derived from the statistical theory for sediment exchange, and the corresponding saturation recovery coefficient are discussed in Chap. 3. For equations for 1-D non-equilibrium sediment transport, there are three types of saturation recovery coefficients. In addition to the saturation recovery coefficients for sediment concentration and for carrying-sediment capacity, or $\tilde{\alpha}$ and α^* in Eq. (5.2), a comprehensive saturation recovery coefficient can be defined as

$$\alpha = \frac{\tilde{\alpha}S}{S - S^*} - \frac{\alpha^* S^*}{S - S^*} \tag{5.4}$$

Apparently, using Eqs. (5.4) and (5.2) is equivalent to Eq. (5.3).

Detailed discussions of $\tilde{\alpha}$, α^* and α are presented in Chap. 6. While the previously recommended values of α, $\alpha = 1$ for the scour situation and $\alpha = 0.25$ for the deposition situation, are empirical, they have been widely used in the mathematic models for sediment transport with reasonable agreement with reality. It is recommended to apply the new findings of $\tilde{\alpha}$, α^* and α, as discussed in Chap. 6, in the models for sediment transport. It is also important to investigate if these empirical saturation recovery coefficients are consistent with the theoretical results, and can still be applied in engineering applications. In fact, the difference between the empirical values and the theoretical results are not significant, resulting in even less impact on the longitudinal variation of the sediment concentration. Based on Tables 6.15–6.22, the theoretical values of the comprehensive saturation recovery coefficient, averaged over the range of $\frac{\omega}{u_*} = 0.001 \sim 2$, are listed in Table 5.1 for particle size $D = 0.01$ ~ 1.0 mm.

A great variation range of sedimentation is listed in Table 5.1, with S/S^* varying from 0.300 to 2.456. As listed in this table, the theoretical values of the recovery coefficient vary between 0.148 and 1.531, with an average of 0.605. This is consistent with the empirical value that varies between 0.25 and 1.0 with an average of 0.5. It is noted that a higher empirical value of the saturation coefficient is set for the scour situation and a lower value for the deposition situation. This is reasonable for $\tilde{\alpha}$ and α^*. But it is inconsistent with the theoretical value for the comprehensive recovery

Table 5.1 Theoretical values for comprehensive saturation recovery coefficient

D (mm)	α		
	S/S*=0.300	S/S*=2.456	Average
0.01	1.493	1.568	1.531
0.025	0.838	0.875	0.857
0.05	0.639	0.658	0.648
0.10	0.488	0.495	0.492
0.25	0.333	0.337	0.335
0.50	0.222	0.225	0.223
1.00	0.147	0.150	0.148
Average	0.587	0.615	0.605

coefficient α. As listed in Table 5.1, the difference in the comprehensive recovery coefficient is negligible for different sedimentation intensity (S/S^*). It is also noted that the theoretical value of α reduces as the particle size increases. This indicates that saturation of finer particles recovers quicker than coarser particles when the incipient motion condition is satisfied. For finer particles with $D = 0.01$ mm ~ 0.05 mm, the average recovery coefficient is $\alpha = 1.01$. For coarser particles with $D = 0.1$ mm ~ 1 mm, $\alpha = 0.300$.

The most important conclusion that can be drawn from the theoretical results is that the comprehensive saturation recovery coefficient shows little variation with the deposition and scour processes. This is a new finding that needs further investigation. This is also consistent with the fact that the scour is typically overestimated by the mathematical models with the empirical saturation recovery coefficient. The details of the saturation recovery coefficient are discussed in Chap. 6.

Equation (5.3) can be written as

$$\frac{d(S - S^*)}{dx} = -\alpha\frac{\omega}{q}\left(S - S^*\right) - \frac{dS^*}{dx}$$

Integrating over x yields

$$S - S^* = \left(S_0 - S_0^*\right)\exp\left(-\frac{\alpha\omega L}{q}\right) - \exp\left(-\frac{\alpha\omega L}{q}\right)\int_0^L \exp\left(\frac{\alpha\omega x}{q}\right)\frac{dS^*}{dx}dx \quad (5.5)$$

where subscript "0" denotes values at the entrance section, the parameters without subscript "0" denote the values at the exit section, and L is the channel length for integration. This equation applies to a general 2-D non-uniform flow. Considering a simple case of 2-D flow with the water depth varying linearly within the channel length L, or

$$h = h_0 + \frac{h_L - h_0}{L}x \quad (5.6)$$

where h_L is the water depth at the exit section. This simplified 2-D flow concept can be used to approximate the characteristics of natural reservoirs. Assuming the carrying-sediment capacity can be generally written as

$$S^* = k\left(\frac{q^3}{gh^4\omega}\right)^m \tag{5.7}$$

then

$$\begin{aligned} S^* &= k\left(\frac{q^3}{gh^4\omega}\right)^m = k\left(\frac{q^3}{gh_0^4\omega}\right)^m \frac{1}{\left[1+\frac{(h_L-h_0)}{h_0}\frac{x}{L}\right]^{4m}} \\ &= S_0^*\left[1-4m\frac{(h_L-h_0)x}{h_0 L}\right] = S_0^* - 4m\frac{(h_L-h_0)x}{h_0 L}S_0^* \end{aligned} \tag{5.8}$$

At $x = L$,

$$S_L^* - S_0^* = -4m\frac{h_L - h_0}{h_0}S_0^* \tag{5.9}$$

Differentiating Eq. (5.8) with x and plugging in Eq. (5.9) yields,

$$\frac{dS^*}{dx} = -4m\frac{h_L - h_0}{h_0}\frac{S_0^*}{L} = -\frac{S_0^* - S_L^*}{L} \tag{5.10}$$

Removing subscript "L" in S_L^*, plugging Eq. (5.10) into Eq. (5.5), and then integrating with x yields

$$S = S^* + (S_0 - S^*)\exp\left(-\frac{\alpha\omega L}{q}\right) + (S_0^* - S^*)\frac{q}{\alpha\omega L}\left[1-\exp\left(-\frac{\alpha\omega x}{q}\right)\right] \tag{5.11}$$

It is noted that $l = q/\omega$ represents the settling distance of a particle from the water surface to the bed in laminar flow. Then Eq. (5.11) can be rewritten as

$$S = S^* + (S_0 - S^*)\exp\left(-\frac{\alpha L}{l}\right) + (S_0^* - S^*)\frac{l}{\alpha L}\left[1-\exp\left(-\frac{\alpha L}{l}\right)\right]$$

Equation (5.10) is the governing equation for the longitudinal variation of sediment concentration in 2-D steady non-uniform flow. It shows the sediment concentration at the exit section depends on the sediment concentration S_0 and carrying-sediment capacity S_0^* at the entrance, carrying-sediment capacity at the exit section S_L^*, relative channel length L/l and the saturation recovery coefficient α. The sediment concentration at the exit section consists of three parts: the carrying-sediment capacity S_L^* at this section, the residual concentration, or $(S_0 - S_0^*)e^{-\frac{\alpha\omega L}{q}}$, of the surplus sediment concentration $S_0 - S_0^*$ at the entrance after decaying over the distance of L/l, and the adjustment term due to the variation of hydraulic factors (or the resulting carrying-sediment capacity). In order to provide a general concept of the magnitudes for these three parts, Table 5.2 lists the computed results of these three terms based on the data measured in Wotousi desilting canal. As listed in this table, the error in the computed sediment concentration is within 10% for six out of the eight sections. For equilibrium sediment transport $S_0^* - S^* = 0$, the last term of the above equation disappears, and this equation becomes the form of [10, 9, 2] formulas. However, as shown in Table 5.2, the discrepancy between the computed sediment concentration and the measured data is even greater if the last term (variation of carrying-sediment capacity) is neglected.

The comparison between the sediment concentration computed with Eq. (5.11) and the measured data is shown in Fig. 5.2 for various data sources. As shown in this figure, the errors in the computed results are generally within 20–30% of the data, and these errors do not increase with the channel lengths. Due to the random factors and errors associated with individual measurements, the sediment concentration averaged over a time period shall be more accurate. It is noted that the same equation for carrying-sediment capacity, or Eq. (5.27), was used in the calculations for different data sets. The adopted value for the saturation recovery coefficient α is a little different from the values recommended before. In these calculations, α was set to 1 for scouring experiments in clear water flume, and to 0.5 for the other three data sources. The comparison shows the difference between the adopted α and the recommended value has negligible impact on the computed sediment concentration. As shown in Fig. 5.2, Eq. (5.11) is also capable of predicting the sediment concentration for a channel scour with clear water. This is because no limitation of sedimentation or scour process is set in the derivation of this equation.

Table 5.2 Comparison between computed sediment concentration and data measured at Wotousi desilting canal

Section	S^*	$(S_0 - S^*)\exp(-\frac{\alpha L}{l})$	$(S_0^* - S^*)\frac{l}{\alpha L}[1 - \exp(-\frac{\alpha L}{l})]$	Computed S	Measured S
No	(kg/m^3)	(kg/m^3)	(kg/m^3)	(kg/m^3)	(kg/m^3)
2	26.2	5.42	..8.22	23.4	30.8
3	23.4	..1.64	2.24	24.0	27.8
4	13.3	0.33	7.57	21.2	23.5
5	9.77	5.64	3.09	18.5	19.2
6	6.97	8.73	2.50	18.2	17.6
7	7.46	9.56	..0.32	16.7	16.7
8	9.33	9.24	..1.77	16.8	14.8
9	24.6	7.06	..14.8	16.9	14.3

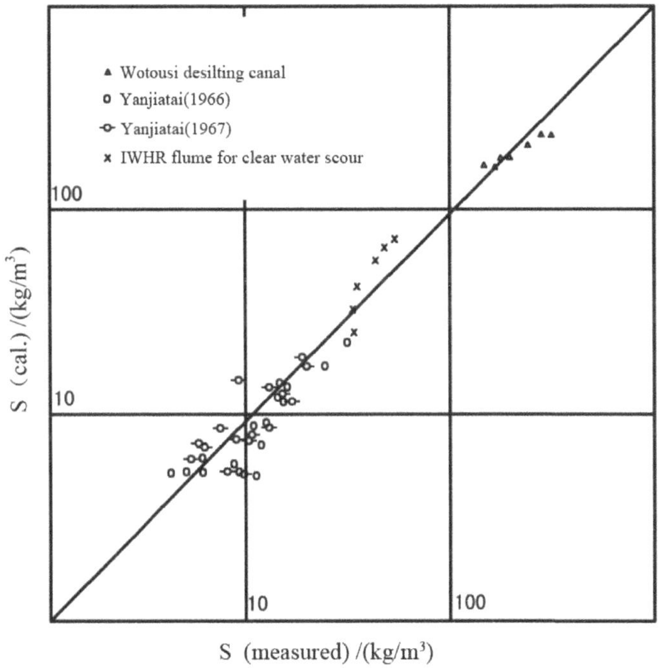

Fig. 5.2 Computed sediment concentration compared with data

5.1.2 Longitudinal Variation of Sediment Concentration for Non-uniform Sediment

Equation (5.11) cannot be generally applied for truly non-uniform sediment mixture. Further investigations are needed in order to determine the conditions under which this equation can be applied and whether this equation can be used to compute the fractional sediment concentration.

In order to better analyze the sediment concentration and carrying-sediment capacity for non-uniform sediment, a new concept of so-called water volume percentage K_l is introduced. K_l is defined as the percentage of the water volume that is required to carry the saturated sediment concentration for the l-th size class. According to the definitions of the size distributions of the suspended load and of the carrying-sediment capacity, or $P_{4.l}$ and $P^*_{4.l}$, respectively, the fractional sediment concentration S_l and the fractional carrying-sediment capacity S^*_l for the l-th size class can be expressed in terms of the total concentration S and total carrying-sediment capacity of the non-uniform sediment mixture, which is

$$S_l = P_{4.l} S \quad (l = 1, 2, \cdots, n) \tag{5.12}$$

$$S_l^* = P_{4.l}^* S^* = P_{4.l} S^* \quad (l = 1, 2, \cdots, n) \tag{5.13}$$

Let's group the sediment mixture according to the particle sizes, and divide the unit water volume into n parts. Each part of the water volume, which accounts K_l of the unit water volume, has the carrying-sediment capacity of S_l^* for the l-th size class. By grouping the sediment concentration for the l-th size class into the corresponding part of the water volume, sediment transport within each water part can be considered as uniform sediment transport. This approach is reasonable based on the fact that flow carrying sediment does not noticeably impact the flow itself and the sediment transport. This is particularly true when the sediment concentration is low. Let $S(\omega_l) = S(l)$ and $S^*(\omega_l) = S^*(l)$ be the concentration and the carrying-sediment capacity, respectively, for the uniform sediment with the settling velocity of ω_l. Assuming the carrying-sediment capacity for the l-th size class is S_l^*, the water volume to transport this size class is Q_l^*, and the carrying-sediment capacity for the uniform sediment with this size is $S^*(l)$, the potential fractional sediment transport rate for this size class is $G_l^* = Q_l S^*(l)$, and the total potential sediment transport rate is $G^* = \sum G_l^*$. Dividing G^* by the total water volume, one obtains

$$S^* = \sum S_l^* = \frac{G^*}{Q} = \sum \frac{G_l^*}{Q} = \sum \frac{Q_l^*}{Q} \frac{G_l^*}{Q_l^*} = \sum K_l S^*(l)$$

where

$$S_l = \frac{G_l}{Q} = \frac{Q_l^*}{Q} \frac{G_l}{Q_l^*} = K_l S(l) \tag{5.14}$$

$$S_l^* = \frac{G_l^*}{Q} = \frac{Q_l^*}{Q} \frac{G_l^*}{Q_l^*} = K_l S^*(l) \tag{5.15}$$

Comparing Eqs. (5.14) and (5.15) with Eqs. (5.12) and (5.13), one obtains

$$P_{4.l} S = K_l S(l) \tag{5.16}$$

$$P_{4.l} S^* = K_l S^*(l) \tag{5.17}$$

Generally $S(l) \neq S$, $S^*(l) \neq S^*$, and thus the water volume percentage K_l does not equal to the size distribution $P_{4.l}^*$ or $P_{4.l}$. K_l can also be considered as the percentage of the energy that is needed to transport the l-th size class sediment so its physical meaning can be better understood.

The differential equation for the longitudinal variation of the non-uniform sediment concentration is derived in the following. As shown in Fig. 5.3, the mass balance for the l-th size class within the control water volume can be expressed as

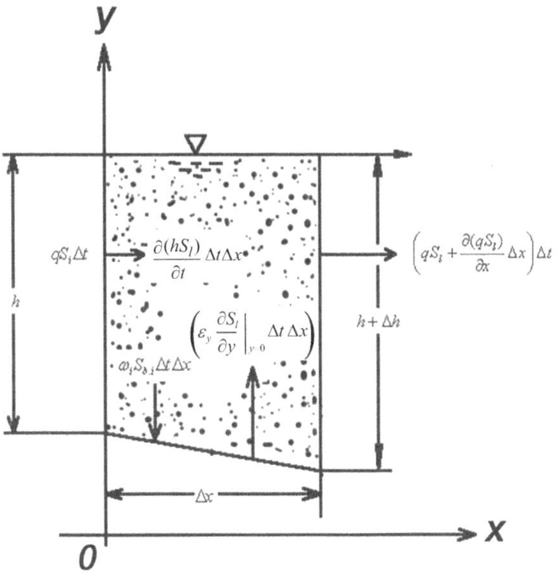

Fig. 5.3 Mass variation of suspended load within a water column for non-equilibrium sediment transport

$$K_l S(l) q \Delta t + \frac{d(K_l S(l) q)}{dx} \Delta x \Delta t + \omega_l K_l S(l) \Delta x \Delta t$$
$$= -\varepsilon_y \frac{\partial}{\partial y} [K_l S(l)]|_{y=0} \Delta x \Delta t + K_l S(l) q \Delta t$$

For steady sediment transport, $\frac{\partial(K_l h S(l))}{\partial t} \Delta t \Delta x$ is neglected. As $\Delta x \to 0$, following the similar derivation as used in Sect. 5.1.1 for uniform sediment, one obtains

$$\frac{d}{dx}[K_l S(l)] = -\frac{\omega_l}{q}\left[\tilde{\alpha} K_l S(l) - \alpha^* K_l S^*(l)\right] \tag{5.18}$$

plugging Eqs. (5.16) and (5.17) in which yields

$$\frac{d}{dx}(P_{4.l} S) = -\frac{\omega_i}{q}\left(\tilde{\alpha}_l P_{4.l} S - \alpha_l^* P_{4.l}^* S^*\right) = -\frac{\tilde{\alpha}_l \omega_i}{q}\left(P_{4.l} S - P_{4.l}^{**} S^{**}\right) \tag{5.19}$$

Assuming $\tilde{\alpha}_l = \alpha_l^* = \alpha_l$, and $P_{4.l}^* \approx P_{4.l}$, then

$$\frac{d(S P_{4.l})}{dx} = -\frac{\alpha_l \omega_l}{q} P_{4.l}\left(S - S^*\right) \tag{5.20}$$

It is noted that $P_{4.l}^* \approx P_{4.l}$ is true in mass scour or deposition condition. It will be illustrated later in Chap. 7 that $P_{4.l}^* = P_{4.l}$ in mass deposition condition (deposition occurs for all size classes), and $P_{4.l}^* \approx P_{4.l}$ in mass scour condition (scour occurs for all size classes).

Summing up Eq. (5.20) for all size classes, and introducing $\omega = \Sigma_l P_{4.l}\omega_l$, one obtains

$$\frac{dS}{dx} = -\frac{\alpha\omega}{q}\left(S - S^*\right) \tag{5.21}$$

Equation (5.21) has the same form as Eq. (5.3). However, ω is a constant in Eq. (5.3), while it is a variable in Eq. (5.21).

It is noted that the previous derivation was made in order to obtain Eqs. (5.19) and (5.20). If using the concept of the sediment exchange intensity, the results, as shown in Eq. (3.92), are the same as Eq. (5.21), but with more in-depth physical meanings.

If the average settling velocity is computed as $\omega = \sum_{l=1}^{n} P_{4.l}\omega_l$, then Eq. (5.3) for uniform sediment can also be used for non-uniform sediment. The calculation of total carrying-sediment capacity for non-uniform sediment mixture will be discussed later. It is assumed that ω does not vary in the longitudinal direction in the derivation of Eq. (5.11), which cannot be guaranteed for non-uniform sediment transport. Therefore, the application of Eq. (5.11) cannot be directly extended to non-uniform sediment. If the relative channel length L/l is small, ω may not show significant variation in the longitudinal direction. For this situation, Eq. (5.11) can be approximately applied to non-uniform sediment, with S_0^* and S^* being computed using the average settling velocity ω_0 at the entrance and the settling velocity ω at the exit section, respectively. In fact, the data used to verify Eq. (5.11) all belongs to non-uniform sediment transport within a relative short channel length (L/l is small).

When L/l changes significantly, which is particularly true for reservoirs, Eq. (5.11) cannot be used for non-uniform sediment transport. Alternatively, the integration of Eq. (5.20) shall be used. It is assumed that the carrying-sediment capacity S^* show linear variation in the longitudinal direction in the derivation of Eq. (5.11). In order to solve Eq. (5.20), let's assume S_l^* vary linearly within the studied reach. Following the derivation procedure used in Sect. 5.1.1 for uniform sediment, introducing

$$l_l = \frac{q}{\omega_l} \tag{5.22}$$

and integrating Eq. (5.20), one obtains

$$S_{L.l} = S_{L.l}^* + \left(S_{0.l} - S_{0.l}^*\right)\exp\left(-\frac{\alpha L}{l_l}\right) + \frac{\left(S_{0.l}^* - S_{L.l}^*\right)l_l}{\alpha L}\left[1 - \exp\left(-\frac{\alpha L}{l_l}\right)\right] \quad (l = 1, 2, \ldots, n) \tag{5.23}$$

Equation (5.23) is the governing equation for the longitudinal variation of the fractional sediment concentration for non-uniform sediment mixture. It is equivalent to Eq. (5.10) with subscript l. Considering Eqs. (5.12) and (5.13), summation of Eq. (5.23) for all size classes yields the total sediment concentration that is

$$S = S^* + \left(S_0 - S_0^*\right)\sum_{l=1}^{n} P_{4.l.0}\exp\left(-\frac{\alpha L}{l_l'}\right) + S_0^*\sum_{l=1}^{n} P_{4.l.0}\frac{l_l'}{\alpha L}\left[1-\exp\left(-\frac{\alpha L}{l_l'}\right)\right] - S^*\sum_{l=1}^{n} P_{4.l}\frac{l_l'}{\alpha L}\left[1-\exp\left(-\frac{\alpha L}{l_l'}\right)\right] \tag{5.24a}$$

where S_0^* and S^* are the total carrying-sediment capacity at the entrance and at exit sections, respectively, and $P_{4.l.0}$ and $P_{4.l}$ are the size distribution (in weight) of the suspended load at the entrance and at the exit sections, respectively.

Equation (5.24a) is under mass deposition or mass scour condition. In more general condition, $P_{4.l}^*$ is different from $P_{4.l}$. Assuming $\tilde{\alpha}_l \approx \alpha_l^* = \alpha$, integrating Eq. (5.20) yields

$$S = S^* + S_0\sum P_{4.l.0}\exp\left(-\frac{\alpha L}{l_l'}\right) - S_0^*\sum P_{4.l.0}^*\exp\left(-\frac{\alpha L}{l_l'}\right) + S_0^*\sum P_{4.l.0}^*\frac{l_l'}{\alpha L}\left[1-\exp\left(-\frac{\alpha L}{l_l'}\right)\right] - S^*\sum P_{4.l}^*\frac{l_l'}{\alpha L}\left[1-\exp\left(-\frac{\alpha L}{l_l'}\right)\right] \tag{5.24b}$$

The sediment concentration computed with Eq. (5.23) compared to the field data is shown in Fig. 5.4 for three data sources. The results computed with Eqs. (5.10) and (5.24a) compared to the field data are listed in Table 5.3 for Diudiuwan desilting basin. As listed in Table 5.3, the sediment concentration computed with Eq. (5.10) and with Eq. (5.24a) does not show significant difference when the relative channel length L/l is small, but show significant difference when L/l is large. For reservoir sedimentation, Eq. (5.11) generally underestimates the sediment concentration, while the results computed with (5.24a) show reasonable agreement with the field data.

- In addition to the sediment concentration S, another unknown of Eq. (5.24a) is the size distribution of suspended load $P_{4.l}$ at the exit section. As discussed later, S^* can only be determined after $P_{4.l}$ is known for non-uniform sediment. However, it is not considered an unknown. Instead, it is considered as a function of S_0, S and $P_{4.l.0}$. Therefore, for given flow parameters and the sediment concentration S_0 and size distribution $P_{4.l.0}$ at the entrance section, the sediment concentration and the size distribution at the exit section cannot be determined merely based on Eq. (5.24a). One more equation system is needed for the size distribution in order to closure the equation system. It should be emphasized that it is not sufficient to just investigate the longitudinal variation of sediment concentration for non-equilibrium transport of non-uniform sediment, even for uniform flow. Previous

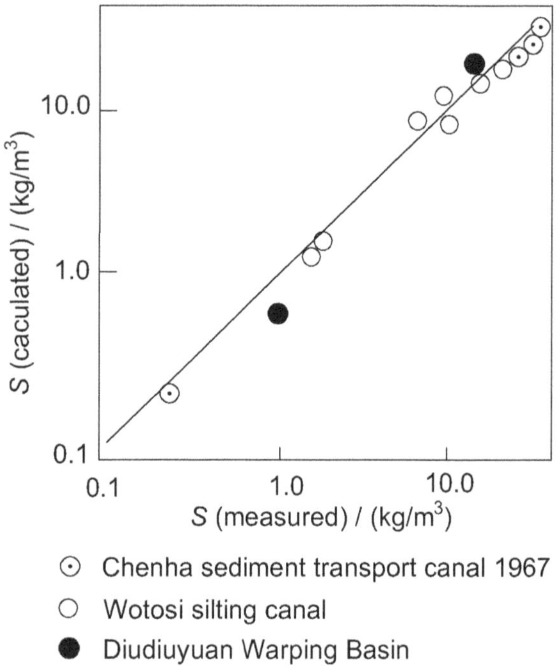

Fig. 5.4 Comparison between computed sediment concentration and field data

Table 5.3 Verification of computed sediment concentration for Diudiuyuan warping basin

<table>
<tr><th rowspan="2">Section No.</th><th rowspan="2">Measured S (kg/m^3)</th><th colspan="2">Computed S (kg/m^3)</th><th rowspan="2">Channel relative length L/l'</th></tr>
<tr><th>Eq. (5.10)</th><th>Eq. (5.24a)</th></tr>
<tr><td>1</td><td>1.66</td><td></td><td></td><td rowspan="2">0.828</td></tr>
<tr><td rowspan="2">2</td><td rowspan="2">1.63</td><td rowspan="2">1.66</td><td rowspan="2">1.65</td></tr>
<tr><td rowspan="2">1.19</td></tr>
<tr><td rowspan="2">7</td><td rowspan="2">1.40</td><td rowspan="2">1.11</td><td rowspan="2">1.23</td></tr>
<tr><td rowspan="2">2.34</td></tr>
<tr><td rowspan="2">9</td><td rowspan="2">0.931</td><td rowspan="2">0.313</td><td rowspan="2">0.600</td></tr>
<tr><td rowspan="2">8.52</td></tr>
<tr><td>11</td><td>0.214</td><td>0.027</td><td>0.198</td></tr>
</table>

studies only considered the non-equilibrium transport of uniform sediment so this problem was not exposed

5.1.3 Carrying-Sediment Capacity for Non-uniform Sediment

The sediment concentration and carrying-sediment capacity discussed in previous sections are for total suspended load that includes both bed material load and wash load. The grain size variation range of the total load is wide. So what's the difference in the carrying-sediment capacity between non-uniform sediment mixture and uniform sediment? Is that reasonable to investigate sediment transport in terms of total load that contradicts with the general concept of dividing sediment into bed material load and wash load? These questions will be discussed later in Chap. 10. This section will focus on the relation of the carrying-sediment capacity between non-uniform and uniform sediment.

Based on Eq. (5.17), the water volume percentage for each size class of the non-uniform sediment mixture can be written as

$$K_l = S^* \frac{P_{4.l}}{S^*(l)}$$

Summing up for all the size classes and considering $\sum_{l=1}^{n} K_l = 1$, one obtains

$$S^* = \frac{1}{\sum_{l=1}^{n} \frac{P_{4.l}}{S^*(l)}} \tag{5.25a}$$

This is the formula for the total carrying-sediment capacity for non-uniform sediment. If the formula for the carrying-sediment capacity $S^*(l)$ of uniform sediment and the size distribution of suspended load can be determined, then the total carrying-sediment capacity for non-uniform sediment can be solved. For example, if Zhang formula (1989) or Eq. (4.43), is used for the carrying-sediment capacity of uniform sediment, or $S^*(l) = k\left(\frac{V^3}{gh\omega_l}\right)^m$, Eq. (5.25a) becomes

$$S^* = \frac{1}{\sum_{l=1}^{n} \frac{P_{4.l}}{k\left(\frac{V^3}{gh\omega_l}\right)^m}} = k\left(\frac{V^3}{gh}\right)^m \frac{1}{\sum_{l=1}^{n} P_{4.l}\omega_l^m} = k\left(\frac{V^3}{gh\omega}\right)^m \tag{5.25b}$$

where $\omega^m = \sum_{l=1}^{n} P_{4.l}\omega_l^m$. If the average settling velocity is defined as

$$\omega = \left(\sum_{l=1}^{n} P_{4.l}\omega_l^m\right)^{\frac{1}{m}} \tag{5.26}$$

the total carrying-sediment capacity for non-uniform sediment can still be determined with the formula for uniform sediment. On the other hand, if the average settling velocity is defined as $\omega = \sum_{l=1}^{n} P_{4.l}\omega_l$, $\omega = \omega(d_{cp})$ or $\omega = \omega(d_{50})$, as used in previous studies for non-uniform sediment transport, the impact of the settling velocity cannot be reasonable represented. This resulted in system errors in the computed carrying-sediment capacity when the size variation range of the sediment mixture is wide. It is also noted that the size distribution of the carrying-sediment capacity, $P^*_{4.l}$, should be used in the computation of the carrying-sediment capacity, while $P_{4.l}$ is used here only under the mass deposition or mass scour condition. Figure 5.5 shows the relation between the sediment concentration S^* and $V^3/(gh\omega)$ for a number of measured data sources. The average settling velocity is calculated with Eq. (5.26) and $m = 0.92$. As shown in this figure, the relation between S^* and $V^3/(gh\omega)$ can be characterized as a linear band. This indicates that Eq. (5.25b) can reasonably predict the total carrying-sediment capacity for non-uniform sediment. For $m = 0.92$, Eq. (5.25b) can be written as

$$S^* = K\frac{V^{2.76}}{h^{0.92}\omega^{0.92}} \tag{5.27}$$

Since the system errors in the measurements are limited, the scattering of data points in Fig. 5.5 is caused by the transort condition that may not in equilibrium and other random factors. In addition, three values of K are found based on these

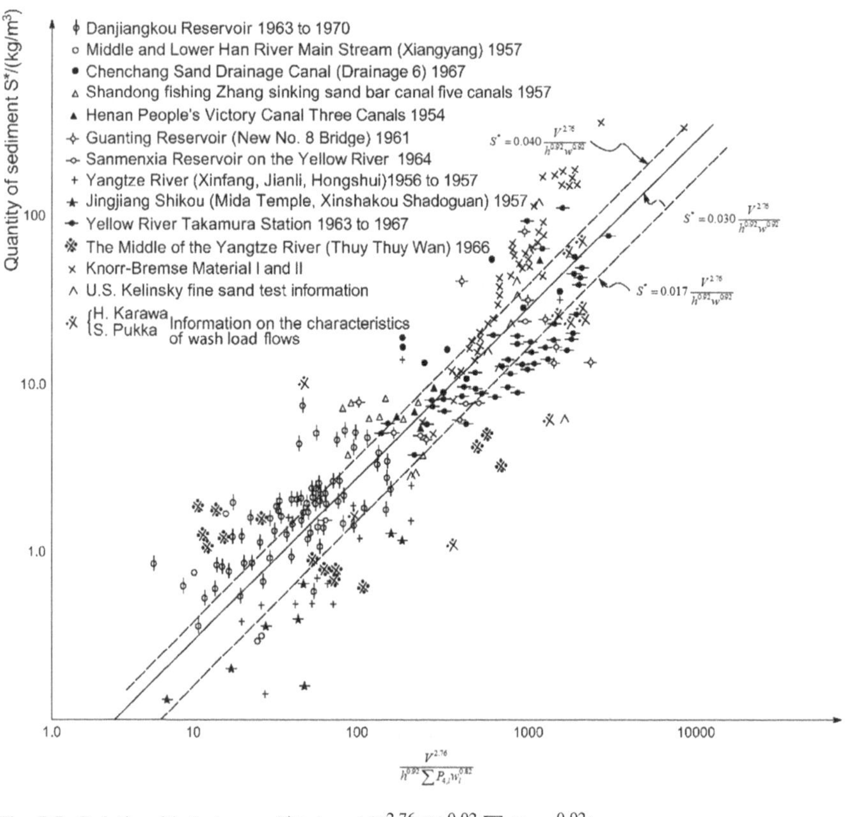

Fig. 5.5 Relationship between $S^*(\omega)$ and $V^{2.76}/(h^{0.92}\sum P_{4.l}\omega_l^{0.92})$

massive data sets: $K = 0.03$ for the average condition of all data sets, and for the downstream reach of the Yellow River and other rivers with high sediment concentration; $K = 0.03 \sim 0.04$ for reservoirs; and $K = 0.017$ for the Yangtze River, the Han River and other rivers in southern China that are approximately under equilibrium condition. Another factor that may contribute to the scattering of data points is the inaccuracy of the adopted formula for uniform sediment transport, which was discussed by Han [6] and in Chap. 4 of this book. It should be emphasized that the influx sediment concentration at the entrance section can be introduced for investigating the unsaturated sediment transport if the sediment concentration is computed with the formulas for non-equilibrium sediment transport. In other words, while the carrying-sediment capacity is calculated with Eq. (5.27), the sediment concentration can be computed with the non-equilibrium approach. By doing that, the errors in the computed results will be significantly reduced, as shown in Figs. 5.2 and 5.4.

5.2 Longitudinal Variation of the Size Distribution of Suspended Load

5.2.1 Sorting of Suspended Load in Mass Deposition Process

As suspended load deposits, coarser particles settle faster than fine particles, and thus the size distribution of suspended load become finer and finer along the flow path, resulting sorting of suspended load. This is clearly shown in Table 5.4, which lists the size distributions of suspended load at various sections along the Wotousi desilting canal. In previous studies, the variation of the size distributions is either described qualitatively [1], or is quantified with a simple approach by using a threshold particle size. All of the particles coarser than this threshold size are assumed to deposit, while all the particles finer than this threshold size do not deposit.

The sorting of suspended load is analyzed as follows. According to Eq. (5.19)

$$\frac{dS_l}{dx} = -\frac{\omega_l}{q}\left(\tilde{\alpha} S - \alpha^* S^*\right) P_{4.l}$$

Multiplying the flow discharge of channel width Bq, and denoting the sediment discharge for the l-th size group $W_l = S_l Bq$, the above equation can be written as

$$\frac{dW_l}{dx} = -\left(\tilde{\alpha} S - \alpha^* S^*\right) B P_{4.l} \omega_l \tag{5.28}$$

Summation for all size groups gives the equation of total sediment discharge, $W = SBq$, which is

$$\frac{dW}{dx} = -\left(\tilde{\alpha} S - \alpha^* S^*\right) B \sum P_i \omega_i = -\left(\tilde{\alpha} S - \alpha^* S^*\right) B \omega \tag{5.29}$$

Table 5.4 Longitudinal variation of the size distribution of suspended load at Wotousi desilting canal

Section No.	S (kg/m^3)	Variation of size distributions of suspended load, $P_{4.l}$						
		0.25–0.15	0.15–0.10	0.10–0.06	0.06–0.03	0.03–0.01	0.01–0.004	< 0.004
1	33.9	0.9	1.5	6.5	17.1	23.9	15.0	35.1
2	30.8	1.0	1.6	5.5	14.0	25.3	17.5	35.1
3	27.8	1.1	0.4	4.4	11.7	26.0	16.5	39.9
4	23.5		0.4	2.5	5.0	18.4	19.7	54.0
5	19.2			0.5	2.5	13.3	19.9	63.8
6	17.6			0.6	1.1	9.5	22.9	65.9
7	16.7				1.2	9.6	20.9	68.3
8	14.8				1.4	4.7	14.2	79.7
9	14.3					5.6	17.5	76.9

where ω is an average settling velocity. Dividing Eq. (5.28) by Eq. (5.29), and considering $P_{4.l} = W_l / W$, one obtains

$$\frac{dW_l}{dW} = \frac{\omega_l}{\omega}\frac{W_l}{W} \tag{5.30}$$

Integrating this equation from W_0 to W yields

$$W_l = W_{0.l} \exp\left(\omega_l \int_{W_0}^{W} \frac{dW}{\omega W}\right)$$

It is noted that the average settling velocity ω is an implicit function of W, and thus the above integral cannot be directly solved. According to the mean value theorem for integrals, it can be written as

$$W_l = W_{0.l}\left(\frac{W}{W_0}\right)^{\frac{\omega_l}{\omega_m}} \tag{5.31}$$

where ω_m is the integral mean value of ω on the interval $[W_0, W]$. Introducing the deposition percentage (or relative sedimentation quantity)

$$\lambda = \frac{W_0 - W}{W_0} = \frac{S_0 - S}{S_0} \tag{5.32}$$

and considering $P_{4.l.0} = W_{0.l}/W_0$, Eq. (5.31) can be rewritten as

$$P_{4.l} = \frac{W_l}{W} = \frac{W_{0.l}}{W_0}\frac{W_0}{W}(1-\lambda)^{\frac{\omega_l}{\omega_m}} = P_{4.l.0}\frac{(1-\lambda)^{\frac{\omega_l}{\omega_m}}}{1-\lambda}$$

or

$$P_{4.l} = P_{4.l.0}\frac{(1-\lambda)^{\frac{\omega_l}{\omega_m}}}{1-\lambda} \quad (l = 1, 2, \cdots, n) \tag{5.33}$$

Summing up this equation for l, and considering $\sum_{l=1}^{n} P_{4.i} = 1$, one obtains

$$\frac{\sum_{l=1}^{n} P_{4.l.0}(1-\lambda)^{\frac{\omega_l}{\omega_m}}}{1-\lambda} = 1 \tag{5.34}$$

For a given λ, let's define a function

$$y = 1 - \frac{1}{1-\lambda}\sum_{l=1}^{n} P_{4.l.0}(1-\lambda)^{\frac{\omega_l}{\omega_m}} = F(\omega_m)$$

With a given λ this function is a monotonically decreasing function with the single value zero at ω_m. $P_{4.l}$ can be solved after ω_m is determined. The physical meaning of ω_m can be interpreted as the effective settling velocity for suspended load deposition when the sediment discharge decreases

For given λ, ω_m can be solved from Eq. (5.34b) using an iteration approach. Instead of this approach ω_m can be solved with introducing a sorting curve of $\lambda_\upsilon \sim \eta_{\upsilon}$., i.e., assuming a series value for $\eta_\upsilon = (1-\lambda_\upsilon)^{\frac{1}{\omega_m}}$, solve Eq. (5.34) for λ_υ. For given λ using interpolation method η can be obtained. Then from Eq. (5.33) $P_{4.l}$ are solved.

Equation (5.33) indicates that the size distribution of suspended load $P_{4.l}$ at the exit section depends on the size distribution $P_{4.l.0}$ at the entrance, the effective settling velocity for deposition ω_m, the settling velocity for this size class ω_l, and the deposition percentage λ. It is not an explicit function of the hydraulic parameters. Instead, the hydraulic parameters impact the size distribution implicitly through the deposition percentage λ. It is noted from Eq. (5.33) that for $\omega_l/\omega_m > 1$, or for coarser sediment, $(1-\lambda)^{\frac{\omega_l}{\omega_m}} < 1-\lambda$ and $P_{4.l} < P_{4.l.0}$; and for $\omega_i/\omega_m/ < 1$, or for finer sediment, $(1-\lambda)^{\frac{\omega_l}{\omega_m}} > 1-\lambda$ and $P_{4.l} > P_{4.l.0}$. The sorting (thinning) of suspended load during deposition as described in Eq. (5.33) is consistent with reality.

Figure 5.6 shows the comparison of the size distribution of suspended load between the field data and the results computed with Eq. (5.33) for a number of sections downstream of the entrance of Wotousi desilting canal. A good agreement between the computed $P_{4.l}$ and the field data can be observed, with errors not exceeding 5%. The errors seem to be system errors, with the percentages generally being underestimated for coarser sediment, or with the finer size distribution being calculated. This is because the horizontal variation of the size distributions is neglected in the derivation based on the vertical 2-D assumption, and thus the coarser sediment would deposit along the entire width of the sections. This is not true in reality. For a truly horizontal 2-D channel, typically only finer sediment enters and deposits in the slow flow zones near the banks. That's why the real size distribution tends to be coarser than the computed results for these areas. In order to consider this fact, a correction exponent m, which is less than 1, is introduced for the ratio of ω_l/ω_m, and Eqs. (5.33) and (5.34) can be written as

$$P_{4.l} = P_{4.l.0}\frac{(1-\lambda)^{\left(\frac{\omega_l}{\omega_m}\right)^m}}{1-\lambda} \tag{5.35}$$

$$\sum P_{4.l} = P_{4.l.0}\frac{(1-\lambda)^{\left(\frac{\omega_l}{\omega_m}\right)^m}}{1-\lambda} = 1 \tag{5.36}$$

Based on field data, m is approximately 0.75 for long and narrow desilting canals and warping basins, and natural rivers; m is approximately 0.5 for reservoirs, and stomach shape warping basins. While m varies between 0.5 and 1.0, its impact to the size distribution is limited, as shown in Table 5.5. Therefore, $m = 1$ can still be used for 2-D analysis.

The size distributions computed with Eq. (5.35) are shown in Figs. 5.7, 5.8, 5.9 and 5.10 for Wotousi desilting canal, Yanjiatai warping basin, Diudiuyan warping basin, and Danjiangkou Reservoir, respectively. As shown in these figures, the errors in the computed $P_{4.l}$ do not exceed 5%, and are not system errors. If averaging over a certain period of time, the computed size distributions at the exit sections show even better agreement with measured data. Tables 5.6 and 5.7 list the size distributions

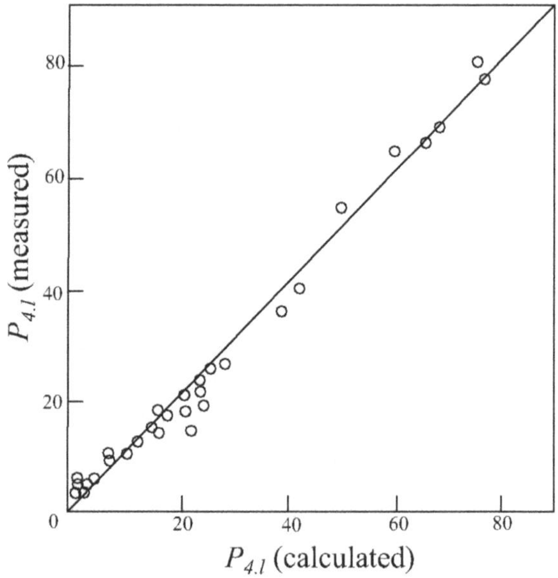

Fig. 5.6 Verification of size distribution of suspended load in Wotousi desilting canal

Table 5.5 Verification of the size distribution of suspended load in Wotousi desilting canal

Section No	S (kg/m^3)	Deposition percentage λ	Size distributions $P_{4.l}$								
			Results		Weight percentage for various size groups						
					0.25–0.15	0.15–0.10	0.10–0.06	0.06–0.03	0.03–0.01	0.01–0.004	< 0.004
1	33.9		Measured data		0.9	1.5	6.5	17.1	23.9	15.0	35.1
2	30.8	0.091	Measured data		1.0	1.6	5.5	14.0	25.3	17.5	35.1
			Computed	m = 1	0.1	0.4	3.8	15.4	25.3	16.4	38.6
				m = 3/4	0.1	0.6	4.7	15.2	24.7	16.2	38.5
				m = 1/2		1.0	5.0	15.4	24.0	16.2	38.4
3	27.8	0.181	Measured data		1.1	0.4	4.4	11.7	26.0	16.5	39.9
			Computed	m = 1		0	1.3	11.8	28.1	18.0	42.8
				m = 3/4		0	2.5	12.2	25.0	17.6	42.7
				m = 1/2		0.4	3.6	13.2	23.0	17.3	42.5
4	23.5	0.307	Measured data		0	0.4	2.5	5.0	18.4	19.7	54.0
			Computed	m = 1			0	4.2	24.6	20.7	50.5
				m = 3/4		0	0.2	6.6	23.6	20.0	49.6
				m = 1/2		0	1.5	9.2	20.5	18.4	50.4
5	19.2	0.434	Measured data			0	0.5	2.5	13.3	19.9	63.8
			Computed	m = 1			0	0.2	15.5	23.3	61.0
				m = 3/4			0	1.1	17.4	22.0	59.5
				m = 1/2		0	0.5	4.8	16.2	18.5	60.0
6	17.6	0.481	Measured data			0	0.6	1.1	9.5	22.9	65.9
			Computed	m = 1				0	10.4	23.8	65.8

(continued)

Table 5.5 (continued)

Section No	S (kg/m^3)	Deposition percentage λ	Size distributions $P_{4.l}$								
			Results		Weight percentage for various size groups						
					0.25–0.15	0.15–0.10	0.10–0.06	0.06–0.03	0.03–0.01	0.01–0.004	< 0.004
				m = 3/4			0	0.3	13.6	22.2	63.9
				m = 1/2		0	0.1	3.0	13.7	18.3	64.9
7	16.7	0.508	Measured data				0	1.2	9.6	20.9	68.3
			Computed	m = 1				0	7.3	23.9	68.8
				m = 3/4				0	11.2	22.0	66.8
				m = 1/2			0	2.3	12.0	17.7	68.0
8	14.8	0.563	Measured data				0	1.4	4.7	14.2	79.7
			Computed	m = 1				0	2.0	22.4	75.6
				m = 3/4				0	6.3	21.1	72.6
				m = 1/2			0	1.3	8.3	16.7	73.7
9	14.3	0.578	Measured data					0	5.6	17.5	76.9
			Computed	m = 1				0	1.1	21.6	77.3
				m = 3/4				0	5.1	20.8	74.1
				m = 1/2			0	0.3	6.7	16.1	76.4

averaged over time for Sanmenxia and Danjiangkou Reservoirs, respectively. The errors in the results are less than 2%.

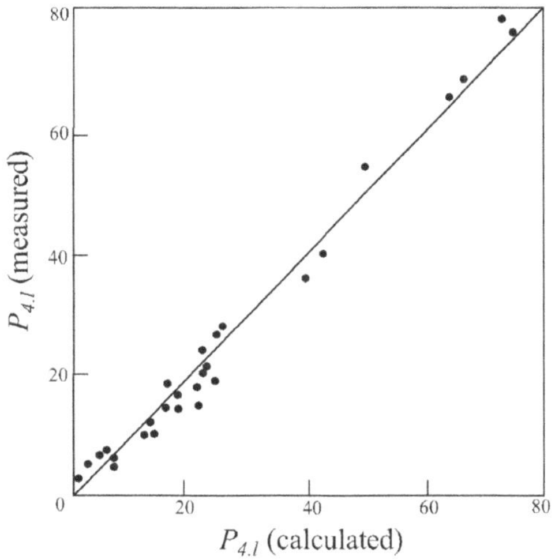

Fig. 5.7 Verification of the Size Distribution of Suspended Load in Wotousi Desilting Canal

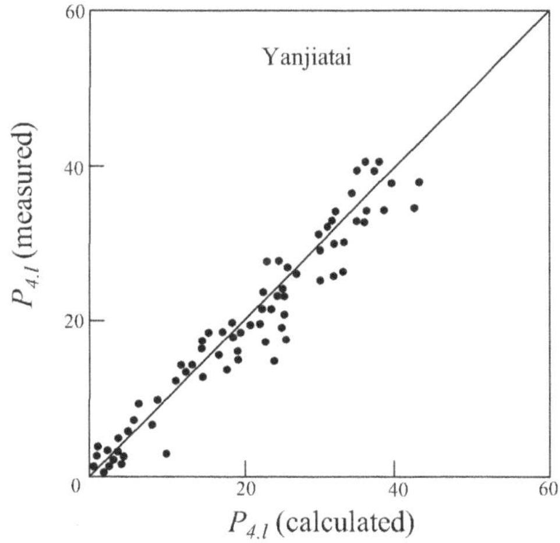

Fig. 5.8 Verification of the Size Distribution of Suspended Load in Yanjiatai Warping Basin

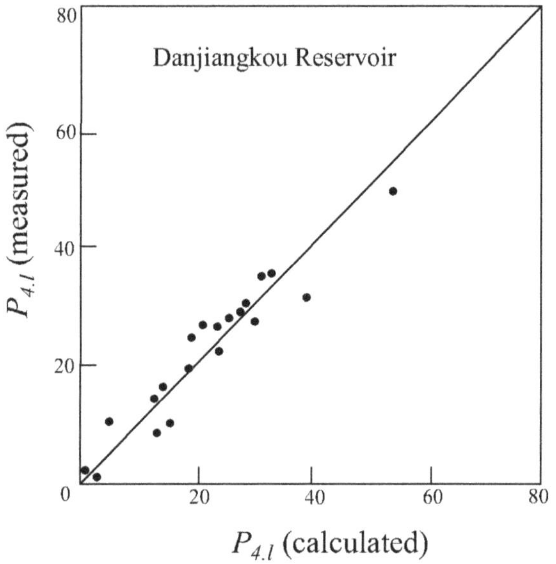

Fig. 5.9 Verification of the Size Distribution of Suspended Load in Danjiangkou Reservoir

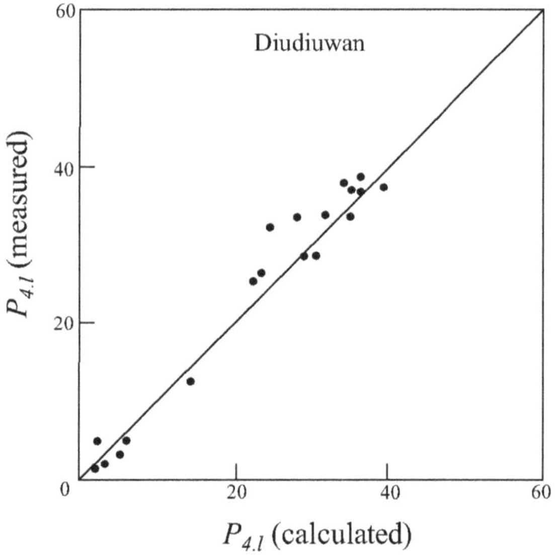

Fig. 5.10 Verification of the Size Distribution of Suspended Load in Diudiuyan Warping Basin

Table 5.6 Variation of the Size Distribution of Suspended Load in Sanmenxia Reservoir between Tongguan and Beicun

Sections	Periods	Sediment ransport quantity (10^8 tons)	Deposition percentage λ	Size distributions for suspended load $P_{4.l}$							
				< 0.005	0.005–0.01	0.01–0.025	0.025–0.05	0.05–0.10	0.10–0.25	0.25–0.50	0.50–1.0
Tongguan	64.3–64.10	23.4		14.0	9.9	17.5	25.4	28.2	4.3	0.6	0.1
Beicun (measured)	64.3–64.10	13.4	0.426	24.5	13.9	21.1	25.3	13.7	1.3	0.2	
Beicun (computed)	64.3–64.10			23.2	15.2	22.7	23.6	14.6	0.7		

Table 5.7 Variation of the size distribution of suspended load in the variable backwater zones of Danjiangkou reservoir

Sections	Periods	Sediment transport quantity (10^6 tons)	Deposition percentage λ	Size distributions for suspended load $P_{4.l}$						
				< 0.01	0.01–0.025	0.025–0.05	0.05–0.10	0.10–0.25	0.25–0.50	0.50–1.0
35.1	70.8.17–10.8	12.05		19.8	20.9	24.8	26.6	6.1	1.8	
24..1 (measured)	70.8.17–10.8	5.89	0.51	34.5	28.7	22.9	13.5	0.4		
24..1 (computed)	70.8.17–10.8			35.6	29.3	22.7	11.2	1.1	0.1	

5.2.2 *Size Distribution of Sediment Supplement and Sorting of Bed Material in Scour Process*

5.2.2.1 Method 1 [3]

When channel scour occurs, the finer sediment is scoured away more than the coarser sediment, resulting in coarsening of bed material. As a result, the particle size of the sediment supplement becomes coarser and the supplement concentration gradually reduces. Sediment supplement is part of the sediment within the water body that is picked up from the bed material. This is a sorting process. Previous studies for bed material coarsening did not have a simple and clear theoretical characterization of this process. The analysis for bed material coarsening presented in this section is for the mass scour condition, and the analysis for general conditions is discussed in Chap. 9.

Based on reservoir scour data, the incipient flow velocity has little influence on sediment transport. At the backwater zone of a reservoir, the flow velocity is sufficiently large so that all suspended load can start moving. In the deep water zone, the sediment can also start suspension because the fine bed material is deposited in this area with lower dry density and loose contact among the particles, and because the bottom flow velocity is relatively large even if the depth-averaged flow velocity is small. Therefore, all the bed material is assumed to be able to start moving in the following scour analysis. The principles of this analysis, with certain corrections, can also be applied to the case with only part of the bed material able to start moving.

Because finer sediment is scoured more than coarser sediment, the instantaneous size distribution of sediment supplement is always finer than the bed material. As discussed later, the bed material gradation $P_{1.l}$ has a simple relation with the size distribution of suspended load $P_{4.l}$, which is described in Eq. (5.65), when the deposition percentage approaches to zero. Let assume this equation also applies to the relation between the instantaneous bed material gradation and the instantaneous size distribution of suspended load supplement in scour condition. This is because the results should be consistent between the case with the deposition rate approaches zero and the case with the scour rate approaches zero. Equation (5.65) can be applied to the instantaneous relation between size distribution of supplement concentration and gradation of bed material. Thus

$$\widetilde{P}_{4.l} = \left(\frac{\omega}{\omega_l}\right)^m P_{1.l} \tag{5.37}$$

where ω is the average settling velocity due to bed material sorting that is determined with Eq. (5.67). Strictly speaking, the size distribution of sediment supplement should be impacted by the flow condition in the scour process. But for a slow progressive scour, the impact of the flow condition is relatively stable, and is implicitly included in $\widetilde{P}_{4.l}$ and $P_{1.l}$, and thus Eq. (5.37) is a reasonable approximation. For an intensive retrogressive scour, bed material sorting rarely occurs, and Eq. (5.37) no longer applies. For this case, the size distribution for sediment supplement can be assumed to be the same as the bed material gradation. The following analysis is for the mass scour condition, or the condition with all size classes being scoured (no deposition for any size class). Let W_l denote the fractional sediment quantity for the l-th size class of sediment supplement, W denote the total quantity of sediment supplement, then

$$\Delta W_l = Q\widetilde{S}_l \Delta t = Q\widetilde{S}\,\Delta t \frac{\omega^m P_{1.l}}{\omega_l^m}$$
$$\Delta W = Q\widetilde{S}\,\Delta t$$
$$\frac{\Delta W_l}{\Delta W} = \frac{\omega^m P_{1.l}}{\omega_l^m} \tag{5.38}$$

where $\widetilde{S}_l$ is the concentration of the sediment supplement. Let V the total bed material quantity to be involved in the scour process, and V_l the quantity of the corresponding l-th size class, the bed material gradation after the scour is

$$P_{1.l} = \frac{V_l - W_l}{V - W} \tag{5.39}$$

Plugging Eq. (5.39) into Eq. (5.38), and letting $\Delta W \rightarrow 0$ yields

$$\frac{dW_l}{dW} = \left(\frac{\omega}{\omega_l}\right)^m \frac{V_l - W_l}{V - W} \tag{5.40}$$

Integrating over the interval [0, W], and introducing the median value of integral of the settling velocity ω_m, over the interval [V, $V - W$], one obtains

$$\frac{V_l - W_l}{V_l} = \left(\frac{V - W}{V}\right)^{\left(\frac{\omega_m}{\omega_l}\right)^m} \tag{5.41}$$

Considering W_l/W is the average size distribution of the eroded sediment $\widetilde{P}_{4.l}$, W/V the scour percentage λ^*, and V_l/V the initial bed material gradation $P_{1.l.0}$ before scour. Then

$$\frac{V_l - W_l}{V_l} = 1 - \frac{W_l}{W}\frac{W}{V}\frac{V}{V_l} = 1 - \frac{\widetilde{P}_{4.l}\lambda^*}{P_{1.l.0}}$$
$$\frac{V - W}{V} = 1 - \lambda^*$$

Plugging them into Eq. (5.41) gives

$$\widetilde{P}_{4.l} = P_{1.l.0}\frac{1 - (1 - \lambda^*)^{\left(\frac{\omega_m}{\omega_l}\right)^m}}{\lambda^*} \tag{5.42}$$

In engineering applications, λ^* can be considered as the ratio of the scour depth $\Delta\tilde{h}$ to the thickness of the bed layer h that is involved in the scour process and $h = (\Delta\tilde{h} + \Delta\tilde{h}_0)$, where $\Delta\tilde{h}_0$ is the thickness of the coarsening layer. Thus, the scour percentage $\lambda^* = \Delta\tilde{h}/(\Delta\tilde{h} + \Delta\tilde{h}_0)$. Here, $\Delta\tilde{h}$ and $\Delta\tilde{h}_0$ measured in *ton/m*^{2} are imaginary depths, and are defined as the mass per unit bed area to avoid conversion between the volume and the mass. Based on the field data analysis for the downstream reach of the Yellow River and the downstream reach of the Danjiankou Reservoir, $\Delta\tilde{h}_0 \approx 1(t/m^2)$. When all erodible sediment is scoured, the scour percentage $\lambda^* = 1$. Equation (5.42) describes how the average size distribution of the sediment supplement vary when the scour percentage varying between 0 and λ^*. As $\lambda^* \to 0$, $\widetilde{P}_{4.l} = \frac{\omega^m}{\omega_l^m} P_{1.l}$, which is consistent with Eq. (5.37); for $\lambda^* = 1$, all bed material is eroded and thus $\widetilde{P}_{4.l} = P_{1.l}$. This is as expected. When scour starts, the size distribution of the sediment supplement is the instantaneous size distribution that satisfy Eq. (5.37). For $\lambda^* = 1$, all bed material is eroded and thus the size distribution of the sediment supplement equals the average bed material gradation. It is also seen from Eq. (5.42) that the size distribution of the sediment supplement become coarser and coarser as λ^* increases from 0 to 1.

Equation (5.42) illustrates the sorting process of the eroded bed material in the scour process, i.e., more finer sediment being eroded than coarser sediment. For coarser sediment $\left(\frac{\omega_m}{\omega_l}\right) < 1$, $\frac{1-(1-\lambda^*)^{\left(\frac{\omega_m}{\omega_l}\right)^m}}{\lambda^*} < \frac{1-(1-\lambda^*)}{\lambda^*} = 1$, and thus $P_{4.l} < P_{1.l}$. For finer sediment $\left(\frac{\omega_m}{\omega_l}\right) > 1$, $\frac{1-(1-\lambda^*)^{\left(\frac{\omega_m}{\omega_l}\right)^m}}{\lambda^*} > 1$, and thus $P_{4.l} > P_{1.l}$. ω_m of Eq. (5.42) is determined with the relation $\sum_{l=1}^{n} P_{4.l} = 1$, or

$$1 = \sum_l P_{1.l.0} \frac{1 - (1 - \lambda^*)^{\left(\frac{\omega_m}{\omega_l}\right)^m}}{\lambda^*} \tag{5.43}$$

The procedure to determine ω_m is referred to that used for Eq. (5.34). The verification of Eq. (5.42) will be discussed later together with Eq. (5.54).

5.2.2.2 Method 2 [4]

This method also defines the fractional sediment concentration and carrying-sediment capacity for the mass scour condition as

$$S_l = P_{41} S = P_{41.0} S_0 + \widetilde{P}_{4.1} \Delta S \tag{5.44}$$

$$S_l^* = P_{4.l}^* S^* = P_{4.l.0}^* S_0^* + \widetilde{P}_{4.l}^* \Delta S^* \tag{5.45}$$

where $P_{4.l}^*$ is the size distribution of the carrying-sediment capacity corresponding to the bed material gradation, $\widetilde{P}_{4.l}$ is the size distribution of the sediment supplement, or the bed material that is picked up. It is obvious that $\widetilde{P}_{4.l} = \widetilde{P}_{4.l}^*$, i.e. the size distribution of the sediment supplement is essentially the size distribution of the carrying-sediment capacity for the eroded sediment, Plugging Eqs. (5.44) and (5.45) into Eq. (5.20) obtains

$$\begin{aligned} \frac{d(P_{4.l}S)}{dx} &= \frac{d\left(\widetilde{P}_{4.l}\Delta S\right)}{dx} = \frac{\alpha\omega_l}{q}\left(P^*_{4.l}\Delta S^* - \widetilde{P}_{4.l}\Delta S\right) \\ &= \frac{\alpha\omega_l}{q}\widetilde{P}_{4.l}\left(\Delta S^* - \Delta S\right) \end{aligned} \tag{5.46}$$

Summation for all size groups gives

$$\frac{d\Delta S}{dx} = \frac{\alpha\omega^*}{q}\left(\Delta S^* - \Delta S\right) \tag{5.47}$$

Dividing Eq. (5.46) by Eq. (5.47) yields

$$\begin{aligned} \frac{d\left(\widetilde{P}_{4.l}\Delta S\right)}{d(\Delta S)} &= \Delta S\frac{d\widetilde{P}_{4.l}}{d(\Delta S)} + \widetilde{P}_{4.l} = \frac{\omega_l}{\omega^*}\widetilde{P}_{4.l} \\ \frac{d\widetilde{P}_{4.l}}{\widetilde{P}_{4.l}} &= \left(\frac{\omega_l}{\omega^*} - 1\right)\frac{d(\Delta S)}{\Delta S} \end{aligned} \tag{5.48}$$

Integration with condition $\Delta S = \Delta S_m$, and $\widetilde{P}_{4.l} = P_{1.l.0}$ gives

$$\ln\frac{\widetilde{P}_{4.l}}{P_{1.l.0}} = \left(\frac{\omega_l}{\omega^*} - 1\right)\ln\frac{\Delta S}{\Delta S_m} \tag{5.49}$$

Considering

$$\frac{\Delta S}{\Delta S_m} = \lambda^* \tag{5.50}$$

Equation (5.49) becomes

$$\widetilde{P}_{4.l} = P_{1.l.0}\lambda^{*\left(\frac{\omega_l}{\omega_m} - 1\right)} \tag{5.51}$$

Similar to Method 2, ω_m is determined with

$$\sum_{l=1}^{n}\widetilde{P}_{4.l} = \sum_{l=1}^{n}P_{1.l.0}\lambda^{*\left(\frac{\omega_l}{\omega_m} - 1\right)} = 1 \tag{5.52}$$

For coarser sediment, $\frac{\omega_l}{\omega_m} - 1 > 0$, $\lambda^{*\left(\frac{\omega_l}{\omega_m}-1\right)} < 1$, and thus $\widetilde{P}_{4.l} < P_{1.l.0}$. For finer sediment, $\frac{\omega_l}{\omega_m} - 1 < 0$, $\lambda^{*\left(\frac{\omega_l}{\omega_m}-1\right)} > 1$, and $\widetilde{P}_{4.l} > P_{1.l.0}$. Equation (5.51) shows the particle size of the sediment supplement are generally finer than the bed material.

5.2.3 *Variation of the Size Distribution of Suspended Load in Mass Scour Condition*

5.2.3.1 Method 1 [3]

The size distribution variation of suspended load are complicated during the scour process. The size distribution of the sediment supplement depends on the bed material gradation, and the size distribution of the sediment from inflow at the entrance also varies. In mass scour condition, it is reasonable to neglect the size distribution variation of the sediment from inflow. This will simplify the calculation of the size distribution of suspended load for reservoir scour, which can then be determined by adding up the size distribution of the sediment inflow and that of the sediment supplement. During a scour process, the sediment concentration for the l-th size class at the exit section is

$$S_l = S_{l.0} + \widetilde{S}_l$$

or

$$P_{4.l}S = P_{4.l.0}S_0 + \widetilde{P}_{4.l}\widetilde{S}$$

Thus

$$P_{4.l} = \frac{S_0}{S}P_{4.l.0} + \widetilde{P}_{4.l}\frac{\widetilde{S}}{S}$$

Considering $\widetilde{S} = S - S_0$, and introducing the deposition percentage as defined in Eq. (5.32), the above equation can be written as

$$P_{4.l} = \frac{P_{4.l.0} - \lambda\widetilde{P}_{4.l}}{1-\lambda} \tag{5.53}$$

plugging Eq. (5.42) in which yields

$$P_{4.l} = \frac{1}{1-\lambda}\left\{P_{4.l.0} - \frac{\lambda}{\lambda^*}P_{1.l.0}\left[1 - (1-\lambda^*)^{\left(\frac{\omega_m}{\omega_l}\right)^m}\right]\right\} \tag{5.54}$$

and the scour percentage λ^* is

$$\lambda^* = \frac{\Delta \tilde{h}}{\Delta \tilde{h} + \Delta \tilde{h}_0} = \frac{Q(S - S_0)\Delta t}{Q(S - S_0)\Delta t + B_k L \Delta \tilde{h}_0} \tag{5.55}$$

where Q is the mean flow discharge during the scour process, Δt the duration of scour, B_k the average width of scour, $\Delta \tilde{h}_0$ the thickness of the bed material that is involved in the scour process but is not picked up, or the thickness of the bed armor layer. It is noted that if considering the bed material coarsening, the sediment concentration and the size distribution of both the sediment supplement and the suspended load at the exit section vary with time during a scour process. This is inconsistent with the assumption of steady non-equilibrium sediment transport. However, the bed material coarsening sometimes occurs quickly. Its impact to the sediment concentration is more significant than the impact of the bed deformation to the flow condition and to the sediment concentration, and thus shall be considered. It is also noted that Eqs. (5.54) and (5.55) can be solved.

Equation (5.54) is verified using the field data collected at the reach between Huayuankou and Gaocun and at the reach between Gaocun and Aishan, both downstream of the Sanmenxia Reservoir before the reconstruction. The results are shown in Fig. 5.11, 5.12, 5.13 and Fig. 5.14, respectively. The size distribution of the suspended load, $P_{4.l}$, at the exit sections (Gaocun and Aishan Stations, respectively) is calculated with Eq. (5.54) based on the size distribution $P_{4.l.0}$ at the entrance sections (Huayuankou and Gaocun Stations, respectively), the average bed gradation $P_{1.l.0}$ prior to scour, the deposition percentage λ, and the scour percentage λ^*. Figures 5.11 and 5.13 show the data that is averaged between 1961 and 1964, while Figs. 5.12 and 5.14 show the data for mean annual value of 1964. A good agreement between the computed results and the data is observed in these figures.

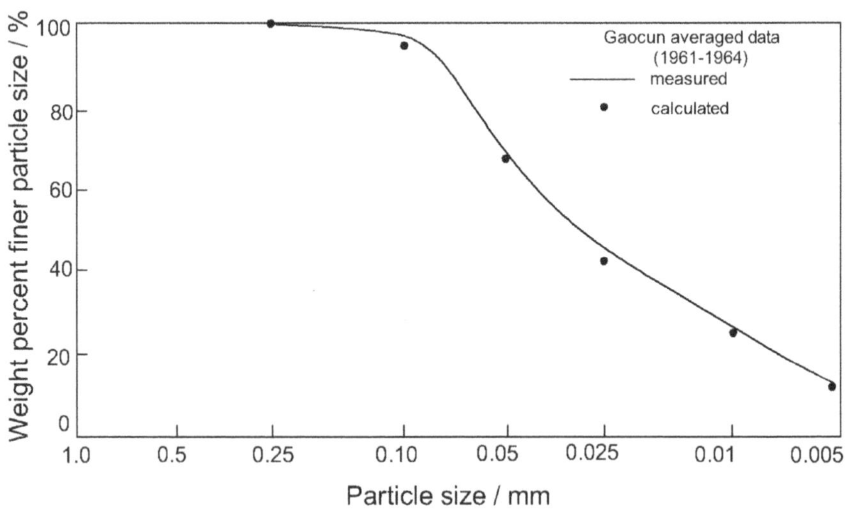

Fig. 5.11 Verification of Eq. (5.54) using 1961–1964 size distribution of suspended load at Gaocun Station

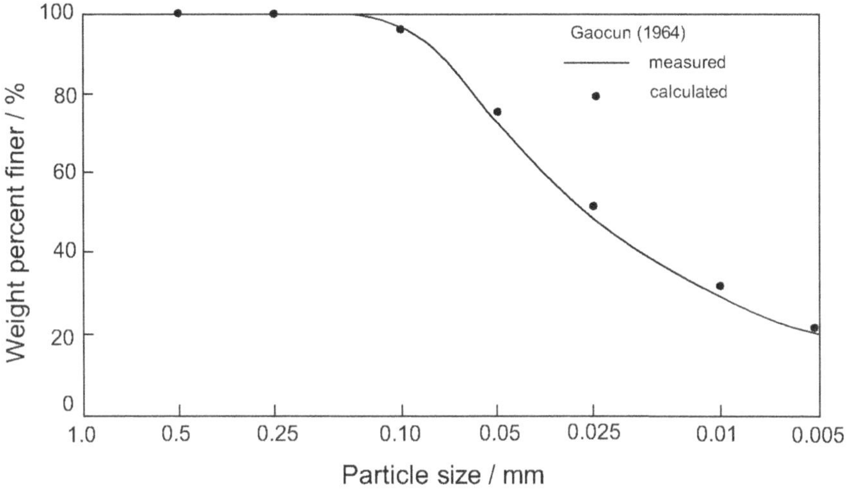

Fig. 5.12 Verification of Eq. (5.54) using 1964 size distribution of suspended load at Gaocun Station

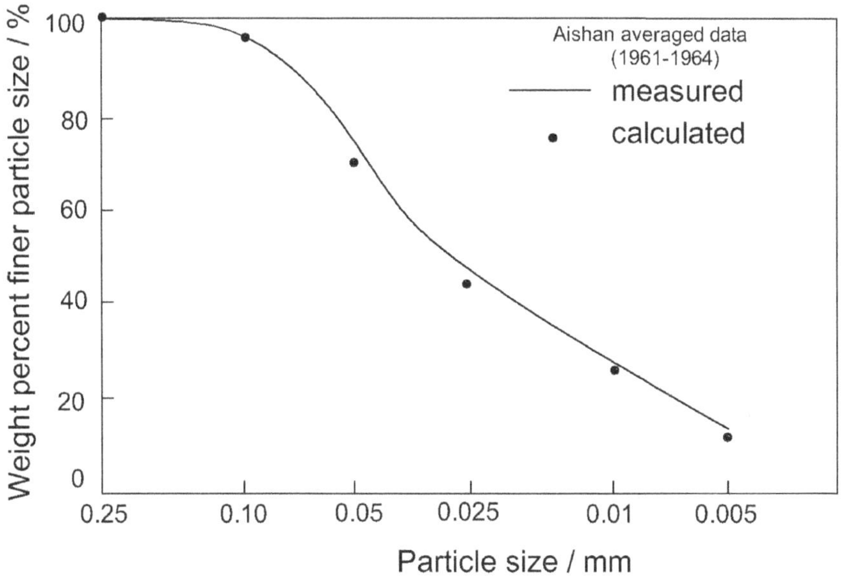

Fig. 5.13 Verification of Eq. (5.54) using 1961–1964 size distribution at Aishan Station

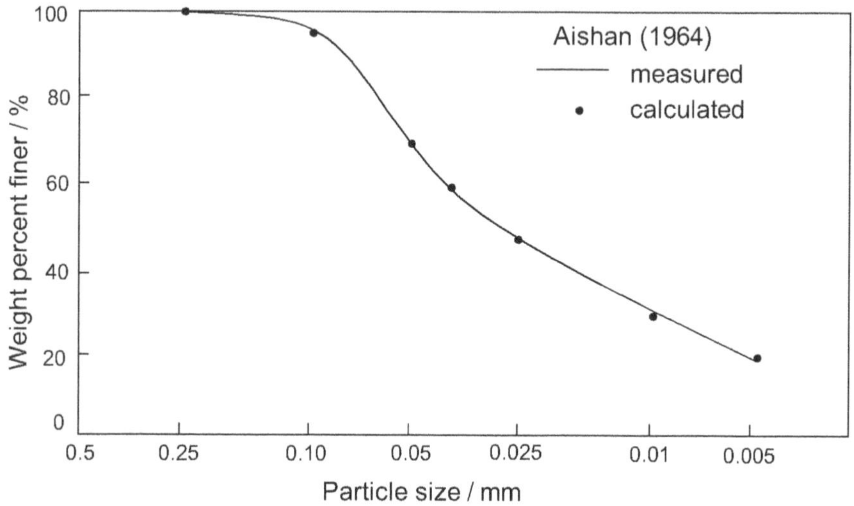

Fig. 5.14 Verification of Eq. (5.54) using 1964 size distribution of suspended load at Aishan Station

For a very small λ^*, or for an instantaneous condition, $\widetilde{P}_{4.l} = (\omega/\omega_l)^m P_{1.l}$, and Eq. (5.53) becomes

$$P_{4.l} = \frac{1}{1-\lambda}\left[P_{4.l.0} - \lambda\left(\frac{\omega}{\omega_l}\right)^m P_{1.l}\right] \tag{5.56}$$

where $P_{1.i}$ is the bed material gradation. In an instantaneous condition, the sediment concentration and the size distribution of the sediment supplement and of the suspended load at the exit section are independent of time.

5.2.3.2 Method 2 [4]

Equation (5.44) can be written as

$$P_{4.l} = \frac{S_0}{S} P_{4.l.0} + \frac{\Delta S}{S}\widetilde{P}_{4.l} \tag{5.57}$$

Summing up over l yields

$$S = S_0 + \Delta S \tag{5.58}$$

So Eq. (5.57) can be rewritten as

$$P_{4.l} = \frac{S_0}{S} P_{4.l.0} + \frac{S - S_0}{S}\widetilde{P}_{4.l} \tag{5.59}$$

Also, Eq. (5.32) can be rewritten as

$$\frac{S}{S_0} = 1 - \lambda \tag{5.60}$$

and thus

$$\frac{S - S_0}{S} = 1 - \frac{S_0}{S} = 1 - \frac{1}{1-\lambda} = -\frac{\lambda}{1-\lambda} \tag{5.61}$$

Plugging Eqs. (5.60) and (5.61) into Eq. (5.59), and considering Eq. (5.51), one obtains

$$P_{4.l} = \frac{P_{4.l.0}}{1-\lambda} - \frac{\lambda}{1-\lambda}\widetilde{P}_{4.l} = \frac{1}{1-\lambda}\left[P_{4.l.0} - \frac{\lambda}{\lambda^*}P_{1.l.0}\lambda^{*\left(\frac{\omega_l}{\omega_m}\right)}\right] \tag{5.62}$$

Equation (5.62) is verified using data collected in the Yellow River downstream of the Sanmenxia Reservoir during the clear water scour periods. The results are shown in Figs. 5.15, 5.16, 5.17 and 5.18, respectively. Figures 5.15 and 5.16 show the results for the exit sections of the reach between Huayuankou and Gaocun and of the reach between Gaocun and Aishan, respectively, with the data averaged between 1961 and 1964. Figures 5.17 and 5.18 shows the data of mean annual value for 1964. A good agreement between the computed results and the data can also be observed in these figures. This indicates that the difference between Method 1 and Method 2 is insignificant for the data collected in the downstream reach of the Yellow River. In fact, the difference between these two methods is insignificant for $\lambda^* < 0.662$, as discussed later in Chap. 7. While these two methods do not show noticeable difference in the computed results, Method 1 has more advantage in theoretical derivation.

5.3 Variation of Bed Material Gradation

Bed material not only impacts the size distribution of suspended load when scour occurs, but also impacts the dry density of sediment in deposition and scour process. Therefore, bed material gradation is an important topic for the study of non-equilibrium sediment transport.

5.3.1 Relation Between Bed Material Gradation and the Size Distribution of Suspended Load in Deposition Process

When sediment deposition occurs, the bed material gradation, or the size distribution of the deposited material, has a simple relation with the size distribution of suspended

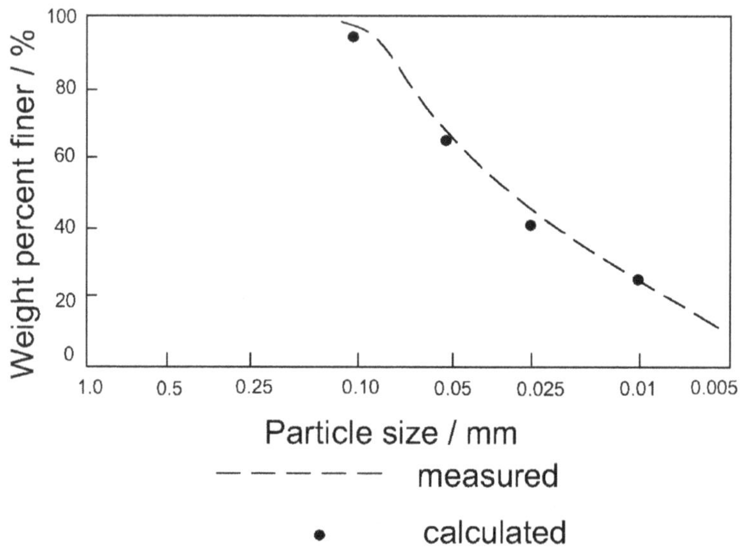

Fig. 5.15 Verification of Eq. (5.62) using 1961–1964 data at Gaocun Station

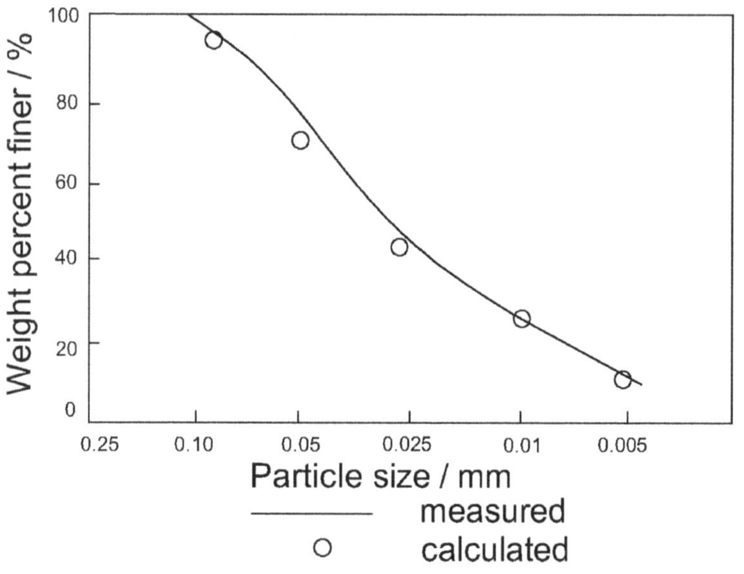

Fig. 5.16 Verification of Eq. (5.62) using 1961–1964 data at Aishan Station

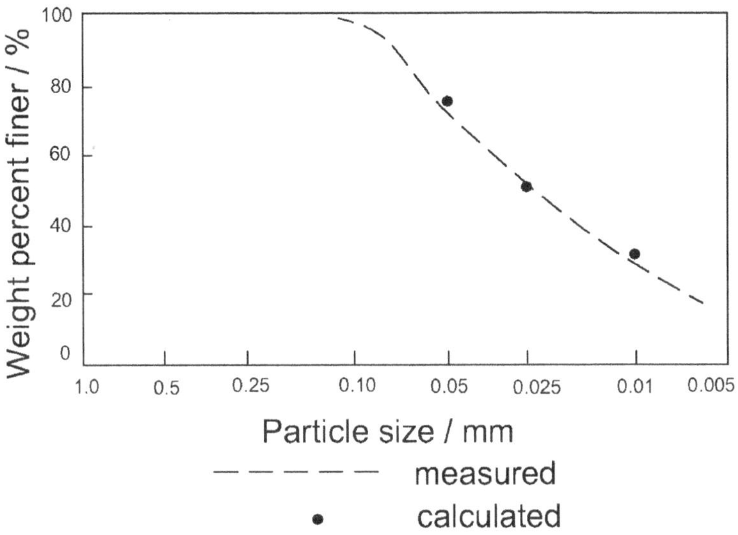

Fig. 5.17 Verification Eq. (5.62) using 1964 data at Gaocun Station

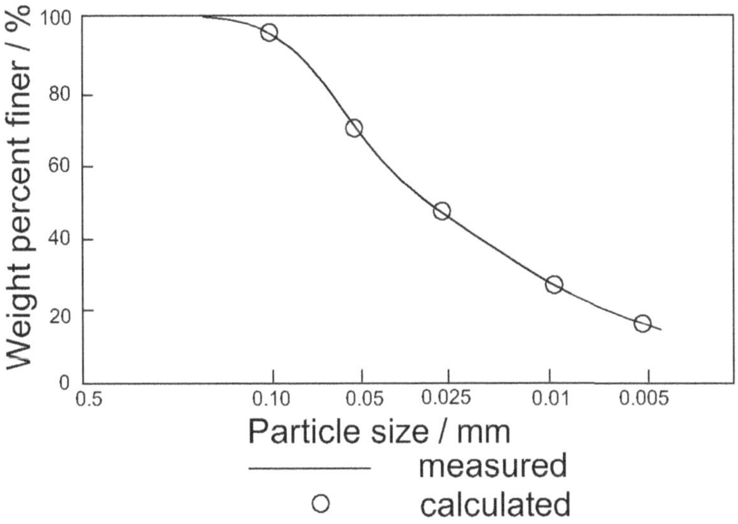

Fig. 5.18 Verification of Eq. (5.62) using 1964 data at Aishan Station

load. According to mass balance, the deposition amount of the l-th size class within a reach equals the change to the sediment transport rate for this size class between the entrance and the exit sections, or

$$V_l = QS_{l.0}\Delta t - QS_l\Delta t = QS_0P_{4.l.0}\Delta t - QSP_{4.l}\Delta t$$

Summation for all size classes yields

$$V = Q(S_o - S)\Delta t$$

Thus

$$P_{1.l} = \frac{V_l}{V} = \frac{S_{l.0}}{S_0 - S} - \frac{S_l}{S_0 - S} = \frac{S_0}{S_0 - S}\left(\frac{S_{l.0}}{S_0} - \frac{S_l}{S_0}\right)$$
$$= \frac{S_0}{S_0 - S}\left(\frac{P_{4.l.0}S_0}{S_0} - \frac{P_{4.l}S}{S_0}\right)$$

Introducing the deposition percentage λ, this equation can be rewritten as

$$P_{1.l} = \frac{P_{4.l.0} - (1-\lambda)P_{4.l}}{\lambda} \tag{5.63}$$

plugging Eq. (5.35) in which, the size distribution of the deposited material can be written as the following equation in terms of the size distribution of sediment inflow at the entrance

$$P_{1.l} = \frac{P_{4.l.0}}{\lambda}\left[1 - (1-\lambda)^{\left(\frac{\omega_l}{\omega_m}\right)^m}\right] \tag{5.64}$$

where ω_m is determined with Eq. (5.36). As shown in Eq. (5.64), the mean gradation of the deposited material $P_{1.l}$ within a reach depends on the size distribution of the suspended load $P_{4.l.0}$ at the entrance, the settling velocity ω_l for this size class, the effective settling velocity for deposition ω_m, and the deposition percentage λ.

Equations (5.63) and (5.64) can be used to determine the deposited material gradation averaged over a given reach, not the relation between bed material gradation and size distribution of suspended load at any sections,, which can be solved by setting λ approach to zero in Eq. (5.64). Equation (5.64) can be expressed in the following form of a power series expansion of λ

$$P_{1.l} = \frac{P_{4.l.0}}{\lambda}\left[1 - \left\{1 - \left(\frac{\omega_l}{\omega_m}\right)^m \lambda + \frac{1}{2!}\left(\frac{\omega_l}{\omega_m}\right)^m\left[\left(\frac{\omega_l}{\omega_m}\right)^m - 1\right]\lambda^2 \cdots\cdots\right\}\right]$$

Let $\lambda \to 0$, and considering $\omega_m \to \omega$, and $P_{4.l} \to P_{4.l.0}$, then

$$P_{1.l} = \frac{\omega_l^m P_{4.l}}{\omega^m} \tag{5.65}$$

where $\omega^m = \sum_{l=1}^{n} P_{4.l}\omega_l^m$ as discussed before. Equation (5.65) indicates that the gradation of the deposited material at a given section has a simple relation with the size distribution of the suspended load. Solving Eq. (5.65) for $P_{4.l}$, one obtains

$$P_{4.l} = \frac{\omega^m}{\omega_l^m} P_{1.l} \tag{5.66}$$

To simplify the calculation, the settling velocity ω^m can be expressed in terms of the gradation of the deposited material, or

$$\omega^m = \frac{1}{\sum_{l=1}^{n} \frac{P_{1.l}}{\omega_l^m}} \tag{5.67}$$

It is noted that ω has a clear physical meaning no mater it is expressed in terms of the size distribution of the suspended load, or in terms of the gradation of the deposited material. As shown in Eq. (5.65), for coarser sediment, $(\omega_l/\omega)^m > 1$, and thus $P_{1.l} > P_{4.l}$; for finer sediment, $(\omega_l/\omega)^m < 1$, and thus $P_{1.l} < P_{4.l}$. As a result, the bed material gradation is always coarser than the suspended load, which is consistent with the reality.

Due to the limitation of the measurement technology, it is difficult to accurately measure the deposited material gradation in the field. Thus, it is hard to strictly verify the formulas for the gradation of the deposited material. However, these formulas seem to be able to predict the overall gradation of the deposited material based on the verification with the filed data of Yanjiatai and Diudiuwan warping basins. Figure 5.19 shows the reach-averaged median particle size D_{50} of the deposited material, compared to the results computed with Eq. (5.64), for Yanjiatai and Diudiuwan warping basins. Figure 5.20 shows the D_{50} measured at various sections compared to the results computed with Eq. (5.65). The computed results overall show reasonable agreement with the data, while discrepancy can be observed. The formula for the deposited material gradation is derived based on the size distribution of suspended load according to mass balance. Therefore, if the formula for the size distribution of suspended load is verified, the formula for the deposited material gradation can also be implicitly verified.

It is noted that the computed gradation of the deposited material is the instantaneous bed material gradation for a given time. The total bed material gradation averaged over a period, which is sometimes required in the computation of reservoir sedimentation, can be determined with these instantaneous gradations.

It should be emphasized that Eq. (5.65) is valid only for $\lambda \to 0$. It is not valid for a general non-equilibrium sediment transport. For this case, Eq. (4.26) shall be used, or $P_{1.l} = \frac{S^* P_{4.l}^*}{S^*(l)} = \frac{\omega_l^m}{\omega^m} P_{4.l}^*$. Also considering $P_{4.l}^* = \widetilde{P}_{4.l}$, one can still obtain the results similar to those discussed before.

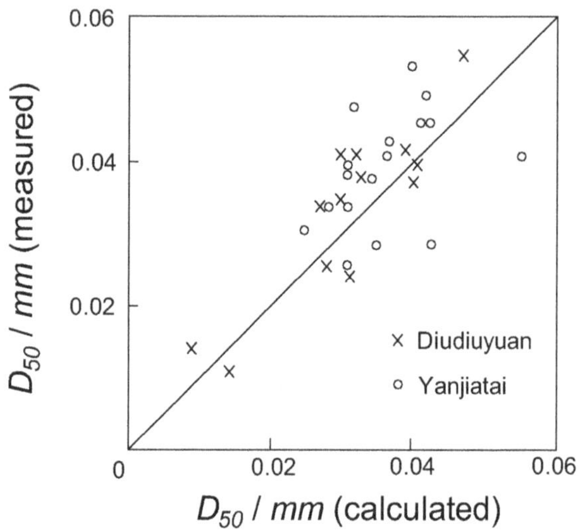

Fig. 5.19 Verification of averaged D_{50} computed with Eq. (5.64) in the reach at Diudiuyuan and Yanjiatai

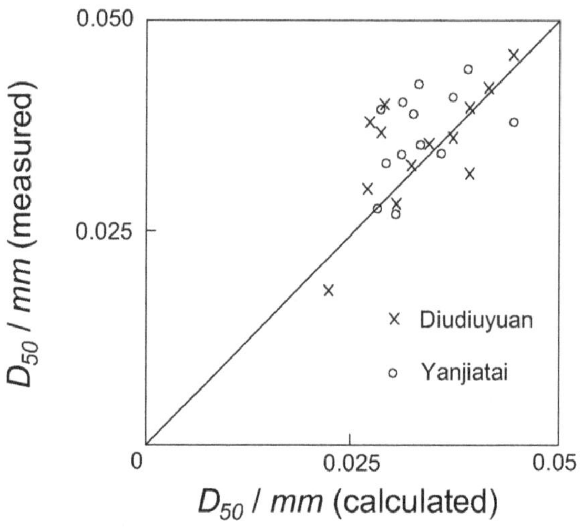

Fig. 5.20 Verification of averaged D_{50} computed with Eq. (5.65) on the sections at Diudiuyuan and Yanjiatai

5.3.2 *Bed Material Coarsening in Scour Process*

5.3.2.1 Method 1 [3]

As discussed before, the bed material becomes coarser and coarser during a scour process. The formula for the sediment supplement from the bed material to the suspended load was also derived. This formula can be slightly revised to obtain the formula for the variation of the bed material gradation. It is noted that $V - W$ is the total bed material amount that still remains on the bed after scour, $V_l - W_l$ is the fraction for the l-th size group. Thus, the bed material gradation can be written as

$$P_{1.l} = \frac{V_l - W_l}{V - W} = \frac{\frac{V_l}{V} - \frac{W_l}{W}\frac{W}{V}}{1 - \frac{W}{V}} = \frac{P_{1.l.0} - \widetilde{P}_{4.l}\lambda^*}{1 - \lambda^*} \tag{5.68}$$

with

$$\frac{W}{V} = \lambda^*$$
$$\frac{V_l}{V} = P_{1.l.0}$$
$$\frac{W_l}{W} = \widetilde{P}_{4.l}$$

Plugging Eq. (5.42) into Eq. (5.68) yields

$$P_{1.l} = \frac{1}{1-\lambda^*}\left\{P_{1.l.0} - \lambda^* \frac{\left[1 - (1-\lambda^*)^{\left(\frac{\omega_m}{\omega_l}\right)^m}\right]}{\lambda^*} P_{1.l.0}\right\} = \frac{P_{1.l.0}(1-\lambda^*)^{\left(\frac{\omega_m}{\omega_l}\right)^m}}{1-\lambda^*} \tag{5.69}$$

where $P_{1.l}$ is the bed material gradation, and ω_m can be determined with the following condition

$$1 = \frac{\sum P_{1.l.0}(1-\lambda^*)^{\left(\frac{\omega_m}{\omega_l}\right)^m}}{1-\lambda^*} \tag{5.70}$$

Similar to those for the size distribution of the sediment supplement, Eqs. (5.69) and (5.70) are also developed based on the assumption of mass erosion (all size groups are able to be start moving).

The comparison of the bed material gradation between the results computed with Eq. (5.69) and the measured data are shown in Figs. 5.21, 5.22, 5.23 and 5.24. The data used in these figures were collected in the lower reaches of the Yellow River between Huayuankou and Gaocun and between Gaocun and Aishan for the same periods as for the data shown in Figs. 5.11, 5.12, 5.13 and 5.14. As shown in these figures, the computed results generally show a reasonable agreement with the data. The percentages for the finer size groups are slightly under-estimated with Eq. (5.69). This is because the sheltering impact of the coarser sediment to the finer sediment, which results in an insufficient sorting for the finer sediment, is not considered in the derivation of Eq. (5.69).

Equation (5.69) indicates that the bed material gradation $P_{1.l}$ during a scour depends on the initial bed material gradation $P_{1.l.0}$, the effective settling velocity ω_m, the settling velocity of the l-th size class ω_l, and the scour percentage λ^*. For

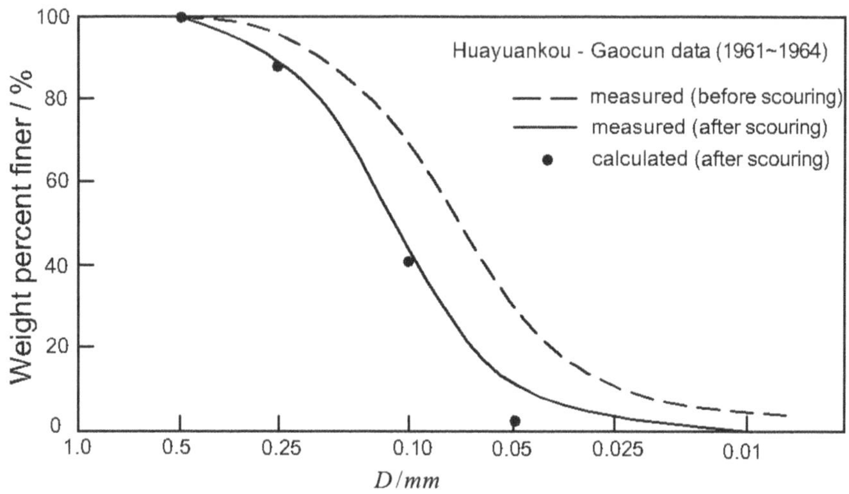

Fig. 5.21 Verification of Eq. (5.69) using 1961–1964 bed material gradation data of the reach between Huayuankou and Gaocun

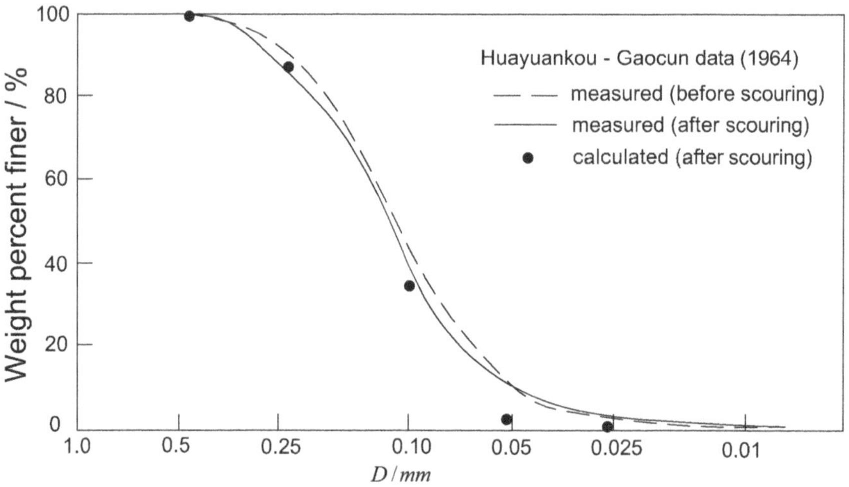

Fig. 5.22 Verification of Eq. (5.69) using 1964 bed material gradation data of the reach between Huayuankou and Gaocun

$(\omega_m/\omega_l)^m < 1$, or for coarser sediment, $P_{1.l} > P_{1.l.0}$; for $(\omega_m/\omega_l)^m > 1$, or for finer sediment, $P_{1.l} < P_{1.l.0}$. In other words, the bed material become coarser as the scour percentage λ^* increases, which is consistent with the reality.

Comparison between Eqs. (5.35) and (5.69) indicates that these two equations has a similar form. The difference is that ω_l is the numerator of the exponent in Eq. (5.35), while it is the denominator of the exponent in Eq. (5.69). The similarity

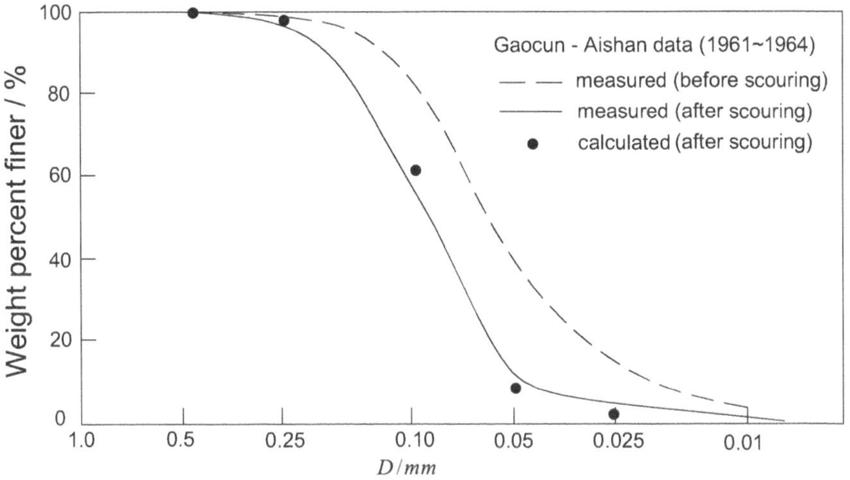

Fig. 5.23 Verification of Eq. (5.69) using 1961–1964 bed material gradation data of the reach between Gaocun and Aishan

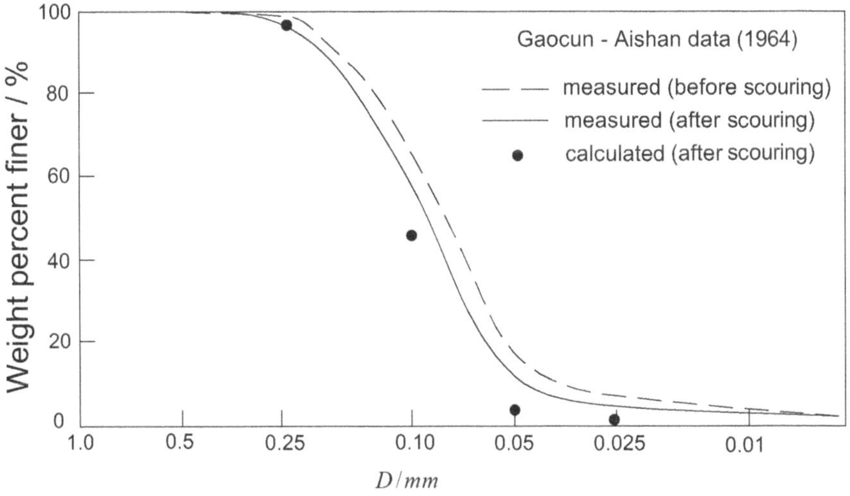

Fig. 5.24 Verification of Eq. (5.69) using 1964 bed material gradation data of the reach between Gaocun and Aishan

of the form for these two formulas represents the common features of the sorting processes for the suspended load and for the bed material. The difference is caused by the different characteristics of the thinning and the coarsening processes.

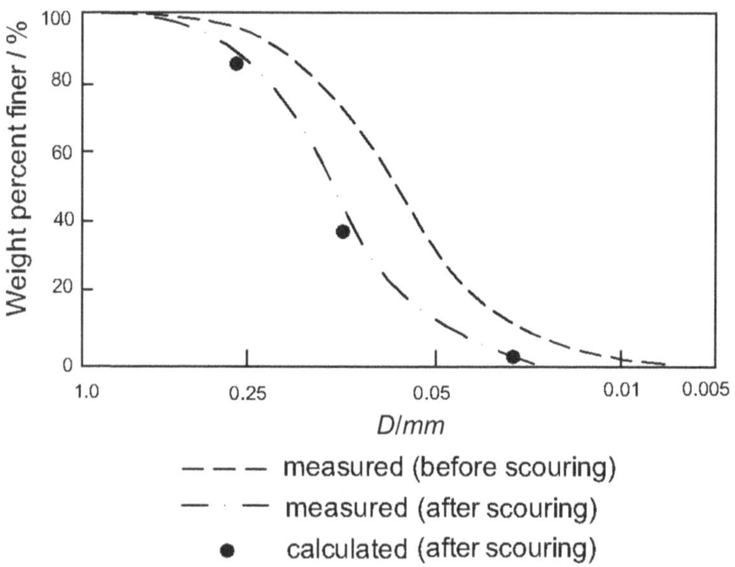

Fig. 5.25 Verification of Eq. (5.71) using 1961–1964 bed material gradation data of the reach between Huayuankou and Gaocun

5.3.2.2 Method 2 [4]

Plugging Eq. (5.51) into (5.68) yields

$$P_{1.l} = \frac{1}{1-\lambda^*}\left[P_{1.l.0} - \lambda^* P_{1.l.0}\lambda^{*\left(\frac{\omega_l}{\omega_m}-1\right)}\right] = P_{1.l.0}\frac{1-\lambda^{*\left(\frac{\omega_l}{\omega_m}\right)}}{1-\lambda^*} \tag{5.71}$$

The formula is verified using the same data set as used for Method 2, and the results are shown in Figs. 5.25, 5.26, 5.27, and 5.28. Figures 5.25 and 5.26 show the results for the reach between Huayuankou and Gaocun with the bed material gradation data averaged between 1961 and 1964 and with the data averaged in 1964, respectively. Figures 5.27 and 5.28 show the results for the reach between Gaocun and Aishan. A good agreement between the computed results and the data can be observed in these figures.

5.4 Solutions to Non-equilibrium Sediment Transport

Sections 5.1–5.3 investigate the longitudinal variations of sediment concentration, the size distribution of suspended load, and the bed material gradation for non-equilibrium sediment transport. These three topics of non-equilibrium sediment

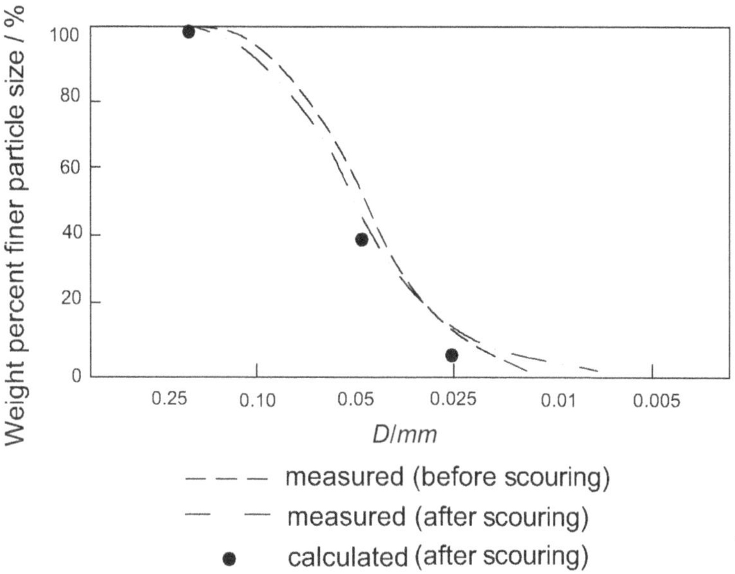

Fig. 5.26 Verification of Eq. (5.71) using 1964 bed material gradation data of the reach between Huayuankou and Gaocun

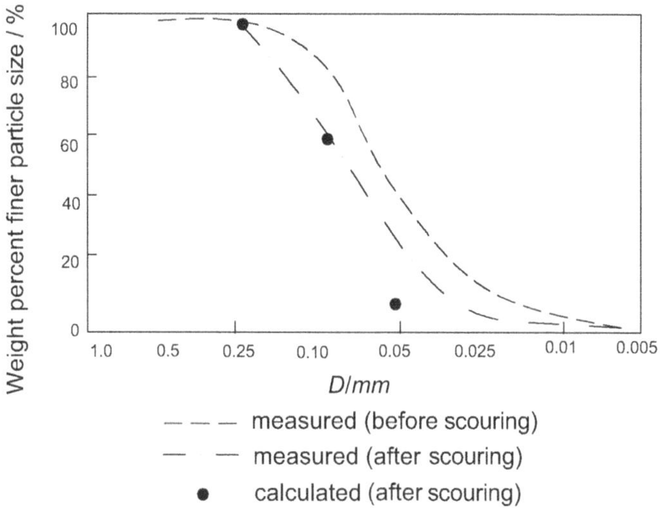

Fig. 5.27 Verification of Eq. (5.69) using 1961–1964 bed materia gradation data of the reach between Gaocun and Aishan

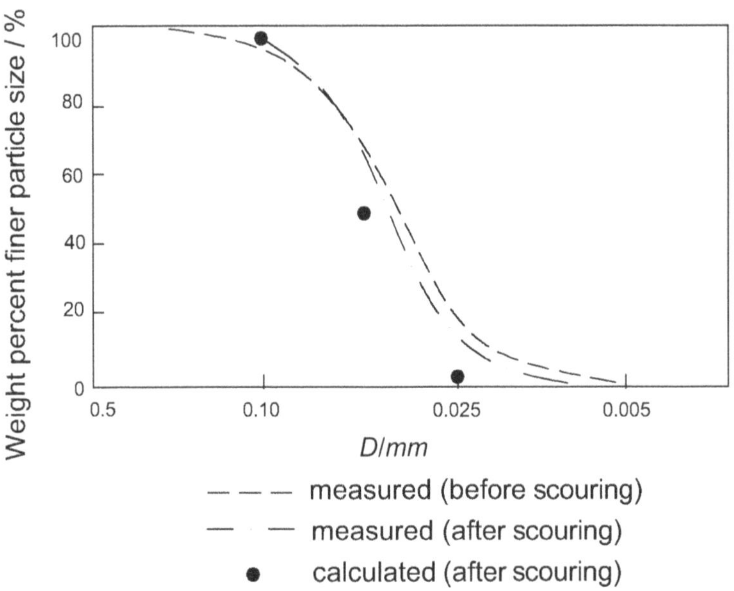

Fig. 5.28 Verification of Eq. (5.71) using 1964 bed material gradation data of the reach between Gaocun and Aishan

transport are correlated. From the mathematical aspect, the parameters for sediment transport cannot be solved individually from the corresponding equations. Instead, they should be determined by solving the system of equations for these parameters. The equation system and the applications are summarized as follows in different conditions:

(1) For deposition of non-uniform sediment, if the reach of interest is not very long, one should solve Eqs. (5.24a) and (5.35), or the following system of equations:

$$\begin{cases} S = S^* + \left(S_0 - S_0^*\right)e^{-\frac{\alpha L}{l'}} + \left(S_0^* - S\right)\frac{l'}{\alpha L}\left(1 - e^{-\frac{\alpha L}{l'}}\right) \\ P_{4.l} = P_{4.l.0}\frac{(1-\lambda)^{\left(\frac{\omega_l}{\omega_m}\right)^m}}{(1-\lambda)} \quad (l = 1, 2, \ldots, n) \end{cases} \tag{5.72}$$

The unknowns include the sediment concentration S and the size distribution of suspended load $P_{4.l}$ at the exit section. The settling distance $l = \omega/q$ with $\omega_0 = \sum_{l=1}^{n} P_{4.l.0}\omega_l$ and $\omega = \sum_{l=1}^{n} P_{4.l}\omega_l$. S_0^* and S^* can be determined using Eq. (5.27) with $P_{4.l.0}$ and $P_{4.l}$, respectively, or using Eq. (5.25) with simple equations for the carrying-sediment capacity of uniform sediment that has good agreement with field data of the subject reach.

(2) For deposition of non-.uniform sediment, if the reach of interest is long, or the longitudinal variations of the sediment concentration and the size distribution of suspended load are significant, one should solve Eqs. (5.24a) and (5.35), or the following system of equations:

$$\begin{cases} S = S^* + \left(S_0 - S_0^*\right) \sum\limits_{l=1}^{n} P_{4.l.0} e^{-\frac{\alpha L}{l_l}} + S_0^* \sum\limits_{l=1}^{n} P_{4.l.0} \frac{l_l}{\alpha L}\left(1 - e^{-\frac{\alpha L}{l_l}}\right) \\ -S^* \sum\limits_{l=1}^{n} P_{4.l} \frac{l_l}{\alpha L}\left(1 - e^{-\frac{\alpha L}{l_l}}\right) \\ P_{4.l} = P_{4.l.0} \frac{(1-\lambda)^{\left(\frac{\omega_l}{\omega_m}\right)^m}}{1-\lambda} \quad (l = 1, 2, \cdots, n) \end{cases} \tag{5.73}$$

The unknowns are S and $P_{4.l}$ at the exit section. The settling distance $l = \omega/q$. S_0^* and S^* can be determined using Eqs. (5.25a), (5.25b) ad (5.26) with $P_{4.l.0}$ and $P_{4.l}$, respectively.

If the entrance section is set at the saturation point, or at $S_0 - S_0^* = 0$, system of Eq. (5.73) can be significantly simplified. This simplification can be applied to the case with coarse sediment being discharged from a reservoir.

(3) For scour of non-uniform sediment, one should solve Eqs. (5.24a), and (5.54) or (5.62), or the following equation system

$$\begin{cases} S = S^* + \left(S_0 - S_0^*\right) \sum\limits_{l=1}^{n} P_{4.l.0} e^{-\frac{\alpha L}{l_l}} + S_0^* \sum\limits_{l=1}^{n} P_{4.l.0} \frac{l_l}{\alpha L}\left(1 - e^{-\frac{\alpha L}{l_l}}\right) \\ -S^* \sum\limits_{l=1}^{n} P_{4.l} \frac{l_l}{\alpha L}\left(1 - e^{-\frac{\alpha L}{l_l}}\right) \\ P_{4.l} = \frac{1}{1-\lambda}\left\{P_{4.l.0} - \frac{\lambda}{\lambda^*} P_{1.l.0}\left[1 - (1-\lambda^*)^{\left(\frac{\omega_m}{\omega_l}\right)^m}\right]\right\} \quad or \\ \widetilde{P}_{4.l} = P_{1.l.0} \frac{1-(1-\lambda^*)^{\left(\frac{\omega_m}{\omega_l}\right)^m}}{\lambda^*} \quad (l = 1, 2, \cdots, n) \end{cases} \tag{5.74}$$

where ω_m is solved with Eq. (5.43) or (5.52).

For an intensive scour (for example, retrogressive scour) without bed material sorting, one should solve Eq. (5.25) and Eq. (5.62) with $\lambda^* = 1$, or the following system of equations

$$\begin{cases} S = S^* + \left(S_0 - S_0^*\right) \sum\limits_{l=1}^{n} P_{4.l.0} e^{-\frac{\alpha L}{l'_l}} + S_0^* \sum\limits_{l=1}^{n} P_{4.l.0} \frac{l'_l}{\alpha L}\left(1 - e^{-\frac{\alpha L}{l'_l}}\right) \\ -S^* \sum\limits_{l=1}^{n} P_{4.l} \frac{l'_l}{\alpha L}\left(1 - e^{-\frac{\alpha L}{l'_l}}\right) \\ P_{4.l} = \frac{1}{1-\lambda}(P_{4.l.0} - \lambda P_{1.l.0}) \quad (l = 1, 2, \cdots, n) \end{cases} \tag{5.75}$$

Iterations are required to solve system of Eqs. (5.72)–(5.75). The iteration procedure is as follows: assume a value of S; solve for λ, λ^*, $P_{4.l}$, ω ω_m, S^* and S; if the computed S equals the assumed S, the solution of S is found; otherwise assume another value of S, repeat the procedure until the computed S equals the assumed S.

(4) After the sediment concentration and the size distribution of suspended load are determined for the exit section, the bed material gradation can be determined with corresponding formulas as discussed in Sect. 5.3.

In Sect. 5.1, the formulas for the longitudinal variation of sediment concentration are verified with field data for known size distribution of the suspended load at the exit section. In Sect. 5.2, the formulas for the size distribution of suspended load are verified with field data for known sediment concentration at the exit section. If both the sediment concentration and the size distribution of the suspended load are unknown for the exit section, how well do the results computed with the above systems of equations agree with the field data? It is our opinion that the error in the sediment concentration computed with the above systems of equations shall be smaller than that computed with formulas in Sect. 5.1. However, it is unclear for the error in the size distribution of suspended load. When solving the above systems of equations, the sediment concentration can be adjusted during the iterations. For example, if the assumed sediment concentration is smaller than the computed result, then the computed ω is smaller, which will result in an increased sediment concentration. The sediment concentration computed with the formulas in Sect. 5.1 with the measured size distribution of suspended load, and that computed with system of Eqs. (5.72) with the size distribution as unknown, are compared with the data collected in Wuotousi desilting canal. The results are listed in Table 5.8. As shown in this table, the results computed with the system of equations show a better agreement with the data than those computed with formulas in Sect. 5.1 with the measured size distribution.

In order to investigate the accuracy of the above system of equations for the long-term reservoir sedimentation, the sediment concentration and the size distribution of suspended load computed with system of Eqs. (5.73) are compared with the field data collected at Danjiangkou Reservoir during various deposition periods in 1970. The data was collected for 53 days. The average sediment concentration at the exit Section is computed with system of Eq. (5.73) for eight periods. The comparison

Table 5.8 Comparison of sediment concentration between computed results and the data measured in Wotousi desilting canal

Section No	Measured S (kg/m^3)	Computed results			
		Based on measured gradation		Based on computed gradation	
		S^* (kg/m^3)	S (kg/m^3)	S^* (kg/m^3)	S (kg/m^3)
1	33.9				
2	30.8	6.27	25.2	16.5	26.3
3	27.8	7.80	22.2	24.7	23.7
4	23.5	5.61	19.5	15.4	20.3
5	19.2	6.18	17.5	10.2	18.3
6	17.6	5.73	16.7	8.73	17.2
7	16.7	6.51	16.1	8.22	16.3
8	14.8	10.0	15.4	11.7	15.7
9	14.3	43.5	16.2	26.3	16.5

Table 5.9 Comparison of Sediment Concentration under Non-equilibrium Condition between Computed Results and the Data Measured in Danjiangkou Reservoir in 1970

Time segment	Days	Sediment concentration at entrance Sect. (35.1) (kg/m^3)	Sediment concentration at exit Sect. (24.1) (kg/m^3)	
			Measured	Computed
1	16	0.066	0.028	0.025
2	1	3.52	0.034*	0.824
3	19	0.469	0.048	0.133
4	3	1.19	0.297	0.490
5	1	4.56	2.29	1.98
6	3	2.83	1.59	1.72
7	4	0.971	0.628	0.521
8	6	0.112	0.014	0.051
Total	53	1.54	0.759	0.794

Note * Peak sediment discharge at exit not measured

of the computed results and the field data is listed in Table 5.9. Except for one period during which the peak sediment discharge was not measured, which resulted in a lower average sediment concentration, the computed results have a reasonable agreement with the data. This is particularly true for the sediment concentration averaged over the 53 days. The computed size distribution of suspended load at the exit Section averaged over the 53 days of data collection, is listed in Table 5.10. As shown in this table, the computed size distribution also shows a reasonable agreement with the data.

5.5 Summary

This chapter investigates three topics of 1-D non-equilibrium sediment transport in steady non-uniform flow. The longitudinal variations of the sediment concentration, the size distribution of suspended load, and the bed material gradation are analyzed, and corresponding governing equations are derived, with certain simplified conditions. Simplifications have been kept as less as possible, and non-uniform flow condition and non-uniform sediment are included in our analysis. The formulas are verified with the filed data. Thus, the results can reasonably represent the reality, particularly for the case with deposition.

The main content of this chapter was accomplished in 1972 when Han worked at Yangtze River Scientific Research Institute of Yangtze Water Resources Commission [12]. The first draft was published in 1973 [3], and the results were later published in *Chinese Science Bulletin* [4] and presented on *the 1st International Symposium on River Sedimentation* [5]. Most parts of this chapter are similar these previous

Table 5.10 Comparison of the size distribution of suspended load under non-equilibrium condition between computed results and the data measured in Danjiangkou reservoir in 1970

Section No.	Days	Sediment Concentration (kg/m^3)	λ	Percentage					
				< 0.01	0.01–0.025	0.025–0.05	0.05–0.10	0.10–0.25	0.25–0.50
35.1	53	1.54		19.8	20.9	24.8	26.6	6.1	1.8
24.1 (Measured)	53	0.759		34.5	28.7	22.9	13.5	0.4	
24.1 (Computed)	53	0.794	0.49	34.5	29.0	23.1	12.3	1.1	

publications. However, a second method is added in this Chapter for scour situation with additional illustration of the saturation recovery coefficient.

References

1. China Institute of Water Resources and Hydropower Research (IWHR, 1959) Study on reservoir sedimentation, Water Resource and Electric Power Press (in Chinese)
2. Dou GR (1963) Suspended sediment movement by tidal currents and bed deformation calculations. J Hydraul Eng 1964(4) (in Chinese)
3. Han QW (1973) Preliminary studies of the non-equilibrium transport of sediment in reservoir. In: Report on sedimentation in reservoir, task group on yellow river sedimentation research, pp 145–168 (in Chinese)
4. Han QW (1979) A Study of the non-equilibrium transportation of non-uniform suspended load. Chin Sci Bull 24(17):804–808 (in Chinese)
5. Han QW (1980) A study on the non-equilibrium transportation of suspended load. In: Proceedings of the international symposium on river sedimentation, vol 2. Guanghua Press, pp 793–801 (in Chinese)
6. Han QW (2003) Reservoir sedimentation. Science Press, Beijing (in Chinese)
7. Han QW, He MM (1987) Mathematical modeling of reservoir sedimentation and erosion. J Sediment Res 1987(3):14–29 (in Chinese)
8. Han QW, Huang YL (1974) Computation methods of sedimentation and erosion in the reservoir and utilization of digital computer. Selected papers on Yangtze water conservancy and hydroelectric power research, pp 45–85 (in Chinese)
9. Karauschev AV (1960) Problems of the dynamics of natural river flows. Hydrometeorological Publishing House
10. Mikhayev NV, Yunevich DL (1959) River channel regulation for land reclamation purpose. Agricultural Publishing House
11. Tsvezizova NV (1965) Calculation method of suspended load settling in turbulent flow. In: Proceedings of river channel evolution. Institute of Water Resources and Hydroelectric Research. SCIENCE Press, Bejing, China
12. Zhang ZT (1988) Study on basic principles of river sediment movement. In: Proceedings of development of river sediment research by Institute of Yangtze water conservancy research, vol 6. Institute of Yangtze Water Conservancy Research Press, Wuhan, pp 23–47

Chapter 6
Saturation Recovery Coefficient

6.1 Governing Equation and Characteristics of Saturation Recovery Coefficient

The sediment saturation recovery coefficient is important for the study on non-equilibrium sediment transport and for the deposition or scour computation in the mathematical models. This chapter presents the in-depth analysis of the saturation recovery coefficient, that has been conducted by the author of this book.

6.1.1 Governing Equation for Saturation Recovery Coefficient

It can be derived from Eq. (3.62) the longitudinal variation of the sediment transport rate per unit width is equal to the bottom boundary value of the diffusion equation,

$$\frac{d(qS_l)}{dx} = -\varepsilon_y \frac{\partial S_l(y)}{\partial y}\Big|_{y=0} + \omega S_{b.l}$$

where S_l is the depth-averaged fractional sediment concentration for the particle size of D_l, $S_l(y)$ and $S_{b.l}$ are the sediment concentrations at the height of y and at the bottom, respectively.

Introducing the sediment transition intensities $\overline{\lambda}_{1.4.l}$ and $\overline{\lambda}_{4.1.l}$, as defined in Eq. (2.99), yields

Q. Han and M. He, *Non-equilibrium Transport of Non-uniform Suspended Load*, Hydroscience and Engineering, https://doi.org/10.1007/978-981-97-1123-9_6

$$
\begin{aligned}
q\frac{dS_l}{dx} &= -\varepsilon_y \frac{\partial S_l(y)}{\partial y}\Big|_{y=0} + \omega S_{b.l} \\
&= (\overline{\lambda}_{1.4.l} - \overline{\lambda}_{4.1.l})\frac{\pi}{6}D_l^3\gamma_s \\
&= -(1-\varepsilon_{0.l})(1-\varepsilon_{4.l})\frac{q}{L_{4.l}}P_{4.l}S - \frac{4m_0}{\pi D^2}\frac{P_{1.l}\beta_l R_{1.l}}{t_{0.4.l}}\frac{\pi}{6}D_l^3\gamma_s \\
&= \frac{2}{3}m_0\gamma_s D_l \frac{P_{1.l}R_{1.l}\beta_l}{t_{0.4.l}} - (1-\varepsilon_{0.l})(1-\varepsilon_{4.l})\frac{qP_{4.l}S}{L_{4.l}}
\end{aligned}
$$

where $P_{1.l}$ is the bed material gradation. Letting $\frac{dS_l}{dx} = 0$, or for strong equilibrium sediment transport, $S = S^*$, $P_{4.l} = P_{4.l}^*$, and $L_{4.l} = L_{4.l}^*$, and plugging Eq. (2.49) for $R_{1.l}$ (probability for sediment to be at rest on the bed surface), one obtains

$$
S_l^* = \frac{2}{3}m_0\gamma_s\Omega_l P_{1.l}\frac{D_l L_{4.l}^*}{T_{0.4.l}q} = P_{1.l}S^*(l) \tag{6.1}
$$

where

$$
S^*(l) = \frac{2}{3}m_0\gamma_s\Omega_l\frac{D_l L_{4.l}^*}{T_{0.4.l}q} \tag{6.2}
$$

is the transport capacity for carrying uniform sediment with the particle size of D_l, and

$$
\Omega_l = \frac{\beta_l}{1-(1-\varepsilon_{1.l})(1-\beta_l)+(1-\varepsilon_{0.l})(1-\varepsilon_{4.l})} \tag{6.3}
$$

is a composite probability. It is noted that $L_{4.l}^*$ is the single-step distance for strong-equilibrium sediment transport, while $L_{4.l}$ the single-step distance for the case without strong equilibrium transport, which depends on the vertical distribution of sediment concentration and the non-equilibrium transport status of sediment transport. It should also be emphasized that $S^*(l)$ is different from S_l^*, $S(l)$ is different from S_l, while the following simplified notion is used for the other parameter

$$
X_l = X(l)
$$

Considering Eq. (4.4), or

$$
P_{4.l}^* S^* = P_{1.l}S^*(l)
$$

Plugging Eq. (3.89) into Eq. (6.2) one obtains

$$
\begin{aligned}
\frac{dS_l}{dx} &= (1-\varepsilon_{0.l})(1-\varepsilon_{4.l})\left(\frac{P_{4.l}^*S^*}{L_{4.l}^*} - \frac{P_{4.l}S}{L_{4.l}}\right) \\
&= \frac{(1-\varepsilon_{0.l})(1-\varepsilon_{4.l})}{\frac{L_{4.l}\omega_l}{q}}\frac{\omega_l}{q}\left(P_{4.l}S^*\frac{L_{4.l}}{L_{4.l}^*} - P_{4.l}S\right) \\
&= \frac{\omega_l}{q}[\alpha_l^* P_{4.l}S^* - \tilde{\alpha}_l P_{4.l}S] \\
&= \frac{\omega_l}{q}\left[\alpha_l^* P_{1.l}^* S^*(l) - \tilde{\alpha}_l P_{4.l}S\right]
\end{aligned}
\tag{6.4}
$$

where

$$
\alpha_l^* = \frac{(1-\varepsilon_{0.l})(1-\varepsilon_{4.l})q}{L_{4.l}^*\omega_l} \tag{6.5}
$$

is the saturation recovery coefficient of carrying-sediment capacity, and

$$
\tilde{\alpha}_l = \frac{(1-\varepsilon_{0.l})(1-\varepsilon_{4.l})q}{L_{4.l}\omega_l} \tag{6.6}
$$

is the saturation recovery coefficient of sediment concentration.

As discussed before, $L_{4.l}^*$ is the single-step suspension distance for strong equilibrium sediment transport. Thus, the corresponding suspension height should be determined using the distribution of sediment concentration under the equilibrium condition. For example, the suspension height can be determined based on the following equation of concentration

$$
S_l(y) = S_b(l)\exp\left(-\frac{\omega y}{\varepsilon_s}\right) = S_b(l)\exp\left(-\frac{6\omega y}{\kappa u_* h}\right) \tag{6.7}
$$

after which $L_{4.l}^*$ can be determined. This will be discussed later.

6.1.2 Properties of Saturation Recovery Coefficient

The distribution of sediment concentration is usually described in the form of numerical results. The accuracy of the boundary condition is also limited. Therefore, it is difficult to quantitatively illustrate the characteristics of the saturation recovery coefficients $\tilde{\alpha}_l$ and α_l^*. With qualitatively theoretical analysis and comparison, the properties of these two coefficients are investigated as follows [2].

Property 1

The saturation recovery coefficients α_l^* and $\tilde{\alpha}_l$, which are defined in Eqs. (6.5) and (6.6), respectively, can be combined into one coefficient α_l as described in the following equation under two conditions:

$$\frac{dS_l}{dx} = -\frac{\alpha_l \omega_l}{q}(P_{4.l}S - P_{4.l}^* S^*) = -\frac{\alpha_l \omega_l}{q}(S_l - S_l^*) \tag{6.8}$$

These two conditions include: (1) for strong equilibrium sediment transport or if the impact of the distribution of sediment concentration on $L_{4.l}$ can be neglected, $L_{4.l} = L_{4.l}^*$, and $\tilde{\alpha}_l = \tilde{\alpha}_L^* = \alpha_l$; ; and/or (2) under a general condition

$$\alpha_l = \tilde{\alpha}_l \frac{S_l}{S_l - S_l^*} - \alpha_l^* \frac{S_l^*}{S_l - S_l^*} \tag{6.9}$$

Or

$$\alpha_l(S_l - S_l^*) = \tilde{\alpha}_l S_l - \alpha_l^* S_l^*$$

where α_l is the composite saturation recovery coefficient. Also, if the difference between the distribution of sediment concentration and the equilibrium profile is negligible, $L_{4.l} \approx L_{4.l}^*$, Eq. (6.8) is also approximately satisfied.

Property 2

$\tilde{\alpha}_l < \alpha_l^*$ for sediment deposition, and $\tilde{\alpha}_l > \alpha_l^*$ for scour.

When sediment deposition occurs, the vertical distribution of sediment concentration is more uniform than the equilibrium profile, the single-step suspension distance $L_{4.l} > L_{4.l}^*$, and thus $\tilde{\alpha}_l < \alpha_l^*$ based on Eqs. (6.5) and (6.6). When scour occurs, the vertical distribution of sediment concentration is less uniform than the equilibrium profile, with sediment concentrating near the bottom, $L_{4.l} < L_{4.l}^*$, and thus $\tilde{\alpha}_l > \alpha_l^*$.

Property 3

For α_l as defined in Eq. (6.9), it is always true that $\alpha_l < \tilde{\alpha}_l$ and $\alpha_l < \alpha_l^*$.

For deposition, $S_l > S_l^*$, $\tilde{\alpha}_l < \alpha_l^*$ based on Property 2. Thus,

$$\alpha_l = \tilde{\alpha}_l \frac{S_l}{S_l - S_l^*} - \alpha_l^* \frac{S_l^*}{S_l - S_l^*} < \alpha_l^* \frac{S_l}{S_l - S_l^*} - \alpha_l^* \frac{S_l^*}{S_l - S_l^*} = \alpha_l^* \tag{6.10}$$

and

$$\alpha_l = \tilde{\alpha}_l \frac{S_l}{S_l - S_l^*} - \alpha_l^* \frac{S_l^*}{S_l - S_l^*} < \tilde{\alpha}_l \frac{S_l}{S_l - S_l^*} - \tilde{\alpha}_l \frac{S_l^*}{S_l - S_l^*} = \tilde{\alpha}_l \tag{6.11}$$

For scour, $S_l < S_l^*$, and $\tilde{\alpha}_l > \tilde{\alpha}_l^*$ based on Property 2. Thus,

$$\alpha_l = \alpha_l^* \frac{S_l^*}{S_l^* - S_l} - \tilde{\alpha}_l \frac{S_l}{S_l^* - S_l} < \alpha_l^* \frac{S_l^*}{S_l^* - S_l} - \alpha_l^* \frac{S_l}{S_l^* - S_l} = \alpha_l^* \tag{6.12}$$

and

$$\alpha_l = \alpha_L^* \frac{S_l^*}{S_l^* - S_l} - \tilde{\alpha}_l \frac{S_l}{S_l^* - S_l} < \tilde{\alpha}_l \frac{S_l^*}{S_l^* - S_l} - \tilde{\alpha}_l \frac{S_l}{S_l^* - S_l} = \tilde{\alpha}_l \tag{6.13}$$

As a result of combining two saturation recovery coefficients into one composite coefficient, the value of the composite coefficient reduces. Even if $\tilde{\alpha}_l$ and α_l^* are both greater than 1, α_l may still be less than 1. Let consider an example with α_l^*=2, and $S_l = 3\,\text{kg/m}^3$. For deposition, assuming $\tilde{\alpha}_l = 1.5$ and $S_l^* = 2.1429$, then

$$\alpha_l = \frac{3 \times 1.5 - 2 \times 2.1429}{3 - 2.1429} = 0.25$$

For Scour, assuming $\tilde{\alpha}_l = 2.5$ and $S_l^* = 4.5$, then

$$\alpha_l = \frac{2.5 \times 3 - 2 \times 4.5}{3 - 4.5} = 1.0$$

It can be seen that $\alpha_l < \tilde{\alpha}_l$ and $\alpha_l < \tilde{\alpha}_l^*$. Based on field data, α_l is far less than 1 for most cases [6, 7], and can be as small as 0.01 [9]. This is consistent with Property 3.

Property 4
Assuming that the probability for the sediment settling on the bed to change to the bed material is 1, then $(1 - \varepsilon_{0.l})(1 - \varepsilon_{4.l}) = 1$. It can be derived from Eqs. (6.5) and (6.6) that

$$\tilde{\alpha}_l = \frac{q}{\omega_l L_{4.l}} = \frac{1}{\frac{\omega_l L_{4.l}}{q}} > 1 \tag{6.14}$$

$$\alpha_l^* = \frac{q}{\omega_l L_{4.l}^*} = \frac{1}{\frac{\omega_l L_{4.l}^*}{q}} > 1 \tag{6.15}$$

Theoretical analysis indicates that the relative single-step distance $\omega_l L_{4.l}^*/q$ is always less than 1 [6, 7], as listed in Table 6.1. Thus, from (6.15) $\alpha_l^* > 1$. The difference of the distribution of sediment concentration between the non-equilibrium and the equilibrium profile is limited, therefore, $\omega_l L_{4.l}/q$ is also much less than 1, and $\tilde{\alpha}_l$ is greater than 1. It is noted that the relative suspension height η_l^* in Table 6.1 is computed with Eq. (4.21).

As illustrated in Property 4, assuming that the event for the sediment settling on the bed to change into the bed material is deterministic rather than with a certain probability, then $\tilde{\alpha}_l$ and α_l^* are both greater than 1 [6, 7]. In fact, in author's earliest derivation of the 1-D governing equation for non-equilibrium sediment transport based on the 2-D diffusion equation, the change of the sediment settling on the bed to the bed material was not considered as a random process, and the concept of exchange intensity was not applied either. As shown in Chap. 5, the governing equation for non-equilibrium transport of uniform sediment is

Table 6.1 Relative single-step suspension distance

$\frac{\omega_l}{u_*}$	Relative suspension height η_l^*	Relative discharge $\frac{q(\eta_l^*)}{q}$	$\frac{\omega_l L_{4,l}^*}{q}$
0.001	1.000	1.000	0.0025
0.01	0.976	0.972	0.0244
0.1	0.574	0.514	0.129
0.5	0.265	0.206	0.262
1.0	0.133	0.0911	0.246
2.0	0.0666	0.0443	0.286

$$\frac{dS}{dx} = -\frac{\omega}{q}(\tilde{\alpha}S - \alpha^* S^*)$$

where

$$\tilde{\alpha} = \frac{S_b}{S} > 1$$

is the ratio of the bottom sediment concentration to the depth-averaged concentration, and

$$\alpha^* = \frac{S_b^*}{S^*} > 1$$

is the ratio of the bottom carrying-sediment capacity to the depth-averaged capacity. It is noted that they are both greater than 1. This explains the puzzle that is hard to understand before: the saturation recovery coefficients are greater than 1 based on theories, but are equal or less than 1 based on field data, and it is even assumed $\alpha = 0.01$ in some mathematical models for the lower [9]. The two reasons attributed to this contradict include: (1) the previous theory with the saturation recovery coefficient greater than 1 is incomplete, which did not consider the probabilistic characteristics for the sediment settling on the bed to change to the bed material; and (2) one composite coefficient α was used to characterize the two saturation recovery coefficients $\tilde{\alpha}$ and α^*.

6.2 Numerical Results for Saturation Recovery Coefficient of Carrying-Sediment Capacity Under Equilibrium Condition

As shown in Eqs. (6.5) and (6.6), the difference in the saturation recovery coefficient between the equilibrium and the non-equilibrium conditions lies in the difference between the single-step distance $L_{4.l}^*$ and $L_{4.l}$, which is caused by the different vertical distributions of sediment concentration. For non-equilibrium condition, there is no analytical solution to the distribution of sediment concentration, which makes it difficult to determine the saturation recovery coefficient. For simplicity, this section focuses on the equilibrium condition.

6.2.1 *Supplement Equations*

6.2.1.1 Single-Step Suspension Distance for Suspended Load

The single-step suspension distance for suspended load is the distance on the bed between a suspended particle's two consecutive contacts with the bed. Within this distance, the particle is first lifted up from the bed, moves with water, and then falls down on the bed. All these motions are impacted by the longitudinal flow velocity $V(y)$, vertical (fluctuating) velocity V_y and the fall velocity of the particle. As discussed in Chap. 4, the horizontal distance of the movement when the particle moves upwards $x_{u.l}$ is

$$x_{u.l} = \frac{1}{\overline{u}_{y.u.i}} \int_0^{h_l^*} V(y)dy = \frac{q(h_i^*)}{\overline{u}_{y.u.l}}$$

and the horizontal distance of the movement when the particle moves downwards is

$$x_{d.l} = \frac{1}{\overline{u}_{y.d.l}} \int_0^{h_i^*} V(y)dy = \frac{q(h_i^*)}{\overline{u}_{y.d.l}}$$

where $\overline{u}_{y.u.l}$ and $\overline{u}_{y.d.l}$ are the average speed for the particle to move upwards and downwards, respectively, and h_l^* is the suspension height, which is also the upper limit of the integration in the above equations. The single-step distance is

$$L_{4.l}^* = x_{u.l} + x_{d.i} = q(h_i^*)\left(\frac{1}{\overline{u}_{y.u.l}} + \frac{1}{\overline{u}_{y.d.l}}\right) \tag{6.16}$$

The mathematical expectation for the speed of the particle's vertical movement is

$$\begin{cases} \overline{u}_{y.u.l} = \frac{\sigma_y}{\sqrt{2\pi}\varepsilon_{4.l}} \exp\left[-\frac{1}{2}\left(\frac{\omega_l}{\sigma_y}\right)^2\right] - \omega_l = \frac{u_*}{\sqrt{2\pi}\varepsilon_{4.l}} \exp\left[-\frac{1}{2}\left(\frac{\omega_l}{\sigma_y}\right)^2\right] - \omega_l \\ \overline{u}_{y.d.l} = \frac{\sigma_y}{\sqrt{2\pi}(1-\varepsilon_{4.l})} \exp\left[-\frac{1}{2}\left(\frac{\omega_l}{\sigma_y}\right)^2\right] + \omega_l \end{cases} \tag{6.17}$$

Thus, the single-step distance can be written as

$$L_{4.l}^* = \frac{q(h_l^*)}{\omega_l}\left\{\left[\frac{1}{\sqrt{2\pi}\varepsilon_{4.l}}\frac{u_*}{\omega_l}\exp\left(-\frac{1}{2}\left(\frac{\omega_l}{u_*}\right)^2\right) - 1\right]^{-1} + \left[\frac{1}{\sqrt{2\pi}(1-\varepsilon_{4.l})}\frac{u_*}{\omega_l}\exp\left(-\frac{1}{2}\left(\frac{\omega_l}{u_*}\right)^2\right) + 1\right]^{-1}\right\} \tag{6.18}$$

In order to verify the theoretical results for the saturation recovery coefficient, a computation method is developed to solve the numerical results for the saturation recovery coefficient for the carrying-sediment capacity. For strong equilibrium condition, $S_l^* = S_l$, $\tilde{\alpha}_l = \alpha_l^*$, and thus the saturation recovery coefficient for the carrying-sediment capacity is the same as the saturation recovery coefficient under the equilibrium condition. In this condition, the distribution of sediment concentration can be represented with the equilibrium profile.

6.2.1.2 Cumulative Flow Discharge on a Suspended Particle

For simplicity, let assume the vertical distribution of sediment concentration follows an exponential distribution

$$S(l) = S_b(l)\exp\left(-\frac{\omega y}{\varepsilon_s}\right) = S_b(l)\exp\left(-\frac{6\omega y}{\kappa u^* h}\right)$$

where κ is von Karman's constant, which has a mean value of 0.4, and the sediment diffusion coefficient is

$$\varepsilon_s = \frac{\kappa}{6}u_* h$$

Assuming the sediment diffusion coefficient ε_s equals the momentum transfer coefficient ε_y, the vertical gradient of flow velocity can be written as

$$\frac{dV}{dy} = \frac{\tau_0}{\rho}\frac{1}{\varepsilon_s}\left(1 - \frac{y}{h}\right) = \frac{u_*^2}{\frac{\kappa}{6}u_* h}\left(1 - \frac{y}{h}\right) = \frac{6u_*}{\kappa h}\left(1 - \frac{y}{h}\right)$$

The bottom shear stress τ_0 is

$$\tau_0 = \rho u_*^2 = \rho g h J$$

where g is the acceleration of gravity, and J the energy slope. Integrating the above equation for dV/dy yields the vertical distribution of the flow velocity that reads

$$V(\eta) = V_b + \frac{3u_*}{\kappa}[1 - (1 - \eta)^2] \tag{6.19}$$

where

$$\eta = \frac{y}{h} \tag{6.20}$$

is the relative depth above the bottom. The depth-averaged flow velocity is

$$\overline{V} = \int_0^1 V d\eta = V_b + \frac{2u_*}{\kappa} \tag{6.21}$$

Thus,

$$V(\eta) = \overline{V} - \frac{2u_*}{\kappa} + \frac{3u_*}{\kappa}[1 - (1 - \eta)^2] = \overline{V} + \frac{u_*}{\kappa} - \frac{3u_*}{\kappa}(1 - \eta)^2 \tag{6.22}$$

The cumulative flow discharge per unit width from the bottom to the suspension height h_l^* is

$$\begin{aligned} q(h_l^*) &= \int_0^{h_l^*} V(y)dy = h \int_0^{\eta_l^*} V(\eta)d\eta \\ &= h\left\{\left(\overline{V} + \frac{u_*}{\kappa}\right)\eta_l^* + \frac{u_*}{\kappa}[(1 - \eta_l^*)^3 - 1]\right\} \\ &= h\overline{V}\left\{\left(1 + \frac{u_*}{\kappa\overline{V}}\right)\eta_l^* + \frac{u_*}{\kappa\overline{V}}[(1 - \eta_l^*)^3 - 1]\right\} = q q_1(\eta_l^*) \end{aligned} \tag{6.23}$$

For $h_l = h$, $\eta_l = 1$ and $q_1(1) = q$ is the flow discharge per unit width

$$q = h\overline{V} \tag{6.24}$$

Therefore, Eq. (6.24) can be written as

$$
\begin{aligned}
q_1(\eta_l^*) &= \frac{q(h_l^*)}{q} = \left(1 + \frac{u_*}{\kappa \overline{V}}\right)\eta_l^* + \frac{u_*}{\kappa \overline{V}}[(1 - \eta_l^*)^3 - 1] \\
&= \left(1 + \frac{u_*}{\kappa \overline{V}}\right)\eta_l^* + \frac{u_*}{\kappa \overline{V}}[-3\eta_l^* + 3\eta_l^{*2} - \eta_l^{*3}] \\
&= \left(1 - \frac{2u_*}{\kappa \overline{V}}\right)\eta_l^* + \frac{3u_*}{\kappa \overline{V}}\eta_l^{*2} - \frac{u_*}{\kappa \overline{V}}\eta_l^{*3}
\end{aligned}
\tag{6.25}
$$

where

$$
\eta_l^* = \frac{h_l^*}{h} \tag{6.26}
$$

is the relative suspension height. It is noted that $u_*/\overline{V}$ varies for different rivers and different flow conditions, and should be determined based on the characteristics of rivers. For simplicity, a constant $u_*/\overline{V}$ is assumed in this analysis. Based on a parabolic distribution of the flow velocity, as shown in Eq. (6.23), and the coefficient of flow velocity at water surface $V_M/\overline{V} = 0.85$ for alluvial rivers, the value of $u_*/\overline{V}$ can be determined. According to Eq. (6.23), the flow velocity at the water surface $(\eta = 1)$ is

$$
V_M = \overline{V} + \frac{u_*}{\kappa} = \frac{\overline{V}}{0.85} \tag{6.27}
$$

Taking $\kappa = 0.4$ yields

$$
\overline{V} = 14.17u_* \tag{6.28}
$$

It is noted that Eq. (6.29) can be used as an approximation when the relation between $\overline{V}$ and u_* is hard to determine. If a more accurate relation between $\overline{V}$ and u_* is needed, Eq. (4.36) is recommended. Plugging Eq. (6.29) into Eq. (6.26) yields

$$
q_1\left(\eta_l^*\right) = \frac{q(h_l^*)}{q} = 0.647\eta_l^* + 0.529\eta_l^{*2} - 0.176\eta_l^{*3} \tag{6.29}
$$

6.2.1.3 Suspension Height

The suspension height is defined as the mathematical expectation of the maximum height that sediment particles can be suspended ([1, 5], also see Eq. (131) in Appendix A). Plugging Eq. (6.7) into Eq. (4.21), one obtains

$$h_l^* = \int_0^h \frac{2yS_b(l)}{hS(l)} \exp\left(-\frac{6\omega_l y}{\kappa u_* h}\right) dy = \int_0^1 \frac{2S_b(l)\eta}{\overline{S}(l)} \exp\left(-\frac{6\omega_l \eta}{\kappa u_*}\right) d\eta$$
$$= \frac{2S_b(l)h}{\overline{S}(l)} \left\{ \left(\frac{\kappa u_*}{6\omega_l}\right)^2 \left[1 - \exp\left(-\frac{6\omega_l}{\kappa u_*}\right)\right] - \frac{\kappa u_*}{6\omega_l} \exp\left(-\frac{6\omega_l}{\kappa u_*}\right) \right\} \tag{6.30}$$

The depth-averaged sediment concentration for uniform sediment with a particle size of D_l is

$$\overline{S}(l) = \frac{1}{h} \int_0^h S(l, y) dy = \int_0^1 S_b(l) \exp\left(-\frac{6\omega_l}{\kappa u_* \eta}\right) d\eta$$
$$= S_b(l) \frac{\kappa u_*}{6\omega_l} \left[1 - \exp\left(-\frac{6\omega_l}{\kappa u_*}\right)\right] = A_0 S_b(l) \tag{6.31}$$

Plugging Eq. (6.32) into Eq. (6.31) yields

$$\eta_l^* = \frac{h^*(l)}{h} = 2\left\{ \frac{\kappa u_*}{6\omega_l} - \exp\left(-\frac{6\omega_l}{\kappa u_*}\right) \left[1 - \exp\left(-\frac{6\omega_l}{\kappa u_*}\right)\right]^{-1} \right\} \tag{6.32}$$

Plugging Eq. (6.29) into Eq. (6.16) yields

$$L_{4.l}^* = \frac{q\left(h_l^*\right)}{\omega_l} \left(\frac{\omega_l}{\overline{u}_{y.4.l}} + \frac{\omega_l}{u_{y.d.l}}\right) = \frac{q}{\omega_l} q_1\left(\eta_l^*\right) \left(\frac{\omega_l}{\overline{u}_{y.4.l}} + \frac{\omega_l}{\overline{u}_{y.d.l}}\right) = \frac{q}{\omega_l} F_l \left(\frac{\omega_l}{u_*}\right) \tag{6.33}$$

As shown in this equation, the single-step distance can be replaced with the so-called settling distance q/ω, or the horizontal distance of movement when a particle settles from the water surface to the bottom in laminar flow, only when $F_l(\omega_l/u_*) = 1$.

6.2.1.4 Formula for Saturation Recovery Coefficient

Plugging Eqs. (6.16), (6.17) and (6.29) into Eq. (6.5), the saturation recovery coefficient for the carrying-sediment capacity can be written as

$$\alpha_l^* = \frac{(1-\varepsilon_{0.l})(1-\varepsilon_{4.l})}{L_{4.l}^* \omega_l} q = \frac{(1-\varepsilon_{0.l})(1-\varepsilon_{4.l})}{0.647\eta_l^* + 0.529\eta_l^{*2} - 0.176\eta_l^{*3}}$$

$$\times \left[\frac{1}{\frac{1}{\sqrt{2\pi}} \frac{u_*}{\varepsilon_{4.l}\omega_l} e^{-\frac{1}{2}(\frac{\omega_l}{u_*})^2} - 1} + \frac{1}{\frac{1}{\sqrt{2\pi}} \frac{u_*}{\omega_l(1-\varepsilon_{4.l})} e^{-\frac{1}{2}(\frac{\omega_l}{u_*})^2} + 1} \right]^{-1}$$

$$= (1-\varepsilon_{0.l})(1-\varepsilon_{4.l}) \left(\frac{\omega_l}{\overline{u}_{y.u.l}} + \frac{\omega_l}{\overline{u}_{y.D.l}} \right)^{-1} \frac{q}{q_l(\eta_l^*)} = (1-\varepsilon_{0.l}) f_l \quad (6.34)$$

where η_l^* is determined with Eq. (6.32).

6.2.2 Numerical Results for Saturation Recovery Coefficient Under Strong Equilibrium Condition

The numerical results for the saturation coefficient α_l^* are calculated in this section for the strong equilibrium condition. Table 6.2 lists the fall velocity ω_l and the threshold bottom flow velocity for sediment to cease motion, $V_{b.c.0.l}$, for various particle sizes.

Table 6.3 list the suspension probability $\varepsilon_{4.l}$, the average upward velocity $\overline{u}_{y.u.l}/\omega_l$ and the downward velocity $\overline{u}_{y.d.l}/\omega_l$ of a particle, the relative suspension height η_l^*, and the cumulative flow discharge from the bottom to the suspension height normalized by the discharge per unit width, or $q(\eta_l^*)/q$, respectively, for a series of ω_l/u_*. In addition, Table 6.3 also lists the results for f_l that is defined as

$$f_l = (1-\varepsilon_{4.l}) \left(\frac{\omega_l}{\overline{u}_{y.u.l}} + \frac{\omega_l}{\overline{u}_{y.d.l}} \right)^{-1} \frac{q}{q(\eta_l^*)} \quad (6.35)$$

and other parameters related to the single-step distance.

Table 6.4 lists the probability for sediment to cease motion $1-\varepsilon_{0.l}$, and the saturation recovery coefficient α_l^* for different particle sizes and different ω_l/u_*.

Table 6.2 Fall velocity and threshold velocity to cease motion

D_l (mm)	ω_l (cm/s)	$V_{b.c.1.0.l}$ (cm/s)	$\frac{V_{b.c.1.0.l}}{\omega_l}$
0.005	0.0023	1.5	652
0.01	0.00919	2.13	232
0.025	0.0574	3.36	58.5
0.05	0.23	4.75	20.7
0.1	0.919	6.73	7.32
0.25	2.68	10.6	3.96
0.5	5.57	15	2.69

Table 6.3 Single-step distance and other parameters

$\frac{\omega_l}{u_*}$	η_l^*	$\frac{q_1(\eta_l)}{q}$	$\varepsilon_{4.l}$	$1-\varepsilon_{4.l}$	$\frac{\omega_l}{\overline{u}_{y.u.l}}$	$\frac{\omega_l}{\overline{u}_{y.d.l}}$	$\left(\frac{\omega_l}{\overline{u}_{y.u.l}}+\frac{\omega_l}{\overline{u}_{y.d.l}}\right)^{-1}$	f_l
0.001	1.000	1.000	0.4996	0.5004	0.00125	0.00125	400	200
0.01	0.976	0.972	0.496	0.504	0.0126	0.0125	39.8	20.6
0.1	0.574	0.514	0.460	0.540	0.131	0.120	3.99	4.19
0.5	0.265	0.206	0.308	0.692	0.778	0.495	0.785	2.64
1	0.133	0.0911	0.159	0.841	1.92	0.777	0.371	3.42
2	0.0666	0.0443	0.0228	0.977	5.48	0.973	0.155	3.42

The ceasing motion probability $1-\varepsilon_{0.l}$ is computed using Eq. (4.5) with $\sigma_x = 0.37\overline{V}_b=2u_*$. The saturation recovery coefficient α_l^* is computed with Eq. (6.35), which is the product of $1-\varepsilon_{0.l}$ as listed in Table 6.4 and f_l as listed in Table 6.3.

The following observations can be made from Table 6.4: (1) For a constant particle size or a constants ω_l, as u_* increases, ω_l/u_* decreases and the probability for sediment to cease motion, or $1-\varepsilon_{0.l}$, deceases and f_l increases. α_l^* is the product of $1-\varepsilon_{0.l}$ and f_l, and thus it is not a monotonous function of ω_l/u_*. (2) Although the variation of α_l^* is complicated, the variation range of α_l^* is between 0.0214 and 5.88,

Table 6.4 Ceasing motion probability and saturation recovery coefficient

D_l (mm)	$\omega_l/u_*=0.001$			$\omega_l/u_*=0.01$			$\omega_l/u_*=0.1$		
	u_* (cm/s)	$1-\varepsilon_{0.l}$	α_l^*	u_* (cm/s)	$1-\varepsilon_{0.l}$	α_l^*	u_* (cm/s)	$1-\varepsilon_{0.l}$	α_l^*
0.005	2.3	0.0089	1.78*	0.23	0.287	5.88	0.023	1	3
0.01	9.19	0.0049	0.980*	0.919	0.0618	1.27	0.0919	1	3
0.025	57.4	0.0038	0.763*	5.74	0.008	0.164*	0.574	0.524	1.57
0.05	230	0.0036	0.722	23	0.0047	0.0964*	2.3	0.0475	0.142*
0.1	919	0.0035	0.703	91.9	0.0043	0.0882	9.19	0.0099	0.0297*
0.25	2680	0.0035	0.703	268	0.0037	0.0759	26.8	0.0062	0.0180*
0.5	5570	0.0035	0.703	557	0.0036	0.0738	55.7	0.0051	0.0153*
D_l (mm)	$\omega_l/u_*=0.5$			$\omega_l/u_*=1.0$			$\omega_l/u_*=2.0$		
	u_* (cm/s)	$1-\varepsilon_{0.l}$	α_l^*	u_* (cm/s)	$1-\varepsilon_{0.l}$	α_l^*	u_* (cm/s)	$1-\varepsilon_{0.l}$	α_l^*
0.005	0.0046	1	2.64	0.0023	1	3.28	0.00115	1	0.334
0.01	0.0184	1	2.64	0.00919	1	3.28	0.0046	1	0.334
0.025	0.115	1	2.64	0.0574	1	3.28	0.0287	1	0.334
0.05	0.46	0.993	2.62	0.23	1	3.28	0.115	1	0.334
0.1	1.83	0.195	0.514*	0.919	0.832	2.73	0.456	1	0.334
0.25	5.36	0.0436	0.115*	2.68	0.235	0.770*	1.34	0.896	2.98*
0.5	11.1	0.0227	0.0599*	5.57	0.0885	0.291*	2.78	0.501	1.67*

which is considered limited. It is also noted that α_l^* can be greater or less than 1, which is consistent with existing literatures. (3) α_l^* is typically greater than 1 when u_* is small. (4) If excluding the uncommon data for $u_* < 1$ cm/s and $u_* > 60$ cm/s, sixteen data points are left in Table 6.4, which are denoted with * in this table, and the mean saturation recovery coefficient averaged over these 16 data points is 0.65, with the maximum value occurring at $\omega_l/u_* = 2$ and $D_l = 0.25$ mm. Considering the suspension just starts at this unrepresentative condition, this data point can be ignored, and the mean saturation recovery coefficient averaged over the remaining 15 data points is 0.5. In practical calculation, α_l^* can be set to 0.5, and this would result in a sufficient accuracy of the results. Or, if a better accuracy is desired for α_l^*, the values listed in Table 6.4 can be used. It is noted that the values listed in Table 6.4 are somewhat different than the results as listed in Tables 6.15, 6.16, 6.17, 6.18, 6.19, 6.20, 6.21 and 6.22 for $S/S_* = 1$. This discrepancy is caused by the different values of ω_l and different accuracies used in the calculations. The results in Tables 6.15, 6.16, 6.17, 6.18, 6.19, 6.20, 6.21 and 6.22 are recommended.

6.2.3 *Opinions on Empirical Coefficient of Saturation Recovery*

As discussed before, the following empirical values were used for the saturation recovery coefficient in many mathematical models for sediment transport: $\alpha = 1.0$ for scour, and $\alpha = 0.25$ for deposition. Our theoretical analysis indicates that the mean value of α is 0.5 for the strong equilibrium condition. Based on this theoretical value, the rationality of the empirical saturation recovery coefficients and the associated problems are described as follows.

The values of the saturation recovery coefficient, 0.25 for deposition, and 1.0 for scour, are approximately consistent with the theoretical value of $\alpha_l^* = 0.5$ for equilibrium condition. It is also noted that the discrepancy in the saturation recovery coefficient has a very limited impact on the computed sediment concentration. The empirical values used for the saturation recovery coefficient in the past mathematical models are considered acceptable.

Since the theoretical values of the saturation recovery coefficient, α^* for equilibrium transport and $\tilde{\alpha}$ for non-equilibrium transport, have been derived, the empirical values of the saturation recovery coefficient are recommended to be replaced with the theoretical values. The difference between α^*, $\tilde{\alpha}$ and the composite saturation recovery coefficient shall also be noted. The composite coefficient is typically used in the mathematical models, and the values can be found in Tables 6.15, 6.16, 6.17, 6.18, 6.19, 6.20, 6.21 and 6.22, or simply in Table 5.1.

6.3 Saturation Recovery Coefficient for Non-equilibrium Sediment Transport

The theoretical saturation recovery coefficient was discussed in Sect. 6.2 for equilibrium sediment transport, and will be derived as follows for the non-equilibrium condition [4] based on the two types of vertical profiles for sediment concentration [3].

6.3.1 *Saturation Recovery Coefficient for Exponent Profile of Sediment Concentration*

The exponent profile for sediment concentrations reads

$$S(l, y) = S_b \exp\left(-\frac{6\omega_l C y}{\kappa u_* h}\right) \tag{6.36}$$

where C represents the saturation condition. $C > 1$ denotes scour condition with a greater gradient of the vertical profile of sediment concentration; $C < 1$ indicates deposition with a smaller gradient of sediment concentration profile; and $C = 1$ represents an equilibrium sediment concentration profile. The bottom sediment concentration for non-equilibrium condition can be written as

$$S_b(l) = \mu S_b^{*\prime}(l) + (1 - \mu) S_b^*(l) \tag{6.37}$$

where μ is a weighting factor, $S_b^*(l)$ the carrying-sediment capacity at the bottom for given flow and sediment conditions, and $S_b^{*'}(l)$ the carrying-sediment capacity at the bottom corresponding to the equilibrium condition with the depth-averaged sediment concentration for the non-equilibrium sediment transport, or $S_b^*(l) = S_b^{*\prime}(l)$ for $S = S^*$. As done in previous section, the relative suspension height is

$$\eta_l = \frac{h_l}{h} = \int_0^1 \frac{2\eta}{S(l)} S(l, \eta) d\eta = \frac{2}{S(l)} \int_0^1 \eta S_b(l) \exp\left(-\frac{6\omega_l C}{\kappa u_*}\eta\right) d\eta$$
$$= \frac{2S_b(l)}{S(l)} \left\{ \left(\frac{\kappa u_*}{6\omega_l C}\right)^2 \left[1 - \exp\left(-\frac{6\omega_l C}{\kappa u_*}\right)\right] - \frac{\kappa u_*}{6\omega_l C} \exp\left(-\frac{6\omega_l C}{\kappa u_*}\right) \right\}$$

and the depth-averaged sediment concentration is

$$S(l) = \int_0^1 S_b(l) \exp\left(-\frac{6\omega_l C \eta}{\kappa u_*}\right) d\eta$$

$$= S_b(l)\frac{\kappa u_*}{6\omega_l C}\left[1-\exp\left(-\frac{6\omega_l C}{\kappa u_*}\right)\right] = A_1 S_b(l) \tag{6.38}$$

Thus, the relative suspension height can be written as

$$\eta_l = 2\left\{\frac{\kappa u_*}{6\omega_l C} - \exp\left(-\frac{6\omega_l C}{\kappa u_*}\right)\left[1-\exp\left(-\frac{6\omega_l C}{\kappa u_*}\right)\right]^{-1}\right\} \tag{6.39}$$

Similar to Eq. (6.30), the flow discharge per unit width that acts on the suspended particles is

$$\frac{q(\eta_l)}{q} = 0.647\eta_l + 0.529\eta_l^2 - 0.176\eta_l^3 \tag{6.40}$$

Similar to Eq. (6.35), the saturation recovery coefficient for non-equilibrium sediment transport is

$$\begin{aligned}
\tilde{\alpha}_l &= \frac{(1-\varepsilon_{0.l})(1-\varepsilon_{4.l})}{L_{4.l}\omega_l}q = \frac{(1-\varepsilon_{0.l})(1-\varepsilon_{4.l})}{0.647\eta_l + 0.529\eta_l^2 - 0.178\eta_l^3} \\
&\quad \times \left\{\left[\frac{u_*}{\sqrt{2\pi}\varepsilon_{4.l}\omega_l}\exp\left(-\frac{1}{2}\left(\frac{\omega_l}{u_*}\right)^2\right) - 1\right]^{-1}\right. \\
&\quad \left. + \left[\frac{1}{\sqrt{2\pi}}\frac{u_*}{(1-\varepsilon_{4.l})\omega_l}\exp\left(-\frac{1}{2}\left(\frac{\omega_l}{u_*}\right)^2\right) + 1\right]^{-1}\right\}^{-1} \\
&= (1-\varepsilon_{0.l})(1-\varepsilon_{4.l})\left(\frac{\omega_l}{u_{4.u.l}} + \frac{\omega_l}{u_{4.d.l}}\right)^{-1}\frac{q}{q(\eta_l)} \\
&= (1-\varepsilon_{0.l})(1-\varepsilon_{4.l})\left(\frac{\omega_l}{u_{4.u.l}} + \frac{\omega_l}{u_{4.d.l}}\right)^{-1}\frac{q}{q(\eta_l^*)}\frac{q(\eta_l^*)}{q(\eta_l)} = \alpha_l^*\frac{q(\eta_l^*)}{q(\eta_l)}
\end{aligned} \tag{6.41}$$

It is noted that the ratio of the saturation recovery coefficient between the non-equilibrium condition and the equilibrium condition equals the ratio of the flow discharge per unit width acting on the suspended particles, or $q(\eta_l^*)/q(\eta_l)$. $q(\eta_l^*)$ and $q(\eta_l)$ can be determined with Eqs. (6.30) and (6.41), respectively. Since the non-equilibrium condition also includes the equilibrium condition with the saturation coefficient $C = 1$, Eqs. (6.39)–(6.41) also apply for the equilibrium condition, with η_l, $L_{4.l}$ and $\tilde{\alpha}_l$ being replaced with η_l^*, $L_{4.l}^*$ and $\tilde{\alpha}_l^*$, respectively.

6.3.2 Saturation Recovery Coefficient for Rouse Profile of Sediment Concentration

The Rouse profile of sediment concentration reads

$$S_l(y) = S_b(l)\left(\frac{h-y}{y}\frac{a}{h-a}\right)^{\frac{C\omega_l}{\kappa u_*}} = S_b(l)\left(\frac{1-\eta}{\eta}\frac{\eta_a}{1-\eta_a}\right)^{\frac{C\omega_l}{\kappa u_*}} \tag{6.42}$$

Assuming an exponent profile for the flow velocity

$$V\left(\frac{y}{h}\right) = V(\eta) = V_M + \frac{u_*}{\kappa}\ln\eta \tag{6.43}$$

the depth-averaged flow velocity is

$$\overline{V} = \int_0^1 \left(V_m + \frac{u_*}{\kappa}\ln\eta\right)d\eta = V_M - \frac{u_*}{\kappa} \tag{6.44}$$

and the flow velocity profile can be written as

$$V(\eta) = \overline{V} + \frac{u_*}{\kappa}(1+\ln\eta) \tag{6.45}$$

Thus, the flow discharge acting on the suspended sediment is

$$\begin{aligned} q(h_l) &= h\int_0^{\eta_l}\left[\overline{V} + \frac{u_*}{\kappa}(1+\ln\eta)\right]d\eta \\ &= h\left[\overline{V}\eta_l + \frac{u_*}{\kappa}\eta_l + \frac{u_*}{\kappa}\eta(\ln\eta - 1)\Big|_0^{\eta_l}\right] \\ &= h\overline{V}\left[\eta_l + \frac{u_*}{\kappa\overline{V}}(\eta_l\ln\eta_l)\right] \end{aligned} \tag{6.46}$$

Assuming $\overline{V} = 0.85V_M$, and plugging it into Eq. (6.45), one obtains $\frac{u_*}{\kappa\overline{V}} = 0.176$. Plugging this relation into Eq. (6.47) yields

$$\frac{q(h_l/h)}{q} = \frac{q(\eta_l)}{q} = \frac{h\overline{V}}{h\overline{V}}(\eta_l + 0.176\eta_l\ln\eta_l) = (\eta_l + 0.176\eta_l\ln\eta_l) \tag{6.47}$$

Similar to the derivation of Eq. (6.38) it can be derived that

$$\eta_l = \frac{2S_b(l)}{S(l)} \int_{\eta_a}^{1} \eta \left(\frac{1-\eta}{\eta} \frac{\eta_a}{1-\eta_a} \right)^{\frac{C\omega_l}{\kappa u_*}} d\eta \tag{6.48}$$

The depth-averaged sediment concentration is

$$S(l) = S_b(l) \int_{0}^{1} \left(\frac{1-\eta}{\eta} \frac{\eta_a}{1-\eta_a} \right)^{\frac{C\omega_l}{\kappa u_*}} d\eta \tag{6.49}$$

Plugging Eq. (6.50) into (6.49) yields

$$\eta_l = 2 \int_{\eta_a}^{1} \eta \left(\frac{1-\eta}{\eta} \frac{\eta_a}{1-\eta_a} \right)^{\frac{C\omega_l}{\kappa u_*}} d\eta \left[\int_{\eta_a}^{1} \left(\frac{1-\eta}{\eta} \frac{\eta_a}{1-\eta_a} \right)^{\frac{C\omega_l}{\kappa u_*}} d\eta \right]^{-1} \tag{6.50}$$

where $\eta_a = a/h$ is a reference location near the bottom, and it is typically 0.05 or less. $q(\eta_l)$ and $q(\eta_l^*)$ can be determined with Eqs. (6.48) and (6.51), or with Eqs. (6.40) and (6.41), while $q(\eta_l^*)$ is computed with $C = 1$. The saturation recovery coefficient for the non-equilibrium and equilibrium conditions, based on the exponent profile or the Rouse profile of sediment concentration, can be determined by plugging $q(\eta_l^*)$ and $q(\eta_l)$, respectively, into Eq. (6.42) or Eq. (6.53). Based on analysis of massive computed results, the difference in the saturation recovery coefficient when using the Rouse profile or the exponent profile was found to be insignificant if $\eta_a = 0.05$ is used as the reference location. For simplicity, the saturation recovery coefficient is derived in the following based on the exponent profile of sediment concentration. The flow velocity distribution is typically assumed to satisfy the parabolic velocity profile or the logarithmic velocity profile. The flow discharge acting on the suspended particles $q(\eta)$ computed with the parabolic velocity profile show insignificant difference from the results with the logarithmic velocity profile. The logarithmic velocity profile is used in the following analysis.

Plugging Eqs. (6.48), (6.18) and the following equation

$$L_{4.l} = \frac{q(\eta_l)}{\omega_l} \left[\frac{\omega_l}{\overline{u}_{4.a.y}} + \frac{\omega_l}{\overline{u}_{4.d.y}} \right] \tag{6.51}$$

into Eq. (6.6), one obtains

$$\tilde{\alpha}_l = \frac{(1-\varepsilon_{0.l})(1-\varepsilon_{4.l})}{L_{4.l}\omega_l}q = \frac{(1-\varepsilon_{0.l})(1-\varepsilon_{4.l})}{\eta_l + 0.176\eta_l \ln \eta_l}$$
$$\times \left[\frac{1}{\frac{1}{\sqrt{2\pi}}\frac{u_*}{\varepsilon_{4.l}\omega_l}e^{-\frac{1}{2}\left(\frac{\omega_l}{u_*}\right)^2} - 1 + \frac{1}{\sqrt{2\pi}}\frac{u_*}{\omega_l(1-\varepsilon_{4.l})}e^{-\frac{1}{2}\left(\frac{\omega_l}{u_*}\right)^2} + 1}\right]^{-1}$$
$$= \alpha_l^* \frac{q(\eta_l^*)}{q(\eta_l)} \tag{6.52}$$

where η_l is determined with Eq. (6.40). It should emphasized that the difference in the saturation recovery coefficient between the non-equilibrium and the equilibrium conditions is caused by $q(\eta)$ and $q(\eta^*)$, the difference between η and η^* is caused by the sediment concentration profiles, while the other factors are not impacted by the sediment concentration profiles.

6.4 Figures and Analysis for the Computed Saturation Recovery Coefficient for Non-Equilibrium Sediment Transport

The saturation recovery coefficients were computed using Eqs. (6.6), (6.40), (6.47), and (6.52) for various sediment grain sizes, sedimentation conditions and flow conditions (u_*). The computed saturation coefficients include $\tilde{\alpha}_l$ for non-equilibrium condition, α_l^* for equilibrium condition, the composite saturation recovery coefficient α_l, as well as the depth-averaged and the bottom saturation recovery coefficients.

It is noted that, two parameters are included in the sediment concentration profiles, not matter for the exponent profile or for the Rouse profile. As shown in Eqs. (6.37) and (6.38), these two parameters are C and μ for the exponent profile. The following can be derived from these two equations

$$\frac{\frac{S}{S^*}}{\mu\frac{S}{S^*} + (1-\mu)} = \frac{\beta}{\mu\beta + (1-\mu)} = \frac{A_0}{A}, \quad B = \frac{1}{C}\frac{1-e^{-\frac{6\omega C}{\kappa u_*}}}{1-e^{-\frac{6\omega}{\kappa u_*}}}$$

where $B = S/S^*$. It is noted that B, A_0/A and C are related. If one parameter is given, the other two can be determined. μ is an independent parameter that need to be specified. Based on preliminary analysis, it is appropriate to set $\mu = 0.2$ or $\mu = 0.5$. A larger value of μ indicates a higher non-saturation status for the bottom sediment concentration. The saturation recovery coefficient for $\mu = 0.5$ is one to three times larger than the value for $\mu = 0.2$, while is less than 1.5 times for most cases. The difference also increases with ω/u_*. The results presented in this section are based on $\mu = 0.2$, which is a representative condition for general rivers.

6.4.1 Computed Saturation Recovery Coefficient for Non-Equilibrium Sediment Transport

Figures 6.1, 6.2, 6.3 and 6.4 show the saturation coefficient $\tilde{\alpha}$ versus ω/u_* computed for the sediment particle size of 0.005 mm, 0.025 mm, 0.1 mm, and 1 mm, respectively, with the variation range of C between 0.25 and 4. The following characteristics of $\tilde{\alpha}$ can be observed from these figures: (1) $\tilde{\alpha}$ generally varies between 0.008 and 20; for $D \le 0.01$ mm, $\tilde{\alpha}$ ranges between 1 and 10; for 0.01 mm $< D \le 1$ mm, $\tilde{\alpha}$ varies between 0.01 and 10, and is less than 1 for most cases. (2) For $\omega/u_* < 0.01 \sim 0.05$, $\tilde{\alpha}$ is independent of the sedimentation state or C, and equals the value in the equilibrium condition. This is because the sediment concentration profile is very uniform for this case and is not impacted by the sedimentation state. (3) For $0.001 < \omega/u_* < 2$, $\tilde{\alpha}$ is less than 1 for most cases. (4) $\tilde{\alpha} < \alpha^*$ when deposition occurs ($C < 1$), $\tilde{\alpha} = \alpha^*$ ($C = 1$) for equilibrium sediment transport, and $\tilde{\alpha} > \alpha^*$ ($C > 1$) when scour occurs. (5) $\tilde{\alpha}$ is independent of the sedimentation state, and is inversely proportional to ω/u_* when ω/u_* is small, which is consistent with the field data of desilting basins [8].

Figures 6.5, 6.6, 6.7, 6.8, 6.9 and 6.10 show the relation between $\tilde{\alpha}$ and the shear velocity u_* for various particle sizes, respectively. Similar trend can be observed for the relation between $\tilde{\alpha}$ and u_*. However, different from Figs. 6.1, 6.2, 6.3 and 6.4, these figures shows: (1) $\tilde{\alpha}$ is independent of the sedimentation state for a small particle size that results in small ω/u_* and uniform sediment concentration distribution, and (2) the impact of the sedimentation state on $\tilde{\alpha}$ becomes more significant when the particle size become coarser, which result in a larger ω/u_*, and more non-uniform distribution of sediment concentration.

It is also noted that the saturation recovery coefficient $\tilde{\alpha}$ for scour ($C > 1$) is larger than that for deposition ($C < 1$), and $\tilde{\alpha}$ increases with the non-saturation coefficient C. This is consistent with the analysis in previous section.

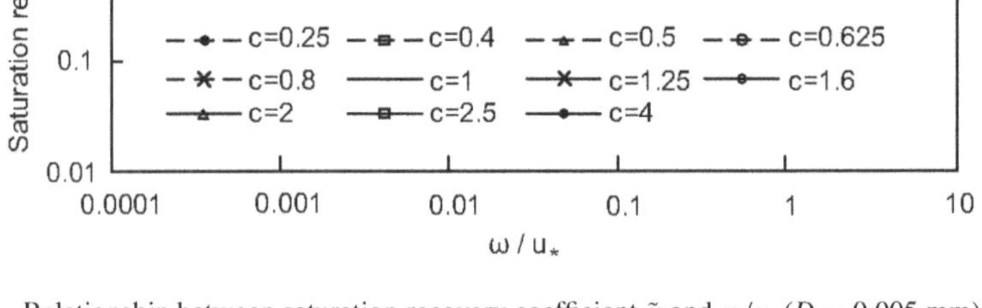

Fig. 6.1 Relationship between saturation recovery coefficient $\tilde{\alpha}$ and $\omega/u_*(D = 0.005$ mm)

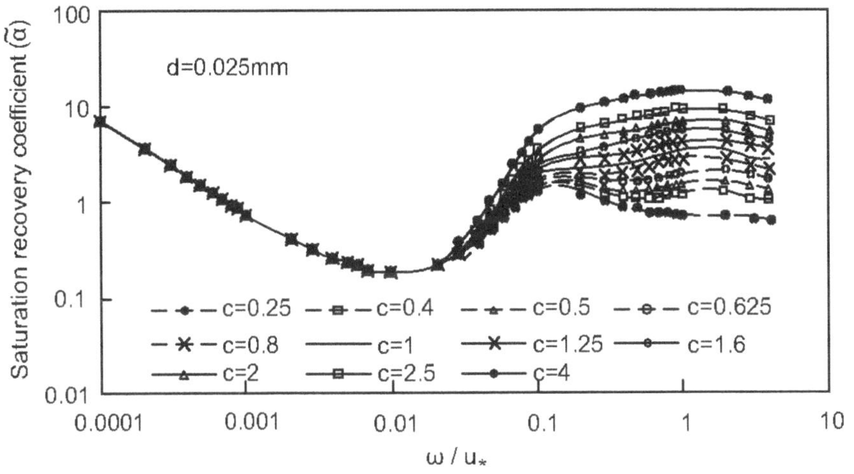

Fig. 6.2 Relationship between saturation recovery coefficient $\tilde{\alpha}$ and $\omega/u_*(D = 0.025\text{ mm})$

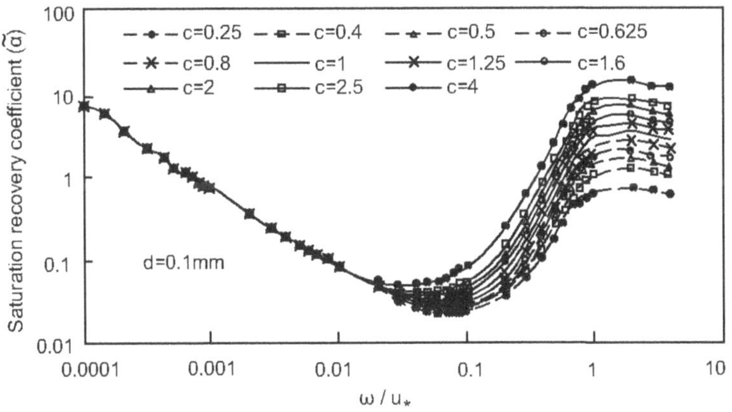

Fig. 6.3 Relationship between saturation recovery coefficient $\tilde{\alpha}$ and ω/u_* $(D = 0.1\text{ mm})$

6.4.2 Average Saturation Recovery Coefficient for Non-equilibrium Sediment Transport

6.4.2.1 Average Saturation Recovery Coefficient

The following equation can be derived from the governing equation for fractional sediment transport:

$$\frac{d(P_{4,l}S)}{dx} = -\frac{1}{q}(\alpha_l\omega_l P_{4,l}S - \alpha_l^*\omega_l P_{4,l}^*S^*)$$

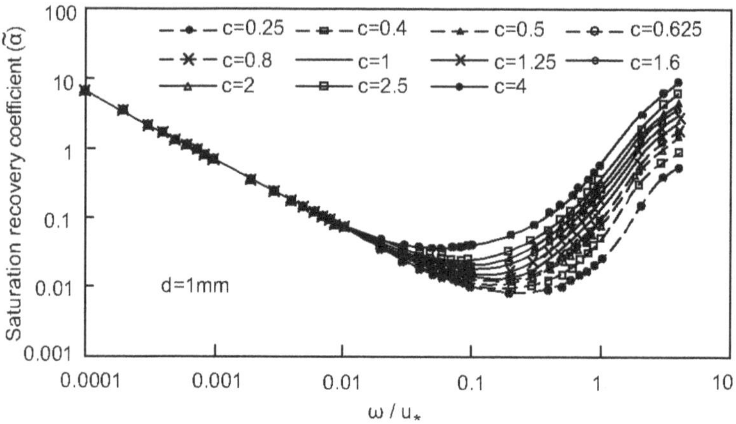

Fig. 6.4 Relationship between saturation recovery coefficient $\tilde{\alpha}$ and $\omega / u_*(D = 1\ \text{mm})$

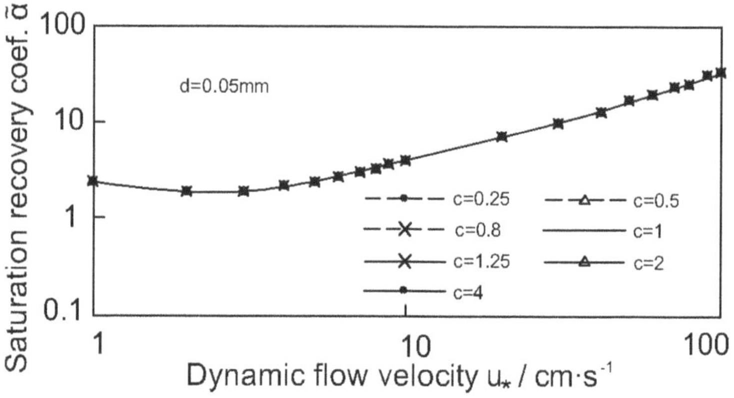

Fig. 6.5 Saturation recovery coefficient $\tilde{\alpha}$ computed ($D = 0.005$ mm)

where $P_{4.l}$ and $P_{4.l}^*$ are the size distributions of sediment concentration and sediment transport capacity, respectively. Summation of this equation for all size groups yields

$$\frac{dS}{dx} = -\frac{1}{q}\left[\sum_{l=1}^{n}(\alpha_l \omega_l P_{4.l})S - \sum_{l=1}^{n}(\alpha_l^* \omega_l P_{4.l}^*)S^*\right] \tag{6.53a}$$

Thus, the governing equation for the total sediment concentration of non-uniform sediment mixture is

$$\frac{dS}{dx} = -\frac{\omega_m}{q}\left(\tilde{\alpha}_m S - \alpha_m^* S^*\right) \tag{6.53b}$$

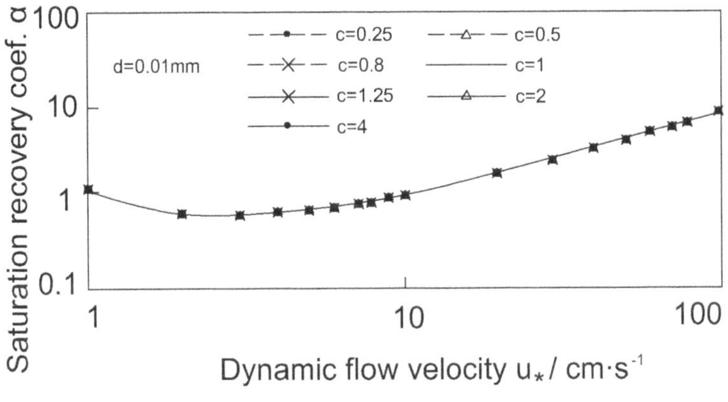

Fig. 6.6 Saturation recovery coefficient $\tilde{\alpha}$ computed ($D = 0.01$ mm)

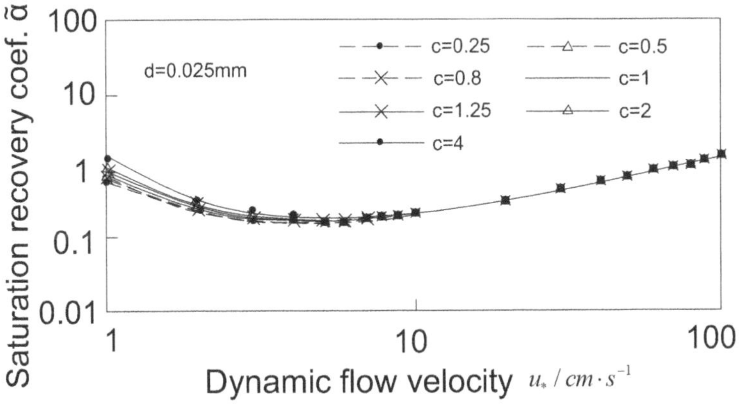

Fig. 6.7 t saturation recovery coefficient $\tilde{\alpha}$ computed ($D = 0.025$ mm)

where ω_m and ω_m^* are the average sediment fall velocity defined as

$$\omega_m = \sum_{l=1}^{n} P_{4.l}\omega_l \tag{6.54a}$$

and

$$\omega_m^* = \sum_{l=1}^{n} P_{4.l}^*\omega_l \tag{6.54b}$$

Comparing Eq. (6.53a) with Eq. (6.53b), the average saturation recovery coefficient of sediment concentration for non-equilibrium transport of non-uniform

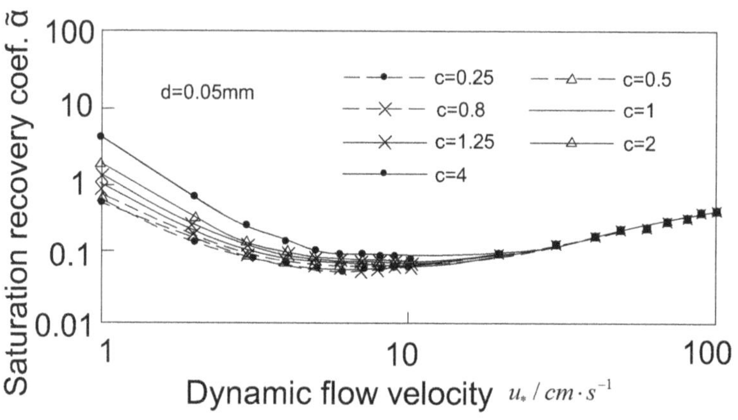

Fig. 6.8 Saturation recovery coefficient $\tilde{\alpha}$ computed ($D = 0.05$ mm)

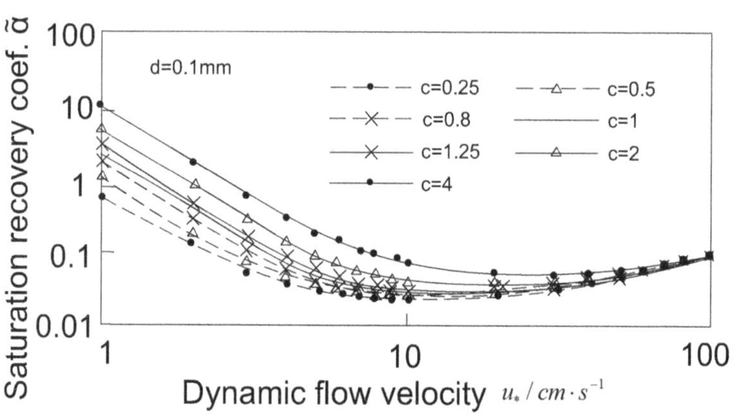

Fig. 6.9 Saturation recovery coefficient $\tilde{\alpha}$ computed ($D = 0.1$ mm)

sediment can be written as

$$\tilde{\alpha}_m = \frac{\sum\limits_{l=1}^{n} (\alpha_l \omega_l P_{4.l})}{\sum\limits_{l=1}^{n} (\omega_l P_{4.l})} = \frac{1}{\omega_m} \sum_{l=1}^{n} (\alpha_l \omega_l P_{4.l}) \tag{6.55a}$$

and the average saturation recovery coefficient of carrying-sediment capacity for equilibrium transport of non-uniform sediment is

$$\alpha_m^* = \frac{1}{\omega_m} \sum_{l=1}^{n} (\alpha_l^* \omega_l P_{4.l}^*) \tag{6.55b}$$

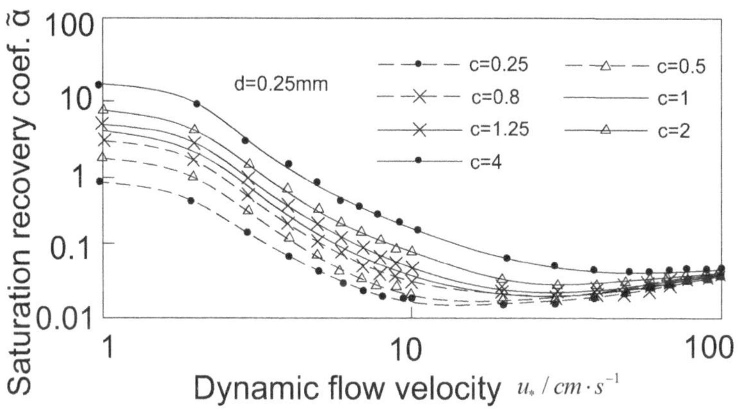

Fig. 6.10 t saturation recovery coefficient $\tilde{\alpha}$ computed ($D = 0.25$ mm)

Equations (6.55a), and (6.55b), indicate that the average saturation recovery coefficient for non-uniform sediment should be computed based on the weighted average of the products of the fall velocities and size distribution of all size groups. Figure 6.11 shows the average saturation recovery coefficient computed based on the size distribution of suspended load, which is averaged over the period between 1987 and 1999, of the downstream reach of the Yellow River. As shown in this figure, the average saturation recovery coefficient for the downstream reach of the Yellow River is less than 1 for most cases. However, when the shear velocity is less than 2 cm/s, the saturation recovery coefficient may exceed 1, and can even be 10 when scour occurs. This is the typical case for desilting basins.

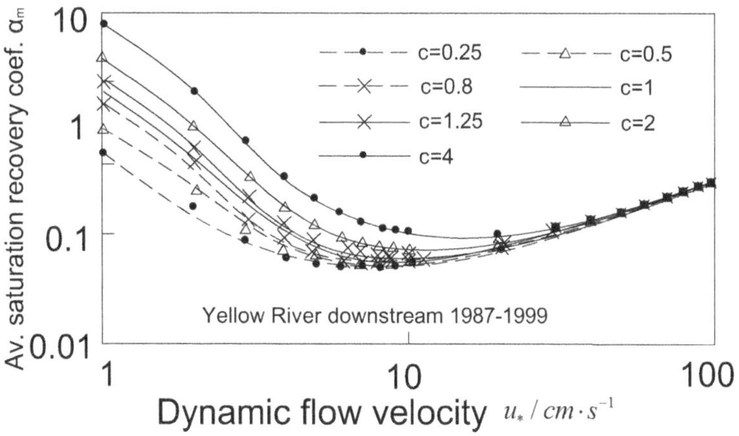

Fig. 6.11 Average sediment saturation recovery coefficient $\tilde{\alpha}_m$ computed for the downstream reach of the Yellow River

6.4.2.2 Average Bottom Saturation Recovery Coefficient

Let introduce the average bottom saturation recovery coefficient $\alpha_{b.m}$, which can be written as a function of the average saturation recovery coefficient with the following relationship

$$\alpha_{b.m} = \frac{\alpha_m}{\alpha_{1.m}} \tag{6.56}$$

where $\alpha_{1.m}$ is the ratio of the bottom sediment concentration to the depth-averaged concentration, which is

$$\alpha_{1.m} = \frac{S_b}{S} = \frac{6C\omega_m}{\kappa u_*}\left[1 - \exp\left(-\frac{6c\omega_m}{\kappa u_*}\right)\right]^{-1} = A_m^{-1} \tag{6.57}$$

Plugging Eq. (6.56) into Eq. (6.57), the bottom saturation recovery coefficient becomes

$$\alpha_{b.m} = \frac{\alpha_m}{\alpha_{1.m}} = \alpha_m\frac{\kappa u_*}{6C\omega_m}\left[1 - \exp\left(-\frac{6c\omega_m}{\kappa u_*}\right)\right] = A_m\alpha_m \tag{6.58}$$

Figure 6.12 shows the average bottom saturation recovery coefficient computed based on the size distribution of suspended load, which is averaged over the period between 1987 and 1999, of the downstream reach of the Yellow River. As shown in this figure, the bottom saturation recover coefficient is greater than 0.1 for the shear velocity less than 4 cm/s or larger than 30 cm/s, and is 0.05–0.1 for typical flow condition in natural rivers with the shear velocity varying between 4 and 30 cm/s.

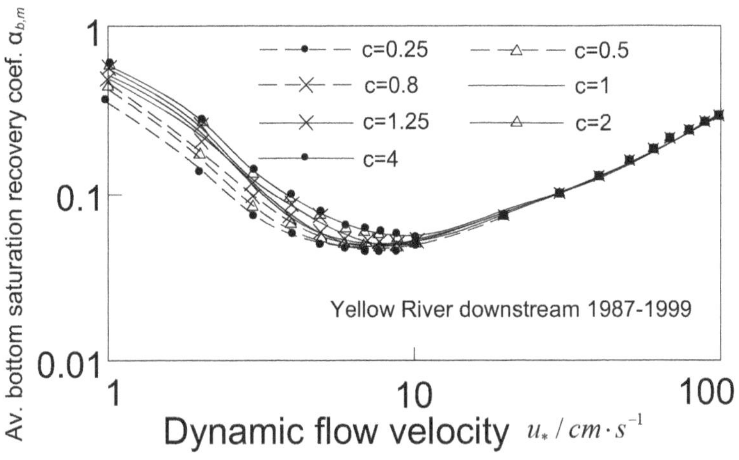

Fig. 6.12 Average bottom sediment saturation recovery coefficient $\alpha_{b.m}$ computed for the downstream reach of the Yellow River

6.4.3 *Composite Saturation Recovery Coefficient for Non-equilibrium Sediment Transport*

The composite saturation recovery coefficient is defined in Eq. (6.8), and its relation with $\tilde{\alpha}$ and α^* is described in Eq. (6.9). The composite saturation recovery coefficient α is generally used in the mathematical models [2], and it represents how fast the sediment concentration approaches the carrying-sediment capacity. The composite saturation recovery coefficient computed for different particle size groups are shown in Figs. 6.13, 6.14, 6.15, 6.16, 6.17 and 6.18, respectively. As show in these figures, the composite saturation recovery coefficient α varies with particle sizes D. As D decreases, α increases, and the sediment concentration reaches saturation quicker. α is almost independent of the non-saturation coefficient C for wash load. For bed material load, α is generally less than 1, and varies with the shear velocity u_* with the minimum value α occurring for a critical u_*. α for deposition ($C < 1$) is slightly greater than that for scour($C > 1$), which implies the sediment concentration reaches saturation quicker for the case with deposition than that the case with scour. This is different from previous concepts, and is also different from our previous recommendation for the empirical saturation recovery coefficient. It is also noted that the composite saturation recovery coefficient varies insignificantly as sedimentation or scour develops, and is approximately independent of C. This will simplify the practical application of the composite saturation recovery coefficient.

The three characteristics of the composite saturation recovery coefficient are summarized as follows: (1) α is less than $\tilde{\alpha}$ and α^*. For example, for $D \geq 0.25$ mm, α is less than 1 and is much less than $\tilde{\alpha}$ and α^*. This is consistent with the theoretical analysis. (2) α varies insignificantly as sedimentation or scour develops, and is approximately independent of the sedimentation status. This will simplify the application of α in the mathematical models. (3) Different from $\tilde{\alpha}$ and α^*, α for scour is less than that for equilibrium condition and for deposition, and α for deposition is

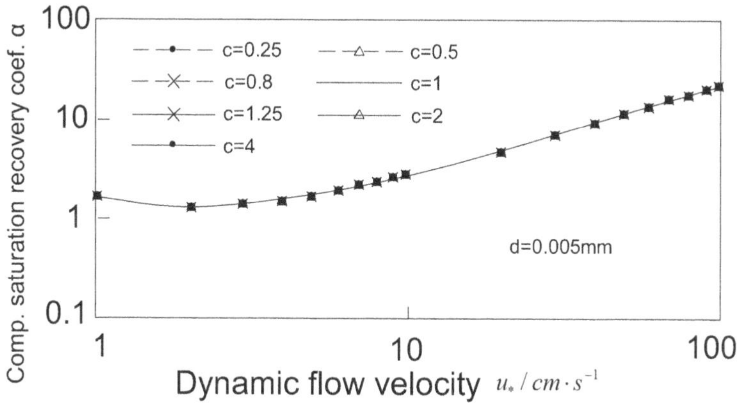

Fig. 6.13 Composite sediment saturation recovery coefficient α computed for $D = 0.005$ mm

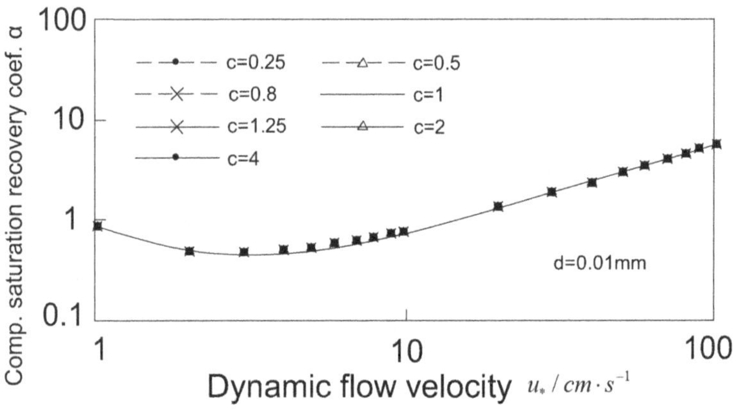

Fig. 6.14 Composite sediment saturation recovery coefficient α computed for $D = 0.01$ mm

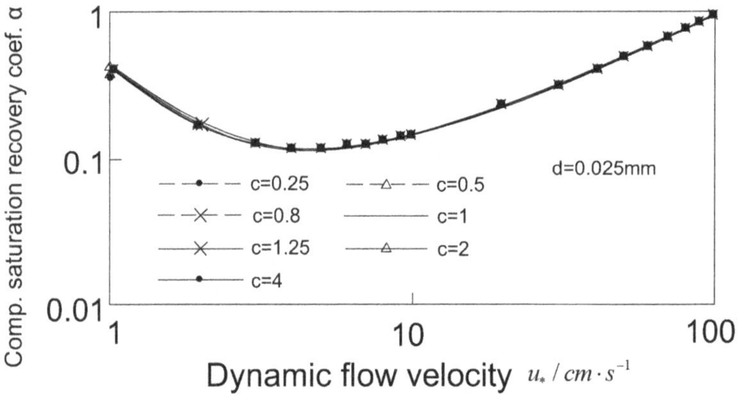

Fig. 6.15 Composite sediment saturation recovery coefficient α computed for $D = 0.25$ mm

larger than that for equilibrium condition and for scour. However, the variation range of α is very limited. Further investigation is needed for the mechanism. This also illustrates the reason for the slow recovery of sediment concentration with limited scour intensity for the channel scour downstream of a reservoir.

6.4.4 Average Composite Saturation Recovery Coefficient for Non-uniform Sediment

The composite saturation recovery coefficient α varies with particle sizes. It is necessary to compute the coefficient that is averaged over all size classes for non-uniform

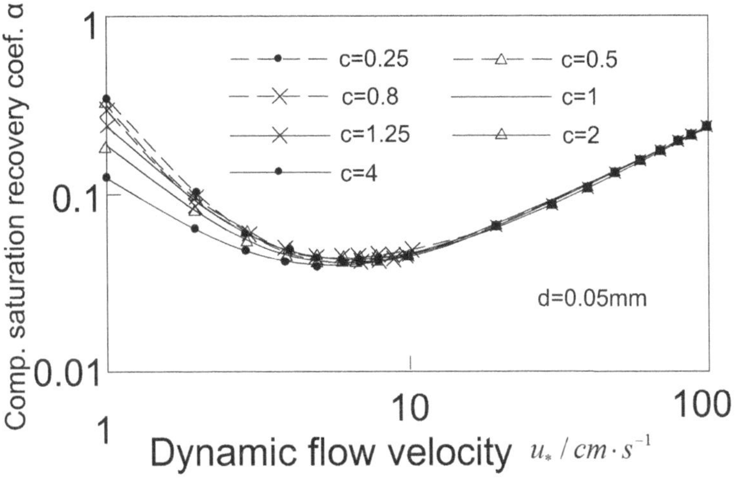

Fig. 6.16 Composite sediment saturation recovery coefficient α computed for $D = 0.05$ mm

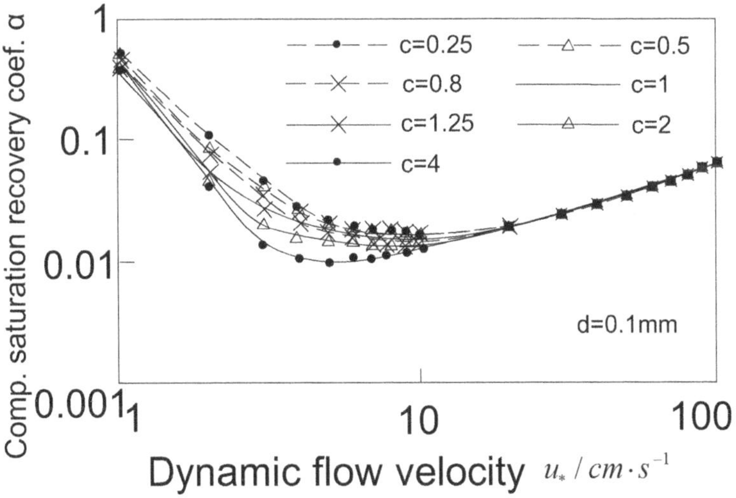

Fig. 6.17 Composite sediment saturation recovery coefficient α computed for $D = 0.1$ mm

sediment mixture. The depth-averaged sediment concentration (or carrying-sediment capacity) for equilibrium sediment transport ($C = 1$) is

$$S^* = S_b^* \frac{\kappa u_*}{6\omega_m} \left[1 - \exp\left(-\frac{6\omega_m}{\kappa u_*} \right) \right] = S_b^* A_m^* \tag{6.59}$$

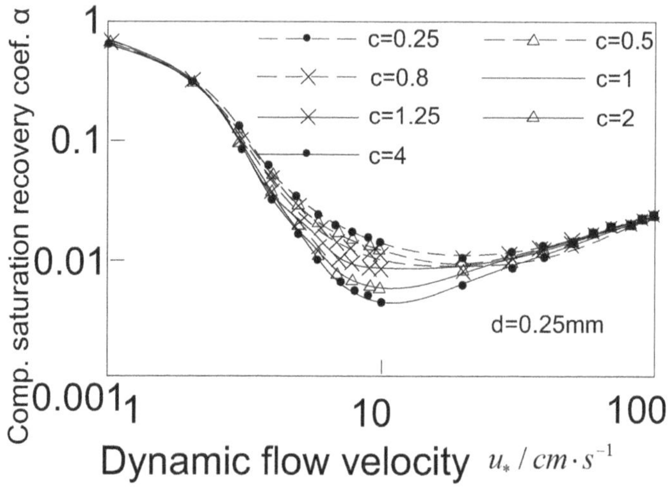

Fig. 6.18 Composite sediment saturation recovery coefficient α computed for $D = 0.25$ mm

and the relation between the sediment concentration and the bottom concentration corresponding to the equilibrium concentration profile is

$$S = S_b' \frac{\kappa u_*}{6\omega_m}\left[1 - \exp\left(-\frac{6\omega_m}{\kappa u_*}\right)\right] = S_b' A_m^* \tag{6.60}$$

Apparently, the real bottom sediment concentration S_b is between S_b' and S_b^*. Assuming

$$S_b = \mu S_b' + (1-\mu) S_b^* \quad (0 \le \mu \le 1) \tag{6.61}$$

$$S = S_b A_m = \left[\mu S_b' + (1-\mu) S_b^*\right] A_m = [\mu S + (1-\mu) S_*] \frac{A_m}{A_m^*} \tag{6.62}$$

where

$$A_m = \frac{\kappa u_*}{6\omega_l C}\left[1 - \exp\left(-\frac{6C\omega_m}{\kappa u_*}\right)\right],$$

$$A_m^* = \frac{\kappa u_*}{6\omega_l}\left[1 - \exp\left(-\frac{6\omega_m}{\kappa u_*}\right)\right]$$

$$\frac{A_m}{A_m^*} = \frac{1}{C}\left[1 - \exp\left(-\frac{6C\omega_m}{\kappa u_*}\right)\right]\left[1 - \exp\left(-\frac{6\omega_m}{\kappa u_*}\right)\right]^{-1}$$

The following non-saturation relationship can be derived from Eq. (6.62)

$$\frac{S}{S^*} = \frac{(1-\mu)A_m/A_m^*}{1-\mu A_m/A_m^*} \tag{6.63}$$

where $\mu = 0.2$. As shown in Eq. (6.63), after μ is determined, the three parameters, S/S^*, C and ω_m/u_*, are correlated. After these three parameters are determined, the saturation recovery coefficients α^* and $\tilde{\alpha}$ can be computed with Eqs. (6.35) and (6.52) for each size group, and the composite saturation recovery coefficient for all size groups can be determined based on the following equation

$$\alpha = \frac{\tilde{\alpha}}{1 - S_*/S} - \frac{\alpha^*}{S/S_* - 1} \tag{6.64}$$

The average composite saturation recovery coefficient can also be determined directly from $\tilde{\alpha}_m$ and α_m^*. In fact, it can be derived from Eq. (6.9) that

$$\alpha_m = \frac{\tilde{\alpha}_m}{1 - S^*/S} + \frac{\alpha_m^*}{S/S^* - 1} \tag{6.65}$$

Figure 6.19 shows the average composite saturation recovery coefficient α_m computed based on the size distribution of suspended load, which is averaged over the period between 1987 and 1999, of the downstream reach of the Yellow River. As shown in this figure, α_m is less than 1. The minimum α_m occurs for the shear velocity of 4 cm/s. The variation range of the minimum α_m for various C is between 0.005 and 0.02 with an average value of 0.01, which is consistent with the empirical saturation recovery coefficient used in the mathematical models for the Yellow River [9].

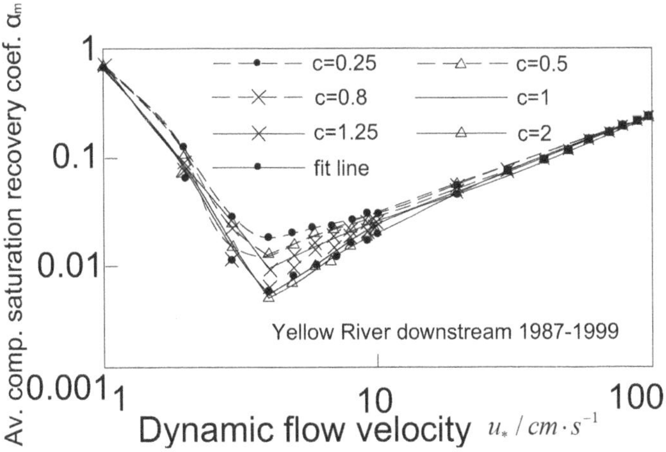

Fig. 6.19 Mean composite sediment saturation recovery coefficient computed for the downstream reach of the Yellow River

Based on the computed results, the following fitting relationship between α_m and u_* is derived for the downstream reach of the Yellow River:

$$\overline{\alpha} = \begin{cases} 0.8u_*^{-3.16} & u_* \leq 4\,\text{cm/s} \\ 0.0025u_* & u_* \geq 4\,\text{cm/s} \end{cases} \tag{6.66}$$

where u_* is in cm/s.

It is noted that simplifications were used in the above calculation of α_m. Alternatively, if the average values of A_m^* and A_m are not computed with the average fall velocity ω_m, the procedure to determine α_m is as follows. The following equation can be derived from Eqs. (3.21) and (6.32)

$$\begin{aligned} S^* &= \sum P_{4.l}^* S^* = \sum P_{1.l} S^*(l) = \sum P_{1.l} S_b^*(l) \frac{\kappa u_*}{6\omega_l}\left[1 - \exp\left(-\frac{6\omega_l}{\kappa u_*}\right)\right] \\ &= \sum P_{1.l} S_b^*(l) A_l^* = \overline{S_b^*}\,\overline{A}^* \end{aligned} \tag{6.67}$$

where

$$A_l^* = \frac{\kappa u_*}{6\omega_l}\left[1 - \exp\left(-\frac{6\omega_l}{\kappa u_*}\right)\right] \tag{6.68}$$

$$\overline{S_b^*} = \sum P_{1.l} S_b^*(l) \tag{6.69}$$

$$\overline{A}^* = \frac{\sum P_{1.l} S_b^*(l) A_l^*}{\sum P_{1.l} S_b^*(l)} \tag{6.70}$$

Similarly,

$$S = \sum P_{1.l} S_b'(l) A_l^* = \overline{S'}_b \overline{A}^* \tag{6.71}$$

Multiplying Eq. (6.39) by $P_{1.l}$, and summing up for all size groups yields

$$\sum P_{1.l} S_b(l) = \mu \sum P_{1.l} S_b'(l) + (1 - \mu) \sum P_{1.l} S_b^*(l)$$

or

$$\overline{S}_b = \mu \overline{S'}_b + (1 - \mu) \overline{S_b^*} \tag{6.72}$$

Thus,

$$S = \sum P_{1.l} S(l) = \sum P_{1.l} S_b(l) A_l = \overline{S}_b \overline{A} \tag{6.73}$$

where

$$A_l = \frac{\kappa u_*}{6\omega_l C}\left[1 - \exp\left(-\frac{6\omega_l C}{\kappa u_*}\right)\right] \tag{6.74}$$

$$\overline{A} = \frac{\sum P_{1.l} S_b(l) A_l}{\sum P_{1.l} S_b(l)} \tag{6.75}$$

Plugging Eqs. (6.69), (6.71), and (6.73) for $\overline{S}_b^*$, $\overline{S}_b'$, and $\overline{S}_b$, respectively, into Eq. (6.72), one obtains

$$\frac{S}{\overline{A}} = \mu \frac{S}{\overline{A}^*} + (1 - \mu)\frac{S^*}{\overline{A}^*}$$

Rearranging this equation yields

$$B = \frac{S}{S^*} = \frac{(1 - \mu)\overline{A}/\overline{A}^*}{1 - \mu \overline{A}/\overline{A}^*} \tag{6.76}$$

It is seen that A^* and $\overline{A}$ are replaced with A_m^* and A_m in previously simplified calculation procedure of α_m.

6.5 Numerical Results for Saturation Recovery Coefficient for Non-equilibrium Sediment Transport ($\mu = 0.5$)

The saturation recovery coefficient was presented in Sect. 6.4 for $\mu = 0.2$, which is a reasonable approximation for sedimentation and scour in natural rivers. However, the non-saturation coefficient C used in the calculation is different from the real non-saturation condition $B = S/S^*$. This is particularly true in the case with scour. For example, for $C = 4$, the corresponding B varies within a wide range. Therefore, it is more reasonable to use B as a factor in the calculation of saturation recovery coefficient. The results presented in this section is based on $\mu = 0.5$, and B varying from 0.3 for scour to 5.0 for deposition. This variation range of non-saturation condition is approximately consistent with the channel scour downstream of a reservoir and the reservoir sedimentation. Table 6.5 lists the relation between the non-saturation coefficient C and $B = S/S^*$ for various ω/u_* with $\mu = 0.5$. Table 6.6 lists the relative suspension heights for various ω/u_* and B.

Table 6.5 Variation of non-saturation coefficient C with ω/u_* and S/S^* ($\mu = 0.5$)

ω/u_*	Non-saturation coefficient C ($\mu = 0.5$)							
	S/S* = 0.3	S/S* = 0.5	S/S* = 0.8	S/S* = 1	S/S* = 1.015	S/S* = 1.078	S/S* = 1.166	S/S* = 2.456
0.001	122.285	59.4497	17.0671	1	0.0099	0	0	0
0.005	25.5845	12.8266	4.2473	1	0.8	0.0111	0	0
0.01	13.5025	7.0008	2.6453	1	0.8987	0.4993	0.0029	0
0.05	3.8832	2.3579	1.3678	1	0.9775	0.8887	0.7787	0.0003
0.1	2.7434	1.8013	1.2136	1	0.987	0.9359	0.8727	0.4317
0.5	2.1679	1.5008	1.1254	1	0.9926	0.9637	0.9285	0.7003
1	2.1667	1.5	1.125	1	0.9926	0.9638	0.9288	0.7036
2	2.1667	1.5	1.125	1	0.9926	0.9638	0.9288	0.7036

Table 6.6 Variation of relative suspension height h_l/h with ω/u_* and S/S^* ($\mu = 0.5$)

ω/u_*	Ceasing motion prob /($\times 10^{-4}$)	Relative upward velocity		Relative average suspension height $\eta_l = h_l/h$ ($\mu = 0.5$)							
		$\overline{U}_{y.u}/\omega$	$\overline{U}_{y.d}/\omega$	S/S* = 0.3	S/S* = 0.5	S/S* = 0.8	S/S* = 1	S/S* = 1.015	S/S* = 1.078	S/S* = 1.166	S/S* = 2.456
0.001	0.5004	797.5216	798.2484	0.71	0.853	0.957	0.998	1	1	1	1
0.005	0.502	159.2141	159.9409	0.698	0.842	0.947	0.988	0.99	1	1	1
0.01	0.504	79.4262	80.153	0.683	0.828	0.934	0.975	0.978	0.988	1	1
0.05	0.5199	15.5997	16.3266	0.572	0.72	0.832	0.876	0.879	0.89	0.903	1
0.1	0.5398	7.6262	8.3533	0.453	0.596	0.712	0.759	0.762	0.773	0.788	0.893
0.5	0.6915	1.2822	2.0183	0.123	0.178	0.237	0.266	0.267	0.275	0.285	0.37
1	0.8413	0.5251	1.2876	0.062	0.089	0.119	0.133	0.134	0.138	0.144	0.189
2	0.9773	0.1866	1.0276	0.031	0.044	0.059	0.067	0.067	0.069	0.072	0.095

Tables 6.7, 6.8, 6.9, 6.10, 6.11, 6.12, 6.13 and 6.14 list the saturation recovery coefficient $\tilde{\alpha}$ in terms of ω/u_* and B computed for $\mu = 0.5$ with various particle sizes D, respectively. The following can be observed from these tables: (1) $\tilde{\alpha}$ does not exceed 10 except for $D = 0.005$ mm and $\omega/u_* = 0.01$. (2) $\tilde{\alpha}$ is still less than 1 for most cases when $D \geq 0.05$ mm. For $D = 0.5$ mm, $\tilde{\alpha}$ is less than 1 except for $\omega/u_* = 2$. (3) The relationship between $\tilde{\alpha}$ and ω/u_* is complicated for a given B. The results indicate that $\tilde{\alpha}$ may have both the maximum and the minimum values for $D = 0.005$ mm and $D = 0.01$ mm, but may only have the minimum value for $D \geq 0.025$ mm. For all ω/u_* and B considered in the calculations, the minimum value of $\tilde{\alpha}$ is 0.012. (4) $\tilde{\alpha}$ for the case with scour is generally 0.3–1 times greater than that for deposition. $\tilde{\alpha}$ for equilibrium condition is between the values for scour and for deposition, while is closer to the value for deposition.

Table 6.7 Sediment saturation recovery coefficient $\tilde{\alpha}$ for $D = 0.005$ mm

ω/u_*	Saturation recovery coefficient $\tilde{\alpha}$								
	S/S* = 0.3	S/S* = 0.5	S/S* = 0.8	S/S* = 1	S/S* = 1.015	S/S* = 1.078	S/S* = 1.166	S/S* = 2.456	S/S* = 5
0.001	2.7507	2.2036	1.9207	1.8298	1.8244	1.8244	1.8244	1.8244	
0.005	9.7846	7.8022	6.7804	6.4526	6.4334	6.3587	6.3587	6.3587	
0.01	24.0789	19.0894	16.5272	15.7070	15.6590	15.4724	15.2464	15.2464	
0.05	8.1331	6.1550	5.1649	4.8536	4.8354	4.7651	4.6801	4.1479	
0.1	5.5883	4.0098	3.2339	2.9942	2.9803	2.9266	2.8619	2.4621	2.2872
0.5	6.2128	4.1496	3.0070	2.6340	2.6122	2.5274	2.4249	1.7884	1.5151
1	7.5119	5.0973	3.7441	3.2947	3.2681	3.1648	3.0393	2.2357	1.8705
2	7.5646	5.1819	3.8431	3.3972	3.3709	3.2683	3.1435	2.3419	1.9743

Table 6.8 Sediment saturation recovery coefficient $\tilde{\alpha}$ for $D = 0.01$ mm

ω/u_*	Saturation recovery coefficient $\tilde{\alpha}$								
	S/S* = 0.3	S/S* = 0.5	S/S* = 0.8	S/S* = 1	S/S* = 1.015	S/S* = 1.078	S/S* = 1.166	S/S* = 2.456	S/S* = 5
0.001	1.4879	1.1919	1.0389	0.9897	0.9868	0.9868	0.9868	0.9868	
0.005	1.1043	0.8806	0.7653	0.7283	0.7261	0.7177	0.7177	0.7177	
0.01	2.1311	1.6895	1.4627	1.3901	1.3859	1.3693	1.3493	1.3493	
0.05	8.1293	6.1521	5.1625	4.8513	4.8332	4.7629	4.6780	4.1460	
0.1	5.5883	4.0098	3.2339	2.9942	2.9803	2.9266	2.8619	2.4621	2.2872
0.5	6.2128	4.1496	3.0070	2.6340	2.6122	2.5274	2.4249	1.7884	1.5151
1	7.5119	5.0973	3.7441	3.2947	3.2681	3.1648	3.0393	2.2357	1.8705
2	7.5646	5.1819	3.8431	3.3972	3.3709	3.2683	3.1435	2.3419	1.9743

Table 6.9 Sediment saturation recovery coefficient $\tilde{\alpha}$ for $D = 0.025$ mm

ω/u_*	Saturation recovery coefficient $\tilde{\alpha}$								
	S/S* = 0.3	S/S* = 0.5	S/S* = 0.8	S/S* = 1	S/S* = 1.015	S/S* = 1.078	S/S* = 1.166	S/S* = 2.456	S/S* = 5
0.001	1.1429	0.9156	0.7980	0.7602	0.7580	0.7580	0.7580	0.7580	
0.005	0.3340	0.2663	0.2314	0.2203	0.2196	0.2171	0.2171	0.2171	
0.01	0.2633	0.2087	0.1807	0.1718	0.1712	0.1692	0.1667	0.1667	
0.05	0.9691	0.7334	0.6154	0.5783	0.5762	0.5678	0.5577	0.4943	
0.1	3.5412	2.5409	2.0493	1.8974	1.8886	1.8545	1.8135	1.5602	1.4494
0.5	6.2128	4.1496	3.0070	2.6340	2.6122	2.5274	2.4249	1.7884	1.5151
1	7.5119	5.0973	3.7441	3.2947	3.2681	3.1648	3.0393	2.2357	1.8705
2	7.5646	5.1819	3.8431	3.3972	3.3709	3.2683	3.1435	2.3419	1.9743

Table 6.10 Sediment saturation recovery coefficient $\tilde{\alpha}$ for $D = 0.05$ mm

ω/u_*	Saturation recovery coefficient $\tilde{\alpha}$								
	S/S* = 0.3	S/S* = 0.5	S/S* = 0.8	S/S* = 1	S/S* = 1.015	S/S* = 1.078	S/S* = 1.166	S/S* = 2.456	S/S* = 5
0.001	1.0777	0.8633	0.7525	0.7169	0.7148	0.7148	0.7148	0.7148	
0.005	0.2509	0.2000	0.1738	0.1654	0.1649	0.1630	0.1630	0.1630	
0.01	0.1513	0.1199	0.1038	0.0987	0.0984	0.0972	0.0958	0.0958	
0.05	0.1244	0.0942	0.0790	0.0742	0.0740	0.0729	0.0716	0.0635	
0.1	0.2913	0.2090	0.1686	0.1561	0.1553	0.1525	0.1492	0.1283	0.1192
0.5	6.1898	4.1342	2.9958	2.6242	2.6025	2.5180	2.4159	1.7817	1.5095
1	7.5119	5.0973	3.7441	3.2947	3.2681	3.1648	3.0393	2.2357	1.8705
2	7.5646	5.1819	3.8431	3.3972	3.3709	3.2683	3.1435	2.3419	1.9743

Table 6.11 Sediment saturation recovery coefficient $\tilde{\alpha}$ for $D = 0.1$ mm

ω/u_*	Saturation recovery coefficient $\tilde{\alpha}$								
	S/S* = 0.3	S/S* = 0.5	S/S* = 0.8	S/S* = 1	S/S* = 1.015	S/S* = 1.078	S/S* = 1.166	S/S* = 2.456	S/S* = 5
0.001	1.0555	0.8455	0.7370	0.7021	0.7000	0.7000	0.7000	0.7000	
0.005	0.2262	0.1804	0.1568	0.1492	0.1487	0.1470	0.1470	0.1470	
0.01	0.1233	0.0978	0.0847	0.0805	0.0802	0.0793	0.0781	0.0781	
0.05	0.0491	0.0372	0.0312	0.0293	0.0292	0.0288	0.0283	0.0250	
0.1	0.0569	0.0408	0.0329	0.0305	0.0303	0.0298	0.0291	0.0251	0.0233
0.5	1.3180	0.8803	0.6379	0.5588	0.5542	0.5362	0.5144	0.3794	0.3214
1	6.4961	4.4080	3.2378	2.8491	2.8262	2.7368	2.6283	1.9334	1.6175
2	7.5646	5.1819	3.8431	3.3972	3.3709	3.2683	3.1435	2.3419	1.9743

Table 6.12 Sediment saturation recovery coefficient $\tilde{\alpha}$ for $D = 0.25$ mm

ω/u_*	Saturation recovery coefficient $\tilde{\alpha}$								
	S/S* = 0.3	S/S* = 0.5	S/S* = 0.8	S/S* = 1	S/S* = 1.015	S/S* = 1.078	S/S* = 1.166	S/S* = 2.456	S/S* = 5
0.001	1.0499	0.8411	0.7331	0.6984	0.6964	0.6964	0.6964	0.6964	
0.005	0.2204	0.1757	0.1527	0.1453	0.1449	0.1432	0.1432	0.1432	
0.01	0.1171	0.0928	0.0804	0.0764	0.0761	0.0752	0.0741	0.0741	
0.05	0.0382	0.0289	0.0242	0.0228	0.0227	0.0224	0.0220	0.0195	
0.1	0.0352	0.0252	0.0204	0.0189	0.0188	0.0184	0.0180	0.0155	0.0144
0.5	0.2916	0.1948	0.1411	0.1236	0.1226	0.1186	0.1138	0.0839	0.0711
1	1.9352	1.3132	0.9646	0.8488	0.8420	0.8153	0.7830	0.5760	0.4819
2	6.9522	4.7624	3.5319	3.1222	3.0980	3.0037	2.8890	2.1523	1.8144

Table 6.13 Sediment saturation recovery coefficient $\tilde{\alpha}$ for $D = 0.5$ mm

ω/u_*	Saturation recovery coefficient $\tilde{\alpha}$								
	S/S* = 0.3	S/S* = 0.5	S/S* = 0.8	S/S* = 1	S/S* = 1.015	S/S* = 1.078	S/S* = 1.166	S/S* = 2.456	S/S* = 5
0.001	1.0479	0.8394	0.7317	0.6970	0.6950	0.6950	0.6950	0.6950	
0.005	0.2182	0.1740	0.1512	0.1439	0.1435	0.1418	0.1418	0.1418	
0.01	0.1148	0.0910	0.0788	0.0749	0.0747	0.0738	0.0727	0.0727	
0.05	0.0347	0.0263	0.0220	0.0207	0.0206	0.0203	0.0200	0.0177	
0.1	0.0292	0.0210	0.0169	0.0157	0.0156	0.0153	0.0150	0.0129	0.0120
0.5	0.1406	0.0939	0.0681	0.0596	0.0591	0.0572	0.0549	0.0405	0.0343
1	0.7217	0.4897	0.3597	0.3165	0.3140	0.3040	0.2920	0.2148	0.1797
2	4.0580	2.7798	2.0616	1.8224	1.8083	1.7532	1.6863	1.2563	1.0591

Table 6.14 Sediment saturation recovery coefficient $\tilde{\alpha}$ for $D = 1.0$ mm

ω/u_*	Saturation recovery coefficient $\tilde{\alpha}$								
	S/S* = 0.3	S/S* = 0.5	S/S* = 0.8	S/S* = 1	S/S* = 1.015	S/S* = 1.078	S/S* = 1.166	S/S* = 2.456	S/S* = 5
0.001	1.0465	0.8384	0.7307	0.6961	0.6941	0.6941	0.6941	0.6941	
0.005	0.2168	0.1729	0.1503	0.1430	0.1426	0.1409	0.1409	0.1409	
0.01	0.1133	0.0899	0.0778	0.0739	0.0737	0.0728	0.0718	0.0718	
0.05	0.0326	0.0247	0.0207	0.0194	0.0194	0.0191	0.0187	0.0166	
0.1	0.0258	0.0185	0.0149	0.0138	0.0138	0.0135	0.0132	0.0114	0.0106
0.5	0.0830	0.0555	0.0402	0.0352	0.0349	0.0338	0.0324	0.0239	0.0202
1	0.3135	0.2127	0.1563	0.1375	0.1364	0.1321	0.1268	0.0933	0.0781
2	1.6876	1.1560	0.8573	0.7579	0.7520	0.7291	0.7013	0.5225	0.4404

Tables 6.15, 6.16, 6.17, 6.18, 6.19, 6.20, 6.21 and 6.22 list the composite saturation recovery coefficient α in terms of ω/u_* and B computed for $\mu = 0.5$ with various particle sizes D, respectively. The following can be observed from these tables: (1) α is less than 10 except for $D = 0.005$ mm and $\omega/u_* = 0.01$, and α is less than 1 for $u_* = 1$–100 cm/s and $D \geq 0.01$ mm. (2) α is less than $\tilde{\alpha}$, as listed in Tables 6.7, 6.8, 6.9 and 6.10, for deposition and for scour. (3) α shows insignificant variations, which is generally within 3–6%, with $B = S/S^*$. Thus, the impact of B on α can be ignored. This will simplify the application of α in the mathematical models, in which α can be selected based on D and ω/u_*. (4) The saturation recovery coefficients are presented in this chapter in the forms of figures for $\mu = 0.2$ and in the forms of tables for $\mu = 0.5$. The results for $\mu = 0.5$ are recommended. (5) α for scour is generally less than that for deposition, while the difference is limited. This is different from our previous recommendations for the empirical saturation recovery coefficients.

Table 6.15 Composite sediment saturation recovery coefficient α for $D = 0.005$ mm

u* (cm/s)	ω/u_*	Composite saturation recovery coefficient α							
		S/S* = 0.3	S/S* = 0.5	S/S* = 0.8	S/S* = 1.015	S/S* = 1.078	S/S* = 1.166	S/S* = 2.456	S/S* = 5
2.211	0.001	1.4351	1.4560	1.4660	1.4690	1.4690	1.4690	1.4690	
0.442	0.005	5.0246	5.1030	5.1411	5.1527	5.1551	5.1551	5.1551	
0.221	0.01	12.1191	12.3246	12.4261	12.4575	12.4639	12.4714	12.4714	
0.0442	0.05	3.4480	3.5521	3.6081	3.6267	3.6306	3.6353	3.6633	
0.0221	0.1	1.8824	1.9786	2.0353	2.0552	2.0594	2.0646	2.0967	2.1105
0.0044	0.5	1.1002	1.1185	1.1421	1.1563	1.1600	1.1649	1.2076	1.2354
0.0022	1	1.4873	1.4920	1.4969	1.4994	1.5000	1.5009	1.5084	1.5144
0.0011	2	1.6112	1.6125	1.6139	1.6147	1.6148	1.6151	1.6171	1.6185

Table 6.16 Composite sediment saturation recovery coefficient α for $D = 0.01$ mm

u* (cm/s)	ω/u_*	Composite saturation recovery coefficient α							
		S/S* = 0.3	S/S* = 0.5	S/S* = 0.8	S/S* = 1.015	S/S* = 1.078	S/S* = 1.166	S/S* = 2.456	S/S* = 5
8.844	0.001	0.7762	0.7875	0.7930	0.7946	0.7946	0.7946	0.7946	
1.769	0.005	0.5671	0.5759	0.5802	0.5816	0.5818	0.5818	0.5818	
0.884	0.01	1.0726	1.0908	1.0997	1.1025	1.1031	1.1038	1.1038	
0.177	0.05	3.4464	3.5505	3.6065	3.6250	3.6290	3.6337	3.6562	
0.0884	0.1	1.8824	1.9786	2.0353	2.0552	2.0594	2.0646	2.0903	2.1105
0.0177	0.5	1.1002	1.1185	1.1421	1.1563	1.1600	1.1649	1.1969	1.2354
0.0088	1	1.4873	1.4920	1.4969	1.4994	1.5000	1.5009	1.5064	1.5144
0.0044	2	1.6112	1.6125	1.6139	1.6147	1.6148	1.6151	1.6166	1.6185

Table 6.17 Composite sediment saturation recovery coefficient α for $D = 0.025$ mm

u* (cm/s)	ω/u_*	Composite saturation recovery coefficient α							
		S/S* = 0.3	S/S* = 0.5	S/S* = 0.8	S/S* = 1.015	S/S* = 1.078	S/S* = 1.166	S/S* = 2.456	S/S* = 5
55.27	0.001	0.5962	0.6049	0.6091	0.6104	0.6104	0.6104	0.6104	
11.06	0.005	0.1715	0.1742	0.1755	0.1759	0.1760	0.1760	0.1760	
5.528	0.01	0.1325	0.1348	0.1359	0.1362	0.1363	0.1364	0.1364	
1.106	0.05	0.4109	0.4233	0.4299	0.4322	0.4326	0.4332	0.4359	
0.553	0.1	1.1928	1.2538	1.2898	1.3023	1.3050	1.3083	1.3246	1.3374
0.111	0.5	1.1002	1.1185	1.1421	1.1563	1.1600	1.1649	1.1969	1.2354
0.0553	1	1.4873	1.4920	1.4969	1.4994	1.5000	1.5009	1.5064	1.5144
0.0276	2	1.6112	1.6125	1.6139	1.6147	1.6148	1.6151	1.6166	1.6185

Table 6.18 Composite sediment saturation recovery coefficient α for $D = 0.05$ mm

u* (cm/s)	ω/u_*	Composite saturation recovery coefficient α							
		S/S* = 0.3	S/S* = 0.5	S/S* = 0.8	S/S* = 1.015	S/S* = 1.078	S/S* = 1.166	S/S* = 2.456	S/S* = 5
221.1	0.001	0.5622	0.5704	0.5744	0.5755	0.5755	0.5755	0.5755	
44.22	0.005	0.1288	0.1308	0.1318	0.1321	0.1322	0.1322	0.1322	
22.11	0.01	0.0762	0.0774	0.0781	0.0783	0.0783	0.0784	0.0784	
4.422	0.05	0.0527	0.0543	0.0552	0.0555	0.0555	0.0556	0.0560	
2.21	0.1	0.0981	0.1031	0.1061	0.1071	0.1073	0.1076	0.1089	0.1100
0.442	0.5	1.0962	1.1143	1.1379	1.1520	1.1557	1.1605	1.1925	1.2308
0.221	1	1.4873	1.4920	1.4969	1.4994	1.5000	1.5009	1.5064	1.5144
0.111	2	1.6112	1.6125	1.6139	1.6147	1.6148	1.6151	1.6166	1.6185

Table 6.19 Composite sediment saturation recovery coefficient α for $D = 0.1$ mm

u* (cm/s)	ω/u_*	Composite saturation recovery coefficient α							
		S/S* = 0.3	S/S* = 0.5	S/S* = 0.8	S/S* = 1.015	S/S* = 1.078	S/S* = 1.166	S/S* = 2.456	S/S* = 5
884.4	0.001	0.5506	0.5587	0.5625	0.5637	0.5637	0.5637	0.5637	
176.9	0.005	0.1162	0.1180	0.1189	0.1191	0.1192	0.1192	0.1192	
88.44	0.01	0.0621	0.0631	0.0637	0.0638	0.0638	0.0639	0.0639	
17.69	0.05	0.0208	0.0214	0.0218	0.0219	0.0219	0.0220	0.0221	
8.844	0.1	0.0192	0.0201	0.0207	0.0209	0.0210	0.0210	0.0213	0.0215
1.768	0.5	0.2334	0.2373	0.2423	0.2453	0.2461	0.2471	0.2539	0.2621
0.884	1	1.2862	1.2902	1.2944	1.2966	1.2972	1.2979	1.3027	1.3096
0.442	2	1.6112	1.6125	1.6139	1.6147	1.6148	1.6151	1.6166	1.6185

Table 6.20 Composite sediment saturation recovery coefficient α for $D = 0.25$ mm

u* (cm/s)	ω/u_*	Composite saturation recovery coefficient α							
		S/S* = 0.3	S/S* = 0.5	S/S* = 0.8	S/S* = 1.015	S/S* = 1.078	S/S* = 1.166	S/S* = 2.456	S/S* = 5
2594	0.001	0.5477	0.5557	0.5596	0.5607	0.5607	0.5607	0.5607	
518	0.005	0.1132	0.1149	0.1158	0.1161	0.1161	0.1161	0.1161	
259	0.01	0.0589	0.0599	0.0604	0.0606	0.0606	0.0606	0.0606	
51.89	0.05	0.0162	0.0167	0.0169	0.0170	0.0170	0.0171	0.0172	
25.94	0.1	0.0119	0.0125	0.0128	0.0129	0.0130	0.0130	0.0132	0.0133
5.19	0.5	0.0516	0.0525	0.0536	0.0543	0.0544	0.0547	0.0562	0.0580
2.59	1	0.3832	0.3844	0.3856	0.3863	0.3864	0.3867	0.3881	0.3901
1.30	2	1.4808	1.4820	1.4833	1.4839	1.4841	1.4843	1.4857	1.4875

Table 6.21 Composite sediment saturation recovery coefficient α for $D = 0.5$ mm

u* (cm/s)	ω/u_*	Composite saturation recovery coefficient α							
		S/S* = 0.3	S/S* = 0.5	S/S* = 0.8	S/S* = 1.015	S/S* = 1.078	S/S* = 1.166	S/S* = 2.456	S/S* = 5
5387	0.001	0.5467	0.5546	0.5585	0.5596	0.5596	0.5596	0.5596	
1077	0.005	0.1121	0.1138	0.1147	0.1149	0.1150	0.1150	0.1150	
538	0.01	0.0578	0.0588	0.0592	0.0594	0.0594	0.0595	0.0595	
107	0.05	0.0147	0.0152	0.0154	0.0155	0.0155	0.0155	0.0156	
53.9	0.1	0.0098	0.0103	0.0106	0.0107	0.0108	0.0108	0.0109	0.0110
10.8	0.5	0.0249	0.0253	0.0259	0.0262	0.0263	0.0264	0.0271	0.0280
5.39	1	0.1429	0.1433	0.1438	0.1440	0.1441	0.1442	0.1447	0.1455
2.69	2	0.8643	0.8650	0.8658	0.8662	0.8663	0.8664	0.8672	0.8682

Table 6.22 Composite sediment saturation recovery coefficient α for $D = 1$ mm

u* (cm/s)	ω/u_*	Composite saturation recovery coefficient α							
		S/S* = 0.3	S/S* = 0.5	S/S* = 0.8	S/S* = 1.015	S/S* = 1.078	S/S* = 1.166	S/S* = 2.456	S/S* = 5
10,972	0.001	0.5460	0.5539	0.5577	0.5589	0.5589	0.5589	0.5589	
2194	0.005	0.1113	0.1131	0.1139	0.1142	0.1142	0.1142	0.1142	
1097	0.01	0.0570	0.0580	0.0585	0.0586	0.0587	0.0587	0.0587	
219	0.05	0.0138	0.0142	0.0145	0.0145	0.0145	0.0146	0.0147	
109.7	0.1	0.0087	0.0091	0.0094	0.0095	0.0095	0.0095	0.0097	0.0097
21.9	0.5	0.0147	0.0149	0.0153	0.0155	0.0155	0.0156	0.0160	0.0165
10.97	1	0.0621	0.0623	0.0625	0.0626	0.0626	0.0626	0.0629	0.0632
5.49	2	0.3594	0.3597	0.3600	0.3602	0.3603	0.3603	0.3606	0.3611

References

1. Han QW (1979) Reservoir sedimentation, vol 1. the Yangtze Valley Planning Office, Hydrology Department
2. Han QW (2005) Size distribution of carrying-sediment capacity. In: Proceedings of 6th national symposium on basic theory of sedimentation. Water Conservancy of Yellow River Press, pp 3–20. (in Chinese)
3. Han, Q.W. (2006), Boundary Condition of Diffusion equation and Coefficient of saturation recovery, *Journal of Changsha University of Science & Technology(Natural Science)*, 3(3), pp. 7–19. (in Chinese)
4. Han QW, Chen XJ (2008) Theoretical method for calculating the saturation recovery coefficient. J Sediment Res 6:8.16. (in Chinese)
5. Han QW, He MM (1984) Stochastic theory of sediment motion. Science Publishing House, Beijing (in Chinese)
6. Han QW, He MM (1997) Study on the two-dimensional equations for non-uniform non-equilibrium suspended sediment transport and the boundary conditions. J Hydraulic Eng 1:1–10 (in Chinese)

7. Han QW, He MM (1997) Preliminary study on saturation recovery coefficients. J Sediment Res 3:33–40 (in Chinese)
8. Yang JY (2005) Study on the recovery saturation coefficient of over-.saturation sediment transport in settling tanks. J Sediment Res 3:42–47 (in Chinese)
9. Yellow River Conservancy Commission (YRCC), Survey and Design Institute (1998) Numerical model for sedimentation of lower Yellow River. (in Chinese)

Chapter 7
Size Distribution for Carrying-Sediment Capacity and Effective Bed Material Gradation

7.1 Introduction

The size distribution for the carrying-sediment capacity typically does not equal the size distribution of sediment concentration except in the case with strong equilibrium or with massive deposition or scouring. Otherwise, based on the following equation for non-equilibrium sediment transport

$$\frac{d(P_{4.l}S)}{dx} = -\frac{\alpha\omega_l}{q}\left(P_{4.l}S - P^*_{4.l}S^*\right) \tag{7.1}$$

the sedimentation condition for each size group would be the same. In the general case, one cannot assume $P^*_{4,l} = P_{4,l}$, or the following relation

$$P^*_{4.l} = P_{1.l.1}\frac{S^*(l)}{S^*(\omega^*_1)} \tag{7.2}$$

which indicates the size distribution for carrying-sediment capacity is depends only on the bed material gradation $P_{1.l.1}$ [6, 12, 13]. For non-equilibrium sediment transport, the bed material gradation is not only correlated to the size distribution of sediment concentration. It is unreasonable to determine the carrying-sediment capacity solely based on the bed material gradation without considering the sediment concentration and its size distribution in the water body. On the other hand, it is also unreasonable to use the size distribution of sediment concentration to represent the size distribution for carrying-sediment capacity. Intuitively, it is more appropriate to include the contributions from both the bed material gradation and the size distribution of sediment concentration in the determination of the size distribution for carrying-sediment capacity. This concept was first developed by Han [6] and He and Han [12, 13], and was further proved by Han [7, 8] based on theoretical analysis. It

Q. Han and M. He, *Non-equilibrium Transport of Non-uniform Suspended Load*, Hydroscience and Engineering, https://doi.org/10.1007/978-981-97-1123-9_7

is obvious that the sediment carried by the water should contribute to the carrying-sediment capacity, and in fact contributes the most if the river is not very long. From the aspect of sediment exchange, the bed material gradation should also contribute to the size distribution for carrying-sediment capacity, particularly in scour process. This contribution can be expressed as

$$P^*_{4.l.1} = \frac{P_{1.l.1} S^*(l)}{S^*} \tag{7.3}$$

It is noted that $P_{1.l.1}$ is not directly used to determine the carrying-sediment capacity. Instead, its contribution is made in the form of Eq. (7.3). Of course, weighting factors should be used when summing up the contribution from the bed material gradation and the contribution from the size distribution of sediment concentration.

For non-equilibrium transport of non-uniform sediment, there are two kinds of bed material gradation. One is the real bed material gradation $P_{1.l.1}$. The other is the gradation during sediment exchange process, which not only depends on the real bed material gradation $P_{1.l.1}$, but is also impacted by the coarser sediment that settles down during the exchange process. This gradation is referred to as the effective bed material gradation $P_{1.l}$. The interval of sediment exchange is much shorter than the time scale of the studied sediment transport problem, and thus the impact of settled sediment should be considered. Similar to the size distribution for carrying-sediment capacity, the effective bed material gradation should be determined with the summation of the real bed material gradation $P_{1.l.1}$ and the size distribution of sediment concentration.

In order to illustrate the importance of introducing the size distribution for carrying-sediment capacity and the effective bed material gradation for non-equilibrium transport of non-uniform suspended load, an example of sediment transport in a pebble-bed stream is discussed in the following for four cases.

Case 1. Clear water flows on pebble bed without suspended load inflow at the entrance. From the concept of dividing suspended load into bed material load and wash load, all the suspended load would be wash load in this case. Since there is no wash load inflow at the entrance, there is no sediment in the water. From the viewpoint of non-equilibrium transport of non-uniform suspended load, the explanation is different. Because the bed material is pebble, ω^* is very large and the carrying-sediment capacity is approximately zero. Thus, no sediment can be picked up from the bed, and the sediment concentration is zero.

Case 2. The concentration of the suspended load that is carried by flow in a pebble-bed stream does not exceed the carrying-sediment capacity $S^*\omega$computed with the average fall velocity ω. In this case, settling of suspended load only occurs during the instantaneous sediment exchange process. For a larger time scale, the settled sediment will not deposits on the bed and will all be transported by flow. From the concept of dividing suspended load into bed material load and wash load, the wash load is not involved in bed evolution, and does not deposit on the bed. From the viewpoint of non-equilibrium transport of non-uniform suspended load, the explanation is different and depends on how to determine the bed material gradation. If using the existing pebble gradation as the bed material gradation, the computed carrying-sediment capacity is zero, which is apparently inconsistent with reality. If the carrying-sediment capacity is zero, suspended load will deposits and the bed material will become finer. The sediment deposited on the bed will then contribute the carrying-sediment capacity that shall be determined with the size distribution of the deposited sediment, or using ω to represent ω^*. Consequently, the deposited sediment will be picked up and carried away by the flow. In other words, suspended sediment will deposit on the pebble-bed to become part of the bed material during the instantaneous sediment change process. However, for a larger time scale, the bed material still consists of the pebble with all suspended load be carried away by the flow. Therefore, from a dynamic aspect, the bed material gradation shall include the sediment deposited on the bed during sediment exchange process, or the effective bed material gradation shall be used.

Case 3. The concentration of the suspended load that is carried by flow in a pebble-bed stream exceeds the carrying-sediment capacity $S^*(\omega)$ computed with the average fall velocity ω, resulting in sediment deposition on the bed. From the concept of dividing suspended load into bed material load and wash load, part of the suspended load is bed material load if considering the fining of bed material after sediment deposits on the bed, and thus sediment deposition occurs. However, it is hard to explain the initial sediment deposition on a gravel-bed, when all suspended load is wash load. From the viewpoint of non-equilibrium transport of non-uniform suspended load, the bed material gradation shall be the size distribution after the suspended load deposits on the bed. As a result, $S^*(\omega) < S$, or the sediment concentration exceeds the carrying-sediment capacity, and sediment deposition occurs.

Case 4. Clear water flows on a bed with medium to fine sediment. In this case, the sediment concentration is zero and its size distribution does not exist. Apparently, the size distribution of suspended load cannot be used to represent the size distribution for carrying-sediment capacity, or it is hard to explain the reason for the initial scour. In this case, the bed material gradation plays a major role for the carrying-sediment capacity.

In conclusion, the size distribution for carrying-sediment capacity depends on both the bed material gradation and the size distribution of suspended load. The suspended load consumes part of the carrying-sediment capacity of the flow, and thus impacts the carrying-sediment capacity. From the sediment exchange aspect, the bed material gradation also impacts the carrying-sediment capacity. As a result, to determine the carrying-sediment capacity instead of the bed material gradation an effective bed material gradation, shall be used, which represents both the dynamic process and the characteristics of the instantaneous sediment exchange. This effective bed material gradation shall include both the bed material gradation and the suspended load that instantaneously deposits on the bed, and thus is closely related to the sediment concentration. Thus, it is more appropriate to use the effective bed material gradation to determine the size distribution of the carrying-sediment capacity for non-equilibrium sediment transport.

In his study for non-equilibrium transport of non-uniform sediment, Han [6] first developed the concept of the size distribution for carrying-sediment capacity, and the effective bed material gradation. This concept was later adopted by other Chinese researchers with various understandings and expressions for the size distribution for carrying-sediment capacity, as reviewed in Han [8] and in Chap. 1 of this book.

It is noted that the bed material gradation was used to sum up the fractional carrying-sediment capacity, and the derived gradation was also referred to as "the size distribution for carrying-sediment capacity" in some studies. However, these studies are for strong equilibrium condition, and "the size distribution for carrying-sediment capacity" is actually the size distribution of sediment concentration, which is different from the size distribution of carrying-sediment capacity for non-equilibrium transport as discussed in this chapter.

7.2 Theoretical Basis and Characteristics for the Size Distribution of Carrying-Sediment Capacity

Theoretically the size distribution for carrying-sediment capacity has the following properties [8]:

Property 1. For flow with medium to low sediment concentration, the fractional carrying-sediment capacity can be approximated with the carrying-sediment capacity of the uniform sediment with the corresponding particle size, or.

$$S_l^* = P_{4.l}^* S^*(\omega^*) = P_{1.l} S^*(l) \tag{7.4}$$

where $P_{1.l}$ is the effective bed material gradation. In strong equilibrium condition,

$$S_l^* = P_{4.l}^* S^*\left(\omega^*\right) = P_{1.l.1} S^*(l) \tag{7.5}$$

where $P_{1.l.1}$ is the actual bed material gradation, $S^*(l) = S^*(\omega_l)$ is the carrying-sediment capacity for the uniform sediment with particle size of D_1. Based on Eq. (7.4.23),

$$S^* P_{4.l}^* = \frac{2}{3} m_0 \gamma_s \frac{\beta_l P_{1.l.1} D_l L_{4.l}^*}{1-(1-\varepsilon_l)(1-\beta_l)+(1-\varepsilon_{0.l})(1-\varepsilon_{4.l})} \frac{1}{q t_{4.0.l}} \tag{7.6a}$$

For $P_{1.l} = P_{4.l}^* = 1$, or for uniform sediment

$$S^*(l) = \frac{2}{3} m_0 \gamma_s \frac{\beta_l D_l L_{4.l}^*}{1-(1-\varepsilon_l)(1-\beta_l)+(1-\varepsilon_{0.l})(1-\varepsilon_{4.l})} \frac{1}{q t_{4.0.l}} \tag{7.6b}$$

Plugging Eq. (7.6) into the previous equation yields

$$P_{4.l.}^{*} S^{*} = S_l^{*} = P_{1.l.1} S^{*}(l)$$

It is noted that the general form of Eq. (7.4) is used here to represent a formula for the carrying-sediment capacity, which is true no matter for equilibrium or non-equilibrium condition. However, the formulas for $P_{4.l}^{*}$ and $P_{1.l}$ are different between equilibrium and non-equilibrium conditions [12, 13]. In strong equilibrium condition, $P_{4.l}^{*} = P_{4.l}$, $P_{1.l} = P_{1.l.1}$ with $P_{1.l.1}$ be the actual bed material gradation, and Eq. (7.5) holds. In a more general condition, $P_{1.l}$ is the effective bed material gradation that is different from $P_{1.l.1}$ and $P_{4.l} \neq P_{4.l}^{*}$. This will be discussed further below.

Property 1 is important for the size distribution of carrying-sediment capacity. It illustrates the components of the carrying-sediment capacity for non-uniform sediment mixture, and its relation to the carrying-sediment capacity of uniform sediment. In order to better understand the reliability of the previous derivation and the mechanism for the carrying-sediment capacity for non-uniform sediment, Eq. (7.4) is proved in the following using a different approach. For steady and strong equilibrium sediment transport

$$\varepsilon_y \frac{dS_l^{*}(x)}{dy} + \omega_l S_l^{*} = 0$$

or

$$\frac{dS_l^{*}(y)}{S_l^{*}(y)} = -\frac{\omega_l}{\varepsilon_y} dy$$

Integrating this equation with $y = 0$, $S_l^{*}(0) = S_{b.l}^{*}$ yields

$$S_l^{*}(y) = S_{bl}^{*} \exp\left(-\int_0^y \frac{\omega_l}{\varepsilon_y} dt\right) \tag{7.7}$$

where $S_{b.l}^{*}$ is the carrying-sediment capacity on the bed for the lth size group of the non-uniform sediment mixture, $S_l^{*}(y)$ is the carrying-sediment capacity for the lth size group at a vertical location of y, and ε_y the sediment diffusion coefficient in the y-direction. The depth-averaged carrying-sediment capacity is

$$S_l^{*} = \frac{1}{h\overline{V}} \int_0^h V S_l^{*}(y) dy = \frac{S_{bl}^{*}}{\overline{V}} \int_0^1 V \exp\left(-\int_0^{\eta} \frac{\omega_l}{\varepsilon_y} dt\right) d\eta \tag{7.8}$$

where h is the water depth, V the flow velocity at y, $\overline{V}$ the depth-averaged flow velocity, and $\eta = y/h$ the relative vertical location. For uniform sediment

$$S^*(l) = \frac{1}{h\overline{V}} \int_0^h V V S_l^*(y) dy = \frac{S_b^*(l)}{\overline{V}} \int_0^1 V \exp\left(- \int_0^\eta \frac{\omega_l}{\varepsilon_y} dt \right) d\eta \tag{7.9}$$

It can be derived from Eqs. (7.8) and (7.9) that

$$\frac{S_l^*}{S^*(l)} = \frac{S_{b.l}^*}{S_b^*(l)} \tag{7.10}$$

For a certain volume with an unit area on the bed surface and with a thickness of a particle size of D_1, the number of particles that are at rest is $N_{1.l} = \frac{P_{1.l.1} m_0}{\frac{\pi}{4} D_l^2} R_{1.l}$ based on Sect. 3.3.4, and the number of suspended particles is $N_{4.l} = \frac{R_{4.l}}{R_{1.l}} N_{1.l}$. As discussed before, let $R_{1.l}$ be the probability for particles at rest on bed, and $R_{4.l}$ the probability for bed material to be suspended. The weight of the suspended sediment within this volume, is

$$W_{b.l} = N_{4.l} \frac{\pi}{6} D_l^3 \gamma_s = \frac{P_{1.l.1} m_0 D_l}{\frac{\pi}{4} D_l^2} R_{1.l} \frac{R_{4.l}}{R_{1.l}} \frac{\pi}{6} D_l^3 \gamma_s = \frac{2}{3} m_0 \gamma_s P_{1.l.1} D_l R_{4.l} \tag{7.11}$$

The weight per unit volume, or the sediment concentration at the bottom is

$$S_{b.l}^* = \frac{W_{b.l}}{D_l} = \frac{2}{3} m_0 \gamma_s P_{1.l.1} R_{4.l} \tag{7.12}$$

Similarly, the bottom concentration of the uniform sediment with the particle size of D_1 is

$$S_b^*(l) = \frac{2}{3} m_0 \gamma_s R_{4.l} \tag{7.13}$$

Plugging Eqs. (7.12) and (7.13) into Eq. (7.10) yields

$$\frac{S_l^*}{S^*(l)} = P_{1.l.1} \tag{7.14}$$

Considering $S_l^* = P_{4.l}^* S^*$, Eq. (7.14) becomes

$$P_{4.l}^* S^* = S_l^* = P_{1.l.1} S^*(l)$$

which is the same as Eq. (7.5) for Property 1. This not only proves Property 1 from a different approach, but also indicates that the water percentage ($K_l = P_{1.l} = P_{1.l.1}$) is constant along the depth. It is noted that the relationship between the carrying-sediment capacity (or sediment concentration) and the bed material gradation that was derived by Li [14] for equilibrium condition can also be represented by Property 1, though a different sediment exchange model was used by Li. According to Li [14], the bottom sediment transport capacity is

$$S_{b.l}^{*} \equiv \overline{V}_l(1 - A_l) = \overline{V} P_{1.l.1}(1 - A_l) \tag{7.15}$$

$$A_l = \omega_l \left[u_* \exp\left(-\frac{\omega_l^2}{2u_*^2}\right) + \omega_l \frac{1}{\sqrt{2\pi}} \int_{-\infty}^{\omega} \exp\left(-\frac{t^2}{2}\right) dt \right]^{-1} \tag{7.16}$$

and the depth-averaged fractional sediment transport capacity is

$$S_l^* = P_{1.l.1}\overline{V}(1 - A_l)\frac{\kappa u_*}{6\omega_l}\left[1 - \exp\left(-\frac{6\omega_l}{\kappa u_*}\right)\right] = P_{1.l.1}S^*(l) \tag{7.17}$$

Hence, the carrying-sediment capacity for the uniform sediment with the particle size of D_1 can be written as

$$S^*(l) = \overline{V}(1 - A_l)\frac{\kappa u_*}{6\omega_l}\left[1 - \exp\left(-\frac{6\omega_l}{\kappa u_*}\right)\right] \tag{7.18}$$

Dividing Eq. (7.17) by Eq, (7.18) yields

$$\frac{S_l^*}{S^*(l)} = P_{1.l.1} \tag{7.19}$$

or

$$P_{4.l}^* S^* = S_l^* = P_{1.l.1} S^*(l)$$

which is the same as Eq. (7.5) for Property 1.

Property 2. The total carrying-sediment capacity for the non-uniform sediment mixture can be written as.

$$S^* = \sum P_{1.l} S^*(l) \tag{7.20}$$

which can be derived with summation of Eq. (7.4) for all size groups. In strong equilibrium condition, $P_{1.l}$ can be replaced with $P_{1.l.1}$, and this equation can also be derived from Eq. (7.5). It can be seen that the total carrying-sediment capacity is the linear summation of the transport capacities $S^*(l)$ for the uniform sediment with corresponding sizes, with the weighting factor of either the effective bed material gradation $P_{1.l}$ or the bed material gradation $P_{1.l.1}$. Also, under strong equilibrium condition, the summation of the transport rate for non-uniform bed load is

$$P_{b.l}^* q_b^* = q_{b.l}^* = P_{1.l.1} q^*(D_l) = P_{1.l.1} q_b^*(l) \tag{7.21}$$

or

$$q_b^* = \sum P_{1.l.1} q_b^*(D_l) = \sum P_{1.l.1} q_b^*(l) \tag{7.22}$$

where $q_{b.l}^*$ is the bed load transport capacity per unit width of lth size group, q_b^* the total bed load transport capacity, $q_b^*(D_l)$ the transport capacity per unit width for the uniform sediment with the particle size of D_1, $P_{1.l.1}$ the bed material gradation, and $P_{b.l}^*$ the size distribution for bed load. The difference between the bed load transport capacity and the transport rate q_b shall be emphasized. The bed load transport is rarely in equilibrium condition. However, the saturation recovers very quickly (Li 1981). Equation (7.22) was first developed by Einstein [2]. This equation was later revised by introducing the parameter $D_l/\overline{D}$ after considering the impact among various size groups, but with the same summation method. For example, Ashida and Michiue [1], and Taizo and Yukio [16] proposed

$$q_b^* = \sum P_{1.l.1} q_b^* \left(D_l, \frac{D_l}{\overline{D}} \right) \tag{7.23}$$

Han (1990) also proved this formula based on the statistical theory for sediment exchange on the bottom. It should be pointed out that some studies used different summation method. For example, using the bed load gradation $P_{T.l}$ in the summation, Locinski et al. [15] proposed

$$q_b^* = \sum P_{b.l}^* q_b^*(D_l) \tag{7.24}$$

which is apparently incorrect. In fact, Eq. (7.24) was derived with summation of the following formula

$$P_{bl}^* q_b^* = P_{bl}^* q_b^*(D_l) \tag{7.25}$$

From Eq. (7.25),

$$q_b^* = q_b^*(D_l) \quad (l = 1, 2, \ldots, n) \tag{7.26}$$

This indicates that the total transport capacity for non uniform bed load equals the transport capacity for the uniform sediment of each size group, which is obviously impossible.

Property 3. In strong equilibrium condition, the total carrying-sediment capacity computed with the size distribution of suspended load equals that computed with the bed material gradation. Because it is in strong equilibrium condition, Eq. (7.5) is satisfied and $P_{4l} = P_{4l}^*$, and thus Eq. (7.5) can be written as.

$$P_{4.l} = P_{1.l.1} \frac{S^*(l)}{S^*} \tag{7.27}$$

Summation over l yields

$$S^* = \sum P_{1.l.1} S^*(l) \tag{7.28}$$

Also, from Eq. (7.5),

$$P_{1.l.1} = P_{4.l} \frac{S^*}{S^*(l)} \tag{7.29}$$

Summation for l yields

$$S^* = \frac{1}{\sum \frac{P^*_{4.l}}{S^*(l)}} \tag{7.30}$$

Since they are both for S^*, Eq. (7.28) should equal Eq. (7.30). Property 3 was first discovered by Han [4]. It is noted that the summation method to determine the total carrying-sediment capacity of suspended load used in the HEC.6 model, which was developed by the U.S. Army Corps of Engineers, is a special case as described in Eq. (7.28) [3]. This method seems to only apply to strong equilibrium sediment transport. It may not be appropriate to directly apply it to a general condition of sediment transport.

If Zhang's formula is used for the carrying-sediment capacity of uniform sediment, or

$$S^*(l) = \kappa \left(\frac{V^3}{gh\omega_l} \right)^m \tag{7.31}$$

then from Eqs. (7.27) and (7.29),

$$S^* = \sum P_{1.l.1} \kappa \left(\frac{V^3}{gh\omega_l} \right)^m = \kappa \left(\frac{V^3}{gh\omega} \right)^m \tag{7.32}$$

$$S^* = \frac{1}{\sum \frac{P_{4.l}}{\kappa \left(\frac{V^3}{gh\omega_l} \right)^m}} = \kappa \left(\frac{V^3}{gh\omega} \right)^m \tag{7.33}$$

and the average fall velocity ω for determining the carrying-sediment capacity is defined as

$$\omega^m = \sum P_{4.l} \omega_l^m = \frac{1}{\sum \frac{P_{1l.1}}{\omega_l^m}} \tag{7.34}$$

It is noted that if the average fall velocity is determined with Eq. (7.34), the formula for the total transport capacity of the non-uniform sediment mixture is the same as the formula for uniform sediment. In a strong equilibrium condition, both the size distribution of suspended load and the bed material gradation can be used to determine the total carrying-sediment capacity as long as Eq. (7.34) is used for the average fall velocity.

This can be used to explain the fact that the average fall velocity was determined using the size distribution of suspended load in some studies for the carrying-sediment capacity, which reads

$$\omega = \sum P_{4l}\omega_l \tag{7.35}$$

or was set to a representative fall velocity. This average fall velocity or the representative fall velocity, such as $\omega(D_{50})$, was introduced intuitively without theoretical derivation. Because $m \approx 1$, the results are approximately consistent with the theoretical relationship as shown in Eq. (7.34). In addition, the bed material gradation was used in some other studies for the summation of the fractional sediment concentration, which is also acceptable. Of course, if the sediment transport is not in strong equilibrium condition, either the bed material gradation or the size distribution of suspended load can not be solely used to determine the total carrying-sediment capacity in order to avoid errors.

Property 4. For strong equilibrium sediment transport, the size distribution of suspended load is correlated with the actual bed material gradation $P_{1.l.1}$. In other words, for a given bed material gradation, the size distribution of suspended load can be determined, and vice versa. Plugging Eq. (7.31) into Eqs. (7.27) and (7.29), respectively, yields.

$$P_{4.l} = P_{1.l.1}\frac{S^*(l)}{S^*(\omega^*)} = \frac{\frac{P_{1.l.1}}{\omega_l^m}}{\sum \frac{P_{1.l.1}}{\omega_l^m}} \tag{7.36}$$

$$P_{1.l.1} = P_{4.l}\frac{S^*(\omega^*)}{S^*(l)} = \frac{P_{4.l}\omega_l^m}{\sum P_{4.l}\omega_l^m} \tag{7.37}$$

Property 5. For strong equilibrium sediment transport, the bed material gradation $P_{1.l.1}$ equals the water percentage K_1 that carries the lth size group of sediment.

Let consider an unit-area water column that carries n groups of sediment with the particle size of D_1, D_2, D_n, respectively. Defining a water body with the water volume of Q_l that carries sediment with the weight of G_l^* and then the sediment concentration equal to $S^*(l)$

$$S^*(l) = \frac{G_l^*}{Q_l}$$

The fractional carrying-sediment capacity

$$S_l^* = \frac{G_l^*}{\sum Q_l} = \frac{G_l^*}{Q_l}\frac{Q_l}{\sum Q_i} = S^*(l)K_l \tag{7.38}$$

where

$$K_l = \frac{Q_l}{\sum Q_l} \tag{7.39}$$

is the water percentage that carries the lth size group of sediment, which was first introduced by [4], and was later verified by [11]. Comparing Eq. (7.38) with Eq. (7.5), one obtains

$$K_l S^*(l) = P_{1.l.1} S^*(l) \tag{7.40}$$

or

$$K_1 = P_{1.l.1} \tag{7.41}$$

Introduction of the concept of the water percentage provide an additional approach for analysis of the non-uniform carrying-sediment capacity.

Property 6. For non-equilibrium sediment transport, with the following relationship between the size distribution for carrying-sediment capacity and the water percentage is still satisfied:

$$P^*_{4.l} S^* = S^*_l = K_l S^*(l) \tag{7.42}$$

However, Eq. (7.5), or

$$S^*_l = P^*_{4.l} S^*(\omega^*) = P_{1.l.1} S^*(l)$$

is generally not true. For non-equilibrium sediment transport, the size distribution for carrying-sediment capacity should be determined based on both the bed material gradation and the size distribution of suspended sediment. Apparently, the size distribution for carrying-sediment capacity generally does not equal the bed material gradation alone or the size distribution of suspended sediment alone.

In fact, the derivation of Eq. (7.38) only used the carrying-sediment capacity and the water percentage without involvement of S and $P_{4.l}$, and thus Eqs. (7.38) and (7.39) still applies for non-equilibrium sediment transport. However, the derivation of Eq. (7.5) is limited for the strong equilibrium condition. In addition, for sediment laden flow on pebble or rock bed, the bed material gradation (the gradation for part of the bed material that can be exchanged with suspended load or can be suspended) is zero, or $S^*(l) = 0$. According to Eq. (7.4)

$$S^*_l = P_{1.l.1} S^*(l) = 0 \tag{7.43}$$

and then

$$S^* = \sum S_l^* = 0 \tag{7.44}$$

which is apparently not true. Therefore, for non-equilibrium sediment transport, the carrying-sediment capacity cannot be determined solely based on the bed material gradation. In fact, for quite some cases, such as the case with the carrying-sediment capacity of suspended load is close to saturation state or is in over-saturation condition (channel deposition occurs), the size distribution for carrying-sediment capacity should be determined mainly from the size distribution of suspended load. This is why the size distribution of suspended load was used in previous studies to determine the carrying-sediment capacity, or was used to represent the size distribution for carrying-sediment capacity.

Property 7. After dividing the non-uniform sediment mixture into various size groups, the fractional carrying-sediment capacity (for each size group) and the total carrying-sediment capacity are.

$$S_l^* = P_{4.l}^* S^*(\omega^*) = K_{1.l} S^*(l) + K_{2.l} S^*(l) \tag{7.45}$$

$$S^* = \sum K_{1l} S^*(l) + \sum K_{2l} S^*(l) = K_l S^*(l) \tag{7.46}$$

$$K_l = K_{1.l} + K_{2.l} \tag{7.47}$$

$$\sum K_{1.l} = K_1 \tag{7.48}$$

$$\sum K_{2.l} = K_2 = (1 - K_1) \tag{7.49}$$

$$K_1 + K_2 = 1 \tag{7.50}$$

where K_1 and K_2 are the water percentages for the two parts of the carrying-sediment capacity that are related to the sediment concentration and the bed material gradation, respectively, and are also represent the percentage of water power, and thus can be summated. The expressions of $K_{1.l}$ and $K_{2.l}$, and their relation with the original bed material gradation and the gradation of the bed material that is converted from the suspended load are discussed later.

The verification of the previous equations is straightforward. The carrying-sediment capacity for each size group, $S^*(l) = S^*(\omega_l)$, multiplying the corresponding water percentage $K_{1.l}$ or $K_{2.l}$ becomes the fractional carrying-sediment capacity. Summation of the fractional carrying-sediment capacities for all size groups is the total carrying-sediment capacity S^*.

Property 8. For non-equilibrium sediment transport, corresponding to the size distribution for carrying-sediment capacity, the effective bed material gradation $P_{1.l}$ also exists according to Eq. (7.4), and equals the water percentage K_l, or.

$$P_{1.l} = K_l \tag{7.51}$$

In fact, based on Properties 6 and Eq. (7.42)

$$P^*_{4.l} S^*(\omega^*) = S^*_l = K_l S^*(l) \tag{7.52}$$

This is not only true in the strong equilibrium condition, but also applies to other general conditions. Also, introducing the effective bed material gradation, which is related to the size distribution for carrying-sediment capacity according to property 1 and Eq. (7.4), one obtains

$$P^*_{4.l} S^*(\omega^*) = P_{1.l} S^*(l) \tag{7.53}$$

Comparing Eq. (7.53) with Eq. (7.52) yields

$$P_{1.l} = K_l \tag{7.54}$$

Property 9. As K_l consists of two terms as shown in Eq. (7.47), the effective bed material gradation should also consist of these two terms. From Eqs. (7.4) and (7.45) yields.

$$P_{1.l} = \frac{P^*_{4.l} S^*(\omega^*)}{S^*(l)} = \frac{K_{1.l} S^*(l)}{S^*(l)} + \frac{K_{2.l} S^*(l)}{S^*(l)} = K_{1.l} + K_{2.l} \tag{7.55}$$

Property 10. If the suspended sediment concentration is fully converted into the carrying-sediment capacity, and all bed material can be suspended, the fractional carrying-sediment capacity, the total carrying-sediment capacity, and the effective bed material gradation can be written as.

$$P^*_{4.l} S^*(\omega^*) = P_{4.l} S + \left[1 - \frac{S}{S^*(\omega)}\right] P^*_{4.l.1} S^*(\omega^*_1) \tag{7.56}$$

$$S^*(\omega^*) = S + \left[1 - \frac{S}{S^*(\omega)}\right] S^*(\omega^*_1) \tag{7.57}$$

$$\begin{aligned} P^*_{4.l} &= P_{4.l} \frac{S}{S^*(\omega^*)} + \left[1 - \frac{S}{S^*(\omega)}\right] \frac{S^*(\omega^*_1)}{S^*(\omega^*)} P^*_{4.l.1} \\ &= \frac{S}{S^*(\omega^*)} P_{4.l} + \left[1 - \frac{S}{S^*(\omega^*)}\right] P^*_{4.l.1} \end{aligned} \tag{7.58}$$

$$\begin{aligned} P_{1.l} &= \frac{P^*_{4.l} S^*(\omega^*)}{S^*(l)} = \frac{P_{4.l} S}{S^*(l)} + \left[1 - \frac{S}{S^*(\omega)}\right] P_{1.l.1} \\ &= \frac{S}{S^*(\omega)} P'_{1.l.1} + \left[1 - \frac{S}{S^*(\omega)}\right] P_{1.l.1} \end{aligned} \tag{7.59}$$

where

$$\sum P_{4.l}^{*} = 1 \tag{7.60}$$

$$\sum P_{1.l} = 1 \tag{7.61}$$

$P_{1.l.1}$is the bed material gradation, $P_{4.l.1}^{*}$is the size distribution of carrying-sediment capacity corresponding to the bed material gradation,

$$S^{*}(\omega_{1}^{*}) = \sum P_{1.l.1} S^{*}(l) \tag{7.62}$$

$$S^{*}(\omega) = \left[\sum \frac{P_{4.l}}{S^{*}(l)}\right]^{-1} \tag{7.63}$$

$$P_{4.l.1}^{*} = \frac{S^{*}(l)}{S^{*}(\omega_{1}^{*})} P_{1.l.1} \tag{7.64}$$

$S^{*}(\omega^{*})$ is the total carrying-sediment capacity that is

$$S^{*}(\omega^{*}) = \sum P_{1.l} S^{*}(l) = \frac{1}{\sum \frac{P_{4.l}^{*}}{S^{*}(l)}} \tag{7.65}$$

and $P_{1.l.1}'$ is the gradation of the material that is deposited out of the suspended load.

$$P_{1.l.1}' = \frac{S^{*}(\omega)}{S^{*}(l)} P_{4.l} \tag{7.66}$$

The derivation of Eq. (7.56) is as follows. Since all suspended load is converted into the carrying-sediment capacity, its contribution to the fractional carrying-sediment capacity is $P_{4.l} S$, which should equal $K_{1.l} S^{*}(l)$ based on Eq. (7.47) or

$$P_{4.l} S = K_{1.l} S^{*}(l) \tag{7.67}$$

Thus

$$K_{1.l} = \frac{P_{4.l} S}{S^{*}(l)} = \frac{\mathrm{S}}{S^{*}(\omega)}\left[\frac{S^{*}(\omega)}{S^{*}(l)} P_{4.l}\right] = \frac{\mathrm{S}}{S^{*}(\omega)} P_{1.l.1}' = K_{1} P_{1.l.1}' \tag{7.68}$$

$$K_{1} = \sum K_{1.l} = \sum \frac{P_{4.l} S}{S^{*}(l)} = \frac{S}{S^{*}(\omega)} \tag{7.69}$$

If the water percentage K_2 for the bed material to convert into the carrying-sediment capacity was 1, its contribution to the carrying-sediment capacity should be $P_{4.l.1}^{*} S^{*}(\omega_{1}^{*})$, which accounts for $K_{2.l} S^{*}(l)$ of the total carrying-sediment capacity.

However, $K_2 = 1 - K_1$. Therefore, the contribution of the bed material to the carrying-sediment capacity is $(1 - K_1)P^*_{4.l.1}S^*(\omega^*_1)$, and the corresponding fractional carrying-sediment capacity can be written as

$$K_{2.l}S^*(l) = (1 - K_1)P_{1.l.1}S^*(l) = (1 - K_1)P^*_{4.l.1}S^*(\omega^*_1) \quad (7.70)$$

plugging Eq. (7.64) in which yields

$$K_{2.l} = \frac{(1 - K_1)P^*_{4.l.1}S^*(\omega^*_1)}{S^*(l)} = K_2 P_{1.l.1} = \left[1 - \frac{S}{S^*(\omega)}\right]P_{1.l.1} \quad (7.71)$$

Combining Eqs. (7.47), (7.70), (7.69) and (7.67), one obtains

$$P^*_{4.l}S^*(\omega^*) = (K_{1.l} + K_{2.l})S^*(l) = P_{4.l}S + \left[1 - \frac{S}{S^*(\omega)}\right]P^*_{4.l.1}S^*(\omega^*_1) \quad (7.72)$$

which is the same as Eq. (7.56). Equation (7.57) can be obtained by summing up Eq. (7.72) for$l$. Equation (7.58) can be derived by dividing Eq. (7.72) by $S^*(\omega*)$. It is noted that the size distribution for carrying-sediment capacity is the summation of $P_{4.l}$ and $P^*_{4.l.1}$ with the weighting factor of K_1 andK_2, respectively. Considering Eq. (7.53), multiplying Eq. (7.58) by $\frac{S^*(\omega*)}{S^*(l)}$ yields Eq. (7.59). It is noted from Eq. (7.59) that the effective bed material gradation consists of two terms, and is the sum of the original bed material gradation $P_{1.l.1}$ and the gradation of the deposited material $P'_{1.l.1}$ with the weighting factors of K_1 and K_2, respectively. Summation of Eq. (7.59) for l yields

$$\begin{aligned} \sum P_{1.l} &= \sum \frac{P_{4.l}S}{S^*(l)} + \sum \left[1 - \frac{S}{S^*(\omega)}\right]\frac{S^*(\omega^*_1)}{S^*(l)}P^*_{4.l.1} \\ &= \frac{S}{S^*(\omega)} + \left[1 - \frac{S}{S^*(\omega)}\right]\sum P_{1.l.1} = 1 \end{aligned} \quad (7.73)$$

Similarly, considering Eq. (7.57), it can be derived from Eq. (7.58) that$\sum P^*_{4.l} = 1$.

It should be emphasized that the only constraint condition for Eqs. (7.56) through (7.59) is

$$K_2 = 1 - \frac{S}{S^*(\omega)} > 0 \quad (7.74)$$

which is actually the condition for scour. If this condition is not satisfied, or in the case with strong equilibrium condition or with massive sediment deposition, Eqs. (7.56)–(7.59) can not be directly applied. In this case, the second terms on the right sides of these equations should be deleted, and the water percentage of the first terms should be changed to 1. In order to change the original water percentage of $\frac{S}{S^*(\omega)}$ to 1, it has to be divided by $\frac{S}{S^*(\omega)}$. As a result, Eqs. (7.68) and (7.69) should be revised accordingly, or

$$K_{1.l} = \frac{\frac{P_{4.l} S}{S^*(l)}}{\frac{S}{S^*(\omega)}} = \frac{P_{4l} S^*(\omega)}{S^*(l)} \tag{7.75}$$

Thus

$$\mathrm{K}_1 = \sum \frac{S^*(\omega) P_{4.l}}{S^*(l)} = \frac{S^*(\omega)}{S^*(\omega)} = 1 \tag{7.76}$$

Also, from Eq. (7.58), one obtains

$$P^*_{4.l} S^*(\omega^*) = P_{4.l} S^*(\omega) \tag{7.77}$$

Summing up for l yields

$$S^*(\omega^*) = S^*(\omega) \tag{7.78}$$

$$P^*_{4.l} = P_{4.l} \tag{7.79}$$

$$P_{1.l} = P_{4.l} \frac{S^*(\omega^*)}{S^*(l)} = \frac{P_{4.l} S^*(\omega)}{S^*(l)} = (\frac{\omega_1}{\omega})^m P_{4.l} = P'_{1.l} \tag{7.80}$$

Equations (7.56)–(7.59) together with Eqs. (7.77)–(7.80) form a general set of governing equations for the size distribution for carrying-sediment capacity. While Eqs. (7.56)–(7.59) describe the case with apparent scour, Eqs. (7.77)–(7.80) apply to the case with apparent sediment deposition. As shown in Eqs. (7.77)–(7.80), in the case with apparent deposition, the size distribution for carrying-sediment capacity equals the size distribution of suspended load, the carrying-sediment capacity is only dependent on the size distribution of suspended load, and the effective bed material gradation $P_{1.l}$ equals the gradation of the deposited material $P'_{1.l}$ which only depends on the size distribution of suspended load and is independent of the bed material gradation. These characteristics were discussed early by Han [6], and by He and Han [13].

It is noted that Property 10 only illustrates a simple relation for the size distribution of carrying-sediment capacity in a special case. More generally, suspended load does not all convert into the carrying-sediment capacity, and only partial bed material is suspended. Discussions for these general conditions were presented in [7, 13], and [8], and will be presented in the following section.

7.3 Size Distribution for Carrying-Sediment Capacity and Effective Bed Material Gradation in a Cross Section

7.3.1 *Fractional Carrying-Sediment Capacity and Water Percentage for Scour and Mild Deposition*

In most cases of sediment transport, including the mild deposition, mild scour, and equilibrium conditions, the sedimentation characteristics are different between coarser and finer sediment. According to Eq. (10.10) of Chap. 10, the threshold particle size D_k is defined as

$$S^*(\omega_1^*) = \sum_{l=1}^{n} P_{1.l.1} S^*(l)$$

$$D_k = f^{-1}(\omega_1^*).$$

where D_k is the inverse function of $\omega_k^* = \omega_1^* = f(D_k)$. If there are coarser particles in the bed material that cannot be suspended, $P_{1.l.1}$ should be replaced with $P_{1.l.1.1} = P_{1.l.1} / \sum_{l=1}^{n} P_{1.l.1}, \quad (l = 1, 2, \ldots, n < N)$, or the gradation of the bed material after excluding the coarser particles $n < l \leq N$ According to the above equation and the definition of the threshold particle size, one obtains

$$S^*(\omega_1^*) = S^*(D_k) = S^*(\omega_k) \tag{7.81}$$

The basis for the determination of the threshold particle size will be discussed in detail in Sect. 10.2.2.

Assume that the bed material consists of N size groups, the suspended load consists of n size groups, and $n < N$. The bed material with a particle size $D_l > D_n$ cannot be suspended. The finer suspended particles with $D_l \leq D_k$ are not involved in the exchange with the bed material (in a longer time scale, not in the instantaneous exchange process). The coarser suspended sediment with $\mathrm{D}_{k+1} \leq \mathrm{D}_l \leq \mathrm{D}_n$ are the particles that take part in the exchange with the bed material, and D_n is the maximum particle size that can be suspended. Let divide the sediment transport capacity into three parts. The first part is the finer portion of the suspended load inflow with the particle size $D_l \leq D_k (l = 1, 2, \ldots, k \leq n)$, which is all counted in the carrying-sediment capacity. The second part is the coarser portion of the suspended load with the particle size $\mathrm{D}_{k+1} \leq \mathrm{D}_l \leq \mathrm{D}_n$, which partially settles on the bed and partially is counted in the carrying-sediment capacity. The third part is contributed from the bed material. If the variation range of the grain size for the bed material is wider than that for the suspended load, it is possible that the coarser bed material with $\mathrm{D}_n < \mathrm{D}_l \leq \mathrm{D}_N$ cannot be suspended, which should be excluded from the contribution to the carrying-sediment capacity.

The carrying-sediment capacity and corresponding size distributions are analyzed as follows. Corresponding to the three parts of carrying-sediment capacity, the water percentage can be divided into three parts, K_1, K_2, and K_3, each carrying the corresponding part of sediment. Each water percentage can be further divided into various parts based on the size groups, which are denoted as $K_{1.l}$, $K_{2.l}$, and $K_{3.l}$. Considering the lth size group for the first part of sediment, which does not take part in the exchange with the bed material from an aspect of a longer time scale, the fractional sediment concentration is completely converted into the fractional carrying-sediment capacity, or $P_{4.l}^* S^* = P_{4.l} S$. Based on Eq. (7.52), one obtains

$$P_{4.l}^* S^* = P_{4.l} S = K_{1.l} S^*(l)$$

or the water percentage

$$K_{1.l} = \frac{S}{S^*(l)} P_{4.l} \quad (l = 1, 2, \ldots, k) \tag{7.82}$$

Summation over l yields

$$K_1 = \sum_{l=1}^{k} K_{1.l} = S \sum_{l=1}^{k} \frac{P_{4.l}}{S^*(l)} = S P_{4.1} \sum_{i=1}^{k} \frac{P_{4.l.1}}{S^*(l)} \tag{7.83}$$

where

$$P_{4.l.1} = \begin{cases} \frac{P_{4.l}}{P_{4.1}} & (l = 1, 2, \ldots, k) \\ 0 & (l > k) \end{cases} \tag{7.84}$$

$$P_{4.1} = \sum_{l=1}^{k} P_{4.l} \tag{7.85}$$

Thus,

$$\sum_{l=1}^{k} \frac{P_{4.l}}{S^*(l)} = P_{4.1} \sum_{l=1}^{k} \frac{P_{4.l.1}}{S^*(l)} = \frac{P_{4.1}}{S^*(\omega_1)}$$

where

$$\frac{1}{S^*(\omega_1)} = \sum_{l=1}^{k} \frac{P_{4.l.1}}{S^*(l)} \tag{7.86}$$

$S^*(\omega_1)$ represents the average carrying-sediment capacity determined by the size distribution $P_{4.l.1}$, which is normalized from $P_{4.l}$ according to Eq. (7.84) and satisfies $\sum_{l=1}^{k} P_{4.l.1} = 1$ for the finer portion of the suspended load, with the corresponding fall velocity of ω_1. Plugging Eq. (7.86) into Eq. (7.83) yields

$$K_1 = \frac{P_{4.1} S}{S^*(\omega_1)} \tag{7.87}$$

Similar to Eq. (7.53), the size distribution of the carrying-sediment capacity that is contributed from the original bed material can be determined with the following equation:

$$P^*_{4.l.1} = P_1 P_{1.l.1.1} S^*(l) \big/ S^*(\omega^*_{1.1}) \tag{7.88}$$

It is noted that $S^*(\omega^*_{1.1})$, rather than $S^*(\omega^*_1)$, is used in the above equations considering part of the coarser bed material, with the size group index size $l > n$, cannot be suspended, and thus should be excluded from the bed material sorting process. Also,

$$S^*(\omega^*_{1.1}) = \sum_{l=1}^{n} P_{1.l.1.1} S^*(l) \tag{7.89}$$

$$P_{1.l.1.1} = \begin{cases} \frac{P_{1.l.1}}{P_1} & (l \le n) \\ 0 & (l > n) \end{cases} \tag{7.90}$$

where

$$P_1 = \sum_{l=1}^{n} P_{1.l.1} \tag{7.91}$$

is the proportion of the bed material that can be suspended. Due to the short time interval adopted in the analysis, it is assumed that the bed material does not change after the coarser suspended sediment particles settle on the bed. As the bed material does, the deposited sediment particles also contribute to the size distribution of carrying-sediment capacity after the sorting process. In other words, the fractional sediment concentration for these coarser particles is $P_{4.l} S$, which will all be converted into the bed material. The fractional carrying-sediment capacity $S \frac{P_{4.l} S^*(l)}{S^*(\omega^*_{1.1})}$ corresponding to the deposited coarser particles should equal $K_{2.l} S^*(l)$, or

$$K_{2,l} S^*(l) = S P_{4,l} \frac{S^*(l)}{S^*\left(\omega^*_{1.1}\right)} \quad (l = k+1, k+2, \ldots, n)$$

Thus,

$$K_{2.l} = P_{4.l} \frac{S}{S^*(\omega^*_{1.1})} = \frac{S P_{4.2} P_{4.l.2}}{S^*(\omega^*_{1.1})} \quad (l = k+1, k+2, \ldots, n) \tag{7.92}$$

Summation over l yields

$$K_2 = \frac{SP_{4.2}}{S^*(\omega_{1.1}^*)} \tag{7.93}$$

It is also noted that

$$P_{4.l.2} = \begin{cases} \frac{P_{4.l}}{P_{4.2}} & (k < l \le n) \\ 0 & (l > n) \end{cases} \tag{7.94}$$

$$P_{4.2} = \sum_{l=k+1}^{n} P_{4.l} \tag{7.95}$$

From Eq. (7.88), the size distribution $P_{4.l.1}^*$ for the carrying-sediment capacity corresponding to the bed material gradation $P_{1.l.1}$ can be written as

$$P_{4.l.1}^* = \frac{S^*(l)}{S^*(\omega_{1.1}^*)} P_{1.l.1} \tag{7.96}$$

As discussed before, $S^*(\omega_{1.1}^*)$ is used here so that only part of the bed material that can be suspended is included in the bed material's contribution to the carrying-sediment capacity. Thus, Eq. (7.96) can be rewritten as

$$P_{4.l.1}^* = \frac{S^*(l)}{S^*(\omega_{1.1}^*)} P_1 P_{1.l.1.1} = P_1 P_{4.l.1.1}^* \tag{7.97}$$

and $P_{4.l.1.1}^*$ is the size distribution of the carrying-sediment capacity corresponding to the bed material gradation $P_{1.l.1}$, or

$$P_{4.l.1.1}^* = \frac{S*(l)}{S*(\omega_{1.1}^*)} P_{1.l.1.1} \tag{7.98}$$

Considering Eqs. (7.87) and (7.93), the water percentage that is required to carry the part of the carrying-sediment capacity corresponding to bed material, if $K_1+K_2 < 1$, is

$$K_3 = 1 - K_1 - K_2 = 1 - \frac{SP_{4.1}}{S^*(\omega_1)} - \frac{SP_{4.2}}{S^*(\omega_{1.1}^*)} \tag{7.99}$$

The fractional carrying-sediment capacity related to the bed material gradation is

$$K_3 S^*(\omega_{1.1}^*) P_{4.l.1}^* = K_{3.l} S^*(l) \tag{7.100}$$

plugging Eq. (7.97) in which yields

$$K_3 S^*(l) P_1 P_{1.l.1.1} = K_{3.l} S^*(l) \tag{7.101}$$

or

$$K_{3.l} = P_1 K_3 P_{1.l.1.1} = P_1 K_3 \frac{S^*(\omega^*_{1.1})}{S^*(l)} P^*_{4.l.1.1} \tag{7.102}$$

Summation for l yields

$$\sum_{l=1}^{n} K_{3.l} = P_1 K_3 \tag{7.103}$$

It is noted that sum of $K_{3.l}$ is less than K_3. This because K_3, as defined in Eq. (7.99), is the remainder of the water percentage after subtracting K_1, the water percentage for the finer portion of suspended load, and subtracting K_2, the water percentage for the coarser portion of suspended load that deposits on the bed and then become part of the carrying-sediment capacity. Part of K_3, or $P_1 K_3$ is to carry the sediment for the part of the bed material that can be suspended. The other part of K_3, or $(1 - P_1)K_3$ corresponds to the part of the bed material that cannot be suspended, and thus is not used, or K_3 is in unsaturated condition. Here, $1 - P_1$ is the percentage of the bed material that cannot be suspended. It is easier to understand the unsaturated condition of the water percentage from the aspect of sediment exchange. During sediment exchange with the bed material, the suspended sediment particles may deposit completely on the bed surface, or partially fall into the gaps among the bed material which consists of coarse bed material. Sediment pickup only occurs on the bed surfaces that are occupied by the size groups of $l \leq n$, and the total percentage of these bed surface areas is $1 - P_1$. As discussed in Han [6] and in He and Han [13], the bed material gradation is also the percentage of the area occupied by the size group on the bed surface.

7.3.2 *Size Distribution for Carrying-Sediment Capacity and Effective Bed Material Gradation [6, 8, 12, 13]*

The fractional sediment transport capacity for the lth size group is the sum of the three parts as discussed in Sect. 7.3.1, or

$$P^*_{4.l} S^*(\omega^*) = (K_{1.l} + K_{2.l} + K_{3.l}) S^*(l)$$

where $S^*(\omega^*)$ is the total sediment transport capacity. Plugging Eqs. (7.82), (7.92) and (7.102) in it yields

$$P_{4.l}^{*}S^{*}(\omega^{*}) = SP_{4.1}P_{4.l.1} + SP_{4.2}P_{4.l.2}\frac{S^{*}(l)}{S^{*}(\omega_{1.1}^{*})} + \left[1 - \frac{P_{4.1}S}{S^{*}(\omega_{1})} - \frac{SP_{4.2}}{S^{*}(\omega_{1.1}^{*})}\right]P_{1}P_{4.l.1.1}^{*}S^{*}(\omega_{1.1}^{*}) \tag{7.104}$$

with the following constraint condition

$$K_1 + K_2 = \frac{P_{4.1}S}{S^{*}(\omega_{1.1}^{*})} + \frac{P_{4.2}S}{S^{*}(\omega_{1.1}^{*})} < 1 \tag{7.105}$$

Summing up over l, Eq. (7.104) becomes

$$S^{*}(\omega^{*}) = SP_{4.1} + SP_{4.2}\frac{S^{*}(\omega_{2}^{*})}{S^{*}(\omega_{1.1}^{*})} + \left[1 - \frac{P_{4.1}S}{S^{*}(\omega_{1})} - \frac{SP_{4.2}}{S^{*}(\omega_{1.1}^{*})}\right]P_{1}S^{*}(\omega_{1.1}^{*}) \tag{7.106}$$

where

$$S^{*}(\omega_{2}^{*}) = \sum_{l=k+1}^{n} P_{4.l.2}S^{*}(l) \tag{7.107}$$

As shown in Eq. (7.106), the carrying-sediment capacity consists of three parts. The first part corresponds to the finer portion of the suspended load, or $SP_{4.1}$, which is completed counted in the sediment transport capacity. The second part corresponds to the coarser portion of the suspended load, or $SP_{4.2}$, which is partially counted in the sediment transport capacity. In fact, for $l \geq k + 1$

$$S^{*}(l) < S^{*}\left(\omega_{1.1}^{*}\right) \tag{7.108}$$

This is because

$$S^{*}(\omega_{1.1}^{*}) = \sum_{l=1}^{n} \frac{P_{1.l.1}}{P_{1}}S^{*}(l) = \sum_{l=1}^{n} P_{1.l.1}S^{*}(l) = S^{*}(\omega_{k}) \tag{7.109}$$

where ω_k is the fall velocity of particle with the threshold size D_k (see Sect. 10.2.2). Also, it can be derived that

$$S^{*}(\omega_{1.1}^{*}) = S^{*}(\omega_{k}) > S^{*}(\omega_{k+1}) \geq S^{*}(\omega_{l}), \quad (l \geq k+1) \tag{7.110}$$

Thus,

$$S^{*}(\omega_{2}^{*}) = \sum_{l=k+1}^{n} P_{4.l.2}S^{*}(l) < S^{*}(\omega_{k+1})\sum_{l=k+1}^{n} P_{4.l.2} = S^{*}(\omega_{k+1}) \tag{7.111}$$

Based on Eqs. (7.109) through (7.111), one obtains

$$S^*(\omega_2^*) < S^*(\omega_{k+1}^*) < S^*(\omega_k) = S^*(\omega_{1.1}^*) \tag{7.112}$$

As shown in Eq. (7.112), $\frac{S^*(\omega_2^*)}{S^*(\omega_{1.1}^*)} < 1$. Based on Eq. (7.106), only part of $SP_{4.2}$ is counted in the carrying-sediment capacity. Furthermore, according to Eq. (7.110), the fractional carrying-sediment capacity exists for all the coarser suspended sediment with the size group index $l \geq k+1$, i.e.

$$SP_{4.2}P_{4.l.2}\frac{S^*(l)}{S^*(\omega_{1.1}^*)} < SP_{4.2}P_{4.l.2} = SP_{4.l} \quad (l = k+1, \ldots, n)$$

which indicates the coarser suspended load only partially contribute to the carrying-sediment capacity. As shown in Eq. (7.106), the third part of the carrying-sediment capacity is related to the bed material. The water percentage for this part is K_3P_1, while $K_3(1-P_1)$ out of this volume percentage is not used, or this water percentage is in unsaturated condition.

The size distribution for carrying-sediment capacity can be determined by dividing Eq. (7.104) by $S^*(\omega*)$, which reads

$$\begin{aligned} P_{4.l}^* = {} & P_{4.1}P_{4.l.1}\frac{S}{S^*(\omega^*)} + P_{4.2}P_{4.l.2}\frac{S}{S^*(\omega^*)}\frac{S^*(l)}{S^*(\omega_{1.1}^*)} \\ & + \left[1 - \frac{P_{4.1}S}{S^*(\omega_1)} - \frac{P_{4.2}S}{S^*(\omega_{1.1}^*)}\right]P_1P_{4.l.1.1}^*\frac{S^*(\omega_{1.1}^*)}{S^*(\omega^*)} \end{aligned} \tag{7.113}$$

Summation for l yields

$$1 - \frac{P_{4.1}S}{S^*(\omega^*)} - \frac{P_{4.2}S}{S^*(\omega^*)}\frac{S^*(\omega_2^*)}{S^*(\omega_{1.1}^*)} = \left[1 - \frac{P_{4.1}S}{S^*(\omega_1)} - \frac{P_{4.2}S}{S^*(\omega_{1.1}^*)}\right]P_1\frac{S^*(\omega_{1.1}^*)}{S^*(\omega^*)} \tag{7.114}$$

Thus Eq. (7.113) can be rewritten as

$$\begin{aligned} P_{4.l}^* = {} & P_{4.1}P_{4.l.1}\frac{S}{S^*(\omega^*)} + P_{4.2}P_{4.l.2}\frac{S}{S^*(\omega^*)}\frac{S^*(l)}{S(\omega_{1.1}^*)} \\ & + \left[1 - \frac{P_{4.1}S}{S^*(\omega^*)} - \frac{P_{4.2}S}{S^*(\omega^*)}\frac{S^*(\omega_2^*)}{S^*(\omega_{1.1}^*)}\right]P_{4.l.1.1}^* \end{aligned} \tag{7.115}$$

Multiplying Eq. (7.113) with $\frac{S^*(\omega^*)}{S^*(l)}$, and considering Eq. (7.53), the effective bed material gradation can be derived, which is

$$P_{1.L} = P_{4.l}^*\frac{S^*(\omega^*)}{S^*(l)} = P_{4.1}\frac{S}{S^*(l)}P_{4.l.1} + P_{4.2}\frac{S}{S^*(\omega_{1.1}^*)}P_{4.l.2}$$

$$+\left[1-\frac{P_{4.1}S}{S^*(\omega_1)}-\frac{P_{4.2}S}{S^*(\omega_{1.1}^*)}\right]P_1P_{1.l.1.1}\quad (l=1,2,\cdots,n) \tag{7.116}$$

As shown in Eqs. (7.113)–(7.116), the size distribution of carrying-sediment capacity and the effective bed material gradation also consists of three parts. It is noted that after considering part of the bed material cannot be suspended, the carrying-sediment capacity of the flow is not completely saturated, or the water percentage is not fully utilized, resulting $\sum_{l=1}^{n} P_{1.l} < 1$. In fact, based on Eq. (7.116),

$$\sum_{l=1}^{n} P_{1.l} = 1-\left[1-\frac{P_{4.l}S}{S^*(\omega_1)}-\frac{P_{4.2}S}{S^*(\omega_{1.1}^*)}\right](1-P_1) < 1 \tag{7.117}$$

For the condition as described in Eq. (7.112), the expression within the square bracket of Eq. (7.117) is less than 1, $1-P_1<1$, and thus $P_1 = \sum_{l=1}^{n} P_{1.l} < 1$. On the other hand, the size distribution for carrying-sediment capacity, as described in Eq. (7.115), is derived by dividing Eq. (7.113) by $S^*(\omega^*)$, and is thus normalized. As a result, $\sum_{l=1}^{n} P_{4.l}^* = 1$.

7.4 Size Distribution for Carrying-Sediment Capacity and Effective Bed Material Gradation for Pure Deposition (No Scour of Bed Material)

For pure deposition, $K_1+K_2>1$, and the third part of the sediment transport capacity does not exist, or no bed material is picked up during the sediment exchange process. This does not include the case for a gravel bed with bed material not erodible. Instead, it refers to the case with the flow being so weak that the sediment inflow exceeds the carrying-sediment capacity of the flow. In fact, in order for the sum of $K_1=\frac{P_{4.1}S}{S^*(\omega_1)}$ and $K_2=\frac{P_{4.2}S}{S^*(\omega_{1.1}^*)}$ to equal 1, $S^*(\omega_1)$ needs to be greater than $S^*(\omega_{1.1}^*)$. This requires the percentage of the finer suspended load (with size group index $l<k$) is small so that less suspended sediment will be completely counted in the sediment transport capacity. As a result, both K_1 and K_1+K_2 reduce.

For $K_3=0$, Eqs. (7.104), (7.106), (7.113), and (7.116) can be simplified as

$$P_{4.l}^*S^*(\omega^*) = P_{4.1}P_{4.l.1}S + P_{4.2}P_{4.l.2}S\frac{S^*(l)}{S^*(\omega_{1.1}^*)} \tag{7.118}$$

$$S^*(\omega^*) = P_{4.1}S + P_{4.2}S\frac{S^*(\omega_2^*)}{S^*(\omega_{1.1}^*)} \tag{7.119}$$

$$P_{4.l}^* = P_{4.1}\frac{S}{S^*(\omega^*)}P_{4.l.1} + \frac{P_{4.2}S}{S^*(\omega^*)}\frac{S^*(l)}{S^*(\omega_{1.1}^*)}P_{4.l.2} \tag{7.120}$$

$$P_{1.l} = \frac{P_{4.1}}{S^*(l)} P_{4.l.1} S + \frac{P_{4.2}}{S^*(\omega_{1.1}^*)} P_{4.l.2} S \tag{7.121}$$

It is noted that the above equations are derived after the threshold particle size D_k is determined, and $K_1 + K_2 > 1$. This threshold particle size D_k should be reduced to make the above equations correct with $K_1 + K_2 = 1$. This reduced threshold particle size is denoted as $D_{k.1}$.

If $K_1 + K_2 = 1$ is satisfied after the adjustments, the following observations can be made from the above equations. Firstly, as $\frac{S^*(\omega_2^*)}{S^*(\omega_{1.1}^*)} < 1$, $S^*(\omega^*) < P_{4.1}S + P_{4.2}S = S$ based on Eq. (7.119), and thus sediment deposition occurs. Secondly, plugging Eq. (7.118) into the governing equation for non-equilibrium sediment transport, one obtains

$$\begin{gathered} \frac{d(P_{4.l}S)}{dx} = -\frac{\alpha\omega_l}{q}(SP_{4.l} - P_{4.1}P_{4.l.1}S) = -\frac{\alpha\omega_l}{q}(P_{4.l}S - P_{4.l}S) = 0 \\ (l = 1, 2, \ldots, k) \end{gathered} \tag{7.122}$$

which indicates no scour or deposition will occur for the finer sediment particles with $D \leq D_{k.1}$. For the coarser sediment particles with $D > D_{k.1}$,

$$\begin{aligned} \frac{d(P_{4.l}S)}{dx} &= -\frac{\alpha\omega_l}{q}\left[SP_{4.l} - P_{4.2}P_{4.l.2}S\frac{S^*(l)}{S^*(\omega_{1.1}^*)}\right] \\ &= -\frac{\alpha\omega_l}{q}SP_{4.l}\left[1 - \frac{S^*(l)}{S^*(\omega_{1.1}^*)}\right] \quad (l = k+1, k+2, \ldots, n) \end{aligned} \tag{7.123}$$

Because$1 > K + 1$, $S^*(l) < S^*\left(\omega_{1.1}^*\right)$, and thus $\frac{dP_{4.1}S}{dx} < 0$, or all size groups of the coarser sediment will deposit. Thirdly, considering Eq. (7.119), summation of Eq. (7.120) over l yields $\sum_{l=1}^{n} P_{4.l}^* = 1$. Fourthly,

$$\sum_{l=1}^{n} P_{1.l} = \frac{P_{4.1}S}{S^*(\omega_1)} + \frac{P_{4.2}S}{S^*\left(\omega_{1.1}^*\right)} = K_1 + K_2 = 1 \tag{7.124}$$

which is obviously correct as the threshold particle size D_k has been adjusted to $D_{k.1}$. Equation (7.124) indicates that the sum of the water percentages for carrying the coarser and the finer sediment is 1.

In order to avoid the iterations involved in the determination of the threshold particle size $D_{k^\circ 1}$ due to $K_1 + K_2' > 1$, an alternate approach can be used. In this approach, K_2 is reduced to K_2' so that $K_1 + K_2' > 1$, or $K_2' = 1 - K_1$ is used to replace the original K_2. Then, Eqs. (7.118)–(7.121) become

$$P_{4.l}^* S^*(\omega^*) = P_{4.1}P_{4.l.1}S + \left[1 - \frac{P_{4.1}S}{S^*(\omega_1)}\right]P_{4.l.2}S^*(l) \tag{7.125}$$

$$S^*(\omega^*) = P_{4.1}S + \left[1 - \frac{P_{4.1}S}{S^*(\omega_1)}\right]S^*(\omega_2^*) \tag{7.126}$$

$$P_{4.l}^* = P_{4.1}\frac{S}{S^*(\omega^*)}P_{4.L.1} + \left[1 - \frac{P_{4.1}S}{S^*(\omega_1)}\right]\frac{S^*(l)}{S^*(\omega^*)}P_{4.l.2} \tag{7.127}$$

$$P_{1.l} = P_{4.1}P_{4.l.1}\frac{S}{S^*(l)} + \left[1 - \frac{P_{4.1}S}{S^*(\omega_1)}\right]P_{4.l.2} \tag{7.128}$$

The following observations can be made from the above equations. Firstly, because $\frac{S^*(\omega_2^*)}{S^*(\omega_{1.1}^*)} < 1$, based on Eqs. (7.124) and (7.126), one obtains

$$\begin{aligned} S^*(\omega^*) &= P_{4.1}S + \left[1 - \frac{P_{4.1}S}{S^*(\omega_1)}\right]S^*(\omega_2^*) \\ &< P_{4.1}S + \frac{P_{4.2}S}{S^*(\omega_{1.1}^*)}S^*(\omega_2^*) < P_{4.1}S + P_{4.2}S = S \end{aligned} \tag{7.129}$$

which indicates the carrying-sediment capacity is less than the inflow sediment concentration. The following relation is used in the derivation of Eq. (7.129):

$$1 - \frac{P_{4.1}S}{S^*(\omega^*)} = 1 - K_1 = K_2' < K_2 = \frac{P_{4.2}S}{S^*(\omega_{1.1}^*)} \tag{7.130}$$

Secondly, considering Eq. (7.126), summation of Eq. (7.127) over l yields

$$\sum P_{4.l}^* = \frac{P_{4.1}S}{S^*(\omega^*)} + \left[1 - \frac{P_{4.1}S}{S^*(\omega_1)}\right]\frac{S^*(\omega_2^*)}{S^*(\omega^*)} = 1 \tag{7.131}$$

Thirdly, summation of Eq. (7.128) over l yields

$$\sum P_{1.l} = \frac{P_{4.1}S}{S^*(\omega_1)} + \left[1 - \frac{P_{4.1}S}{S^*(\omega_1)}\right] = K_1 + K_2' = 1 \tag{7.132}$$

which indicates that the sum of the water percentages for carrying the finer and the coarser suspended sediment is 1, or $K_1 + K_2' = 1$.

7.5 General Expressions for the Size Distribution for Carrying-Sediment Capacity and Effective Bed Material Gradation

Based on the results in the scour and mild deposition condition and in the pure deposition condition, which were discussed in Sects. 7.3 and 7.4, respectively, the formulas for the fractional carrying-sediment capacity, the size distribution for carrying-sediment capacity, and the effective bed material gradation are summarized as follows under the general condition:

$$P_{4.l}^{*}S^{*}(\omega^{*})=\begin{cases}P_{4.1}P_{4.l.1}S+P_{4.2}P_{4.l.2}S\frac{S^{*}(l)}{S^{*}(\omega_{1.1}^{*})}\\ \quad+\left[1-\frac{P_{4.1}S}{S^{*}(\omega_{1})}-\frac{P_{4.2}S}{S^{*}(\omega_{1.1}^{*})}\right]P_{1}P_{4.l.1.1}^{*}S^{*}(\omega_{1.1}^{*})\\ \quad\left(l=1,2,\ldots,n;\quad \frac{P_{4.1}S}{S^{*}(\omega_{1})}+\frac{P_{4.2}S}{S^{*}(\omega_{1.1}^{*})}<1\right)\\ P_{4.1}P_{4.l.1}S+[1-\frac{P_{4.1}S}{S^{*}(\omega_{1})}]P_{4.l.2}S^{*}(l)\\ \quad(l=1,2\ldots.n\ ;\quad Otherwise)\end{cases}\tag{7.133}$$

$$S^{*}(\omega^{*})=\begin{cases}P_{4.1}S+P_{4.2}S\frac{S^{*}(\omega_{2}^{*})}{S^{*}(\omega_{1.1}^{*})}\\ \quad+\left[1-\frac{P_{4.1}S}{S^{*}(\omega_{1})}-\frac{P_{4.2}S}{S^{*}(\omega_{1.1}^{*})}\right]P_{1}S^{*}(\omega_{1.1}^{*})\\ \quad(l=1,2,\ldots,n;\quad \frac{P_{4.1}S}{S^{*}(\omega_{1})}+\frac{P_{4.2}S}{S^{*}(\omega_{1.1}^{*})}<1)\\ P_{4.1}S+[1-\frac{P_{4.1}S}{S^{*}(\omega_{1})}]S^{*}(\omega_{2}^{*})\\ \quad(l=1,2,\ldots,n;\quad Otherwise)\end{cases}\tag{7.134}$$

$$P_{4.l}^{*}=\begin{cases}P_{4.1}P_{4.l.1}\frac{S}{S^{*}(\omega^{*})}+P_{4.2}P_{4.l.2}\frac{S}{S^{*}(\omega^{*})}\frac{S^{*}(l)}{S^{*}(\omega_{1.1}^{*})}\\ \quad+\left[1-\frac{P_{4.1}S}{S^{*}(\omega_{1})}-\frac{P_{4.2}S}{S^{*}(\omega_{1.1}^{*})}\right]P_{1}P_{4.l.1.1}^{*}\frac{S^{*}(\omega_{1.1}^{*})}{S^{*}(\omega^{*})}\\ \quad(l=1,2,\ldots,n;\quad \frac{P_{4.1}S}{S^{*}(\omega_{1})}+\frac{P_{4.2}S}{S^{*}(\omega_{1.1}^{*})}<1)\\ P_{4.1}P_{4.l.1}\frac{S}{S^{*}(\omega*)}+\left[1-\frac{P_{4.1}S}{S^{*}(\omega_{1})}\right]\frac{S^{*}(l)}{S^{*}(\omega_{1.1}^{*})}P_{4.l.2}\\ \quad(l=1,2,\ldots,n;\quad Otherwise)\end{cases}\tag{7.135}$$

$$P_{1.l}=\begin{cases}P_{4.1}P_{4.l.1}\frac{S}{S^{*}(l)}+P_{4.2}P_{4.l.2}\frac{S}{S^{*}(\omega_{1.1}^{*})}\\ \quad+\left[1-\frac{P_{4.1}S}{S^{*}(\omega_{1})}-\frac{P_{4.2}S}{S^{*}(\omega_{1.1}^{*})}\right]P_{1}P_{1.l.1.1}\\ \quad\left(l=1,2,\ldots,n;\quad \frac{P_{4.1}S}{S^{*}(\omega_{1.1}^{*})}+\frac{P_{4.2}S}{S^{*}(\omega_{1.1}^{*})}<1\right)\\ P_{4.1}P_{4.l.1}\frac{S}{S^{*}(l)}+\left[1-\frac{P_{4.1}S}{S^{*}(\omega_{1})}\right]P_{4.l.2}\\ \quad(l=1,2\ldots.n\ ;\quad Otherwise)\end{cases}\tag{7.136}$$

where $P_{4.l.1.1}^{*}$ is determined with Eq. (7.98). These equations can be simplified in special cases, which will be discussed in the following section.

7.6 Size Distribution for Carrying-Sediment Capacity and Effective Bed Material Gradation in Special Cases

7.6.1 *Case 1: Suspended Load Inflow is Partially Counted in Carrying-Sediment Capacity and All Bed Material Can Be Suspended*

For this case the percentage of the coarser bed material with the size group index $l > n$ is zero, or all bed material belongs to the size groups $l \le n$. Thus, $S^*(\omega_{1.1}^*) = S^*(\omega_1^*)$, $P_1 = 1$, $P_{1l.1.1} = P_{1l.1}$, $P_{4.l.1.1}^* = P_{4.l.1}^*$, and Eqs. (7.133)–(7.136) can be written as

$$P_{4.l}^* S^*(\omega^*) = \begin{cases} P_{4.1} P_{4.l.1} S + P_{4.2} S \frac{S^*(l)}{S^*(\omega_1^*)} P_{4.l.2} + \left[1 - \frac{P_{4.1} S}{S^*(\omega_1)} - \frac{P_{4.2} S}{S^*(\omega_1^*)}\right] P_{4.l.1}^* S^*(\omega_1^*), \\ \quad (l = 1, 2, \ldots, n\,; \quad \frac{P_{4.1} S}{S^*(\omega_1)} + \frac{P_{4.2} S}{S^*(\omega_1^*)} < 1) \\ P_{4.1} P_{4.l.1} + \left[1 - \frac{P_{4.1} S}{S^*(\omega_1)}\right] P_{4.l.2} S^*(l), \quad (l = 1, 2, \ldots, n; \quad Otherwise) \end{cases} \tag{7.137}$$

$$S^*(\omega^*) = \begin{cases} P_{4.1} S + P_{4.2} S \frac{S^*(\omega_2^*)}{S^*(\omega_1^*)} + \left[1 - \frac{P_{4.1} S}{S^*(\omega_1)} - \frac{P_{4.2} S}{S^*(\omega_1^*)}\right] S^*(\omega_1^*), \\ \quad (l = 1, 2, \ldots, n; \quad \frac{P_{4.1} S}{S^*(\omega_1)} + \frac{P_{4.2} S}{S^*(\omega_1^*)} < 1) \\ P_{4.1} S + \left[1 - \frac{P_{4.1} S}{S^*(\omega_1)}\right] S^*(\omega_2^*), \quad (l = 1, 2, \ldots, n; \quad Otherwise) \end{cases} \tag{7.138}$$

$$P_{4.l}^* = \begin{cases} P_{4.1} P_{4.l.1} \frac{S}{S^*(\omega^*)} + P_{4.2} P_{4.l.2} \frac{S}{S^*(\omega^*)} \frac{S^*(l)}{S^*(\omega_1^*)} \\ \quad + \left[1 - \frac{P_{4.1} S}{S^*(\omega_1)} - \frac{P_{4.2} S}{S^*(\omega_1^*)}\right] P_{4.l.1}^* \frac{S^*(\omega_1^*)}{S^*(\omega *)} \\ \quad (l = 1, 2, \ldots, n; \quad \frac{P_{4.1} S}{S^*(\omega_1)} + \frac{P_{4.2} S}{S^*(\omega_1^*)} < 1) \\ P_{4.1} P_{4.l.1} \frac{S}{S^*(\omega^*)} + \left[1 - \frac{P_{4.1} S}{S^*(\omega_1)}\right] \frac{S^*(l)}{S^*(\omega_1^*)} P_{4.l.2} \\ \quad (l = 1, 2, \ldots, n; \quad Otherwise) \end{cases} \tag{7.139}$$

$$P_{1.l} = \begin{cases} P_{4.1} P_{4.l.1} \frac{S}{S^*(L)} + P_{4.2} P_{4.l.2} \frac{S}{S^*(\omega_1^*)} \\ \quad + \left[1 - \frac{P_{4.1} S}{S^*(\omega_1)} - \frac{P_{4.2} S}{S^*(\omega_1^*)}\right] P_{1.l.1.1} \\ \quad \left(l = 1, 2, \ldots, n; \quad \frac{P_{4.1} S}{S^*(\omega_1)} + \frac{P_{4.2} S}{S^*(\omega_2)} < 1\right) \\ P_{4.1} P_{4.l.1} \frac{S}{S^*(l)} + \left[1 - \frac{P_{4.1} S}{S^*(\omega_1)}\right] P_{4.l.2} \\ \quad (l = 1, 2, \ldots, n; \quad Otherwsie) \end{cases} \tag{7.140}$$

where $P_{4l.1}^*$ is determined with Eq. (7.65).

7.6.2 Case 2: All Suspended Load Inflow is Counted in Carrying-Sediment Capacity and Bed Material Can Be Partially Suspended

This case typically occurs on a mixed sand and gravel bed, or on a mixed coarse sand and pebble bed. Because all suspended load is included in the carrying-sediment capacity, only scour will occur with no sediment deposition. Thus, the carrying-sediment capacity and the corresponding size distribution caused by the pure deposition do not exist. Only if the suspended load is partially counted in the sediment transport capacity then sediment deposition will occur. In this case, $k = n$, $P_{4.1} = 1$, $P_{4.2} = 0$, $P_{4l.1} = P_{4l}$, $S^*(\omega_1) = S^*(\omega)$, the water percentage to carry the finer sediment $\frac{S}{S^*(\omega)} < 1$, and the following equations can be derived from Eqs. (7.133)–(7.136):

$$P_{4.l}^* S^*(\omega^*) = P_{4.l} S + \left[1 - \frac{S}{S^*(\omega)}\right] P_1 P_{4.l.1.1}^* S^*(\omega_{1.1}^*) \quad (l = 1, 2, \ldots, n) \tag{7.141}$$

$$S^*(\omega^*) = S + \left[1 - \frac{S}{S^*(\omega)}\right] P_1 S^*(\omega_{1.1}^*) \quad (l = 1, 2, \ldots, n) \tag{7.142}$$

$$\begin{aligned} P_{4.l}^* &= P_{4.l} \frac{S}{S^*(\omega^*)} + \left[1 - \frac{S}{S^*(\omega)}\right] P_1 P_{4.l.1}^* \frac{S^*(\omega_{1.1}^*)}{S^*(\omega^*)} \\ &= P_{4.l} \frac{S}{S^*(\omega^*)} + \left[1 - \frac{S}{S^*(\omega^*)}\right] P_{4.l.1}^* \quad (l = 1, 2, \ldots, n) \end{aligned} \tag{7.143}$$

$$\begin{aligned} P_{1.l} &= P_{4.l.1} \frac{S}{S^*(l)} + \left[1 - \frac{S}{S^*(\omega)}\right] P_1 P_{1.l.1.1} \\ &= P_{4.l} \frac{S}{S^*(L)} + \left[1 - \frac{S}{S^*(\omega^*)}\right] P_{1.l.1.1} \frac{S^*(\omega^*)}{S^*(\omega_{1.1}^*)}, \quad (l = 1, 2, \ldots, n) \end{aligned} \tag{7.144}$$

where

$$S^*(\omega) = \sum_{l=1}^{n} \frac{P_{4.l}}{S^*(l)} \tag{7.145}$$

The following properties can be derived from the above equations: (1) $S^*(\omega^*) > S$, and thus only scour will occur. This is actually a prerequisite condition for Eq. (7.141). (2) $\sum P_{4.l}^* = 1$. (3)

$$\sum P_{1.l} = \frac{S}{S^*(\omega)} + \left[1 - \frac{S}{S^*(\omega)}\right] P_1 = P_1 + (1 - P_1) \frac{S}{S^*(\omega)} < 1$$

7.6.3 Case 3: All Suspended Load Inflow is Counted in Carrying-Sediment Capacity and All Bed Material Can Be Suspended

This case typically occurs on a sand bed with fine sediment inflow. Because, all suspended load inflow is converted into the carrying-sediment capacity, the sediment concentration will increase. In this case, $\frac{S}{S^*(\omega)} < 1$, bed scour will occur, $P_{4.1} = 1$, $P_{4.2} = 0$, $P_{4.l.1} = P_{4.l}$, $S^*(\omega^*_{1.1}) = S^*(\omega^*_1)$, $S^*(\omega_1) = S^*(\omega)$ $P_{1.l.1.1} = P_{1.l.1}$, $P^*_{4.l.1.1} = P^*_{4.l.1}$, and Eqs. (7.133) through (7.136) can be simplified as follows

$$P^*_{4.l}S^*(\omega^*) = P_{4.l}S + \left[1 - \frac{S}{S^*(\omega)}\right]P^*_{4.l.1}S^*(\omega^*_1), \quad \left(\frac{S}{S^*(\omega)} < 1; \quad l = 1, 2, \ldots, n\right) \tag{7.146}$$

$$S^*(\omega^*) = S + \left[1 - \frac{S}{S^*(\omega)}\right]S^*(\omega^*_1) \quad \left(\frac{S}{S^*(\omega)} < 1; \quad l = 1, 2, \ldots, n\right) \tag{7.147}$$

$$\begin{aligned} P^*_{4.l} &= P_{4.l}\frac{S}{S^*(\omega^*)} + \left[1 - \frac{S}{S^*(\omega)}\right]P^*_{4.l.1}\frac{S^*(\omega^*_1)}{S^*(\omega_*)} \\ &= P_{4.l}\frac{S}{S^*(\omega^*)} + \left[1 - \frac{S}{S^*(\omega*)}\right]P^*_{4.l.1}, \quad \left(\frac{S}{S^*(\omega^*)} < 1; \quad l = 1, 2, 3, \ldots, n\right) \end{aligned} \tag{7.148}$$

$$P_{1.l} = P_{4.l}\frac{S}{S^*(l)} + \left[1 - \frac{S}{S^*(\omega)}\right]P_{1.l.1} = \frac{P_{4.l}S}{S^*(l)} + \left[1 - \frac{S}{S^*(\omega^*)}\right]P_{1.l.1}\frac{S^*(\omega^*)}{S^*(\omega^*_1)} \tag{7.149}$$

$$\sum P_{1.l} = \frac{S}{S^*(\omega)} + \left[1 - \frac{S}{S^*(\omega)}\right] = 1 \tag{7.150}$$

The properties in this case are the same as those in Case 2 that are discussed in Sect. 7.6.2.

In addition, for $\frac{S}{S^*(\omega)} = 1$, the above equations become

$$S^*(\omega^*) = S \tag{7.151}$$

$$S^*_l = P^*_{4.l}S^*(\omega^*) = P_{4.l}S = S_l \tag{7.152}$$

$$P^*_{4l} = P_{4l} \tag{7.153}$$

$$P_{1.l} = P^*_{4.l}\frac{S^*}{S^*(l)} = P_{4.l}\frac{S}{S^*(l)} \tag{7.154}$$

This equation set is in the strong equilibrium condition. Equations (7.151) through (7.153) are consistent with the definition of the strong equilibrium as described in Eq. (6.3). It is noted that, in this strong equilibrium condition, the effective bed material gradation is exactly the gradation of the instantaneously deposited sediment during the exchange between the suspended load and the bed material, or $P_{4.l}S/S^*(l)$, which is also called the gradation of the deposited material as the deposition percentage approaches zero. Also, Eq. (7.154) is consistent with the general relation between the effective bed material gradation and the size distribution of the carrying-sediment capacity. Equations (7.151)–(7.154) do not directly address the original bed material gradation $P_{1.l.1}$. However, the gradation of the deposited material is the same as the original bed material gradation in this case, and thus

$$P_{1.l.1} = P'_{1.l.1} = P_{1.l} \tag{7.155}$$

7.6.4 Case 4: All Bed Material Can Be Suspended, $P_1 = 1$ and $P_{4.l.2} = P_{1.l.1}$

In this case, the coarser particles of the suspended load all deposit on the bed, and the gradation of the deposited material $P_{4.l.2}$ is the same as the original bed material gradation. The first formula of Eq. (7.133) then becomes

$$\begin{aligned} P^*_{4.l}S^*(\omega^*) &= P_{4.1}P_{4.l.1}S + \left[1 - \frac{P_{4.1}S}{S^*(\omega_1)}\right]P^*_{4.l.1}S^*(\omega^*_1) \\ &\quad + \left[P_{4.2}P_{4.l.2}S\frac{S^*(l)}{S^*(\omega^*_1)} - P_{4.2}SP^*_{4.l.1}\right] \\ &= P_{4.1}P_{4.l.1}S + \left[1 - \frac{P_{4.1}S}{S^*(\omega_1)}\right]P^*_{4.l.1}S^*(\omega^*_1) \end{aligned} \tag{7.156}$$

after utilizing Eq. (7.96) with $P_{4.l.2} = P_{1.l.1}$, which leads to

$$P_{4.2}P_{4.l.2}\frac{S^*(l)}{S^*(\omega^*_1)} = P_{4.l}P_{1.l.1}\frac{S^*(l)}{S^*(\omega^*_1)} = P_{4.l}P^*_{4.l.1} \tag{7.157}$$

The equations $P_1 = 1$, $P_{4.l.2} = P_{1.l.1}$, $P^*_{4l,1,1} = P^*_{4l,1}$, and $S^*(\omega^*_{1.1}) = S^*(\omega^*_1)$ are also used to obtain Eq. (7.156). Then the following equations can be derived:

$$S^*(\omega^*) = S + \left[1 - \frac{P_{4.1}S}{S^*(\omega_1)}\right]S^*(\omega^*_1) \tag{7.158}$$

$$P^*_{4.l} = P_{4.1}P_{4.l.1}\frac{S}{S^*(\omega*)} + \left[1 - \frac{P_{4.1}S}{S^*(\omega_1)}\right]P^*_{4.l.1}\frac{S^*(\omega^*_1)}{S^*(\omega^*)}$$

$$= P_{4.1} P_{4.l.1} \frac{S}{S^*(\omega_1^*)} + \left[1 - \frac{P_{4.1} S}{S^*(\omega_1)}\right] P^*_{4.l.1} \tag{7.159}$$

$$P_{1.l} = P^*_{4.l1} \frac{S^*(\omega^*)}{S^*(l)} = P_{4.1} P_{4.l.1} \frac{S}{S^*(l)} + \left[1 - \frac{P_{4.1} S}{S^*(\omega_1)}\right] P_{1.l.1} \tag{7.160}$$

Equations (7.158)–(7.160) indicate that all wash load is converted to the carrying-sediment capacity, with the remainder carrying-sediment capacity being contributed from the bed material. Compared to the traditional approach of dividing suspended load into wash load and bed material load with the carrying-sediment capacity being considered only for the bed material load, the carrying-sediment capacity is considered in our analysis for both bed material load and wash load. The following comments are made from our analysis: (1) The constraint condition for the above equations, or$P_{4l.2} = P_{11.1}$, is strict. (2) In previous studies, the finer sediment is simply excluded from the carrying-sediment capacity, and only the carrying-sediment capacity $S^*(\omega_1^*)$ for the bed material load is included. This is not accurate. (3) Another prerequisite condition for the above equations is $P_{4.1} S / S^*(\omega_1) < 1$, or the concentration of the wash load is less than the corresponding carrying-sediment capacity. This condition is generally satisfied under a wide range of sedimentation conditions.

7.6.5 *Case 5: For Pure Deposition with Suspended Load Influx Being Partially Counted in Carrying-Sediment Capacity*

This case includes two situations: one with all finer particles being counted in the carrying-sediment capacity and the coarser particles being partially included in the transport capacity; the other with both finer and coarser particles being partially included in the carrying-sediment capacity. The first situation was already discussed in Sect. 7.4, and the corresponding carrying-sediment capacity and size distributions are described in Eqs. (7.125)–(7.128). This section focuses on the second situation.

For this case, $P_{4.1} = 0$, $P_{4.2} = 1$, $P_{4.l.2} = P_{4.l}$, and $P_{4.l} = P_{1.l.1}$. Also, because the bed material consists of the sediment particles that deposit out of the suspended load, all the bed material can be suspended and $S^*(\omega^*_{1.1}) = S^*(\omega_1^*)$. Based on Eq. (7.5), one obtains

$$P^*_{4.l} S^*(\omega^*) = P_{4.l} S^*(l) \tag{7.161}$$

If the carrying-sediment capacity $S^*(\omega*)$ consists of three parts, $K_{2,l}$ can be determined with Eq. (7.92). However, only the second part of $S^*(\omega*)$ exists for this case. Therefore, $K_{2,l}$ needs to be adjusted by introducing a correction factor β so that $K_2 = 1$, or

$$K_{2.l} = \frac{SP_{4.l}\beta}{S^*(\omega_1^*)} \tag{7.162}$$

$$1 = K_2 = \frac{S\beta}{S^*(\omega_1^*)} \tag{7.163}$$

Thus,

$$\beta = \frac{S^*(\omega_1^*)}{S}$$

Plugging into Eq. (7.162) yields

$$K_{2,l} = \frac{SP_{4,l}\beta}{S^*(\omega_1^*)} = \frac{P_{4,l}S}{S^*(\omega_1^*)}\frac{S^*(\omega_1^*)}{S} = P_{4,l}$$

Thus,

$$P_{4.l}^*S^*(\omega*) = K_{2.l}S^*(l) = P_{4.l}S^*(l)$$

which is the same as Eq. (7.161). Based on the above equation,

$$S^*(\omega^*) = S^*(\omega_{2.2}^*) \tag{7.164}$$

$$P_{4.l}^* = P_{4.l}\frac{S^*(l)}{S^*(\omega^*)} \tag{7.165}$$

$$P_{1.l} = P_{4.l}^*\frac{S^*(\omega*)}{S^*(l)} = P_{4.l}\frac{S^*(\omega*)}{S^*(l)}\frac{S^*(l)}{S^*(\omega^*)} = P_{4.l} = P_{1.l.1} \tag{7.166}$$

where

$$S^*(\omega_{2.2}^*) = \sum_{l=1}^{n} P_{4.l}S^*(l) = \sum_{l=1}^{n} P_{1.l.1}S^*(l) \tag{7.167}$$

It is noted that the relation $P_{4.l} = P_{1.l.1}$ is not directly used in the derivation of Eqs. (7.164)–(7.166) from Eq. (7.5). But it was assumed in Eq. (7.5). Based on this relation

$$S^*(\omega^*) = \sum_{l=1}^{n} P_{4.l}S^*(l) = \sum_{l=1}^{n} P_{1.l.1}S^*(l) = S^*(\omega_1^*) \tag{7.168}$$

Then, based on Eqs. (7.165), (7.166), and (7.36), one obtains

$$P_{4.l}^* = P_{4.l}\frac{S^*(l)}{S^*(\omega*)} = P_{4.l}\left(\frac{\omega^*}{\omega_l}\right)^m = P_{1.l.1}\left(\frac{\omega^*}{\omega_l}\right)^m \tag{7.169}$$

$$P_{1.l.1} = P_{4.l} = P_{4.l}^*(\frac{\omega_l}{\omega*})^m \tag{7.170}$$

As shown in Eq. (7.165), for the coarse particles with $\omega^* < \omega_l, P_{4,l}^* < P_{4.l}$; for the fine particles with $\omega^* > \omega$, $P_{4,l}^* > P_{4.l}$. This indicates the carrying-sediment capacity has a higher percentage of fine sediment than the sediment concentration. This is because by assuming both fine and coarse sediment particles are partially converted into the carrying-sediment capacity, more fine particles are converted into the carrying-sediment capacity than coarse particles do. This is unreasonable. The reason is that the graduation of bed material, $P_{1.l.1}$ is not equal the size distribution of suspended load, $P_{4.l}$(see Sect. 7.7.1.2).

7.7 Size Distributions for Carrying-Sediment Capacity and Suspended Sediment at the Exit Section of a Reach with Significant Scour or Sedimentation

The relationship of the instantaneous size distribution for carrying-sediment capacity and effective bed material gradation at a section with the corresponding concentration and size distribution of suspended load and the bed material gradation was discussed in Sects. 7.2–7.6. In this section, we will investigate the size distribution for carrying-sediment capacity and the bed material gradation within a reach during a period of time. The focus will be the size distribution for carrying-sediment capacity and the variation of the bed material gradation after channel sedimentation has occurred for a certain period of time for given concentration and size distribution of suspended load at the entrance section and initial bed material gradation of the reach. For this case, the variations of the size distribution of suspended load and of the bed material gradation with sedimentation quantity should be considered.

For simplicity, our analysis of the impact of channel sedimentation focuses on the variations of the bed material gradation and of the suspended load size distribution, and their impacts to the size distribution for carrying-sediment capacity at the exit section and to the effective bed material gradation at the end of study period for the cases with massive deposition and with massive scour. For the other cases, most of them are with mild erosion or mild deposition. For these cases, the impact of channel sedimentation is negligible, and the results for the size distribution for carrying-sediment capacity and for the effective bed material gradation, which were previously derived for a section, can be approximately used for a short reach.

7.7.1 Size Distribution for Carrying-Sediment Capacity, Bed Material Gradation, and Size Distribution of Suspended Sediment for the Case with No Scour of Bed Material

For this case, the sediment influx exceeds the carrying-sediment capacity, and thus only sediment deposition, without scour, occurs on the sand bed. There are two situations: one is that all finer sediment particles and only part of the coarser particles change to the carrying-sediment capacity; the other is that all sediment particles deposit, or in massive deposition condition.

7.7.1.1 Case 1: All Finer Sediment Particles and Part of Coarser Particles of the Suspended Load Influx Change to Carrying-Sediment Capacity

Because all finer sediment particles change to carrying-sediment capacity, they do not deposit in a short reach, and thus the quantity of the finer sediment at the exit section remains the same as the influx. However, this is not true for coarser particles. In this section, the variation of the size distribution for suspended load at the exit section is analyzed first, followed by the variations of the corresponding size distribution for carrying-sediment capacity and of the effective bed material gradation. For massive sediment deposition, the size distribution of the coarser part of the suspended load, $P_{4.l,2}$, can be used to represent the bed material gradation, and thus,

$$S^*(\omega_1^*) = \sum_{l=k+1}^{n} P_{4.l.2} S^*(l) \tag{7.171}$$

For the coarser suspended load, Eq. (7.123) can be rewritten as

$$\frac{d(P_{4.l}S)}{dx} = \frac{d(P_{4.2}P_{4.l.2}S)}{dx} = -\frac{\alpha}{q} S P_{4.2} - \left[\omega_l P_{4.l.2} - \frac{\omega_l S^*(l)}{S^*(\omega_1^*)} P_{4.l.2}\right] \tag{7.172}$$

Utilizing Eq. (7.31) for the carrying-sediment capacity, one obtains

$$\omega_l S^*(l) P_{4.l.2} = \kappa \left(\frac{V^3}{gh\omega_l}\right)^m \omega_l P_{4.l.2} \tag{7.173}$$

$$\sum_{l=k+1}^{n} \omega_l S^*(l) P_{4.l.2} = \kappa \left(\frac{V^3}{gh}\right)^m \sum_{l=K+1}^{n} \omega_l^{1-m} P_{4.l.2} = \kappa \left(\frac{V^3}{gh}\right)^m \omega_{2.1}^{1-m} = \omega_{2.1} S^*(\omega_{2.1}) \tag{7.174}$$

where

$$\omega_{2.1} = \left[\sum_{l=k+1}^{n} P_{4.l.2} \omega_l^{1-m} \right]^{\frac{1}{1-m}} \tag{7.175}$$

Considering $P_{4.2}S = S_2$, summation of Eq. (7.172) for $l = k + 1$ through $l = n$ yields

$$\frac{dS_2}{dx} = -\frac{\alpha}{q} S_2 - \left[\omega_2 - \frac{S^*(\omega_{2.1})}{S^*(\omega_1^*)} \omega_{2.1} \right] \tag{7.176}$$

where ω_2 is

$$\omega_2 = \sum_{l=k+1}^{n} P_{4.l.2} \omega_l \tag{7.177}$$

Assigning $m = 1$, Eqs. (7.172) and (7.175) can be significantly simplified, and Eq. (7.172) can be written as

$$\frac{d(P_{4.l.2}S_2)}{dx} = -\frac{\alpha S_2 P_{4.l.2}}{q} \left[\omega_l - \frac{\omega_1^* S^*(\omega_1^*)}{S^*(\omega_1^*)} \right] = -\frac{\alpha S_2 P_{4.l.2}}{q} \left(\omega_l - \omega_1^* \right) \tag{7.178}$$

after utilizing

$$\omega_l S^*(l) = \omega_l \kappa \left(\frac{V^3}{gh\omega_l} \right) = \omega_1^* S^*(\omega_1^*) \tag{7.179}$$

Summation of Eq. (7.178) for l yields

$$\frac{dS_2}{dx} = -\frac{\alpha S_2}{q} \left(\omega_2 - \omega_1^* \right) \tag{7.180}$$

Dividing Eq. (7.180) by Eq. (7.178) gives

$$\frac{d(P_{4.l.2}S_2)}{dS_2} = P_{4.l.2} + S_2 \frac{dP_{4.l.2}}{dS_2} = \frac{\omega_l - \omega_1^*}{\omega_2 - \omega_1^*} P_{4.l.2}$$

or

$$S_2 \frac{d(P_{4.l.2})}{dS_2} = \left(\frac{\omega_l - \omega_1^*}{\omega_2 - \omega_1^*} - 1 \right) P_{4.l.2} = \frac{\omega_l - \omega_2}{\omega_2 - \omega_1^*} P_{4.l.2} \tag{7.181}$$

Integrating within the interval between $P_{4.l.2.0}S_{2.0}$ and $P_{4.2}S_2$, one obtains

$$\ln \frac{P_{4.l.2}}{P_{4.l.2.0}} = \frac{\omega_l - \omega_{2.m}}{\omega_{2.m} - \omega_1^*} \ln \frac{P_{4.2}S_2}{P_{4.2.0}S_{2.0}}$$

or

$$P_{4.l.2} = P_{4.l.2.0}\left(\frac{P_{4.2}S_2}{P_{4.2.0}S_{2.0}}\right)^{\frac{\omega_l-\omega_{2.m}}{\omega_{2.m}-\omega_1^*}} = P_{4.2.l.0}(1-\lambda_2)^{\frac{\omega_l-\omega_{2.m}}{\omega_{2.m}-\omega_1^*}} \quad (k+1 \le l \le n) \tag{7.182}$$

where ω_{2m} is a median value of ω_2 within the interval between $P_{4.2.0}S_0$ and $P_{4.2}S$, which can be determined with

$$\sum_{l=k+1}^{n} P_{4.l.2} = P_{4.l.2.0}(1-\lambda_2)^{\frac{\omega_l-\omega_{2.m}}{\omega_{2.m}-\omega_1^*}} = 1 \tag{7.183}$$

and

$$\lambda_2 = \frac{P_{4.2.0}S_0 - P_{4.2}S}{P_{4.2.0}S_0} \tag{7.184}$$

is the deposition percentage. The impact of sedimentation on the size distribution of suspended load at the exit section is considered in Eq. (7.182) for a relatively long reach with significant sediment deposition.

In order to illustrate the characteristics associated with Eq. (7.182), we first prove the following relationship between ω_1^*, as described in Eq. (7.171), and ω_2, which is defined in Eq. (7.177)

$$\omega_2 = \sum_{l=k+1}^{n} P_{4.l.2}\omega_l > \frac{1}{\sum_{l=k+1}^{n} \frac{P_{4.l.2}}{\omega_l}} = \omega_1^* \tag{7.185}$$

Or, equivalently

$$\sum_{l=1}^{n} P_{4.l}\omega_l > \frac{1}{\sum_{l=1}^{n} \frac{P_{4.l}}{\omega_l}} \tag{7.186}$$

where

$$\sum_{l=1}^{n} P_{4.l} = 1 \tag{7.187}$$

We need to prove

$$\sum_{l=1}^{n} P_{4.l}\omega_l = \sum_{l=1}^{n} k_l P_{4.l}\omega_1 > \frac{1}{\sum\limits_{l=1}^{n} \frac{P_{4.l}}{k_l\omega_1}} = \frac{\omega_1}{\sum\limits_{l=1}^{n} \frac{P_{4.l}}{k_l}} \tag{7.188}$$

where

$$\omega_l = k_l\omega_1 \tag{7.189}$$

From Eq. (7.188) we need to prove

$$\sum_{l=1}^{n} P_{4.l}k_l \times \sum_{l=1}^{n} \frac{P_{4.l}}{k_l} > 1 \tag{7.190}$$

The left side of the greater-than symbol in Eq. (7.190) can be expressed as

$$\sum_{l=1}^{n} P_{4.l}^2 + \sum_{l=1}^{n}\sum_{r=l+1}^{n} P_{4.l}P_{4.r}\left(\frac{k_r}{k_l} + \frac{k_l}{k_r}\right) > 1 \tag{7.191}$$

Since

$$\frac{k_r}{k_l} + \frac{k_l}{k_r} > 2 \tag{7.192}$$

Equation (7.191) turns to

$$\begin{aligned} &\sum_{l=1}^{n} P_{4.l}^2 + \sum_{l=1}^{n}\sum_{r=l+1}^{n} P_{4.l}P_{4.r}\left(\frac{k_r}{k_l} + \frac{k_l}{k_r}\right) \\ &> \sum_{l=1}^{n} P_{4.l}^2 + \sum_{l=1}^{n}\sum_{r=l+1}^{n} (2P_{4.l}P_{4.r}) = \left(\sum_{l=1}^{n} P_{4.l}\right)^2 = 1 \end{aligned} \tag{7.193}$$

This proves Eq. (7.185). It is noted that ω_{2m} in Eq. (7.182) is a median value between $\omega_{2.0}$ corresponding to $P_{4.1.2.0}$ at the entrance section and ω_2 corresponding to $P_{4.l.2}$ at the exit section. Since $\omega_2 > \omega_1^*$, the following relation is true:

$$\omega_2 > \omega_{2.m} > \omega_1^* \tag{7.194}$$

Hence, according to Eq. (7.182), for the coarser particles with $\omega_l > \omega_{2.m}, x = \frac{\omega_l - \omega_{2m}}{\omega_{2m} - \omega_1^*} > 0$, $(1 - \lambda_2)^x < 1$, and thus $P_{4.1.2} < P_{4.1.2.0}$. For the finer particles with $\omega_l < \omega_{2.m}$, $x < 0$, $(1 - \lambda)^x > 1$, and thus $P_{4.1.2} > P_{4.1.2.0}$. This indicates that the percentages for the coarser particles reduce as deposition continues, and this reduction increases with λ, which is true during the deposition process.

The size distribution of suspended load at the exit section is derived as follows. The fractional sediment concentration is

$$SP_{4l} = P_{4l.1.0}P_{4.1.0}S_0 + \mathrm{P}_{4.2}P_{4l.2}S \tag{7.195}$$

$$S = P_{4.1.0}S_0 + P_{4.2}S = P_{4.1}S + P_{4.2}S \tag{7.196}$$

Assuming the fractional sediment concentration remains the same for the finer sediment, i.e., no deposition or scour for the finer sediment, then $P_{4.1.0}S_0 = P_{4.1}S$. Considering Eq. (7.182), $P_{4,l}$ can also be written as

$$\begin{aligned} P_{4.l} &= P_{4.l.1.0}P_{4.1.0}\frac{S_0}{S} + P_{4.l.2}P_{4.2} = \frac{P_{4.l.1.0}P_{4.1.0}}{1-\lambda} + P_{4.l.2}P_{4.2} \\ &= \frac{P_{4.l.1.0}P_{4.1.0}}{1-\lambda} + P_{4.2}P_{4.l.2.0}(1-\lambda_2)^{\frac{\omega_l-\omega_{2.m}}{\omega_{2.m}-\omega_1^*}} \end{aligned} \tag{7.197}$$

where

$$\lambda = \frac{S_0 - S}{S_0} \tag{7.198}$$

Summation of Eq. (7.197) over l, one obtains

$$P_{4.2} = 1 - P_{4.1.0}\frac{S_0}{S} = 1 - \frac{P_{4.1.0}}{1-\lambda} \tag{7.199}$$

Considering Eq. (7.198), Eq. (7.184) can be written as

$$\lambda_2 = 1 - \frac{P_{4.2}S}{P_{4.2.0}S_0} = 1 - (1-\lambda)\frac{P_{4.2}}{P_{4.2.0}} \tag{7.200}$$

Similar to Eqs. (7.125)–(7.128), the carrying-sediment capacity and its size distribution at the exit section, and the effective bed material gradation can be written as

$$P_{4.l}^*S^*(\omega^*) = P_{4.1.0}P_{4.l.1.0}S_0 + \left[1 - \frac{P_{4.1.0}S_0}{S^*(\omega_{1.0})}\right]P_{4.l.2}S^*(l) \tag{7.201}$$

$$S^*(\omega^*) = P_{4.1.0}S_0 + \left[1 - \frac{P_{4.1.0}S_0}{S^*(\omega_{1.0})}\right]S^*(\omega_2^*) \tag{7.202}$$

$$P_{4.l}^* = P_{4.1.0}\frac{S_0}{S^*(\omega^*)}P_{4.l.1.0} + \left[1 - \frac{P_{4.1.0}S_0}{S^*(\omega_{1.0})}\right]\frac{S^*(l)}{S^*(\omega^*)}P_{4.l.2} \tag{7.203}$$

$$P_{1.l} = P_{4.1.0}P_{4.l.1.0}\frac{S_0}{S^*(l)} + \left[1 - \frac{P_{4.1.0}S_0}{S^*(\omega_{1.0})}\right]P_{4.l.2} \tag{7.204}$$

where the average fall velocity is determined based on the finer sediment at the entrance section

$$S^*(\omega_{1.0}) = \sum_{l=1}^{k} P_{4.l.1.0}\omega_l \tag{7.205}$$

and $S^*(\omega_2^*)$ is determined with Eq. (7.107). As shown in these equations, the carrying-sediment capacity and its size distribution at the exit section consist of two parts: one is related to the concentration and the size distribution for the finer sediment at the entrance section; the other is related to the size distribution of suspended load at the exit section, which also reflects the impact of the bed material gradation. It is noted that the formulas derived in this section are based on the assumption that Eq. (7.123) is true. The threshold particle size, which is used to divide the sediment into the finer part and the coarser part, is adjusted to satisfy $K_1 + K_2 = 1$ in this equation. Hence, this condition must be satisfied when applying equations of this section. Otherwise, the results may be incorrect. Also, if Eq. (7.125) is used to replace Eq. (7.118) in the derivation of Eq. (7.123), the simple form as shown in Eq. (7.123) can not be derived.

It is noted that the sediment concentration at the exit section in Eqs. (7.196) and (7.197) is unknown, which can be determined with the governing equation for non-equilibrium sediment transport. This can be done by solving coupled Eqs. (7.197), (7.202) and (7.203) and the equation for non-equilibrium sediment transport for $S^*(\omega^*)$, $P_{4.l}^*$, $P_{4.l}$ and S.

7.7.1.2 Case 2: Massive Deposition

Different from the previous case with all finer sediment changing to the carrying-sediment capacity, deposition occurs for the sediment of all size classes for the case with massive deposition. This is also different from the case with pure deposition (see Sect. 7.4). This non-equilibrium sediment transport for this case was already discussed in Chap. 5. In this section, the formulas for the size distribution for carrying-sediment capacity, the effective bed material gradation, and the size distribution of suspended load are derived based on the general concepts for non-equilibrium sediment transport.

The general equation for non-equilibrium sediment transport reads

$$\frac{d(P_{4.l}S)}{dx} = -\frac{\alpha\omega_l}{q}\left[P_{4.l}S - P_{4.l}^* S^*(\omega^*)\right] \tag{7.206}$$

Summation over l yields

$$\frac{dS}{dx} = -\frac{\alpha}{q}\left[\omega S - \omega^* S^*(\omega^*)\right] \tag{7.207}$$

where

$$\omega = \sum_{l=1}^{n} P_{4.l}\omega_l \tag{7.208}$$

$$\omega^* = \sum_{l=1}^{n} P_{4.l}^*\omega_l \tag{7.209}$$

Comparing Eq. (7.206) with Eq. (7.207), one obtains

$$\frac{d(P_{4.l}S)}{dS} = \frac{\omega_l\left[P_{4.l}S - P_{4.l}^*S^*(\omega^*)\right]}{\omega S - \omega^* S^*(\omega^*)} \tag{7.210}$$

after utilizing

$$\frac{d(P_{4.l}S)}{dS} = \frac{d(P_{4.l}S)}{dx}\Big/\frac{dS}{dx} \tag{7.211}$$

which is the proportion of the weight of the material deposited out of the suspended load for the lth size class to the total deposited weight, or the size gradation of the deposited material $P'_{1,l,1}$. For massive deposition, it is also the bed material gradation $P_{1,l,1}$. Hence,

$$\begin{aligned} P'_{1.l.1} = P_{1.l.1} &= \frac{d(P_{4.l}S)}{dS} = \frac{\omega_l\left[P_{4.l}S - P_{4.l}^*S^*(\omega^*)\right]}{\omega S - \omega^* S^*(\omega^*)} \\ &= \frac{\omega S}{\omega S - \omega^* S^*(\omega^*)}\frac{\omega_l}{\omega}P_{4.l} - \frac{\omega_l P'_{1.l.1}S^*(l)}{\omega S - \omega^* S^*(\omega^*)} \end{aligned} \tag{7.212}$$

where

$$P_{4.l}^*S^*\left(\omega^*\right) = P'_{1.l.1}S^*(l) \tag{7.213}$$

is determined with Eq. (7.40) after assuming the water percentage K_1 is only a function of the deposited material gradation (that also represents the impact of the size distribution of suspended load) for the case with the bed material consisting of deposited sediment, or $K_l = P'_{1,l,1}$. Also, utilizing Eq. (7.31) with $m = 1$ for the carrying-sediment capacity, one obtains

$$\omega_l S^*(l) = \omega^* S^*(\omega^*) \tag{7.214}$$

Then, Eq. (7.214) can be rewritten as

$$P'_{1.l.1}\left[1 + \frac{\omega_l S^*(l)}{\omega S - \omega^* S^*(\omega^*)}\right] = P'_{1.l.1}\left[1 + \frac{\omega^* S^*(\omega^*)}{\omega S - \omega^* S^*(\omega^*)}\right]$$

$$= P'_{1.l.1}\left[\frac{\omega S}{\omega S - \omega^* S^*(\omega^*)}\right] = \left[\frac{\omega S}{\omega S - \omega^* S^*(\omega^*)}\right]\frac{\omega_l}{\omega} P_{4.l} \tag{7.215}$$

or

$$P_{1.l} = P'_{1.l.1} = \frac{\omega_l}{\omega} P_{4.l} \tag{7.216}$$

where

$$\omega = \sum_{l=1}^{n} P_{4l}\omega_l = \frac{1}{\sum \frac{P'_{1.l.1}}{\omega_l}} \tag{7.217}$$

which is consistent with Eq. (7.208). Considering Eq. (7.53), the following can be derived from Eq. (7.213):

$$P_{1.l} S^*(l) = P^*_{4.l} S^*\left(\omega^*\right) = P'_{1.l.1} S^*(l) = P_{4.l}\frac{\omega_l}{\omega} S^*(\omega_l) = P_{4.l} S^*(\omega) \tag{7.218}$$

Summation for l yields

$$S^*(\omega^*) = S(\omega) \tag{7.219}$$

$$\omega = \omega^* \tag{7.220}$$

Also, from Eqs. (7.218) and (7.219)

$$P^*_{4.l} = P_{4.} \tag{7.221}$$

and

$$P_{1.l} = P'_{1.l.1} \tag{7.222}$$

which is an important characteristics for the massive deposition condition. It is noted that Eqs. (7.219)–(7.222) were already derived in Sect. 7.2, as shown in Eqs. (7.77)–(7.80). Here, these equations are derived again with a different approach. The purpose is to further illustrate the characteristics of the size distribution for carrying-sediment capacity and of the bed material gradation for the massive deposition condition. It is also noted that $P^*_{4l} = P_{4.l}$ was introduced as an assumption in Chap. 5, while Eq. (7.221) is derived theoretically. Equations (7.220)–(7.222) not only illustrate the characteristics for massive deposition, but also significantly simplify the analysis.

The characteristics of the suspended load sorting are illustrated in the following for the case with massive deposition. Considering

$$\frac{d(P_{4.l}S)}{dS} = P_{4.l} + \frac{SdP_{4.l}}{dS} = P'_{1.l.1} = \frac{P_{4.l}\omega_l}{\omega} \tag{7.223}$$

then

$$S\frac{d(P_{4.l})}{dS} = \left(\frac{\omega_l}{\omega} - 1\right)P_{4.l} \tag{7.224}$$

Integration of Eq. (7.224) yields the formula for the size distribution variations of suspended load during the sediment deposition process [9]

$$P_{4.l} = P_{4.l.0}(1-\lambda)^{\left(\frac{\omega_l}{\omega_m}-1\right)} \tag{7.225}$$

where $P_{4.l.0}$ is the size distribution of suspended load at the exit section, ω_m is determined with $\sum P_{4l} = 1$,

$$\lambda = \frac{S_0 - S}{S_0} \tag{7.226}$$

is the deposition percentage. Equation (7.225) illustrates the coarsening and fining of suspended sediment during the deposition process. For the finer sediment with $\omega_l < \omega_m, P_{4.l} > P_{4.l.0}$; for the coarser sediment with $\omega_l > \omega_m, P_{4.l} < P_{4.l.0}$.

7.7.2 *Size Distribution for Carrying-Sediment Capacity, Bed Material Gradation, and Size Distribution of Suspended Sediment for the Case with Massive Scour*

For massive scour condition, scour occurs for all size classes, and no deposition occurs for any size class. For this case, all suspended load influx changes to the carrying-sediment capacity, and in general all the bed material can be suspended. While the exchange between suspended load and the bed material sill occurs, no sediment deposition occurs in a longer time scale. During the massive scour process, the bed material gradation varies rapidly, and its impact on the size distribution for carrying-sediment capacity also varies. Hence, the impact of the scour rate or the scour percentage on the size distribution for carrying-sediment capacity should be considered. Also, the size distribution of suspended load has a relatively simple form of expression under certain simplified conditions.

For a scour within a short period of time, as discussed in Sect. 7.6.3, all suspended load changes to the carrying-sediment capacity, all bed material can be suspended, and $\frac{S}{S^*(\omega)} < 1$ that indicate scour occurs. Within this short period of time, the bed material gradation remains unchanged, and Eqs. (7.146)–(7.149) also apply to the case with massive scour. However, after considering the coarsening of bed material, the size distribution for carrying-sediment capacity at the exit section and the effective bed material gradation both vary. Two sets of results for the coarsening of the bed material gradation $P_{1.l.1}$ and the corresponding size distribution for carrying-sediment capacity $P^*_{1.l.1}$ were derived with two different methods. One method was developed by [4], and the other was discussed in [5]. These two methods were briefly discussed in Chap. 5. Starting from general concepts for the size distribution for carrying-sediment capacity, the bed material coarsening, the size distribution for carrying-sediment capacity, and the effective bed material gradation were analyzed in the following based on these two methods with a more strict derivation. The difference between these two methods is also discussed.

7.7.2.1 Method 2 [5]

Plugging Eq. (7.146) into the governing equation for non-equilibrium sediment transport, one obtains

$$\frac{d(P_{4.l}S)}{dx} = -\frac{\alpha\omega_l}{q}\left(P_{4.l}S - P^*_{4.l}S^*\right) = \frac{\alpha\omega_l}{q}\left[1 - \frac{S}{S^*(\omega)}\right]P^*_{4.l.1}S^*(\omega^*) \quad (7.227)$$

Since $\frac{S}{S^*(\omega)} < 1$, then $\frac{d(P_{4.l}S)}{dx} > 0$, or scour occurs for each size class. This is the basic characteristic for massive scour.

The variations of the bed material gradation and of the carrying-sediment capacity in a relatively longer reach or within a longer time interval are analyzed in the following. For this case, the impact of bed material coarsening needs to be considered. As discussed in Han [5], Han et al. (1982), and Han [8], the following relationship is true for the exit section:

$$P_{4.l}S = P_{4.l.0}S_0 + \Delta P_{4.l}\Delta S \quad (7.228)$$

where S_0 and $P_{4.l.0}$ is the concentration and the size distribution, respectively, of the suspended load at the entrance section of a reach, ΔS is the net sediment concentration eroded from the bed, and $\tilde{P}_{4.l}$ is the size distribution of the eroded sediment, or the size distribution of the sediment supplement. Plugging Eq. (7.234) into Eq. (7.233) yields

$$\frac{d(\Delta P_{4,l}\Delta S)}{dx} = \frac{\alpha\omega_l}{q}\left[1 - \frac{S}{S^*(\omega)}\right]S^*(\omega^*_1)P^*_{4,l,1} \quad (7.229)$$

Set the size distribution of the eroded sediment equal the size distribution for the carrying-sediment capacity corresponding to the bed material gradation, or

$$\Delta P_{4.l} = P^*_{4.l.1} \quad (7.230)$$

This is because the eroded sediment consists of the finer particles sorted out of the bed material. The relationship between the size distribution for carrying-sediment capacity, $P^*_{4.1}$, and the bed material gradation, $P_{1.l.1}$ also applies to the size distribution of the eroded sediment. Plugging Eq. (7.230) into (7.229), one obtains

$$\frac{d\left(P^*_{4.l.1}\Delta S\right)}{dx}=\frac{\alpha\omega_l}{q}\left[1-\frac{S}{S^*(\omega)}\right]S^*\left(\omega^*_1\right)P^*_{4.l.1} \tag{7.231}$$

Summation for l yields

$$\frac{d(\Delta S)}{dx}=\frac{\alpha\omega^*_1}{q}\left[1-\frac{S}{S^*(\omega)}\right]S^*\left(\omega^*_1\right) \tag{7.232}$$

after utilizing

$$\sum_{l=1}^{n}P^*_{4.l.1}\omega_l=\sum_{l=1}^{n}\frac{S^*(l)}{S^*\left(\omega^*_1\right)}P_{1.l.1}\omega_l=\sum_{l=1}^{n}\frac{\omega^*_1}{\omega_l}P_{1.l.1}\omega_l=\omega^*_1 \tag{7.233}$$

which can be derived by considering Eq. (7.31) with $m=1$. It is noted that ω^*_1 is the same as the definition in Eq. (7.63). Dividing Eq. (7.231) by Eq. (7.232),

$$\frac{d\left(P^*_{4.l.1}\Delta S\right)}{d\Delta S}=\Delta S\frac{dP^*_{4.l.1}}{d(\Delta S)}+P^*_{4.l.1}=\frac{\omega_l}{\omega^*_1}P^*_{4.l.1} \tag{7.234}$$

or

$$\frac{dP^*_{4.l.1}}{P^*_{4.l.1}}=\left(\frac{\omega_l}{\omega^*_1}-1\right)\frac{d(\Delta S)}{\Delta S} \tag{7.235}$$

Integrating this equation with the condition: $\Delta S=\Delta S$, $P^*_{4.l.1}=P_{4.l.1}$, and $\Delta S=\Delta S_M$, $P^*_{4.l.1}=P_{4.l.0}$, one obtains

$$\ln\frac{P_{1.l.1.0}}{P^*_{4.l.1}}=\int_{\Delta S}^{\Delta S_m}\left(\frac{\omega_l}{\omega^*_1}-1\right)\frac{d(\Delta S)}{\Delta S}=\left(\frac{\omega_l}{\omega^*_{1.m}}-1\right)\ln\frac{\Delta S_M}{\Delta S} \tag{7.236}$$

or

$$P^*_{4.l.1}=P_{1.l.1.0}\left(\frac{\Delta S}{\Delta S_M}\right)^{\left(\omega_l/\omega^*_{1.m}-1\right)} \tag{7.237}$$

where ΔS_m is the sediment concentration if all erodible bed material is scoured. For this case, the size distribution of the eroded sediment is the same as the bed material gradation $P_{1,1,0}$. Define the scour percentage

$$\lambda*=\frac{\Delta S}{\Delta S_M}=\begin{cases}\frac{\Delta G}{G}=\frac{\Delta G}{\Delta G+\Delta G_0}=\frac{\Delta h}{\Delta h+\Delta h_0} & (\Delta h+\Delta h_0<\Delta h_m)\\ \frac{\Delta G}{G_m}=\frac{\Delta h}{\Delta h_m} & (\Delta h+\Delta h_0>\Delta h_m)\end{cases} \tag{7.238}$$

where G is the total quantity of the sediment involved in the bed material sorting process, ΔG is the quantity of the eroded sediment, ΔG_0 is the quantity of the sediment that is involved in the sorting process but is not eroded, ΔG_m is the quantity of the erodible bed material, and Δh, Δh_0, and Δh_m is the corresponding scour depth. Since it is very rare for all erodible bed material to be totally scoured away, the first formula of Eq. (7.238) is generally used. Hence,

$$P^*_{4.l.1} = \left(\frac{\Delta S}{\Delta S_M}\right)^{(\omega_l/\omega^*_{1.m}-1)} = P_{1.l.1.0}(\lambda^*)^{(\omega_l/\omega^*_{1.m}-1)} = \Delta P_{4.l} \tag{7.239}$$

where $\omega^*_{1.m}$ is a median value of ω^*_1 for the interval of $\lambda^* = 0 \sim \lambda^*$. For a given λ^*, it is determined with

$$\sum P^*_{4.l.1} = \sum P_{1.l.1.0}(\lambda^*)^{(\omega_\lambda/\omega^*_{1.m}-1)} = 1 \tag{7.240}$$

It is noted that Eq. (7.239) is the same as the formula that was presented in [5], which reads

$$\Delta P_{4.l} = P_{1.l.1.0}(\lambda^*)^{(\omega_\lambda/\omega^*_{1.m}-1)} \tag{7.241}$$

The coarsening process of the bed material gradation $P_{1.1.1}$ for the case with massive scour is analyzed in the following. Let ΔG be the total quantity of the eroded sediment, ΔG_1 be the fractional quantity of the eroded sediment for the lth size class, G and G_1 is the total quantity and the fractional quantity, respectively, of the sediment that is involved in the scour process. The bed material gradation after the scour is

$$P_{1.l.1} = \frac{G_l - \Delta G_l}{G - \Delta G} = \frac{\frac{G_l}{G} - \frac{\Delta G_l}{\Delta G}\frac{\Delta G}{G}}{1 - \frac{\Delta G}{G}} = \frac{P_{1.l.1.0} - \lambda * \Delta P_{4.l}}{1 - \lambda *} \tag{7.242}$$

where $\lambda^* = \frac{\Delta G}{G}$ is the scour percentage. As previously discussed,

$$\frac{\Delta G_l}{G} = \Delta P_{4.l} \tag{7.243}$$

is the size distribution of the eroded sediment, and

$$\frac{G_l}{G} = P_{1.l.1.0} \tag{7.244}$$

is the initial bed material gradation. Plugging Eq. (7.241) into Eq. (7.242), one obtains

$$P_{1.l.1} = \frac{1}{1-\lambda^*}\left[P_{1.l.1.0} - \lambda * P_{1.l.1.0}\left(\lambda^*\right)^{\left(\omega_l/\omega^*_{1.m}-1\right)}\right] = \frac{P_{1.l.1.0}\left[1-\left(\lambda^*\right)^{\left(\omega_l/\omega^*_{1.m}\right)}\right]}{1-\lambda^*} \tag{7.245}$$

For the coarser sediment with $\omega_l > \omega^*_{1,m}$, $(\lambda^*)^{\omega_l/\omega^*_{1.m}} < \lambda^*$, and $P_{1.l.1} > P_{1.l.1.0}$. For the finer sediment with $\omega_l < \omega^*_{1,m}$, $(\lambda^*)^{\omega_l/\omega^*_{1.m}} > \lambda^*$, and $P_{1.l.1} < P_{1.l.1.0}$. It is seen that Eq. (7.245) describes a coarsening process. Equations (7.245) and (7.241) are the same as the relationships that were presented in [5].

The size distribution for the carrying-sediment capacity at the exit section, after considering the impact of the bed material coarsening, is analyzed in the following. Similar to Eqs. (7.201)–(7.204), one obtains

$$P^*_{4.l}S^*(\omega^*) = P_{4.l.0}S_0 + \left[1 - \frac{S_0}{S^*(\omega)}\right]P_{1.l.1}S^*(\omega^*_l) \tag{7.246}$$

$$S^*(\omega^*) = S_0 + \left[1 - \frac{S_0}{S^*(\omega)}\right]S^*(\omega^*_1) \tag{7.247}$$

$$\begin{aligned} P^*_{4.l} &= P_{4.l.0}\frac{S_0}{S^*(\omega^*)} + \left[1 - \frac{S_0}{S^*(\omega)}\right]P^*_{4.l.1}\frac{S^*(\omega^*_1)}{S^*(\omega^*)} \\ &= P_{4.l.0}\frac{S_0}{S^*(\omega^*)} + \left[1 - \frac{S_0}{S^*(\omega^*)}\right]P^*_{4.l.1} \end{aligned} \tag{7.248}$$

$$\begin{aligned} P_{1.l} &= P_{4.l.0}\frac{S_0}{S^*(l)} + \left[1 - \frac{S_0}{S^*(\omega)}\right]P_{1.l.1} \\ &= P'_{1.l.1}\frac{S_0}{S^*(\omega)} + \left[1 - \frac{S_0}{S^*(\omega)}\right]P_{1.l.1} \end{aligned} \tag{7.249}$$

where $P^*_{4.l.1}$ at the exit section is determined with Eq. (7.239), and $P_{1.l.1}$ within a reach is determined with Eq. (7.245), $P'_{1.l.1}$ is the gradation of the deposited material corresponding to $P_{4.l.1}$ as $\lambda \to 0$, which is

$$P'_{1.l.1} = \frac{P_{4.l.0}S^*(\omega)}{S^*(l)} \tag{7.250}$$

$S^*\left(\omega^*_1\right)$ is determined with the corresponding equations using $P^*_{4.l.1}$ at the exit section, and $S^*(\omega)$ is determined with the corresponding equations using $P_{4.l.0}$ at the entrance section. As shown in Eqs. (7.248) and (7.249), for massive scour, the size distribution of carrying-sediment capacity and the effective bed material gradation consist of the two parts that is related to the size distribution of suspended load and the bed material gradation, respectively.

The simplified expressions for the size distribution for carrying-sediment capacity are investigated in the following for the massive scour condition. Plugging Eq. (7.241) into Eq. (7.228), the size distribution variation of the suspended load for massive scour condition is

$$P_{4.l} = \frac{P_{4.l.0}S_0}{S} + \frac{(S - S_0)}{S} P_{1.l.1.0}(\lambda^*)^{(\omega_l/\omega^*_{1.m} - 1)}$$
$$= \frac{1}{1-\lambda}\left[P_{4.l.0} - \frac{\lambda}{\lambda^*} P_{1.l.1.0}(\lambda^*)^{\omega_l/\omega^*_{1.m}}\right] \tag{7.251}$$

This was developed by Han [5], which was referred to as the first mode for the size distribution of suspended load at the exit section for massive scour condition. Plugging $P^*_{4,l.1}$, as presented in Eq. (7.239), into Eq. (7.248)

$$P^*_{4.l} = \frac{S_0}{S^*(\omega^*)}\left\{P_{4.l.0} + \left[1 - \frac{S^*(\omega^*)}{S_0}\right]\frac{P_{1.l.1.0}}{\lambda^*}(\lambda^*)^{\omega_l/\omega^*_{1.m}}\right\} \tag{7.252}$$

Set

$$S = \beta_1 S^*(\omega^*) \tag{7.253}$$

Then

$$\frac{S_0}{S} = \frac{S_0}{\beta_1 S^*(\omega^*)} \tag{7.254}$$

$$\frac{S - S_0}{S} = 1 - \frac{S_0}{\beta_1 S^*(\omega^*)} = -\frac{S_0}{\beta_1 S^*(\omega^*)}\left[1 - \frac{\beta_1 S^*(\omega^*)}{S_0}\right] \tag{7.255}$$

Considering Eq. (7.240), Eq. (7.251) can be written as

$$P_{4.l} = \frac{S_0}{\beta_1 S^*(\omega^*)} P_{4.l.0} + \left[1 - \frac{S_0}{\beta_1 S^*(\omega^*)}\right] P^*_{4.l.1} \tag{7.256}$$

It is seen that only for $\beta_1 = 1$, Eq. (7.251), or Eq. (7.256), is the same as the right side of Eq. (7.252). Also,

$$\frac{S_0}{S^*(\omega^*)} = \frac{S_0}{S} = \frac{1}{1-\lambda}$$
$$1 - \frac{S^*(\omega^*)}{S_0} = 1 - \frac{S}{S_0} = \lambda \tag{7.257}$$

Hence, Eq. (7.252) becomes

$$P^*_{4.l} = \frac{1}{1-\lambda}\left[P_{4.l.0} - P_{1.l.1.0}\frac{\lambda}{\lambda^*}(\lambda^*)^{\omega_l/\omega^*_m}\right] \tag{7.258}$$

Comparing with Eq. (7.252), it is concluded that

$$P_{4,1} = P_{4,1}^* \tag{7.259}$$

In fact, this only occurs when the scour is under a weak equilibrium condition, $S = S^*(\omega^*)$, which is rare. However, $\beta_1 \approx 1$, and for quite some cases. Comparing the size distribution for suspended sediment for the massive erosion condition, which is presented in Eq. (7.257)6, with the size distribution for carrying-sediment capacity, which is listed in Eq. (7.249)8, β_1 only results in different weighting factors for $P_{4.l.0}$ and $P_{4.l.1}^*$. If β_1 is close to 1 (for example, for $0.8 < \beta_1 < 1$), considering $S < S^*(\omega^*)$ for massive scour, the difference between $P_{4.l}$ and $P_{4.l}^*$ is limited. Setting $\beta_1 = 0.8$, Eq, (7.257)6 becomes

$$P_{4.l} = 1.25\frac{S_0}{S^*(\omega^*)}P_{4.l.0} + \left[1 - 1.25\frac{S_0}{S^*(\omega^*)}\right]P_{4.l.1}^* \tag{7.260}$$

For a general condition, the difference between $P_{4.l}$, as presented in Eq. (7.256), and $P_{4.l}^*$, as determined in Eq. (7.248) is much less than the difference between β_1 and 1. Hence, for $\beta_1 \geq 0.8$,

$$P_{4.l}^* \approx P_{4.2} \tag{7.261}$$

This will significantly simplify the analysis for non-equilibrium sediment transport by directly applying Eq. (7.261) together with Eq. (7.251), without introducing the size distribution for carrying-sediment capacity and the effective bed material gradation. This simplification was addressed, without derivation, in [12, 13]). It is worthy to point that $P_{4.l}^*$ is slightly smaller than it is if $P_{4.l}^*$ is replaced with $P_{4.l}$, considering $P_{4.l.1}^*$ is larger than $P_{4,l,0}$.

If the difference between Eqs. (7.251) and (7.252) is significant, and $P_{4.l}^*$ is significantly different from $P_{4,2}$, Eqs. (7.248) and (7.249) should be used to determine the size distribution for carrying-sediment capacity at the exit section, followed by determining $P_{4.l}$ with the governing equation for non-equilibrium sediment transport.

7.7.2.2 Method 1 [4]

For a reach with the length of x, setting the initial quantity of the bed material that is involved in erosion is W_0, and the bed material left on the bed after the erosion is W, the corresponding fractional bed material quantities are

$$W_{0.l} = P_{11.1.0}W_0 \tag{7.262}$$

$$W_l = P_{1l.1}W \tag{7.263}$$

The ratio $\frac{\Delta W_l}{\Delta W}$ represents the variation of the bed material gradation, or the gradation of the material that is eroded from the bed and then becomes part of the carrying-sediment capacity, and

$$P_{4.l.1}^{*}=\lim_{\Delta W\to 0}\frac{\Delta W_l}{\Delta W}=\frac{dW_l}{dW}=\frac{d(P_{1.l.1}W)}{dW}=P_{1.l.1}+W\frac{dP_{1.l.1}}{dW} \tag{7.264}$$

Also, as discussed before,

$$P_{4.l.1}^{*}=\frac{S^{*}(l)}{S^{*}(\omega_1^{*})}P_{1.l.1}=\left(\frac{\omega_1^{*}}{\omega_l}\right)^{m}P_{1.l.1} \tag{7.265}$$

Plugging Eq. (7.265) into Eq. (7.264) yields

$$\frac{dP_{1.l.1}}{P_{1.l.1}}=\left[\left(\frac{\omega_1^{*}}{\omega_l}\right)^{m}-1\right]\frac{dW}{W} \tag{7.266}$$

Integrating with the condition: $W=W_0$, $P_{1.l.1}=P_{1.l.1.0}$, and utilizing the mean value theorem for integrals, [4] obtained

$$\begin{aligned}P_{1.l.1}&=P_{1.l.1.0}\left(\frac{W}{W_0}\right)^{\left(\omega_{1.m}^{*}/\omega_l\right)^{m}-1}=P_{1.l.1.0}\left(\frac{W_0-\Delta W}{W_0}\right)^{\left(\omega_{1.m}^{*}/\omega_l\right)^{m}-1}\\&=P_{1.l.1.0}(1-\lambda^{*})^{\left(\omega_{1.m}^{*}/\omega_l\right)^{m}-1}\end{aligned} \tag{7.267}$$

where ΔW is the weight of the material that is eroded from the bed within the reach with the length of x,

$$\lambda^{*}=\frac{\Delta W}{W_0} \tag{7.268}$$

is the scour percentage, $\omega_{1,m}^{*}$ is a median value between $\omega_{1,0}^{*}$ and $\omega_{1,x}^{*}$, which are determined by

$$S^{*}\left(\omega_{1.0}^{*}\right)=\sum_{l=k+1}^{n}P_{1.l.1.0}S^{*}(l) \tag{7.269}$$

$$S^{*}(\omega_{1.x}^{*})=\sum_{l=k+1}^{n}P_{1.l.1}S^{*}(l) \tag{7.270}$$

In fact, $\omega_{1,m}^{*}$ can also be determined with

$$\sum_{l=k+1}^{n}P_{1.l.1}=\sum_{l=k+1}^{n}P_{1.l.1.0}\left(1-\lambda^{*}\right)^{\left(\frac{\omega_{1.m}^{*}}{\omega_l}\right)^{m}-1}=1 \tag{7.271}$$

Hence, the size distribution for the carrying-sediment capacity related to the bed material, or the size distribution of the eroded material, is

$$P^*_{4.l.1} = \frac{W_{0.l} - W_l}{W_0 - W} = \frac{\frac{W_{0,l}-W_l}{W_0-W}\frac{W}{W_0}}{1 - \frac{W}{W_0}} = \frac{P_{1.l.1.0} - P_{1.l.1}(1-\lambda^*)}{\lambda^*} \tag{7.272}$$

Plugging Eq. (7.267) into Eq. (7.272), [10] obtained

$$\begin{aligned} P^*_{4.l.1} &= \frac{1}{\lambda^*}\Big[P_{1.l.1.0} - P_{1.l.1.0}(1-\lambda^*)^{(\omega^*_{1.m}/\omega_l)^m - 1}(1-\lambda^*)\Big] \\ &= \frac{P_{1.l.1.0}}{\lambda^*}\Big[1 - (1-\lambda^*)^{(\omega^*_{1.m}/\omega_l)^m}\Big] - \Delta P_{4.l} \end{aligned} \tag{7.273}$$

This is the size distribution for carrying-sediment capacity at the exit section that is related to the bed material in a reach with the length of x. Equations (7.267) and (7.273) were developed by Han [4], which were referred to as the first mode for the bed material coarsening and corresponding size distribution of the eroded material ($\Delta P_{4,l}$).

The size distribution for carrying-sediment capacity at the exit section, after considering bed material coarsening, is discussed as follows. According to Eqs. (7.146) and (7.149)

$$\begin{aligned} P^*_{4.l} S^*(\omega^*) &= P_{4.l.0} S_0 + \left[1 - \frac{S_0}{S^*(\omega)}\right] P^*_{4.l.1} S^*(\omega^*_1) \\ &= P_{4.l.0} S_0 + \frac{P_{1.l.1.0}}{\lambda^*}\Big[1 - (1-\lambda^*)^{\omega^*_{1.m}/\omega_l - 1}\Big]\left[1 - \frac{S_0}{S^*(\omega)}\right](l) \end{aligned} \tag{7.274}$$

$$S^*(\omega^*) = S_0 + \left[1 - \frac{S_0}{S^*(\omega)}\right] S^*(\omega^*_1) \tag{7.275}$$

$$\begin{aligned} P^*_{4.l} &= P_{4.l.0}\frac{S_0}{S^*(\omega^*)} + \left[1 - \frac{S_0}{S^*(\omega)}\right]\frac{S^*(\omega^*_1)}{S^*(\omega^*)} P^*_{4.l.1} \\ &= P_{4.l.0}\frac{S_0}{S^*(\omega^*)} + \left[1 - \frac{S_0}{S^*(\omega^*)}\right] P^*_{4.l.1} \\ &= P_{4.l.0}\frac{S_0}{S^*(\omega^*)} + \left[1 - \frac{S_0}{S^*(\omega^*)}\right] \frac{P_{1.l.1.0}}{\lambda^*}\Big[1 - (1-\lambda^*)^{\omega^*_{1.m}/\omega_l - 1}\Big] \end{aligned} \tag{7.276}$$

$$\begin{aligned} P_{1,1} &= P_{4.l.0}\frac{S_0}{S^*(l)} + \left[1 - \frac{S}{S^*(\omega)}\right] P_{1.l.1} \\ &= \frac{S_0}{S^*(\omega)} P'_{1.l.1} + \left[1 - \frac{S}{S^*(\omega)}\right] P_{1.l.1} \end{aligned} \tag{7.277}$$

Similar to Eqs. (7.228), (7.230) and (7.272), the fractional sediment concentration for massive scour condition can be derived, which is

$$
\begin{aligned}
P_{4.l}^{*}S &= P_{4.l.0}S_0 + (S + S_0) - P_{4.l.1}^{*} \\
&= P_{4.l.0}S_0 + (S + S_0) - \frac{P_{1.l.1.0}}{\lambda^{*}}\left[1 - (1 - \lambda^{*})^{\omega_{1.m}^{*}/\omega_l}\right]
\end{aligned}
$$

or

$$
\begin{aligned}
P_{4.l} &= P_{4.l.0}\left(\frac{S_0}{S} + \frac{S - S_0}{S}\right)P_{4.l.1}^{*} \\
&= P_{4.l.0}\frac{S_0}{S} + \left(\frac{S - S_0}{S}\right)\frac{P_{1.l.1.0}}{\lambda^{*}}\left[1 - (1 - \lambda^{*})^{(\omega_{1.1.m}^{*}/\omega_l)^m}\right] \\
&= \frac{1}{1-\lambda}\left\{P_{4.l.0} - \frac{\lambda}{\lambda^{*}}P_{1.l.1.0}\left[1 - (1 - \lambda)^{(\omega_{1.1.m}^{*}/\omega_l)^m}\right]\right\}
\end{aligned} \tag{7.278}
$$

This is the formula that was developed by Han [4] for the size distribution of suspended load at the exit section for the massive scour condition, which was referred to as the first mode results.

Similar to our previous discussion for Eqs. (7.252) and (7.249), for $0.8 < \frac{S}{S^{*}(\omega^{*})} = \beta_1 < 1$, the right side of Eq. (7.276) is approximately the same as Eq. (7.278), resulting in $P_{4.l}^{*} = P_{4,l}$.

The difference between Eq. (7.267)

$$
P_{1.l.1} = P_{1.l.1.0}\frac{(1 - \lambda^{*})^{(\omega_{1.m}^{*}/\omega_l)^m}}{1 - \lambda^{*}}
$$

and Eq. (7.245)

$$
P_{1.l.1} = P_{1.l.1.0}\frac{1 - (\lambda^{*})^{\omega_{1.m}^{*}/\omega_l}}{1 - \lambda^{*}}
$$

or the difference between the first method and the second method for bed material coarsening, is analyzed in the following. Firstly, both equations were derived based on the same assumption (massive scour) with reasonable theoretical basis. The second mode involves the governing equation for non-equilibrium sediment transport. Secondly, both equations were verified with the same data sets, which included the bed material gradation data collected in a downstream reach of the Yellow River between Huayuankou Station and Aishan Station during the clear water erosion periods from 1961 to 1964 after completion of the Sanmenxia Reservoir [5, 10], as discussed in Chap. 5. Thirdly, the difference in the computed results between these two equations for the scour percentage $\lambda^{*} < 0.662$ is listed in Table 7.1. In the calculations, the suspended load is divided into eight classes, with the same initial bed material gradation $P_{1.l.1.0} = 1/8 = 0.125$. As listed in this table, for $\lambda^{*} < 0.662$, the absolute difference between these two equations does not exceed 0.03 for any size class, which is within the accuracy requirement for engineering practice. Fourthly, both equations show $P_{1.l.1} \rightarrow P_{1.l.1.0}$ for $\lambda^{*} \rightarrow 0$, which is reasonable. However, Eq. (7.273) shows the limit of $P_{4.l.1}^{*}$ exists for $\lambda^{*} \rightarrow 0$, which is the function value at

$\lambda^* = 0$; while Eq. (7.240) shows the limit of $P^*_{4.l.1}$ does not exist for $\lambda^* \to 0$. Theoretically, the value of $P^*_{4.l.1}$ can be determined for small λ^* based on $\sum_{l=1}^{n} P^*_{4.l.1} = 1$. In conclusion, $P_{1,l,1}$ and $P^*_{4.l.1}$ can be determined with either the first method or the second method in engineering calculation for the massive scour condition. However, the first method seems to be more appropriate for theoretical analysis that needs to include the case for $\lambda^* \to 0$.

The results listed Table 7.1 are computed for a synthetic case with $P_{1.l.1.0} = 1/8 = 0.125$. For the first method, Eq. (7.267) becomes

$$P_{1.l.1} = P_{1.l.1.0}\left(1 - \lambda^*\right)^{(\omega^*_{1.m}/\omega_l - 1)} = 0.125\frac{\eta^{1/\omega_l}}{1 - \lambda^*}$$

$$\lambda^* = 1 - \sum_{l=1}^{8} P_{1.l.1.0}\eta^{1/\omega_l} = 1 - 0.125\sum_{l=1}^{8} \eta^{1/\omega_l}$$

Assuming various μ, the corresponding $P_{1.l.1}$ and λ^* are determined based on the above equations. For the second method, Eq. (7.246) can be written as

$$P_{1.l.1} = P_{1.l.1.0}\frac{1 - (\lambda^*)^{\omega_l/\omega^*_{1.m}}}{1 - \lambda^*} = \frac{0.125}{1 - \lambda^*}(1 - \eta^{\omega_l})$$

$$\lambda^* = 0.125\sum_{l=1}^{8} \eta^{\omega_l}$$

Assuming various η, the corresponding $P_{1.l.1}$ and λ^* are determined based on the above equations.

7.8 Size Distribution for Carrying-Sediment Capacity and Effective Bed Material Gradation at the Exit Section of a Relatively Long Reach for the Case with Significant Scour or Deposition

The size distribution for carrying-sediment capacity and the effective bed material gradation as discussed in Sects. 7.4 and 7.5 are for the same section. In other words, the relationships of the size distribution for carrying-sediment capacity and the effective bed material gradation to the size distribution of suspended load and the bed material gradation were investigated in those two sections for the same cross section. Considering the impact of channel sedimentation on the bed material gradation, the size distribution for carrying-sediment capacity at the exit section, the bed material gradation, and the size distribution of suspended load were analyzed in Sect. 7.7. For a relatively long reach, the influx sediment concentration decays along the flow

Table 7.1 Difference in computed bed material gradation between the first method, Eq. (7.267), and the second method, Eq. (7.246)

λ^*		Bed material gradation $P_{1,l,1}$								$\sum P_{1,l,1}$
	D (mm)	< 0.005	0.009–0.01	0.01–0.025	0.025–0.05	0.05–0.10	0.10–0.25	0.25–0.50	0.50–1.00	
	ω (cm/s)	0.000575	0.00536	0.0279	0.0982	0.536	1.723	4.033	8.280	
0.1526	Method 1	0.0166	0.0989	0.1470	0.1475	0.1475	0.1475	0.1475	0.1475	1.0000
	Method 2	0.0196	0.1069	0.1384	0.1449	0.1474	0.1476	0.1476	0.1476	1.0000
0.2299	Method 1	0.0000	0.0637	0.1350	0.1542	0.1608	0.1618	0.1622	0.1623	1.0000
	Method 2	0.0054	0.0432	0.1413	0.1589	0.1628	0.1628	0.1628	0.1628	1.0000
0.4892	Method 1	0.0000	0.0000	0.0051	0.0837	0.2010	0.2302	0.2384	0.2416	1.0000
	Method 2	0.0004	0.0037	0.0187	0.0591	0.1860	0.2416	0.2452	0.2453	1.0000
0.6618	Method 1	0.0000	0.0000	0.0000	0.0003	0.1014	0.2472	0.3112	0.3399	1.0000
	Method 2	0.0001	0.0011	0.0059	0.0199	0.0947	0.2196	0.3155	0.3432	1.0000
0.8310	Method 1	0.0000	0.0000	0.0000	0.0000	0.0027	0.1301	0.3520	0.5152	1.0000
	Method 2	0.0001	0.0007	0.0035	0.0123	0.0631	0.1774	0.3400	0.4029	1.0000

path due the sediment exchange. The corresponding impact on the size distribution for carrying-sediment capacity at the exit section is discussed in this section.

7.8.1 Carrying-Sediment Capacity at the Exit Section Corresponding to the Sediment Influx at the Entrance Section

In order to analyze the mechanisms for the size distribution variation for carrying-sediment capacity within a reach, two topics should be investigated first. One is how the suspended load influx changes to the carrying-sediment capacity at the exit section after being transport within a reach. The other is how the variation of the bed material gradation, which is caused by channel sedimentation, impacts the carrying-sediment capacity at the exit section. The coarser particles of the suspended load influx constantly change to the bed material during the sediment exchange process, and the concentrations of these coarser size classes decay along the flow path. At the same time, the proportion of the carrying-sediment capacity that is related to the bed material increases. For simplicity, the suspended load is divided into two parts. The finer part of the suspended load consists of the particles that completely change to the carrying-sediment capacity, and the coarser part consist of the particles that partially changes to the transport capacity. After depositing on the bed within a distance of x, the coarser suspended particles remaining in the water body change to the carrying-sediment capacity, which is

$$S_0 P_{4.2.0} P_{4.l.2.0} \exp\left(-\frac{\alpha\omega_l x}{q}\right)$$

This is the remainder of the fractional suspended sediment concentration after depositing on the bed to become the bed material. This remainder part then completely changes to the carrying-sediment capacity at the exit section. Also, it is assumed that all bed material can be suspended, and thus can change to the carrying-sediment capacity. Equation (7.88), the quantity of the coarser particles that change to the carrying-sediment capacity at the exit section is

$$S_0 P_{4.2.0} P_{4.l.2.0} \exp\left(-\frac{\alpha\omega_l x}{q}\right) = K_{2.l} S^*(l) \tag{7.279}$$

The water percentage for these coarser particles is

$$K_{2.l} = \frac{S_0}{S^*(l)} P_{4.2.0} P_{4.l.2.0} \exp\left(-\frac{\alpha\omega_l x}{q}\right) \tag{7.280}$$

$$K_2 = S_0 P_{4.2.0} \sum_{l=k+1}^{n} \frac{K_{x.2} P_{4.l.2.0} \exp\left(-\frac{\alpha\omega_l x}{q}\right)}{S^*(l) K_{x.2}}$$
$$= S_0 P_{4.2.0} \sum_{l=k+1}^{n} \frac{K_{x.2}}{S^*(l)} P_{4.l.2} = \frac{P_{4.2.0} S_0 K_{x.2}}{S^*(\omega_2)} \quad (7.281)$$

with

$$\frac{1}{S^*(\omega_2)} = \sum_{l=k+1}^{n} \frac{P_{4.l.2.0} \exp\left(-\frac{\alpha\omega_l x}{q}\right)}{S^*(l) \sum_{l=1}^{n} P_{4.l.0} exp\left(-\frac{\alpha\omega_l x}{q}\right)} = \sum_{l=k+1}^{n} \frac{P_{4.l.2}}{S^*(l)} \quad (7.282)$$

$$P_{4.l.2} = \frac{P_{4.l.2.0} \exp\left(-\frac{\alpha\omega_l x}{q}\right)}{\sum_{l=k+1}^{n} P_{4.l.2.0} \exp\left(-\frac{\alpha\omega_l x}{q}\right)} = \frac{P_{4.l.2.0} \exp\left(-\frac{\alpha\omega_l x}{q}\right)}{K_{x.2}} \quad (7.283)$$

$$K_{x.2} = \sum_{l=1}^{n} P_{4.l.2.0} \exp\left(-\frac{\alpha\omega_l x}{q}\right) = \frac{K_2 S^*(\omega_2)}{P_{4.l.0} S_0} \quad (7.284)$$

where the subscript "0" denotes the value at the entrance section, $P_{4.l.2x}$ is the normalized size distribution for the coarser part of suspended load at the exit section that is left after the sediment exchange with the distance of x, $K_{x.2}$ is the product of the decay coefficient for the concentration of the coarser sediment to change to the carrying-sediment capacity at the exit section and the corresponding water percentage, and $S^*(\omega_2)$ is the carrying-sediment capacity for the coarser size classes at the exit section.

All finer particles of the suspended load influx change to the carrying-sediment capacity at the exit section. Hence,

$$S_0 P_{4.1.0} P_{4.l.1.0} = K_{1.l} S^*(l) \quad (7.285)$$

or

$$K_{1.l} = \frac{P_{4.1.0} P_{4.l.1.0} S_0}{S^*(l)} \quad (7.286)$$

$$K_1 = \sum_{l=1}^{n} P_{4.1.0} S_0 \frac{P_{4.l.1.0}}{S^*(l)} = \frac{S_0 P_{4.1.0}}{S^*(\omega_{1.0})} \quad (7.287)$$

where

$$P_{4.1.0} = \sum_{l=1}^{k} P_{4.l.0} \quad (7.288)$$

$$P_{4.l.1.0} = \frac{P_{4.l.0}}{P_{4.1.0}} \quad (1 < l \leq k) \tag{7.289}$$

$$\frac{1}{S^*(\omega_{1.0})} = \sum_{l=k+1}^{n} \frac{P_{1.l.1.0}}{S^*(l)} \tag{7.290}$$

7.8.2 *Carrying-Sediment Capacity at the Exit Section Related to the Bed Material Gradation After Bed Material Coarsening*

Adopting the first method for bed material coarsening, the variation of the bed material gradation after bed material coarsening can be determined with Eq. (7.267). The corresponding $P^*_{4.l.1}$ can be computed with Eq. (7.273). Then, the fractional and total carrying-sediment capacities, the size distribution for carrying-sediment capacity, and the bed material gradation can be determined with Eqs. (7.274) through (7.277), respectively. This was already discussed in Sect. 7.7.2.2.

For a general condition, the bed material should include the coarser particles that deposit on the bed during the sediment exchange between the suspended load and the bed material within the time interval of Δt, which is $\sum_{l=k+1}^{n} P_{4.2.0} P_{4.l.2.0} S_0 \left[1 - \exp\left(-\frac{\alpha \omega_l x}{q}\right)\right]$. This is particularly true if Δt is very large, or if the coarser part of the suspended load is finer than the bed material. For a very small Δt, the bed material sorting process may exclude the particles deposited within this time step. Instead, these deposited particles can be included in the bed material for the next time step, with the sorting process being described with Eqs. (7.267) and (7.273). In other words, $P_{1.l.1}$ and $P_{1.l.1.0}$ in Eqs. (7.267) and (7.273) can be understood as the initial bed material gradation at each time step if Δt is very small. On the other hand, if Δt is very large, the bed material gradation should include the quantity and the size distribution of the coarser particles that deposit on the bed. Assuming the flow rate is Q in a reach with the length of x, the weight of sediment for the lth (coarser) size class that deposits on the bed within Δt is

$$\Delta G_l = P_{4.2.0} P_{4.l.2.0} S_0 \left[1 - \exp\left(-\frac{\alpha \omega_l x}{q}\right)\right] \tag{7.291}$$

The gradation of the deposited material is

$$P'_{1.l.1} = \frac{\Delta G_l}{\sum_{l=k+1}^{l=n} \Delta G_l} = \frac{P_{4.l.2.0}\left[1 - \exp\left(-\frac{\alpha \omega_l x}{q}\right)\right]}{1 - K_{x.2}} \tag{7.292}$$

The thickness of the deposited material is

$$\Delta h_1 = \frac{\sum_{l=k+1}^{n} \Delta G_l Q \Delta t}{\gamma_s' B x} = \frac{P_{4.2.0} S_0 (1 - K_{x.2}) Q \Delta t}{\gamma_s' B x} \tag{7.293}$$

The fractional thickness of the deposited material is

$$\Delta h_{1.l} = \frac{\Delta G_l Q \Delta t}{\gamma_s' B x} = P'_{1.l.1} \Delta h_1 \tag{7.294}$$

The gradation of the bed material that participates in the scour and sorting process is

$$P'_{1.l.1.0} = \begin{cases} \frac{P'_{1.l.1.} \Delta h_1 + P_{1.l.1.0}(\Delta h_2 - \Delta h_1 + \Delta h_0)}{\Delta h_2 + \Delta h_0} & (k + 1 \le l \le n, and \Delta h_2 \ge 0) \\ \frac{P_{1.l.1.0}(\Delta h_2 - \Delta h_1 + \Delta h_0)}{\Delta h_2 + \Delta h_0} & (l \le k) \end{cases} \tag{7.295}$$

where Δh_0 is the thickness of the bed material that participates in the scour and sorting process but is not scoured away, or the depth of armor layer, and Δh_2 is the sum of the deposition depth of the coarser suspended sediment, h_1, and the scour depth of the bed material, Δh, or

$$\Delta h_2 = \Delta h + \Delta h_1$$

which is referred to as the net sedimentation depth. The magnitudes of Δh_2 and Δh represent the intensity of sedimentation and sediment exchange between suspended load and the bed material. Based on Eq. (7.295),

$$\begin{aligned} \sum_{l=1}^{n} P'_{1.l.1.0} &= \sum_{l=1}^{k} P_{1.l.1.0} \left(\frac{\Delta h_2 + \Delta h_0 - \Delta h_1}{\Delta h_2 + \Delta h_0} \right) + \sum_{l=k+1}^{n} \frac{P'_{1.l.1} \Delta h_1 + P_{1.l.1.0}(\Delta h_2 + \Delta h_0 - \Delta h_1)}{\Delta h_2 + \Delta h_0} \\ &= \frac{\Delta h_2 + \Delta h_0 - \Delta h_1}{\Delta h_2 + \Delta h_0} \left(\sum_{l=1}^{k} P_{1.l.1.0} + \sum_{l=k+1}^{k} P_{1.l.1.0} \right) + \frac{\Delta h_1}{\Delta h_2 + \Delta h_0} = 1 \end{aligned}$$

Also,

$$\Delta h = \frac{(S - S_0) Q \Delta t}{\gamma_s' B x} \tag{7.296}$$

For $\Delta h_2 < 0$, or $\Delta h < -\Delta h_1$, pure deposition occurs. For this case, the third terms in the formulas for the carrying-sediment capacity, the size distribution for carrying-sediment capacity, and the bed material gradation disappear, or the coarser particles depositing on the bed can not be suspended and thus do not contribute to the carrying-sediment capacity. For $0 \le \Delta h_2 \le \Delta h_1$, the coarser particles depositing on the bed are not be fully suspended ($\Delta h_1 \ge -\Delta h$), resulting in channel deposition ($\Delta h_2 \ge 0$). However, the eroded sediment also includes part of the bed material, and thus bed material gradation varies. For this case, the third terms in these formulas should be included. For $\Delta h_2 \ge \Delta h_1$, channel scour occurs.

7.8.3 Carrying-Sediment Capacity and Corresponding Size Distribution at the Exit Section After Considering the Impacts of Sediment Exchange Between Suspended Load and Bed Material and Bed Material Coarsening

There are three cases: (1) case 1: all finer suspended load influx completely changes to the carrying-sediment capacity at the exit section, and the coarser suspended load partially changes to the carrying-sediment capacity; (2) case 2: all suspended load influx partially changes to the carrying-sediment capacity at the section; and (3) case 3: all suspended load influx completely changes to the carrying-sediment capacity at the exit section. For case 3, if $K_3 > 0$ and the bed material is erodible, massive scour occurs, which was discussed before; if $K_3 > 0$ and the bed material can not be suspended, this case is equivalent to that for non-equilibrium sediment transport on a gravel-bed, which will be analyzed in Sect. 7.9. Case 2 will be discussed in Sect. 7.8.4. This section focuses on case 1.

Considering Eqs. (7.287) and (7.281), the water percentage for the part of the carrying-sediment capacity that is related to the bed material is

$$K_3 = 1 - K_1 - K_2 = 1 - \frac{S_0 P_{4.1.0}}{S^*(\omega_{1.0})} - \frac{P_{4.2.0} S_0 K_{x.2}}{S^*(\omega_x)} \tag{7.297}$$

For $\Delta h_2 < 0$, $K_1 + K_2 > 1$ $K_3 < 0$, and the carrying-sediment capacity consists of two parts. The water percentage for the part of the carrying-sediment capacity at the exit section, which is related to the coarser particles of the suspended load influx after the exchange with the bed material, is

$$K_2 = 1 - K_1 = 1 - \frac{S_0 P_{4.1.0}}{S^*(\omega_{1.0})} \tag{7.298}$$

Hence, the fractional carrying-sediment capacity and other parameters at the exit section can be written as

$$P^*_{4.l} S^*(\omega^*) = \begin{cases} (K_{1.l} + K_{2.l} + K_{3.l}) S^*(l) = P_{4.1.0} P_{4.l.1.0} S_0 + P_{4.2.0} P_{4.l.2.0} S_0 \exp\left(-\frac{\alpha \omega_l x}{q}\right) & \\ \quad + \left[1 - \frac{S_0 P_{4.1.0}}{S^*(\omega_{1.0})} - \frac{S_0 P_{4.2.0} K_{x.2}}{S^*(\omega_2)}\right] P^*_{4.l.1} S^*(\omega^*_1) & (K_3 > 0) \\ (K_{1.l} + K_{2.l}) S^*(l) = P_{4.1.0} P_{4.l.1.0} S_0 & \\ \quad + \left[1 - \frac{S_0 P_{4.1.0}}{S^*(\omega_{1.0})}\right] P_{4.2.0} P_{4.l.2.0} S_0 \exp\left(-\frac{\alpha \omega_l x}{q}\right) & (K_3 \le 0) \end{cases} \tag{7.299}$$

$$S^*(\omega^*) = \begin{cases} P_{4.1.0}S_0 + P_{4.2.0}S_0K_{x.2} + \left[1 - \dfrac{S_0P_{4.1.0}}{S^*(\omega_{1.0})} - \dfrac{S_0P_{4.2.0}K_{x.2}}{S^*(\omega_2)}\right]S^*(\omega_1^*) \\ \qquad (K_3 > 0) \\ P_{4.1.0}S_0 + \left[1 - \dfrac{S_0P_{4.1.0}}{S^*(\omega_{1.0})}\right]P_{4.2.0}S_0K_{x.2} \qquad (K_3 \leq 0) \end{cases} \tag{7.300}$$

$$P_{4.l}^* = \begin{cases} \dfrac{P_{4.1.0}P_{4.l.1.0}S_0}{S^*(\omega^*)} + \dfrac{P_{4.2.0}P_{4.l.2.0}S_0K_{x.2}}{S^*(\omega_1)} \\ +\left[1 - \dfrac{P_{4.1.0}S_0}{S^*(\omega_{1.0})} - \dfrac{P_{4.2.0}S_0K_{x.2}}{S^*(\omega_2)}\right]P_{4.l.1}^*\dfrac{S^*(\omega_1^*)}{S^*(\omega^*)} \qquad (K_3 > 0) \\ \dfrac{P_{4.1.0}P_{4.l.1.0}S_0}{S^*(\omega^*)} + \left[1 - \dfrac{P_{4.1.0}S_0}{S^*(\omega_1)}\right]\dfrac{P_{4.2.0}P_{4.l.2.0}S_0K_{x.2}}{S^*(\omega^*)} \qquad (K_3 \leq 0) \end{cases} \tag{7.301}$$

$$P_{1.l} = \begin{cases} \dfrac{P_{4.1.0}P_{4.l.1.0}S_0}{S^*(l)} + \dfrac{P_{4.2.0}P_{4.l.2.0}S_0K_{x.2}}{S^*(l)} \\ +\left[1 - \dfrac{P_{4.1.0}S_0}{S^*(\omega_{1.0})} - \dfrac{P_{4.2.0}S_0K_{x.2}}{S^*(\omega_2)}\right]P_{4.l.1}^*\dfrac{S^*(\omega_1^*)}{S^*(l)} \qquad (K_3 > 0) \\ \dfrac{P_{4.1.0}P_{4.l.1.0}S_0}{S^*(l)} + \left[1 - \dfrac{P_{4.1.0}S_0}{S^*(\omega_1)}\right]\dfrac{P_{4.2.0}P_{4.l.2.0}S_0K_{x.2}}{S^*(l)} \qquad (K_3 \leq 0) \end{cases} \tag{7.302}$$

where $P_{4.l.1}^*$ is determined with Eq. (7.273). Based on Eq. (7.300), one obtains

$$\left[1 - \frac{S_0P_{4.1.0}}{S^*(\omega_{1.0})} - \frac{S_0P_{4.2.0}K_{x.2}}{S^*(\omega_2)}\right]\frac{S^*(\omega_1^*)}{S^*(\omega^*)} = 1 - \frac{S_0P_{4.1.0}}{S^*(\omega^*)} - \frac{S_0P_{4.2.0}K_{x.2}}{S^*(\omega^*)} \quad (K_3 < 0) \tag{7.303}$$

$$\left[1 - \frac{S_0P_{4.1.0}}{S^*(\omega_{1.0})}\right]\frac{S_0P_{4.2.0}K_{x.n}}{S^*(\omega^*)} = 1 - \frac{S_0P_{4.1.0}}{S^*(\omega^*)} \quad (K_3 > 0) \tag{7.304}$$

Utilizing Eqs. (7.303) and (7.304), Eqs. (7.299) through (7.302) can be rewritten in different forms, if needed. Also, considering Eqs. ((7.303) and (7.304), it can be derived from Eqs. (7.301) and (7.302), respectively, that $P_{4.l}^* = 1$ and $\sum P_{1.l} = 1$.

7.8.4 Sediment Transport Capacity and Corresponding Size Distribution for the Case with All Suspended Load Influx Partially Changing to Carrying-Sediment Capacity

The fractional quantity of the suspended load influx that changes to the carrying-sediment capacity at the exit section for the lth size class is

$$S_0 P_{4.l.0} \exp\left(-\frac{\alpha\omega_l x}{q}\right) = K_{1.l} S^*(l) \tag{7.305}$$

or

$$K_1 = \sum_{l=1}^{n} K_{1.l} = \sum_{l=1}^{n} \frac{S_0 P_{4.l.0}}{S^*(l)} \exp\left(-\frac{\alpha\omega_l x}{q}\right) = \frac{S_0 K_x}{S^*(\omega)} \tag{7.306}$$

with

$$\frac{1}{S^*(\omega)} = \sum_{l=1}^{n} \frac{1}{S^*(l)} \frac{P_{4.l.0} \exp\left(-\frac{\alpha\omega_l x}{q}\right)}{\sum P_{4.l.0} \exp\left(-\frac{\alpha\omega_l x}{q}\right)} = \sum_{l=1}^{n} \frac{P_{4.l}}{S^*(l)} \tag{7.307}$$

where $P_{4.l}$ is the size distribution of suspended load at the exit section, and

$$K_x = \sum_{l=1}^{n} P_{4.l.0} \exp\left(-\frac{\alpha\omega_l x}{q}\right) \tag{7.308}$$

The water percentage for the part of the carrying-sediment capacity, which is related the bed material including the coarser particles settling out of the suspended load, is

$$K_2 = 1 - K_1 = 1 - \frac{S_0 P_{4.l.0} K_x}{S^*(\omega)} \tag{7.309}$$

Hence, the fractional carrying-sediment capacity and other parameters at the exit section can be written as

$$P_{4.l}^* S^*(\omega^*) = (K_{1.l} + K_{2.l}) S^*(l) = P_{4.l.0} S_0 \exp\left(-\frac{\alpha\omega_l x}{q}\right) + \left[1 - \frac{S_0 K_x}{S^*(\omega)}\right] P_{4.l.1}^* S^*(\omega_1^*) \tag{7.310}$$

$$S^*(\omega^*) = S_0 K_x + \left[1 - \frac{S_0 K_x}{S^*(\omega)}\right] S^*(\omega_1^*) \tag{7.311}$$

$$P_{4.l}^* = \frac{P_{4.1.0} S_0}{S^*(\omega^*)} \exp\left(-\frac{\alpha \omega_l x}{q}\right) + \left[1 - \frac{S_0 K_x}{S^*(\omega)}\right] P_{4.l.1}^* \frac{S^*(\omega_1^*)}{S^*(\omega^*)} \tag{7.312}$$

$$P_{1.l} = \frac{P_{4.l.0} S_0}{S^*(l)} \exp\left(-\frac{\alpha \omega_l x}{q}\right) + \left[1 - \frac{S_0 K_x}{S^*(\omega)}\right] P_{4.l.1}^* \frac{S^*(\omega_1^*)}{S^*(l)} \tag{7.313}$$

These equations apply to the case with $K_1 < 1$, or $K_2 > 0$.

For $K_2 \leq 0$, sediment transport is characterized by massive deposition. For this case, the original bed material can not be eroded, and all suspended load can partially change to the carrying-sediment capacity. However, $K_1 \neq 1$ according to Eq. (7.306). In order to satisfy $K_1 = 1$, the following correction to $K_{1.l}$ in Eq. (7.305) is made

$$K_{1.l} = \frac{S_0 P_{4.l.0} \exp\left(-\frac{\alpha \omega_l x}{q}\right)}{S^*(l)} \left[\frac{S_0 K_x}{S^*(l)}\right]^{-1} = \frac{S^*(\omega) P_{4.l.0} \exp\left(-\frac{\alpha \omega_l x}{q}\right)}{S^*(l) K_x} = \frac{S^*(\omega)}{S^*(l)} P_{4.l} \tag{7.314}$$

which leads to

$$K_1 = \sum_{l=1}^{n} K_{1.l} = S^*(\omega) \sum_{l=1}^{n} \frac{P_{4.l}}{S^*(l)} = 1 \tag{7.315}$$

Hence,

$$P_{4.l}^* S^*(\omega*) = K_{1.l} S^*(l) = P_{4.l} S^*(\omega) \tag{7.316}$$

$$S^*(\omega^*) = S^*(\omega) \tag{7.317}$$

$$P_{4.l}^* = P_{4.l} \tag{7.318}$$

$$P_{1.l} = P_{4.l}^* \frac{S^*(\omega*)}{S^*(l)} = P_{4.l} \frac{S^*(\omega)}{S^*(l)} = P'_{1.l.1} \tag{7.319}$$

It is noted that Eqs. (7.316)–(7.319) are the same as Eqs. (7.224), (7.225), (7.227), and (7.228) that were previously derived in Sect. 7.7.1.2 for the case with massive deposition.

7.8.5 Carrying-Sediment Capacity and Corresponding Size Distribution for Reach Length and Scour Percentage Approaching Zero

If the reach length $x \to 0$,

$$K_{2.l} = \frac{S_0 P_{4.l.2} P_{4.2}}{S^*(l)} \quad (l = 1, 2, \ldots, K) \tag{7.320}$$

$$K_2 = \frac{S_0 P_{4.2}}{S^*(\omega_2)} \tag{7.321}$$

$$K_{1.l} = \frac{S_0 P_{4.1.2} P_{4.l.1.0}}{S^*(l)} \quad (l = k + 1, \ldots, n) \tag{7.322}$$

$$K_1 = \frac{S_0 P_{4.2.0}}{S^*(\omega_1)} \tag{7.323}$$

$$\begin{aligned} K_3 = 1 - K_1 - K_2 &= 1 - \frac{S_0 P_{4.1.0} K_x}{S^*(\omega_2)} - \frac{S_0 P_{4.2.0}}{S^*(\omega_1)} \\ &= 1 - S_0\left[\sum_{l=1}^{K} \frac{P_{4.1.0} P_{4.l.1.0}}{S^*(l)} + \sum_{l=k+1}^{n} \frac{P_{4.2.0} P_{4.l.2.0}}{S^*(l)}\right] \\ &= 1 - S_0 \sum_{l=1}^{n} \frac{P_{4.l}}{S^*(l)} = \frac{S_0}{S^*(\omega)} \end{aligned} \tag{7.324}$$

As $x \to 0, P_{1.l.1} \to P_{1.l.1.0}$. Base on Eq. (7.273),

$$\begin{aligned} P^*_{4.l.1} &= \lim_{\lambda^* \to 0} \frac{P_{1.l.1.0}}{\lambda^*}\left[1 - (1 - \lambda^*)^{(\omega^*_{1.m}/\omega_l)^m}\right] \\ &= \lim_{\lambda^* \to 0} \frac{P_{1.l.1.0}}{\lambda^*}\left\{1 - \left[1 - \left(\frac{\omega^*_{1.m}}{\omega_l}\right)^m \lambda^* + \left(\frac{\omega^*_{1.m}}{\omega_l}\right)^m \left(\left(\frac{\omega^*_{1.m}}{\omega_l}\right)^m - 1\right]\frac{\lambda^*}{2!} + \cdots\right)\right]\right\} \\ &= \lim_{\lambda^* \to 0} P_{1.l.1.0}\left\{\left(\frac{\omega^*_{1.m}}{\omega_l}\right)^m - \left(\frac{\omega^*_{1.m}}{\omega_l}\right)^m\left[\left(\frac{\omega^*_{1.m}}{\omega_l}\right)^m - 1\right]\frac{\lambda^*}{2!} + \cdots\right\} \\ &= P_{1.l.1}\left(\frac{\omega^*_1}{\omega_l}\right)^m = P_{1.l.1}\frac{S^*(l)}{S^*(\omega^*_1)} \end{aligned} \tag{7.325}$$

Also, for $x \to 0$ and $\lambda^* \to 0$, $S \to S_0$ and $P_{4.l} \to P_{4.l.0}$. Therefore, for $K_3 > 0$, $x \to 0$ and $\lambda^* \to 0$, Eqs. (7.299)–(7.302) can be written as

$$
\begin{aligned}
P_{4.l}^* S^*(\omega^*) &= P_{4.1.0} P_{4.l.1.0} S_0 + P_{4.2.0} P_{4.l.2.0} S_0 + \left[1 - \frac{S_0}{S^*(\omega)}\right] P_{4.l.1}^* S^*(\omega_1^*) \\
&= P_{4.l} S + \left[1 - \frac{S}{S^*(\omega)}\right] P_{4.l.1}^* S^*(\omega_1^*)
\end{aligned}
\tag{7.326}
$$

$$
S^*(\omega^*) = S + \left[1 - \frac{S}{S^*(\omega)}\right] S^*(\omega_1^*)
\tag{7.327}
$$

$$
P_{4.l}^* = \frac{P_{4.l} S}{S^*(\omega^*)} + \left[1 - \frac{S}{S^*(\omega)}\right] P_{4.l.1}^* \frac{S^*(\omega_1^*)}{S^*(\omega^*)}
\tag{7.328}
$$

$$
P_{1.l} = + \frac{P_{4.l} S}{S^*(l)} + \left[1 - \frac{S_0}{S^*(\omega)}\right] P_{1.l.1}
\tag{7.329}
$$

Equations (7.326)–(7.329) are the same as Eqs. (7.146)–(7.149), which were previously derived in Sect. 7.6.3 for the case with all suspended load influx being counted in carrying-sediment capacity and all bed material suspendable.

7.9 Example Applications of the Size Distribution for Carrying-Sediment Capacity

The basic results that were obtained in the previous sections for the size distribution for carrying-sediment capacity and the effective bed material gradation have been widely applied and verified, particularly in the mathematical models for sediment transport (Han 1987). These results can apply to general conditions, illustrate mechanisms and develop new concepts for sediment transport, and have reasonable agreement with reality. As shown in Chap. 3 and in this chapter, the water percentage is a useful parameter to illustrate the different transport mechanisms for coarser and finer sediment particles. This section discusses two important applications of the size distribution for carrying-sediment capacity.

7.9.1 Suspended Load Transport in Gravel-Bed Rivers and Carrying-Sediment Capacity

For a gravel-bed river, if $S < S(\omega^*)$, the suspended load influx can be completely counted in the carrying-sediment capacity, while the bed material cannot be suspended. For this case, $P_1 = 0$, and thus Eqs. (7.142)–(7.144) can be simplified as

$$S^*\left(\omega^*\right) = S \tag{7.330}$$

$$P_{4.l}^* = P_{4.l}\frac{S}{S^*(\omega^*)} = P_{4.l} \tag{7.331}$$

$$P_{1.l} = P_{4.l}\frac{S}{S^*(l)} = P_{4.l}\frac{S^*(\omega^*)}{S^*(l)} = P'_{1.l.1} \tag{7.332}$$

For this case, the carrying-sediment capacity equals the sediment concentration at the suspended load influx, the size distribution for carrying-sediment capacity equals the size distribution of the suspended load influx, and thus the concentration and the size distribution of suspended load remain constant along the flow path. However, only the water percentage of $\frac{S}{S^*(\omega)}$ is used to transport the suspended load, while the remainder $1 - \frac{S}{S^*(\omega)}$ is unsaturated. Based on the formulas for carrying-sediment capacity and corresponding size distribution, the concentration and the size distribution of suspended load remain constant along the flow path. From the aspect of water percentage, part of the water percentage is unused and thus the sediment transport is unsaturated. If using the concept of wash load, the sediment influx can be considered as wash load and thus does not participate in channel evolution. However, this explanation is less clear and profound than the explanation based on the formulas for carrying-sediment capacity and corresponding size distribution together with the concept of water percentage. Also, according to Eq. (7.332), the effective bed material gradation $P_{1.l}$ is the same as the gradation of the deposited material $P'_{1.l.1}$ for $\lambda = 0$. This is consistent with Eq. (5.65), as expected.

For the case with $1 - \frac{S}{S^*(\omega)} < 0$, or $S > S^*(\omega)$, since $P_1 = 0$, Eqs. (7.330)–(7.332) still apply, while the carrying-sediment is characterized by massive deposition. For this case, the fractional quantity of the suspended load influx that directly changes to the carrying-sediment capacity for the lth size class is $P_{4.l}S$, and the fractional carrying-sediment capacity is

$$P_{4.l}S = K_{1.l}S^*(l)$$

Thus,

$$K_{1.l} = \frac{P_{4.l}S}{S^*(l)} \tag{7.333}$$

$$K_1 = \sum_{l=1}^{n} K_{1.l} = \frac{S}{S^*(\omega^*)}$$

Since the total water percentage $K_1 \neq 1$, the following adjustment is made for $K_{1.l}$

$$K_{1.l} = \frac{P_{4.l}S}{S^*(l)}\left[\frac{S}{S^*(\omega)}\right]^{-1} = P_{4.l}\frac{S^*(\omega)}{S^*(l)} \tag{7.334}$$

Now

$$K_1 = \sum_{l=1}^{n} K_{1.l} = \sum_{l=1}^{n} P_{4.l}\frac{S^*(\omega)}{S^*(l)} = 1 \tag{7.335}$$

Hence,

$$P^*_{4.l}S^*(\omega^*) = K_{1.l}S^*(l) = P_{4.l}S^*(\omega) \tag{7.336}$$

$$S^*\left(\omega^*\right) = S^*(\omega) \tag{7.337}$$

$$P^*_{4.l} = P_4 \tag{7.338}$$

$$P_{1.l} = P_{4.l}\frac{S^*(\omega)}{S^*(l)} = P'_{1.l.1} \tag{7.339}$$

Plugging Eq. (7.336) into the governing equation for non-equilibrium sediment transport yields $\frac{d(P_{4.l}S)}{dx} = -\frac{\alpha\omega_l}{q}P_{4.l}[S - S^*(\omega)] < 0$, which indicates sediment deposition occurs and the "wash load" participates in channel evolution. This is as expected.

This example indicates that utilizing the formulas for the carrying-sediment capacity and corresponding size distribution, combining with the governing equation for non-equilibrium sediment transport, the concept of carrying-sediment capacity can be extended to the gravel-bed or the hard-bed river, and can be used to explain the phenomena of suspended load transport in a gravel-bed river.

7.9.2 Variation and Recovery of Sediment Concentration in the Reach Downstream of a Reservoir During Clear Water Discharge Period

The recovery of sediment concentration in the downstream reach of a reservoir during a clear water discharge period was briefly discussed in Chap. 5, and will be analyzed in detail in Chap. 10. As discussed in Chap. 10, the size distributions of the suspended load and for the carrying-sediment capacity become finer due to the sediment exchange along the flow path. As a result, the carrying-sediment capacity increases with the flow path and approximately uniform scour occurs within a long distance downstream of a reservoir. This scour mechanism will be theoretically analyzed and will be illustrated in Chap. 10 with a large amount of field data. This section focuses on the variation of carrying-sediment capacity during the scour process.

During the scour process in a downstream reach of a reservoir, the carrying-sediment capacity varies and mostly decreases compared with that before the reservoir is operated. This decrease may be significant under certain conditions. In the past, the decrease in the carrying-sediment capacity as scour evolves is attributed to the decrease in the longitudinal slope, increases to the flow area and the Manning's roughness coefficient, and coarsening of the bed material. The variations of these parameters can be represented in the water depth h, the flow velocity V and the average fall velocity ω of sediment concentration, according to the analysis. The relationship for the carrying-sediment capacity should remain the same. However, based on the field data, this relationship also varies during a scour process if the average fall velocity ω of sediment concentration is used. This variation is caused by the size distribution variation for carrying-sediment capacity.

For the case with the suspended load completely changing to the carrying-sediment capacity and the bed material being completely suspendable, the carrying-sediment capacity and its size distribution are described in Eqs. (7.147) and (7.148), respectively, which read

$$S^* = S + \left[1 - \frac{S}{S^*(\omega)}\right] S^*(\omega_1^*) \tag{7.147}$$

$$P_{4.l}^* = P_{4.l} \frac{S}{S^*(\omega^*)} + \left[1 - \frac{S}{S^*(\omega^*)}\right] P_{4.l.1}^* \tag{7.148}$$

Diving Eq. (7.147) by $\kappa\left(\frac{V^3}{gh}\right)^{0.92}$, one obtains

$$\frac{1}{\omega^{*m}} = \frac{S}{S(\omega)}\frac{1}{\omega^m} + \left[1 - \frac{S}{S(\omega)}\right]\frac{1}{\omega_1^{*m}} \tag{7.340}$$

As shown in Eq. (7.340), the term involving the fall velocity for the carrying-sediment capacity, $1/\omega^{*m}$, is a weighted sum of the term with the average fall velocity associated with the suspended sediment concentration, $1/\omega^m$, and the term with the average fall velocity corresponding to the bed material gradation, $1/\omega_1^{*m}$. The average fall velocity of the bed material, ω_1^*, is typically larger than the average fall velocity of the suspended load, ω. Therefore, $1/\omega^{*m} < 1/\omega$, $S^*(\omega^*) < S^*(\omega)$, and the carrying-sediment capacity decreases as scour evolves. Other factors can also cause the decrease in carrying-sediment capacity, such as the probability for the incipient suspension of bed material less than the suspension probability for the particles that are already suspended in the water, and the difference in the saturation recovery coefficients between equilibrium and non-equilibrium conditions. The multiple values of the carrying-sediment capacity caused by these two factors will be discussed in Chap. 11. The difference between the size distribution for carrying-sediment capacity and the size distribution of the sediment concentration, which was used in the past, is discussed in the following.

The difference in the carrying-sediment capacity when utilizing the size distribution for carrying-sediment capacity versus using the size distribution of the sediment concentration is analyzed as follows with the field data. Figure 7.1 shows the relationship of the carrying-sediment capacity at the Huayuankou Station that is downstream of the Sanmenxia Reservoir during scour periods. The x-axis is $S^*(\omega)$, which is the carrying-sediment capacity computed based on size distribution of the suspended load. The y-axis is S which is the sediment concentration. As shown in this figure, $S^*(\omega)$ is apparently larger than S, and the relationship of the carrying-sediment capacity is scatted, which is also significantly different from the relationship (solid line) for the equilibrium condition. Therefore, if the relationship $P_{4,l}^* = P_{4,l}$, which applies to the massive deposition condition and the equilibrium condition, is used to compute the carrying-sediment capacity, or $S^*(\omega^*) = S^*(\omega)$ is used, it is hard to explain the variation of the carrying-sediment capacity.

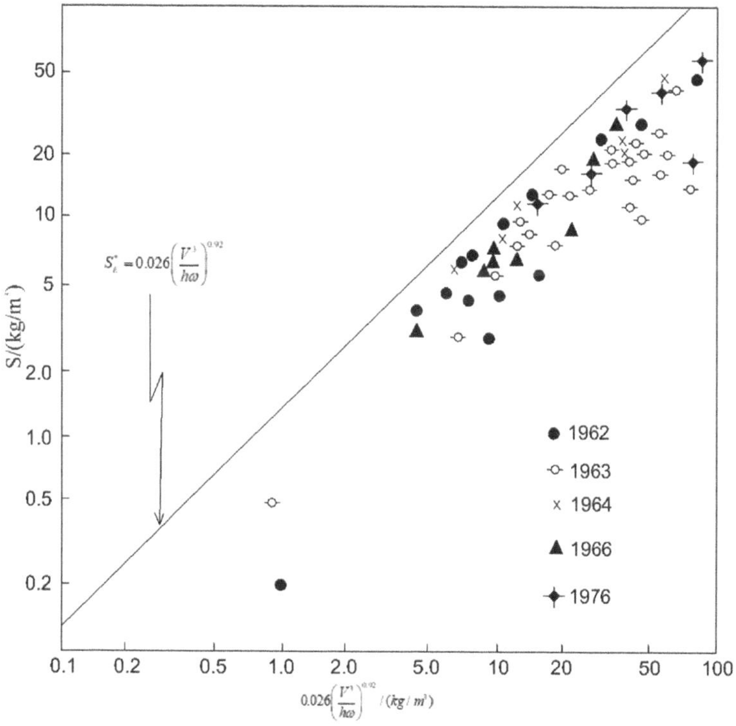

Fig. 7.1 Comparison between S and $S^*(\omega)$ during Scour Periods at Huayuankou Station of the Yellow River

Figure 7.2 shows the comparison between S and $S^*(\omega)$ based on the same field data as used in Fig. 7.1. $S^*(\omega)$ is computed with Eqs. (7.147) and (7.148) that include the size distribution for carrying-sediment capacity. As shown in this figure, the relationship is less scatted, and it is get closer to the relationship (solid line) for the equilibrium condition. This indicates that it is necessary to include Eqs. (7.147) and (7.148) for the investigation of the channel scour downstream of a reservoir. For channel scour, $S < S^*(\omega^*)$. Therefore, the computed carrying-sediment capacity $S^*(\omega)$ is below the relationship (solid line in Fig. 7.2) for the equilibrium condition, which is also related to the two factors that were discussed before for the multiple values of the sediment transport capacity. This will be further discussed in Chap. 11.

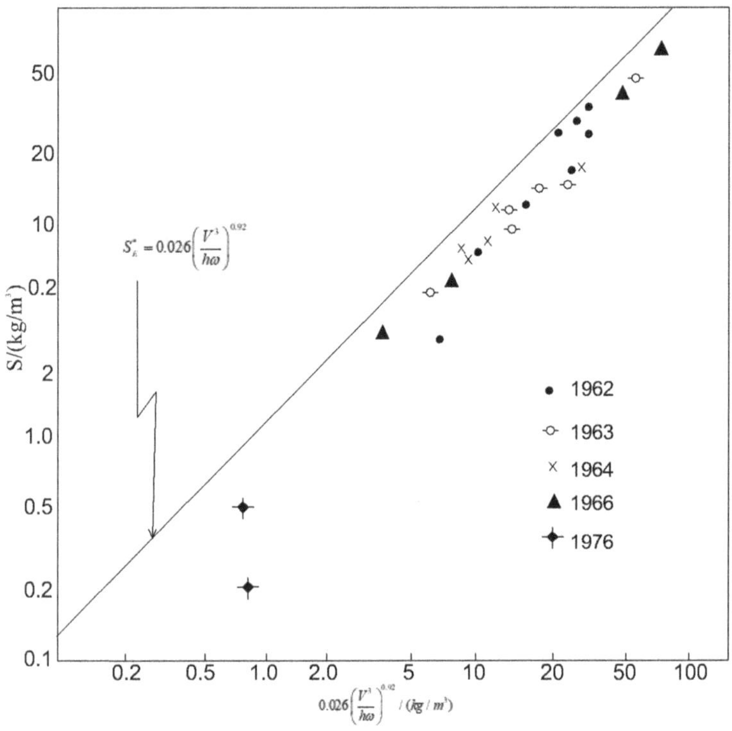

Fig. 7.2 Comparison between S and $S^*(\omega)$ during Scour Periods at Huayuankou Station of the Yellow River

Figures 7.3 and 7.4 show computed $S^*(\omega)$ and $S^*(\omega)$, respectively, compared to S based on the field data collected at the Xiangyang Station that is downstream of the Danjiankou Reservoir during multiple scour periods. Similar conclusion can be made.

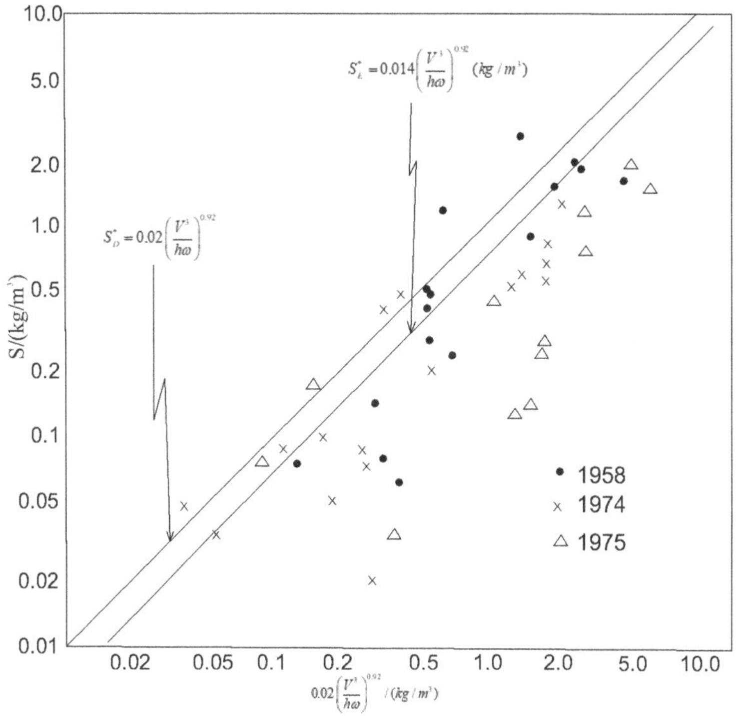

Fig. 7.3 Comparison between S and $S^*(\omega)$ during Scour Periods at Xiangyang Station of the Han River

Based on these comparisons, it is concluded that the relationship for the sediment transport capacity does not change during a scour process. The average fall velocity ω, which is calculated based on the size distribution of suspended load, is not a representative fall velocity to be used in determining the carrying-sediment capacity. The representative fall velocity ω^* shall be determined based on the size distribution for carrying-sediment capacity, and the resulting relation for carrying-sediment capacity is more stable.

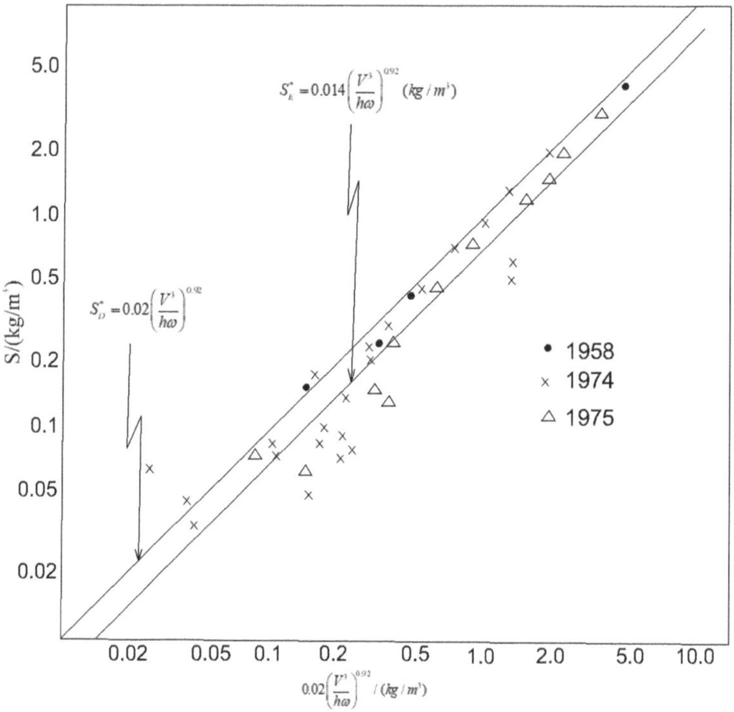

Fig. 7.4 Comparison between S and $S^*(\omega)$ during Scour Periods at Xiangyang Station of the Han River

References

1. Ashida K, Michiue M (1972) Study on hydraulic resistence and bed load transport rate in alluvial streams. Proc Jpn Soc Civ Eng 206:59–69 (In Japanese)
2. Einstein HA (1950) The bed-load function for sediment transportation in open channel flow. Tech Bull 1026 (USDA. Soil Conservation Service)
3. Feldman A (1981) HEC models for water resources, system simulation, theory and experiences. Academic Press, New York
4. Han QW (1973) Preliminary studies for the non-equilibrium transport of sediment in reservoirs, report on sedimentation in reservoir, Task Group on Yellow River Sedimentation Research, pp 145–168 (in Chinese)
5. Han QW (1979) A study of the non-equilibrium transportation of non-uniform suspended load. Chin Sci Bull 24(17):804–808 (in Chinese)
6. Han QW (1982) Reservoir sedimentation, vol 5. The Yangtze Valley Planning Office, Hydrology Department
7. Han QW (2003) Reservoir sedimentation. Science Press, Beijing (in Chinese)
8. Han QW (2005) Size distribution of carrying: sediment capacity. In: Proceedings of 6th national symposium on basic theory of sedimentation. Water Conservancy of Yellow River Press, pp 3–20 (in Chinese)
9. Han QW, He MM (1981) Statistical characters of sediment exchange. J Hydraul Eng 1:10–22 (in Chinese)

10. Han QW, He MM (1987) Mathematical modeling of reservoir sedimentation and fluvial process. J Sediment Res 3:14–29 (in Chinese)
11. Han QW, He MM (1997) Study on the two dimensional equations for non-uniform non-equilibrium suspended sediment transport and the boundary conditions. J Hydraul Eng 1:1–10 (in Chinese)
12. He MM, Han QW (1989) Concepts about grain size distribution of carrying capacity and effective bed material. J Hydraul Eng 3:7–16 (in Chinese)
13. He MM, Han QW (1990) The determination of the size distributions of carrying capacity and effective bed material. J Hydraul Eng 3:l–12 (in Chinese)
14. Li YT (1987) Preliminary study on bed material gradation under equilibrium transport. J Sediment Res 1:82–87 (in Chinese)
15. Rocinski KI (1967) Specific discharge of bed-load sediment. Hydrometeorological Publishing House, Leningrad
16. Taizo H, Yukio O (1978) Calculation formula of mixed sand with gravel and flow discharge. Report Summary, 5th annual Conference on Advanced research, Kanto Branch. 81–82

Chapter 8
Numerical Results for Vertical 2-D Sediment Diffusion Equation and Approximate Vertical Distribution of Concentration in Non-equilibrium Condition

8.1 Numerical Results for Vertical 2-D Sediment Diffusion Equation

8.1.1 Bottom Boundary Condition [1–5]

Neglecting sediment diffusion in the longitudinal direction, the vertical two-dimensional (2-D) equation for steady non-equilibrium transport of uniform sediment is

$$\frac{\partial[V_x S(l)]}{\partial x} + \frac{\partial\left[V_y S(l)\right]}{\partial y} = \frac{\partial}{\partial y}\left[\varepsilon_y \frac{\partial S(l)}{\partial y} + \omega_l S(l)\right] \tag{8.1}$$

where $S(l) = S(x, y, l)$ is the concentration of the uniform sediment of the l-th size group at the location (x,y), ω_l the sediment fall velocity, ε_y the vertical (in y-direction) sediment diffusion equation, and V_x and V_y are the flow velocity components in the longitudinal direction (x-direction) and in the vertical direction, respectively. Integrating Eq. (8.1) over the water depth yields

$$q\frac{dS(l)}{dx} = \int\limits_{h_1(x)}^{h_2(x)} \left\{\frac{\partial}{\partial x}[V_x S(l)] + \frac{\partial}{\partial y}\left[V_y S(l)\right]\right\}dy = -\varepsilon_y \frac{\partial S(l)}{\partial y}\Big|_{h_1(x)+a} - \omega_l S_b(l)$$

where $h_1(x)$ is the channel bottom profile, $h_2(x)$ the water surface profile, $h_1(x)+a$ a location that is above the bottom by a small vertical distance of a, and $S_b(l)$ the sediment concentration at the bottom. Setting the x-axis to be the channel bottom profile $h_1(x)$, then the above equation can be written as

Q. Han and M. He, *Non-equilibrium Transport of Non-uniform Suspended Load*, Hydroscience and Engineering, https://doi.org/10.1007/978-981-97-1123-9_8

$$q\frac{d\overline{S}(l)}{dx} = -\varepsilon_y \frac{\partial S(l)}{\partial y}\Big|_{y=a} - \omega_l S_b(l) \tag{8.2}$$

Based on the statistical theory for sediment exchange on the bed [4, 5], the longitudinal variation of the depth-averaged sediment concentration can be described by Eqs. (3.87) and (3.92). Combining these two equations,

$$\begin{aligned} q\frac{d\overline{S}(l)}{dx} &= -\varepsilon_y \frac{\partial S(l)}{\partial y}\Bigg|_{y=a} - \omega_l S_b(l) = (\lambda_{1.4.l} - \lambda_{4.1.l})\frac{\pi}{6} D_l^3 \gamma_s \\ &= \omega_l\left[\alpha_l^* \overline{S}^*(l) - \tilde{\alpha}_l \overline{S}(l)\right] \end{aligned} \tag{8.3}$$

where $\overline{S}(l)$ is the depth-averaged sediment concentration, $\overline{S}^*(l)$ the depth-averaged carrying-sediment capacity, and α_l^* and $\tilde{\alpha}_l$ are the saturation recover coefficients under the equilibrium and the non-equilibrium conditions that are determined with Eqs. (3.90) and (3.91), respectively.

If the unsteady term need to be included, considering

$$\int_0^h \frac{\partial S(l)}{\partial t} dy = \frac{\partial}{\partial t}\int_0^h S(l) dy = \frac{\partial\left[\overline{S}(l)h\right]}{\partial t} \tag{8.4}$$

Equation (8.3) should be replaced with

$$\frac{\partial\left[\overline{S}(l)h\right]}{\partial t} + q\frac{\partial S(l)}{\partial x} = -\varepsilon_y \frac{\partial S(l)}{\partial y} - \omega_l S_b(l) = \omega_l\left[\alpha_l^* S^*(l) - \tilde{\alpha}_l S(l)\right] \tag{8.5}$$

This is not only a boundary condition, but also the governing equation for bed change. As discussed in Chap. 3, Eq. (8.5) can also be written as the following equation for channel sedimentation [6]

$$\frac{\partial\left(\gamma_s' y_l\right)}{\partial \tau} + \omega_l\left(\alpha_l^* P_{4.l}^* S^* - \tilde{\alpha} P_4 S\right) = 0 \tag{8.6}$$

As the boundary condition used in the statistical theory, Eq. (8.3) has the following characteristics. (1) This boundary condition was developed based on the sediment exchange intensity with the statistical theory, considering various factors. The equations for the carrying-sediment capacity for non-uniform sediment developed based this boundary condition can illustrate the mechanisms for sediment transport, apply to comprehensive cases, and have a reasonable agreement with reality. As discussed in Chap. 4, this also shows the advantage and reliability of the sediment exchange intensity based on the statistical theory. (2) Considering the difficulty in determining the parameters at the bottom, the depth-averaged parameters are used in the proposed boundary condition, Eq. (8.3). The one-dimensional (1-D) equation for non-equilibrium sediment transport was also used in the derivation. As the 1-D equation for non-equilibrium sediment transport, Eq. (8.3) has been widely applied to the Yellow River, the Yangtze River, and many other rivers and reservoirs. The sedimentation quantity determined with this equation agrees well with the field data. Therefore, using Eq. (8.3) as a boundary condition has a good theoretical basis with reliability. (3) The unknown of the governing equation, or the sediment concentration, is also included in the boundary equation. Hence, trial-and-error methods are needed to solve the governing equation with this boundary condition. In other words, the boundary condition is unknown, while the equation it should satisfy is defined. This condition is different from the previous study but more reliable in practice.

8.1.2 Methods for 2-D Sediment Diffusion Equation

For simplicity, the steady uniform flow condition and uniform sediment are considered in this analysis. For non-uniform sediment mixture, the vertical distribution of the total sediment concentration can be determined by summing up the results of each size group.

Setting the x-axis to be the channel bottom profile towards the downstream direction and the y direction to be normal to the bottom, the 2-D sediment diffusion equation for the steady uniform condition is

$$q\frac{\partial S(l)}{\partial x} = -\varepsilon_y \frac{\partial S(l)}{\partial y}\Big|_{y=a} - \omega_l S_b(l) \tag{8.7}$$

The distribution of sediment concentration at the entrance section ($x = 0$) is assumed to be the Rouse profile multiplying by a correction factor B_0

$$S(0, y) = B_0 S_{b.l}^{*}\left(\frac{1-\eta}{\eta}\frac{\eta_a}{1-\eta_a}\right)^{\frac{\omega_l}{\kappa u_*}} \tag{8.8}$$

where $\eta = y/h$, $B_0 > 1$ for deposition, and $B_0 < 1$ foe scour. This distribution is only one example for the boundary condition at x=0. In the following analysis, various distributions will be assigned to $S(0, y, l)$. Because of no sediment exchange at the water surface, the boundary condition at the water surface is

$$\varepsilon\frac{\partial S_l}{\partial y}\Big|_{y=0} + \omega_l S_h = 0 \tag{8.9}$$

The boundary condition at the bottom is described in Eq. (8.3). Hence, the 2-D sediment diffusion equation, Eq. (8.7), need to be solved together with the boundary conditions as described in Eqs. (8.3), (8.8) and (8.9).

Other parameters are also involved in solving the sediment diffusion equation, such as

$$\tilde{\alpha}_l = \frac{(1-\varepsilon_{0.l})(1-\varepsilon_{4.l})q}{L_{4.l}\omega_l}$$
$$\tilde{\alpha}_l^* = \frac{(1-\varepsilon_{0.l})(1-\varepsilon_{4.l})q}{L_{4.l}^*\omega_l}$$
$$L_{4.l}^* = \frac{q\left(\eta_l^*\right)}{\omega_l}\left[\left\{\frac{1}{\sqrt{2\pi}}\frac{u_*}{\omega_l\varepsilon_{4.l}}e^{-\frac{1}{2}\left(\frac{\omega_l}{u_*}\right)^2} - 1\right\}^{-1} + \left\{\frac{1}{\sqrt{2\pi}}\frac{u_*}{(1-\varepsilon_{4.l})\omega_l}e^{-\frac{1}{2}\left(\frac{\omega_l}{u_*}\right)^2} - 1\right\}^{-1}\right] \tag{8.10}$$

$$L_{4.l} = \frac{q(\eta_l)}{\omega_l}\left[\left\{\frac{1}{\sqrt{2\pi}}\frac{u_*}{\omega_l\varepsilon_{4.l}}e^{-\frac{1}{2}\left(\frac{\omega_l}{u_*}\right)^2} - 1\right\}^{-1} + \left\{\frac{1}{\sqrt{2\pi}}\frac{u_*}{(1-\varepsilon_{4.l})\omega_l}e^{-\frac{1}{2}\left(\frac{\omega_l}{u_*}\right)^2} - 1\right\}^{-1}\right] \tag{8.11}$$

$$\eta_l^* = \frac{h_l^*}{h} = 2\int_0^1 \frac{S^*(\eta, l)}{S_l^*}\eta d\eta \tag{8.12}$$

$$\eta_l = \frac{h_l}{h} = 2\int_0^1 \frac{S(\eta, l)}{S}\eta d\eta \tag{8.13}$$

where $q(\eta_l)$ and $q\left(\eta_l^*\right)$ depend on the vertical distribution of flow velocity for given η_l and η_l^*, and η_l and η_l^* depend on the vertical distributions of sediment concentration and carrying-sediment capacity. It is noted that while the sediment concentration and the carrying-sediment capacity for the l-th size group are denoted as $S(l)$ and $S^*(l)$, respectively, the subscript l is used to denote the other parameters for the l-th size group, or $X(l)$ is simply denoted as X_l for the other parameters. The depth-averaged sediment concentration for the l-th size group is determined with

$$\overline{S}(l) = \int_0^1 S(\eta, l)d\eta \tag{8.14}$$

The corresponding $\bar{S}^*(l)$ is determined with the formula for carrying-sediment capacity.

The following two trial-and-error methods can be used to solve the diffusion equation.

Method 1. The procedure of this method is as follows: assuming $S(l)$ at $x = \Delta x_1$, solve equation for $S(y, l)$; compute $\overline{S}(l)$ and $\overline{S}^*(l)$, the corresponding relative suspension height η_l and η_l^*, and the flow discharge on suspended particles per unit width $q(\eta_l)$ and $q(\eta_l^*)$; determine $\tilde{\alpha}_l$ and α_l^* with Eqs. (6.34) and (6.52), respectively; then compute $\varepsilon \frac{\partial S(l)}{\partial y}\big|_{y=a}$ and $\omega_l S_b(l)$; compare $-\varepsilon_y \frac{\partial S(l)}{\partial y}\Big|_{y=a} - \omega_l S_b(l)$ with $\omega_l \left[\alpha_l^* \overline{S}^*(l) - \tilde{\alpha}_l \overline{S}(l)\right]$; if they are the same, then $S(y,l)$ is the solution; otherwise, assume another $S(y,l)$ and repeat the above steps until the solution is found. In order to simplify the calculation, the vertical distribution of sediment concentration at the previous section can be used to determine $\overline{\alpha}_l$ and α_l^* for the study section. For uniform sediment, the subscript l can be ignored in these parameters for simplicity of expressions. It is noted that total $\overline{S}$ and $\overline{S}^*$ for non-uniform sediment can be determined by summing up $\overline{S}(l)$ and $\overline{S}^*(l)$ for uniform sediment for known size distributions of suspended load and the carrying-sediment capacity need to be determined. For non-uniform sediment transport, $P_{1.l.1}$ and $P_{4.l}(0)$ at the entrance section are known. Considering Eq. (7.14), the size distribution for carrying-sediment capacity can be determined with

$$P_{4.l}^* = \frac{P_{1.l.1} S_b^*(l)}{S_b^*} = \frac{P_{1.l.1} S_b^*(l)}{\sum P_{1.l.1} S_b^*(l)} \tag{8.15}$$

or directly using

$$P_{4.l}^* = \frac{P_{1.l.1} S^*(l)}{S^*} = \frac{P_{1.l.1} S^*(l)}{\sum P_{1.l.1} S^*(l)} \tag{8.16}$$

It is noted that the numerical results listed in this chapter is only for uniform sediment.

Method 2. Since the sediment concentration has insignificant impact on the saturation recovery coefficients, $\tilde{\alpha}_l$ and α_l^* that were previously determined in Chap. 6 based on the vertical distribution of sediment concentration for the non-equilibrium condition can be directly applied here. The trial-and-error procedure is as follows: assuming $S_b(l)$, solve for $S(y, l)$; compute $\overline{S}(l)$ and $\overline{S}^*(l)$; look up $\tilde{\alpha}_l$ and α_l^* for $\beta = \overline{S}(l)/\overline{S}^*(l)$, ω_l and u_* based on results presented in Chap. 6; then compute $\varepsilon \frac{\partial S(l)}{\partial y}\Big|_{y=a}$ and $\omega_l S_b(l)$; compare $-\varepsilon_y \frac{\partial S(l)}{\partial y}\Big|_{y=a} - \omega_l S_b(l)$ with $\omega_l \left[\alpha_l^* \overline{S}^*(l) - \tilde{\alpha}_l \overline{S}(l)\right]$; if they are the same, then $S(y, l)$ is the solution; otherwise, assume another $S_b(l)$ and repeat the above steps until the solution is found. It is noted that β_0 is $\overline{S}(l)/\overline{S}^*(l)$ at the entrance section, while β is $\overline{S}(l)/\overline{S}^*(l)$ for the study section.

8.1.3 Numerical Results for Vertical 2-D Sediment Diffusion Equation

The numerical results presented in this section were computed with Method 2. The computation was conducted for two cases: $\tilde{\alpha}_l \neq \alpha_l^*$, and $\tilde{\alpha}_l = \alpha_l^*$. The computation for the later case was actually conducted with the composite saturation recovery coefficient α. The numerical results in various conditions [4, 7, 8] are presented in this section. For simplicity of presentations, the sediment concentration $\overline{S}(l)$ for the l-th size group is simply denoted as S.

Tables 8.1 and 8.2 list the results of the vertical size distributions of sediment concentration in the scour and deposition conditions, respectively, with $\tilde{\alpha}_l = \alpha_l^*$. Tables 8.3 and 8.4 list the results computed with $\tilde{\alpha}_l \neq \alpha_l^*$. In addition to the numerical results computed with Method 2, the vertical distribution of sediment concentration in the equilibrium condition, and the Rouse distribution in the non-equilibrium condition, as discussed in Sect. 8.2, are also listed in these tables. The depth-averaged sediment concentration for the two theoretical distributions is the same as that determined from the numerical results. The following observations ae made from the results:

(1) The vertical distribution of sediment concentration computed from the 2-D sediment diffusion equation in the non-equilibrium condition ($\beta = S/S^* \neq 1$) are significantly different from the results in the equilibrium condition. The difference increases as the difference between β and 1 increases. Only for 0.9 $< \beta <$ 1.1, the distribution of sediment concentration can be approximated by the distribution in the equilibrium condition.

(2) If the depth-averaged sediment concentration is the same, For the case of sediment deposition, the computed sediment concentration in the non-equilibrium condition is higher than that in the equilibrium condition at the upper part of the water, but is lower at the deeper part of the water, particularly at the bottom. This can be observed in Tables 8.2 and 8.4. For scour, compared to that in equilibrium condition, the computed sediment concentration in the non-equilibrium condition is lower than that in the equilibrium condition for the upper part of the water, but is higher for the deeper part of the water, particularly at the bottom. This is shown in Tables 8.1 and 8.3. Hence, compared to the equilibrium condition, the vertical gradient of sediment concentration in the non-equilibrium condition decreases for sediment deposition, but increases for scour.

(3) In both sediment deposition and scour conditions, the computed vertical distribution of sediment concentration approaches the distribution in equilibrium condition after a sufficiently long distance.

(4) By comparing to the distribution in the equilibrium condition, it is concluded that the numerical results computed in the non-equilibrium condition are reliable and reasonable. This implies the boundary condition and the trial-and-error method developed for the 2-D sediment diffusion equation are appropriate.

Table 8.1 Numerical results of longitudinal variation of vertical distribution of sediment concentration for the case of scour

$\beta_0 = 0.5$ (Scour) $D = 0.025$ mm ($\omega = 0$ 0.053 cm/s), $\omega/u_* = 0.1$, $S^* = 0.469$

y/h	$S/S_b{}^*$								
	$x = 0.5$			$x = 3.0$			$x = 4.5$		
	Difference[a]	Equi-dist.[b]	Rouse dist.[c]	Difference[a]	Equi-dist.[b]	Rouse dist.[c]	Difference[a]	Equi dist.[b]	Rouse dist.[c]
0.0	0	0	0	0	0	0	0	0	0
0.05	0.135	0.153	0.128	0.161	0.178	0.160	0.180	0.193	0.183
0.1	0.168	0.184	0.161	0.200	0.214	0.197	0.223	0.223	0.224
0.2	0.209	0.226	0.205	0.248	0.262	0.247	0.276	0.285	0.277
0.3	0.242	0.258	0.240	0.286	0.300	0.286	0.318	0.326	0.320
0.4	0.273	0.289	0.275	0.323	0.335	0.326	0.357	0.364	0.360
0.5	0.307	0.319	0.310	0.362	0.371	0.366	0.398	0.403	0.400
0.6	0.346	0.353	0.350	0.406	0.411	0.409	0.444	0.446	0.445
0.7	0.396	0.395	0.400	0.462	0.459	0.463	0.501	0.498	0.501
0.8	0.468	0.452	0.471	0.540	0.525	0.539	0.581	0.570	0.578
0.9	0.601	0.553	0.600	0.681	0.643	0.677	0.722	0.698	0.717
0.95	0.750	0.667	0.750	0.835	0.775	0.835	0.874	0.842	0.874
β	0.693	0.693	0.693	0.806	0.806	0.806	0.874	0.874	0.874
S	0.325	0.325	0.325	0.378	0.378	0.378	0.410	0.410	0.410

Notes: [a]Difference—computed with trial-and-error method 2 as discussed in Sect. 8.1.2

[b]Equi-Dist—based on vertical distribution of sediment concentration in equilibrium condition

[c]Rouse-Dist—based on Rouse's vertical distribution of sediment concentration in non-equilibrium condition

Table 8.2 Numerical results of longitudinal variation of vertical distribution of sediment concentration for the case of deposition

$\beta_0 = 5$ (Deposition), $D = 0.025$ mm ($\omega = 0\ 0.053$ cm/s), $\omega/u_* = 0.5$, $S^* = 0.0829$

y/h	$S/S_b{}^*$								
	$x = 1.0$			$x = 1.5$			$x = 3.5$		
	Trial–error[a]	Equi-dist.[b]	Rouse dist.[c]	Trial–error[a]	Equi-dist.[b]	Rouse dist.[c]	Trial–error[a]	Equi-dist.[b]	Rouse dist.[c]
1.00	0	0	0	0	0	0	0	0	0
0.95	0.0018	0.0011	0.0028	0.0013	0.0008	0.0018	0.0006	0.0006	0.0008
0.90	0.0048	0.0029	0.0061	0.0035	0.0022	0.0040	0.0016	0.0015	0.0019
0.80	0.0132	0.0079	0.0146	0.0094	0.0059	0.0099	0.0046	0.0043	0.0049
0.70	0.0252	0.0155	0.0258	0.0178	0.0116	0.0178	0.0091	0.0083	0.0094
0.60	0.0419	0.0268	0.0414	0.0296	0.202	0.290	0.0158	0.0145	0.0161
0.50	0.0660	0.0446	0.0638	0.0467	0.0335	0.0453	0.0261	0.0241	0.0262
0.40	0.103	0.0740	0.0982	0.0729	0.0556	0.0707	0.0431	0.0400	0.0426
0.30	0.165	0.129	0.157	0.118	0.0967	0.115	0.0742	0.0694	0.0726
0.20	0.290	0.252	0.279	0.211	0.190	0.208	0.144	0.136	0.139
0.10	0.675	0.695	0.663	0.511	0.523	0.506	0.381	0.376	0.369
0.05	1.47	1.77	1.47	1.16	1.33	1.15	0.906	0.955	0.906
β	2.30	2.30	2.27	1.72	1.72	1.71	1.09	1.24	1.09
S	0.191	0.191	0.188	0.143	0.143	0.142	0.103	0.103	0.101

Notes: [a]Trial–error—computed with trial-and-error method 2 as discussed in Sect. 8.1.2
[b]Equi-dist—based on vertical distribution of sediment concentration for equilibrium condition
[c]Rouse-dist—based on Rouse's vertical distribution of sediment concentration for non-equilibrium condition

Table 8.3 Numerical results of longitudinal variation of vertical distribution of sediment concentration for the case of scour

$\beta_0 = 5$ (Scour), $D = 0.025$ mm ($\omega = 0\ 0.053$ cm/s), $\omega/u_* = 0.1$, $S^* = 0.469$

y/h	$S/S_b{}^*$								
	$x = 1.0$			$x = 3.0$			$x = 6.0$		
	Difference[a]	Equi-dist.[b]	Rouse dist.[c]	Trial–error[a]	Equi-dist.[b]	Rouse dist.[c]	Trial–error[a]	Equi-dist.[b]	Rouse dist.[c]
1.00	0	0	0	0	0	0	0	0	0
0.95	0.127	0.145	0.126	0.164	0.184	0.169	0.195	0.204	0.202
0.90	0.158	0.175	0.157	0.204	0.217	0.207	0.242	0.246	0.245
0.80	0.196	0.214	0.200	0.254	0.268	0.259	0.298	0.301	0.300
0.70	0.227	0.245	0.235	0.293	0.306	0.300	0.342	0.345	0.344
0.60	0.257	0.274	0.268	0.331	0.342	0.339	0.383	0.385	0.385
0.50	0.289	0.303	0.303	0.370	0.379	0.379	0.425	0.426	0.426
0.40	0.326	0.335	0.342	0.415	0.419	0.423	0.472	0.472	0.472
0.30	0.374	0.374	0.389	0.472	0.468	0.478	0.529	0.527	0.528
0.20	0.445	0.428	0.458	0.551	0.536	0.553	0.608	0.603	0.605
0.10	0.577	0.524	0.582	0.693	0.656	0.691	0.747	0.738	0.743
0.05	0.728	0.632	0.728	0.848	0.791	0.848	0.897	0.890	0.897
β	0.66	0.66	0.68	0.82	0.82	0.83	0.92	0.92	0.93
S	0.308	0.308	0.317	0.386	0.386	0.391	0.434	0.434	0.434

Notes: [a]Trial–Error—computed with trial-and-error method 2 as discussed in Sect. 8.1.2
[b]Equi-Dist—based on vertical distribution of sediment concentration for equilibrium condition
[c]Rouse-Dist—based on Rouse's vertical distribution of sediment concentration for non-equilibrium condition

Table 8.4 Numerical results of longitudinal variation of vertical distribution of sediment concentration for the case of deposition

$\beta_0 = 2.0$ (Deposition), $D = 0.025$ mm ($\omega = 0\ 0.053$ cm/s), $\omega/u_* = 0.1$, $S^* = 0.469$

y/h	$S/S_b{}^*$								
	$x = 1.0$			$x = 3.0$			$x = 6.0$		
	Difference[a]	Equi-dist.[b]	Rouse dist.[c]	Trial–error[a]	Equi-dist.[b]	Rouse dist.[c]	Trial–error[a]	Equi-dist.[b]	Rouse dist.[c]
1.00	0	0	0	0	0	0	0	0	0
0.95	0.415	0.348	0.467	0.384	0.467	0.328	0.255	0.246	0.268
0.90	0.510	0.463	0.536	0.403	0.362	0.411	0.314	0.291	0.316
0.80	0.627	0.567	0.621	0.490	0.443	0.480	0.384	0.364	0.378
0.70	0.705	0.650	0.687	0.552	0.507	0.533	0.435	0.416	0.426
0.60	0.777	0.725	0.745	0.606	0.567	0.580	0.481	0.465	0.470
0.50	0.846	0.802	0.802	0.657	0.627	0.627	0.527	0.514	0.514
0.40	0.917	0.888	0.864	0.711	0.694	0.678	0.576	0.569	0.562
0.30	0.997	0.992	0.938	0.771	0.775	0.739	0.634	0.636	0.620
0.20	1.09	1.13	1.03	0.848	0.887	0.817	0.712	0.727	0.698
0.10	1.24	1.39	1.20	0.975	1.08	0.957	0.845	0.890	0.835
0.05	1.38	1.68	1.38	1.11	1.31	1.11	0.985	1.07	0.985
β	1.74	1.75	1.69	1.36	1.36	1.32	1.11	1.11	1.10
S	0.817	0.823	0.791	0.638	0.638	0.620	0.523	0.523	0.515

Notes: [a]Trial–error—computed with trial-and-error method 2 as discussed in Sect. 8.1.2
[b]Equi-dist.—based on vertical distribution of sediment concentration for equilibrium condition
[c]Rouse-dist.—based on Rouse's vertical distribution of sediment concentration for non-equilibrium condition

For further investigate, the computed vertical distributions of sediment concentration are shown in Figs. 8.1, 8.2, 8.3, 8.4, 8.5, 8.6, 8.7, 8.8, 8.9, 8.10, 8.11, 8.12, 8.13 and 8.14. The values of B_0, particle size D (with fall velocity ω), ω/u_*, α^* and $\tilde{\alpha}$ are listed in each figure. The x-axis is the relative sediment concentration $S(y)/S_b^*$. The y-axis is the relative vertical location *y/h*. Each solid line represents the numerical results of the vertical distribution of sediment concentration computed for a section with a relative distance of $X = x/h$ from the entrance section. The theoretical results of the Rouse's distribution are represented by symbols. Figs. 8.1, 8.2, 8.3, 8.4, 8.5, 8.6, 8.7, 8.8, 8.9 and 8.10 show the results for the case with . $\tilde{\alpha}_l = \alpha_l^*$. Figures 8.11, 8.12, 8.13 and 8.14 show the results for the case with $\tilde{\alpha}_l \neq \alpha_l^*$. As shown in these figures, the numerical results agree well with the theoretical results except for the case of deposition with $B_0 = 5 \sim 8$.

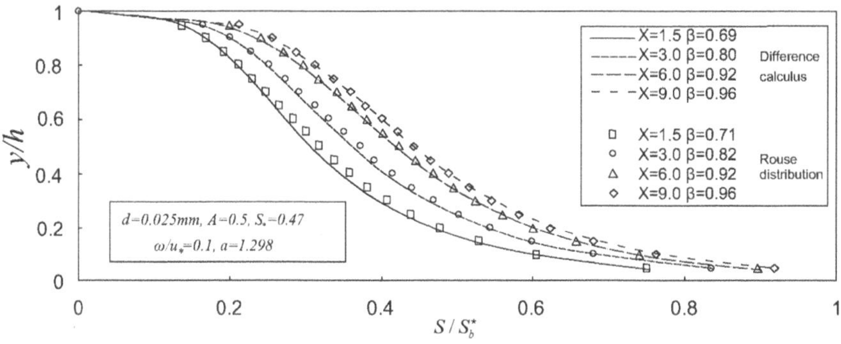

Fig. 8.1 Computed longitudinal variation of vertical distribution of sediment concentration for the case of scour ($D = 0.025$ mm, $\beta_0 = 0.5$, $\omega/u_* = 0.1$)

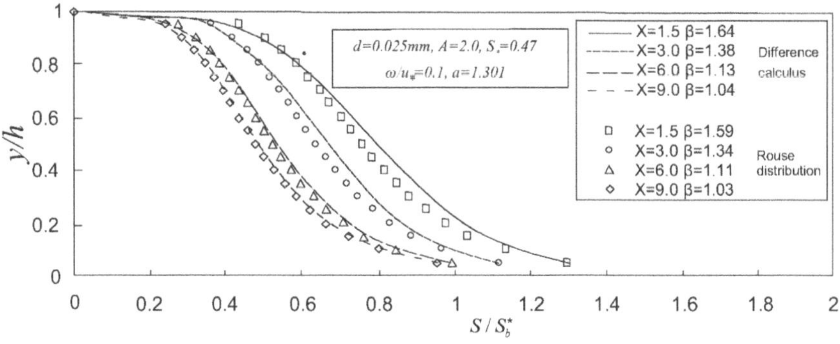

Fig. 8.2 Computed longitudinal variation of vertical distribution of sediment concentration for the case of deposition ($D = 0.025$ mm, $\beta_0 = 2$, $\omega/u_* = 0.1$)

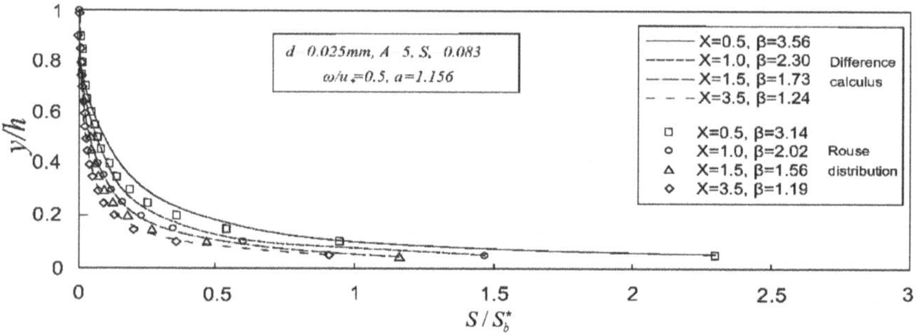

Fig. 8.3 Computed longitudinal variation of vertical distribution of sediment concentration for the case of deposition ($D = 0.025$ mm, $\beta_0 = 5$, $\omega/u_* = 0.5$)

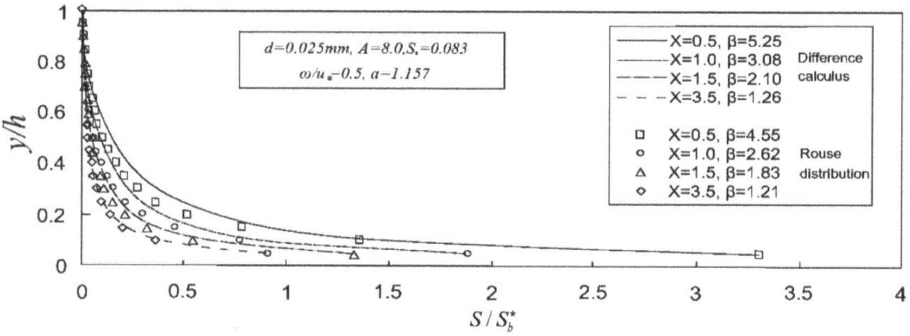

Fig. 8.4 Computed longitudinal variation of vertical distribution of sediment concentration for the case of deposition ($D = 0.025$ mm, $\beta_0 = 8.0$, $\omega/u_* = 0.5$)

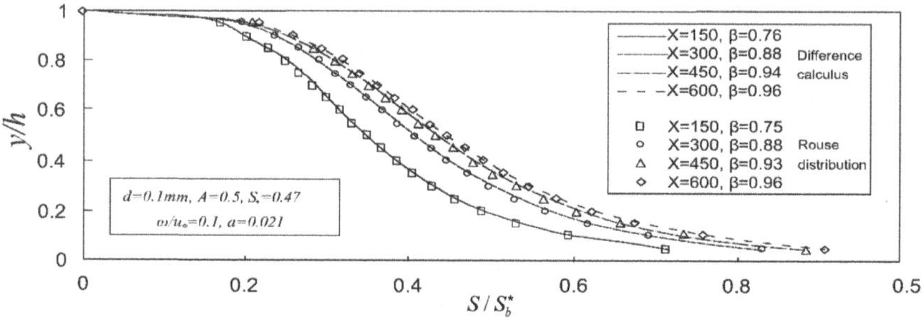

Fig. 8.5 Computed longitudinal variation of vertical distribution of sediment concentration for the case of scour ($D = 0.1$ mm, $B_0 = 0.5$, $\omega/u_* = 0.1$)

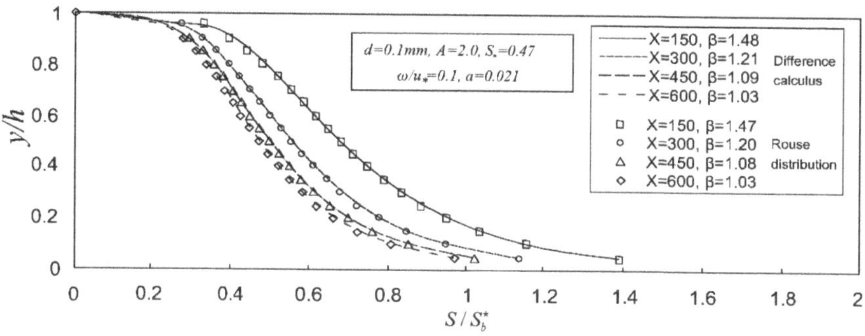

Fig. 8.6 Computed longitudinal variation of vertical distribution of sediment concentration for the case of deposition ($D = 0.1$ mm, $B_0 = 2.0$, $\omega/u* = 0.1$)

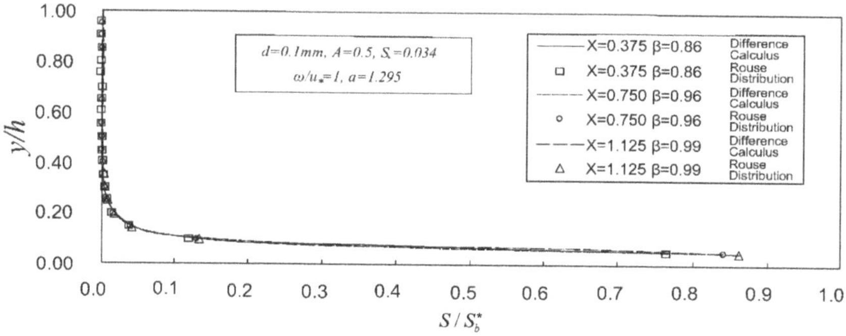

Fig. 8.7 Computed longitudinal variation of vertical distribution of sediment concentration for the case of scour ($D = 0.1$ mm, $B_0 = 0.5$, $\omega/u* = 1$)

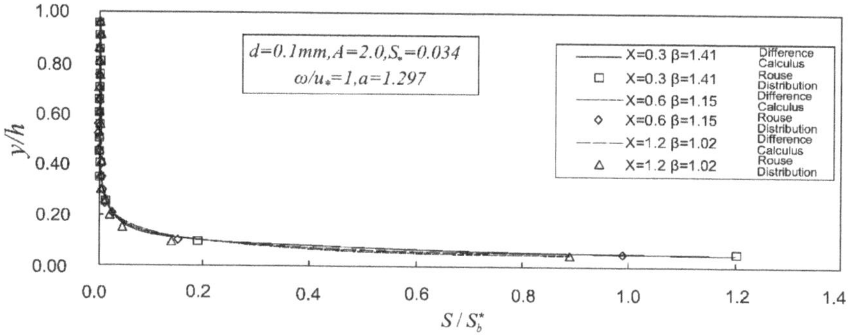

Fig. 8.8 Computed longitudinal variation of vertical distribution of sediment concentration for the case of deposition ($D = 0.1$ mm, $B_0 = 2.0$, $\omega/u* = 1$)

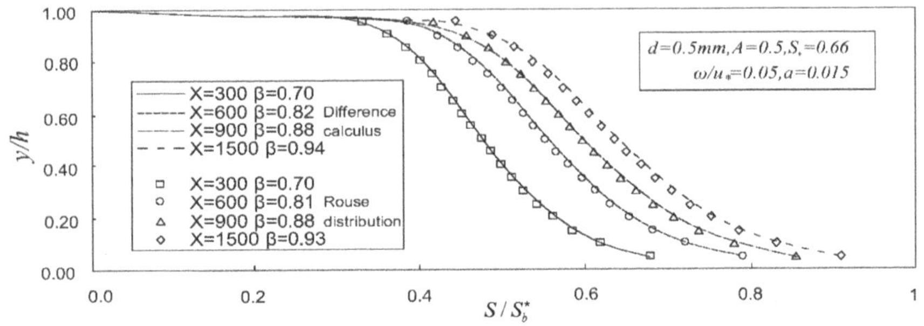

Fig. 8.9 Computed longitudinal variation of vertical distribution of sediment concentration for the case of scour ($D = 0.5$ mm, $B_0 = 0.5$, $\omega/u_* = 0.05$)

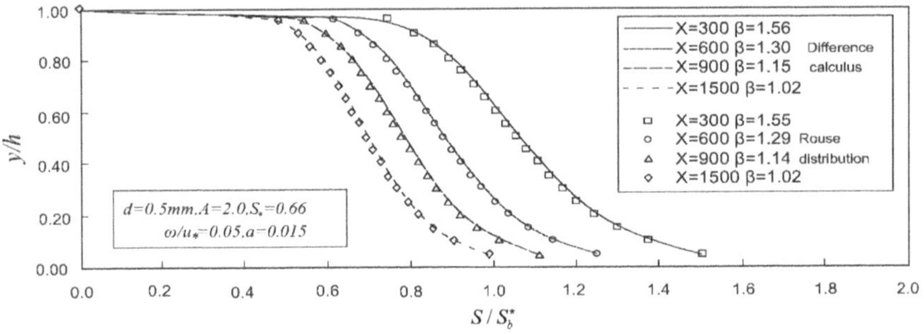

Fig. 8.10 Computed longitudinal variation of vertical distribution of sediment concentration for the case of deposition ($D = 0.5$ mm, $B_0 = 2.0$, $\omega/u_* = 0.05$)

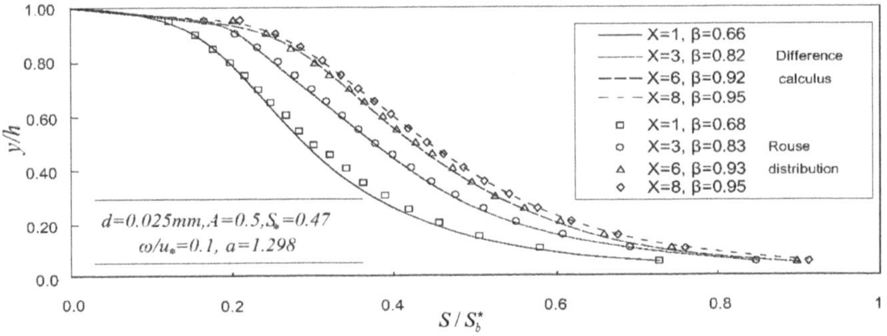

Fig. 8.11 Computed longitudinal variation of vertical distribution of sediment concentration for the case of scour ($D = 0.025$ mm, $B_0 = 0.5$, $\omega/u_* = 0.1$)

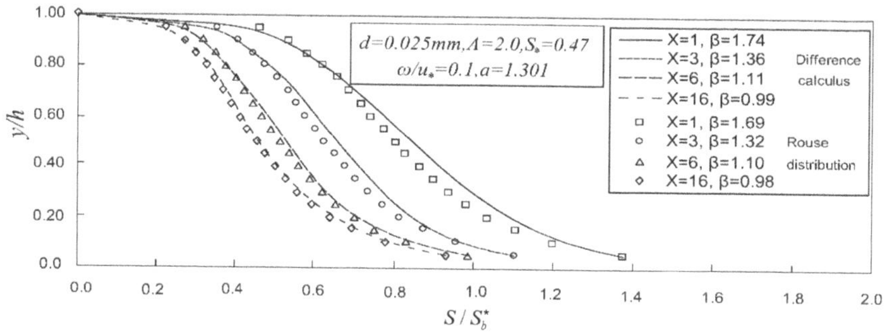

Fig. 8.12 Computed longitudinal variation of vertical distribution of sediment concentration for the case of deposition ($D = 0.025$ mm, $B_0 = 2.0$, $\omega/u_* = 0.1$)

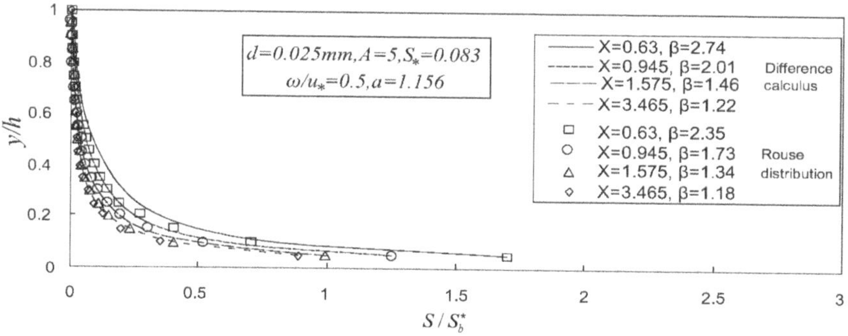

Fig. 8.13 Computed longitudinal variation of vertical distribution of sediment concentration for the case of scour ($D = 0.025$ mm, $B_0 = 5.0$, $\omega/u_* = 0.5$)

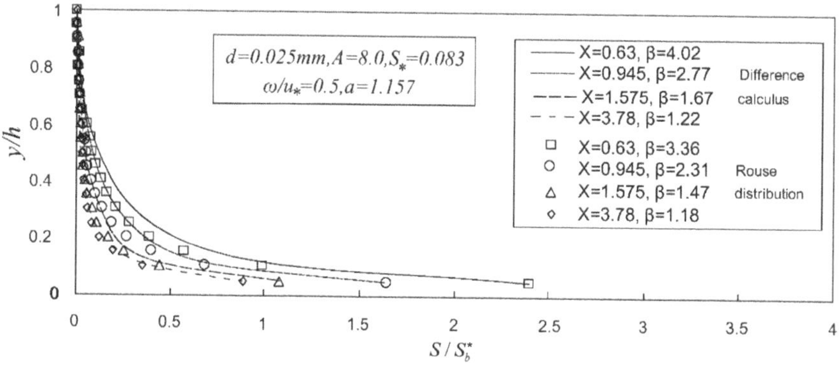

Fig. 8.14 Computed longitudinal variation of vertical distribution of sediment concentration for the case of deposition ($D = 0.025$ mm, $B_0 = 8.0$, $\omega/u_* = 0.5$)

8.2 Approximate Theoretical Solutions of the Vertical Distribution of Sediment Concentration for Non-equilibrium Sediment Transport

8.2.1 Theoretical Derivation

8.2.1.1 Exponential Vertical Distribution of Sediment Concentration

For steady equilibrium sediment transport in uniform flow, the diffusion equation for suspended load transport is

$$\varepsilon_y \frac{dS}{dy} + \omega S = 0 \tag{8.17}$$

where S is the sediment concentration, ω the sediment fall velocity, and ε_y the sediment diffusion coefficient in the vertical direction. The vertical profile of the flow velocity can be written as

$$\frac{du}{dy} = \frac{u_*}{\kappa y} \tag{8.18}$$

and the vertical profile of the shear stress is

$$\tau = \tau_0\left(1 - \frac{y}{h}\right) = \rho\varepsilon_m \frac{du}{dy} \tag{8.19}$$

where τ is the shear stress, τ_0 the shear stress on the bed, and ε_m the turbulent diffusion coefficient. Assuming the sediment diffusion coefficient ε_y equals the turbulent coefficient ε_m, one obtains

$$\varepsilon_y = \varepsilon_m = \frac{\tau}{\rho\frac{du}{dy}} = \frac{\frac{\tau_0}{\rho}\left(1 - \frac{y}{h}\right)}{\frac{du}{dy}} = \frac{u_*^2}{\frac{u_*}{\kappa y}}\left(1 - \frac{y}{h}\right) = \kappa u_*\left(1 - \frac{y}{h}\right)y \tag{8.20}$$

where κ is the von Karman constant. For simplicity, assume ε_y is constant over the water depth, and equals the depth-averaged value

$$\overline{\varepsilon}_y = \frac{1}{h}\int_0^h \varepsilon_y dy = \int_0^h \kappa u_*\left(1 - \frac{y}{h}\right)\frac{y}{h}dy = \frac{1}{6}\kappa u_* h \tag{8.21}$$

Plugging Eq. (8.21) into Eq. (8.17),

$$\frac{dS}{dy} = -\frac{\omega S}{\overline{\varepsilon}_y} = -\frac{6\omega S}{\kappa u_* h} \tag{8.22}$$

Integrating Eq. (8.22) gives the following exponential vertical profile of sediment concentration for equilibrium sediment transport

$$S = S_b \exp\left(-\frac{6\omega\eta}{\kappa u_*}\right) \tag{8.23}$$

where S_b is the sediment concentration at the bottom, η the vertical location, normalized by the water depth, above the bottom.

The diffusion equation for non-equilibrium transport of suspended load can be written as

$$\varepsilon_y \frac{dS}{dy} + \omega S = q_s \tag{8.24}$$

or

$$\frac{dS}{dy} + \frac{\omega S}{\varepsilon_y} = \frac{dS}{dy} + \frac{6\omega S}{\kappa u_* h} = \frac{q_s}{\varepsilon_y} = \delta_s \tag{8.25}$$

where q_s is the sedimentation quantity.

As shown in Eq. (8.22), the vertical gradient of sediment concentration for equilibrium sediment transport depends on $\frac{6\omega S}{\kappa u_* h}$, which represents the ratio between the gravitational settling ωS and the turbulent diffusion ε_y. For over-saturated sediment transport, δ_s is related to these two forces, or $\delta_s \propto \frac{6\omega S}{\kappa u_* h}$. In addition, δ_s is also impacted by the saturation coefficient $\overline{S}/\overline{S}^*$. Set

$$\delta_s = \frac{6\omega S}{\kappa u_* h} f\left(\frac{\overline{S}}{\overline{S}^*}\right) \tag{8.26}$$

where $f\left(\frac{\overline{S}}{\overline{S}^*}\right)$ is a function of $\frac{\overline{S}}{\overline{S}^*}$, the ratio between the depth-averaged sediment concentration and the depth-averaged carrying-sediment capacity, which is independent of η. Plugging Eq. (8.26) into Eq. (8.25) yields

$$\frac{dS}{dy} = -\frac{6\omega S}{\kappa u_* h}\left[1 - f\left(\frac{\overline{S}}{\overline{S}^*}\right)\right] = -\frac{6\omega SC}{\kappa u_* h} \tag{8.27}$$

where $C = 1 - f\left(\frac{\overline{S}}{\overline{S}^*}\right)$ is the un-saturation factor, which equals 1 for $\frac{\overline{S}}{\overline{S}^*} = 1$, larger than 1 for $\frac{\overline{S}}{\overline{S}^*} < 1$ (scour), and less than 1 for $\frac{\overline{S}}{\overline{S}^*} > 1$ (deposition).

It is noted that $\overline{S}/\overline{S}^*$ only varies slightly with the distribution of sediment concentration, and thus $\overline{S}/\overline{S}^*$ is considered to be approximately independent of y. With this approximation, integrating Eq. (8.27) with the bottom boundary condition $S = S_b$ at $\eta = 0$, yields

$$\ln\frac{S}{S_b} = -\frac{6\omega C}{\kappa u_*}\frac{y}{h} = -\frac{6\omega C}{\kappa u_*}\eta$$

Then the following exponential vertical distribution of sediment concentration for non-equilibrium sediment transport is obtained

$$S = S_b \exp\left(-\frac{6c\omega}{\kappa u_*}\eta\right) \tag{8.28}$$

Compared to the distribution for equilibrium sediment transport, the un-saturation factor C is included in the distribution for non-equilibrium sediment transport.

8.2.1.2 The Rouse Distribution of Sediment Concentration

Instead of setting the sediment diffusion coefficient to be the depth-averaged value, ε_y can be related to the turbulent diffusion coefficient ε_m with the following relationship

$$\varepsilon_y = \beta_0\varepsilon_m = \beta_0\kappa u_*\left(1 - \frac{y}{h}\right)y \tag{8.29}$$

For β_0=1,

$$\varepsilon_y = \kappa u_*\left(1 - \frac{y}{h}\right)y \tag{8.30}$$

Plugging Eq. (8.30) in Eq. (8.17), one obtains

$$\frac{dS}{dy} = -\frac{\omega S}{\varepsilon_y} = -\frac{\omega S}{\kappa u_*\left(1 - \frac{y}{h}\right)y} \tag{8.31}$$

or

$$\frac{dS}{S} = -\frac{\omega}{\kappa u_*}\frac{dy}{\left(1 - \frac{y}{h}\right)y} = -\frac{\omega}{\kappa u_*}\frac{d\eta}{(1 - \eta)\eta}$$

Integrating this equation

$$\int_a^\eta \frac{dS}{S} = \int_a^\eta -\frac{\omega}{\kappa u_*}\frac{d\eta}{(1-\eta)\eta} = -\frac{\omega}{\kappa u_*}\int_a^\eta \left(\frac{1}{(1-\eta)} + \frac{1}{\eta}\right)d\eta$$

yields the Rouse distribution of sediment concentration for equilibrium sediment transport, which reads

$$S = S_a\left(\frac{1-\eta}{\eta}\frac{a}{1-a}\right)^{\frac{\omega}{\kappa u_*}} \quad (8.32)$$

where η is a relative vertical location y/h above the bottom, and S_a the sediment concentration at a reference location with a relative height of a above the bottom, which is typically set to $a = 0.05$.

The sediment diffusion equation for non-equilibrium sediment transport is

$$\varepsilon_y\frac{dS}{dy} + \omega S = q_s \quad (8.33)$$

or

$$\frac{dS}{dy} + \frac{\omega S}{\varepsilon_y} = \frac{dS}{dy} + \frac{\omega S}{\kappa u_*\left(1-\frac{y}{h}\right)y} = \frac{q_s}{\varepsilon_y} = \delta_s \quad (8.34)$$

where q_s is the sedimentation quantity.

As shown in Eq. (8.31), the vertical gradient of sediment concentration is a function of the ratio between the gravitational settling ωS and the turbulent diffusion ε_y for equilibrium sediment transport. This is also be true for δ_z in the non-equilibrium condition, or

$$\delta_s \propto \frac{\omega S}{\kappa u_*\left(1-\frac{y}{h}\right)y}$$

In addition, δ_s is also impacted by the saturation coefficient $\overline{S}/\overline{S}^*$. Hence, set

$$\delta_s = \frac{\omega S}{\kappa u_*\left(1-\frac{y}{h}\right)y} f\left(\frac{\overline{S}}{\overline{S}^*}\right) \quad (8.35)$$

where $f(\overline{S}/\overline{S}^*)$ is a function of $\overline{S}/\overline{S}^*$, the ratio between the depth-averaged sediment concentration and the depth-averaged carrying-sediment capacity, which is independent of η. Plugging Eq. (8.35) into Eq. (8.34) yields

$$\frac{dS}{dy} = -\frac{\omega S}{\kappa u_*\left(1-\frac{y}{h}\right)y}\left[1 - f\left(\frac{\overline{S}}{\overline{S}^*}\right)\right] = -\frac{\omega SC}{\kappa u_*\left(1-\frac{y}{h}\right)y} \quad (8.36)$$

where $C = 1 - f(\overline{S}/\overline{S}^*)$ is the un-saturation factor, which equals 1 for $\overline{S}/\overline{S}^* = 1$, is larger than 1 for $\overline{S}/\overline{S}^* < 1$ (scour), and is less than 1 for $\overline{S}/\overline{S}^* > 1$ (deposition).

As discussed in Sect. 8.2.1.1, $\overline{S}/\overline{S}^*$ is considered to be approximately independent of y. With this approximation, integrating Eq. (8.36) with the bottom boundary condition $S = S_b$ at $\eta = 0$, yields

$$\int_a^\eta \frac{dS}{S} = \int_a^\eta -\frac{C\omega}{\kappa u_*}\frac{d\eta}{(1-\eta)\eta} = -\frac{C\omega}{\kappa u_*}\int_a^\eta \left(\frac{1}{(1-\eta)} + \frac{1}{\eta}\right)d\eta$$

Then the following Rouse-type distribution of sediment concentration for non-equilibrium sediment transport is obtained

$$S = S_b\left(\frac{1-\eta}{\eta}\frac{a}{1-a}\right)^{\frac{C\omega}{\kappa u_*}} \tag{8.37}$$

Compared to the distribution for equilibrium condition, the un-saturation factor C is included in the distribution for non-equilibrium sediment transport.

The determination of the bottom sediment concentration, S_b, is discussed in the next section.

8.2.2 Bottom Sediment Concentration for Non-equilibrium Sediment Transport

The vertical distribution of sediment concentration S for the case of scour is illustrated in Fig. 8.15. Curve A represents the non-equilibrium vertical distribution of S for the case of scour. Curve B represents the equilibrium vertical distribution, with the depth-averaged sediment concentration $\overline{S}$ being the same as that for curve A, or the distribution of carrying-sediment capacity for $\overline{S} = \overline{S}^{*'}$. Curve C represents the vertical distribution of carrying-sediment capacity S^*. For the case of scour, $S^* > S^{*'}$. For given hydraulic parameters, the sediment concentration and its vertical distribution approach to the carrying-sediment capacity and its distribution, or curve A approaches curve C, as the scour develops, and thus $S_b \le S_b^*$. As the equilibrium distribution, curve B is a special case of curve $S_b \le S_b^*$ A, and the gradient of curve A is milder than that of curve B. For scour, relatively more sediment concentrates on the bottom than the equilibrium distribution, and thus $S_b \ge S_b^{*'}$. Hence, $S_b^* \ge S_b \ge S_b^{*'}$ for the case of scour.

The vertical distribution of sediment concentration S for the case of deposition is illustrated in Fig. 8.16. Curve A represents the non-equilibrium vertical distribution of S for the case of deposition. Curve B represents the equilibrium vertical distribution, with the depth-averaged sediment concentration $\overline{S}$ being the same as that for curve A, or the distribution of sediment transport capacity for $\overline{S} = \overline{S}^*$. Curve C represents

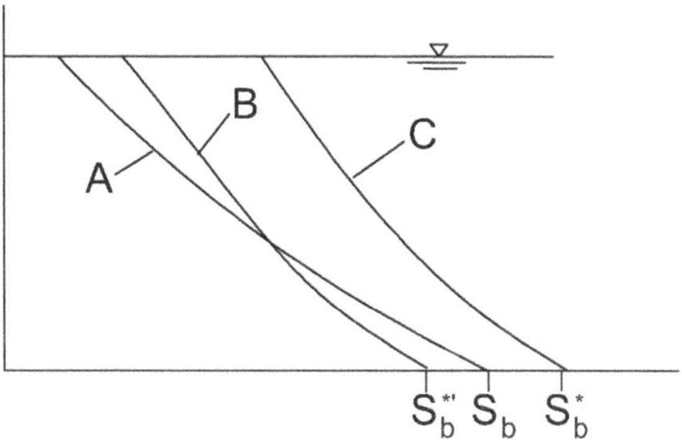

Fig. 8.15 Vertical distribution of sediment concentration for scour

the vertical distribution of sediment transport capacity S^*. For the case of deposition, $S^* < S^{*'}$. For given hydraulic parameters, the sediment concentration and its vertical distribution approach to the carrying-sediment capacity and its distribution, or curve *A* approaches curve C, as the deposition develops, and thus $S_b \geq S_b^*$. As the equilibrium distribution, curve *B* is a special case of curve *A*, and the gradient of Curve *A* is steeper than that of Curve *B*. For deposition, relatively more sediment remains on the upper portion of the water than the equilibrium distribution, and thus $S_b \leq S_b^{*'}$. Hence, $S_b^* \leq S_b \leq S_b^{*'}$ for the case of deposition.

In general, the bottom sediment concentration for non-equilibrium sediment transport can be written as

$$S_b = \mu S_b^{*'} + (1 - \mu) S_b^* \tag{8.38}$$

where μ is a weighting factor, and $0 \leq \mu \leq 1$. Based on the exponential vertical distribution of carrying-sediment capacity, the depth-averaged carrying-sediment

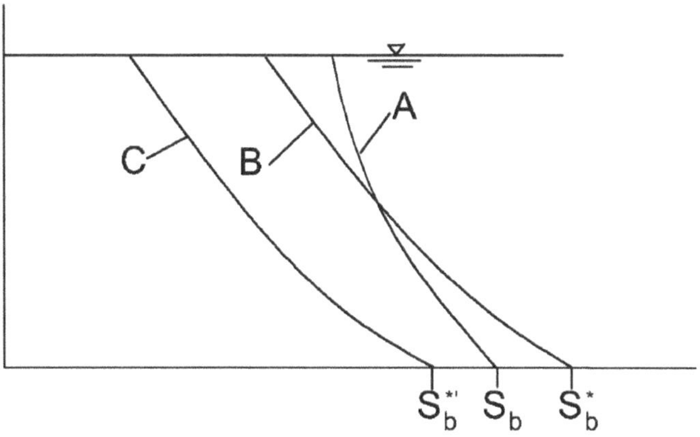

Fig. 8.16 Vertical distribution of sediment concentration for deposition

capacity is

$$\bar{S}^* = \frac{1}{h}\int_0^h S^*(y)dy = \int_0^1 S_b^* \exp\left(-\frac{6\omega\eta}{\kappa u_*}\right)d\eta$$
$$= S_b^*\frac{\kappa u_*}{6\omega}\left[1-\exp\left(-\frac{6\omega}{\kappa u_*}\right)\right] = S_b^* A_0 \quad (8.39)$$

where

$$A_0 = \frac{\kappa u_*}{6\omega}\left[1-\exp\left(-\frac{6\omega}{\kappa u_*}\right)\right] \quad (8.40)$$

and

$$\overline{S} = S_b^{*'} A_0$$

Hence, the bottom sediment concentration can be written as

$$S_b = \left[\mu\overline{S} + (1-\mu)\overline{S}^*\right]\frac{1}{A_0} \quad (8.41)$$

Based on the Rouse distribution, the depth-averaged sediment transport capacity is

$$\overline{S}^* = \frac{1}{h}\int_0^n S^*(y)dy = \int_0^1 S_b^*\left(\frac{1-\eta}{\eta}\frac{a}{1-a}\right)^{\frac{\omega}{\kappa u_*}} d\eta = S_b^* A_0 \quad (8.42)$$

where

$$A_0 = \int_a^1 \left(\frac{1-\eta}{\eta}\frac{a}{1-a}\right)^{\frac{\omega}{\kappa u_*}} d\eta \quad (8.43)$$

Similarly

$$\overline{S} = S_b^{*'} A_0$$

Hence, the bottom sediment concentration can be written as

$$S_b = \left[\mu\overline{S} + (1-\mu)\overline{S}^*\right]\frac{1}{A_0} \quad (8.44)$$

8.2.3 Vertical Distribution of Sediment Concentration for Non-equilibrium Sediment Transport

Plugging Eq. (8.44) in Eq. (8.37), the following Rouse-type distribution of sediment concentration for non-equilibrium sediment transport is obtained

$$S(\eta) = \left[\mu\overline{S} + (1-\mu)\overline{S}^*\right]\frac{1}{A_0}\left(\frac{1-\eta}{\eta}\frac{a}{1-a}\right)^{\frac{C\omega}{\kappa u_*}} \tag{8.45}$$

where A_0 is determined with Eq. (8.43). The depth-averaged sediment concentration

$$\overline{S} = \int_a^1 S_b\left(\frac{1-\eta}{\eta}\frac{a}{1-a}\right)^{\frac{C\omega}{\kappa u_*}} d\eta = S_b A \tag{8.46}$$

with

$$A = \int_a^1 \left(\frac{1-\eta}{\eta}\frac{a}{1-a}\right)^{\frac{C\omega}{\kappa u_*}} d\eta$$

Plugging Eq. (8.44) into Eq. (8.46) yields

$$\overline{S} = \left[\mu\overline{S} + (1-\mu)\overline{S}^*\right]\frac{A}{A_0} = \left[\mu\overline{S} + (1-\mu)\overline{S}^*\right]\frac{\int_a^1 \left(\frac{1-\eta}{\eta}\frac{a}{1-a}\right)^{\frac{C\omega}{\kappa u_*}} d\eta}{\int_a^1 \left(\frac{1-\eta}{\eta}\frac{a}{1-a}\right)^{\frac{\omega}{\kappa u_*}} d\eta} \tag{8.47}$$

or

$$\frac{\frac{\overline{S}}{\overline{S}^*}}{\mu\frac{\overline{S}}{\overline{S}^*} + (1-\mu)} = \frac{A}{A_0} = \frac{\int_a^1 \left(\frac{1-\eta}{\eta}\frac{a}{1-a}\right)^{\frac{C\omega}{\kappa u_*}} d\eta}{\int_a^1 \left(\frac{1-\eta}{\eta}\frac{a}{1-a}\right)^{\frac{\omega}{\kappa u_*}} d\eta}$$

It can be rewritten as

$$\frac{\beta}{\mu\beta + (1-\mu)} = \frac{A}{A_0} \tag{8.48}$$

where $\beta = \overline{S}/\overline{S}^*$ is the saturation coefficient for non-equilibrium sediment transport.

Plugging Eq. (8.44) in Eq. (8.28), the following exponential vertical distribution of sediment concentration for non-equilibrium sediment transport is obtained

$$S(\eta) = \left[\mu\overline{S} + (1-\mu)\overline{S}^*\right]\frac{1}{A_0}\exp\left(-\frac{6C\omega}{\kappa u_*}\eta\right) \tag{8.49}$$

where A_0 is determined with Eq. (8.40). The depth-averaged sediment concentration

$$\overline{S} = \int_0^1 S_b\exp\left(-\frac{6C\omega}{\kappa u_*}\eta\right)d\eta = S_b\frac{\kappa u_*}{6C\omega}\left[1-\exp\left(-\frac{6C\omega}{\kappa u_*}\right)\right] = S_b A \tag{8.50}$$

with

$$A = \frac{\kappa u_*}{6C\omega}\left[1-\exp\left(-\frac{6C\omega}{\kappa u_*}\right)\right] \tag{8.51}$$

Plugging Eq. (8.44) in Eq. (8.50) yields

$$\overline{S} = \left[\mu\overline{S} + (1-\mu)\overline{S}^*\right]\frac{A}{A_0} = \left[\mu\overline{S} + (1-\mu)\overline{S}^*\right]\frac{1}{C}\frac{1-\exp\left(-\frac{6C\omega}{\kappa u_*}\right)}{1-\exp\left(-\frac{6\omega}{\kappa u_*}\right)} \tag{8.52}$$

or

$$\frac{\frac{\overline{S}}{\overline{S}^*}}{\mu\frac{\overline{S}}{\overline{S}^*} + (1-\mu)} = \frac{A}{A_0} = \frac{1}{C}\frac{1-\exp\left(-\frac{6C\omega}{\kappa u_*}\right)}{1-\exp\left(-\frac{6\omega}{\kappa u_*}\right)}$$

It can be rewritten as

$$\frac{\beta}{\mu\beta + (1-\mu)} = \frac{1}{C}\frac{1-\exp\left(-\frac{6C\omega}{\kappa u_*}\right)}{1-\exp\left(-\frac{6\omega}{\kappa u_*}\right)} \tag{8.53}$$

where $\beta = S/S*$ is the saturation coefficient for non-equilibrium sediment transport.

As shown in Eq. (8.53), the un-saturation factor C can be determined for given μ, B, and ω/u_*, and the distribution of sediment concentration can then be determined for non-equilibrium sediment transport. For $\mu = 0.5$, Eq. (8.53) becomes

$$\frac{2\beta}{1+\beta} = \frac{1}{C}\frac{1-\exp\left(-\frac{6C\omega}{\kappa u_*}\right)}{1-\exp\left(-\frac{6\omega}{\kappa u_*}\right)} \tag{8.54}$$

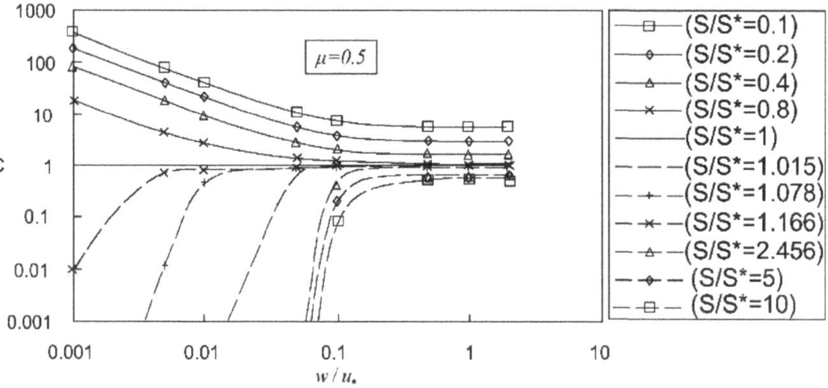

Fig. 8.17 Computed results of non-equilibrium factor C for $\mu = 0.5$

The un-saturation factor C is not only a function of ω/μ_*, but also depends on $\overline{S}/\overline{S}^*$ and μ. This is reasonable because $\overline{S}/\overline{S}^*$ and μ are the parameters to represent the degree of non-equilibrium. Figure 8.17 shows the numerical results of the un-saturation factor C for μ=0.5. As shown in this figure, C decreases as $\overline{S}/\overline{S}^*$ increases; for the case of scour, $\overline{S}/\overline{S}^* > 1$, C is great than 1, and C decreases as ω/μ_* increases; and for the case of deposition, $\overline{S}/\overline{S}^* < 1$, C is less than 1, and C increases as ω/μ_* increases. This is consistent with the engineering experience.

For $C \to 0$, the vertical distribution of sediment concentration can approach uniform for suspended load with any particle size (or ω). This uniform distribution can occur for high-concentration flows, even when the suspended particles are coarse. For $C \to 0$, $\exp\left(-\frac{6C\omega}{\kappa u_*}\right)$ can be written in the expanded form

$$\exp\left(-\frac{6C\omega}{\kappa u_*}\right) = 1 - \frac{6\omega}{\kappa u_*}C + \frac{1}{2!}\left(\frac{6\omega}{\kappa u_*}\right)^2 C^2 - \frac{1}{3!}\left(\frac{6\omega}{\kappa u_*}\right)^3 C^3 + \cdots$$

Plugging in Eq. (8.53) yields

$$\begin{aligned}
&\frac{\beta}{\mu\beta + (1-\mu)} \\
&= \frac{1}{C}\frac{1}{1-\exp\left(-\frac{6\omega}{\kappa u_*}\right)}\left[\frac{6\omega}{\kappa u_*}C - \frac{1}{2!}\left(\frac{6\omega}{\kappa u_*}\right)^2 C^2 + \frac{1}{3!}\left(\frac{6\omega}{\kappa u_*}\right)^3 C^3 - \cdots\right] \\
&= \frac{1}{1-\exp\left(-\frac{6\omega}{\kappa u_*}\right)}\left[\frac{6\omega}{\kappa u_*} - \frac{1}{2!}\left(\frac{6\omega}{\kappa u_*}\right)^2 C + \frac{1}{3!}\left(\frac{6\omega}{\kappa u_*}\right)^3 C^2 - \cdots\right]
\end{aligned}$$

$$\frac{\beta}{\mu\beta + (1-\mu)} = \frac{1}{1-e^{-\frac{6\omega}{\kappa u_*}}}\left[\frac{6\omega}{\kappa u_*} - \frac{1}{2!}\left(\frac{6\omega}{\kappa u_*}\right)^2 c + \frac{1}{3!}\left(\frac{6\omega}{\kappa u_*}\right)^3 c^2 - \ldots\right]$$

For $C \to 0$, neglecting the higher order terms of C, one obtains

$$\frac{\beta}{\mu\beta + (1-\mu)} = \frac{\frac{6\omega}{\kappa u_*}}{1 - \exp\left(-\frac{6\omega}{\kappa u_*}\right)}$$

or

$$\beta = \frac{\overline{S}}{\overline{S}^*} = \frac{(1-\mu)\frac{6\omega}{\kappa u_*}}{1 - \exp\left(-\frac{6\omega}{\kappa u_*}\right) - \mu\frac{6\omega}{\kappa u_*}} \tag{8.55}$$

For various ω/u_*, the saturation coefficient $\beta = \overline{S}/\overline{S}^*$ for the distribution of sediment concentration to become uniform can be determined with Eq. (8.55), and the numerical results are listed in Table 8.5. As listed in this table, the vertical distribution of sediment concentration can be adjusted towards the uniform distribution through the over-saturated sediment transport ($\overline{S}/\overline{S}^* > 1$) during deposition. For finer sediment particles with higher flow velocity, ω/μ_* is smaller, and the saturation coefficient for the distribution of sediment concentration to become uniform is smaller. On the other hand, for coarser sediment particles with lower flow velocity, ω/μ_* is larger, and the saturation coefficient for the uniform distribution is greater. For a small ω/μ_*, the distribution of sediment concentration is already close to the uniform distribution, and thus the increase to the saturation coefficient $\overline{S}/\overline{S}^*$ for the uniform distribution is limited. On the other hand, for a very large ω/μ_*, the distribution is far away from the uniform distribution, and thus the saturation coefficient needs to be increased significantly in order for the distribution to become uniform. For examples, for $\omega/\mu_* = 0.001$, the saturation coefficient $\overline{S}/\overline{S}^*$ only need to be slightly increased to 1.015 for the distribution of sediment concentration to become uniform; while for $\omega/\mu_* = 0.05$, $\overline{S}/\overline{S}^* = 2.456$, or the sediment concentration needs to be 2.456 times of the carrying-sediment transport capacity, in order for the distribution to become uniform.

Table 8.5 Saturation coefficient (S/S^*) required for uniform distribution of sediment concentration under various flow and sediment conditions (ω/u_*)

ω/u_*	0.001	0.005	0.01	0.05
S/S^*	1.015	1.078	1.166	2.456

8.3 Comparison of the Vertical Distribution of Sediment Concentration Between Theoretical and Numerical Results for Non-equilibrium Sediment Transport

As discussed before, Figs. 8.1, 8.2, 8.3, 8.4, 8.5, 8.6, 8.7, 8.8, 8.9, 8.10, 8.11, 8.12, 8.13 and 8.14 show the numerical results of the vertical 2-D sediment diffusion equation compared to the Rouse-type distribution, as described in Eq. (8.45), for non-equilibrium sediment transport. As shown in these figures, Eq. (8.45) in general agrees well with the numerical results. Except for the cases shown in Figs. 8.3, 8.4, 8.13, and 8.14, in which the numerical results show relatively larger discrepancy from Eq. (8.45), the numerical results show good agreement with Eq. (8.55) for the other ten cases. Figures 8.1, 8.2, 8.3, 8.4, 8.5, 8.6, 8.7, 8.8, 8.9 and 8.10 show the results for the case with $\tilde{\alpha}_l \neq \alpha_l^*$, and Figs. 8.11, 8.12, 8.13 and 8.14 show the results for the case with $\tilde{\alpha}_l \neq \alpha_l^*$. The numerical results show reasonably agreement with the theoretical distribution for both $\tilde{\alpha}_l \neq \alpha_l^*$ and $\tilde{\alpha}_l \neq \alpha_l^*$.

In conclusion, the approximate theoretical distribution of sediment concentration developed in this chapter for non-equilibrium sediment transport has theoretical basis and been proved to be reliable. It can be used in the theoretical analysis and engineering sedimentation applications.

References

1. Han QW (1973) Preliminary studies for the non-equilibrium transport of sediment in reservoirs. In: Report on Sedimentation in Reservoir, Task Group on Yellow River Sedimentation Research (in Chinese)
2. Han QW (1981) Statistical characters of sediment exchange. J Hydraulic Eng (1):10–22 (in Chinese)
3. Han QW (2006) Boundary condition of diffusion equation and coefficient of saturation recovery. J Changsha Univ Sci Tech Nat Sci 3(3):7–19 (in Chinese)
4. Han QW, Chen XJ, Xue XC (2010) On the vertical distribution of sediment concentration in non-equilibrium transportation. J Adv Water Sci (4):512–523 (in Chinese)
5. Han QW, He MM (1984) Stochastic theory of sediment motion. Science Publishing House, Beijing (in Chinese)
6. Han QW, Huang YL (1974) Methods of computation of sedimentation and erosion in the reservoir and utilization of digital computer. Yangtze Institute of Water Resources Research 1:45–85 (in Chinese)
7. Han QW, Xue XC (2010) Numerical results for sediment diffusion equation (1). Technical Report, China Institute of Water Resources and Hydropower Research, Department of Sediment Research (in Chinese)
8. Han QW, Xue XC (2010) Numerical results for sediment diffusion equation (2). Technical Report, China Institute of Water Resources and Hydropower Research, Department of Sediment Research (in Chinese)

Chapter 9
Bed Material Coarsening

Bed material coarsening is a common phenomenon in rivers. It occurs not only in the downstream channel of a reservoir during the clear water release, but also in a natural stream or the variable back water zone of a reservoir when scour takes place. Bed material coarsening not only occurs during channel scour. It may also occur when the channel approaches to equilibrium or even during a mild channel accretion. Of course, the extent of bed material coarsening is different for various cases. Bed material coarsening impacts sediment transport, sedimentation quantity and channel roughness. Thus, it is a critical topic for the investigation of channel scour downstream of a reservoir, and is important for sedimentation calculation and sediment transport.

The main contents of this chapter consists of: (1) review of the existing studies on bed material coarsening, including the advances and problems of these studies; (2) a thorough illustration of the mechanisms for bed material coarsening based on the field data and the statistical theory for sediment exchange on the bed, which is a breakthrough of the general concept that scour is the sole cause for bed material coarsening; (3) introducing bed material coarsening caused by the exchange between the coarser and finer sediment particles that generally occurs in streams, which is proved to be the most general mechanism for bed material coarsening; and (4) illustration of the bed material coarsening caused by bed load erosion, by suspended load erosion, and by the exchanges between bed load and bed material and between suspended load and bed material, armoring layer thickness and scour depth limit.

9.1 Literature Review

The existing studies on bed material coarsening covers the following topics: the minimum particle size for coarsening, the relation between the coarsened layer composition and the sedimentation percentage, the thickness of the coarsened layer,

Q. Han and M. He, *Non-equilibrium Transport of Non-uniform Suspended Load*, Hydroscience and Engineering, https://doi.org/10.1007/978-981-97-1123-9_9

and the proportion of the finer sediment particles that fill the voids between the coarser particles.

9.1.1 Minimum Particle Size for Coarsening

The minimum particle size for coarsening is generally considered to be controlled by the critical flow velocity or shear stress for sediment entrainment, or the entrainment probability. For example, Harrison (1950) investigated the mechanism for bed material coarsening based on the flume experiments of the scour on a pebble-sand bed. Harrison used the flow intensity factor as the criterion for bed material coarsening

$$\phi_* = C\frac{\gamma_s - \gamma}{\gamma}\frac{D}{R_b' J} \tag{9.1}$$

where γ_s and γ are the specific weights for the sediment particle and the water, respectively, D the particle diameter, J the slope of the water surface, R_b' the hydraulic radius corresponding to the particle resistance, and C a coefficient to reflect the impact of particles. Equation (9.1) can be rewritten as the ratio between the critical shear stress for sediment entrainment τ_c and the shear stress of the flow, or

$$1 = \frac{C(\gamma_s - \gamma)D/\phi_*}{\gamma R_b' J} = \frac{\tau_c}{\tau_0} \tag{9.2}$$

For $\phi_* = 27$–40, the corresponding $\frac{\tau_c}{(\gamma_s - \gamma)D} = 0.037$–$0.025$, which is close to the minimum dimensionless shear stress of the Shields diagram. Thus, the particle with the diameter of D that is determined with the above relation cannot be moved, or the minimum particle size of the coarsened layer that cannot be entrained by the flow is

$$D_0 = \frac{27}{C}\frac{\gamma}{\gamma_s - \gamma}R_b' J \tag{9.3}$$

Based on the formula for the entrainment flow velocity

$$V_c = 5.4aD_0^{0.36}h^{0.14} \tag{9.4}$$

Xie et al. [19, 20] derived the minimum particle size of the coarsened layer that is

$$D_0 = \left[\frac{q}{5.4ah^{1.14}}\right]^{2.78} \tag{9.5}$$

where q is the flow discharge per unit width, h the water depth, and a a coefficient that is less than 1.

Based on the formula for the critical flow velocity for the entrainment of non-uniform sediment

$$V_s = 0.786\sqrt{\frac{\gamma_s - \gamma}{\gamma} g\left(2.5m\overline{D} + D_0\right)}\left(\frac{h}{D_{90}}\right)^{\frac{1}{6}} \quad (9.6)$$

Qin and Wang [18] obtained the minimum particle size for coarsening

$$D_0 = \frac{\gamma}{0.618(\gamma_s - \gamma)} V^2 \left(\frac{D_{90}}{h}\right)^{\frac{1}{3}} - 2.5m\overline{D} \quad (9.7)$$

where $\overline{D}$ is the average particle size,

$$m = 0.75 - \frac{0.65}{2 + \eta} \quad (9.8)$$

$$\eta = \frac{D_{60}}{D_{10}} \quad (9.9)$$

Han [6, 7] and Han et al. [12, 15] considered that the basic mechanism for bed material coarsening is the exchange between bed material and suspended load and the exchange between bed material and bed load. Bed material coarsening is not just dependent on the critical flow velocity for entrainment or the entrainment probability. Instead, it involves a series of hydraulic and sediment parameters. This presents a more comprehensive illustration for the bed material coarsening caused by scour. The coarsening is not determined by a single particle size but is an non-uniform variation of bed material size distribution. Han proposed a new concept of the bed material coarsening associated with sediment exchange. Based on the concept of sediment exchange intensity, Han thoroughly illustrates the mechanisms for bed material coarsening and first demonstrated that the coarsening is not only caused by scouring but generally by sediment exchange, which are detailed in this chapter.

9.1.2 Scour Depth and Coarsened Layer Thickness

Bed material coarsening generally occurs when bed is eroded, reflecting the sorting mechanism of the flow. Both the coarser and the finer bed material are involved in the coarsening process, while the coarsened layer is formed with the coarser particles that are left on the bed after the erosion. As bed material coarsening develops, the top layer on the bed, or the armor layer, will be comprised of the coarser particles that are hard to be eroded.

Xie [19, 20], Yin [22], Qin and Wang [18], and Yang [21] assumed all the particles that are finer than the minimum size for coarsening are eroded away, and the coarsened

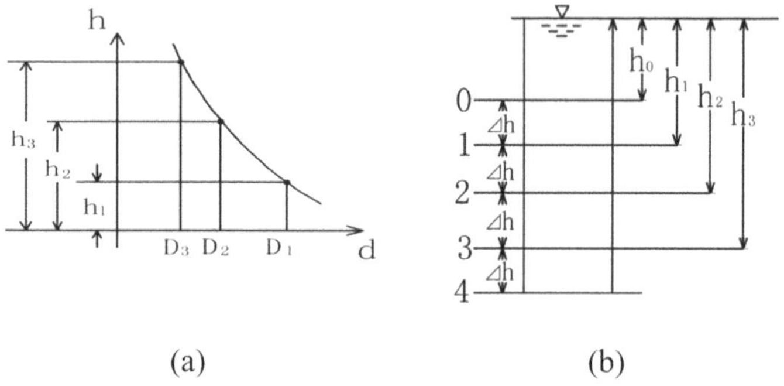

Fig. 9.1 Diagram of bed material coarsening

layer is comprised of the coarser particles. This assumption was made to simplify the computation of the coarsened layer thickness. As an example, Xie's method [19, 20] for determining the scour depth is illustrated in Fig. 9.1, and is described as follows:

(1) For $h_2 = h_1 + \Delta h$, compute the minimum particle size for coarsening D_2 according to Eq. (9.5) and the corresponding weight percent $P_{1\text{-}2}$ for the particles coarser than D_2, then the volume of the coarser particles within the layer between lines 1 and 2, as shown in Fig. 9.1a and b, is

$$\Delta h(1 - \phi)P_{1-2}$$

where ϕ is the porosity rate of this layer.

(2) Similarly, for the layer between lines 2 and 3, the corresponding water depth is $h_1 + 2\Delta h$, the minimum particle size for coarsening is $D_{1\text{-}3}$, and the volume of the coarser particles is

$$\Delta h(1 - \phi)P_{1-3}$$

(3) When the bed material coarsening process stops, the total volume of the coarser particles in the armor layer is

$$\sum_{1-2}^{1-n} \Delta h(1 - \phi)P_{1-l} = \Delta h(1 - \phi)\sum_{1-2}^{1-n} P_{1-l} \tag{9.10}$$

where n is the total number of sub-layers in the armor layer when coarsening stops.

(4) This total volume can also be expressed as

$$\Delta y(1 - \phi_*)P_*$$

where Δy is the thickness of the armor layer, ϕ_* the porosity rate of the armor layer, and P_* the proportion of the coarsened layer within the active layer. These two expressions are equivalent, which leads to

$$\sum_{1-2}^{1-n} P_{1.l} = \frac{\Delta y}{\Delta h}\frac{(1-\phi_*)}{(1-\phi)} P_* \tag{9.11}$$

While Δh is assigned, Δy, ϕ, P_*, and $\sum_{1-2}^{1-n} P_{1.l}$ can be determined step by step with the trial and error method. It is noted that as the scour depth increases, the minimum particle size for coarsening decreases, resulting in a less scour in the deeper sub-layer. In other words, the particles in the upper sub-layer are coarser than those in the deeper sub-layer. The coarser particles on the upper layer are not eroded away when coarsening occurs in the deeper sub-layer. Hence, these coarser particles of the upper layer do not need to be added in the sorting process of the deeper layer in the calculation of the minimum particle size for coarsening and corresponding percent for the deeper layer. This is actually a necessary condition for the derivation of Eq. (9.10). One problem of Xie's method is that based on the assumption, the sediment particles of the top layer between lines 0 and 1 would be completely eroded because of $D_0 = D_M$. In reality, part of the coarser particles of this top layer would be left on the bed. This suggests that bed material coarsening should be considered starting from the top layer between lines 0 and 1.

If the maximum scour depth in the active layer is limited, Eq. (9.11) can be written as

$$P_0 = \frac{\Delta y}{\Delta h'}\frac{(1-\phi_*)}{(1-\phi)} P_* \tag{9.12}$$

where P_0 is the percentage of the particles coarser than D_0 that is the minimum particle size for coarsening at $h_0 + \Delta h$. It is noted that $\Delta h'$ is the depth of the bed material that is involved in scour, and thus the actual depth of the bed material eroded away is

$$\Delta h = \Delta h' - \Delta y \tag{9.13}$$

Plugging Eq. (9.12) into Eq. (9.13)

$$\Delta h = \Delta y\left[\frac{(1-\phi_*)}{(1-\phi)}\frac{P_*}{P_0} - 1\right] \tag{9.14}$$

When the scour completes, setting $P_* = 1$, and neglecting the difference in the porosity rate, then Eq. (9.14)

$$\Delta h = \Delta y\left[\frac{1}{P_0} - 1\right] \tag{9.15}$$

Equation (9.15) was proposed by Yin [22], and was also recommended by U. S. Bureau of Reclamation. Yang [21] also suggested to directly use Eq. (9.15) for the case with narrow size distribution. Qin and Wang [18] also proposed a formula for computing the scour depth layer by layer, in which the coarser particles left in the upper layer is added to the deeper layer in the calculation of scour for the deeper layer.

The formulas discussed above establish the relation between the scour depth and the thickness of the coarsened layer. If the thickness of the armor is known, then the scour depth can be determined.

As shown in Fig. 9.2, Yin [22] proposed the thickness of the coarsened layer, h_0, was equal to D_{65} of the coarsened layer. D_{65} is the particle size corresponding to the percent finer of 65%. It was further assumed that the proportion of the coarser particles in the coarsened layer was $P_l = 0.65 = 65\%$, and the percentage of the finer particles that can be entrained but are hided between the coarser particles was $P_s = 35\%$. The thickness of the coarsened layer can be directly determined based on the initial bed material gradation with the following relation

$$P_A = \frac{P_s}{P_l} P_0 \tag{9.16}$$

where P_0 is the percent of the particles in the initial bed material that are coarser than the minimum particle size for coarsening D_0, or the percent of the bed material that are not movable; P_A the percent of the particles in the initial bed material that are coarser than the minimum particle size D_{m} of coarsened layer thickness h_A. Equation (9.16)

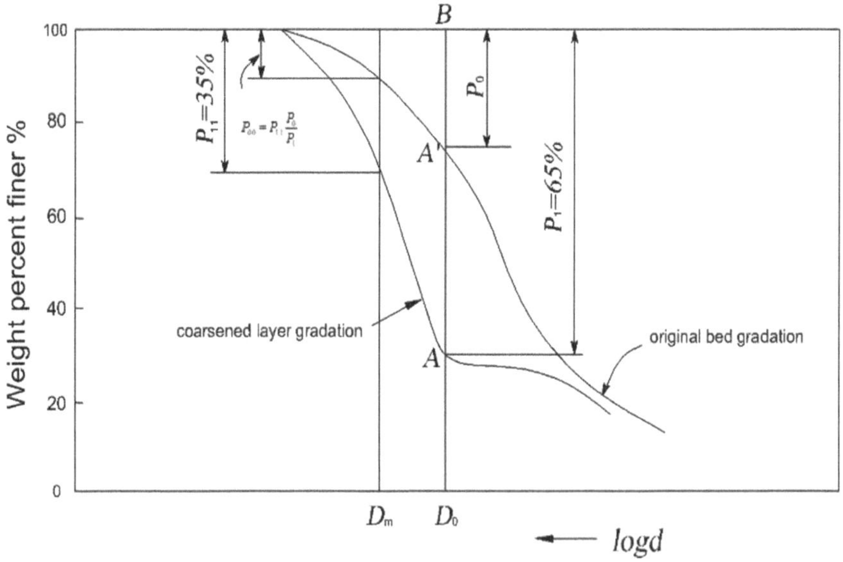

Fig. 9.2 Diagram for determination of $P_{0.l}$ and D_m

is essentially to scale down D_{65} of the coarsened layer with the scale factor of P_0/P_1. This is based on the assumption that scour only occurs for the particles that are finer than D_0.

For the pebble-sand bed, Xie [20] proposed the thickness of the armor layer was equal to thickness of the entire coarsened layer, in which the particle size corresponding to the following percent coarser of the armor

$$P_0 = 0.43P_{1-n} \approx 0.5P_{1-n} \tag{9.17}$$

where $P_{1\sim n}$ is the percent of the particles in the second armor sub-layer. U. S. Bureau of Reclamation proposed

$$h_0 = 3D_0 \tag{9.18}$$

where D_0 is the minimum particle size for coarsening that is determined based on the Shields diagram.

Based on a lot of field data of bed material sorting and composition after scour [6, 7], Han et al. [12, 15] proposed the following empirical values for the thickness of the coarsened layer: D_M, or D_{95} for the pebble bed and for the pebble-sand bed, 0.1–0.3 m for the coarser sand or-gravel bed, 0.2–0.4 m for the coarser sand bed, and 0.7 m for the fine sand bed.

9.1.3 Existing Methods for Calculating Composition of the Coarsened Layer

Assuming the percent of the coarser particles in the coarsened layer is $P_1 = 0.65$, Yin [22] scaled up the percent of the coarser particles using the ratio of P_1 to P_0, which is the percent of the coarser particles in the initial bed material, as the scale factor

$$P_l = \frac{P_1}{P_0} P_{0.l} \tag{9.19}$$

where $P_{0.l}$ and P_l are the accumulative size distribution of the bed material before and after the scour of the coarsened layer, respectively.

Qin and Wang [18] also considered there are finer particles hiding between the coarser particles considered there are finer particles hiding between the coarser particles in the coarsened layer, and the proportion of these finer particles is related to the degree of the non-uniformity of the coarser particles. Assuming the quantity of the finer particles is $P_0\frac{D_0}{D_m}$, the total sediment quantity of the coarsened layer is $(1 - P_0) + P_0\frac{D_0}{D_m}$, and thus

$$P_1 = \begin{cases} \dfrac{\frac{D_0}{D_m} P_{0.l}}{(1-P_0)+P_0\frac{D_0}{D_m}} & (D < D_0) \\ 1 - \dfrac{1-P_{0.l}}{(1-P_0)+P_0\frac{D_0}{D_m}} & (D \geq D_0) \end{cases} \quad (9.20)$$

Assuming the flow shear stress τ follows a normal distribution, Gessler [2, 3] obtained the probability for sediment not to be entrained

$$q = P(\tau < \tau_c) = \frac{1}{\sqrt{2\pi}\sigma_\tau} \int_{-\infty}^{\tau_c} e^{-\frac{(\tau-\tau)^2}{2\sigma_z^2}} d\tau$$

$$= \frac{1}{\sqrt{2\pi}} \int_{-\infty}^{\frac{\tau_c-\bar{\tau}}{\sigma_z}} e^{-\frac{t^2}{2}} d\tau \quad (9.21)$$

Gessler further assumed the size distribution of the bed material that were left on the bed after scour, or the size distribution of the coarsened layer, was proportional to q and the initial size distribution $P_0(D)$, i.e.,

$$P(D) = CqP_0(D) \quad (9.22)$$

where C is determined by

$$\int_{D_m}^{D_M} CP(D)dD = 1 \quad (9.23)$$

D_m and D_M are the maximum and minimum particle size, respectively. Hence, the particle size distribution of the coarsened layer is

$$P(D) = \frac{\int_{D_m}^{D} qP_0(D)dD}{\int_{D_m}^{D_M} qP_0(D)dD} \quad (9.24)$$

Han [6, 7] and Han et al. [12, 15] derived the variation of the bed material gradation after scour, the initial bed material gradation and scour percentage determined by the composition of the coarsened layer, and the ratio of the particle size (fall velocity) to the average particle size (average fall velocity). These will be discussed in detail in this chapter.

9.1.4 Methods for Calculating Bed Material Coarsening Associated with Scour in Mathematical Models

The variation of the bed material composition can also be computed with the formulas for sediment deposition or scour in the mathematical models. Examples include the one-dimensional (1-D) model developed by Han [4, 5], in which the formulas for the variation of bed material gradation and the relationship between the bed material gradation and the size distribution of suspended load were included for the case of massive deposition or scour and for the equilibrium condition; the HEC-6 model [1]; and the model developed by Li [17], in which the relationship between the bed material gradation and the size distribution of suspended load was derived for the strong equilibrium condition. There are also many other similar mathematical models that include the variation of bed material composition associated with scour. However, most of the models only provide numerical results, which are hard to be used to illustrate the mechanisms for bed material coarsening. Also, different methods are used in these models to determine the carrying-sediment capacities for suspended load and for bed load, and the deposition or scour quantity associated with the suspended load and/or the bed load transport. Thus, the results for the variation of bed material gradation, including coarsening, are not well accepted. In some mathematical models, the transport rate of bed load is computed based on the uniform sediment assumption, without considering the variation of bed material gradation. In most studies on bed material coarsening, the impacts of non-uniform sediment on the sediment transport rate and on the entrainment flow velocity are generally included by introducing some correction coefficients.

For a general non-equilibrium sediment transport, He and Han [16] proposed the concept of the size distribution for carrying-sediment capacity, and derived its general relations to the size distribution of suspended load, and to the bed material gradation. These relations apply to various sediment transport conditions, including massive scour or deposition, mild scour and deposition, and equilibrium transport.

9.1.5 Evaluation of Existing Studies on Bed Material Coarsening

The existing studies on bed material coarsening, such as Xie and Xie et al. [19, 20], Qin and Wang [18], Yin [22], and Yang [21], include scour as the main cause for bed material coarsening. They also provide methods to quantify bed material coarsening, the composition of the coarsened layer, and the scour depth, and thus provide supports for the sediment engineering studies. However, these studies lack thorough illustration of the mechanisms and the factors that impact bed material coarsening, provide few reasonable formulas or quantitative relations for coarsening that well agree with the reality, and thus are not sufficient for the description of the actual bed material coarsening. Some subjects related to bed material coarsening are discussed as follows.

1. Bed material coarsening is not solely controlled by the critical flow velocity for sediment entrainment, the corresponding entrainment shear stress, or the entrainment probability. For example, if the bed material is composed of coarser, medium, and finer sediment particles, all the particles can be entrained. However, bed material coarsening still occurs during bed scour in reality. For this case, a less amount of coarser bed material is eroded than the finer bed material. This kind of sediment transport can lasts for a long time, which is independent of the critical condition for sediment entrainment.
2. The time-averaged entrainment flow velocity and the corresponding shear stress are not deterministic, while the instantaneous entrainment velocity is deterministic. Due to the fluctuation of flow velocity, a small amount of the particles that are coarser than the minimum particle size for coarsening still move during scour, and the chance for movement become less for coarser particles. As a result, the assumption that the ratio of the coarser particles remains constant in the coarsened layer, as used by Xie et al. [20], Qin and Wang [18], and Yin [22], and the calculation of the corresponding size distribution lack theoretical basis. Gesseler [2, 3] calculated the composition of the coarsened layer using the probability for sediment not to be entrained, which would reflect the variation of the ratios among the coarser particles in the coarsened layer during scour. However, as discussed later in this chapter, the scour and the resulting amount of coarser particles in the coarsened layer are not only dependent on the no entrainment probability.
3. There is still no reliable method for determining the thickness of the coarsened layer. Alternatively, it can be estimated based on the empirical data of the field investigations. It is difficult to determine the coarsened layer thickness directly based on the height of ripples on a sand bed as Qin and Wang [18] suggested. This is because the real bed surface is three-dimensional. Bed ripples do not occur along the entire cross-section. Instead, ripples may occur in the main channel sometimes, and on the floodplain sometimes. Particularly, the higher flow discharge is typically considered when the bed material coarsening is computed based on the entrainment flow velocity, while the bed ripples already disappear under this high flow discharge. As pointed out by Qin and Wang [18], the heights of bed ripples can be as high as 4–5 m in the Yangzi River, which is not a representative value for the coarsened layer thickness, and which is just an extreme condition for a local reach. Of course, the bed ripple can stir up sediment particles, which enables the bed material within a certain depth below bed surface to take part in the sorting process during scour, and thus is one of the main factors to impact the thickness of the coarsened layer. However, considering the average condition of bed ripples that may or may not occur, the thickness of the coarsened layer is selected to be much less than the heights of the bed ripples, as did in Xie and Xie et al. [19, 20], Han [6, 7], and Han et al. [15].

It should be emphasized that how fast the coarsened layer is formed should also depend on the thickness of the active coarsened layer (the layer disturbed in the coarsened process). Let consider two cases with the active coarsened layer thickness $h_0 = 5.0\,\text{m}$, and $h_0 = 1.0\,\text{m}$, respectively. For both cases, assume the scour depth is

1 m, and the percent of the particles with $D > D_0$ in the initial bed material, or P_0, is 35%. After a scour of 1 m on the bed, the percent of the particles with $D > D_0$ in the bed material is $0.35 \times \frac{6m}{5m} = 0.42 = 42\%$ for the first case, without formation of the coarsened layer yet, and is $0.35 \times \frac{6m}{3m} = 0.70 = 70\%$ for the second case, with sufficient coarser material to form the armor layer. This indicates that for a thicker active layer on the bed, the speed of bed material coarsening is slower, resulting in a slower formation of the armor layer.

4. Most of the existing studies lack in-depth investigations of the mechanisms for bed material coarsening, and thus do not provide systematic theories to profoundly and comprehensively describe the phenomenon of bed material coarsening. Except for Han [6, 7] and Han et al. [15], the other studies considered bed material coarsening was caused only by channel scour. In fact, the exchange between bed material and suspended load is the most general cause for bed material coarsening. In most mathematical models, the erosion of bed material is only a result of the deposition or scour of suspended load, and is not described with any specific equations, which is incomplete.

9.2 Phenomena and Mechanisms for Bed Material Coarsening

As discussed before, most descriptions and expressions for bed material coarsening are simplistic. They mainly focus on the coarsening cause by scour, and only considered the impact of the critical condition for sediment entrainment. Hence, it is necessary to conduct a comprehensive and in-depth investigation of the phenomena and mechanisms for bed material coarsening. The results presented in Han [6, 7] and in Han et al. [15] as well as the recent studies conducted by the author are detailed as follows.

9.2.1 *Six Phenomena for Bed Material Coarsening*

1. Sand on the Top Layer Eroded with Pebble Exposed

For this case, the initial bed material gradation is discrete with the depth. The top layer on the bed is composed of sand, while the deeper layer is composed of pebble. After the sand of the top layer is eroded, the underneath pebble is exposed to form the coarsened layer. Figure 9.3 shows the comparison of the bed material gradations on the floodplain at Huangjiagang in the downstream channel of the Danjiangkou Reservoir before and after the operation of this reservoir. Before construction of this reservoir, the bed material is mainly composed of medium to coarse sand, as shown by Curve A of this figure for 1959. After operation of this reservoir, the sand is eroded away, left the underneath pebble exposed, and the bed material is abruptly

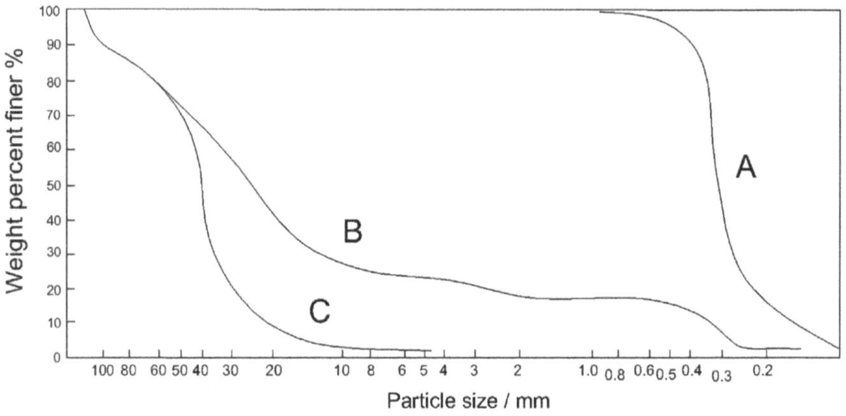

Fig. 9.3 Comparison of bed material gradations on the floodplain at Huangjiagang in the downstream channel of the Danjiangkou Reservoir before and after reservoir operation

coarsened, as shown by Curve B for 1974. As bed material coarsening develops, the bed material gradation is shown by Curve C.

2. Bed Material Coarsening on Pebble Beds

When scour occurs on a pebble bed, the smaller pebble is moved as bed load downstream, and the coarser pebble remains on the bed and finally forms the armor layer. Curves A and B in Fig. 9.4 show the compositions of the pebble on a bed before and after the scour, respectively. As shown in these two curves, the finer particles with the diameter $D < 10\,\text{mm}$ are almost completely eroded away, and massive scour also occurs for the medium coarse particles with $D = 10$–50 mm. Figure 9.5 shows the difference in compositions after scour between the armor layer on the surface and the deeper layers underneath. The bed material in the deeper layers is comprised of a great amount of the medium coarse particles with $D = 10$–50 mm and the finer particles with $D < 10\,\text{mm}$ that fill the voids among the coarser particles. When these particles are exposed on the bed surface, they will be eroded away, with the coarser particles left on the bed to form the armor layer. This type of bed material coarsening not only occurs in the downstream channel of a reservoir, but is also found in natural streams without sediment deposition, by comparing the bed material compositions between the surface layer and underneath layers. The thickness of the coarsened layer as shown in Fig. 9.5 is approximately 0.1 m, and the bed material composition in the layer between 0.1 and 0.4 m deep is evidently finer. Also, the comparison of the bed material gradations between the surface layer and the underneath layers, as shown in Fig. 9.5, is similar to the difference between curve A and curve B in Fig. 9.4.

3. Bed Material Coarsening on Pebble-Sand Beds

For this case, the sand with $D < 2\,\text{mm}$ and the gravel with $D = 2$–10 mm are eroded away, resulting in the pebble being exposed. As this process continues, an

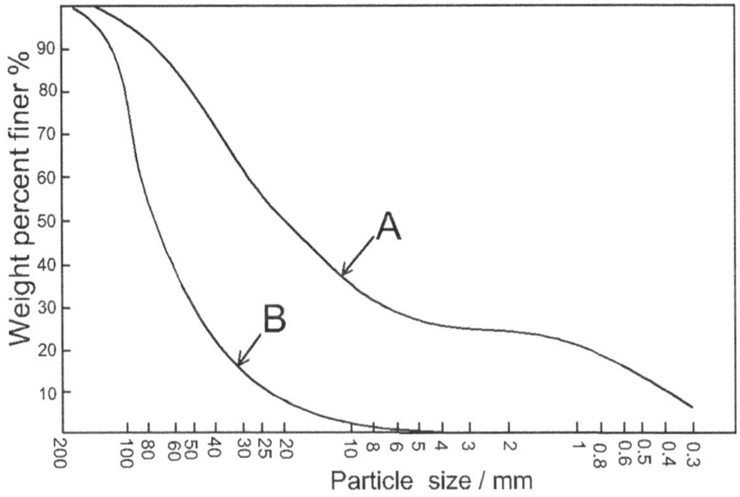

Fig. 9.4 Comparison of the size distributions of pebble on the bed before and after scour

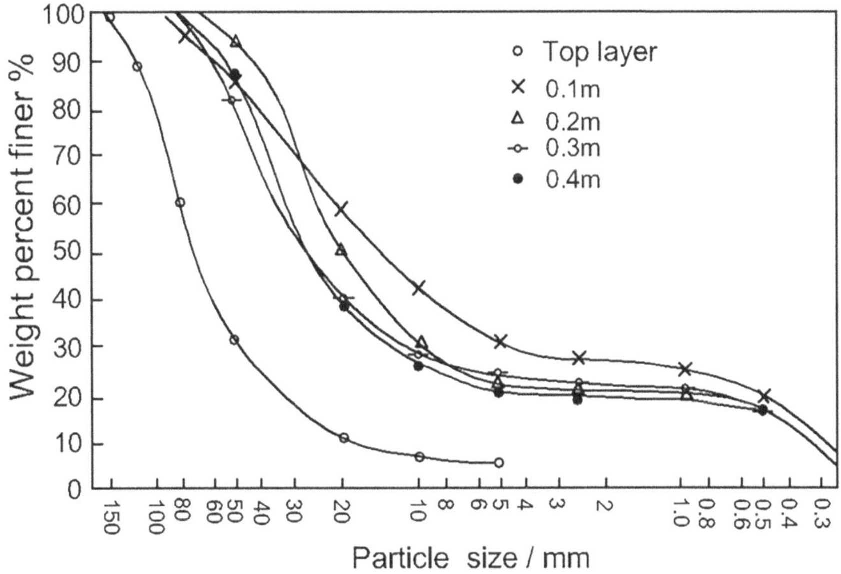

Fig. 9.5 Comparison of the size distributions of bed material between the top layer and deep layer of Han River

armor layer, which is mainly composed of pebbles, is finally formed, as shown in Fig. 9.6. Curve A in this figure shows the bed material gradation of a pebble-sand bed before the construction of a reservoir, which can be characterized as a "recliner-type" distribution. The original bed material is mainly composed of the sand with $D < 1$ mm and the pebble with $D > 10$ mm, with a very small proportion of the coarse sand and pebble with $D = 1$–10 mm. The composition of the top bed layer after scour is shown in curve C. As shown in this curve, the sand and pebble with $D < 10$ mm are almost completely eroded away, and a significant amount of the small pebble with $D = 10$–50 mm is also eroded. It is noted that curve *C* is the composition of the top bed layer, or the gradation of the armor layer. The bed material gradation of the deeper layers underneath the armor layer is shown in curve *B*, which is finer than the armor layer. This indicates that bed material coarsening on a pebble-sand bed typically occurs only at the top layer on the bed. The armor layer shown in Fig. 9.6 is similar to that as shown in Fig. 9.4. However, the quantities of bed material that are eroded away before the armor layers formed are significantly different between these two cases, with a greater amount of bed material being eroded in the case as shown in Fig. 9.6.

4. Bed Material Coarsening on Sand Beds

An example of bed material coarsening on a sand bed is shown in Fig. 9.7. This figure shows the variation of the bed material compositions averaged within the reach between Xiangyang and Yicheng of the Han River. As shown in this figure,

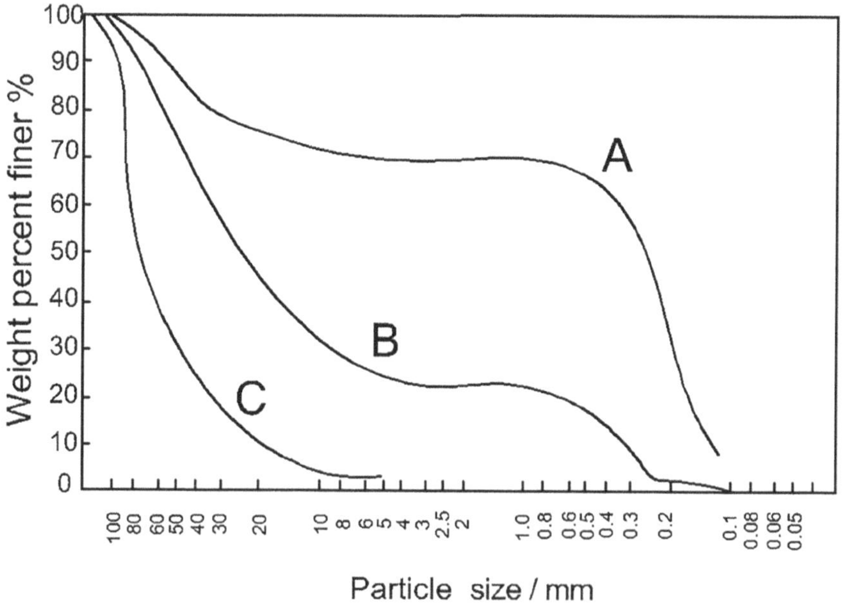

Fig. 9.6 Diagram for bed material coarsening on a pebble-sand bed

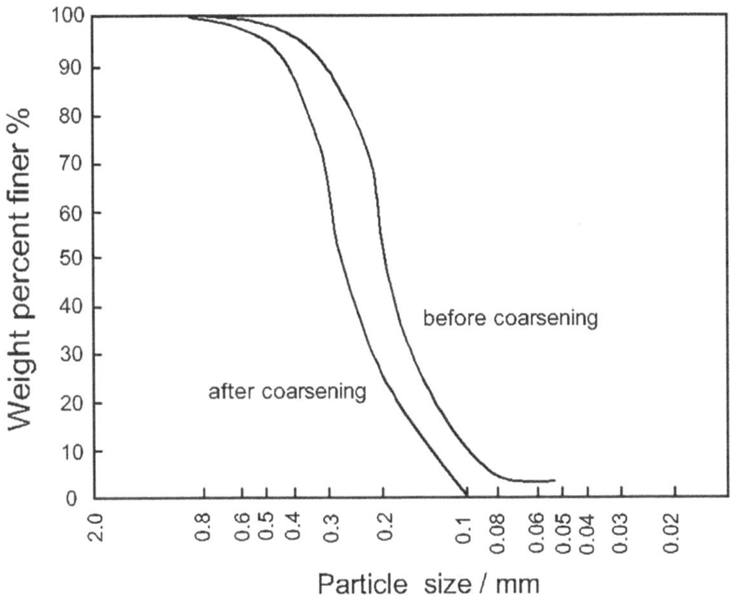

Fig. 9.7 Comparison of bed material gradations on a medium to fine sand bed before and after bed material coarsening

noticeable bed material coarsening occurred in this reach, and a significant amount of the finer bed material with $D < 0.2$–0.3 mm was eroded away.

5. Bed Material Coarsening Caused by Deposition of the Gravel and Pebble from the Bed of the Upper Reach

After the median to fine sand on the bed is eroded away during the clear-water scour periods, the small gravel and pebble of the upper reach are also moved by the water towards the lower reach in the form of bed load. During the movement, the bed load constantly contacts with the bed. As the bed load is transported to the lower reach, it may become the bed material of this reach. It may then stop movement for a sufficiently long time if it is significantly coarser than the local bed material. As sufficient gravel and pebble accrete, bed material coarsening will occur and a coarsened layer will be finally formed in this lower reach. Figure 9.8 shows the bed material gradation in the left branch of the Han River at Taipingdian. As shown in curve A, the bed material before the construction of the Danjiangkou Reservoir is composed of sand and a small amount of gravel, without coarser material with $D > 10$ mm. After a long-term clear-water scour downstream this reservoir, the pebble on the upper reach bed was continuously moved to the lower reach, and finally deposited in the left branch at Taipingdian due to the impact of local channel condition. The bed material gradation after the long-term scour is shown in curve B of Fig. 9.8. Comparing curves A and B, an obvious bed material coarsening can

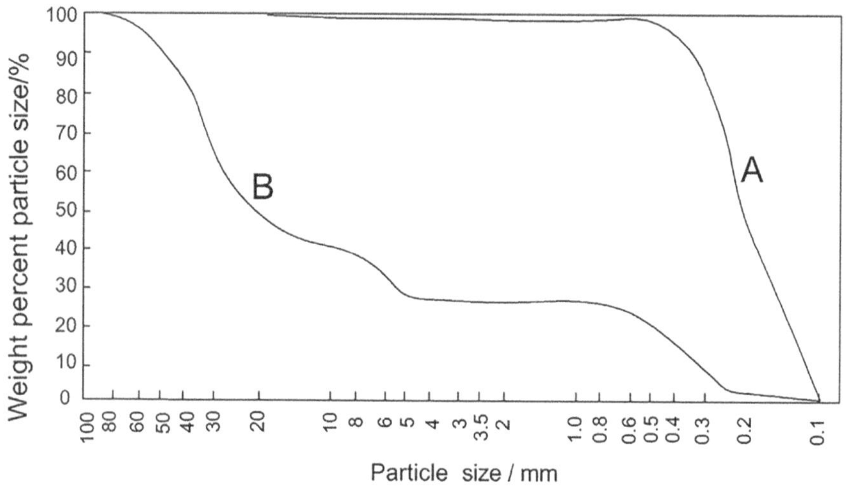

Fig. 9.8 Diagram for coarsening associated from sediment exchange for Taipingdian station of the Han River

be observed. Due to the pebble deposition, the navigation condition in this reach is worsened and the local dredging is needed to maintain this navigation channel.

A similar example can be found from the variation of the bed material composition at the Tiexie station at the downstream reach of the Yellow River during the clear-water release period of the Sanmenxia Reservoir. The original bed material at this Tiexie reach is composed of medium to fine sand, and based on the geological survey, the gravel and pebble are buried approximately 15 m below the bed surface. The depth of the clear-water scour is only a couple of meters, which should not expose the gravel and pebble that are buried 15 m deep. However, due to the continuously movement of the gravel and pebbles from the upper reach, a small amount of gravel and even some pebble have been found in this reach, and the maximum diameter of these particles is 59.5 mm. This accelerates the bed material coarsening in this reach. Also, as shown in Fig. 9.8, the coarse sand and gravel with $D > 1$ mm were also found in the Xiangyang-Yicheng reach of the Han River after the operation of the Danjiangkou Reservoir. These coarser particles should also be moved from the upper reach to this reach in the form of bed load.

6. Bed Material Coarsening Caused by the Deposition of the Coarser Suspended Load from the Upper Reach

During a scour, the suspended load that is eroded from the bed is compatible with the local bed material composition. However, when this suspended load is transported to the lower reach, it may become incompatible with the bed material, resulting in deposition of the coarser suspended load and erosion of local finer bed material in this lower reach. As a result, the suspended load becomes finer, and the bed material become coarser in this lower reach. Table 9.1 lists the difference in the annual fractional sediment transport quantities in 1974 between the Huangzhuang station

Table 9.1 Fractional sedimentation quantities in 1974 within Huangzhuang–Xiantao reach of the Han River

Diameter (mm)	< 0.01	0.01–0.025	0.025–0.05	0.05–0.1	0.1–0.25	0.25–0.5	Total
Quantity(10^4 tons)	– 287	– 268	– 190	– 160	20	132	– 753

and the downstream Xiantao station of the Han River [9], or the annual fractional sedimentation quantities in the reach between Huangzhuang and Xiantao. As listed in this table, the total scour quantity in this reach was 7.53 million tons in 1974. However, while 9.05 million tons of finer sediment with $D < 0.1$ mm was scoured, 1.52 million tons of coarser sediment with $D > 0.1$ mm deposited in this reach. This sediment exchange caused the bed material to become coarser in this reach. The bed material coarsening in this reach was not only caused by the scour of this reach, resulting in a decreased proportion for finer particles and in an increased proportion for coarser particles in the bed material. It was mainly caused by the deposition of the 1.52×10^6 tons of the coarser particles that came from the upstream. This type of bed material coarsening can be referred to as the coarsening caused by sediment exchange. Because of this sediment exchange, while the scour in the reach between Huangzhuang and Maliang was insignificant before 1980, bed material coarsening can still be observed, as shown in Fig. 9.9.

It is noted that the bed material coarsening caused by sediment exchange not only occurs during a scour. It may also occur under the approximate equilibrium condition or during a deposition [9]. As an example, the data measured in 1975 at the Huangzhang to Xiantao stations of the Han River that are located downstream of the Danjiangkou Reservoir is analyzed as follows.

During the 1975 August flood event, the tributaries of the Han River near Xiangyang, including the South River and the Tangbai River, brought a significant amount of sediment into the Han River. The annual sediment transport quantity at Huangzhuang was 58.90 million tons. The contribution from the tributaries was 29.70 million tons, the sediment release from the reservoir was 1.69 million tons, and the quantity of sediment eroded from the channel upstream of Huangzhang was 27.51 million tons, resulting in a mild deposition in the amount of 0.76 million tons. Based on the data of the fractional sedimentation quantities in this reach, as shown in Table 9.2, exchange occurred between the coarser and the finer sediment particles. While 8.88 million tons of finer sediment with $D < 0.05$ mm was scoured, 9.64 million tons of coarser sediment with $D > 0.05$ mm deposited. While a mild deposition occurred in the reach, bed material coarsening still occurred due to the exchange between the coarser and the finer sediment particles. This reach is approximately 217 km long, and the main channel is approximately 300 m wide. Assuming a dry density of 1.3 tons/m^3 for the bed material, the deposited coarser sediment in the amount of 9.64 million tons is equivalent to 0.134 m in depth. It is noted that 27.51 million tons out of the 58.90 million tons of suspended load that passed the Huangzhuang station was the coarser sediment eroded from the channel upstream of

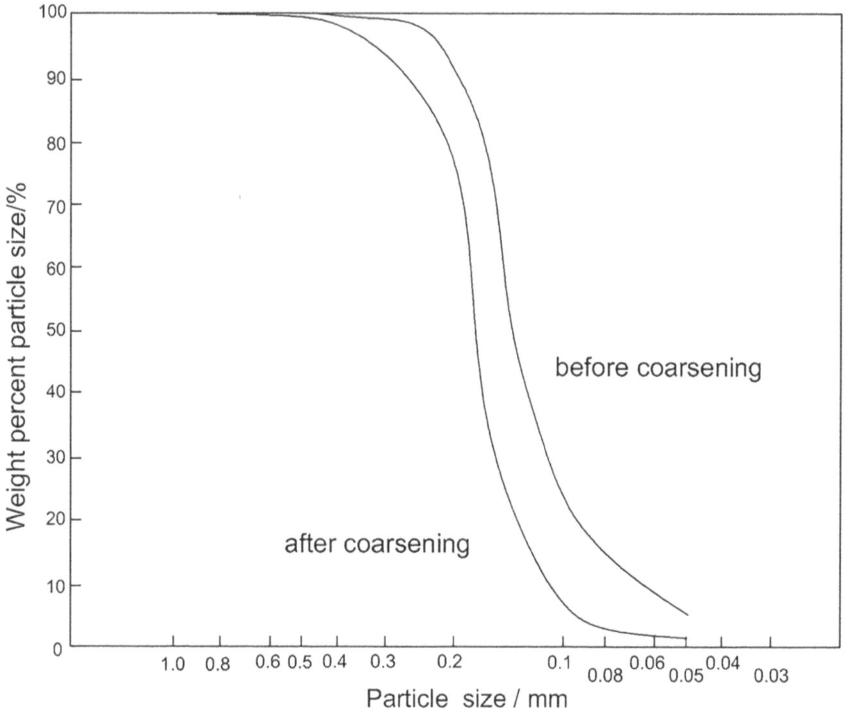

Fig. 9.9 Comparison of bed material gradations on a medium to fine sand bed before and after the coarsening caused by sediment exchange

this station. When the coarser sediment was transported to the lower reach that had finer bed material, the relatively coarser suspended load and the finer bed material were incompatible, resulting intensive sediment exchange. Hence, if the coarse sediment from the upstream is transported to a reach that has relatively finer bed material and that is with a mild deposition, bed material coarsening will occur as the result of the exchange between the coarser and the finer sediment particles.

As another example of this type of bed material coarsening, Table 9.3 lists the fractional sedimentation quantities in 1971 within the reach between Aishan and Lijin of the Yellow River that is downstream of the Sanmenxia Reservoir [9]. A

Table 9.2 Fractional sedimentation quantities in 1975 within Huangzhuang–Xiantao reach of the Han River

Diameter (mm)	< 0.01	0.01–0.025	0.025–0.05	0.05–0.1	0.1–0.25	0.25–0.5	Total
Quantity (10^6 tons)	− 1.63	− 2.49	− 4.76	0.74	4.67	4.23	0.76

Table 9.3 Fractional sedimentation quantities during 1971 flood season in Aishan–Lijin reach of the Yellow River

Diameter(mm)	< 0.005	0.005–0.01	0.01–0.025	0.025–0.05	0.05–0.1	0.1–0.25	Total
Quantity(10^6 tons)	– 4.4	– 3.9	– 38.3	– 20.5	134.5	13.6	81

significant deposition occurred within this reach, with the net deposition of approximately 81 million tons. While 67.1 million tons of finer sediment with $D < 0.05$ mm was scoured, 148.1 million tons of coarser sediment with $D > 0.05$ mm deposited. The length of this reach is approximately 262 km, with the channel width of approximately 500 m. Assuming sediment was deposited within the main channel only, the deposition depth would be 1.10 m, and thus the bed material was almost composed of the coarser sediment with $D > 0.05$ mm only. Of course, sediment deposition is distributed non-uniformly within the stream, and thus the actual bed material coarsening may be less severe as estimated.

Bed material coarsening is an exception of the general conditions that would occur during a mild deposition. More generally, bed material become finer during a deposition. Bed material coarsening may occur during a mild deposition. However, it typically does not last for a long time or occur in a large portion of a river. This type of bed material coarsening will adjust the transport capacity, and alleviate the magnitude of scour or deposition along the flow path.

It is noted that these six types of bed material coarsening, as discussed above, are not isolated from each other. Instead, the formation of bed material coarsening in reach is typically composed of several types of coarsening. These six types of coarsening are differentiated based on phenomena. Based on the characteristics of sediment transport that cause bed material coarsening, it can be divided into three types of bed material coarsening. They include the coarsening caused by scour in the form of suspended load, the coarsening caused by scour in the form of bed load, and the coarsening caused by sediment exchange. Among the six phenomena for bed material coarsening, the first and the fourth phenomena, and the initial stage of the third phenomenon are caused by the scour of the bed material in the form of suspended load. The second phenomenon, the second stage of the third phenomenon, and the final stage of the first phenomenon are formed through the scour of bed material in the form of bed load. The coarsening through sediment exchange is caused by the exchange between the coarser sediment particles from the upstream and the finer particles on the local bed. The fifth phenomenon for bed material coarsening is caused by the exchange between the coarser bed load from the upstream and the local finer bed material. The sixth phenomenon is caused by the exchange between the coarser suspended load from the upstream and the local finer bed material. The mechanisms of these three types of bed material coarsening will be analyzed in this chapter.

Also, Fig. 9.10 shows the bed material gradations before and after bed material coarsening at central bar downstream of Xincheng of the Han River. The thickness of the coarsened layer on this sand bed was shallower than 0.2 m. Bed material

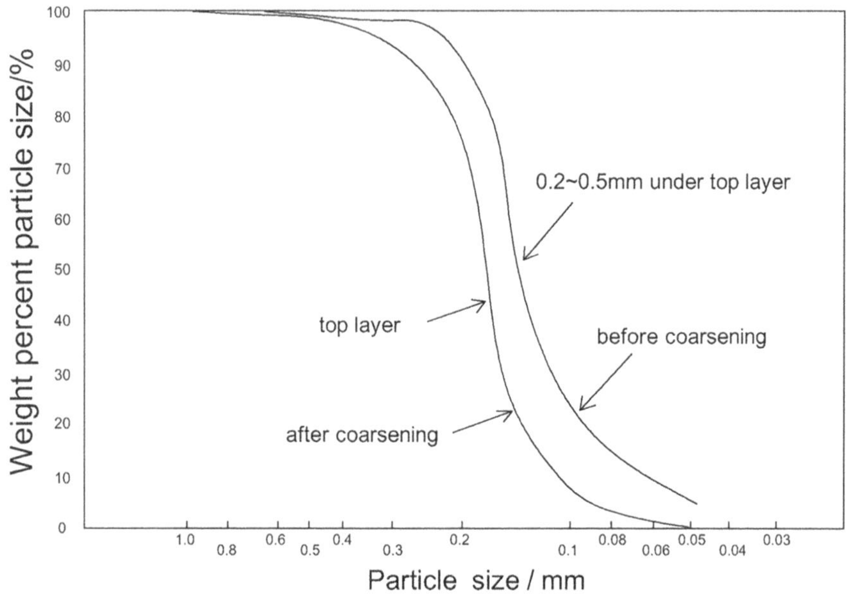

Fig. 9.10 Bed material gradations at central bar downstream of Xincheng of the Han River

coarsening did not occur in the layer between 0.2 and 0.5 m, resulting in a finer bed material composition in this deeper layer.

9.2.2 Mechanisms for Bed Material Coarsening

The six phenomena of bed material coarsening are described in the previous section. The general mechanisms for bed material coarsening are analyzed as follows.

Over the years, bed material coarsening is considered to be controlled by the relative magnitudes of the critical flow velocity for sediment entrainment and of the actual flow velocity. If the entrainment flow velocity for a given bed material particle is larger than the flow velocity, this particle will stop moving and remain on the bed. Otherwise, it will be eroded away. However, the entrainment flow velocity is not the absolutely criterion to determine whether a sediment particle move or stop moving. Based on field experiments and sediment transport theories, when sediment entrainment occurs on the bed surface, individual or a few coarser particles always move [13]. In other words, it is not completely consistent with the reality to use the difference between the entrainment velocity and the actual flow velocity to determine the bed material coarsening. Also, even the bed material can be completely entrained, the gradation of the material remaining on the bed is still coarser than the composition of the material that is eroded away. As the scour continues, the sediment particles left on the bed will become coarser and coarser. For the case with the bed material being partially entrained, not all particles that are finer than the critical grain size for entrainment are eroded away. The composition of the eroded material is still finer than the corresponding portion of the original bed material, resulting in coarsening of bed material. It is only an approximation, not a correct illustration for the mechanisms for scour of bed load, to assume all particles finer than the entrainment particle size, which is determined with the formulas for sediment entrainment, are eroded away. As for the movement of the fine particles that fill the voids among the coarser particles it is not controlled by the entrainment flow velocity.

In reality, for the bed with non-uniform sediment, the critical flow velocities for various movement states, such as the entrainment velocity and the critical velocity for sediment to cease motion, and the corresponding transition probabilities are different between the coarser and the finer sediment particles. Also, the single-step distance and the time for sediment to be at rest on the bed (sediment resting time) are also different [12]. As a result, the proportion of moving sediment vary with size classes, and the composition of the transported sediment is different from the bed material gradation. The composition of the transported sediment is finer than the bed material. If the sediment inflow is insufficient, scour will occur on the bed, with a greater amount of finer sediment particles being eroded than the coarser particles, resulting in bed material coarsening. Also, even when the total sediment transport is approximately equilibrium, if the particles that fall on the bed are coarser than the original bed material, the exchange between the coarser and the finer particles will occur, also resulting in bed material coarsening. In general, the exchange between the coarser and the finer sediment particles is the basic mechanism for bed material coarsening. This is true for all three types of bed material coarsening that are caused by the scour of suspended load, the scour of bed load, and the sediment exchange, respectively. In fact, these three types of bed material coarsening are not absolutely independent from each other. Instead, they are closely related. They are differentiated in this chapter just for the purpose of highlighting the different characteristics associated with each type of bed material coarsening.

In order to further illustrate the mechanisms for bed material coarsening, analysis is conducted in the following based on the theory for sediment exchange on the bed that was developed by the author. Assuming the sediment particles within the bed surface layer have three states of movement: at rest, moving in the form of bed load, and moving in the form of suspended load. The bed surface layer is defined as the layer from the bed upward to the maximum hopping height of the bed load. The movement of bed load include rolling and hopping. Rolling and hopping were differentiated based on the movement mechanisms, resulting in four states for sediment movement [12]. To be consistent with the general practice, the rolling particles and the hopping particles are combined and are referred to as bed load, with the motion parameters being determined with the combined results of the rolling state and of the hopping state [14, 15]. In total, there are six kinds of intensities for sediment exchange on the bed. The numbers of sediment particles that change from the resting state (i.e., from the bed material) to bed load and to suspended load, respectively, on a unit area of the bed surface per unit time, or the intensities for the transitions from the resting state to the bed load movement and to the suspended load movement, respectively, are

$$\lambda_{1.b.l} = n_{0.l}\frac{\beta_{1.b.l}}{t_{0.b.l}} = n_{0.l}\frac{(1-\beta_l)\varepsilon_{1.l}}{t_{0.b.l}} \tag{9.25}$$

$$\lambda_{1.4.l} = n_{0.l}\frac{\beta_{1.4.l}}{t_{0.4.l}} = n_{0.l}\frac{\beta_l}{t_{0.4.l}} \tag{9.26}$$

where $\beta_{i.j.l}$ is the transition probability for the sediment of the lth size class to change from state i to state j; the subscripts "1" denotes the rest state, "b" denotes the bed load movement, "4" denotes the suspended load movement, and l is the index of sediment size class; $\varepsilon_{1.l}$ the probability for the sediment to change from the rest state to rolling; β_l the probability for sediment to change from the rest state to suspension, or the incipient suspension probability; $(1-\beta_l)\varepsilon_{1.l}$ the probability for the sediment to change form the rest state to rolling but without being suspended, or the probability to change to bed load; $t_{0.b.l}$ and $t_{0.4.l}$ are the time for the sediment, which changes from the rest state to bed load and to suspended load, respectively, to stay off the bed at a height equivalent to the sediment particle diameter, since only after this time period the particle would leave the bed then the other particles can be moved, n_0 the total number of sediment particles that are at rest on the bed, and the number of particles in the lth size class that are at rest on a unit area of the bed surface is

$$n_{0.l} = n_0 R_{1.l} P_{1.l} = n_l R_{1.l} = \frac{m_0 R_{1.l} P_{1.l}}{\frac{\pi}{4} D_l^2} \tag{9.27}$$

where $P_{1.l}$ is the bed material gradation, $R_{1.l}$ the probability for the sediment of the lth size class to be at rest on the bed [12], which is

$$R_{1.l} = \frac{(1-\varepsilon_{0.l})(1-\varepsilon_{4.l})}{1+(1-\varepsilon_{0.l})(1-\varepsilon_{4.l})-(1-\varepsilon_{1.l})(1-\beta_l)} = \frac{(1-\varepsilon_{0.l})(1-\varepsilon_{4.l})}{\Omega_0} \tag{9.28}$$

For coarser sediment particles, neglecting the cohesive force and the additional pressure of film water, then $\varepsilon_{1.l} = \varepsilon_{0.l}$, $\beta_l = \varepsilon_{4.l}$, and thus $\Omega_0 \approx 1$. It is noted that $n_0 R_{1.l}$ is the number of particles in the lth size class that are at rest on a unit area of the bed surface, while the number of particles in the lth size class per unit area of the bed surface is $n_l = \frac{m_0 P_{1.l}}{\frac{\pi}{4} D_l^2}$ in which m_0 is the stationary sediment compaction coefficient. As proved by Han and He [12], the proportion of the area on the bed surface that is occupied by a given size class is equivalent to the weight percent of this size class.

The number of particles that change from bed load to bed material per unit area per unit time is

$$\lambda_{b.1.l} = \beta_{b.1.l} K_{b.l} \mu_{b.l} U_{b.l} = (1-\varepsilon_{0.l})(1-\varepsilon_{4.l}) K_{b.l} \mu_{b.l} U_{b.l} \tag{9.29}$$

and the number of particles that change from suspended load to bed material is

$$\lambda_{4.1.l} = \beta_{4.1.l} K_{4.l} \mu_{4.l} U_{4.l} = (1 - \varepsilon_{0.l})(1 - \varepsilon_{4.l}) K_{4.l} \mu_{4.l} U_{b.l} \tag{9.30}$$

where $\varepsilon_{0.l}$ is the probability for the sediment not to cease movement, $\varepsilon_{4.l}$ the suspension probability, $(1 - \varepsilon_{0.l})$ the cease-movement probability; $K_{b.l}$ and $K_{4.l}$ the number of particles within a water column with a unit area base that move as bed load and as suspended load, respectively; $\mu_{b.l} = 1/L_{b.l}$ and $\mu_{4.l} = 1/L_{4.l}$, where $L_{b.l}$ and $L_{4.l}$ the single-step distances for bed load and for suspended load, respectively; $U_{b.l}$ and $U_{4.l}$ the average velocities for bed load and for suspended load, respectively; $\mu_{b.l} U_{b.l} = U_{b.l}/L_{b.l} = 1/t_{b.l}$, and $\mu_{4.l} U_{4.l} = U_{4.l}/L_{4.l} = 1/t_{4.l}$, in which $t_{b.l}$ and $t_{4.l}$ the single-step travel time for bed load and for suspended load, respectively. As shown in Eq. (9.29), after the $K_{b.l}$ bed load particles within a water column with a unit area base travel a single-step distance in a time interval of $t_{b.l}$, a portion of these bed load particles will stop movement with a probability of $\beta_{b.1.l}$, or with the amount of $K_{b.l}/t_{b.l} = \lambda_{b.1.t}$, and will then change to bed material. Therefore, $\lambda_{b.1.l}$ is also called the number of bed load particles within a water column with a unit area base that change to bed material per unit time. A similar meaning can be observed from Eq. (9.30) for suspended load.

Based on Eqs. (9.25), (9.26), (9.29), and (9.30), the number of particles in the lth size class that deposit on the bed within a time interval of Δt is

$$I_{b.n.j} \Delta t = \left[\beta_{b.1.l} K_{b.l} \mu_{b.l} U_{b.l} + \beta_{4.1.l} K_{4.l} \mu_{4.l} U_{4.l} - \frac{\beta_{1.b.l} n_{0.l}}{t_{0.b.l}} - \frac{\beta_{1.4.l} n_{0.l}}{t_{0.4.l}} \right] \Delta t \tag{9.31}$$

These deposited particles cause the variation of the bed material gradation.

It can be seen that the variation of the bed material composition is very complicated. It depends on the transition probabilities $\beta_{b.1.l}$, $\beta_{4.1.l}$, $\beta_{b.1.l}$, and $\beta_{1.4.l}$, the number of particles that move as bed load $K_{b.l}$, the number of particles that move as suspended load $K_{4.l}$, the number of particles that are at rest on the bed surface $n_{0.l}$, the entrainment time $t_{0.b.l}$ and $t_{0.4.l}$, the single-step distances $L_{b.l} = 1/\mu_{b.l}$ and $L_{4.l} = 1/\mu_{4.l}$, and the average sing-step velocities $U_{b.l}$ and $U_{4.l}$. It is noted that

$$K_{b.l} U_{b.l} = q_{b.n.l} = \frac{q_{b.l}}{\gamma_s \frac{\pi}{6} D_l^3} \tag{9.32}$$

$$K_{4.l} U_{b.l} = q_{s.n.l} = \frac{q_{s.l}}{\gamma_s \frac{\pi}{6} D_l^3} \tag{9.33}$$

where $q_{b.l}$ and $q_{s.l}$ are the fractional transport rates for the lth size class in weight, per unit width for bed load and for suspended load, respectively, $q_{b.n.l}$ and $q_{s.n.l}$ the fractional transport rates, in number of particles, per unit width for bed load and for suspended load, respectively. As $K_{b.l}$ is the number of bed load particles of the lth size class within a water column with a unit area base, $K_{b.l} U_{b.l} = K_{b.l} L_{b.l}/t_{b.l}$ is the number of particles within a water column with a bottom surface area of 1 $1 \times L_{b.l}$ that move per unit time, which is the fractional bed load transport rate, in number of particles, per unit width, or $q_{b.n.l}$. A similar reasoning can be made for the suspended load transport rate per unit width.

Plugging Eqs. (9.29), (9.30), (9.32), and (9.33) in Eq. (9.31), and multiplying the weight of a single particle that is $\gamma_s \frac{\pi}{6} D_l^3$, one obtains

$$\begin{aligned} I_{b.l}\Delta t = & \left[\beta_{b.1.l} q_{b.l}\mu_{b.l} + \beta_{4.1.l} q_{4.l}\mu_{4.l} - \frac{2}{3} m_0 \gamma_s D_l \frac{R_{1.l} P_{1.l} \varepsilon_{1.l}(1-\beta_l)}{t_{0.b.l}} \right. \\ & \left. - \frac{2}{3} m_0 \gamma_s D_l \frac{R_{1.l} P_{1.l} \beta_l}{t_{0.4.l}} \right] \Delta t \\ = & \left[(1-\varepsilon_{0.l})(1-\varepsilon_{4.l})(P_{b.l} q_b \mu_{b.l} + P_{4.l} S q \mu_{4.l}) \Delta t \right. \\ & \left. - \frac{2}{3} m_0 \gamma_s D_l R_{1.l} P_{1.l} \left(\frac{\varepsilon_{1.l}(1-\beta_l)}{t_{0.b.l}} + \frac{\beta_l}{t_{0.4.l}}\right) \right] \Delta t \end{aligned} \tag{9.34}$$

where γ_s' is the specific weight of a single sediment particle, $P_{b.l}$ the size distribution of bed load, $P_{4.l}$ the size distribution of suspended load, q_b the bed load transport rate per unit width, q the flow discharge per unit width, and S the sediment concentration. Equation (9.34) represents the net fractional sedimentation quantity of the lth size class on a unit area of the bed surface within Δt. A positive quantity denotes deposition, and a negative quantity denotes scour. For $I_{b.l} = 0$, a weak equilibrium condition is achieved. For this case, while the quantity and the composition of the sediment on the bed remain constant, a non-equilibrium exchange may still occur between the bed load and the suspended load. For a strong equilibrium exchange between the bed load and the bed material,

$$\lambda_{b.1.l} = \lambda_{1.b.l} \tag{9.35}$$

To simplify the analysis, let's assign $\Omega_0 \approx 1$ in Eq. (9.28), which is generally acceptable for bed load. For $\Omega_0 \neq 1$, a similar conclusion can be derived, with a more in-depth illustration for a few aspects. Considering the terms of Eq. (9.34) that are related to $\lambda_{b.1.1}$ and $\lambda_{1.b.l}$, and plugging Eq. (9.28) for $R_{1.l}$ into Eq. (9.34) yield

$$(1-\varepsilon_{1.l})(1-\varepsilon_{4.l})\left[\mu_{b.l} P_{b.l}^* q_b^* - \frac{2}{3} m_0 \gamma_s D_l \varepsilon_{1.l}(1-\beta_l)\frac{P_{1.l}}{t_{0.b.l}}\right] = 0 \tag{9.36}$$

$$P_{b.l}^* q_b^* = \frac{2}{3} m_0 \gamma_s D_l \frac{\varepsilon_{1.l}(1-\beta_l) P_{1.l}}{\mu_{b.l} t_{0.b.l}} \tag{9.37}$$

where q_b^* and $P_{b.l}^*$ are the bed load transport rate and corresponding size distribution, respectively, when the exchange between bed load and bed material is in a strong equilibrium condition. Hence, q_b^* is referred to as the bed load transport capacity, and $P_{b.l}^*$ is called the size distribution for bed load transport capacity. Similarly, when a strong equilibrium exchange occurs between suspended load and bed material,

$$\lambda_{4.1.l} = \lambda_{1.4.l}$$

Considering the terms of Eq. (9.34) that are related to $\lambda_{4.1.l}$ and $\lambda_{4.1.l}$, then

$$(1-\varepsilon_{1.l})(1-\varepsilon_{4.l})\left(\mu_{4.l}qP_{4.l}^{*}S^{*}-\frac{2}{3}m_{0}\gamma_{s}D_{l}\beta_{l}\frac{P_{1.l}}{t_{0.4.l}}\right)=0 \tag{9.38}$$

or,

$$P_{4.l}^{*}S^{*}=\frac{2}{3}m_{0}\gamma_{s}D_{l}\frac{\beta_{l}P_{1.l}}{\mu_{4.l}t_{0.4.l}q} \tag{9.39}$$

where S^* and $P_{4.l}^*$ are the suspended load concentration and the corresponding size distribution, respectively, when the exchange between suspended load and bed material is in a strong equilibrium condition. They are called the capacity for carrying suspended load and the size distribution for the capacity for carrying suspended load, respectively. Plugging Eqs. (9.28), (9.37), (9.39) in Eq. (9.34) yields

$$\begin{aligned}I_{b.l}\Delta t=&\big[(1-\varepsilon_{0.l})(1-\varepsilon_{4.l})\mu_{b.l}\big(P_{b.l}q_{b}-P_{b.l}^{*}q_{b}^{*}\big)+\\&(1-\varepsilon_{0.l})(1-\varepsilon_{4.l})\mu_{4.l}q\big(P_{4.l}S-P_{4.l}^{*}S^{*}\big)\big]\Delta t\end{aligned} \tag{9.40}$$

If the bed material is too coarse to be suspended, or only the bed load exists, Eq. (9.40) is simplified to

$$I_{b.l}\Delta t=(1-\varepsilon_{0.l})(1-\varepsilon_{4.l})\mu_{b.l}\big(P_{b.l}q_{b}-P_{b.l}^{*}q_{b}^{*}\big)\Delta t \tag{9.41}$$

If the bed material is so fine that all the particles are suspended load after being entrained, or the proportion of bed load is negligible, then

$$I_{b.l}\Delta t=(1-\varepsilon_{0.l})(1-\varepsilon_{4.l})\mu_{4.l}q\big(P_{4.l}S-P_{4.l}^{*}S^{*}\big)\Delta t \tag{9.42}$$

The following complicated mechanisms for bed material coarsening can be observed from the above equations:

(1) The bed material coarsening is impacted by various hydraulic and sediment factors. As shown in Eqs. (9.37), (9.39) and (9.40), the variation of bed material depends on the probability for the sediment not to be entrained $(1-\varepsilon_{1.l})$, the cease-movement probability $1-\varepsilon_{0.l}$, the probability for the sediment not to be suspended $(1-\varepsilon_{4.l})$, the single-step distance for bed load and for suspended load $\mu_{b.l}$, $\mu_{4.l}$, respectively, the flow discharge per unit width q, capacity for carrying suspended load S^*, the bed load transport capacity q_b^*, the corresponding size distributions, the probability for bed material to remain rest $R_{1.l}$, bed material gradation, and the entrainment time. In summary, bed material coarsening depends on four exchange intensities: $\lambda_{1.b.l}$, $\lambda_{1.4.l}$, $\lambda_{b.1.l}$, and $\lambda_{4.1.l}$, or depends on the fractional sedimentation quantities.

(2) Generally, the exchange between bed load and bed material, and the exchange between suspended load and bed material both in the strong equilibrium condition, or

$$\left.\begin{aligned} q_b &= q_b^* \\ P_{b.l} &= P_{b.l}^* \\ S &= S^* \\ P_{4.l} &= P_{4.l}^* \end{aligned}\right\} \tag{9.43}$$

cannot be satisfied in reality [11]. As long as sediment moves and sedimentation occurs, the bed material gradation will vary.

(3) If the bed material is so coarse that no suspended load movement occurs, $\varepsilon_{4.l} = 0$. From Eqs. (9.41) and (9.37)

$$I_{b.l} = (1 - \varepsilon_{0.l})\mu_{b.l}\left(P_{b.l}q_b - P_{b.l}^* q_b^*\right) = (1 - \varepsilon_{0.l})\left[\mu_{b.l}P_{b.l}q_b - \frac{2}{3}m_0\gamma_s D_l \frac{\varepsilon_{1.l}P_{1.l}}{t_{0.b.l}}\right]$$

It is seen that if scour occurs for all size classes, the entrainment probability $\varepsilon_{1.l}$ varies more significantly with size classes than $D_l/t_{0.b.l}$ does. In fact, the variation of $\varepsilon_{1.l}$ can be up to several orders of magnitude for various size classes. As particle size decreases, $\varepsilon_{1.l}$ increases. This indicates that a greater amount of finer particles will be eroded than coarser particles, resulting in bed material coarsening. For an extreme case with clear water discharge, the bed load transport rate q_b is zero, and

$$I_{b.l} = -\frac{2}{3}m_0\gamma_s D_l \frac{\varepsilon_{1.l}(1 - \varepsilon_{0.l})P_{1.l}}{t_{0.b.l}}$$

Hence, a channel scour must occur for this case, with a greater amount of finer particles being eroded than coarser particles. This bed material coarsening is caused by the scour of bed load, as shown in Figs. 9.4 and 9.5. It is noted that even for a very coarse particle, $\varepsilon_{1.l}$ may not equal zero, and thus it may be eroded. On the other hand, a finer particles may not be eroded. This is because

$$1 - \varepsilon_{1.l} = \frac{1}{\sqrt{2\pi}} \int_{-\infty}^{\frac{V_{b.c.1.l} - \overline{V}_b}{\sigma}} \exp\left(-\frac{t^2}{2}\right)dt \tag{9.44}$$

which is not zero even for the case with $-\infty < \overline{V_b} < V_{b.c.1.l}$. Here, $\overline{V_b}$ is the time-averaged bottom flow velocity, and $V_{b.c.1.l}$ is the instantaneous entrainment velocity for the lth size class. Of course, $V_{b.c.1.l}$ is greater than $\overline{V}_b$ for most of the time. As shown in Eq. (9.44), $-\varepsilon_{1.l}-\varepsilon_{1.l}$ does not equal 1 as long as $V_{b.c.1.l} < \infty$. This indicates that the entrainment probability cannot be zero even for a very coarse particle.

(4) For a weak equilibrium movement of pure coarse bed load, $q_b = q_b^*$, but the bed material gradation may still vary. It can be derived from Eq. (9.41) that

$$I_{b.l} = \sum_{l=1}^{l=n} (1 - \varepsilon_{0.l})\mu_{b.l}\left(P_{b.l} - P^*_{b.l}\right) = 0 \tag{9.45}$$

If the sediment inflow is short of finer particles, or $P_{b.l} < P^*_{b.l}$, then $I_{b.l} < 0$, indicating a scour of the finer particles. If coarser particles are over-saturated in the sediment inflow, or $P_{b.l} > P^*_{b.l}$, then $I_{b.l} > 0$, indicting a deposition of these coarser particles. It is true for the case $I_{b.l} = 0$ or the difference between $I_{b.l}$ and zero can be neglected. If the coarser particles do not occur in the bed material, then $P^*_{b.l} = 0$, and thus $I_{b.l}$ is always positive. Hence, these coarser particles will deposit on the bed, resulting in the maximum particle size of the bed material being coarser than the original bed material, as shown in Fig. 9.8. These two cases both belong to the typical bed material coarsening caused by the exchange between bed load and bed material.

(5) A similar analysis can be performed for the bed material coarsening caused by the exchange between suspended load and bed material. For example, if the particles of the sediment influx are sufficiently fine so that the bed load barely exist, the following relation can be derived from Eqs. (9.42) and (9.39):

$$\begin{aligned} I_{b.l} &= (1 - \varepsilon_{0.l})(1 - \varepsilon_{4.l})\mu_{4.l}q\left(P_{4.l}S - P^*_{4.l}S^*\right) \\ &= (1 - \varepsilon_{0.l})(1 - \varepsilon_{4.l})\left(\mu_{4.l}qP_{4.l}S - \frac{2}{3}m_0\gamma_s D_l \frac{\varepsilon_{4.l}P_{1.l}}{t_{0.4.l}}\right) \end{aligned} \tag{9.46}$$

Similar to the scour of bed load, as the particle size of the suspended load decreases, $\varepsilon_{4.l}$ increases, resulting in a greater percentage of the eroded material. As more amount of finer sediment is eroded than the coarser sediment, bed material coarsening occurs. This is true no matter $I_{b.l}$ being negative for all size groups, or only for finer particles. This is the typical bed material coarsening caused by the scour of suspended load, as shown in Fig. 9.7.

The bed material coarsening caused by the exchange between suspended load and bed material can also be illustrated with Eq. (9.46). For a weak equilibrium suspended load transport, $S^* = S$, and thus Eq. (9.46) can be written as

$$I_{b.l} = (1 - \varepsilon_{0.l})(1 - \varepsilon_{4.l})\mu_{4.l}qS\left(P_{4.l} - P^*_{4.l}\right) \tag{9.47}$$

If the coarser particles are over-saturated in the sediment inflow, $P_{4.l} > P^*_{4.l}$, $I_{b.l} > 0$, and deposition occurs for these coarser particles. Correspondingly, the finer particles are relatively insufficient in the sediment inflow $P_{4.l} < P^*_{4.l}$, $I_{b.l} < 0$, and scour occurs for these finer particles. This is the bed material coarsening caused by the exchange between suspended load and bed material, which is characterized by the deposition of coarser particles and the scour of finer particles.

In order to better understand the deposition or scour of suspended sediment and the variation of bed material composition, Let's link Eq. (9.42) with the equation for non-equilibrium sediment transport. In fact, $I_{b.l}$ of Eq. (9.42) represents the quantity

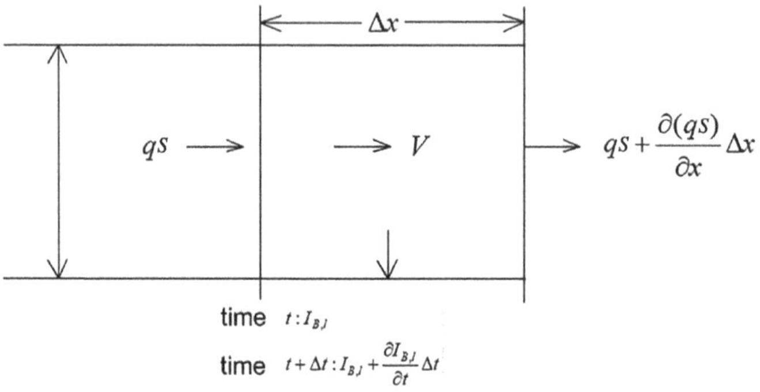

Fig. 9.11 Diagram of suspended load scour

of the sediment that deposits on a unit area of the bed surface per unit time. As shown in Fig. 9.11, the sediment transport quantity at the entrance section is $qP_{4.l}\Delta t$, and the sediment transport quantity becomes $\left(qP_{4.l}S + \frac{\partial(qP_{4.l}S)}{\partial x}\Delta x\right)\Delta t$ after the distance of Δx, and thus the quantity for sediment depositing on the bed is

$$qS\Delta t - (qS + \frac{\partial(qP_{4.l}S)}{\partial x}\Delta x)\Delta t = -\frac{\partial(qP_{4.l}S)}{\partial x}\Delta x\Delta t$$

Also, the quantity of sediment deposited within Δt can be expressed as $\left[(I_{b.l} + \frac{\partial(I_{b.l})}{\partial t}\Delta t) - I_{b.l}\right]\Delta x$. These two expressions are equivalent, and thus

$$-\frac{\partial(qP_{4.l}S)}{\partial x}\Delta x\Delta t = \frac{\partial I_{b.l}}{\partial t}\Delta x\Delta t$$

or,

$$\frac{\partial(qP_{4.l}S)}{\partial x} = -\frac{\partial I_{b.l}}{\partial t}$$

For a sufficiently small Δt, the temporal variation term can be neglected. Or, taking the mean value of $I_{b.l}$ within Δt, then [5]:

$$\begin{aligned}\frac{d(qP_{4.l}S)}{dx} &= (1-\varepsilon_{0.l})(1-\varepsilon_{4.l})\mu_{4.l}\left(-P_{4.l}S + P_{4.l}^{*}S^{*}\right)\\ &= \frac{\alpha_l\omega_l}{q}\left(P_{4.l}^{*}S^{*} - P_{4.l}S\right)\end{aligned} \tag{9.48}$$

where

$$\alpha_l = (1-\varepsilon_{0.l})(1-\varepsilon_{4.l})\mu_{4.l}\frac{q}{\omega_l} \tag{9.49}$$

is called the saturation recovery coefficient. For a more detailed analysis, let use the Eq. (7.137) [8, 16] to determine the size distribution for carrying-sediment capacity:

$$P_{4.l}^{*}S^{*}(\omega^{*}) = P_{4.l}P_{4.1.l}S + P_{4.2}P_{4.l.2}S\frac{S^{*}(l)}{S^{*}(\omega_{1}^{*})} + \left[1 - \frac{P_{4.l}S}{S^{*}(\omega_{1})} - \frac{P_{4.2}S}{S^{*}(\omega_{1}^{*})}\right]P_{4.l.1}^{*}S^{*}(\omega_{1}^{*}) \tag{9.50}$$

The condition of $1 - \frac{P_{4.l}S}{S^{*}(\omega_{1})} - \frac{P_{4.2}S}{S^{*}(\omega_{1}^{*})} > 0$ must be satisfied for this case. Otherwise, setting $1 - \frac{P_{4.l}S}{S^{*}(\omega_{1})} - \frac{P_{4.2}S}{S^{*}(\omega_{1}^{*})} = 0$, the bed material cannot be eroded, and thus the fractional carrying-sediment capacity will decrease along the flow path, resulting in sediment deposition. Hence, Eq. (9.50) only applies to the case of scour and the case of mild deposition. The parameters in Eq. (9.50) are defined as follows:

$$P_{4.1} = \sum_{l=1}^{k} P_{4.l} \tag{9.51}$$

$$P_{4.2} = \sum_{l=k+1}^{n} P_{4.l} \tag{9.52}$$

$$P_{4.l.1} = \begin{cases} 0\,(l \geq k+1) \\ \dfrac{P_{4.l}}{P_{4.1}}\,(l < k) \end{cases} \tag{9.53}$$

$$P_{4.l.2} = \begin{cases} 0\,(l \geq k+1) \\ \dfrac{P_{4.l}}{P_{4.2}}\,(l < k) \end{cases} \tag{9.54}$$

in which k is the index of the coarsest size class of the fine particles that does not take part in the exchange with the bed material.

$$\omega_{1} = \left(\sum P_{4.l.1}\omega_{l}^{m}\right)^{\frac{1}{m}} \tag{9.55}$$

ω^{*} is the sediment fall velocity used in the formula for carrying-sediment capacity

$$S^{*}(\omega^{*}) = \kappa\left(\frac{V^{3}}{gh\omega^{*}}\right)^{m} \tag{9.56}$$

which is determined with

$$\omega^{*} = \left(\sum P_{4.l}^{*}\omega_{l}^{m}\right)^{\frac{1}{m}} \tag{9.57}$$

For non-equilibrium suspended load transport, $P_{1.l.1}$ is the actual bed material gradation, and $P_{1.l}$ is the effective bed material gradation. Also,

$$P^*_{4.l.1} = \frac{S^*(\omega_l)}{S^*(\omega_1^*)} P_{1.l.1} \tag{9.58}$$

where

$$S^*(\omega_1^*) = \sum P_{1.l.1} S^*(\omega_l) = \kappa \left(\frac{V^3}{gh\omega_1^*} \right)^m \tag{9.59}$$

$$\omega_1^* = \left(\sum P_{1.l.1} \omega_l^{-m} \right)^{\frac{1}{m}} \tag{9.60}$$

The physical meaning of Eq. (9.50) is described as follows. Based on some analyses [5, 8, 16], the suspended load transport is very complicated when the transport is non-equilibrium. The carrying-sediment capacity does not only depend on the bed material gradation, and cannot be simply determined with Eq. (9.39). Or, if this equation is used, $P_{1.l}$ should be the effective bed material gradation. Under a general condition, the carrying-sediment capacity $S^*(\omega^*)$ and the fractional carrying-sediment capacity $P^*_{4.l} S^*$ not only depend on the bed material composition, but are also significantly impacted by the sediment inflow $P_{4.l} S$. It is obvious that the sediment already suspended in the water should be a part of the carrying-sediment capacity. Hence, the fractional carrying-sediment capacity is comprised of three components. The first component is the carrying-sediment capacity $P_{4.1} P_{4.1.l} S$ that is associated with the finer suspended particles ($l \leq k$). The second component is $P_{4.2} P_{4.l.2} S$ that is associated with the coarser suspended particles ($l \geq k + 1$). Here, $P_{4.1}$ and $P_{4.2}$ are the weight percentages for suspended particles with $D_l \leq D_k$ and with $D_l > D_k$, respectively. The quantity for the coarse particles that deposit on the bed and then are eroded is $P_{4.l.2} \frac{S^*(l)}{S^*(\omega_1^*)}$, in which $\frac{S^*(l)}{S^*(\omega_1^*)} = \frac{\omega_1^{*m}}{\omega_l^m} < 1$, indicating the coarse particles that deposit on the bed will only be partially eroded. The equation $\left(\frac{\omega_1^*}{\omega_l} \right)^m < 1$ exists, because ω_1 is the average sediment fall velocity while ω_l is the fall velocity for the coarser particles. The third component of the carrying-sediment capacity is a part of the carrying-sediment capacity $P^*_{4.l.1} S^*(\omega_1^*)$ that is associated with the particles eroded from the bed. Due to the impacts of the first two components, the third component of carrying-sediment capacity should multiply $\left[1 - \frac{P_{4.l} S}{S^*(\omega_1)} - \frac{P_{4.2} S}{S^*(\omega_1^*)} \right]$ so that the total water percentage will not exceed 100%, or the power expended for sediment transport will not exceed the available power of the flow.

Rewriting Eq. (9.50) as

$$P_{4.l}^{*}S^{*}(\omega^{*}) - P_{4.l}S = P_{4.2}P_{4.l.2}S\left[\frac{S^{*}(l)}{S^{*}(\omega_1^{*})} - 1\right] + \left[1 - \frac{P_{4.1}S}{S^{*}(\omega_1)} - \frac{P_{4.2}S}{S^{*}(\omega_1^{*})}\right]P_{4.l.1}^{*}S^{*}(\omega_1^{*}) \tag{9.61}$$

and then plugging which in Eq. (9.48) yields

$$\frac{d(P_{4.l}S)}{dx} = -\frac{\alpha_l\omega_l}{q}\left[1 - \frac{S^{*}(l)}{S^{*}(\omega_1^{*})}\right]P_{4.2}P_{4.l.2}S + \frac{\alpha_l\omega_l}{q}\left[1 - \frac{P_{4.1}S}{S^{*}(\omega_1)} - \frac{P_{4.2}S}{S^{*}(\omega_1^{*})}\right]P_{4.l.1}^{*}S^{*}(\omega_1^{*}) \tag{9.62}$$

As shown in this equation, the finer suspended particles ($l \le k$) do not directly impact the variation of the fractional capacity for carrying suspended load because they do not take part in the exchange with the bed material. However, the existence of these finer particles and their quantity still impact the quantity of the exchange between the coarser suspended particles and the bed material, and impact the quantity of the particles that are eroded from the bed. Either from the theoretical viewpoint or based on the analysis of the field data, these fine particles are equivalent to the wash load. This clarifies the theoretical meaning of "wash load", not only confirming that wash load does not take part in the exchange with bed material, but also emphasizing that the concentration of wash load does impact the exchange between the coarser suspended sediment and the bed material. It is further proved that the carrying-sediment capacity also exists for wash load.

Let further analyze Eq. (9.62). For the finer suspended particles in the size class of $l \le k$, they are completely transported through the reach during the scour process or the mild scour and mild deposition process. However, the finer bed material of these size classes is typically hidden in the voids among the coarser particles, and they can be possibly eroded only after the coarser particles are scoured. Actually, fir this case $P_{4.l.2} = 0$, Eq. (9.62) is positive, indicating a scour of bed material. While $P_{4.l.2} \neq 0$, the coarser suspended particles, part of them will deposit because $-\frac{\alpha_l\omega_l}{q}\left[1 - \frac{S^{*}(l)}{S^{*}(\omega_1^{*})}\right]P_{4.2}P_{4.l.2}S$ is negative. Hence, Eq. (9.62) illustrates in detail the bed material coarsening associated with the exchange between suspended load and bed material. During this sediment exchange, the coarser suspended particles deposit on the bed, and the finer particles on the bed are picked up, resulting in the coarsening of bed material. For a special case with $\frac{d(P_{4.l}S)}{dx} = 0$, no net scour or net deposition occurs. For this case, the coarser suspended particles still deposit on the bed, and the finer bed material must be eroded in the amount that is equivalent to the deposition quantity of the coarser particles, resulting in the bed material coarsening solely caused by the exchange between suspended load and bed material, as shown in Fig. 9.9. It is noted that the derived formula for the fractional carrying-sediment capacity also includes the case of mild deposition, and thus the bed material coarsening for this case can also be described with Eq. (9.62). The examples are illustrated in Tables 9.2 and 9.3.

9.3 Bed Material Coarsening Caused by Bed Load Erosion

9.3.1 Relationship Between Bed Load Composition and Bed Material Gradation

Separating the formulation for the non-uniform bed load transport rate into three zones based on the flow intensities, Han [7] developed a complete set of formulas for the pebble, the gravel, and the sandy bed load. The general formula for the transport capacity of the lth size group of the non-uniform bed load is

$$q_b^*(l) = \kappa \gamma_s D\omega_{1.l}\left(\frac{D_l}{\overline{D}}\right)^{m_2}\left(\frac{V_b}{\omega_{1.l}}\right)^{m_1}$$

where γ_s is the specific weight of sediment, D_l the diameter of the lth size group, $\overline{D}$ the average particle size, m_1, m_2 and κ are the parameters that vary with the sediment particle size and the flow intensity. Han [11] also utilized the following formulation:

$$q = \frac{1}{n}H^{\frac{5}{3}}J^{\frac{1}{2}}$$

where q is the flow discharge per unit width, J the energy slope. The Manning's roughness coefficient is

$$n = \frac{H^{\frac{1}{6}}}{6.5\sqrt{g}}\left(\frac{H}{\overline{D}}\right)^{-\frac{1}{4+\lg\frac{H}{\overline{D}}}} = \frac{H^{\frac{1}{6}-m_3}\overline{D}^{m_3}}{6.5\sqrt{g}}$$

where g is the acceleration of gravity, and

$$m_3 = \frac{1}{4+\lg\frac{H}{\overline{D}}}$$

The water depth is

$$H = \left(\frac{q\overline{D}^{m_3}}{6.5\sqrt{gJ}}\right)^{\frac{2}{3+2m_3}}$$

The bottom flow velocity is

$$V_b = 3.73u_*$$

Also,

$$\omega_{1.l} = \sqrt{\frac{4}{3C_x}\frac{\gamma_s - \gamma}{\gamma} g D_l} = \sqrt{3.33\frac{\gamma_s - \gamma}{\gamma} g D_l}$$

where γ is the specific gravity of water. Using these relations to replace V_b and $\omega_{1.l}$, the formula for the bed load transport rate per unit width can be rewritten as [11]

$$q_b^*(l) = K\gamma_s\left(\frac{\gamma}{\gamma_s - \gamma}\right)^{\frac{m_1-1}{2}} \times \frac{q^{\frac{m_1}{3+2m_3}} J^{\frac{m_1}{2}-\frac{m_1}{6+4m_3}}}{g^{\frac{m_1}{6+4m_3}-\frac{1}{2}} \overline{D}^{m_2-\frac{m_1 m_3}{3+2m_3}} D_l^{\frac{m_1}{2}-1.5-m_2}} \tag{9.63}$$

It should be emphasized that the most important part of the study on bed material coarsening is for the characteristics of the armor layer that finally forms. For this case, the composition of the coarser particles is relatively uniform with a narrow particle size distribution, and the roughness coefficient is also stable. Hence, the impact of the average particle size on the roughness can be neglected, and the average particle size in the formula for the roughness coefficient can be approximated by the particle size of a given size class, or

$$n = \frac{H^{\frac{1}{6}}}{6.5\sqrt{g}}\left(\frac{H}{\overline{D}}\right)^{-\frac{1}{4+\lg\frac{H}{D}}} = \frac{H^{\frac{1}{6}-m_3} D_l^{m_3}}{6.5\sqrt{g}} \tag{9.64}$$

It is noted that

$$\frac{V_b}{\omega_{1.l}} = \frac{3.73u_*}{\sqrt{3.33\frac{\gamma_s-\gamma}{\gamma} g D_l}} = \frac{3.73\sqrt{gHJ}}{\sqrt{3.33\frac{\gamma_s-\gamma}{\gamma} g D_l}} = 2.04\sqrt{\frac{\gamma}{\gamma_s - \gamma}}\sqrt{\frac{HJ}{D_l}} \tag{9.65}$$

The flow discharge per unit width can be rewritten as

$$q = \frac{1}{n} H^{\frac{5}{3}} J^{\frac{1}{2}} = \frac{6.5\sqrt{g} H^{\frac{5}{3}} J^{\frac{1}{2}}}{H^{\frac{1}{6}-m_3} D_l^{m_3}} = 6.5\sqrt{g}\frac{H^{\frac{3}{2}+m_3} J^{\frac{1}{2}}}{D_l^{m_3}} \tag{9.66}$$

or

$$H = \left(\frac{q D_l^{m_3}}{6.5\sqrt{gJ}}\right)^{\frac{2}{3+2m_2}} \tag{9.67}$$

Plugging Eq. (9.67) in Eq. (9.65) yields

$$\begin{aligned} \frac{V_b}{\omega_{1.l}} &= 2.04\sqrt{\frac{\gamma}{\gamma_s - \gamma}}\left(\frac{q D_l^{m_3}}{6.5\sqrt{gJ}}\right)^{\frac{1}{3+2m_3}}\sqrt{\frac{J}{D_l}} \\ &= 2.04\sqrt{\frac{\gamma}{\gamma_s - \gamma}}\frac{q^{\frac{1}{3+2m_3}} J^{\frac{1}{2}-\frac{1}{6+4m_3}}}{6.5^{\frac{1}{3+2m_3}} g^{\frac{1}{6+4m_3}} D_l^{\frac{1}{2}-\frac{m_3}{3+2m_3}}} \end{aligned} \tag{9.68}$$

Therefore,

$$q_b^*(l) = K\gamma_s\left(\frac{\gamma}{\gamma_s-\gamma}\right)^{\frac{m_1-1}{2}} \times \frac{q^{\frac{m_1}{3+2m_1}} J^{\frac{m_1}{2}-\frac{m_1}{6+4m_3}}}{g^{\frac{m_1}{6+4m_3}-\frac{1}{2}}\overline{D}^{m_2} D_l^{\frac{m_1}{2}-\frac{m_1 m_3}{3+2m_3}-1.5-m_2}} \tag{9.69}$$

where

$$K = 2.04^{m_1}\sqrt{3.33} \times 6.5^{-\frac{m_1}{3+2m_1}}\kappa$$

For $V_b/\omega_{1.l} < 0.70$, which is approximately equivalent the condition with pebble bed load movement in natural streams, $m_1 = 10$, $m_2 = 1$, $m_3 = 1/6$, and thus

$$q_b^*(l) = K\gamma_s\left(\frac{\gamma}{\gamma_s-\gamma}\right)^{4.5} \times \frac{q^3 J^{3.5}}{g\overline{D}D_l^2} \tag{9.70}$$

For $0.70 < V_b/\omega_{1.l} < 1.15$, which is approximately in the range of gravel bed load movement, $m_1 = 6$, $m_2 = 10/21$, $m_3 = 1/7$, and thus

$$q_b^*(l) = K\gamma_s\left(\frac{\gamma}{\gamma_s-\gamma}\right)^{25} \times \frac{q^{\frac{42}{23}} J^{\frac{48}{23}}}{g^{\frac{19}{46}}\overline{D}^{\frac{10}{21}} D_l^{\frac{737}{566}}} \tag{9.71}$$

For $0.90 < V_b/\omega_{1.l} < 2.00$, $m_1 = 5$, $m_2 \approx 7/26$, $m_3 = 1/8$, and thus

$$q_b^*(l) = K\gamma_s\left(\frac{\gamma}{\gamma_s-\gamma}\right)^{2} \times \frac{q^{\frac{20}{13}} J^{\frac{45}{26}}}{g^{\frac{7}{26}}\overline{D}^{\frac{7}{26}} D_l^{\frac{7}{13}}} \tag{9.72}$$

For $V_b/\omega_{1.l} > 2.0$, $m_1 = 4$, $m_2 = 0$, $m_3 = 1/8$, and thus

$$q_b^*(l) = K\gamma_s\left(\frac{\gamma}{\gamma_s-\gamma}\right)^{1.5} \times \frac{q^{\frac{16}{13}} J^{\frac{18}{13}}}{g^{\frac{3}{26}} D_l^{\frac{9}{26}}} \tag{9.73}$$

Utilizing the relation

$$P_{b.l}^* q_b^* = P_{1.l} q_b^*(l)$$

where $P_{1.l}$ is the bed material gradation hereafter, not the effective bed material gradation. The relationships between the size distribution for the bed load transport capacity and the bed material gradation can be derived from Eqs. (9.70) through (9.73) for various ranges of $V_b/\omega_{1.l}$, respectively. These relationships are

$$P_{b.l}^* = \frac{\frac{P_{1.l}}{D_l^2}}{\sum_{l=1}^{n}\frac{P_{1.l}}{D_l^2}} = P_{1.l}\frac{\overline{D}_1^2}{D_l^2} \tag{9.74}$$

$$P_{b.l}^{*} = \frac{\frac{P_{1.l}}{D_l^{\frac{737}{566}}}}{\sum_{l=1}^{n} \frac{P_{1.l}}{D_l^{\frac{737}{566}}}} \approx \frac{P_{1.l} D_l^{-0.76}}{\sum_{l=1}^{n} P_{1.l} D_l^{-0.76}} = P_{1.l} \frac{\overline{D_l^{\frac{19}{25}}}}{D_l^{\frac{19}{25}}} \tag{9.75}$$

$$P_{b.l}^{*} = \frac{\frac{P_{1.l}}{D_l^{\frac{7}{13}}}}{\sum_{l=1}^{n} \frac{P_{1.l}}{D_l^{\frac{7}{13}}}} = P_{1.l} \frac{\overline{D_3^{\frac{7}{13}}}}{D_l^{\frac{7}{13}}}.. \tag{9.76}$$

$$P_{b.l}^{*} = \frac{\frac{P_{1.l}}{D_l^{\frac{9}{26}}}}{\sum_{l=1}^{n} \frac{P_{1.l}}{D_l^{\frac{9}{26}}}} = P_{1.l} \frac{\overline{D_4^{\frac{9}{26}}}}{D_l^{\frac{9}{26}}} \tag{9.77}$$

where the average particle sizes are

$$\overline{D_1^2} = \left[\sum \frac{P_{1.l}}{D_l^2}\right]^{-1} \tag{9.78}$$

$$\overline{D_l^{\frac{19}{25}}} = \left(\sum_1 \frac{P_{1.l}}{D_l^{\frac{19}{25}}}\right)^{-1} \tag{9.79}$$

$$\overline{D_1^{\frac{7}{13}}} = \left[\sum \frac{P_{1.l}}{D_l^{\frac{7}{13}}}\right]^{-1} \tag{9.80}$$

$$\overline{D_1^{\frac{9}{26}}} = \left[\sum \frac{P_{1.l}}{D_l^{\frac{9}{26}}}\right]^{-1} \tag{9.81}$$

Or, the general form of these relationships can be written as

$$P_{b.l}^{*} = P_{1.l} \frac{\overline{D_K^{\upsilon}}}{D_l^{\upsilon}} \tag{9.82}$$

$$\overline{D_K^{\upsilon}} = \left[\sum \frac{P_{1.l}}{D_l^{\upsilon}}\right]^{-1} \tag{9.83}$$

where $K = 1, 2, 3, 4$ and $\nu = 2, \ \frac{19}{25}, \ \frac{7}{13}, \ \frac{9}{26}$.

As shown in Eqs. (9.74) through (9.77), because roughness coefficient n is considered to be a function of D_l instead of $\overline{D}$, the resultant sediment particle sorting as described in these equations occurs faster than the case in which n is assumed to be dependent of $\overline{D}$ [8]. During the period between the start of bed material coarsening and the final formation of the armor layer, the coarse particles can be scoured only during large flood events. As a result, the transport rates of these coarse particles are very small, or V_b/ω_l is very small, and v should be larger. While Eqs. (9.74) through (9.77) are approximate, the larger v for coarser particles will cause the resultant coarsening process to have a better agreement with the reality.

9.3.2 Bed Material Coarsening During Scour

The bed material layer is typically thick in an alluvial stream, and the composition of bed material varies with depth. Thus, the bed material gradation is always referenced to a certain depth of bed material. Within this depth, the bed material can possibly exchange with bed load and suspended load. Because of the rolling of sand ripples and the local bed deformation, the bed material that takes part in the exchange is not limited to the surface layer on the bed except for a pebble bed without pebble ripples. After the scour, a layer of coarser particles typically forms. These particles take part in the flow's sorting process, but are not scoured away. This layer is called the coarsened layer. Apparently, the bed material of this coarsened layer and the layer above which is involved in the sorting process. It is reasonable based on the coarsened layer to specify the active bed material layer thickness, which should equal the thickness of the coarsened layer adding the scour depth. The thickness of the coarsened layer depends on the bed material composition. As scour proceeds, the bed material become coarser and coarser until the armor layer forms. For the case of sediment deposition, the exchange between the coarser and finer particles is typically limited to the surface layer on the bed, and this thickness of the coarsened layer can be used to represent the depth of the active bed material layer. The governing equation for bed material coarsening and the solutions are derived as follows [6, 7, 15].

Let assume a scour and resultant bed material coarsening occur in a reach that is Δx long within a time interval of Δt. The depth of the coarsening layer is h_0, the scour depth by the end of t is h, the scour depth within Δt is Δh, and the thickness of bed material that participates the scour is $h_0 + h_E$, in which h_E is the scour depth at the end of the scour. Then, the weight of the bed material in the lth size group before the scour at time t is

$$P_{1.l}(t)\gamma_s'(t)(h_0 + h_E - h)B\Delta x \tag{9.84}$$

where B is the bed width. The weight of the bed material in the lth size group at $t + \Delta t$ is

$$P_{1.l}(t + \Delta t)\gamma_s'(t + \Delta t)(h_0 + h_E - h - \Delta h)B\Delta x \tag{9.85}$$

Assuming the weight of the eroded bed material is $g_b(t + \alpha_1 \Delta t)$, the fractional weight of the lth size class is

$$g_b(t + \alpha_1 \Delta t)P_{b.l}^*(t + \alpha_2 t)B\Delta t \tag{9.86}$$

The variation of bed material weight should equal the weight of eroded sediment, or

$$\begin{aligned}&\left[P_{1.l}(t+\Delta t)\gamma_s'(t+\Delta t)(h_0+h_E-h-\Delta h)-P_{1.l}(t)\gamma_s'(t)(h_0+h_E-h)\right]B\Delta x\\&\quad=-P_{b.l}^*(t+\alpha_2\Delta t)\times g_b(t+\alpha_1\Delta t)B\Delta t\end{aligned}\tag{9.87}$$

Summation for l yields

$$\begin{aligned}&\gamma_s'(t+\Delta t)(h_0+h_E-h_1-\Delta h)-\gamma_s'(t)(h_0+h_E-h)B\Delta x\\&\quad=-g_b(t+\alpha_1\Delta t)B\Delta t\end{aligned}\tag{9.88}$$

Substituting Eq. (9.88) for $-g_b(t+\alpha_1\Delta t)B\Delta t$ in Eq. (9.87), one obtains

$$\begin{aligned}&\left[P_{1.l}(t+\Delta t)\gamma_s'(t+\Delta t)(h_0+h_E-h-\Delta h)-P_{1.l}(t)\gamma_s'(t)(h_0+h_E-h)\right]B\Delta x\\&=P_{b.l}^*(t+\alpha_2\Delta t)\left[\gamma_s'(t+\Delta t)(h_0+h_E-h-\Delta h)-\gamma_s'(t)(h_0+h_E-h)\right]\end{aligned}$$

Dividing which by Δt gives

$$\begin{aligned}&\frac{P_{1.l}(t+\Delta t)\gamma_s'(t+\Delta t)-P_{1.l}(t)\gamma_s'(t)}{\Delta t}(h_0+h_E-h)B\Delta x\\&\quad=P_{1.l}(t+\Delta t)\gamma_s'(t+\Delta t)\frac{\Delta h}{\Delta t}-P_{b.l}^*(t+\alpha_2\Delta t)\frac{\gamma_s'(t+\Delta t)-\gamma_s'(t)}{\Delta t}(h_0+h_E-h)\\&\quad+P_{b.l}^*\gamma_s'(t+\Delta t)\frac{\Delta h}{\Delta t}=0\end{aligned}$$

Let $\Delta t\to 0$, then

$$\begin{aligned}&\frac{d(P_{1.1}\gamma_s')}{dt}(h_0+h_E-h)+\gamma_s'(P_{b.l}^*-P_{1.l})\frac{dh}{dt}-P_{b.l}^*\frac{d\gamma_s'}{dt}(h_0+h_E-h)\\&=\frac{dP_{1.l}}{dt}\gamma_s'(h_0+h_E-h)+P_{1.l}\frac{d\gamma'}{dt}(h_0+h_E-h)\\&\quad-P_{b.l}^*\frac{d\gamma_s'}{dt}(h_0+h_E-h)-1-\gamma_s'(P_{b.l}^*-P_{1.l})\frac{dh}{dt}\\&=\frac{dP_{1.l}}{dt}\gamma_s'(h_0+h_E-h)+\gamma_s'(P_{b.l}^*-P_{1.l})\frac{dh}{dt}-(P_{b.l}^*-P_{1.l})\frac{d\gamma_s'}{dt}(h_0+h_E-h)=0\end{aligned}\tag{9.89}$$

This complicated equation can be simplified. Note that

$$\begin{aligned}\frac{P_{b.l}^*-P_{1.l}}{h_0+h_E-h}\frac{d\left[\gamma_s'(h_0+h_E-h)\right]}{dt}&=\left(P_{b.l}^*-P_{1.l}\right)\frac{d\gamma_s'}{dt}\\&\quad-\frac{\left(P_{b.l}^*-P_{1.l}\right)\gamma_s'}{h_0+h_E-h}\frac{dh}{dt}\end{aligned}\tag{9.90}$$

Plugging Eq. (9.90) in Eq. (9.89) yields

$$\frac{dP_{1.l}}{dt}=\left(\frac{P_{b.l}^{*}-P_{1.l}}{\gamma_s'}\right)\frac{d\gamma_s'}{dt}-\frac{\left(P_{b.l}^{*}-P_{1.l}\right)}{h_0+h_E-h}\frac{dh}{dt}$$
$$=\frac{(P_{b.l}-P_{1.l})}{\gamma_s'(h_0+h_E-h)}\frac{d\left[\gamma_s'(h_0+h_E-h)\right]}{dt} \tag{9.91}$$

The weight of the bed material that is left after scour at time t is

$$W=(h_0+h_E-h)B\Delta x\gamma_s' \tag{9.92}$$

The weight of the bed material that is involved in scour is

$$W_0=(h_0+h_E)B\Delta x\gamma_s' \tag{9.93}$$

Plugging Eq. (9.92) into Eq. (9.91) yields [10]

$$\frac{dP_{1.l}}{dt}=\frac{\left(P_{b.l}^{*}-P_{1.l}\right)}{W}\frac{dW}{dt} \tag{9.94}$$

which is the differential equation for the variation of bed material gradation during a scour. Plugging Eqs. (9.74) through (9.77), respectively, in Eq. (9.94), a general formulation can be derived

$$\frac{dP_{1.l}}{dt}=\frac{P_{1.l}\left(\frac{\overline{D}}{D_l}\right)^{v}-P_{1.l}}{W}\frac{dW}{dt}=\frac{\left[\left(\frac{\overline{D}}{D_l}\right)^{v}-1\right]P_{1.l}}{W}\frac{dW}{dt} \tag{9.95}$$

in which v is 2, 9/25, 7/13, and 9/26, respectively. At the beginning of the time interval, $W=W_0$, and $P_{1.l}=P_{1.l.0}$. Integrating Eq. (9.95) within this interval leads to

$$\int_{P_{1.l.0}}^{P_{1.l}}\frac{dP_{1.l}}{P_{1.l}}=\int_{W_0}^{W}\left[\left(\frac{\overline{D}}{D_l}\right)^{v}-1\right]\frac{dW}{W}=\left[\left(\frac{\overline{D_m}}{D_l}\right)^{v}-1\right]\int_{W_0}^{W}\frac{dW}{W}$$

where $\overline{D}_m$ is a median value of $\overline{D}$ based on the mean value theorems for integrals. Hence,

$$\ln\frac{P_{11}}{P_{1.l.0}}=\left[\left(\frac{\overline{D}_m}{D_l}\right)-1\right]\ln\frac{W}{W_0}$$

Or

$$P_{1.l}=P_{1.l.0}\left(\frac{W}{W_0}\right)^{\left(\frac{\overline{D}_m}{D_l}\right)^{V}-1} \tag{9.96}$$

It is noted that W_0 is the quantity of the bed material that takes part in the scour within a reach that has a length of Δx and a width of B, and $W_0 - W$ is the quantity of the bed material that is eroded away. Apparently,

$$\frac{W_0 - W}{W} = \frac{Bh\gamma_s'\Delta x}{B(h_E + h_0)\gamma_s'\Delta x} = \frac{Bh\gamma_s'\Delta x}{B(h + h_0)\gamma_s'\Delta x} = \lambda^* \quad (9.97a)$$

where h is the thickness of eroded bed material, h_0 the thickness of the coarsened layer, $h_0 + h$ the thickness of the bed material layer that is involved in the sorting, and λ^* the scour percentage that represents the scour intensity, or the ratio between the quantity of eroded sediment and the quantity of the bed material that is involved in the scour process. It is noted that the thickness of the bed material involved in scour can be changed from $h_0 + h_E$, as used in Eq. (9.93), to $h + h_0$. In the previous description of the bed material coarsening process, h varies with time while h_E is constant, and $h = h_E$ when the scour stops. This is also a boundary condition for the integration, which is $h = h_E$ at $W = W$. In fact, W_0 is the weight of the bed material involved in scour, which should equal the sum of the eroded sediment quantity and the sediment quantity left in the coarsened layer, or $h_0 + h$. Hence, the sediment involved in scour should be all of the bed material that remains in the layer of $h_0 + h_E$ when the scour stops, not only the sediment that participates in the erosion during the scour process. Assigning $h_E = h$, $\Delta h = h(t+\Delta t) - h(t)$, $W_0 = (h_0 + h)B\gamma_s'\Delta x$ and $W = h_0 B\gamma_s'\Delta x$, Eqs. (9.84) through (9.96) are still correct. However, the scour percentage becomes

$$\lambda^* = \frac{W_0 - W}{W_0} = \frac{h}{h + h_0} \quad (9.97b)$$

Equation (9.97b) seems consistent with Eq. (9.97a). However, Eq. (9.97a) is for the case with constant $P_{1.l.0}$, while Eq. (9.97b) is for the case with $P_{1.l.0}$ varying with h. For the case with a very thin bed material layer, Eq. (9.97a) can be used. Otherwise, Eq. (9.97b) should be applied. Plugging Eq. (9.97b) in Eq. (9.96) gives

$$P_{1.l} = P_{1l.0}\left(1 - \lambda^*\right)^{\left(\frac{\overline{D}_m}{D_l}\right)^V - 1} \quad (9.98)$$

For a general condition, it is difficult to determine the median particle size $\overline{D}_m$ with the mean value theorem for integrals. However, for the cases included in our analysis, the sum of the weight percent for all size classes should equal 1. For given λ^*, D_1, v, and $P_{11.0}$, $\overline{D}_m$ can be determined with the following relation

$$1 = \sum_{l=1}^{n} P_{1.l} = \sum_{l=1}^{n} P_{1.l.0}(1 - \lambda^*)^{(\frac{\overline{D}_m}{D_l})^v - 1} \quad (9.99)$$

Once $\overline{D}_m$ is determined, $P_{1.l}$ can be calculated.

It is noted that a special consideration should be applied to the depth of the active bed material layer and the coarsened layer thickness if the erodible depth of the bed material is limited, the bed material can be completely eroded away, or no bed material is erodible (as e.g., for a pebble bed). For these cases, assuming the maximum erodible depth is $\tilde{h}_M$, the scour depth, the coarsened layer thickness, and the scour percentage are

$$h\gamma_s' = \tilde{h} = \begin{cases} \tilde{h} & (\tilde{h} + \tilde{h}_0 < \tilde{h}_M) \\ \tilde{h} & (\tilde{h} + \tilde{h}_0 \geq \tilde{h}_M \geq \tilde{h}) \\ \tilde{h}_M & (\tilde{h} > \tilde{h}_M) \end{cases} \tag{9.100}$$

$$h_0\gamma_s' = \tilde{h}_0 = \begin{cases} \tilde{h}_0 & (\tilde{h} + \tilde{h}_0 < \tilde{h}_M) \\ \tilde{h}_M - \tilde{h} & (\tilde{h} + \tilde{h}_0 \geq \tilde{h}_M \geq \tilde{h}) \\ \tilde{h}_M & (\tilde{h} > \tilde{h}_M) \end{cases} \tag{9.101}$$

$$\lambda^* = \begin{cases} \frac{\tilde{h}}{\tilde{h}+\tilde{h}_0} & (\tilde{h} + \tilde{h}_0 < \tilde{h}_M) \\ \frac{\tilde{h}}{\tilde{h}_M} & (\tilde{h} + \tilde{h}_0 \geq \tilde{h}_M \geq \tilde{h}) \\ 1 & (\tilde{h} > \tilde{h}_M) \end{cases} \tag{9.102}$$

where $\tilde{h}$, $\tilde{h}_0$, and $\tilde{h}_M$ are the virtual scour depth, the virtual thickness of the coarsened layer, and the virtual maximum erodible depth of bed material, respectively. They are expressed as the weight per unit area on the bed to avoid the conversions associated from the various γ_s'. If the dry density varies insignificantly or is hard to determine, in Eq. (9.102) $\tilde{h} + h$ can be used. It has to be emphasized that if $\tilde{h} = \tilde{h}_M$, $\lambda^* = 1$ cannot be taken as the second term of Eq. (9.102), otherwise according to Eq. (9.98) all $P_{1.l} = 0$. For this case. It is assumed $\lambda^* = 0.99$, or all $P_{1.l} = 0$ except the largest size group $P_{1.n} = 1$, thus to keep the continuous variation of the gradation of bed material.

The characteristics demonstrated in Eq. (9.98) are analyzed as follows. For the bed material particles that are coarser than the average particle size, $\overline{D}_m$, $\frac{\overline{D}_m}{D_l} < 1$, $\cdot\left(\frac{\overline{D}_m}{D_l}\right)^{\upsilon} - 1 < 0$, and thus

$$(1-\lambda^*)\left(\frac{\overline{D}_m}{D_l}\right)^{\upsilon} - 1 = \frac{1}{(1-\lambda^*)\left|\left(\frac{\overline{D}_m}{D_l}\right)^{\upsilon} - 1\right|} > 1$$

Based on Eq. (9.98), $P_{1,l} > P_{1.l.0}$, or the percent of these coarser particles increases. On the other hands, for the bed material particles that are finer than $\overline{D}_m$, $\frac{\overline{D}_m}{D_l} > 1$, $\left(\frac{\overline{D}_m}{D_l}\right)^{\upsilon} - 1 > 0$, $(1-\lambda^*)\left(\frac{\overline{D}_m}{D_l}\right)^{\upsilon} < 1$, and thus $P_{1.l} < P_{1.l.0}$ according to Eq. (9.98), indicating the percent of these finer particles decrease. These percent variations of the coarser and finer particles both lead to bed material coarsening.

For the case with various λ^*, the following method can be used to determine the relationship between $P_{1.l}$ and λ^*. Firstly, assume

$$\eta = (1 - \lambda^*)^{\overline{D}_m^{\upsilon}} \tag{9.103}$$

Then set η a certain value and calculate

$$\eta^{\frac{1}{D_l^{\upsilon}}} = (1 - \lambda^*)^{(\frac{\overline{D}_m}{D_l})^{\upsilon}} \tag{9.104}$$

$$P'_{1.l} = P_{1.l.0}(1 - \lambda^*)^{(\frac{\overline{D}_m}{D_l})^{\upsilon}} \tag{9.105}$$

$$\lambda^* = 1 - \sum P'_{1.l} = 1 - \sum P_{1.l.0}(1 - \lambda^*)^{(\frac{\overline{D}_m}{D_l})^{\upsilon}} \tag{9.106}$$

$$P_{1.l} = \frac{P'_{1.l}}{1 - \lambda^*} = \frac{P'_{1.l}}{\sum P'_{1.l}} = \frac{P_{1.l.0}(1 - \lambda^*)^{(\frac{\overline{D}_m}{D_l})^{\upsilon}}}{\sum P_{1.l.0}(1 - \lambda^*)^{(\frac{\overline{D}_m}{D_l})^{\upsilon}}} \tag{9.107}$$

and the relation of $P_{1.l} \sim \lambda^*$ can be developed. The $P_{1.l}$ for a given λ^* can be determined based on this relation of $P_{1.l} \sim \lambda^*$ using iteration or interpolation method.

As an example, Table 9.4 lists the computed results of the bed material coarsening occurred on a pebble-sand bed. This table lists the particle size range D'_1 for each size class, the mean particle size D_1 of each size group, the initial weight percent $P_{1.1.0}(D_l)$, and the computed weight percent $P_{1.l}$ and cumulative percent $\sum_{l=1}^{k} P_{1.l}$ after the scour assuming $\eta = 0.5$, $\upsilon = 2$, and the corresponding scour percentage $\lambda^* = 0.671$. By comparing $P_{1.l}$ and $P_{1.l.0}$, obvious bed material coarsening can be observed for $\lambda^* = 0.671$. This can be detailed follows: (1) For the three size classes with $D < 1$ mm, $P_{1.l} < P_{1.l.0}$; for the other seven size classes with $D > 1$ mm, $P_{1.l} > P_{1.l.0}$. (2) For the three size classes with $D > 50$ mm, the absolute quantities barely change, which can be observed from the fact that $P_{1.l.0}\eta^{\frac{1}{D_l^{\upsilon}}}$ is the same as $P_{1.l,0}$. $P_{1.l}$ is greater than $P_{1.l,0}$ because the sum of $P_{1.l.0}\eta^{\frac{1}{D_l^{\upsilon}}}$ is less than 1 after the scour. In other words, after the finer particles are eroded away, the total quantity of the bed material decreases, and thus the percent of the coarser particles increases. (3) For the four size classes with 1 mm $< D <$ 50 mm, the absolute quantities after the scour decreases, however, $P_{1.l} > P_{1..0}$, or the weight percent still increases after the scour, that means the bed material is coarsened.

As shown in Eq. (9.98), the unit of D_l and $\overline{D}_m$ does not impact the results. However, $\overline{D}_m$ is an implicit parameter in Eqs. (9.103) through (9.107). Change to the unit of D_l will cause a significant change to $\eta^{D_l^{-1}}$ but which will not impact the relationship of $P_{1.l} \sim \lambda^*$. In fact assuming the particle size is D_l in one unit, and changes to $K D_1$ in another unit, the following relation can be derived based on these equations:

$$P'_{1.l} = P_{1.l.0}\eta^{(K D_l)^{\upsilon}} = P_{1.l.0}\Big[(1 - \lambda^*)^{K^{-\upsilon}}\Big]\left(\frac{D_m}{D_l}\right)^{\upsilon} \tag{9.108}$$

Setting

Table 9.4 Computed bed material coarsening

D_l'	D_1	$P_{1.l.0}$	$\sum P_{1J.0}$	$P_{1.l.0}\eta^{\frac{1}{\omega_l^v}}$	$P_{1.l}$	$\sum P_{1j}$
0.10–0.25	0.175	0.46	0.460	0	0	0
0.25–0.50	0.375	0.14	0.600	0.0010	0.003	0.003
0.50–1.00	0.75	0.1	0.700	0.0292	0.089	0.092
1.00–5.00	3.00	0.005	0.705	0.0046	0.014	0.106
5.00–10.0	7.50	0.015	0.720	0.0148	0.045	0.151
10.0–20.0	15.0	0.03	0.750	0.0299	0.091	0.242
20.0–50.0	35.0	0.13	0.880	0.1299	0.394	0.636
50.0–75.0	62.5	0.08	0.960	0.0800	0.243	0.879
75.0–100.0	87.5	0.02	0.980	0.020	0.060	0.939
100.0–125.0	125	0.02	1.000	0.020	0.061	1.000
Σ				0.3294	1.000	

Note $v = 2$, $\eta = 0.5$, D in mm, and $\lambda^* = 0.671$

$$\left(1 - \lambda^*\right)^{K-v} = \left(1 - \tilde{\lambda}^*\right) \tag{9.109}$$

then

$$\sum P_{1.l}'\left(\tilde{\lambda}^*\right) = \sum P_{1.l.0}\left(1 - \tilde{\lambda}^*\right)\left(\frac{D_m}{D_l}\right)^v$$

$$P_{1.l}\left(\tilde{\lambda}^*\right) = \frac{P_{11.0}\left(1 - \tilde{\lambda}^*\right)\left(\frac{D_m}{D_l}\right)^v}{\sum P_{1..0}\left(1 - \tilde{\lambda}\right)\left(\frac{D_m}{D_l}\right)^v} \tag{9.110}$$

Equation (9.110) is equivalent to Eq. (9.107). After changing the unit of D_1, the corresponding λ^* and $P_{1.l}$ are different. For the same λ^*, both equations yields the same value of $P_{1.l}$.

9.3.3 Bed Material Coarsening Process

1. Scour and Bed Material Coarsening Process on A Pebble or A Pebble-Sand Bed

As an example, Fig. 9.12 and Table 9.5 show the variation of the bed material gradation in a reach between Huangjiagang and Guanghua of the Han River downstream of the Danjiangkou Reservoir [6, 7]. The percent finer (or the cumulative weight percent passing) of various size classes is shown in Fig. 9.12, and the weight percent is listed in Table 9.5. The initial bed material gradation $P_{1.l.0}$ indicates that the original bed material is composed of pebble and sand. The post-scour bed material gradation in

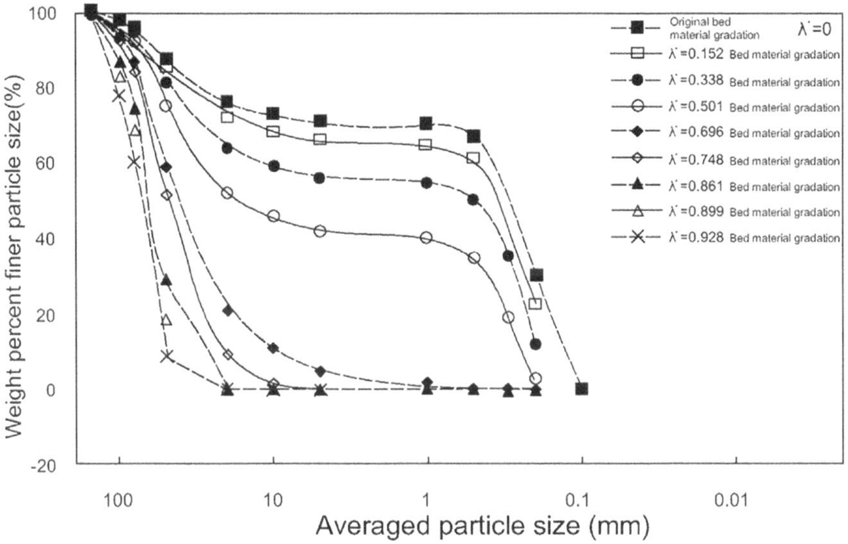

Fig. 9.12 Variation of bed material gradation showing the bed material coarsening process (reach between Huangjiagang and Guanghau, Han River)

1978 indicates that the bed material in this downstream reach became much coarser due to the channel scour after the reservoir operation. A significant amount of the median to fine sand with $D < 1$ mm was eroded away. The coarser particles with $D = 10{-}50$ mm were also massively scoured. The bed material gradation and the gradation for the median to coarse particles (excluding the fine particles filling the voids among these coarse particles) were computed for various λ^* using Eqs. (9.98) and (9.99). Since the bed material include pebble, υ was set to 2 in the computation.

As listed in Table 9.5, since the scour percentage λ^* increase, the bed material composition become coarser and coarser. Two bold lines drawn in this table divide the results into three zones for each λ^*, except for $\lambda^* \geq 0.861$ the two bold lines merged. For the size classes above the upper bold line, $P_{1,l}$ decreases rapidly due to the massive sediment scour for these size classes. For the size classes below the lower bold line, the particles are barely eroded (less than 5%); however, the total bed material quantity decreases, and thus the percent of these size classes increases, or $P_{1,l}$ increases. For the size classes between these two bold lines, a limited scour occurs, and $P_{1,l}$ still increases due to the combined effect of the scour for these sizes classes and the decrease in the total bed material quantity. As discussed later, the armor layer will be formed for $\lambda^* = 0.910$.

2. Analysis and Verification of Scour and Bed Material Coarsening on A Sand Bed

Table 9.6 and Fig. 9.13 show the variation of the bed material gradation on a sand bed of the reach between Huangjiagang and Guanghua in the lower Han River [6, 7]. The field data collected in 1959, before the operation of the Dangjiangkou Reservoir, and the data collected in 1974, after the operation of this reservoir, are listed in Table 9.6

Table 9.5 Computed variation of bed material gradation showing the bed material coarsening process on a pebble-sand bed (Reach between Huangjiagang and Guanghau, Han River)

D_l (mm)	$\overline{D}_l$ (mm)	$P_{1.l.0}\lambda^* = 0$	Computed $P_{1.l}(\lambda^*)$								
			$\lambda^* = 0.0178$	$\lambda^* = 0.152$	$\lambda^* = 0.338$	$\lambda^* = 0.501$	$\lambda^* = 0.696$	$\lambda^* = 0.748$	$\lambda^* = 0.861$	$\lambda^* = 0.899$	$\lambda^* = 0.928$
0.10	0	0	0	0	0	0	0	0	0	0	0
0.20	0.15	0.300	0.292	0.227	0.120	0.027	0	0	0	0	0
0.30	0.25	0.250	0.250	0.251	0.234	0.164	0	0	0	0	0
0.50	0.40	0.120	0.121	0.133	0.150	0.155	0	0	0	0	0
1.00	0.75	0.030	0.030	0.035	0.043	0.053	0.017	0	0	0	0
5.0	3.00	0.010	0.010	0.012	0.015	0.020	0.029	0	0	0	0
10.0	7.50	0.020	0.020	0.024	0.030	0.040	0.064	0.013	0	0	0
20.0	15.0	0.030	0.031	0.035	0.046	0.060	0.098	0.076	0	0	0
50.0	35.0	0.115	0.117	0.135	0.174	0.231	0.378	0.422	0.292	0.184	0.087
80.0	65.0	0.085	0.087	0.100	0.128	0.170	0.280	0.331	0.452	0.495	0.511
100	90.0	0.020	0.021	0.024	0.030	0.040	0.066	0.079	0.123	0.150	0.180
153	125	0.020	0.021	0.024	0.030	0.040	0.066	0.079	0.133	0.171	0.222

Table 9.6 Bed material coarsening on a sand bed

Diameter (mm)	Percent finer			
	Measured (1959)	Computed		Measured (1974)
		$\upsilon = 9/26$	$\upsilon = 7/13$	
0.025				
0.05	0.045	0.009	0.002	0.001
0.10	0.100	0.031	0.013	0.030
0.25	0.740	0.583	0.475	0.100
0.50	0.930	0.855	0.795	0.760
1.00	1.000	1.000	1.000	1.000

and is shown in Fig. 9.13. During this period from 1959 to 1974, the average scour depth was approximately 2.53 m. Assuming the coarsened layer thickness was 0.4 m, the scour depth is equivalent to 6.325 layers, and the depth of the sand involved in the scour process is equivalent to 7.325 layers, and the scour percentage $\lambda^* = 0.863$. Assuming $\upsilon = 9/26 = 0.346$ and $\upsilon = 7/13 = 0.538$, respectively, the bed material gradation after the scour (with$\lambda^* = 0.863$) was computed using Eqs. (9.103) through (9.106). The results are listed in Table 9.6 and are shown in Fig. 9.13. As shown in Fig. 9.13, the bed material gradation computed based on bed load scour has a great discrepancy from the field data, no matter for $\nu = 7/13$ or for $\nu = 9/26$. This indicates the bed material coarsening process develops very slowly if it is caused by the scour of bed load. Particularly, for a weak flow condition ($\nu = 9/26$), this coarsening process is extremely slow. This is consistent with the field observation that the size distribution of bed load is similar to the bed material gradation if massive bed load movement occurs on a sand bed. Our analysis provides a theoretical proof of this phenomenon.

As discussed later, bed material coarsening on a sand bed is mainly caused by the sorting process during the scour of suspended load.

3. Impact of υ on Bed Material Coarsening

lists the bed material gradation after the coarsening process computed with $\upsilon = 5/6$, 2, and 3, respectively. For the smallλ^*, if the difference in λ^* is small ($\lambda^* = 0.211$ vs. 0.244, and $\lambda^* = 0.480$ vs. 0.485), the difference in the computed weight percent $P_{1.l}$ is very small for most size classes except for the first one to the third class. This is particularly true for the coarser size classes that are barely eroded. However, for the large $\lambda^*(\lambda^* = 0.860 - 0.854)$, only coarser particles remain on the bed, and the impact of υ on $P_{1.l}$ is more obvious. The coarsening process develops faster for a larger υ (Table 9.7).

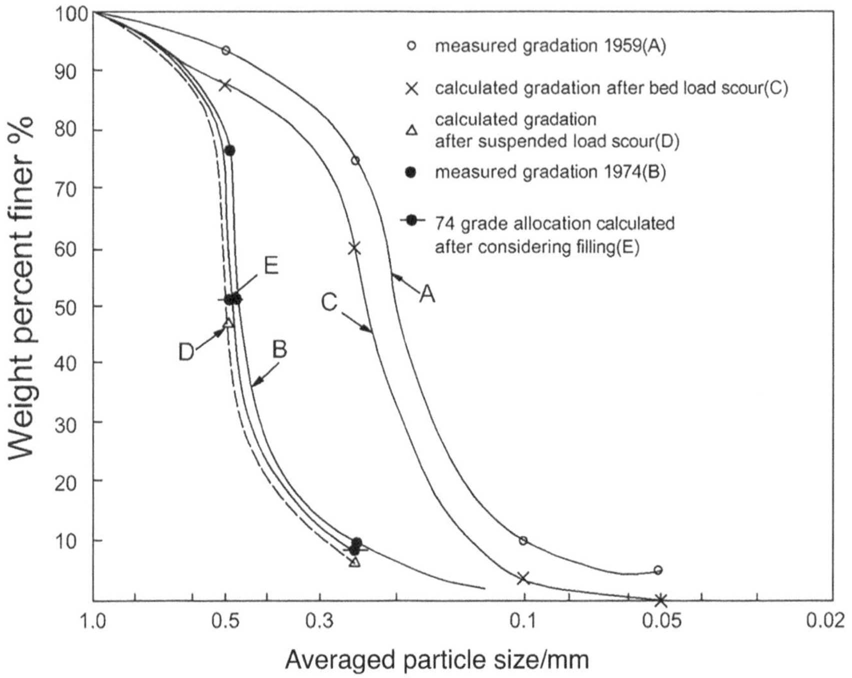

Fig. 9.13 Comparison of bed material gradations in the bed coarsened layer

Table 9.7 Impact of υ on the bed material gradation

Initial gradation		$\upsilon =$ 5/6	$\upsilon = 2$	$\upsilon =$ 5/6	$\upsilon = 2$	Initial gradation		$\upsilon =$ 5/6	$\upsilon = 2$	$\upsilon = 3$
D (mm)	$P_{1.l.0}$	$\lambda^* =$ 0.211	$\lambda^* =$ 0.244	$\lambda^* =$ 0.480	$\lambda^* =$ 0.485	D (mm)	$P_{1.l.0}$	$\lambda^* =$ 0.864	$\lambda^* =$ 0.861	$\lambda^* =$ 0.860
0.1–0.25	0.46	0.371	0.315	0.193	0.084	0.1–0.20	0.30			
0.25–0.50	0.14	0.140	0.160	0.120	0.162	0.20–0.30	0.25			
0.50–1.0	0.10	0.111	0.128	0.122	0.171	0.30–0.50	0.12			
1.0–5.0	0.005	0.006	0.007	0.008	0.010	0.50–1.0	0.03			
5.0–10	0.015	0.019	0.020	0.027	0.029	1.0–5.0	0.01			
10–20	0.030	0.038	0.040	0.056	0.058	5.0–10	0.02	0.007		
20–50	0.130	0.164	0.172	0.246	0.253	10–20	0.03	0.040	0.001	
50–75	0.08	0.101	0.106	0.152	0.155	20–50	0.115	0.366	0.292	0.220
75–100	0.02	0.025	0.026	0.038	0.039	50–75	0.085	0.377	0.453	0.489
100–153	0.02	0.025	0.026	0.038	0.039	75–100	0.020	0.100	0.120	0.140
						100–153	0.020	0.110	0.132	0.151

9.4 Bed Load Armor Layer, Maximum Scour Depth and Bed Material Gradation After Scour

The maximum scour depth is defined as the cumulative scour depth when the erosion finally stops after a long-term bed scour and bed material coarsening process for a reach with clear-water inflow and given flow hydrograph and bed material composition. However, the flow condition still varies and the bed material composition also show a great spatial variation. Particularly, the sediment movement has a random behavior. As a result, a threshold condition for absolutely no more scour does not exit, and the so-called maximum scour depth is only relative. As pointed out before, sediment entrainment or not is actually referred to a certain quantity of sediment transport rate, which has a constraint nature. On the other hand, based on the field data, the channel bed does change little or barely change after a certain period of scour. From this aspect, the maximum scour depth does exist in reality. More importantly, introducing the concept of the maximum scour depth will simplify the analyses of some problems. Hence, the concept of the maximum scour depth is introduced in the following with supplementary condition. With this condition, the sediment transport rate will be very small instead of being zero after the maximum scour depth is reached. This small amount of sediment transport rate is still able to cause the bed change, but can be neglected if a lower precision is applied. This will make the concept of the maximum scour depth more precise.

It is noted that the above definition of the maximum scour depth for the case with clear-water inflow can also be applied to the case with certain sediment inflow that still induces channel scour. However, because this "certain" sediment inflow is practically hard to control and to be satisfied, the maximum scour depth is difficult to be specific, and there is no practical meaning to study the maximum scour depth for this case. Hence, the maximum scour depth is defined in this chapter for the case with clear-water inflow.

The armor layer is the coarsened bed material layer that finally forms when the coarsening process reaches its limit, and thus is closely related to the maximum scour depth. The coarsening process reaches its limits when the bed material of the coarsened layer can no longer be eroded for a given flow hydrograph. In fact, for a bed downstream of a reservoir that experiences clear-water scour, the coarsened layer when scour reaches the maximum scour depth is the armor layer that is resistant to further scour. Hence, the maximum scour depth and the armor layer are the two aspects of the clear-water scour problem. However, this is not true if the maximum scour depth is defined for the case with certain sediment inflow. For a median to fine sand bed that is downstream of a reservoir, the channel hydraulic parameters will become weaker after a certain period of scour. If the sediment discharge from the reservoir increases, the scour will stop and the bed will reach dynamic equilibrium. Although the maximum scour depth is reached at this point of time, the sediment particles on the bed may still be entrained, move as bed load, hop, be suspended, or exchange with sediment inflow from the upstream, resulting in local sediment deposition or scour. Hence, the coarsened layer on the bed will not be resistant to further scour, and cannot be treated as an armor layer. Also, for the median to fine sand bed, the size distribution of the bed material is narrow. Therefore, the difference in the bed material gradation before and after the coarsening process and the resulting difference in the critical entrainment velocity are less obvious than the pebble-sand bed. Also, the critical entrainment velocity for the median to fine sand is very low. As a result, the armor layer is typically difficult to form on the bed with median to fine sand.

For the clear-water scour, the maximum scour depth is controlled by the formation of the armor layer. Based on the field data, the formation of the armor layer involves three subjects: (1) the thickness of the armor layer, (2) the grain size and the gradation of the coarser particles that control the stability of the armor layer, and (3) the proportion and the size range of the finer particles that possibly fill the voids among the coarser particles in the armor layer no matter how thin this armor layer is. These finer particles exist in the armor layer because of the shadowing effect of the coarser particles, and are hardly impact by the flow. Therefore, the proportion and size range of these finer particles are not determined by the flow condition and should be investigated with consideration of other factors. These three subjects are analyzed as follows.

9.4.1 Thickness of Coarsened Layer and Armor Layer

During the bed material coarsening process, the sediment particles that are involved in the sorting process and that are not eroded away form a layer on the bed surface. This layer will prevent the bed material from further scour and is thus called the armor layer. If there is only layer by layer scour without sand ripple movement and local sedimentation, the thickness of the final armor layer will be equivalent to the diameter of a coarse sediment particle. However, if there is sand ripple movement, formation of bed load delta, or local bed change, the bed material will be constantly disturbed. As a result, a continuous sorting will occur during the scour, and the thickness of the armor layer is much greater than the diameter of a coarse sediment particle. It is difficult to theoretically derive general expressions for the thickness of the armor layer and for the thickness for the coarsened layer. Instead, they can be empirically determined for various cases, as discussed in the following.

1. Armor Layer Thickness on A Pebble or A Pebble-Sand Bed

During the clear-water scour in a pebble bed downstream of a reservoir, the movement of pebble is weak under a general flow condition. As a result, it is impossible for pebble ripples to form and for the local sedimentation to occur extensively on the bed. It is reasonable to use the diameter of a coarser sediment particle to represent the thickness of the coarsened layer, as done in previous studies [6, 7]. This can also be observed from the field data that is shown in Fig. 9.5. As shown in this figure, the bed material composition of the surface layer, the thickness of which is equivalent to the maximum diameter of the bed material, is coarser than the other locations that are

0.1–0.4 m deep. This indicates that the coarsening only occurs in this surface layer. What grain size should be used to represent the thickness of the coarsened layer? Considering sediment sorting, the coarsened layer should be thick enough to contain the coarsest particle, and thus the thickness of the coarsened layer should be the maximum particle size D_M. However, considering the maximum particle size of the bed material is usually instable, the coarsened layer thickness can also be assumed to equal D_{90} or D_{95}. It is noted that the coarsening also be partially caused by the scour of bed load, it is more general to use D_M as the coarsened layer thickness. Apparently, the composition of the coarsened layer varies during the scour, and thus the coarsened layer thickness also varies. In the following analysis, the thickness of the coarsened layer that finally forms, or the armor layer thickness, is assigned to be D_{90} or D_M of the armor layer.

2. Coarsened Layer Thickness on A Sand Bed

Sand dunes are typically developed on a sand bed. Also, local deltas are usually formed on the bed downstream of a reservoir during the clear-water scour period. According to the field data of the Han River downstream the Danjiangkou Reservoir, the heights of the dunes are in the order of decimeters, mostly 2–4 dm; and the heights of local deltas vary significantly, from decimeters to 2.3 m. However, the local deltas and sand dunes do not form extensively on the bed, and thus it is hard to determine the thickness of the disturbed layer, or the thickness of the coarsened layer, purely based on the heights of the dunes or of the deltas. Alternatively, the coarsened layer thickness can be approximately estimated by analyzing the field bed material samples. Table 9.8 lists the field data of bed material samples at various depth on the sand bed of the Han River, from which an obvious coarsening can be observed for the surface layer. The comparison of representative bed material gradations is shown in Fig. 9.10. Based on Table 9.8, the coarsened layer thickness of the Han River is approximately 0.37 m in the reach upstream of Huangzhuang, and is 0.30 m for the downstream reach. Considering the variation range, the coarsened layer thickness on the sand bed of Han River is approximately 0.2–0.5 m, which can be used to represent the condition for the bed comprised of relatively coarse sand. For the wandering Yellow River with bed material consisting of relatively fine sand, the coarsened layer thickness was found to be approximately 0.8–1.0 m based on the results computed with the formula for bed material coarsening associated from the suspended load exchange [15]. These values can be used as the reference for determining the coarsened layer thickness on a sand bed. From the perspective of the mechanism for sediment disturbance and sorting process, the coarsened layer thickness is approximately consistent with the armor layer thickness.

3. Coarsened Layer Thickness on A Gravel Bed

The bed material on a gravel bed is usually composed of the gravels with the size of 0.5–10 mm and coarse sand. The entrainment velocity for these particles are higher than that for a sand bed. However, the sand dunes on a gravel bed are usually hard to disappear during high flow, and thus these sand dunes last longer than those on a sand bed. Also, these particles on a gravel bed typically move as bed load, the bed

Table 9.8 Coarsened layer thickness on the medium to coarse sand beds [13]

Reach	Sample location	Bed material description	Selected armor layer thickness (m)
Cihe-Huangzhuang Han River	Scour 37	Coarser in the top layer, and is finer at the depth of 0.5 m	0.3
	Scour 38	Slightly coarser in the top layer than the depth of 0.4 m	0.4
	Zhimatan	Coarse in the top layer, and apparently finer at the depth of 0.4–0.5 m	0.2
	Gongjia-jixinzhou	Slightly finer in the top layer than the depth of 0.4–0.5 m	0.5
	Scour 43	Slightly coarser in the top layer, slightly finer at the depth of 0.2 m, and in the middle at the depth of 0.3 m. But the differences are small	0.3
	Tangbai River mouth	Coarser at the depth of 0.2 m, and apparently finer at the depth of 0.4 m	0.3
	Mean		0.37
Huangzhuang-Makou Han river	Shizitan	Coarser in the top 0.2–0.3 m layer, and finer at the depth of 0.5–0.6 m	0.4
	Zhouwei	Finer in the top layer, and coarser at the depth of 0.1–0.3 m	0.3
	Zhoutou	Coarser in the top layer, and finer at the depth of 0.5 m	0.3
	Yaojiaji	Coarser in the top layer and at the depth of 0.3 m, finer at the depth of 0.1 m	0.3
	Jingangtan	Coarser in the top layer, and finer at the depths of 0.2 m, 0.3 m, and 0.4	0.1
	Dujiatan	Medium size in the top layer, coarser at the depth of 0.2–0.3 m, and finer at the depth of 0.6 m	0.45
	Caobali	Medium size in the top layer, and finer at the depth of 0.3–0.4 m	0.2
	Xia-Baigutan	Approximately the same between the top layer through the depth of 0.5 m	0.5
	Xincheng-xiaxintan	Coarser at the top layer, and finer at the depth of 0.2–0.5 m	0.1
	Xintan	Coarser in the top layer, and finer at the depth of 0.4 m	0.25

(continued)

Table 9.8 (continued)

Reach	Sample location	Bed material description	Selected armor layer thickness (m)
	Zekoe-shangxntan	Finer in the top layer, and coarser at the depth of 0.2–0.4 m	0.4
	Makou-biantan	Coarser in the top layer, and finer at the depth of 0.3–0.7 m	0.2
	Mean		0.30

is relatively flat, and the flow velocity usually has a relatively uniform distribution within a cross-section. As a result, sand dunes usually form more extensively on a gravel bed. According to the field data for a number of streams, the height of the sand dune is mostly 0.1–0.3 m. Without more field data, the coarsened layer thickness and the armor layer thickness can be selected with reference to this range of dune height.

9.4.2 Gradation of the Coarser Particles in Armor Layer and the Maximum Scour Depth

With the equation for the bed material coarsening during the scour of bed load, or Eq. (9.98), the bed material gradation can be determined for various values of the scour percentage λ^*. Finally, as $\lambda^* = \lambda_M^*$, when the coarsening process reaches its limit, the results are the bed material gradation or the bed material gradation of the armor layer. As discussed before, there are always some fine particles filling the voids among the coarse particles. Therefore, the bed material gradation computed with Eq. (9.98) for $\lambda^* = \lambda_M^*$ is only for the coarse particles in the armor layer, and is not the actual bed material gradation that should also include the fine particles that fill the voids. According to Eq. (9.98), the maximum scour depth h_M needs to be known in order to compute the scour percentage λ_M^*. It is noted that Eqs. (9.98) and (9.99) form a set of $m + 1$ equations for m size classes, but with $m + 2$ unknowns that include $P_{1.l}$, λ^*, and h_M. Hence, one more equation is required to closure this equation set. This equation can be derived based on the condition that the coarse particles in the armor layer will no longer be entrained.

For non-uniform sediment, Han and He [13] developed the following equation for the entrainment (depth-averaged) velocity for the lth size class:

$$V_{c.1.l} = 0.268\omega_{1.l}\psi f^{-1}\left(\frac{D}{\overline{D}}, \lambda_{q_{b.l}}\right) \tag{9.111}$$

where $\overline{D}$ is the mean particle size that can also be replaced with D_{50},

$$\psi = \psi\left(\frac{H}{\overline{D}}\right) = \frac{\overline{V}}{u_*} = 6.5\left(\frac{H}{\overline{D}}\right)^{\frac{1}{4+\lg\frac{H}{\overline{D}}}} \tag{9.112}$$

$$\omega_{1.l} = \sqrt{53.9D_l + \frac{3\times10^{-7}}{D_l}(1+0.85H)} \approx \sqrt{53.9D_l} \tag{9.113}$$

and $\lambda_{q_{b.l}}$ is the dimensionless transport rate for the lth size class when the entrainment of this size class occurs. It is noted that Eq. (9.113) is in SI units.

For a pebble bed, the dimensionless transport rate for sediment entrainment criterion should be that for non-uniform sediment, which is $\lambda_{q_{b.l}} = 0.3\times10^{-6}$. Hence,

$$\lambda_{q_{b.l}} = \frac{q_{b.l}}{P_{1.l}\gamma_s D_l \omega_{1.l}} = f\left(\frac{\overline{V}_{b.c.1..l}}{\omega_{1.l}}, \frac{D_l}{\overline{D}}\right) = 0.3\times10^{-6} \tag{9.114}$$

where $\overline{V}_{b.c.1..l}$ is the entrainment velocity expressed by the velocity at bed for the lth size class. The numerical results of the function $f\left(\frac{\overline{V}_{b.c.1..l}}{\omega_{1.l}}, \frac{D_l}{\overline{D}}\right)$ is listed in Table 9.9 [13], which is computed based on the theoretical relationship for low transport rate. For a given $\frac{D_l}{\overline{D}}$, $\frac{\overline{V}_{b.c.1..l}}{\omega_{1.l}} = f^{-1}\left(\frac{D_l}{\overline{D}}, \lambda_{q_{b.l}} = 0.3\times10^{-6}\right)$ can be found from Table 9.9. Then plugging in Eq. (9.111), the entrainment velocity $V_{b.c.l}$ for the lth size class. It is noted that the entrainment velocity has a clear physical meaning here. When the flow velocity equal this entrainment velocity, a certain amount of bed load will be transported while the transport rate is very small. Different from previous formulas, the entrainment velocity is defined here using the sediment transport rate as the quantified indicator.

For the bed material consisting of gravel and sand, because the bed material composition is within a relatively narrow range, the entrainment velocity can be defined using the critical dimensionless transport rate for entrainment of uniform sediment as the criterion, or

Table 9.9 Relation between $\frac{\overline{V}_{b.c.1..l}}{\omega_{1.l}}$ and $\frac{D_l}{\overline{D}}$ for $\lambda_{q_{b.l}} = 0.3\times10^{-6}$

$\frac{D_l}{\overline{D}}$	$\frac{\overline{V}_b}{\omega_{1.l}}$	$\frac{D_l}{\overline{D}}$	$\frac{\overline{V}_b}{\omega_{1.l}}$
0.25	0.304	2.0	0.266
0.50	0.293	3.0	0.262
0.9	0.286	4.0	0.260
1.00	0.282	5.0	0.247
1.10	0.279	7.46	0.235
1.30	0.274	10.0	0.222
1.50	0.271		

$$\lambda_{q_{b.l}} = \frac{q_b}{\gamma_s D_l \omega_{1.l}} = 0.219 \times 10^{-3} \tag{9.115}$$

The corresponding dimensionless entrainment velocity at the bottom is

$$\frac{V_{b.c.1..l}}{\omega_{1.l}} = f^{-1}\left(\lambda_{q_{b.l}} = 0.219 \times 10^{-3}\right) = 0.433 \tag{9.116}$$

Based on this value of $\frac{V_{b.c.1..l}}{\omega_{1.l}}$, the depth-averaged entrainment velocity can be determined with Eq. (9.111), or

$$V_{c.1.l} = 0.268\omega_{1.l}\psi \times 0.433 = 0.116\omega_{1.l}\psi$$

which is based on SI units. If the bed material gradation is known, $\frac{V_{b.c.1..l}}{\omega_{1.l}}$ corresponding to a $\frac{D_l}{D}$ can be found from the table. After determining $\omega_{1.l}$ and ψ, the entrainment velocity $V_{c.1..l}$ can be computed with Eq. (9.111) for each size class. Comparing the flow velocity with the entrainment velocity $V_{c.1..l}$, the size classes that can no longer be entrained under this flow condition, or the particles that finally form the armor layer, can be determined. The gradations of these coarse particles can also be determined. In order to test whether the above method can reasonably determine the composition of the coarse particles in the armor layer, the gradation of the top pebble layer in 1979 within the reach between Huangjiagang and Guanghua of the Han River, which is downstream of the Danjiangkou Reservoir, was used to compute the entrainment velocity. The data and results are listed in Table 9.10.

The above computation of the entrainment velocity is conducted for known bed material gradation and known hydraulic parameters. However, under the general condition in reality, these parameters are usually unknown, and iterations are needed to determine the entrainment velocity. The iteration procedure is as follows:

(1) After the armor layer thickness h_0 is determined, assume a value for the maximum scour depth h_M, and then compute the bed material gradation as

Table 9.10 Bed material gradation of the top gravel layer in Huanghuagang–Guanghua Reach

Diameter D (mm)	$\sum P_{1.1}$	
	Measured (1978)	Computed ($\lambda^* = 0.861$)
1	0	
5	0.050	0
10	0.070	0
20	0.100	0
50	0.310	0.292
80	0.620	0.744
100	0.890	0.867
153	1.00	1.000

the coarsening process reach its limit, or the gradation of the coarse particles in the armor layer.

(2) Based on the scour depth, the roughness coefficient after bed material coarsening, compute the water surface profile and other hydraulic parameters including the water depth and the flow velocity. Based on the gradation of the coarse particles and the water depth, compute the entrainment velocities for various size class. Then, compare the flow velocity with the entrainment velocities to identify which size classes cannot be entrained.

(3) Compare the diameter of the coarsest size class that can be entrained of the h_M layer to the diameter of the finest size class of the computed composition of the final coarsened bed material in the armor layer. If they are equal, the iteration is ended, and the assumed h_M is the actual maximum scour depth. Otherwise, assume another h_M and repeat previous steps. If all particles in the final coarsened layer cannot be entrained, the coarsening process is over-predicted, λ_M^* is over-estimated, and thus h_M should be reduced for a given h_0. On the other hand, if all the size classes in the final coarsened layer can be entrained, λ_M^* is under-estimated, and h_M should be increased.

9.4.3 Percent of Filling Particles in Armor Layer and Bed Material Gradation After Scour

As discussed before, in the coarsened layer and the armor layer, in addition to the coarse particles that control the stability of these layers, there are certain amount of finer particles that fill the voids among these coarse particles. The existence of these finer particles is almost independent of the hydraulic factors. In previous analyses, this void filling phenomenon was considered by assuming an empirical percent for the filling particles. The existence of these finer particles depends on whether they can fill the voids among the coarse particles. This will limit the filling particles to a certain particle size range and to a certain quantity, and make it possible to theoretically analyze the composition and the percent of the filling particles with certain simplifications, instead of purely using the empirical values. In the following analysis, the percent and the composition of the filling particles are investigated for two types of bed material. One is comprised of gravel and pebble, and the other is comprised of coarse and fine sand. During the process of bed material coarsening, finer particles also fill the voids in the coarsened layer. However, these particles play insignificant role in bed stability and bed material sorting. Hence, the filling of these finer particles is not usually considered for a general coarsened layer. In the analyses hereafter, the finer particle filling is only considered for the armor layer unless otherwise specified.

A special analysis of the percent and composition of the filling particles is conducted in the following for the armor layer on the pebble bed. For this case, the thickness of the armor layer is equivalent to the size of the coarsest particle on the bed, or the armor layer only includes the first layer on the bed surface. Compared

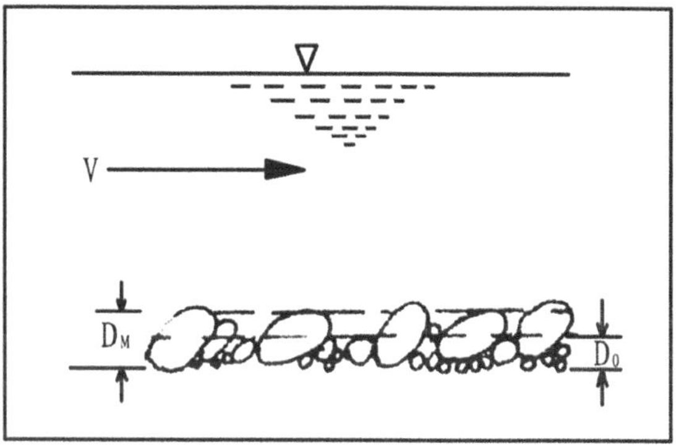

Fig. 9.14 Diagram for the coarser and finer particles in the armor layer

to the layers underneath, this armor layer has a relative small compaction factor (approximately 0.4). Also, the fine particles only fill the lower portion of this layer, which is different from the other layers.

Figure 9.14 shows the diagram of the armor layer that consists of the coarse and fine particles. In this diagram, D_M is the maximum particle size, and D_0 is the minimum particle size of the coarse bed material in the armor layer. Of course, D_0 is also the maximum particle size of the filling particles, and can be used to represent the depth of filling in the armor layer.

The percent finer of the filling particles, or P[D $< D_0$], is defined as the cumulative weight percent of the filling particles with D $< D_0$. Assuming the static compaction factor is 0.4 for the coarse particles, the weight of the coarse particles in the armor layer per unit area is

$$G_c = 0.4 D_M \gamma_s \tag{9.117}$$

where γ_s is the specific weight for the gravel particles that is approximately 2.65 tons/m^3. In the filling sub-layer of the armor layer, the volume of the coarse particles per unit area is 0.4 D_0, and the volume of the filling particles is 0.6 D_0. The grain sizes of these filling particles vary within a wide range, and generally can be as fine as the diameter of suspended load. Han and He [13] performed a specific analysis for the filling of fine particles in the voids between the coarse particles, and proved that the particles that can possibly fill the voids between the coarse particles should be one order of magnitude finer than these coarse particles. Hence, for a pebble bed or for a pebble-sand bed, if the pebble is very coarse (the maximum particle size greater than 300–500 mm), the possible filling particles include the two size groups with 30 mm $> D_1 >$ 10 mm, and with 10 mm $> D_2 >$ 1 mm, respectively. If the pebble is relatively fine (the maximum particle size is less than 150–200 mm), the filling particles are those with 20 mm $> D_1 >$ 5 mm and with 5 mm $> D_1 >$ 0.5 mm.

Let $\gamma'_{s.1}$ and $\gamma'_{s.2}$ be the specific dry weight of the filling particles of the two size classes. The weight of the filling particles with the size of D_1 (first size class) is

$$G_1 = 0.6D_0\gamma'_{s.1} + 0.4D_0\left(1 - \frac{\gamma'_{s.M}}{\gamma_s}\right)\gamma'_{s.1} \tag{9.118}$$

The particles with the size of D_1 and with the specific dry weight of $\gamma'_{s.1}$ will fill the sub-layer that does not contain pebble and that has an equivalent thickness of 0.6 D_0, and the weight of these particles per unit area is $0.6D_0\ \gamma'_{s.1}$ in this sub-layer. Besides, the layer has void rate between pebble particles of size D_0, which is filled with the particles of size D_1 with the weight of $0.4D_0\left(1 - \frac{\gamma'_{s.M}}{\gamma_s}\right)\gamma'_{s.1}$.

The filling particles with the grain size of D_2 only fill the voids between the particles with the size of D_1. The volume of these voids is

$$\left[0.6D_0 + 0.4D_0\left(1 - \frac{\gamma'_{s.M}}{\gamma_s}\right)\right]\left(1 - \frac{\gamma'_{s.1}}{\gamma_s}\right)$$

where $\left(1 - \frac{\gamma'_{s.1}}{\gamma_s}\right)$ is the void ratio of the particles with the size of D_1. The weight of the filling size class of D_2 is

$$\left[0.6D_0 + 0.4D_0\left(1 - \frac{\gamma'_{s.M}}{\gamma_s}\right)\right]\left(1 - \frac{\gamma'_{s.1}}{\gamma_s}\right)\gamma'_{s.2} \tag{9.119}$$

The percent of the filling particles with the grain size of D_1 (in the first size class) is

$$\begin{aligned} P_1(D_1 < D \le D_0) &= \frac{G_1}{G_0 + G_1 + G_2} \\ &= \frac{0.6D_0\gamma'_{s.1}}{0.4\gamma_s D_M + 0.6\gamma'_{s.1}D_0 + 0.6\gamma'_{s.2}\left(1 - \frac{\gamma'_{s.1}}{\gamma_s}\right)D_0} \\ &= \frac{1.5\frac{D_0}{D_M}\gamma'_{s.1}}{\gamma_s + 1.5\left(\gamma'_{s.1} + \gamma'_{s.2} - \frac{\gamma'_{s.1}\gamma'_{s.2}}{\gamma_s}\right)\frac{D_0}{D_M}} \end{aligned} \tag{9.120}$$

The percent of the filling particles with the grain size of D_2 (in the second size class) is

$$\begin{aligned} P_2(D_2 < D \le D_1) &= \frac{G_1}{G_0 + G_1 + G_2} \\ &= \frac{0.6D_0\left(1 - \frac{\gamma'_{s.1}}{\gamma_s}\right)\gamma'_{s.2}}{0.4\gamma_s D_M + 0.6\gamma'_{s.1}D_0 + 0.6\gamma'_{s.2}\left(1 - \frac{\gamma'_{s.1}}{\gamma_s}\right)D_0} \\ &= \frac{1.5\frac{D_0}{D_M}\left(1 - \frac{\gamma'_{s.1}}{\gamma_s}\right)\gamma'_{s.2}}{\gamma_s + 1.5\left(\gamma'_{s.1} + \gamma'_{s.2} - \frac{\gamma'_{s.1}\gamma'_{s.2}}{\gamma_s}\right)\frac{D_0}{D_M}} \end{aligned}$$

$$P_2(D_2) = \frac{G_2}{G_0 + G_1 + G_2} = \frac{\left[1.5 + \left(1 - \frac{\gamma'_{s.M}}{\gamma_s}\right)\right]\left[\left(1 - \frac{\gamma'_{s.1}}{\gamma_s}\right)\gamma'_{s.2}\frac{D_0}{D_M}\right]}{\gamma_s + \left[1.5 + \left(1 - \frac{\gamma'_{s.M}}{\gamma_s}\right)\right]\left[\gamma'_{s.1} + \left(1 - \frac{\gamma'_{s.1}}{\gamma_s}\right)\gamma'_{s.2}\right]\frac{D_0}{D_M}} \quad (9.121)$$

The percent of all filling particles is

$$P_0 = P_1 + P_2 = \frac{G_1}{G_0 + G_1 + G_2} = \frac{1.5\frac{D_0}{D_M}\left(\gamma'_{s.1} + \gamma'_{s.2} - \gamma'_{s.1}\gamma'_{s.2}\right)}{\gamma_s + 1.5\left(\gamma'_{s.1} + \gamma'_{s.2} - \frac{\gamma'_{s.1}\gamma'_{s.2}}{\gamma_s}\right)\frac{D_0}{D_M}}$$

$$P_0 = P_1 + P_2 = \frac{\left[1.5 + \left(1 - \frac{\gamma'_{s.M}}{\gamma_s}\right)\right]\left[\gamma'_{s.1} + \left(1 - \frac{\gamma'_{s.1}}{\gamma_s}\right)\gamma'_{s.2}\right]\frac{D_0}{D_M}}{\gamma_s + \left[1.5 + \left(1 - \frac{\gamma'_{s.M}}{\gamma_s}\right)\right]\left[\gamma'_{s.1} + \left(1 - \frac{\gamma'_{s.1}}{\gamma_s}\right)\gamma'_{s.2}\right]\frac{D_0}{D_M}} \quad (9.122)$$

The specific dry density can be determined with [13]

$$\gamma'_s = 1.89 - 0.483e^{-0.095\frac{D-D_c}{D_c}} \quad (9.123)$$

where D is in mm, the reference particle size $D_c = 1$ mm, and the dry density is in ton/m^3.

The percent finer for all size classes of the bed material is

$$\sum_{k=1}^{l} P_{1.k} = \begin{cases} P_2 & (D_2 < D_l \leq D_1) \\ P_1 + P_2 & (D_1 < D_l \leq D_0) \\ \sum_{k=1}^{l} P'_{i.k}(1 - P_1 - P_2) + P_1 + P_2 & (D_0 < D_l) \end{cases} \quad (9.124)$$

where $P'_{1.l}$ is the percent finer of the coarse particles without considering the filling particles, or the percent that is directly computed with the formula for bed material coarsening.

9.4.4 Verification of Bed Material Coarsening Caused by Scour of Bed Load

This section focuses on the verification of the computed bed material gradation of the armor layer using the field data collected in 1978 within the reach between Huangjiagang and Guanghua of the Han River, which had experienced channel scour since the operation of the upstream Danjiangkou Reservoir. The bed material coarsening of this pebble-sand bed between 1959 and 1978 was illustrated in Table 9.5 and in Fig. 9.11. As listed in Table 9.5, the gradation of the coarse particles in the armor layer computed with $\lambda^* = 0.861$ agrees with the variation trend of the field data collected in 1978. As listed in Table 9.10, the percent finer $\sum P_{1.l}$ computed for the coarse particles (coarser than 50 mm) show a good agreement with the data. It is noted that these results were computed without considering the filling of finer particles. The results computed for $\lambda^* = 0.861$ also show certain discrepancy from the field data that was found to correspond to $\lambda^* = 0.910$.

1. Minimum Particle Size for Coarsening

The average hydraulic parameters are listed in Table 9.11 for this reach between Huangjiagang and Guanghua of the Han River in 1978, when this reach has become equilibrium. This table also list the entrainment velocity that was computed with Eqs. (9.111) through (9.114). A high discharge of 10,000 m^3/s, which would occur for less than 5 days within a year, was used in the computation. Assuming $\overline{D} \approx 80$ mm, $\overline{D}/D_l$ is 0.2 and 0.625 for the particles with $D = 20$ mm and $D = 50$ mm, respectively. Using the entrainment criterion of $\lambda_{q_b} = 0.3 \times 10^{-6}$, the corresponding relative entrainment velocity at the bottom $V_{b.c.1.l}/\omega_{1.l}$ is 0.304 and 0.291. Based on the entrainment velocity, the particles with $D = 20$ mm can be entrained; the particles with $D = 50$ mm cannot be entrained while a small amount of particles are already transported. The bed load transport for these two sizes can also be calculated based on Table 9.11. Assuming the channel width is 500 m, the channel length is 20 km, and $P_{1.l} \approx 0.1$, the daily transport rate for the particles with $D = 20$ mm is

$$Q_b = 500 \text{ m } \times \text{ } 86400 \text{ s } \times 0.0165 \text{ g/(m} \cdot \text{s)} = 0.713 \text{ t/d}$$

The daily transport rate for the particles with $D = 50$ mm is

$$Q_b = 2.82t/d$$

With these transport rates, the annual scour depth would be only 9.82×10^{-6} m and 3.88×10^{-5} m, respectively. The scour depths would be far less than 1 mm even if the scour occurs for 100 year. As listed in Table 9.11, with the same entrainment

Table 9.11 Computed entrainment condition for pebble

D (mm)	Hydraulic parameters				Entrainment computation ($\lambda q_b = 0.3 \times 10^{-6}$)			
	Flow discharge (m^3/s)	Water depth (m)	Flow velocity (m/s)	ψ	$\frac{V_{b.c.1.l}}{\omega_{1.l}}$	$\omega_{1.l}$ (m/s)	$\overline{V}_c$ (m/s)	q_b (g/(m/s))
20	10,000	4.08	1.75	15.98	0.304	1.04	1.35	0.0165
50	10,000	4.08	1.75	13.68	0.291	1.64	1.75	0.0652

criterion, the transport rate for the particles with $D = 50$ mm, which are not considered to be entrained, is higher than the transport rate for the particles $D = 20$ mm, which are considered to be entrained. This is because the entrainment criterion is a relative value, which varies with the particle size and the hydraulic parameters. The transport rate and the scour depth discussed above are the results associated with this criterion, not the results for the same water depth and flow velocity. In fact, as listed in Table 9.11, the bottom flow velocity $V_{b.l} = 0.304 \times 1.04 = 0.316\,\text{m/s}$ and the depth-averaged velocity is 1.35 m/s for the particles with $D = 20$ mm. However, for $D = 50$ mm, the bottom velocity $V_{b.l} = 0.477\,\text{m/s}$, and the depth-average velocity is 1.75 m/s. Due to this significant difference in the flow velocities, the transport rate for $D = 50$ mm is even greater than that for $D = 20$ mm. The larger bottom velocity is due to the larger exposure degree of coarser particles.

Since the particles with $D = 50$ mm cannot be entrained but the particles with $D = 20$ mm can be entrained, the minimum particle size for coarsening D_0 should be between 20 and 50 mm. It is noted that the transport rate for particles with $D = 20$ mm is still very small, which is only 162 tons per year for a flow discharge of 10,000 m^3/s with a flow velocity of 1.75 m/s and with $V_{b.l}/\omega_{1.l} = 0.395$. Also, the particles with the size of 20–50 mm are too coarse to fill the voids between gravel. Hence, the minimum particle size for coarsening can be set to $D_0 = 20\,\text{mm}$.

2. Coarsened Layer Thickness and Calculation of Bed Material Sorting

Let analyze the coarsened layer gradation for the reach between Huangjiagang and Guanghua of the Han River. The basic data of this reach is listed in Table 9.5. The average channel scour depth was 2.53 m. The first stage of scour occurred mainly for coarse to median sand, and the later stage of scour was for pebble. The coarsened layer thickness is different between the coarse to median sand and the pebble. According to the previous analysis, the coarsened layer thickness is approximately 0.40 m for the coarse to median sand, and is equivalent to the maximum particle size for the pebble. For this case, the coarsened layer thickness for the pebble was $h_0 = D_M = 0.153\,\text{m}$. During the first stage of scour, for example, for $\lambda^* \leq 0.748$, the pebble with $D > 20\,\text{mm}$ was barely eroded, and the scour of particles with the size of 10 ~ 20 mm was also insignificant. Therefore, a total of 4 scoured layers, which is approximately equivalent to case for $\lambda^* = 0.748$, is used as the boundary between the scour of the coarse to median sand and the scour of pebble. Within the total scour depth of 2.53 m, the thickness of the four scoured layers of the coarse to median sand was $4 \times h_0 = 1.6\,\text{m}$, and the remaining thickness for pebble scour was 0.93 m. For a coarsened layer thickness of $h_0 = D_M = 0.153\,\text{m}$, the scour of pebble occurred for approximately 6.08 layers. Based on the number of scoured layers, the final scour percentage

$$\lambda_M^* = \frac{h}{h + h_0} = \frac{4 + 6.08}{4 + 6.08 + 1} = \frac{10.08}{11.69} = 0.910.$$

It is noted that the scour depth is calculated here using the number of layers. The dry density is of no importance and thus is not considered.

The gradation of the final coarsened layer corresponding to $\lambda_M^* = 0.910$ is calculated by interpolating the gradations for $\lambda^* = 0.899$ and $\lambda^* = 0.928$. The results are listed in Table 9.12 and are shown in Fig. 9.15. These results can also be directly solved with Eq. (9.98) using the iteration method. The calculation is based on $\upsilon = 2$. As shown in Table 9.12 and Fig. 9.15, the computed gradation for the coarsened layer shows a similar trend as the measured gradation for the armor layer, even without including the fine particles filling the voids.

Table 9.12 Gradation of pebble armor Layer

D (mm)	Computed percent finer for coarsened layer	Percent finer for armor layer(including finer particles filling the voids)	
		Computed	Measured
0.5	0	0	0
1.0	0	0	0
5.0	0	0.0344	0.050
20	0	0.1411	0.100
50	0.147	0.267	0.300
80	0.649	0.699	0.620
100	0.810	0.837	0.890
153	100	1.00	1.000

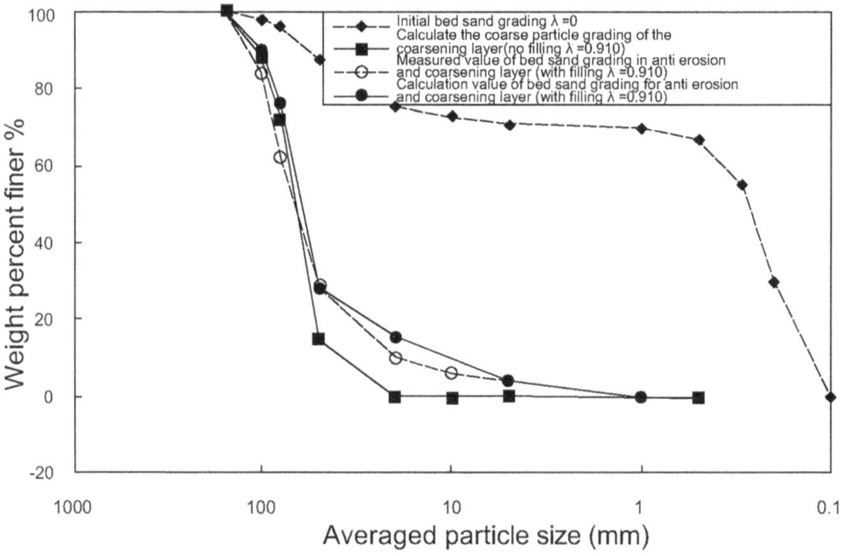

Fig. 9.15 Computed coarsened layer gradation

3. Percent of Filling Particles

As discussed before, the first group of the particles that fill the voids between the coarse particles has a particle size range of 5–20 mm. According to Eq. (9.123), the mean dry density $\gamma'_{s.1} = 1.68\ \text{t/m}^3$. The second group of filling particles has the size of 0.5–5 mm, with the mean dry density $\gamma'_{s,2} = 1.48\,\text{t/m}^3$. The dry density of pebble $\gamma'_{S.M}$ is 1.89 t/m³. For $D_0 = 20\,\text{mm}$ and $D_M = 153\,\text{mm}$, the gradation of the coarser bed material can be computed with the third expression in Eq. (9.124) for $\lambda^* = 0.910$, as listed in the third column of Table 9.12 for the four sizes that are coarser than 20 mm. Then, based on Eq. (9.120)

$$\begin{aligned} P_1(D_1) &= P(5\,\text{mm} < D \le 20\,\text{mm}) \\ &= \frac{\left[1.5 + \left(1 - \frac{\gamma'_{S.M}}{\gamma_s}\right)\right]\gamma'_{s.1}\frac{D_0}{D_M}}{2.65 + \left[1.5 + \left(1 - \frac{\gamma'_{S.M}}{\gamma_s}\right)\right]\left[\gamma'_{s.1} + \left(1 - \frac{\gamma'_{S.1}}{\gamma'_s}\right)\gamma'_{s.2}\right]\frac{D_0}{D_M}} \\ &= \frac{0.392}{2.65 + 0.392 + 0.127} = 0.124 \end{aligned}$$

and based on Eq. (9.121)

$$\begin{aligned} P_2(D_2) &= P(0.5\,\text{mm} < D \le 5\,\text{mm}) \\ &= \frac{\left[1.5 + \left(1 - \frac{\gamma'_{S.M}}{\gamma_s}\right)\right]\left[\left(1 - \frac{\gamma'_{S.1}}{\gamma_s}\right)\gamma'_{s.2}\frac{D_0}{D_M}\right]}{2.65 + 0.392 + 0.127} \\ &= \frac{0.127}{3.169} = 0.041 \end{aligned}$$

According to Eq. (9.122), the percent of all filling particles $P(D_0) = 0.165$. For $\lambda^* = 0.910$ and the given gradation $P'_{1.l}$ of the coarser particles, the percent finer of all particle size classes in the armor layer can be computed with Eq. (9.124), which is

$$\sum_{k=1}^{l} P_{1.k} = \begin{cases} 0.041 & (0.5\,\text{mm} < D_l \le 5\,\text{mm};\ l = 1) \\ 0.165 & (5\,\text{mm} < D_l \le 20\,\text{mm};\ l = 2) \\ 0.165 + P'_{1.l}(1 - 0.165) & (20\,\text{mm} < D_l;\ l > 2) \end{cases}$$

As shown in Table 9.12 and Fig. 9.15, after considering the finer particles that fill the voids between the coarse particles, the computed gradation of the armor layer shows a good agreement with the data. A greater discrepancy between the computed results and the data is found for the particle sizes of 50–80 mm. The percent of these size classes increases dramatically, and the slope of the cumulative percent curve is too steep. As a result, a greater absolute error is found for these size classes, while this error is not obviously observed from the figure.

9.5 Bed Material Coarsening Caused by Scour of Suspended Load

As discussed before, bed material coarsening caused by the suspended load movement can also be divided into two cases. The first case is the coarsening caused by the massive scour of suspended load. For this case, all size classes experience scour, without deposition, and thus no exchange occurs between the suspended load and bed material. This is more general than the definition of clear-water scour, while this also simplify the analysis. The second case is the coarsening caused by the exchange between the suspended load and the bed material. This case also include the general scour, in which the size classes may have different sedimentation patterns. This section focuses on the coarsening caused by the massive scour of suspended load.

Considering as one of the important topics for the non-equilibrium transport of non-uniform suspended load, Han [4, 5] conducted a specific analysis for the bed material coarsening caused by the massive scour of suspended load, and described the process of bed material coarsening during the scour. Han also derived the two modes and the corresponding equations for the variation of bed material gradation during the massive scour process. The results are detailed in Sect. 5.3.2. The following analysis is conducted for one mode, which is described with Eq. (5.69), and then based on which consider the filling of the finer particles. Equation (5.69) can be written as

$$P_{1.l} = P_{1.l.0}\left(1 - \lambda^*\right)^{\left[\left(\frac{\omega_m}{\omega_l}\right)^{\upsilon} - 1\right]} \tag{9.125}$$

where λ^* is the scour percentage, ω_m the median fall velocity, υ the sorting factor that typically varies between 0.5 and 1. It is suggested that $\lambda^* = 0.863$ for rivers, and $\upsilon = 0.5 \sim 3/4$ for reservoirs. From Eq. (7.267) there is

$$P_{1.l} = P_{1.l.0}\left[\frac{W}{W_0}\right]^{\left(\frac{\omega_m}{\omega_l}\right)^{\upsilon} - 1} = P_{1.l.0}(1 - \lambda^*)^{\left(\frac{\omega_m}{\omega_l}\right)^{\upsilon} - 1} \tag{9.126}$$

where the scour percentage is

$$\lambda^* = \frac{\gamma_s' h}{\gamma_s' h + \gamma_{s.0}' h_0} \approx \frac{h}{h_0 + h} \tag{9.127}$$

As shown in Sect. 7, Eq. (7.265)

$$P_{4.l}^* = \left(\frac{\omega_m}{\omega_l}\right)^{\upsilon} P_{1.l} \tag{9.128}$$

is used to derive Eq. (9.125). For the case of massive scour the scour occurs for all size classes, while the scour quantities are different. Hence, the scour percentage λ^* can also be expressed in terms of the sediment transport quantities, or

$$\lambda^* = \frac{(G_s - G_{s.0})\Delta t}{\gamma_s' h_0 \Delta L B + (G_s - G_{s.0})\Delta t} = \frac{(QS - Q_0 S_0)\Delta t}{\gamma_s' h_0 \Delta L B + (QS - Q_0 S_0)\Delta t} \tag{9.129}$$

where Q is the flow discharge, S the sediment concentration, and G the sediment transport rate, the subscript "0" denotes the value at the entrance section, a parameter without the subscript "0" denotes the value at the exit section, Δt is the time of scour, B the erodible width of the bed material, and ΔL the channel length

It is noted that Eq. (9.128) is similar to the formula for the coarsening associated from bed load scour, or. Equation (9.82), with $\left(\frac{D_m}{D_l}\right)^{\upsilon}$ being replaced with $\left(\frac{\omega_m}{\omega_l}\right)^{\upsilon}$.

Using the formula for the coarsening associated from the bed load scour, the gradation of the coarsened layer was computed based on the field data collected on a sand bed within the reach between Huangjiagang and Guanghua of the Han River that is downstream of the Dangjiangkou Reservoir. The results are listed in Table 9.6 and are shown in curve C in Fig. 9.13. As shown in Fig. 9.13, curve C (computed results) shows a great discrepancy from curve B (field data), indicating the bed material coarsening on a sand bed cannot be explained as the results of bed load scour. For a sand bed, all bed material can be entrained. There is an intensive exchange between the bed material and the sediment in the water body, and their compositions do not show big difference. As a result, the bed material coarsening develops very slowly during a scour.

According to the formula for the coarsening associated from the suspended load scour, or Eq. (9.126), the gradation of the coarse particles in the coarsened layer was computed for $\lambda^* = 0.863$. The result is shown in curve D of Fig. 9.13, which shows a good agreement with the data, or curve B. The computed gradation for all size classes in the coarsened layer, after considering the finer particles that fill the voids between the coarse particles, is shown in curve E of Fig. 9.13. Curve E shows an even better agreement with the data (curve B).

Equation (9.125) with $\upsilon = 2/3$ is verified using the bed material data collected during 1961 to 1964, the period of impoundment and flood detention operations of the Sanmenxia Reservoir, in the reaches between Huayuankou and Gaocun and between Gaocun and Aishan of the lower Yellow River. The comparison of the computed results and the data are shown in Figs. 9.16 and 9.17 for these two reaches, respectively. The solid dots in these figures are the computed results without considering the filling of finer particles, which show an overall reasonable agreement with the data. Because the filling of finer particles is not included, the percent for $D \leq 0.05$ mm is slightly underestimated. After considering the impact of the filling, the computed percent for $D \leq 0.05$mm (open dots in these figures) shows a better agreement with the data.

The following method can be used to consider the filling of finer particles in the coarsened layer that is caused by the scour of suspended load. For the scour of

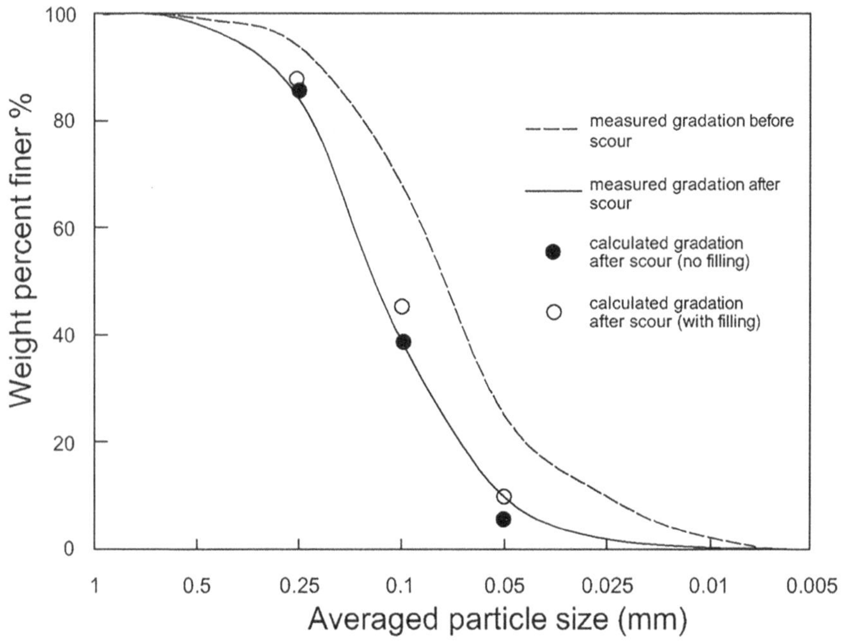

Fig. 9.16 Verification of Eqs. (9.125) and (9.124) (Huayuankou to Gaocun Reach, Yellow River, China)

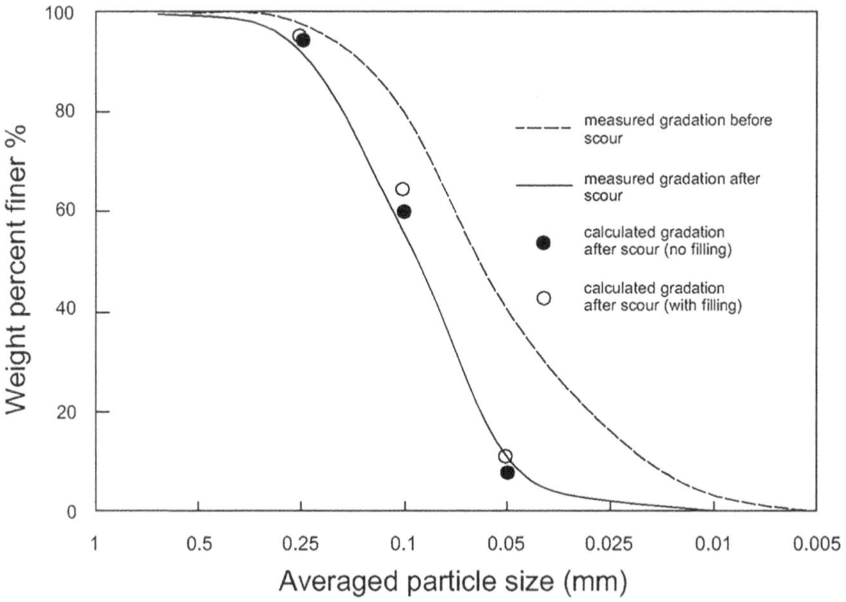

Fig. 9.17 Verification of Eqs. (9.125) and (9.124) (Gaocun to Aishan Reach, Yellow River, China)

suspended load, all the particles can be eroded, and thus the concept of the minimum particle size for coarsening, D_0, does not apply for this case. However, from the perspective of bed material exchange, a threshold particle size still exists [8, 9]. Under the condition of mild scour and mild deposition, the particles coarser than this threshold size will deposit, and the particles finer than this size will be eroded. This threshold particle size D_k is a function of the bed material gradation, or

$$D_k = f^{-1}(\omega_1^*) \tag{9.130}$$

The mean fall velocity of the bed material ω_1^* is defined as

$$S^*\left(\omega_1^*\right) = \sum_{l=1}^{n} P_{1.l.1} S^*(l) \tag{9.131}$$

in which $P_{1.l.1}$ is the bed material gradation. The computed D_k based on this method is approximately equivalent to the threshold particle size to separate the wash load from the bed material load for a bed with median to fine sand [8]. Table 9.13 lists the values of D_k and the weight percent for particles finer than D_k for a number of reaches, consisting of median to fine sand, of the Yangtze River. As listed in this table, the percent for particles finer than D_k is 5–10%. This is consistent with the criteria proposed by Einstein and Zhang [19] to separate the wash load from the bed material load. From the aspect of the mechanism for the suspended load and bed material exchange, the particles finer than D_k cannot stay on the bed and should be completely scoured away. Hence, these finer particles that fall on the bed must fill in the voids between the coarse particles and be sheltered by these coarse particles in order to not be scoured away. In other words, these finer particles only exist in the voids between the coarse particles on the bed.

The percent of filling particles depends on the thickness of filling. Apparently, the thickness of the coarse particles is the thickness of the entire coarsened layer, h_0. There are two situations for the thickness of the fine particles. If the bed has constant supply of fine sediment, such as sediment influx from tributaries, bank collapses, and scour of high floodplain, the filling of fine particles is more complete, and thus the thickness of the fine particles is only slight less than h_0. If the channel has experienced a long-term scour without supply of fine sediment, the thickness of fine particles is much less than h_0, and it is typically $h_0/2$ or less. Following a similar derivation as used in Sect. 9.4.3, the weight of the coarse particles in the coarsened layer with

Table 9.13 Threshold particle size Dk and corresponding percent finer for a number of reached in the Middle Yangtze River, China

Reaches	D_k (mm)	Percent finer
ChenjiaZhuang–Haoxue	0.200	6
Haoxue–Ouchikou	0.080	8
Ouchikou–Jianli	0.052	5
Jianli–Chenglingji	0.018	10

a thickness of h_0 is $h_0\gamma'_{s.M}$, the volume of the fine particles that fill the voids is $Kh_0\left(1-\frac{\gamma'_{s.M}}{\gamma_s}\right)$, and the corresponding weight is $Kh_0\left(1-\frac{\gamma'_{s.M}}{\gamma_s}\right)\gamma'_{s.1}$. Here, K is the percent of the voids that is filled, and $K \le 1$. The percent of the filling particles is

$$P_0(D < D_k) = \frac{Kh_0(1-\frac{\gamma'_{s.M}}{\gamma_s})\gamma'_{s.1}}{h_0\gamma'_{s.M} + Kh_0(1-\frac{\gamma'_{s.M}}{\gamma_s})\gamma'_{s.1}} = \frac{K(1-\frac{\gamma'_{s.M}}{\gamma_s})\gamma'_{s.1}}{\gamma'_{s.M} + K(1-\frac{\gamma'_{s.M}}{\gamma_s})\gamma'_{s.1}} \tag{9.132}$$

where K = 1/2–1/5. According to Eq. (9.132), the dry density $\gamma'_{S.M}$ is 1.39 t/m^3 for the coarse particles with 1 mm $> D > D_k$. For $D < D_K$, the dry density $\gamma'_{s.1} = 1.1\text{t/m}^3$. Also, $\gamma_s = 2.65\text{t/m}^3$. Based on these parameters, $P(D_0) = 0.158$ for $K = 1/2$; $P(D_0) = 0.110$ for $K = 1/3$; $P(D_0) = 0.086$ for $K = 1/4$; and $P(D_0) = 0.077$ for $K = 1/5$. It is noted that the percent of filling particles shows a narrow variation range, and thus its impact to the bed material gradation is very limited. Considering the empirical method of separating wash load from bed material load, it is reasonable to assume $P(D < D_k) = 0.05 - 0.10$. If the scour occurs in a long period, the bed material is constant stirred up because of the movement of sand ripples and the local sedimentation, it is recommended to set K to 1/3–1/5. Once the percent of the filling particles is determined, the bed material of the coarsened layer can be computed with Eq. (9.124).

Based on the field data of the sand beds, as shown in Figs. 9.13, 9.16, and 9.17, the following describes how to determine the threshold particle size D_k and the percent of filling particles $P_0(D < D_k)$. Figure 9.13 shows the data for suspended load scour in the reach between Huangjiagang and Guanghua in the lower Han River. According to Han [8, 9], D_k is 0.25 mm for the reach upstream of Huangzhuang in the Han River. Assuming $K = 1/4$, then $P_0 = 0.09$. Based on the gradations of the coarse particles, the gradation of all size classes in the coarsened layer can be computed with Eq. (9.124). The result is shown in curve E of Fig. 9.13, which agrees well with the data (curve B). For the data shown in Figs. 9.16 and 9.17, D_k is approximately 0.05 mm based on Table 9.13, and the $P(D < D_k) = 0.09$ for $K = 1/4$. As shown in these two figures, after considering the filling particles, the computed percent for $D \le 0.05$ mm shows a better agreement with the data, while the percent for $D \le 0.1$ mm shows a greater discrepancy from the data. This indicates the adjustment to include the filling particles is not practically meaningful if the coarse particles in the coarsened layer are so fine that it is difficult for the filling of even finer particles.

9.6 Bed Material Coarsening Caused by Sediment Exchange

9.6.1 Bed Material Coarsening Caused by Exchange with Suspended Load

The concept and the mechanism of bed material coarsening caused by exchange with suspended load is discussed in Sect. 9.2. For simplicity, it is assumed that the deposition of the coarse suspended sediment particles and the scour of bed material occur independently. These two process are combined through the variation of the bed material composition. As discussed before, the coarser suspended particles, in the amount of $P_{4.2}SP_{4.1.2}$, will deposit when they are exchange with the bed material. The deposition quantity within a distance of Δx should be proportional to $P_{4.2}SP_{4.1.2}$, the fall velocity ω_l, and Δx, and be inversely proportional to the flow discharge per unit width. Comparing the governing equation for 1-D non-equilibrium sediment transport [5] with the formulas for the size distribution of carrying-sediment capacity [8, 10], the derivative of the deposition quantity can be written as

$$\frac{d(P_{4.l}P_{4.l.2}S)}{dx} = -\frac{\alpha(P_{4.l}P_{4.l.2}S)\omega_l}{q} \quad (l \geq k+1) \tag{9.133}$$

where α is the saturation recovery coefficient. Integrating this equation from x = 0 to x with the boundary conditions: at $x = 0$, $P_{4.2}P_{4.l.2}S = P_{4.2.0}P_{4.l.2.0}S_0$, one obtains

$$\ln\frac{P_{4.2}SP_{4.l.2}}{P_{4.2.0}SP_{4.l.2.0}} = -\frac{\alpha\omega_l}{q}x \tag{9.134}$$

or

$$P_{4.2}P_{4.l.2}S = P_{4.2.0}P_{4.l.2.0}S_0\exp\left(-\frac{\alpha\omega_l x}{q}\right)$$

in which α can be set to 0.25. The weight of the coarse particles in the lth size class that deposit within the distance of x is

$$\begin{aligned}\Delta G_l &= (P_{4.2.0}S_0P_{4.l.2.0} - P_{4.2}SP_{4.l.2})Q\Delta t\\ &= P_{4.2.0}S_0P_{4.l.2.0}\left[1-\exp\left(-\frac{\alpha\omega_l x}{q}\right)\right]Q\Delta t\end{aligned} \tag{9.135}$$

The total deposition quantity for all size groups is

$$\Delta G = \sum_l \Delta G_l = P_{4.2.0}S_0(1-K_{x.2})Q\Delta t \tag{9.136}$$

where

$$K_{x.2} = \sum_{l} P_{4.l.2.0} \exp\left(-\frac{\alpha\omega_l x}{q}\right)\Delta G_l \tag{9.137}$$

From Eqs. (9.135) and (9.136), the weight percent of the coarse particles that deposit on the bed can be determined, which is

$$P_{1.l.1.D} = \frac{\Delta G_l}{\sum_l \Delta G_l} = \frac{P_{4.l.2.0}\left[1 - \exp\left(-\frac{\alpha\omega_l x}{q}\right)\right]}{\sum P_{4.l.2.0}\left[1 - \exp\left(-\frac{\alpha\omega_l x}{q}\right)\right]} = \frac{P_{4.l.2.0}\left[1 - \exp\left(-\frac{\alpha\omega_l x}{q}\right)\right]}{1 - K_{x.2}} \tag{9.138}$$

The virtual depth of total deposition is

$$\tilde{h}_D = \frac{P_{4.2.0} S_0 (1 - K_{x.2}) Q \Delta t}{Bx} = \frac{\Delta G}{Bx} \tag{9.139}$$

The virtual depth of scour $\tilde{h}_E$ depends on the sediment inflow and outflow of a given reach and the deposition quantity of the coarse suspended particles. Apparently

$$\frac{\sum (P_{4.l} S - P_{4.l.0} S_0) Q \Delta t}{Bx} = \tilde{h}_{\mathbb{E}} - \tilde{h}_0 \tag{9.140}$$

$$\tilde{h}_E = \tilde{h}_0 + \sum (P_{4.l} S - P_{4.l.0} S_0) \frac{Q \Delta t}{Bx} = \tilde{h}_0 + \tilde{h}_D \tag{9.141}$$

As shown in Eq. (9.141), if $\tilde{h}_E - \tilde{h}_0 > 0$, scour occurs for the entire reach; if $\tilde{h}_E - \tilde{h}_0 < 0$, deposition occurs; and if $\tilde{h}_{\mathbb{E}} - \tilde{h}_0 = 0$, the channel is equilibrium.

The bed material coarsening during the scour of bed material can be computed with Eq. (9.125), which is developed for the case of massive scour,

$$P_{1.l.1.E} = P_{1.l.0}\left(1 - \lambda^*\right)^{\left[\left(\frac{\omega_m}{\omega_l}\right)^{\upsilon} - 1\right]}$$

Assuming the virtual scour depth $\tilde{h}_E > 0$, the scour percentage is

$$\lambda* = \frac{\tilde{h}_E}{\tilde{h}_E + \tilde{h}_0} \tag{9.142}$$

where $\tilde{h}_0$ is the virtual thickness of the coarsened layer, and $\tilde{h}_0 = 0.4\ \text{t/m}^2$ can be adopted. After the scour and the deposition of the coarse suspended particles, the gradation of the ultimate bed material is

$$P_{1.l.1} = \begin{cases} \frac{P_{1.l.1.D}\tilde{h}_D + P_{1.l.1.E}\tilde{h}_0}{\tilde{h}_D + \tilde{h}_0} & (l \geq k+1, \quad \tilde{h}_E > 0) \\ \frac{P_{1.l.1.E}\tilde{h}_0}{\tilde{h}_D + \tilde{h}_0} & (l < k \quad \tilde{h}_E > 0) \end{cases} \tag{9.143}$$

where $\tilde{h}_0$ is the thickness of the coarsened layer, $P_{1.l.1.E}$ is the bed material gradation after the pure scour.

An example is given in the following to illustrate the procedure for the calculation of bed material coarsening caused by the exchange with suspended load. Assume the channel is 50 km long and 1000 m wide, with a flow discharge of 5000 m^3/s and with a scour period of one year. The sediment concentration at the entrance is $S_0 = 0.2603\,\text{kg/m}^3$, and the size distribution of suspended load and the bed material gradation are listed in Table 9.14. With the bed material gradation data,

$$\omega_1^{*-0.92} = \sum P_{4.l.0}\omega_l^{-0.92} = 1.11$$

or

$$\omega_1^{*0.92} = (0.898\,\text{cm/s})^{0.92}$$

Hence, $D_k = f^{-1}\left(\omega_1^{*0.92}\right) = 0.104\,\text{mm} \approx 0.1\,\text{mm}$. There are three size classes in the sediment inflow with $D > D_k$. They are 0.10–0.25 mm, 0.25–0.5 mm, and 0.5–1.0 mm, and the percent is $P_{4.2} = 0.24$ Based on Eq. (9.137)

$$K_{2.x} = \sum_{l=3}^{5} P_{4.l.2.0} e^{-\frac{\alpha\omega_C x}{q}} = 0$$

Based on Eq. (9.138)

$$P'_{1.l.1.D} = \frac{P_{4.l.2.0}\left[1 - \exp\left(-\frac{\alpha\omega_l x}{q}\right)\right]}{1 - K_{x.2}}$$

From Eq. (9.135),

$$\Delta G = \sum_{l=4}^{6} P_{4.2.0} S_0 P_{4.l.2.0} Q\Delta t = 0.24 S_0 Q \Delta t = 9.85 \times 10^6 t$$

The annual transport quantity of suspended load is $S_0 Q \Delta t = 41.04 \times 10^6 t$, and the annual sediment inflow of the coarse particles with $D > 0.1$ mm is $0.24 \times 41.04 \times 10^6 t = 9.85 \times 10^6 t$. Assume all coarse particles deposit ($K_{2.x} = 0$), the fractional weight of these three size classes are listed in Table 9.14. Assuming the channel has no net scour or deposition, or $\tilde{h}_E = 9.85 \times 10^6 t/(5000\,\text{m} \times 1000\,\text{m}) = 0.197\,\text{t/m}^2$, the scour percentage computed with Eq. (9.142) is $\lambda^* = 0.330$. The computed bed material gradation after the coarsening ($P_{1.l.1.E}$) is listed in Table 9.14.

Table 9.14 Computation of the exchange between coarser and finer sediment particles

Size class No.	D_1 (mm)	Initial grain size distribution			$P_{1.l.1.E}$ ($\lambda^* = 0.330$)	Original bed material weight (10^6 tons)			Deposited coarser sediment		Bed material after scour or deposition		Net Sedimentation weight (10^6 tons)	Sediment transport at exit section	
		$P_{4.1.0}$	$P_{1.l1.0}$	$\omega_l^{0.92}\left(\text{cm}^{0.92}\right)$		Participate in scour	Scoured	Remaining on bed	Weight (10^6 tons)	$P'_{1.l.l.D}$	Weight (10^6 tons)	$P_{1.l.1}$		Weight (10^4 tons)	$P_{4.l}$
1	0.01–0.025	0.10	0											4.104	0.100
2	0.025–0.05	0.18	0.05	0.118	0	1.492	− 1.492	0					− 1.492	8.879	0.216
3	0.05–0.10	0.48	0.15	0.589	0.094	4.478	− 2.598	1.88			1.880	0.063	− 2.598	22.297	0.543
4	0.10–0.25	0.14	0.65	1.650	0.712	19.402	− 5.162	14.24	5.746	0.583	19.986	0.670	+ 0.584	5.162	0.126
5	0.25–0.50	0.07	0.15	3.607	0.194	4.478	− 0.598	3.88	2.873	0.292	6.753	0.226	+ 2.275	0.598	0.015
6	0.50–1.00	0.03							1.231	0.125	1.231	0.041	+ 1.231	0	0
Total						29.85	− 9.85	20.00	9.85		29.85	1.00	0	41.040	1.000

The scour quantity corresponding to $\tilde{h}_E$ is $G_{S.E} = 0.197\,\text{t/m}^2 \times 1000\,\text{m} \times 50 \times 1000\,\text{m} = 9.87 \times 10^6\text{t}$. The fractional quantities of sediment erosion (the virtual scour depths corresponding to the pebble with equivalent sizes) are listed in Table 9.14 for various size classes. The fractional quantities of the sediment that take part in the scour, which are

$$W_l = (h_0 + h_E) \times 1000\,\text{m} \times 5000\,\text{m} \times P_{1.l.1.0} = 29.85 \times 10^6 t \times P_{1.l.1.0}$$

and the fractional quantities of the sediment that take part in scour but is not scoured away are also listed in this table. Also listed in this table are the fractional net sedimentation quantities for different size classes. The scour quantity of the finer particles with $D < 0.1$ mm is 4.09×10^6 t, which is equivalent to the deposition quantity of the coarser particles with $D > 0.1$ mm, resulting in not net deposition or scour on the bed. This is a typical example for the exchange between the coarse and fine sediment particles in an alluvial stream. Under the simultaneous actions of the scour of the finer bed material and the deposition of the coarser suspended sediment particles, bed material coarsening develops quickly. In fact, based on Eq. (9.11), the computed post-scour depth of the bed material is $\tilde{h}_E = \tilde{h}_0 = 0.197 + 0.400 = 0.597\text{t/m}^2$ (equivalently, 29.85×10^6 t), and the corresponding bed material gradation is $P_{1.l.1}$, as listed in Table 9.14. Comparing $P_{1.l.1}$ with $P_{1.l.1.0}$, a rapid bed material coarsening can be observed, with $\overline{D}$ increasing from 0.183 to 0.237 mm. The weight percent of the particles coarser than 0.10 mm increases from 80 to 93.5%, the percent of particles coarser than 0.25 mm increases from 15 to 26.7%. In particular, the bed material after the coarsening include the coarse particles with the diameter of 0.5–1 mm, and these particles do not exist in the original bed material. It is noted that the bed material discussed here only includes the sediment within the active layer (0.597 t/m^2) that has been disturbed, and does not include the layers underneath this active layer. Corresponding to bed material coarsening, the suspended load composition at the exit section ($P_{4.l}$) is apparently finer than that at the entrance section ($P_{4.l.0}$). The percent of the coarse particles with $D > 0.1$ mm decreases from 24% at the entrance section to 14.1% at the exit section. The average fall velocity $\overline{\omega^{0.92}}$ decreases from 0.0989 to 0.0597, or a 39.6% decrease. For the same hydraulic condition, the carrying-sediment capacity will increase by 39.6% from the entrance section to the exit section.

9.6.2 *Bed Material Coarsening Caused by Exchange with Bed Load*

The mechanism for bed material coarsening associated with the exchange with bed load was discussed in previous sections based on theoretical analysis and the field data. As bed load move on the bed, it constantly exchange with the bed material. Hence, after assuming all bed load from the upstream deposits on the bed and mixes with bed material, the scour of bed load can be computed, and the bed material coarsening associated from the exchange with bed load can be determined.

Let assume the fractional bed load transport rate is $P_{b.l.o}q_{b.0}$ at the entrance of a reach, and is $P_{b.l}q_b$ at the exit section, the channel width is B, the channel length is x, and the initial bed material gradation is $P_{1.l.1.0}$. After the bed load inflow mixes with the bed material, the gradation of the mixture, or the effective bed material gradation at the beginning of scour is

$$P_{1.l.0} = \begin{cases} \frac{\tilde{h}_D P_{b.l.0} + P_{1.l.1.0}(\tilde{h}_0 - \tilde{h}_D)}{\tilde{h}_0} & (\tilde{h}_D < \tilde{h}_0) \\ P_{b.l.0} & (\tilde{h}_D > \tilde{h}_0) \end{cases} \tag{9.144}$$

The virtual depth for the deposition of the bed load inflow is

$$\tilde{h}_D = \frac{q_b B_b \Delta t}{B \Delta x} \tag{9.145}$$

where B_b is the width of bed load movement. For a sand bed, $B_b \approx B$. For a pebble/gravel bed, $B_b < B$. It can be determined based on the field data. If no data is available, B_b can be empirically set to 0.7–0.9B. In numerical models, $\tilde{h}_D < \tilde{h}_0$ is typical true for each computation segment of a reach. As a result, the first expression of Eq. (9.144) is always adopted. The case with $\tilde{h}_D > \tilde{h}_0$ is only used in the long-term estimate.

After the effective bed material gradation $P_{11.0}$ is determined, the bed material sorting can be computed with Eq. (9.98). Then, the gradation of the coarsened bed material layer, $P_{1.l}$, and the composition of the scoured sediment, or the composition of bed load at the exit section, $P_{b.l}$ can be determined,

$$\begin{aligned} P_{b.l} &= \frac{P_{11.0}(\tilde{h}_D + \tilde{h}_0) - P_{1.l}(1 - \lambda^*)(\tilde{h}_D + \tilde{h}_0)}{(\tilde{h}_D + \tilde{h}_0)} \\ &= P_{11.0}\left[1 - (1 - \lambda^*)\left(\frac{D_m}{D_l}\right)^v\right] = P_{1.l.0}\left[1 - (1 - \lambda^*)P_{1.l}\right] \end{aligned} \tag{9.146}$$

where $P_{1.l}$ is determined with Eq. (9.98).

References

1. Feldman AD (1981) HEC models for water resources system simulation, theory and experience. The Hydraulic Engineering Center, Davis, CA
2. Gesseler J (1971) Critical shear stress for sediment mixture. In: Proceedings of 14th International Association of Hydraulic Research, vol 3
3. Gesseler J (1971) Beginning and creasing of sediment motion, vol 1. River Mechanics
4. Han QW (1973) Preliminary studies for the non-equilibrium transport of sediment in reservoirs. Report on Sedimentation in Reservoir, Task Group on Yellow River Sedimentation Research, pp 145–168 (in Chinese)

5. Han QW (1979) A study of the non-equilibrium transportation of non-uniform suspended load. China Sci Bull 24(17):804–808 (in Chinese)
6. Han QW (1982) Reservoir sedimentation, vol 5. Yangtze Valley Planning Office, Hydrology Department, Oct (in Chinese)
7. Han QW (2003) Reservoir sedimentation. Science Press, Beijing, China (in Chinese)
8. Han QW (2005) Size distribution of carrying sediment capacity. In: Proc of 6th National Symp on BasicTheory of Sedimentation. Water Conservancy of Yellow River Press, pp 3–20 (in Chinese)
9. Han QW (2007) Exchange between coarse and fine particles during deposition and erosion along river. In: Proceedings of the tenth international symposium on river sedimentation, Moscow, Russia, pp 139–144
10. Han QW (2009) In-depth study on non-equilibrium transport of non-uniform suspended load Technical Report, China Institute of Water Resources and Hydropower Research (IWHR) (in Chinese)
11. Han QW, He MM (1981) Statistical characters of sediment exchange. J Hydraulic Eng 1:9–22 (in Chinese)
12. Han QW, He MM (1984) Stochastic theory of sediment motion. Science Publishing House, Beijing, China (in Chinese)
13. Han QW, He MM (1999a) Study on state probabilities and the ratio of bed load to suspended load. J Hydraulic Eng 30(10):7–16 (in Chinese)
14. Han QW, He MM (1999b) Characteristics of incipient sediment motion and incipient velocity. Science Press, Beijing, China (in Chinese)
15. Han QW, Xiang XL, Wang YC (1983) A study on armoring of bed material. In: Proc. of 2nd Int. Symp. on River Sedimentation. Water Resource and Electric Power Press, pp 366–375
16. He MM, Han QW (1990) The determination of the size distributions of carrying capacity and effective bed material. J Hydraulic Eng 3:7–16 (in Chinese)
17. Li YT (1987) Preliminary study on bed material load gradation in equilibrium transport during deposition and erosion. J Sediment Res 1:82–87 (in Chinese)
18. Qin RY, Wang CH (1996) Theory and application of bed load movement in river. China Railway Publishing House, Beijing, China (in Chinese)
19. Wuhan Institute of Hydraulic and Electric Engineering (WIHEE) (1982) River sedimentation engineering. Hydraulic and Electric Press, China (in Chinese)
20. Xie JH (1959) Coarsening calculation for river bed. J Wuhan Inst Hydraulic Electr Eng 2 (in Chinese)
21. Yang CT (1996) Sediment transport theory and practice. The McGraw-Hill Companies, Inc.
22. Yin XL (1963) River bed coarsening during erosion by clean-water inflow. J Hydraulic Eng 11 (in Chinese)

Chapter 10
Exchange Between Coarser and Finer Sediment and Comparison Between Bed Material Load and Wash Load

10.1 Exchange Between Coarser and Finer Sediment Particles

10.1.1 Field Data

Most studies consider the sedimentation properties for the particles with different sizes are the same, and assume either deposition or scour occurs for all size classes, with various sedimentation quantities. In most mathematical model, the sedimentation quantity is related to the difference between the sediment concentration and the carrying-sediment capacity, $S–S^*$, for uniform sediment, and to $P_{4.l}S–P_{4.l}S^*$ for non-uniform sediment. By doing this, the sediment properties are assumed to be the same for various size classes, which is a simplification of the comprehensive phenomena of sedimentation.

Tables 10.1, 10.2, 10.3 and 10.4 list the fractional sedimentation quantities for various size classes that were measured in the reaches between Huangjiagang and Xiangyang, and between Xiangyang and Huangzhuang of the Han River downstream of the Danjiangkou Reservoir, the variable backwater zone of the Danjiangkou Reservoir, the reach between Xinchang and Jiangli of the Yangtze River, a downstream reach of the Yellow River between Aishan and Lijin, and the middle and lower reach of the Yangtze River downstream of the Three Georges Reservoir, respectively. These data sets were compiled from previous literatures and field data. [4, 16, 18, 19, 21].

Q. Han and M. He, *Non-equilibrium Transport of Non-uniform Suspended Load*, Hydroscience and Engineering, https://doi.org/10.1007/978-981-97-1123-9_10

Table 10.1 Field data of fractional sedimentation quantities

Reach	Fractional sedimentation quantity (10^4 t)								Total quantity (10^4 t)			Transport rate at exit (10^4 t)
	Year	$D < 0.01$	0.01 ~ 0.025	0.025 ~ 0.050	0.05 ~ 0.10	0.10 ~ 0.25	0.25 ~ 0.50	0.50 ~ 1.00	Net	Scour	Deposition	
Han River, Xiangyang–Huangzhuang	1974	− 35	− 38	− 282	− 532	− 536	− 42	+ 9	− 1456	− 1465	+ 9	
	1975	− 458	− 891	− 631	− 674	− 364	+ 692	+ 909	− 2236	− 3018	+ 782	
	1976	− 96	− 72	− 94	− 291	− 396	− 43	+ 10	− 981	− 991	+ 10	
	1977	− 29	− 98	+ 37	− 134	− 253	+ 7.8	+ 11	− 459	− 515	+ 56	
	1978	− 24	− 42	− 62	− 117	− 96	+ 15	+ 4	− 322	− 342	+ 20	
	1979	− 127	− 64	− 31	− 222	− 339	+ 35	+ 4	− 743	− 782	+ 38	
	1980	− 167	− 230	− 248	− 258	− 274	+ 27	+ 17	− 1132	− 1177	+ 45	
	1981	− 125	− 154	− 243	− 341	− 343	+ 103	+ 33	− 1071	− 1206	+ 135	
	1982	− 96	− 56	− 159	− 291	− 386	− 96	+ 13	− 1071	− 1084	+ 13	
	1983	− 184	− 324	− 633	− 1315	− 619	+ 1337	+ 8	− 2884	− 3025	+ 141	
	1984	− 313	− 239	− 348	− 395	− 450	+ 78	+ 16	− 1651	− 1745	+ 94	
Han River, Huangz huang–Xiantao	1974	− 287	− 268	− 190	− 160	+ 20	+ 132		− 753	− 905	+ 152	
	1975	− 163	− 249	− 476	+ 74	+ 467	+ 423		+ 76	− 888	+ 964	5890
	1976	− 51	− 84	− 95	− 15	+ 80	+ 116		− 48	− 244	+ 196	
	1977	− 28	+ 107	− 23	− 154	+ 30	+ 66		− 2	− 204	+ 202	
	1978	− 37	− 34	− 29	− 61	− 38	+ 44		− 155	− 199	+ 44	
	1979	− 27	− 64	74	− 105	+ 90	+ 53		+ 21	− 196	+ 217	
	1974–1979	− 593	− 591	− 888	− 421	+ 649	+ 833		861	− 2636	+ 1775	
	1980	− 479	− 248	− 289	− 691	− 279	+ 183	+ 17	− 1786	− 1987	+ 200	
	1981	− 334	− 312	− 380	− 817	− 496	+ 210	+ 32	− 2097	− 2339	+ 242	

(continued)

Table 10.1 (continued)

Reach	Fractional sedimentation quantity (10^4 t)								Total quantity (10^4 t)			Transport rate at exit (10^4 t)
	Year	$D < 0.01$	0.01 ~ 0.025	0.025 ~ 0.050	0.05 ~ 0.10	0.10 ~ 0.25	0.25 ~ 0.50	0.50 ~ 1.00	Net	Scour	Deposition	
	1982	– 422	– 213	– 262	– 688	– 392	+ 102	+ 13	– 1863	– 1978	+ 115	
	1983	– 957	– 888	– 1137	– 1421	– 962	+ 362	+ 55	– 4948	– 5365	+ 417	
	1984	– 578	– 459	– 660	– 1133	– 578	+ 300	+ 19	– 3088	– 3407	+ 318	
	1980–1984	– 2771	– 2120	– 2727	– 4751	– 2707	+ 1156	+ 136	– 13,783	– 15,075	+ 1292	
Variable Backwater Zone, Danjiangkou Reservoir Baihe-Youfanggou	1980	– 92	+ 190	– 405	– 18	+ 145	+ 305	+ 290	+ 415	– 515	+ 930	4400
	1981	– 25	+ 32	+ 182	+ 341	+ 40	+ 456	+ 143	+ 1169	– 25	+ 1194	7730
	1983	– 577	– 431	– 153	+ 174	+ 519	+ 1026	+ 603	+ 1161	– 1161	+ 2322	9330
	1980–1983	– 694	– 209	– 376	+ 497	+ 704	+ 1787	+ 1036	+ 2745	– 1701	+ 4446	7490
	1984	84	– 138	– 280	– 303	– 247	291	242	– 351	– 968	+ 617	2190
	1985	– 182	– 23	– 34	– 34	– 51	62	107	– 155	– 324	+ 169	5440
	1987	– 320	– 186	– 129	25	64	– 47	171	– 422	– 682	+ 260	826
	1984–1987	– 418	– 347	– 443	– 312	– 234	+ 306	+ 520	– 928	– 1974	+ 1046	

Notes "+" denotes deposition; "–" denotes scour

Table 10.2 Fractional sedimentation quantities of Jingjian Reach, Yangtze River in 1979

Reach	Month	Difference in sediment concentration[a] (kg/m^3)		
		Total	$D < 0.1$ mm	$D > 0.1$ mm
Xinchang to Jianli Yangtze River	1	– 0.030	– 0.069	+ 0.039
	2	– 0.010	– 0.074	+ 0.064
	3	– 0.020	– 0.081	+ 0.061
	4	– 0.020	– 0.098	+ 0.078

[a]Sediment concentration at Xinchange Station (upstream) minus that at Janli Station (downstream)

10.1.2 Analysis of Field Data

The following characteristics for non-equilibrium sediment transport (mild scour and deposition) can be observed from the filed data listed in Tables 10.1, 10.2, 10.3 and 10.4.

(1) All the data sets show that the sedimentation patterns are different between the coarser sediment particles and the finer particles. While deposition occurred for the coarser particles, scour occurred for the finer particles. This is a result of the exchange between the coarser particles of the suspended load and the finer particles of the bed material. Let's use the reach between Huangzhuang and Xiantao of the Han River as an example. While 1.52 million tons of coarser suspended sediment with the particle size $D > 0.1$ mm deposited on the bed, 9.05 million tons of finer bed material with $D > 0.1$ mm was eroded, resulting in a net scour of 7.53 million tons in 1974. In 1975, 9.64 million tons of coarser suspended load with $D > 0.05$ mm deposited on the bed, with 8.88 million tons of finer bed material with $D > 0.05$ mm being scoured, resulting in a net deposition of 0.76 million tons. Comparing this net deposition quantity with the annual sediment transport amount of 58.9 million tons in 1975 at Huangzhuang, sediment transport in this reach is close to equilibrium. As a result of the exchange between the coarser suspended load and the median to fine bed material, the bed material gradation became coarser, while the suspended load became finer.

(2) Regular pattern can be observed for the scour of finer sediment and the deposition of coarser particles. Except for a few data sets with small sedimentation quantities and thus with poorer data accuracy, most data sets show that deposition typically occurred for all size classes that were coarser than a particles size, while scour occurred for all size classes that were finer than this particle size. For the reach between Xiangyang and Huangzhuang of the Han River, only four data sets, which accounted for approximately 5.2% of the 11 years of data sets between 1974 and 1984, showed exception. All the other data sets showed that scour occurred for all size classes that were coarser than a threshold particle size, while deposition occurred for all size classes finer than this threshold particle size. The threshold particle size might be different for different years. For the

Table 10.3 Fractional sedimentation quantities for a downstream reach between Aishan and Lijin of the Yellow River

Reach	Time	Fractional sedimentation quantity (10^4 t)			Total quantity (10^4t)			Critical particle size (mm)
		$D < 0.025$	0.025 ~ 0.05	$D > 0.05$	Net	Deposition	Scour	
Aishan–Lijin	1950.7 ~ 1960.6 Non-flood season	− 1660	+ 2060	+ 6800	+ 7200	+ 8860	− 1660	0.025
Aishan–Lijin	1964.11 ~ 1973.10 Non-flood season	− 2100	+ 2900	+ 10,800	+ 11,600	+ 13,700	− 2100	0.025
Aishan–Lijin	1971 Flood season	− 4660	− 2050	+ 14,810	+ 8100	+ 14,810	− 6710	0.05

Notes "+" denotes deposition; "−" denotes scour

Table 10.4 Fractional sedimentation quantities for the reach dow nstream of the three gorges dam after the operation

Reach	Time	Fractional sedimentation quantity[a] (10^4 t)									Quantity[a] (10^4 m^3)
		$D < 0.004$	0.004 ~ 0.008	0.008 ~ 0.016	0.016 ~ 0.031	0.031 ~ 0.062	0.062 ~ 0.125	0.125 ~ 0.250	0.250 ~ 1.00	Total	
Yichangto–Zhicheng	2003–2007	− 239	1176	− 107	− 41	− 1400	− 93	− 1380	− 2076	− 4161	− 9,315
Zhicheng–Shashi	2003–2007	− 937	− 171	− 394	− 407	− 425	− 986	− 5890	+ 643	− 8567	− 9934
Shashi–Jianli	2003–2007	− 102	+ 553	+ 457	− 49	− 826	− 2134	− 4389	− 419	− 6910	− 12,970
Jianli–Luoshan	2003–2007	285	− 1876	− 911	+ 185	+ 1739	− 921	+ 1889	+ 821	+ 1210	− 263
Luoshan–Hankou	2003–2007	− 1049	− 684	− 265	− 292	+ 315	+ 601	+ 2402	+ 363	+ 1392	− 6675
Hankou–Jiujiang	2004–2007	− 7	− 373	− 728	− 1741	− 1477	+ 434	+ 5620	+ 988	+ 3953	− 3312
Jiujiang–Datong	2004–2007	− 2410	− 2076	− 736	− 2720	− 140	− 34	+ 29	− 179	− 8266	− 27,088

[a]"+" denotes deposition; "−" denotes scour

reach between Huangzhuang and Xiantao of the Han River, only three data sets, which accounted for approximately 4.2% of the 11 years of data sets between 1974 and 1984, showed exception. As shown in Table 10.2, for the reach between Xinchang and Jianli of the Yangtze River, no exception was found in the data sets for the four dry months of 1979, with the threshold particle size of 0.10 mm. As listed in Table 10.3, for a downstream reach of the Yellow River between Aishan and Lijin, deposition occurred for the particles coarser than 0.025 mm, and scour occurred for the particles finer than 0.025 mm during two dry seasons in 1950–1960, and in 1964–1973, respectively. For the flood season of 1971, due to the strong flow condition, the threshold particle size was 0.05 mm. The bed material of the variable backwater zone of the Danjiangkou Reservoir consists of gravel with sand, and is short of fine sediment. The scour of the bed material was thus limited. While the data sets for an individual year might show some exceptions, the 3-year-averaged results indeed showed the regular pattern, as listed in Table 10.1.

Table 10.4 lists the fractional sedimentation quantities for various size classes for the reach downstream of the Three Georges Reservoir in 2003 through 2007 after the operation of this reservoir. The short reach between Yichang and Zhicheng was characterized as a pebblel-sand bed with significant scour. All particles that were finer than 1 mm and that could be suspended were eroded away. For the reaches between Shashi and Jianli, and between Jiujiang and Datong, the bed material consisted of fine sediment. Hence, significant scour occurred in these two reaches and the scour occurred for all size classes with $D \leq 1$ mm. For other reaches between Zhicheng and Shashi, between Jianli and Luoshan, between Luoshan and Hankou, and between Hankou and Jiujiang, scour occurred for the finer sediment, while deposition occurred for the coarser sediment. The threshold particle size for scour or deposition gradually decreased towards downstream. The scour in the reaches between Yichang and Zhicheng, between Shashi and Jianli, and between Jiujiang and Datong is close to the massive scour condition, and thus scour occurred for all size classes.

3. The proportion of the coarser particles of the suspended load influx, which are exchanged with the finer bed material, and the ratio of the deposition quantity of the coarser particles to the scour quantity of the finer particles varied significantly. These relative exchange quantities depended on whether the channel were scoured, deposited, or were approximately at equilibrium. The deposition quantity of the coarser sediment when channel deposition occurred, or the scour quantity of the finer sediment when channel scour occurred generally accounted more than 10% of the sediment influx. On the other hands, the scour quantity of the finer sediment in the occurrence of channel deposition, and the deposition quantity of the coarser sediment in the occurrence of channel scour are much smaller than the sediment influx.
4. The threshold particle size for scour or deposition showed insignificant variations for each reach, and in general only varied by an order of one class size. For example, the threshold particle size was generally 0.25 mm for the reach between Xiangyang and Huangzhuang of the Han River except for a few years

with a threshold particle size of 0.5 mm. For the reach between Huangzhuang and Xiantao of the Han River, the threshold particle size was 0.10 mm during the initial scour periods, and then increased to 0.25 mm because of the bed material coarsening. The threshold particle size was 0.1 mm for the reach between Xinchang and Jianli of the Yangtze River, and was 0.025–0.05 mm at downstream reach of the Yellow River between Aishan and Lijin.

The threshold particle size is closely related to the bed material gradation and the channel sedimentation pattern. The threshold particle size is coarser if the bed material coarser, and vice versa. It is also coarser when channel scour occurred, and is finer when channel deposition occurred. For example, the bed material in the reach between Xiangyang and Huangzhuang is coarser than the reach between Huangzhuang and Xiantao. As a result, the threshold particle size was generally 0.25 mm for the former reach, which was coarser than 0.1 mm for the later reach during the initial scour periods. Also, for the reach between Huangzhuang and Xiantao, the threshold particle size was 0.1 mm for 5 years out of the 6-year period between 1974 and 1979, except for 1978 when the threshold particle size was 0.25 mm. Using 0.1 mm as the threshold particle size for scour or deposition, the total deposition quantity of the sediment coarser than this threshold particle size was computed to be 14.82 million tons for the 6-year period of 1974–1979, and the total scour quantity of the sediment finer than this threshold particle size was 24.91 million tons. The resulting net scour quantity was 10.09 million tons for 1974–1979, or 1.68 million tons per year. These 6 years were thus considered as a period of initial scour. Due to the bed material coarsening as scour developed, the threshold particle size increased to 0.25 mm for 1980–1984. The total deposition quantity of the sediment coarser than this threshold particle size was 12.92 million tons for this 5-year period between 1980 and 1984, and the total scour quantity of the sediment finer than this threshold particle size was 150.75 million tons. The resulting net scour quantity was 137.83 million tons for 1980–1984, or 27.57 million tons per year. Significant channel scour occurred in these 5 years. It is noted that deposition was not found for the coarser particle size class of 0.5–1.0 mm between 1974 and 1979 for the reach between Huangzhuang and Xiantao, indicating the sediment influx from the upper reach barely had this size class during this period. However, the deposition quantity for this size class was estimated to be 1.36 million tons between 1980 and 1984. This indicated not only the occurrence of the bed material coarsening in the upper reach (Xiangyang to Huangzhuang), which allowed the coarser particles with $D > 0.5$ mm to entering this reach, but also the bed material coarsening within this reach. The variation of the threshold particle size with channel sedimentation pattern can also be observed from the 1974–1975 data for the reach between Huangzhuang and Xiantao of the Han River. Channel scour occurred in 1974, and the threshold particle size was 0.1 mm. Due to the 785 flood event, the sediment influx of this reach was significantly increased in 1975, and the channel is close to equilibrium condition. As a result, the threshold particle size decreased to 0.05 mm in 1975. For the variable back-water zone of the Danjiangkou Reservoir, the sedimentation magnitudes were relatively large, and thus the threshold particle size varied as much as by two order of class size. The threshold particle size was 0.05 mm for the 3-year deposition period of 1980–1983, but was 0.25 mm for the 3-year scour period of 1984–1987.

5. The deposition quantity of coarser sediment is insignificant when significant channel scour occurred. The scour quantity of finer sediment was also insignificant when significant channel deposition occurred. The former phenomenon could be observed from the field data for the reach between Xiangyang and Huangzhuang in 1974, 1976, and 1982. The later phenomenon was demonstrated by the 1980 and 1981 data collected at the variable back-water zone of the Danjiangkou Reservoir. These two cases were practically with the massive scour or massive deposition condition.
6. The threshold particle size for scour or deposition is similar to the critical particle size D_0 that is traditionally used differentiate bed material load and wash load. As typically did in the past, defining D_0 as the particle size corresponding to the cumulative percent finer of 5–10%, D_0 is approximately 0.1 mm for the Jingjiang Reach of the Yangtze River, 0.025–0.05 mm for the Han River downstream of Xiangyang, and 0.025–0.05 mm for the Yellow River. These values are close to the threshold particle size for scour or deposition. The relationship between the threshold particle size D_1 for scour or deposition and D_0 as well as its determination will be further analyzed in the following section.

10.2 Mechanism for the Coarser and Finer Sediment Exchange

10.2.1 Theoretical Background for the Coarser and Finer Sediment Exchange

The governing equation for the 1-D non-equilibrium transport of non-uniform sediment [5, 6] reads

$$\begin{aligned}\frac{d(P_{4.l}S)}{dx} &= \frac{dS_l}{dx} = -\frac{\omega_l}{q}\left(\tilde{\alpha}S_l - \alpha^* S_l\right) \approx -\frac{\alpha\omega_l}{q}(S_l - S_l^*)\\ &= -\frac{\alpha\omega P_{4.l}}{q}\left(S - S^*\right)\end{aligned}$$

If the relation after the approximately equal sign of this equation is adopted, the sedimentation patterns of different size classes are the same, which is defined as the massive scour or massive deposition condition. However, considering $\tilde{\alpha} \neq \alpha^*$, then the sedimentation patterns are different for different size classes. Based on this concept, and more field data available, Han (EGSE 2008) first divided the non-equilibrium suspended load transport into three cases: missive deposition, massive scour, and mild scour and deposition. Massive deposition refers to the case of significant sediment deposition with deposition occurring for all size class and without scour of the original bed material from a cumulative effect aspect. Massive scour refers to the case of significant scour with scour occurring for all size classes and without deposition of suspended load. Mild scour and deposition refers to the case with different sedimentation patterns for various size classes. While the massive scour and massive deposition are the extreme condition, non-equilibrium sediment transport is generally characterized by the mild scour or erosion condition. Hence, the study for the carrying-sediment capacity and the corresponding size distribution should focus on the mild scour or erosion condition. The massive scour or massive deposition can be considered as the extreme condition of the case of mild scour and deposition, with simpler expression of the size distribution for carrying-sediment capacity. For massive deposition, $P^*_{4.l} = P_{4.l}$. For massive scour, $P^*_{4.l} \approx P_{4.l}$. For these two extreme cases, $P^*_{4.l}$ can be replaced with $P_{4.l}$. Starting from the difference in the sedimentation patterns between the coarser and finer sediment, the characteristics of the mild scour or erosion condition is analyzed as follows.

For the case of mild scour or erosion, or for a general non-equilibrium transport of non-uniform sediment, introducing the size distribution for carrying-sediment capacity [8, 15, 16], $P^*_{4.l}$, the governing equation for the concentration of the l-th size class [5] is

$$\frac{dS_l}{dx} = \frac{d(P_{4.l}S)}{dx} = -\frac{\omega}{q}\left(\tilde{\alpha} P_{4.l}S - \alpha^* P^*_{4.l}S^*\right) \tag{10.1}$$

where $P_{4.l}$ is the size distribution of suspended load. This equation is the same as Eq. (3.92).

The phenomenon for the coarser and finer sediment exchange not only assists in illustrating the sediment exchange mechanism for non-equilibrium sediment transport, but also provides support for the theoretical study for sediment transport. During sediment transport, the suspended particles settle down to the bed, and part of the deposited particles is eroded from the bed. Compared to finer sediment particles, more coarser particles deposit on the bed. Because coarser particles are difficult to get entrained with a lower suspension height and a limited suspension space, less deposited coarser particles are re-suspended, resulting in a net deposition of coarser sediment particles on the bed. On the other hand, less finer particles deposit on the bed and a great portion of the deposited finer particles are re-suspended from the bed. During a channel scour, more finer sediment particles are eroded from bed material than the coarser particles. Combining these two processes, only a smaller portion of the finer sediment particles deposit on the bed, while scour occurs for a greater portion of the finer particles. At the same time, the increase to the carrying-sediment capacity for the finer sediment also limit the scour intensity of the coarser particles. As a result, the overall sediment pattern is characterized by the scour of the finer particles with the deposition of the coarser particles.

This is the basic characteristics for the case of mild scour or erosion, which are also shown in Tables 10.1, 10.2, 10.3 and 10.4. As discussed before, the fractional sedimentation quantities for various size classes, and the threshold particle size for scour or deposition is impacted by the overall sedimentation intensity.

Starting from the theoretical relation, or the equations for non-equilibrium sediment and for the size distribution for carrying-sediment capacity, the basic characteristics for mild erosion and deposition are further proved in the following. Based on Eq. (7.133), the equation for the fractional carrying-sediment capacity for the mild scour and deposition condition is

$$\begin{aligned} P_{4.l}^{*} S^{*}\left(\omega^{*}\right) = {} & P_{4.1} P_{4.l.1} S + P_{4.2} P_{4.l.2} S \frac{S^{*}(\omega_l)}{S^{*}\left(\omega_1^{*}\right)} \\ & + \left[1 - \frac{P_{4.1} S}{S^{*}(\omega_1)} - \frac{P_{4.2} S}{S^{*}\left(\omega_1^{*}\right)}\right] P_{4.l.1}^{*} S^{*}\left(\omega_1^{*}\right) \end{aligned} \tag{10.2}$$

This equation applies to the case with all bed material being suspensible and with the suspended load partially changing to the carrying-sediment capacity. Based on this equation, for $1 \leq l \leq k$, or for the finer sediment particles, one obtains

$$\begin{aligned} & \frac{\tilde{\alpha}_l}{\alpha_l^{*}} P_{4.l} S - P_{4.l}^{*} S^{*}(\omega^{*}) = K_0 P_{4.l} S - P_{4.l}^{*} S^{*}(\omega^{*}) \\ & = (K_0 - 1) P_{4.l} S - \left[1 - \frac{P_{4.1} S}{S^{*}(\omega_1)} - \frac{P_{4.2} S}{S^{*}\left(\omega_1^{*}\right)}\right] P_{4.l.1} S^{*}\left(\omega_1^{*}\right) \end{aligned} \tag{10.3}$$

where $K_0 = \tilde{\alpha}_l / \alpha_l^{*}$. For $k + 1 \leq l \leq n$, or for the coarser particles,

$$\begin{aligned} K_0 P_{4.l} S - P_{4.l}^{*} S^{*}\left(\omega^{*}\right) = {} & P_{4.l} S \left[K_0 - \frac{S^{*}(\omega_l)}{S^{*}\left(\omega_1^{*}\right)}\right] \\ & - \left[1 - \frac{P_{4.1} S}{S^{*}(\omega_1)} - \frac{P_{4.2} S}{S^{*}\left(\omega_1^{*}\right)}\right] P_{4.l.1} S^{*}\left(\omega_1^{*}\right) \end{aligned} \tag{10.4}$$

Plugging Eqs. (10.3) and (10.4) in Eq. (10.1), respectively, for the finer particles $(1 \leq l \leq k)$, one obtains

$$\begin{aligned} \frac{d(P_{4.l} S)}{dx} & = -\frac{\omega_l}{q} \alpha_l^{*} \left(K_0 P_{4.l} S - P_{4.l}^{*} S^{*}\right) \\ & = \frac{\omega_l}{q} \alpha_l^{*} \left\{ -P_{4.l} S (K_0 - 1) + \left[1 - \frac{P_{4.1} S}{S^{*}(\omega_1)} - \frac{P_{4.2} S}{S^{*}\left(\omega_1^{*}\right)}\right] P_{4.l.1}^{*} S^{*}\left(\omega_1^{*}\right) \right\} \end{aligned} \tag{10.5}$$

and for the coarser particles $(k + 1 \leq l \leq n)$.

$$\frac{d(P_{4.l}S)}{dx} = \frac{\omega_l}{q}\alpha_l^*\left\{-P_{4.l}S\left[K_0 - \frac{S^*(\omega_l)}{S^*(\omega_1^*)}\right]\right.$$
$$\left. + \left[1 - \frac{P_{4.1}S}{S^*(\omega_1)} - \frac{P_{4.2}S}{S^*(\omega_1^*)}\right]P_{4.l.1}^*S^*(\omega_1^*)\right\}, \tag{10.6}$$

For deposition, $K_0 = \tilde{\alpha}_l/\alpha_l^* < 1$. For scour, $K_0 > 1$. For the case of mild scour and deposition, K_0 is close to 1, or $K_0 - 1 \approx 0$. Hence, the first term on the right side of Eq. (10.5) is approximately zero, or the finer particles of the sediment influx approximately remain unchanged along the flow path. However, the finer particles of the bed material are picked up. As a result, $\frac{d(P_{4.l}S)}{dx} > 0$, and a net scour occurs for the finer sediment. For the coarser particles of the sediment influx, $\omega_l > \omega_1^*$, and $\frac{S^*(\omega_l)}{S^*(\omega_1^*)} = \left(\frac{\omega_1^*}{\omega_l}\right)^m$ is much less than 1. Also, $K_0 \approx 1$. Hence, $K_0 - \frac{S^*(\omega_l)}{S_l(\omega_1^*)} > 0$, and deposition occurs for the coarser particles of the sediment influx. Although some coarser particles of the bed material are eroded, the erosion is very limited under the mild scour and deposition condition. As a result, the first term on the right side of Eq. (10.6) dominates, and $\frac{d(P_{4.l}S)}{dx} < 0$, or net deposition occurs for the coarser sediment. It is noted that the value of k, which is the threshold size group index to differentiate the finer and coarser sediment, may not be constant. If $l = k$ does not satisfy

$$\left[K_0 - \frac{S^*(\omega_l)}{S^*(\omega_1^*)}\right]P_{4.l}S > \left[1 - \frac{P_{4.1}S}{S^*(\omega_1)} - \frac{P_{4.2}S}{S^*(\omega_1^*)}\right]P_{4.l.1}^*S^*(\omega_1^*) \tag{10.7}$$

then another value of k can be found in the range between $k + 1$ and n to satisfy Eq. (10.7). Because

$$\frac{S^*(\omega_l)}{S^*(\omega_1^*)} = \left(\frac{\omega_1^*}{\omega_1^*}\right) \tag{10.8}$$

as k increases, ω_l is much greater than ω_1^*, $\frac{S^*(\omega_l)}{S^*(\omega_1^*)}$ reduces rapidly, and $K_0 - \frac{S^*(\omega_l)}{S^*(\omega_1^*)}$ increases rapidly. As a result, the weighting coefficient of $P_{4.l}S$ increases quickly until satisfying Eq. (10.7). This explains why the threshold particle size for scour or deposition, D_k, varies within a channel.

10.2.2 Theoretical Analysis of the Threshold Particle Size Between Scour and Deposition

Can the coarser and finer sediment exchange be linked to the differentiation between bed material load and wash load as adopted in the traditional sediment studies? The answer is yes. By doing that, the mechanism of the threshold particle size can be further illustrated. This is also a prerequisite for the following analysis. However, rather than adopting the empirical methods that are traditionally used to differentiate between bed material load and wash load (as e.g., assuming the wash load is the particles finer by weight than 5 ~ 10% of the total load), the threshold particle size is determined in the following from the aspect of sediment exchange.

Let divide the sediment mixture into n size classes, and assume each size class can be suspended. According to the general relationship between the size distribution for carrying-sediment capacity and the bed material gradation for strong equilibrium sediment transport [5]

$$P_{4.l.1}^{*} S^{*}\left(\omega_{1}^{*}\right) = P_{1.l.1} S^{*}(l) \tag{10.9}$$

where ω_1^* is the average fall velocity for carrying-sediment capacity that is determined with the bed material gradation; $P_{1.l.1}$ the bed material gradation, or the weight percent of the bed material for the size class of D_l, $P_{4.l.1}^*$ the size distribution for carrying-sediment capacity corresponding to the bed material gradation. The third subscripts of $P_{1.l.1}$ and $P_{4.l.1}^*$ are used to differentiate them from the effective bed material gradation $P_{1.l}$ and the corresponding size distribution for carrying-sediment capacity $P_{4.l}^*$, respectively. The left side of Eq. (10.9) can be considered as the fractional carrying-sediment capacity for non-uniform sediment, and the right side as the carrying-sediment capacity provided by the bed material for this size class. They equal each other in the strong equilibrium condition. Summation of Eq. (10.9) for all size groups yields

$$S^{*}\left(\omega_{1}^{*}\right) = \sum_{l=1}^{n} P_{1.l.1} S^{*}(l) \tag{10.10}$$

As shown in Eq. (10.9), for $\omega_l < \omega_1^*$, or for the finer particles, $S*(l) > S*(\omega_1^*)$, and thus $P_{4.l.1}^* > P_{1.l.1}$. In other words, when the non-uniform suspended sediment influx and the bed material reach a strong equilibrium condition, because of the wide variation range of the fall velocities for various size classes (up to several orders of magnitude), the finer particles with ω_l much less than ω_1^* make up a larger proportion of the carrying-sediment capacity. On the other hand, the proportion is smaller for the coarser particles.

Define

$$D_k = f^{-1}\left(\omega_1^*\right) \tag{10.11}$$

where f^{-1} is the inverse function of $\omega = f(D)$. The finer particles with $D < D_k$ seem not to participate the exchange with the bed material, or are not involved in bed evolution from a cumulative effect aspect. Specifically, these finer particles do not change to the bed material or deposit on the bed. This is because these finer particles, even if existing in the bed material, will be immediately eroded away. As illustrated in previous equations, the finer particles with $l \leq k$ barely deposit on the bed.

For confirmation, the following will further illustrate, using intuition, the finer particles do not deposit from a cumulative effect aspect. For non-equilibrium sediment transport, let assume the finer suspended particles all deposit on the bed during the sediment exchange process, and become the bed material with gradation $P_{1.l.1} = P_{4.l}$. The quantity of the deposited finer particles is $SQP_{4.l}\Delta t$, the corresponding size distribution for sediment transport is $P^*_{4.l.1}$, and the quantity counted in the sediment transport capacity is $SQP^*_{4.l.1}\Delta t$. As discussed before, $P^*_{4.l.1} > P_{1.l.1} = P_{4.l}$ for the finer particles. The quantity of the l-th size class that is eroded and counted in the carrying-sediment capacity is larger than the deposition quantity of the suspended load for the l-th size class. This is impossible. Thus, the l-th size class of the suspended load influx should all be counted in the sediment transport capacity, and these finer particles do not exchange with the bed material from a cumulative effect aspect. Hence, the threshold particle size D_k has the same property as the threshold particle size that is generally used to differentiate bed material load and wash load. Han [8] proposed to use D_k to differentiate bed material load and wash load, and validated this approach using the field data collected in the Jingjiang Reach of the Yangtze River, the Han River downstream of the Danjiangkou Reservoir, and the lower reach of the Yellow River. Han [8] concluded that the capacity for carrying suspended load and non-equilibrium sediment transport could be analyzed without differentiating bed material load and wash load, and at the same time the different transport patterns between the coarser and finer sediment could be illustrated [6]. The comparison between bed material load and wash load and the differentiating criteria will be discussed in Sects. 10.5 and 10.6.

To further illustrate the fact that the finer particles with $D < D_k$ of the suspended load influx do not deposit or at least barely deposit on the bed, and to verify D_k is indeed equivalent to the threshold particle size to differentiate bed material load and wash load, the typical bed material data collected in the Yangtze River and the Han River is analyzed in the following, with D_k being computed with Eqs. (10.10) and (10.11) [12]. If the bed material contains coarse particles that cannot be suspended, such as coarse sand with $D > D_{n+1}$, pebble, and gravel, these coarse particles are excluded in the bed material gradation, and the adjusted gradation for the bed material then can be suspended is

$$P_{1.l.1.1} = \frac{P_{1.l.1}}{\sum_{l=1}^{n} P_{1.l.1}} \quad (1 \le l \le n < N) \tag{10.12}$$

The procedure to compute D_k is as follows: after excluding the gradation for the coarse size classes of $n + 1 \sim N$ for coarse sand, pebble, and gravel, calculate $P_{1.l.1.1}$ with Eq. (10.12); compute corresponding $\omega^*_{1.1}$ with Eq. (10.10); replacing ω^*_1 with $\omega^*_{1.1}$, determine D_k with Eq. (10.11). The computed results of D_k are listed in Table 10.5.

The following can be observed from Table 10.5:

(1) D_k is approximately 0.25 ~ 0.36 mm for a pebble bed, 0.20–0.25 mm for a pebble-sand bed, 0.10–0.20 mm for a bed consisting of median to fine sand

Table 10.5 Typical bed material gradation and computed D_k for the Yangtze River and Han River

Reach	Type	Percent finer												D_k (mm)	Percent finer for D_k
		< 0.005	0.01	0.025	0.05	0.1	0.25	0.5	1.0	5	10	50	100		
Yichang–Yidu	Gravel, sand					1.7	29.1	44.7	50.7	58.1	65.5	90.2	100	0.213	22
Yidu–Songzikou	Gravel, sand					3.6	44.2	64.2	69.3	80.1	87.7	95.1	100	0.193	29
Songzikou–Chenjiawan	Gravel, sand					4.2	52.5	75.8	79.0	82.1	85.2	96.8	100	0.190	33
Chenjiawan–Haoxue	Median to fine sand	3.1	4.1	6.2	8.3	13.5	79.7	99.4	100					0.02	6
Haoxue–Ouchikou	Median to fine sand		0.6	2.0	3.8	10.5	88.2	100						0.080	8
Ouchikou–Jianli	Median to fine sand	0.2	1.7	3.0	4.8	10.7	90.4	100						0.052	5
Jianli–Chenglingji	Median to fine sand	3.7	6.2	10.8	14.9	22.0	92.4	99.9	100					0.018	10
Huanghuagang–Guanghua	Pebble					0.2	2.5	26.4	35.6	38.5	42.3	78.0	100	0.360	13
Guanghua–Taipingdian	Pebblel, sand				0.7	1.5	14.9	73.7	82.0	82.7	83.3	85.9	100	0.225	16
Taipingdian–Xiangyan	Pebble, sand				0.3	0.9	30.7	78.7	83.5	84.0	84.6	99.4	100	0.235	28

(continued)

Table 10.5 (continued)

Reach	Type	Percent finer												D_k (mm)	Percent finer for D_k
		< 0.005	0.01	0.025	0.05	0.1	0.25	0.5	1.0	5	10	50	100		
Xiangyang–Yicheng	Median to fine sand, pebble				1.0	4.0	56.0	91.9	94.4	94.9	95.6	100		0.181	32
Yicheng–Huangzhuang	Median to fine sand, pebble				2.8	9.8	77.3	98.6	99.0	99.1	99.2	100		0.102	11
Huangzhuang–Zekou	Median to fine sand				4.1	10.8	91.1	99.6	100					0.118	29
Zekou–Xiangtao	Median to fine sand				5.4	23.6	93.1	99.6	100					0.10	24

together with a small amount of pebble and coarse sand, and 0.02–0.12 mm for a bed with median to fine sand. The threshold particle size D_k is coarser as the bed material become coarser. However, the variation range of D_k is very limited. The computed results of D_k are consistent with the empirical particle sizes for differentiating bed material load and wash load, and they are the same for the alluvial channels with the bed material consisting of median to fine sand.

(2) According to Table 10.1, the threshold particle size of the suspended sediment that does not participate the exchange with the bed material was approximately 0.25 mm for the reach between Xiangyang and Huangzhuang of the Han River in 1974–1984, and was 0.10 mm for the reach between Huangzhuang and Xiantao in 1974–1979. This is similar to the computed D_k as listed in Table 10.5, which is 0.102–0.181 mm for the reach between Xiangyang and Huangzhuang, is approximately 0.200 mm for a pebble-sand bed reach between Yichang and Songzikou of the Yangtze River, and is 0.10–0.118 mm for the reach between Huangzhuang and Xiantao of the Han River. The threshold particle size for differentiating bed material load and wash load is slightly larger than D_k for the reach between Xiangyang and Huangzhuang because some gravel and pebble exist in this reach. Hence, it is reasonable to consider the finer suspended sediment with $D < D_k$ does not deposit on the bed to become the bed material.

(3) The proportion of the finer particles with $D < D_k$ to the total bed material, P_k varies significantly, and is a function of the bed material gradation. For a pebble bed, a pebble-sand bed, and a gravel-sand bed, P_k is generally 13 ~ 33%, instead of 5–10%. This is because for a pebble bed or a pebble-sand bed, approximately 35% of the median to fine sediment, most with the particle size smaller than D_k, is stored in the voids among coarser particles, and does not present in the bed surface layer. This results in a larger proportion for the finer bed material with $D < D_k$. Also, for a bed with median to fine sediment that has been scoured for a long period, only a small amount of finer particles with $D < 0.05$ mm exist on the bed due to the bed material sorting. This also results in an increase to D_k and the proportion of the finer sediment with $D < D_k$. Only for the bed with median to fine sediment that is approximately in equilibrium condition, such as the reach between Chenjiawan and Jianli of the Yangtze River, the proportion of the finer sediment with $D < D_k$ is 5–10% of the total bed material. This is within the range of the threshold particle sizes proposed by Zhang and Einstein. It is concluded that D_k can be used as the threshold particle size for more general cases.

(4) As listed in Table 10.5, for the bed with median to fine sediment, such as the reaches between Chenjiawan and Chenglingji, the proportion of the finer particles with $D < D_k$ is 5–10%, with an average of 7%, of the total bed material. This is consistent with the traditional criteria for differentiating bed material load and wash load, which used the particle size with the percent finer of 5–10%.

10.3 Impacts of the Coarser and Finer Sediment Exchange on Sediment Transport and Channel Evolution

The concept of the coarser and finer sediment exchange can illustrate the complex mechanisms for sediment transport and channel evolution. Based on this concept and corresponding study results, the phenomena and mechanisms for some important topics [8, 9, 12] are analyzed in the following.

10.3.1 Long-Distance Scour in the Reach Downstream of a Reservoir

Based on the diffusion equation, the distance of scour (or distance for sediment concentration to recover the saturation) downstream of the Sanmenxia Reservoir on the Yellow River was estimated to be approximately 800 m in the preliminary feasibility study. After completion of this reservoir, channel scour was observed for the entire downstream reach all the way to the mouth of the river, which is more than 800 km long, when the flow charge exceeded 3000 m^3/s. Some studies attributed this to the sediment exchange between the floodplain and the main channel. When the sediment-laden flow goes up to the floodplain, sediment deposit on the floodplain. As the flow retreats back to the main channel, scour occurs. This is apparently one reason, but is not the main reason. For a flow discharge of 2000 m^3/s that is conveyed only within the main channel, is the scour distance significantly shorter? The answer is no. However, this can be explained with the concept of the coarser and finer sediment exchange. When the suspended load from the upstream reach exchanges with the bed material of a given reach, the size distribution of suspended load becomes finer, and thus the carrying-sediment capacity increases and scour occurs. If the flow parameters remain constant, the exchange between the coarser and finer sediment continues as long as the bed material gradation become finer along the flow path, which is typically true for an alluvial stream. This will result in the size distribution of suspended load becoming finer, increase the carrying-sediment capacity, and cause long distance scour along the flow path.

The variation of sediment concentration along the flow path can be written as [6]

$$\frac{dS_l}{dx} = \frac{d(P_{4.l}S)}{dx} = -\frac{\alpha_l \omega_l}{q}\left(P_{4.l}S - P_{4.l}^{*}S^{*}\right) \tag{10.13}$$

Based on Eq. (5.24b), the fractional sediment concentration is

$$\begin{aligned} P_{4.l}S = {} & P_{4.l}^{*}S^{*} + S_0 P_{4.l.0}\exp\left(-\tfrac{\alpha_l\omega_l L}{q}\right) - S_0^{*}P_{4.l.0}\exp\left(-\tfrac{\alpha_l\omega_l L}{q}\right) \\ & + S_0^{*}P_{4.l.0}^{*}\tfrac{q}{\alpha_l\omega_l L}\left[1-\exp\left(-\tfrac{\alpha_l\omega_l L}{q}\right)\right] - S^{*}P_{4.l}^{*}\tfrac{q}{\alpha_l\omega_l L}\left[1-\exp\left(-\tfrac{\alpha_l\omega_l L}{q}\right)\right] \end{aligned} \tag{10.14}$$

where the subscript “0” denotes the value at the entrance section, and the parameter without the subscript “0” denotes the value at the exit section. Assuming sediment transport at the entrance section is in strong equilibrium condition ($S_0^* = S_0$, and $P_{4.l.0}^* = P_{4.l.0}$) and L is sufficiently large so that $\exp\left(-\frac{\alpha_l \omega_l L}{q}\right) \to 0$, Eq. (10.14) can be simplified as

$$P_{4.l}S = P_{4.l}^* S^* + \left(S_0^* P_{4.l.0}^* - S^* P_{4.l}^*\right)\frac{q}{\alpha_l \omega_l x} \tag{10.15}$$

where L is replaced with x. Plugging Eq. (10.15) into Eq. (10.13) yields

$$\frac{dS_l}{dx} = -\left(S_0^* P_{4.l.0}^* - S^* P_{4.l}^*\right)\frac{1}{x} \tag{10.16}$$

Assuming V^3/h remains constant along the flow path, and thus

$$S^*(l) = \kappa\left(\frac{V^3}{gh\omega_l}\right)^m \tag{10.17}$$

is also a constant. Plugging $P_{4.l}^* S^* = P_{1.l.1} S(l)$, Eq. (10.16) can be written as

$$\frac{dS_l}{dx} = S^*(l)\frac{P_{1.l.1} - P_{1.l.1.0}}{x} \tag{10.18}$$

where $P_{1.l.1.0}$ and $P_{1.l.1}$ are the bed material gradations at the entrance and the exit sections, respectively. Summation of Eq. (10.18) for l, one obtains

$$\frac{dS}{dx} = \frac{S^* - S_0^*}{x} \tag{10.19}$$

Let assume the bed material becomes finer, with the gradation varies linearly along the flow path, the effective bed material gradation also shows linear variation along the flow path, or

$$P_{1.l} = P_{1.l.0} + \frac{P_{1.l.L} - P_{1.l.0}}{L}x \tag{10.20}$$

where L is the reach length. With these assumptions, the following conclusions can be made:

(1) $\frac{1}{\omega^{*m}}$ shows linear variation along the flow path. Diving Eq. (10.20) by ω_l^m

$$\frac{1}{\omega^{*m}} = \sum_{l=1}^{n}\frac{P_{1.l}}{\omega_l^m} = \sum_{l=1}^{n}\frac{P_{1.l.0}}{\omega_l^m} + \left(\sum_{l=1}^{n}\frac{P_{1.l.L}}{\omega_l^m} - \sum_{l=1}^{n}\frac{P_{1.l.0}}{\omega_l^m}\right)\frac{x}{L}$$

$$= \frac{1}{\omega_0^{*m}} + \left(\frac{1}{\omega_l^{*m}} - \frac{1}{\omega_0^{*m}}\right)\frac{x}{L} \tag{10.21}$$

(2) The total carrying-sediment capacity increases linearly along the flow path. Multiplying Eq. (10.21) by $\kappa\left(\frac{V^3}{gh}\right)^m$ yields

$$S^* = S_0^* + \left(S_L^* - S_0^*\right)\frac{x}{L} \tag{10.22}$$

(3) The gradient of the fractional sediment concentration, $\frac{dS_l}{dx}$, varies linearly along the flow path. Plugging Eq. (10.20) into Eq. (10.18), one obtains

$$\frac{dS_l}{dx} = S^*(l)\frac{P_{1.l.L} - P_{1.l.0}}{L} \tag{10.23}$$

(4) The fractional carrying-sediment capacity varies linearly along the flow path. Integrating Eq. (10.23) gives

$$S_l = S_{l.0} + S^*(l)\frac{P_{1.l.L} - P_{1.l.0}}{L}x = S_{L.0} + \frac{S_L^* - S_0^*}{L}x$$

(5) As the bed material becomes finer along the flow path, $P_{1.l.L} < P_{1.l.0}$ for the coarser particles. Based on Eq. (10.23), $\frac{dS_l}{dx} < 0$, or deposition occurs for the coarser particles. For the finer bed material particles, $P_{1.l.L} > P_{1.l.0}$. Based on Eq. (10.23), scour occurs for these finer particles.

In summary, assuming the bed material varies linearly and the flow parameters remain constant along the flow path, the mechanisms of the long-distance channel scour downstream of a reservoir, which can extend to the mouth of the channel for some cases, are illustrated. The variations of corresponding parameters are also demonstrated.

10.3.2 Another Form of Bed Material Coarsening Through the Coarser and Finer Sediment Exchange

Bed material coarsening is traditionally considered to be caused by channel scour. More finer particles are scoured than coarser particles, and thus the bed material becomes coarser after scour occurs for a certain period of time. However, from the aspect of the coarser and finer sediment exchange, bed material coarsening not only occurs during scour, but also occurs during equilibrium sediment transport or during a deposition process, as discussed in Chap. 9.

As listed in Table 10.1, the scour quantity for the finer sediment with $d \leq 0.05$ mm was approximately 8.88×10^6 tons in 1975 within the reach between Huangzhuang and Xiantao of the Han River, while the deposition quantity was 9.64×10^6 tons for the coarser sediment with $d > 0.05$ mm. The deposition quantity was close to the scour quantity, resulting in a weak equilibrium channel condition. However, obvious bed material coarsening still occurred in the main channel. The length of this reach is approximately 217 km, and the width of the main channel is approximately 300 m. Assuming the dry density of the sediment is 1.3 ton/m^3. The equivalent deposition depth of the deposited coarser particles in the amount of 9.64×10^6 tons was approximately 0.134 m. The bed material coarsening for the case of mild deposition can be observed in the field data collected in 1971 within the reach between Aishan and Lijin of the Yellow River. As listed in Table 10.3, the net deposition was 81×10^6 tons within this reach in 1971, the scour quantity was 67.1×10^6 tons for the finer sediment with $d \leq 0.05$ mm, and the deposition quantity was 148.1×10^6 tons for the coarser sediment with $d > 0.05$ mm. The reach length between Aishan and Lijin is 262 km, and the channel is approximately 500 m wide. If all coarser sediment particles deposit in the main channel, the average deposition depth was 1.10 m. Therefore, most bed material are coarser than 0.05 mm, resulting in obvious bed material coarsening in the main channel.

The previous examples illustrate that bed material coarsening can occur for the case of mild erosion and deposition, equilibrium, or deposition. Because of the bed material coarsening through the coarser and finer sediment exchange, the bed material gradation for the streams with median to fine sand bed, such as the Yangtze River and the Yellow River, is relatively stable with a narrow variation range of grain size such as 0.025–0.25 mm or 0.05–0.5 mm.

An idealized model for suspended load coarsening through the coarser and finer sediment exchange was demonstrated in Table 9.14 for equilibrium sediment transport. The process of the sediment exchange was illustrated in detailed in this table.

10.3.3 Change to Carrying-Sediment Capacity Caused by the Coarser and Finer Sediment Exchange

Exchange between the coarser and finer sediment particles changes the carrying-sediment capacity of the flow. The scour condition without considering the coarser and finer sediment exchangemay changes to the deposition pattern. This can be illustrated with the 1974–1979 field data measured in the reach between Huangzhuang and Xiantao of the Han River. As listed in Table 10.1, the scour quantity was 26.36×10^6 tons for the finer sediment with $d \leq 0.1$ mm, and deposition quantity was 17.75×10^6 tons for the coarser sediment with $d > 0.1$ mm. For $d \leq 0.1$ mm, $\omega^{0.92} = 0.0022$ with ω being the average fall velocity. For $d > 0.1$ mm, $\omega^{0.92} = 0.0398$. Thus, the ratio of the carrying-sediment capacity between the finer and the coarser sediment is

$$\frac{S^*_{\text{fine}}}{S^*_{\text{coarse}}} = \frac{0.0398}{0.0022} = 18.1$$

This indicates for a deposition of 1 ton of the coarser sediment, 18.1 tons of the finer sediment will be scoured. In the equilibrium or approximately equilibrium condition, the deposition of the coarser sediment in the amount of 1.46×10^6 tons is equivalent to the scour of the finer sediment in the amount of 26.36×10^6 tons. Hence, if no exchange occurred between the coarser and the finer sediment particles, and no finer sediment was eroded, then the quantity of the deposited coarser sediment would be 17.75×10^6 tons-1.46×10^6 tons $= 16.29 \times 10^6$ tons, and the sedimentation pattern of the channel would change from scour to deposition.

During 1980–1984, the deposition quantity of the coarser sediment was 12.92×10^6 tons for the reach between Hungzhuang and Xiantao, while the scour of the finer sediment was 150.75×10^6 tons. Based on the ratio of $\omega^{0.92}$, or the ratio of the carrying-sediment capacity, the scour quantity of the finer sediment, which is equivalent to the deposition of the coarser sediment in the amount of 12.92×10^6 tons, is $\frac{0.0573}{0.00733} \times 12.92 \times 10^6$ tons $= 101.02 \times 10^6$ tons. If no coarser sediment deposited, the scour quantity of the finer sediment would only be 150.75×10^6 tons $-$ 101.02×10^6 tons $= 49.73 \times 10^6$ tons. Hence, it is concluded that the coarser and finer sediment exchange tends to adjust the size distribution of sediment, increase the carrying-sediment capacity, and alleviate channel deposition.

10.4 Definition and Characteristics of Water Percentage

The concept of water percentage was developed by Han [5]. As an important tool, it has been widely applied to the investigations of non-uniform suspended load transport [7, 9, 10, 11, 13]. The concept of water percentage was briefly addressed in previous chapters, and is analyzed in detailed as follows.

10.4.1 Definition of Water Percentage

Let assume the sediment concentration is not very high so that the impacts between the sediment particles and the impact of the sediment on the flow are negligible. This is also a basic assumption for the advection–diffusion equation for the sediment transport. This section focuses on the water percentage for the carrying-sediment capacity. For strong equilibrium sediment transport, or

$$\begin{cases} S = S^* \\ P_{4.l} = P^*_{4.l} \end{cases}$$

where S is the sediment concentration, S^* the sediment transport capacity, $P_{4.l}$ and $P^*_{4.l}$ are the size distributions for suspended load and for carrying-sediment capacity, respectively. For this case, the water percentage can be directly introduced for the

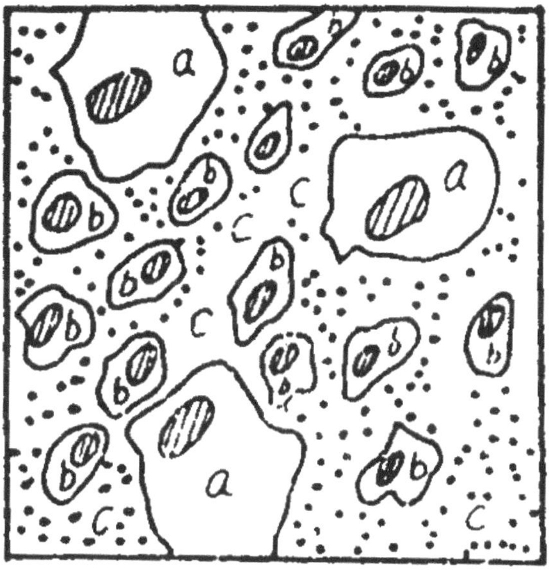

Fig. 10.1 Diagram for sediment transport capacity within a unit water volume

sediment concentration. For a unit-area water column, assume it carries n groups of sediment, with the particle size of $D_1, D_2, \ldots, D_n$, respectively. The corresponding weight of each size group is $G_1, G_2, \ldots, G_n$, respectively. Set a water body around each sediment particle according to $S^*(l)$. Figure 10.1 shows a diagram for the case with three groups of sediment: a, b, and c. The water bodies surrounding the sediment particles of groups a and b are drawn in this figure, while the remaining part of the water column belong to sediment group c.

The water body surrounding a sediment particle is chosen so that the carrying-sediment capacity of this water body is exactly the same as $S^*(l)$ for this size group. Summing up the water bodies for all particles within one size group, the total water quantity for this size group is Q_l that satisfy

$$S^*(l) = \frac{G_l}{Q_l}$$

or

$$Q_l = \frac{G_l}{S^*(l)} \tag{10.24}$$

Alternatively, consider a water volume of 1 m^3, and set the fractional weight of each sediment size group that is carried in this water volume to be $g_1, g_2, \ldots, g_n$. As done before, divide the water volume into different parts, with each part carrying the sediment of one size class. Setting the water volume part to carry the l-th size group is q_l, then $S^*(l) = g_l/q_l$, which is the same as that derived before. As shown in Eq. (10.24), Q_l is the water quantity required to carry the sediment concentration that equals the carrying-sediment capacity, $S^*(l)$, of the uniform sediment with the particle size of D_l. The water percentage is defined as

$$K_l = \frac{Q_l}{\sum_{l=1}^{n} Q_l} = \frac{Q_l}{Q} \tag{10.25}$$

which represents the proportion of the water quantity that is required to carry the size group of D_l in the amount of the carrying-sediment capacity for the uniform sediment with the particle size of D_l. Obviously, $\sum K_l = 1$. However $Q = \sum_{l=1}^{n} Q_l$ has to be proved. In fact, from

$$S_l^* = \frac{G_l}{Q} = P_{1.l} S^*(l) = P_{1.l} \frac{G_l}{Q_l}$$

where $P_{1.l}$ is the effective gradation of bed material, there is $P_{1.l} = \frac{Q_l}{Q}$, Since $\sum P_{1.l} = 1$, then $Q = \sum_{l=1}^{n} Q_l$ and $K_l = P_{1.l}$, i.e. the water Percentage is the effective gradation of bed material,

10.4.2 Properties of Water Percentage [10]

(1) The water percentage can be used to sum up the carrying-sediment capacity for uniform sediment to determine the carrying-sediment capacity for the non-uniform sediment mixture.

The fractional carrying-sediment capacity for the non-uniform sediment mixture is

$$S_l^* = P_{4.l}^* S^* \tag{10.26}$$

where S_l^* is the fractional sediment transport capacity for the l-th size class of the non-uniform sediment mixture, $P_{4.l}^*$ the size distribution for the carrying-sediment capacity, S^* the total carrying-sediment capacity for the non-uniform sediment mixture. According to the definition of the fractional carrying-sediment capacity, and considering Eqs. (10.24) and (10.25), one obtains

$$S_l^* = \frac{G_l}{\sum_{l=1}^{n} Q_l} = \frac{G_l}{Q_l} \frac{Q_l}{\sum_l^n Q_l} = S^*(l) K_l \tag{10.27}$$

Thus,

$$S^*(\omega^*) = S^* = \sum_{l=1}^{n} S_l^* = \sum_{l=1}^{n} K_l S^*(l) \tag{10.28}$$

Equation (10.27) has a clear physical meaning: the fractional carrying-sediment capacity for the l-th size class equals the carrying-sediment capacity for the uniform sediment with this particle size multiplying the water percentage for this size class Eq. (10.28) is the formula to determine the sediment transport for the non-uniform sediment mixture by summing up the carrying-sediment capacity for uniform sediment with the water percentage. It also illustrate the structure of the carrying-sediment capacity for non-uniform sediment, which can be determined by summing up the carrying-sediment capacity for uniform sediment. This significantly simplifies the analysis of the carrying-sediment capacity for non-uniform sediment. Comparing Eq. (10.27) with Eq. (10.26) gives

$$P_{4.l}^* S^*(\omega^*) = S_l^* = K_l S^*(l) \tag{10.29}$$

In the strong equilibrium condition, $P_{4.l}^* = P_{4.l}$, $S^*(\omega^*) = S$, and Eq. (10.29) becomes

$$P_{4.l} S = P_{4.l} S^*(\omega^*) = K_l S^*(l) \tag{10.30}$$

or

$$K_l = \frac{P_{4.l}^* S^*}{S^*(l)} = \frac{P_{4.l} S}{S^*(l)} \tag{10.31}$$

It is noted that $S^*(\omega^*)$ is simply denoted as S^* hereafter, unless otherwise specified. The water percentage can be determined with Eq. (10.31). Summation of Eq. (10.29) for all size classes yields

$$1 = \sum_{l=1}^{n} \frac{S^* P_{4.l}^*}{S^*(l)} \tag{10.32}$$

Using Zhang's formula for the carrying-sediment capacity,

$$S^*(l) = \kappa \left(\frac{V^3}{gh\omega_l} \right)^m$$

and then plugging into Eq. (10.32) gives

$$S^* = \frac{1}{\sum_{l=1}^{n} \frac{P_{4.l}^*}{S^*(l)}} = \kappa\left(\frac{V^3}{gh}\right)^m \frac{1}{\sum_{l=1}^{n} P_{4.l}^* \omega_l^m} = \kappa\left(\frac{V^3}{gh\omega^*}\right)^m = \kappa\left(\frac{V^3}{gh\omega}\right)^m \tag{10.33}$$

where

$$\omega^{*m} = \sum_{l=1}^{n} P_{4.l}^* \omega_l^m \tag{10.34}$$

$$\omega^m = \sum_{l=1}^{n} P_{4.l} \omega_l^m \tag{10.35}$$

Plugging Eq. (10.33) and Zhang's formula for $S^*(l)$ into Eq. (10.31)

$$K_l = \frac{P_{4.l}^* S*}{S^*(l)} = \frac{P_{4.l}^* \omega_l^m}{\omega^{*m}} = \frac{P_{4.l}^* \omega_l^m}{\sum_{l=1}^{n} P_{4.l}^* \omega_l^m} = \frac{P_{4.l} \omega_l^m}{\sum_{l=1}^{n} P_{4.l} \omega_l^m} \tag{10.36}$$

It is noted that the terms after the last equal sign of Eqs. (10.31), (10.33), and (10.36) involve the size distribution of suspended load, and are only true under the strong equilibrium condition.

(2) For the strong equilibrium condition, the water percentage equals the bed material gradation.

Based on the statistical theory for sediment exchange on the bottom, He and Han [13, 15] derived the fractional carrying-sediment capacity S_l^* for equilibrium transportation that is

$$P_{4.l}^* S^* = S_l^* = \frac{2}{3} m_0 \gamma_s \Omega_l \frac{P_{1.l.1} D_l L_{4.l}^*}{q t_{4.0.l}}$$

where $P_{1.l.1}$ is the gradation of bed material. For uniform sediment

$$S^*(l) = \frac{2}{3} m_0 \gamma_s \Omega_l \frac{D_l L_{4.l}^*}{q t_{4.0.l}}$$

Hence,

$$P_{4.l}^* S^* = P_{1.l.1} S^*(l) \tag{10.37}$$

Comparing Eq. (10.37) with Eq. (10.29)

$$P_{1.l.1} S^*(l) = P_{4.l}^* S^* = K_l S^*(l)$$

Thus

$$K_l = P_{1.l.1} \tag{10.38}$$

As mentioned above the water percentage equals the effective bed material gradation. For the strong equilibrium condition, $P_{1.l} = P_{1.l.1}$, and thus the water percentage equals the bed material gradation [7, 13, 15].

(3) The water percentage also represents the energy percentage.

For the studied unit water column, the kinetic energy can be represented by $\rho V^2 Q_l$, and the potential energy by $\gamma\ Q_l J$. The flow velocity V and the water surface slope J do not vary with Q_l. The density (ρ) and the specific weight (γ) of water are constant. Hence, the kinetic energy and the potential energy only vary with Q_l, and the water percentage also represents the energy percentage.

10.5 Utilizing Water Percentage to Integrate the Carrying-Sediment Capacities for Bed Material Load and Wash Load

In the earlier studies for the non-uniform suspended load transport in an alluvial stream, the concept of bed material load and wash load was introduced to simplify the analysis. The bed material load considered to be the great majority of the bed material, up to 90–95% [1, 20], take part in channel evolution, and is closely dependent of the flow. Coming from the watershed, the wash load is almost independent of the flow condition, and can be completely transported through the reach. The relationship between the flow discharge and the sediment transport rate, which were measured in flume experiments, is shown in Fig. 10.2 for various grain size classes [2]. As shown in this figure, the transport rate, q, for the coarser sediment with $D > 0.01$ mm shows close correlation with the flow discharge, Q, while the transport rate for the finer sediment with $D \leq 0.01$ mm, or for the wash load, show no correlation with the flow discharge. Hence, it is a popular approach to exclude the wash load, which comes from the watershed, barely exchanges with the bed material and only spends a small amount of flow energy, and for the analysis of carrying-sediment capacity, only consider the carrying-sediment capacity for the bed material load.

On the other hand, some studies questioned the differentiation between bed material load and wash load. Shan [17] analyzed the carrying-sediment capacity for the coarser and the finer sediment together, without differentiating bed material load and wash load. Chein [1] considered it was necessary to separate wash load from bed material load, but a general law should apply to both bed material load and wash load, and the wash load could not be computed as done for the bed material load as the wash load come from the watershed. In fact, if the sediment concentration is not too high (less than 50–100 kg/m^3) and the sediment particle is not too fine (coarser than

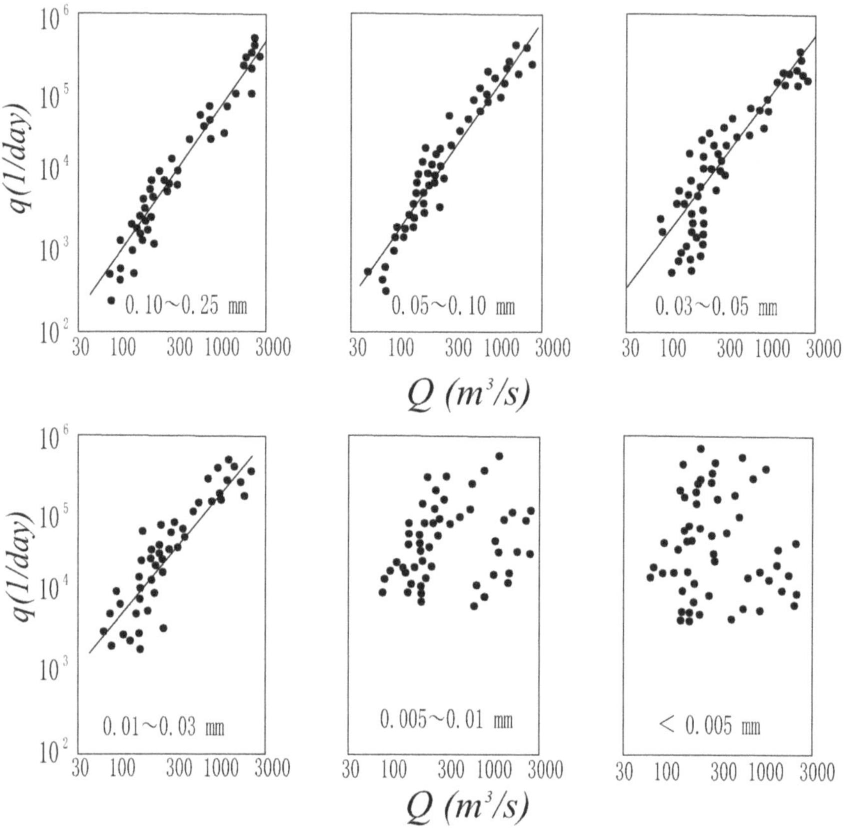

Fig. 10.2 Relationship between fractional sediment transport rate and flow charge on the same River

0.001 mm) so that the assumptions for the sediment diffusion equation are satisfied, then, theoretically, it is unnecessary to differentiate between bed material load and wash load [12, 14]. In most mathematical models for sediment transport, bed material load and wash load are not differentiated for simplicity. This is particularly true for the mathematical models for reservoir sedimentation [5]. It is assumed in these models that both bed material load and wash load follow a common law of transport, without theoretical proof and basis, while the model results can demonstrate the sedimentation patterns for both the coarser and the finer sediment. In fact, this assumption can be theoretically proved with the concept of water percentage [12, 14]. The theoretical proof, together with the field data collected at the Xinchang station of the Yangtze River and at the Gaocun station of the Yellow River, are discussed as follows.

Table 10.6 lists two sets of water percentage [14] measured at the Xinchang station of the Yangtze River during two events of weak equilibrium sediment transport, respectively. According to the formula for carrying-sediment capacity that is generally used for the Jingjiang reach of the Yangtze Rover,

$$S^* = \kappa\left(\frac{V^3}{gh\omega}\right)^m = K\left(\frac{V^3}{h\omega}\right)^{0.92} = 0.017\left(\frac{V^3}{h\omega}\right)^{0.92} \tag{10.39}$$

where $K = \kappa/g^m$, and this formula is in metric (SI) units. As listed in Table 10.6, the computed carrying-sediment capacity S^* is 0.210 kg/m^3 and 2.78 kg/m^3, respectively, which show good agreement with the measured sediment concentration of 0.200 kg/m^3 and 2.51 kg/m^3, respectively. The flow velocity and the water depth during these two events are similar, resulting in a negligible difference (1.5%) in the hydraulic parameter V^3/h. However, the measured sediment concentration shows significant difference between these two events, with the carrying-sediment capacity satisfying the same governing equation, or Eq. (10.39). Because the size distribution of suspended sediment is not completely consistent with the bed material gradation, the sediment transport is not under the strong equilibrium condition during these two events. Instead, they belong to weak equilibrium sediment transport with the carrying-sediment capacity S^* approximately equal to the sediment concentration S.

Based on the exchange of sediment it can be explained that with almost the same flow condition why exists large difference between the sediment concentrations of these two events. During the 3/17/1964 event, with a low concentration, the wash load needed supplement from the bed material. However, wash load barely existed in the bed material. As a result, as the supplement from the bed material, the bed material load is scoured until a weak equilibrium condition was achieved. Opposite to the 3/17/1964 event, the wash load had a high concentration during the 6/23/1963 event, and part of the suspended load needed to deposit on the bed in order to achieve a weak equilibrium condition. As the bed material has a larger fall velocity, the deposited material on the bed consisted of the bed material load. This indicates that the carrying-sediment capacity for bed material load and for wash load is a united

Table 10.6 Bed material load and wash load measured at the Xinchang Station of the Yangtze River

No	Date	Water depth H (m)	Flow velocity V (m/s)	Total sediment concentration S (kg/m^3)	Wash load concentration S_w (kg/m^3)	Bed material load concentration S_b (kg/m^3)	$\frac{V^{2.76}}{h^{0.92}}$	$\frac{V^{2.76}}{h^{0.92}\sum P_{4,l}\omega_l^{0.92}}$
1	3/17/1961	4.14	0.954	0.200	0.075	0.125	0.238	0.210
2	6/23/1963	5.15	1.06	2.57	2.51	0.06	0.245	2.78

whole. Depending on the inflow of wash load, the concentration of bed material load is adjusted so that the total sediment concentration will approach the equilibrium condition. Compared to the wash load, the bed material load is coarser, and expends more flow energy. By adjusting the concentration of bed material load, the adjustment of sediment concentration is achieved more quickly with the least bed change. Hence, it lacks of theoretical basis to simply exclude wash load from the carrying-sediment capacity, considering it does not expend flow energy, and only consider the carrying-sediment capacity for bed material load.

Using the data listed in Table 10.6 as an example, the water percentage and its application are discussed as follows. For a weak equilibrium condition, $S^*(\omega) = S$. As an approximation, set $P^*_{4.l} = P_{4.l}$. Then,

$$K_l S^*(l) = P_{4.l} S^*(\omega)$$

Also noting $m = 0.92$ in Eq. (10.39), Eq. (10.36) can be written as

$$K_l = \frac{P_{4.l} S^*(\omega)}{S^*(l)} = \frac{P_{4.l}\omega_l^{0.92}}{\omega^{0.92}} = \frac{P_{4.l}\omega_l^{0.92}}{\sum_{l=1}^{n} P_{4.l}\omega_l^{0.92}} \tag{10.40}$$

According to previous studies, the threshold particle size to differentiate bed material load and wash load is $D_k = 0.1$ mm for the Jingjiang reach of the Yangtze River. The sediment with size class index $1 \le l \le k$ is wash load, and the sediment with $k+1 \le l \le n$ is bed material load. Summing up Eq. (10.40) for size classes for bed material load and for wash load, respectively, the water percentages to carry the bed material load and wash load, or K_b and K_w, are

$$K_b = \sum_{l=k+1}^{n} K_l = \frac{\sum_{l=k+1}^{n} P_{4.l}\omega_l^{0.92}}{\sum_{l=1}^{n} P_{4.l}\omega_l^{0.92}} \tag{10.41}$$

$$K_W = \sum_{l=1}^{k} K_l = \frac{\sum_{l=1}^{k} P_{4.l}\omega_l^{0.92}}{\sum_{l=1}^{n} P_{4.l}\omega_l^{0.92}} \tag{10.42}$$

Also, based on Eq. (10.40)

$$\frac{P_{4.l} S^*(\omega)}{K_l} = S^*(l) = \kappa\left(\frac{V^3}{gh\omega_l}\right)^{0.92} = K\left(\frac{V^3}{h\omega_l}\right)^{0.92} \tag{10.43}$$

where $S^*(l)$ is the carrying-sediment capacity for the uniform sediment with the particle size of D_l. Based on Eq. (10.27), and utilizing Eqs. (10.41) and (10.43), the ratio of the total carrying-sediment capacity for bed material load ($D_{k+1} \sim D_n$) to the water percentage carrying the bed material load, or S^*_b/K_b, can be written as

$$\frac{S_b^*}{K_b}=\frac{\sum_{l=k+1}^{n}S_i^*}{K_b}=\frac{\sum_{l=k+1}^{n}K_lS^*(l)}{K_b}=\frac{\frac{\sum_{l=k+1}^{n}P_{4.l}\omega_l^{0.92}K\left(\frac{V^3}{h\omega_l}\right)^{0.92}}{\sum_{l=1}^{n}P_{4.l}\omega_l^{0.92}}}{\frac{\sum_{l=k+1}^{n}P_{4.l}\omega_l^{0.92}}{\sum_{l=1}^{n}P_{4.l}\omega_l^{0.92}}}$$
$$=\frac{\sum_{l=k+1}^{n}P_{4.l}K\left(\frac{V^3}{h}\right)^{0.92}}{\sum_{l=k+1}^{n}P_{4.l}\omega_l^{0.92}}=K\left(\frac{V^3}{h}\right)^{0.92}\frac{1}{\sum_{l=k+1}^{n}P_{4.l.b}\omega_l^{0.92}}$$
$$=K\left(\frac{V^3}{h\omega_b}\right)^{0.92}=S^*(b) \tag{10.44}$$

As shown in (10.39), $K=0.017$ for the Jingjiang reach. Then Eq. (10.44) becomes

$$\frac{S_b^*}{K_b}=0.017\left(\frac{V^3}{h\omega_b}\right)^{0.92} \tag{10.45}$$

where

$$\omega_b^{0.92}=\sum_{l=k+1}^{n}P_{4.l.b}\omega_l^{0.92}=\frac{\sum_{l=k+1}^{n}P_{4.l}\omega_l^{0.92}}{\sum_{l=k+1}^{n}P_{4.l}}$$
$$=\frac{\sum_{l=k+1}^{n}P_{4.l}\omega_l^{0.92}}{P_{4.b}}=\sum_{l=k+1}^{n}P_{4.l.b}\omega_l^{0.92} \tag{10.46}$$

and $P_{4.l.b}$ is the normalized size distribution for bed material load, or

$$P_{4.l.b}=\frac{P_{4.l}}{\sum_{l=k+1}^{n}P_{4.l}}\quad(l=k+1,k+2,\ldots,n) \tag{10.47}$$

Similarly, for wash load

$$\frac{S_W^*}{K_w}=0.017\left(\frac{V^3}{h\omega_w}\right)^{0.92}=S^*(w) \tag{10.48}$$

$$\omega_w^{0.92}=\sum_{l=1}^{k}P_{4.l.w}\omega_l^{0.92}=\frac{\sum_{l=1}^{k}P_{4.l}\omega_l^{0.92}}{\sum_{l=1}^{k}P_{4.l}}=\frac{\sum_{l=1}^{k}P_{4.l}\omega_l^{0.92}}{P_{4.w}} \tag{10.49}$$

$$P_{4.l.w}=\frac{P_{4.l}}{\sum_{l=k+1}^{n}P_{4.l}}\quad(l=1,2,\ldots,k) \tag{10.50}$$

where S_w^* is the carrying-sediment capacity for wash load, $P_{4.l.w}$ is the normalized size distribution for wash load.

For total load,

$$\omega^{0.92} = \sum_{l=1}^{n} P_{4.l}\omega_l^{0.92} \tag{10.51}$$

and

$$\frac{S^*}{\sum_{l=1}^{n} K_l} = S^* = 0.017\left(\frac{V^3}{h\omega}\right)^{0.92} \tag{10.52}$$

The ratio of the carrying-sediment capacity to the corresponding water percentage for bed material load, wash load, and total load can be generally written as

$$\frac{S_R^*}{K_R} = 0.017\left(\frac{V^3}{h\omega_R}\right)^{0.92} = S^*(R) \tag{10.53}$$

where R can be b, w, or c, representing bed material load, wash load, and total load, respectively. For total load, $K_c = 1$, $\omega_c = \omega$. Equation (10.53) indicates that a general unified carrying-sediment capacity law can apply to bed material load, wash load, and total load of suspended sediment. Table 10.7 lists the computed results of S_b^*, S_w^*, $S_c^* = S^*$, $\omega_c = \omega$, ω_b, ω_w, $S_b^*/K_b = S^*(b)$, $S_w^*/K_w = S^*(w)$, and $S^*/K_c = S^*$ for the Jingjiang reach of the Yangtze River during the 6/23/1963 and the 3/17/1964 events. As listed in this table, the computed carrying-sediment capacities for bed material load $S_b^* = K_b S^*(b)$, wash load $S_w^* = K_w S^*(w)$, and total load S^* show good agreement with the measured sediment concentrations. Also, comparing the ratios of the measured carrying-sediment capacities to the corresponding water percentages, i.e., $\frac{S_W}{K_W}$, $\frac{S_b}{K_b}$, and S, with the computed $S^*(R) = 0.017\left(\frac{V^3}{h\omega_R}\right)^{0.92}$, a good agreement can be found. On the other hand, if the water percentage is not considered, and directly use

$$S_b^* = 0.017\left(\frac{V^3}{h\omega_b}\right)^{0.92}$$

$$S_w^* = 0.017\left(\frac{V^3}{h\omega_W}\right)^{0.92}$$

to determine the carrying-sediment capacity for bed material load and wash load, respectively, then the computed results show significant discrepancy from the data. For example, the computed S_w^* for wash load is 0.777 kg/m^3 for the 3/17/1964 event, while the measured sediment concentration was 0.075 kg/m^3. For the 6/23/1963 event, the computed S_w^* is 5.25 kg/m^3, compared to a measured sediment concentration of 2.51 kg/m^3. Hence, without considering the water percentage, the computed carrying-sediment capacity for wash load show significant difference from the data.

Table 10.7 Carrying-sediment capacities for wash load and bed material load

Item	3/17/1964			6/23/1963		
	Wash load	Bed material load	Total load	Wash load	Bed material load	Total load
$\left(\frac{V^3}{h}\right)^{0.92}$	0.238	0.238	0.238	0.245	0.245	0.245
Measured S_w, S_b, S (kg/m^3)	0.075	0.125	0.200	2.51	0.0600	2.57
Measured $P_{4.w}$, $P_{4.b}$, $\sum P_{4.l}$	0.372	0.628	1.000	0.977	0.023	1.000
K_w, K_b, K_c	0.101	0.899	1.000	0.521	0.479	1.000
$\omega_w^{0.92}$, $\omega_b^{0.92}$, $\omega^{0.92}$	0.00521	0.0276	0.0192	0.000793	0.0309	0.00150
S_w/K_w, S_b/K_b, S	0.743	0.139	0.200	4.82	0.125	2.57
$S^*(R) = 0.017\left(V^3/h\omega_R\right)^{0.92}$	0.777	0.147	0.211	5.25	0.135	2.78
$\left(V^3/h\omega_R\right)^{0.92}$	45.7	8.62	12.4	309	7.93	163
S_W^*, S_b^*, S^* (kg/m^3) Cal.by S_l^*	0.078	0.132	0.211	2.74	0.065	2.78

Figure 10.3 lists the relationships for $\frac{S_b^*}{K_b}$ versus $\left(\frac{V^3}{h\omega_b}\right)^{0.92}$, $\frac{S_W^*}{K_W}$ versus $\left(\frac{V^3}{h\omega_w}\right)^{0.92}$, and S^* versus $\left(\frac{V^3}{h\omega}\right)^{0.92}$, which can be approximated by a linear regression line. This indicates that the carrying-sediment capacities for wash load, bed material load, and total load satisfy a same law, as presented in Eq. (10.53), after considering the water percentage. As a comparison, the relationships of $S_b^* \sim \left(\frac{V^3}{h\omega_b}\right)^{0.92}$, and $S_W^* \sim \left(\frac{V^3}{h\omega_w}\right)^{0.92}$ are also shown in Fig. 10.3, with no reasonable correlation being found, particularly for wash load. This indicates that simply establishing a relationship between the carrying-sediment capacity for bed material load, or for wash load, and the hydraulic parameters, without including the water percentage, cannot represent the law for sediment transport. It is theoretically questionable and can practically cause a great error to investigate the carrying-sediment capacity of bed material load as a function of $\left(\frac{V^3}{h\omega_b}\right)^m$ without including the water percentage, as generally did in many studies. As shown in Fig. 10.3, a correlation does not exist between S_b^* and $\left(\frac{V^3}{h\omega_b}\right)^{0.92}$, or between S_w^* and $\left(\frac{V^3}{h\omega_w}\right)^{0.92}$. Without considering the water percentage, the carrying-sediment capacity for bed material load or for wash load does not follow the same law of sediment transport as the total load does.

Figure 10.4 shows the relationship between S_R^*/K_R and $\left(V^3/h\omega_R\right)^{0.92}$ for bed material load, wash load, and total load, which were computed based on the 3-year data collected at the Xinchang station of the Yangtze River. As shown in this figure,

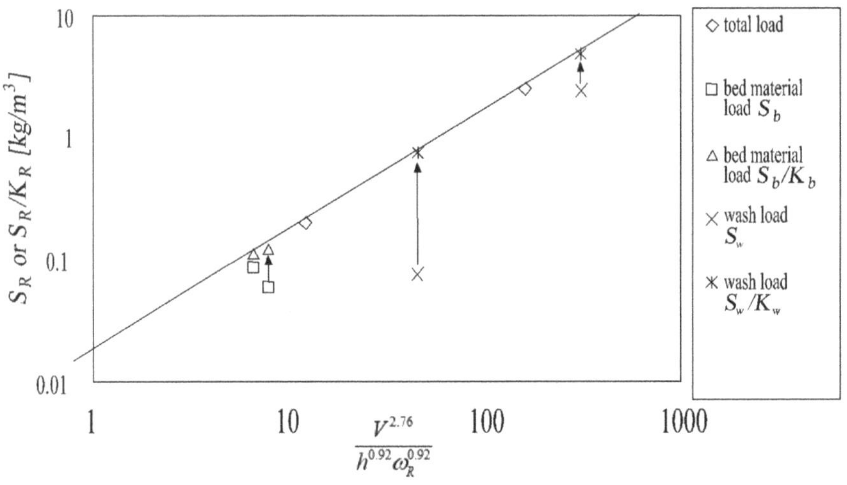

Fig. 10.3 Relationship between S_R^*/K_R, or S_R, and $\left(V^3/gh\omega\right)^{0.92}$ for wash load, bed material load and total load

a linear regression line, which reads $S(R) = S_R^*/K_R = 0.017\left(\frac{V^3}{h\omega_R}\right)^{0.92}$, can be used to approximate these data points. This proves Eqs. (10.45), (10.48), and (10.52), and also proves that the carrying-sediment capacities for bed material load, wash load, and total load follows the same law or formula, after including the impact of the water percentage. Figure 10.5 shows the relationship between S_R^*/K_R and $\left(V^3/h\omega_R\right)^{0.92}$ based on the data measured at the Gaocun station of the Yellow River during the downstream channel erosion period that occurred after the initial operation of the Sanmenxia Reservoir. As shown in this figure, after including the water percentage, the transport capacities for bed material load, wash load, and total load follow a same law, which reads $S_R^*/K_R = 0.025\left(\frac{V^3}{h\omega_R}\right)^{0.92}$.

Different from analyzing the carrying-sediment capacities for bed material load and for wash load, a more general approach is to divide suspended load into a series of size groups and then analyze the fractional carrying-sediment capacity S_l^* for each size group. According to Eq. (10.29), based on which, the carrying-sediment capacity $S^*(l)$ for the uniform sediment can be converted from the fractional carrying-sediment capacity S_l^*. Because of this relation, a generalized equation for the carrying-sediment capacity can be developed for the uniform sediment with different particle sizes. Based on the field data collected at the Xinchang station, Fig. 10.6 shows the relationship between the ratios of the fractional sediment carrying-sediment capacity to the corresponding water percentage, or S_l^*/K_l, and $\left(V^3/h\omega_l\right)^{0.92}$ for various size classes. As shown in this figure, the data points can be approximated by a linear regression line, indicating all size classes follows a same law for carrying-sediment capacity.

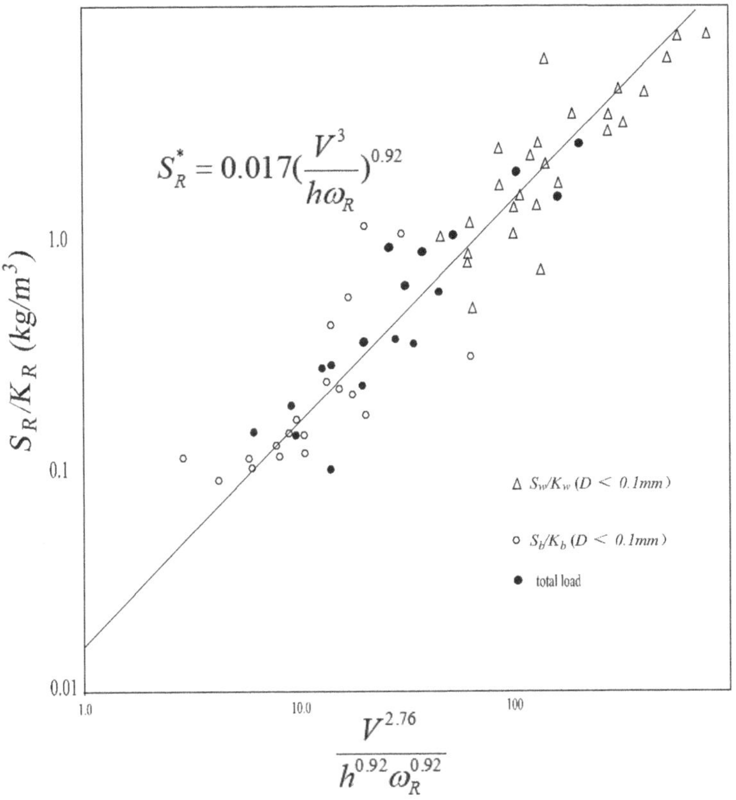

Fig. 10.4 Relationship between S_R^*/K_R and $\left(V^3/h\omega_R\right)^{0.92}$ for wash load, bed material load and total load at the Xinchang Station of the Yangtze River

$$\frac{S_l^*}{K_l} = S^*(l) = 0.017\left(\frac{V^3}{h\omega_l}\right)^m \tag{10.54}$$

The analysis in this section describes the characteristics of bed material load and wash load and their similarities and differences. More importantly, it illustrates the physical meaning of the water percentage, which is proved to be an efficient tool for analyzing the carrying-sediment capacity for non-uniform sediment. It should be emphasized that since a generalized law or formula can be applied to the carrying-sediment capacities for wash load, bed material load, and total load, it is unnecessary to differentiate between bed material load and wash load as long as a proper method is used to determine the carrying-sediment capacity for non-uniform sediment. If the wash load needs to be separated from the bed material load, the variation of the water percentage should be considered instead of simply excluding the concentration of wash load.

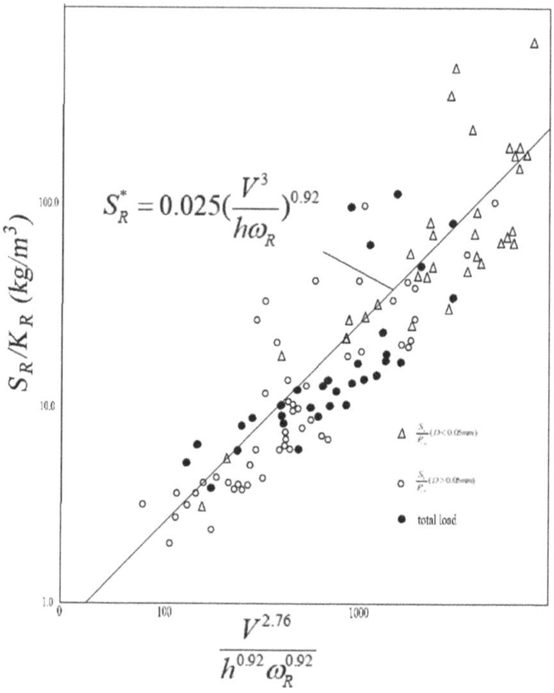

Fig. 10.5 Relationship between S_R^*/K_R and $(V^3/h\omega_R)^{0.92}$ for wash load, bed material load and total load at the Gaocun station of the Yellow River

10.6 Discussion of the Differentiation Between Bed Material Load and Wash Load

As discussed in Sect. 10.5, it is unnecessary to differentiate between bed material load and wash load [14]. However, as the first approximation for the analysis of non-uniform sediment transport, differentiation between bed material load and wash load can be used to illustrate some phenomena for sediment transport, and has played an important role in the analyses of river sediment transport and channel evolution. After the study results are available for the non-uniform carrying-sediment capacity and the non-equilibrium transport of non-uniform sediment, the second approximation was obtained for non-uniform sediment transport. The various transport patterns for the bed material load and wash load, and their similarity and differences can be better illustrated with the carrying-sediment capacity for non-uniform sediment and the law for non-equilibrium sediment transport. It is also proved that the differentiation between bed material load and wash load is theoretically questionable. However, if the differentiation is needed, what's the correct approach? This question is discussed as follows.

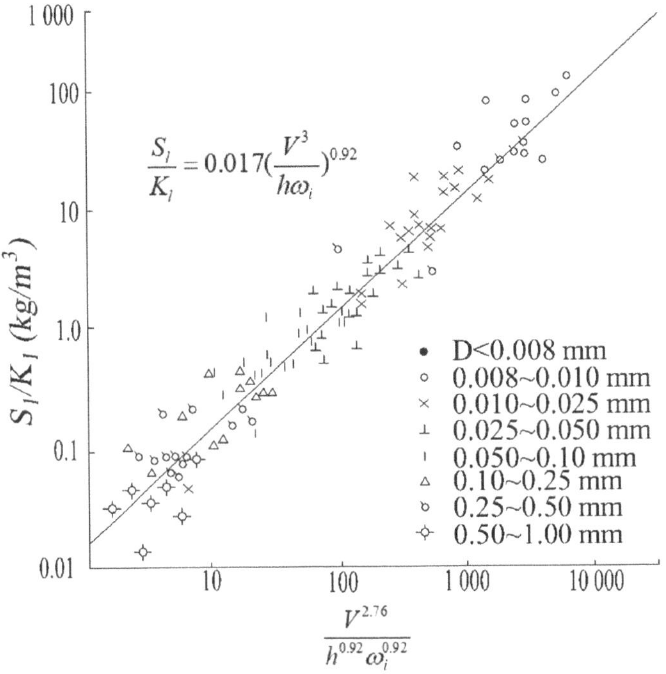

Fig. 10.6 Relationship between S_l^*/K_l and $\left(V^3/h\omega_l\right)^{0.92}$ for the Xinchang Station of the Yangtze River

10.6.1 It is Questionable to Consider that Bed Material Load is Usually Saturated in Alluvial Streams While Wash Load is Unsaturated

The bed material can serve as a sufficient supplement for the bed material load in the water body. Hence, if the bed material load from the upstream is insufficient, the bed material will quickly provide a supplement; if the bed material load from the upstream is over-saturated, part of the bed material load will deposit on the bed and the concentration of bed material load will be adjusted to become consistent with the carrying-sediment capacity. However, this is not true for wash load. If the wash load from the upstream is insufficient, the bed cannot or can barely provide any supplement; if the wash load from the upstream is over-saturated, it barely deposits on the bed and thus the concentration of wash load is hard to be adjusted to be consistent with the carrying-sediment capacity. While these inferences sound reasonable, they are not sufficient to prove that the bed material load is saturated or almost saturated while the wash load is unsaturated.

In fact, the bed material load and wash load form a unity whole for suspended load, they are either both saturated or both unsaturated. This is because the water percentages for bed material load and for wash load, or K_b and K_w can be adjusted, with the sum equal to 1. If the wash load influx is over-saturated and the bed material is sub-saturated, the wash load mixed by the turbulent flow will acquire more water percentage and then decrease K_b until both sub-saturated. In contrary if the inflow wash load is sub-saturated and the bed material is over-saturated, mixed by the turbulent flow the bed material load will obtain more water percentage and then decrease K_w until both over-saturated. At the same time adjusted by the scour and deposition both wash load and bed material load will turn to equilibrium. The 'adjustion' mainly caused by the scour or deposition of coarser particles because the finer particles are hard to deposit and there is few of them at the bed. This can be seen in Table 10.6. As listed in this table, the value of $(V^3/h)^{0.92}$ was approximately the same for the 3/17/1964 event and the 6/23/1963 event, which was 0.238 for the former event, and was 0.245 for the later event. However, the total sediment concentration showed significant difference, which was 0.200 kg/m^3 for the 3/17/1964 event, and was 2.57 kg/m^3 for the 6/23/1963 event that is 12.8 times higher than the 3/17/1964 event. The sediment concentration was very close to the corresponding carrying-sediment capacity for these two events, which was 0.210 kg/m^3 and 2.78 kg/m^3, respectively. This indicates that both bed material load and wash load were approximately saturated for these two events. More importantly, Table 10.6 also shows how the water percentage varies with the wash load influx. During the 3/17/1964 event, the concentration of wash load was very high (2.51 kg/m^3) with a high water percentage for wash load ($K_w = 0.521$). This resulted in a lower water percentage to carry bed material load and a low concentration of bed material load (0.06 kg/m^3). During the 6/23/1963 event, the concentration of wash load was very low (0.075 kg/m^3) with a low water percentage for wash load ($K_w = 0.101$). As a result, the water percentage for bed material load was much higher and the concentration of bed material load was also higher (0.125 kg/m^3). These examples show how the total sediment concentration is adjusted to achieve the equilibrium condition by adjusting the proportion of the bed material load and the water percentage.

Sediment transport is defined as a strong equilibrium transport if the following conditions are satisfied: $S = S^*$ and $P_{4.l} = P^*_{4.l}$, or $S_l = S^*_l = P_{1.l.1}S^*(l) = K_{1.l}S^*(l)$. For an alluvial stream, if the wash load influx is irregular, while the water percentage can be adjusted accordingly, the bed material gradation is adjusted very slowly. As a result, while $S^*_l = K_{1.l}S^*(l)$ is satisfied, $S^*_l = P_{1.l.1}S^*(l)$, or Eq. (10.38) is not true. Sediment transport for this case is not in a strong equilibrium condition, but in a weak equilibrium condition with both the concentration of bed material load and the concentration of wash load are unsaturated. If the wash load influx is steady and is compatible with the bed material gradation, a strong equilibrium sediment transport can occur, with the wash load, bed material load and total load all in equilibrium condition.

For $S \neq S^*$, both the bed material load and wash load are unsaturated, either are sub-saturated or are over-saturated. It is impossible for the concentration of bed material load is saturated while the concentration of wash load is unsaturated. This is because the water percentage can be adjusted constantly until the bed material load and wash load achieve a same degree of saturation.

10.6.2 Similarity and Difference Between the Coarser and Finer Sediment Exchange and the Concept of Bed Material Load and Wash Load

The exchange between the coarser and finer sediment and the concept of bed material load and wash load are two aspects of a same subject. The study for the coarser and finer sediment exchange focuses on the dynamic process and the impact to channel evolution. The concept of bed material load and wash load is used to illustrate the different characteristics of sediment transport and carrying-sediment capacity between the bed material load and the wash load. The characteristics of the coarser and finer sediment exchange are a supplement to and an in-depth illustration of the concept of bed material load and wash load. Firstly, based on the field data, the coarser sediment as defined in the study of the coarser and finer sediment exchange is equivalent to the bed material load, and the finer sediment is equivalent to the wash load. The threshold particle size to differentiate the coarser and fine sediment can be theoretically determined with Eq. (10.11), instead of being empirically estimated. The bed material gradation not only reflects the size of the bed material. More importantly, it reflects the cumulative impact of the flow condition on the bed for alluvial streams, and is the "memory" of the sedimentation characteristics for various size classes. Secondly, the exchange process between the coarser and finer sediment indicates that the finer sediment can be eroded, or supplemented, from the bed, while the coarser sediment deposits on the bed in a reach with a mild erosion and deposition. This sediment exchange increases the carrying-sediment capacity along the flow path so as to be compatible with the weakening of the flow condition. This is not illustrated in the concept of bed material load and wash load. Thirdly, the threshold particle size to differentiate the coarser and finer sediment varies significantly for various reaches and for various sedimentation patterns. This is an improvement to the threshold particle size used to differentiate bed material load and wash load, which is relatively constant for various reaches with different sedimentation patterns. For a reach with strong flow and with coarser bed material, the threshold particle size to differentiate the coarser and finer sediment is significantly larger than the threshold particle size between bed material load and wash load that was estimated in most studies. The threshold particle size between the coarser and finer sediment has a significant impact to the size distribution for carrying-sediment capacity, as discussed in Chap. 9. After utilizing the characteristics and the law of the coarser and finer sediment exchange, the concept of bed material load and wash load is no longer necessary.

10.6.3 Carrying-Sediment Capacity for Wash Load Can Be Computed

Wash load is generally considered to come from the watershed, with the quantity hard to be determined. This might be true if only the source of wash load is considered. However, since the bed material load and wash load form a united whole of suspended load, the carrying-sediment capacity for wash load S_w^* can be computed after determining the transport capacity $S^*(b)$ and water percentage K_b for bed material load, and $S^*(w)$ for given flow condition and sediment characteristics. In fact, for given flow condition and bed material gradation $P_{1.l.1}$, the computation procedure is as follows: determine the threshold particle size with Eq. (10.11); compute the water percentages in the strong equilibrium condition with $K_b = \sum_{l=k+1}^{n} P_{1.l.1}$ and $K_W = \sum_{l=1}^{k} P_{1.l.1}$; calculate $S^*(b)$ and $S^*(w)$ based on the formulas for carrying-sediment capacity; and determine the fractional concentration of bed material load $S_b^* = K_b S^*(b)$, the fractional concentration of wash load $S_w^* = K_w S^*(w)$, and the carrying-sediment capacity of total load. It is noted that the proportion of wash load in the bed material is very small, and is hard to be determined. The size distribution of suspended load can be corrected with Eqs. (10.41) and (10.42). If sediment transport is not in strong equilibrium condition, the known conditions are not sufficient enough for determining S_w^* or S_b^*. For this case, additional conditions are needed. For the case with $S \neq S^*$, the approach for non-equilibrium transport of non-uniform sediment is needed to determine S, S_b, and S_w.

Chein [1, 2] considered the wash load and bed material load follow the same law of sediment transport, which is confirmed by our analysis as described in this chapter. Due to the concern that the quantity of wash load was hard to compute, Chien concurred with the concept of wash load and bed material load. As discussed in this chapter, the quantity of wash load can be determined and it is no longer necessary to differentiate between wash load and bed material load.

10.6.4 Application of the Concept of Bed Material Load and Wash Load to Sediment Transport in Reservoirs

The concept of bed material load and wash load has been used for sediment transport in reservoirs [3, 7], and can be used to explain some phenomena of reservoir sedimentation. For example, with an appropriate threshold particle size, the concept of bed material load and wash load can be used to explain the fact that the bed material load typically deposits on the delta of a reservoir, while the wash load is laden by the density flow downstream of the delta. It is also true that no sedimentation occurs for wash load in some reservoirs with limited height of backwater. However, these phenomena can be better explained with the theory of non-equilibrium transport of non-uniform sediment, which yields more in-depth and more accurate results.

Table 10.8 lists the variation of fractional sediment concentration along the flow path measured in the Wotousi desilting canal. As listed in this table, the fractional concentration of the fine size class with $d < 0.004$ mm shows negligible variation along the channel. Using 0.004 mm as the threshold particle size for wash load, it is true that no deposition occurs for wash load. However, the threshold particle size for wash load becomes 0.01 mm if only considering the reach that is within 850 m from the entrance. Hence, if the wash load in a reservoir is defined as the sediment that does not deposit on the bed, the threshold particle size to differentiate bed material load and wash load varies, and a formulation or method is needed to determine this threshold particle size. Tables 5.2 and 5.5 list the concentration and the size distribution, respectively, of the total suspended load that were computed based on the carrying-sediment capacity for non-uniform sediment and on the approach of non-equilibrium sediment transport. The computed results show good agreement with the data, including the fine particles with $d < 0.004$ mm that did not deposit on the bed. It can be concluded that the sediment concentration, including S_b and S_w, and the size distribution of suspended load can be explained and quantified without utilizing the concept of bed material load and wash load.

The fact that the bed material load typically deposits on the delta of a reservoir while the wash load is laden by the density flow downstream of the delta was explained based on the different properties for the coarser and finer particles [3]. According to the concept of bed material load and wash load, the wash load can be completely transported over the delta without deposition, the plunging of density flow needs finer particles, and thus the wash load is carried with the density flow that occurs downstream of the delta. Figure 10.7 shows the size distribution of suspended load influx of the Guanting Reservoir, and the measured and computed size distributions of the material deposited on the delta and of the sediment that remained in the water after passing the delta. A deposition ratio of $\lambda = 0.70$ on the delta is used in the calculation for the sorting of size distributions. As shown in this figure, the computed size distributions show good agreement with the data. It is concluded that the deposition of bed material load on the delta of a reservoir with wash load being laden by density flow can be better explained and described with the theory of the coarser and finer sediment sorting during deposition, which is more reasonable and convincing than the concept of bed material load and wash load.

10.6.5 Reasonable Method to Differentiate Between Bed Material Load and Wash Load

As discussed above, the more comprehensive and more accurate results can be obtained for suspended load transport with the theory of non-equilibrium transport of non-uniform sediment, and it is no longer necessary to differentiate between bed material load and wash load. The different forms of transport for bed material load and wash load, and the characteristics of the exchange between the coarser and

Table 10.8 Fractional sediment concentration measured at Wotousi desilting canal

Section No.	Distance from entrance (m)	Sediment concentration (kg/m^3)	Fraction sediment concentration (kg/m^3)							
			< 0.004	< 0.001	0.004 ~ 0.010	0.010 ~ 0.030	0.03 ~ 0.06	0.06 ~ 0.10	0.10 ~ 0.15	0.15 ~ 0.25
1	50	33.9	11.9	17.0	5.1	8.1	5.8	2.2	0.5	0.3
2	250	30.8	10.8	16.2	5.4	7.8	4.3	1.7	0.5	0.3
3	450	27.8	10.9	15.4	4.5	7.1	3.2	1.2	0.1	0.3
4	650	23.5	12.9	17.6	4.7	4.4	1.2	0.6	0.1	
5	850	19.2	12.5	15.5	3.9	2.6	0.5	0.1		
6	1050	17.6	11.3	15.1	4.1	1.7	0.2	0.1		
7	1250	16.7	11.4	14.9	3.5	1.6	0.2			
8	1450	14.8	11.8	13.9	2.1	0.7	0.2			
9	1650	14.3	11.0	13.5	2.5	0.8				

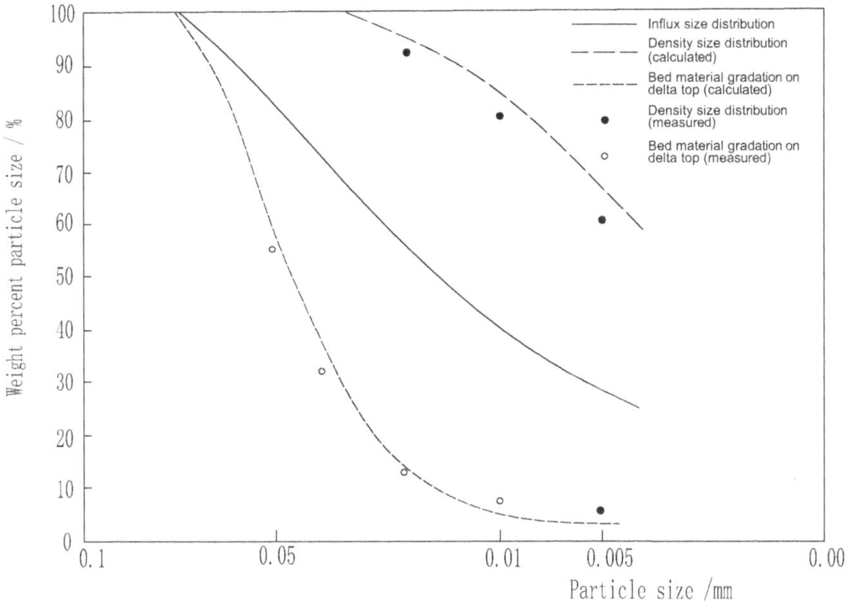

Fig. 10.7 Size distributions of the material deposited on the Delta of the Guanting reservoir and of the sediment laden by density flow

finer sediment should be recognized and utilized. From the aspect of analyzing the carrying-sediment capacity, however, the differentiation between bed material load and wash load is unnecessary. Also, from the mechanics aspect, there is no reason to separate the wash load from the bed material load either. For example, the fine particles with $d < 0.001$ mm are mostly subject to Brownian motion, and are no longer transported solely by turbulence. For this case, it is necessary to differentiate these extremely fine particles from the coarser sediment.

It is unreasonable to use $\omega < VJ$ as the criterion to define wash load, which was proposed by Bagnold with the theory of automatic suspension. Based on Bagnold's hypothesis, for the fine particles with the fall velocity $\omega < VJ$, where V is the flow velocity and J is the slope of the energy grade, the inertial energy gained by these particles as they move with the flow from a higher elevation to a lower elevation is equal to or higher than the suspension energy due to the flow turbulence. As a result, these fine particles, or wash load, are automatically suspended. Apparently, the hypothesis might be right only if the sediment particles move along the paths that are parallel to the channel slope. In reality, when sediment particles are transported by the flow, the particles move up and down repeatedly, and may even move from the water surface down to the bottom, or vice versa. Hence, compared to the energy for suspension, and the inertial energy variation due to the particles' movement, the inertia energy caused by the slope of the flow energy grade is insignificant.

If it is needed to differentiate between bed material load and wash load in the study of the carrying-sediment capacity, the following consideration are recommended: (1) The fractional carrying-sediment capacity for bed material load and wash load, or S_b^* and S_W^*, should be distinguished from the corresponding transport capacity $S^*(b)$ and $S^*(w)$ for uniform sediment, which are related through the water percentage. The carrying-sediment capacity of the uniform sediment corresponding to the bed material load and the wash load, or $S^*(b) = S_b^*/K_b$ and $S^*(W) = S_W^*/K_W$, follows the same law or formulation as the total suspended load. (2) If only the transport capacity S_b^* for bed material load is of interest, it is not recommended to simply exclude the transport capacity S_w^* of wash load, and only investigate the relation between S_b^* and related hydraulic parameters, as done in the past. (3) The threshold particle size between bed material load and wash load can be replaced with the threshold particle size to differential the finer sediment from the coarser sediment, which is determined with Eqs. (10.10) and (10.11)

References

1. Chien N (1957) Notes on the concept of bed material lad and wash load. J Hydraul Eng (1) (in Chinese)
2. Chien N, Wan ZH (1983) Mechanics of sediment transport. Science Publishing House, Beijing (in Chinese)
3. China Institute of Water Resources and Hydropower Research (IWHR) (1958) Study on reservoir sedimentation. Water Resource and Electric Power Press, Beijing (in Chinese)
4. Experts Group of Sedimentation Engineering (EGSE) of Three Gorges Reservoir, Chinese Academy of Engineering (2008) Appraisal report of sedimentation engineering of Three Gorges Reservoir
5. Han QW (1973) Preliminary studies for the non-equilibrium transport of sediment in reservoirs. In: Report on sedimentation in reservoir, task group on Yellow River sedimentation research. pp 145–168 (in Chinese)
6. Han QW (1979) A study of the non-equilibrium transportation of non-uniform suspended load. China Sci Bull 24(17):804–808 (in Chinese)
7. Han QW (2003) Reservoir sedimentation. Science Publishing House, Beijing (in Chinese)
8. Han QW (2005) Size distribution of carrying sediment capacity. In: Proc. of 6th National Symp. on Basic Theory of Sedimentation. Water Conservancy of Yellow River Press, pp 3–20 (in Chinese)
9. Han QW (2007a) Exchange between coarse and fine particles during deposition and erosion along river. In: Proceedings of the tenth international symposium on river sedimentation, Moscow, Russia, pp 139–144
10. Han QW (2007b) Concept of water percentage and its application in sediment transport of non-uniform suspended load. J. Progress Water Sci 18(5):633–644
11. Han QW (2009) In-depth study on non-equilibrium transport of non-uniform suspended load. Technical Report, China Institute of Water Resources and Hydropower Research (IWHR). (in Chinese)
12. Han QW, He MM, Wang YC (1989) A discussion on distinction between wash load and bed material load. In: Proceedings of the 4th international symposium on river sedimentation, vol. 1. China Ocean Press, pp 506–513
13. Han QW, He MM (1997) Study on the two-dimensional equations for non-equilibrium transport of non-uniform suspended load and the boundary conditions. J Hydraulic Eng 1:1–10 (in Chinese)

14. Han QW, Wang YC (1980) Discussion on separation of bed material load and wash load. J. Yangtze River 1980(6):47–55 (in Chinese)
15. He MM, Han QW (1989) Concepts about grain size distribution of carrying capacity and effective bed material. J Hydraulic Eng 3:7–16 (in Chinese)
16. He MM, Han QW (1990) The determination of the size distributions of carrying capacity and effective bed material. J Hydraulic Eng (3):l–12 (in Chinese)
17. Sah YQ (1965) Introduction to sediment dynamics. Industry Press, Bejing, pp 1–302 (in Chinese)
18. Tong ZJ (1989) Analysis of recent channel deformation in the middle reach of the Han River and discussion on channel evolution. In: Selected Papers on Fluvial Process Downstream of Danjioangkou Reservoir (Continued). The Yangtze Valley Planning Office, Hydrology Department, pp 16–36 (in Chinese)
19. Wang YC (1982) Preliminary analysis of channel coarsening downstream of the Danjiangkou Reservoir. In: Selected papers on fluvial process downstream of Danjioangkou Reservoir. Yangtze Valley Planning Office, Hydrology Department, pp 47–54 (in Chinese)
20. Zhang RJ (1961) River mechanics. China Industry Press, Beijing (in Chinses)
21. Xiang XL (1982) Preliminary analysis of channel coarsening downstream of the Danjiangkou Reservoir. In: Selected papers on fluvial process downstream of Danjioangkou Reservoir. Yangtze Valley Planning Office, Hydrology Department, pp 56–64 (in Chinese)

Chapter 11
Multi-valued Nature of Carrying-Sediment Capacity

11.1 Literature Review

There are different opinions on whether the carrying-sediment capacity is single-valued or is multi-valued. In most studies, the carrying-sediment capacity is considered to be equal to the sediment concentration under the equilibrium condition, and thus is a function of the flow and sediment conditions, and is single-valued. However, the carrying-sediment capacity is considered multi-valued in some studies.

Figure 11.1 shows the relationship between S and V that was developed by Rosinskee and Kuzmin [18], where V is the average flow velocity, and S the average sediment concentration of the coarser particles of the bed material, or the sediment concentration for the bed material load. S can be considered as the carrying- sediment capacity in equilibrium condition. As shown in Fig. 11.1, the relationship between S and V is a wide band that is bounded by two curves. The upper curve was considered to represent the relationship between S and V during sediment deposition, and the lower curve to represent the relationship during scour. As long as the flow condition and the sediment concentration are within these two boundary curves, which represent the threshold conditions for deposition and scour, respectively, sedimentation would not occur along the channel. Without considering the sediment fall velocity and the water depth, which is an implicit function of the flow velocity, the derivation by Rosinskee and Kuzmin [18] was not very strict. However, the derived relationship between S and V, which is a wide band, clearly illustrates the multi-valued nature of the carrying-sediment capacity.

Sha [20] considered the carrying-sediment capacity had multiple values, as illustrated in Fig. 11.2. In this figure, $S_D^* = f(V_1)$ is called the carrying- sediment capacity with no deposition, which is the threshold carrying- sediment capacity for no deposition to occur; $S_E^* = f(V_1)$ is the transport capacity with no scour, which is the threshold carrying-sediment capacity for no scour to occur; $S_0^* = f(V_1)$ is the (normal) equilibrium carrying-sediment capacity, and V_1 is called the specific flow velocity that is defined as $V_1 = V/R^{0.2}$ with V being the average flow velocity It is

Q. Han and M. He, *Non-equilibrium Transport of Non-uniform Suspended Load*, Hydroscience and Engineering, https://doi.org/10.1007/978-981-97-1123-9_11

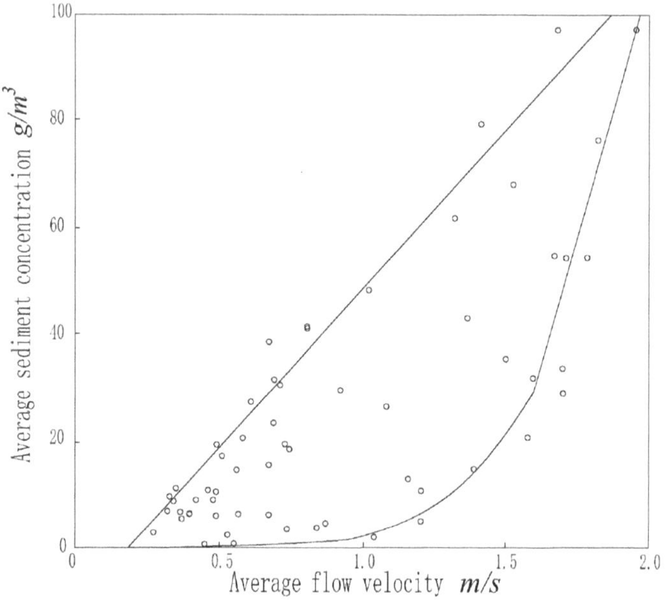

Fig. 11.1 Relationship between sediment concentration and average flow velocity [18]

noted that V is equivalent to the flow velocity corresponding to the hydraulic radius R of 1 m. Sha concluded that no sediment deposition or scour would occur as long as the sediment concentration satisfies $S_E^* \leq S \leq S_D^*$, scour would occur for $S < S_E^*$, and deposition would occur for $S > S_D^*$. The sediment pattern versus the sediment saturation and equilibrium condition are summarized in Table 11.1.

As shown in Fig. 11.2 and supplemented in Table 11.1, Sha defined a range of sediment concentration between S_E^* and S_D^*, within which no sediment deposition or scour would occur. Hence $S = S_E^*$ can be considered as the threshold sediment concentration with no noticeable scour, and $S = S_D^*$ as the threshold sediment concentration with no noticeable deposition. Sha considered the multi-valued nature of carrying-sediment capacity is controlled by the specific velocity V_1, and the threshold specific velocity for incipient sediment transport $V_{0.1}$ is related to the three threshold specific velocities: the entrainment specific velocity $V_{K.1}$, the threshold specific velocity for sediment suspension $V_{S.1}$, and the threshold specific velocity for ceasing movement $V_{H.1}$. The quantified relation for these three specific velocities is shown in Fig. 11.3.

As shown in Fig. 11.3, for particles with $D < 0.04$ mm, $V_{S.1}$ is the least, and $V_{H.1}$, is greater than $V_{S.1}$. Hence, Sha defined $V_{0.1} = V_{H.1}$ for this case, i.e., the threshold specific velocity for incipient sediment transport equal to the threshold specific velocity for sediment to cease movement,. This indicates that the sediment already laden by the water would be suspended as long as the specific flow velocity is higher than the threshold specific velocity for sediment to cease movement. For particles with $D > 0.04$ mm, $V_{S.1}$ is greater than $V_{H.1}$, and Sha considered $V_{0.1}$ could

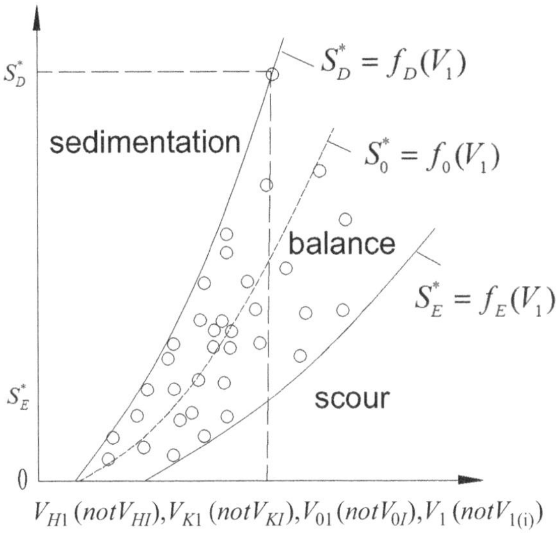

Fig. 11.2 Relationship between carrying-sediment capacity and specific flow velocity [20]

Table 11.1 Relationship between saturation and sedimentation [20]

Concentration versus capacity	Saturation degree	Equilibrium	Stability	Trend	Probability for no deposition P (%)
$S > S_D^*$	Over-saturation	Non-equilibrium	Unstable	Deposition	0.01 ~ 1
$S = S_D^*$	High-saturation	No-deposition equilibrium	Temporally stable	Insignificant deposition	1 ~ 15.9
$S = S^*$	Median-saturation	Normal equilibrium	Stable	No scour and no deposition	15.9 ~ 84.1
$S = S_E^*$	Low-saturation	No-scour equilibrium	Temporally stable	Insignificant scour	84.1 ~ 99.0
$S < S_E^*$	Unsaturation	Non-equilibrium	Unstable	Scour	99.0 ~ 99.99

be close to $V_{S.1}$, and the sediment particles would be suspended when the specific flow velocity exceeds the threshold specific velocity for sediment suspension.

Sha also determined the multiple values of the carrying-sediment capacity using empirical methods. A frequency curve was developed by Sha for the carrying-sediment coefficient K, which varies between 33 and 3390, based on field data. Ranking K from greatest to least, or from deposition to scour, the probability for no sediment deposition to occur was determined, as shown in the last column of Table 11.1. The carrying-sediment coefficient corresponding to the probability is $K = 440$ for the threshold high-saturation condition, $K = 91$ for the threshold low-saturation condition, and $K = 200$ for the median saturation condition. The variation

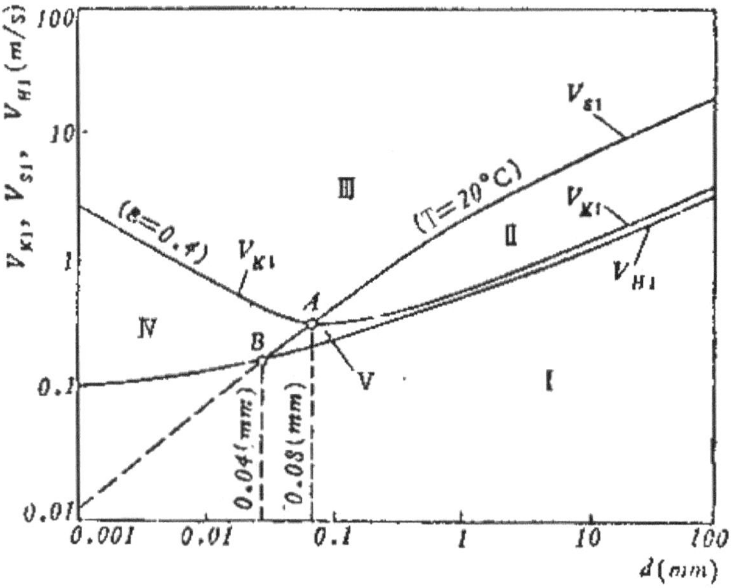

Fig. 11.3 Relationship between specific velocities and sediment particle size [20]

range of K is considered limited. The values of K for the high-saturation condition and for the low-saturation are approximately 2.2 times, and 0.455 times, respectively, of the value for the median saturation condition.

Hou [16] also considered that the carrying-sediment capacity has dual values, which were caused by the three threshold flow velocities, i.e., the dual values because of the entrainment velocity and the threshold velocity for sediment to cease movement, and the dual values because of the threshold velocity for sediment suspension and the threshold velocity for sediment to cease movement. Based on his formulation for carrying-sediment capacity, and utilizing [20] formulations for the three threshold flow velocities, Hou developed a dual-valued relationship for carrying-sediment capacity, as shown in Fig. 11.4. The upper curve in this figure is a convex curve and the lower curve can be approximated with a straight line. The difference between these two curves decreases as the specific velocity V_1 increases, and these two curves eventually merge when V_1 is sufficiently high. It is noted that the maximum difference between these two curves exceeds the range determined by Sha [20]. In Hou's formulation for carrying-sediment capacity, the entrainment probability was considered to be $1/2 \leq \eta \leq 1$, which is questionable. However, this does not impact his conclusion on the dual-valued nature of carrying- sediment capacity.

Based on Sha [20] results, Cao et al. [1] further analyzed the multi-valued nature of carrying-sediment capacity. Introducing

$$S_V = K \frac{\gamma}{\gamma_s - \gamma} \frac{(V - V_0)^3}{g R \omega_m}$$

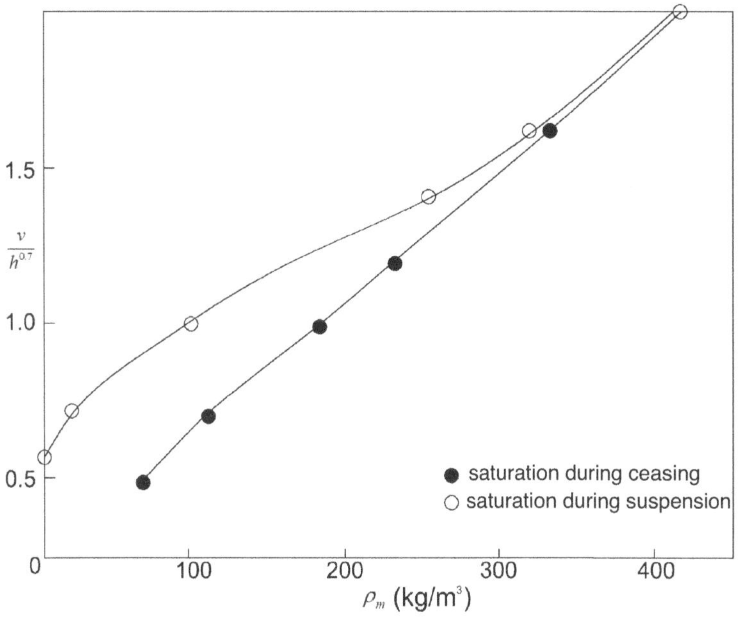

Fig. 11.4 Dual values of sediment transport capacity [16]

where S_V is the sediment concentration by volume, V_0 the threshold flow velocity for incipient sediment transport, andω_m the sediment fall velocity. Different from Sha [20], Cao proposed a different formulation for the threshold specific velocity for incipient sediment transport. As shown in Fig. 11.3, for suspended load with $D < 0.08$ mm, $V_{H.1} < V_{0.1} < V_{K.1}$, the sediment transport is controlled by the entrainment velocity, or $V_{0.1} = V_{K.1}$. For sediment with $D > 0.08$ mm, $V_{S.1} > V_{K.1}$, and thus $V_{0.1} = V_{S.1}$. Using $D = 0.08$ mm as the threshold particle size, Cao did not consider the case with $V_{S.1} < V_{K.1}$ for $D < 0.08$ mm, which is only true for the suspended sediment that is picked up from the bed. On the other hand, the case with $V_{S.1} < V_{K.1}$ in Sha [20] analysis is only true for the suspended sediment that is already laden by the water.

Based on the experimental and field data, Cao developed dual curves for the relationship between the carrying-sediment capacity and the flow parameters, as shown in Fig. 11.5. The corresponding equation for these two curves is

$$S_V^* = \begin{cases} S_{V.D}^* = 0.0005\frac{\gamma}{\gamma_S-\gamma}\frac{(V-V_0)^3}{gh\omega_m} \\ S_{V.E}^* = 0.0002\frac{\gamma}{\gamma_S-\gamma}\frac{(V-V_0)^3}{gh\omega_m} \end{cases} \quad (11.1)$$

where $S_{V.D}^*$ and $S_{V.E}^*$ are the carrying-sediment capacities for deposition and for scour, respectively. The carrying- sediment capacities for deposition and for scour

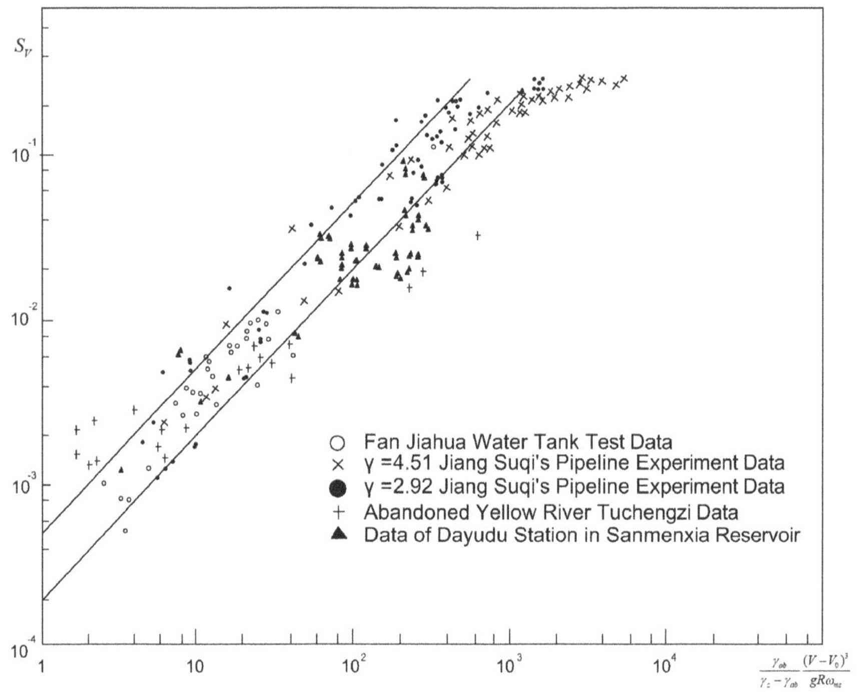

Fig. 11.5 Envelop for carrying-sediment capacity [1]

are approximately 1.43 times and 0.571 times, respectively, of the mean carrying-sediment capacity. The difference in the carrying-sediment capacity between deposition and scour is slightly less than the variation range derived by Sha [20].

It is noted that whether the multi-valued carrying-sediment capacity should be bounded by the two straight lines as derived by Cao et al. [1] or by the curves derived by Sha [20] are questionable. As shown in Fig. 11.5 Eq. (11.1) is not available for high concentration.

Han [7] considered the carrying-sediment capacity is multi-valued, which is caused by the multiple values of the threshold flow velocities. The carrying-sediment capacity discussed here is not in the strong equilibrium condition. Based on the statistical theory for the sediment exchange on the bed, Han derived the formula for the fractional carrying-sediment capacity, or Eq. (4.23), which has a solid theoretical foundation with a general application. This formula reads

$$S_l^* = P_{4.l}^* S^* = \frac{2}{3} m_0 \gamma_s P_{1.l} \Omega_{1.l} \frac{D_l}{\omega_l t_{0.4.l}} \frac{L_{4.l}^* \omega_l}{q} = P_{1.l} S^*(l)$$

with the so-called composite probability

$$\Omega_{1.l} = \frac{\beta_l}{1 + (1 - \varepsilon_{0.l})(1 - \varepsilon_{4.l}) - (1 - \varepsilon_{1.l})(1 - \beta_l)} \tag{11.2}$$

where β_l is the incipient suspension probability, or the probability for the sediment to be picked up from the bed by the flow and then be suspended in the water, $\varepsilon_{4.l}$ the suspension probability, $\varepsilon_{1.l}$ the entrainment probability, $\varepsilon_{0.l}$ the probability for sediment not to cease movement, and $(1 - \varepsilon_{0.l})$ the probability for sediment to cease movement. These probabilities are related to the corresponding threshold flow velocities. It is noted that except for the composite probability, the formula for the fractional carrying-sediment capacity does not include any other parameters that are related to the threshold conditions, velocities and probabilities. Hence, the multi-valued nature of the carrying-sediment capacity can be represented by the composite probability Ω.

In fact, Eq. (11.2) was derived based on the exchange between the suspended sediment and the bed material. The incipient sediment suspension probability represents the condition for bed material to be picked up from the bed, assuming the cohesive force and the additional pressure of film water act on the bed material. However, for the coarser sediment particles such as $D > 0.25 \sim 0.5$ mm, these two forces are negligible. Thus, these coarser particles can be picked up as long as the suspension condition, $V_{y'} > \omega$, is satisfied. For the sediment particles that are already suspended in the water, including the particles that settle down on the bed and then re-suspended before being adherent to the bed material, are controlled only by the threshold velocity for suspension and the suspension probability. Hence, the actual sediment suspension consists of two scenarios. The incipient suspension and suspension of the fine sediment particles that are adherent to the bed material are controlled by the threshold condition for incipient suspension and the incipient suspension probability β. The determination of β was discussed in Chaps. 4 and 6. The suspension of the coarser sediment is controlled only by the suspension condition, $V_{b.y'} > \omega$, and the suspension probability ε_4. As a result, the composite probability also have two parts. One is expressed by Eq. (11.2), and the other is for the case with $\beta_l = \varepsilon_{4.l}$ and $\varepsilon_{1.l} = \varepsilon_{0.l}$, which is

$$\Omega_{0.l} = \frac{\varepsilon_{4.l}}{1 + (1 - \varepsilon_{0.l})(1 - \varepsilon_{4.l}) - (1 - \varepsilon_{0.l})(1 - \varepsilon_{4.l})} = \varepsilon_{4.l} \tag{11.3}$$

Then average composite probability can be written as

$$\Omega_l = \mu_l \Omega_{0.l} + (1 - \mu_l)\Omega_{1.l} \tag{11.4}$$

where $\Omega_{1.l}$ represents the composite probability for the sediment to be picked up from the bed, as defined in Eq. (11.2).

Replacing Ω_l in Eq. (4.23) with Ω_l as defined in Eq. (11.4), the fractional carrying-sediment capacity is

$$S_l^* = P_{4.l}^* S^*$$

$$
\begin{aligned}
&= [(1-\mu_l)\Omega_{1.l} + \mu_l \Omega_{0.l}] \frac{2}{3} m_0 \gamma_s P_{1.l.1} \frac{D_l}{\omega_l t_{0.4.l}} \frac{L^*_{4.l}\omega_l}{q} \\
&= (1-\mu_l) S^*_{1.l} + \mu_l S^*_{0.l} \\
&= \left[(1-\mu_l)\frac{\Omega_{1.l}}{\Omega_{0.l}} + \mu_l\right] S^*_{0.l} = \kappa_l S^*_{0.l} = \kappa_l P_{1.l.1} S^*_0(l)
\end{aligned}
\tag{11.5}
$$

where

$$
S^*_{1.l} = \frac{2}{3} m_0 \gamma_s P_{1.l.1} \Omega_{1.l} \frac{D_l}{\omega_l t_{0.4.l}} \frac{L^*_{4.l}\omega_l}{q} = P_{1.l.1} S^*_1(l) \tag{11.6}
$$

is called the carrying-sediment capacity under the scour equilibrium condition,

$$
S^*_{0.l} = \frac{2}{3} m_0 \gamma_s P_{1.l.1} \Omega_{0.l} \frac{D_l}{\omega_l t_{0.4.l}} \frac{L^*_{4.l}\omega_l}{q} = P_{1.l.1} S^*_0(l) \tag{11.7}
$$

is called the carrying-sediment capacity under the deposition equilibrium condition, and

$$
\frac{S^*_{1.l}}{S^*_{0.l}} = \frac{\Omega_{1.l}}{\Omega_{0.l}} = \frac{S^*_1(l)}{S^*_0(l)} \tag{11.8}
$$

is always less than 1. The correction factor for the multi-valued carrying-sediment capacity is

$$
\kappa_l = \mu_l + (1-\mu_l)\frac{\Omega_{1.l}}{\Omega_{0.l}} \tag{11.9}
$$

which can also be written as

$$
\kappa_l = \frac{S^*_l}{S^*_{0.l}} = \frac{S^*_1(l)}{S^*_0(l)} \tag{11.10}
$$

Based on the conditions that are similar to the Yellow River ($u_* = 0.05$ m/s, $h = 2$ m) and that are similar to the Yangtze River ($u_* = 0.05$ m/s, $h = 10$ m), the correction coefficient κ_l for various size classes are computed for $\mu = 1/3$, $\mu = 1/2$, and $\mu = 2/3$. The results are listed in Table 11.2, in which $\Omega_{0.l}$ is approximated by $\varepsilon_{4.l}$. In this table, μ denotes that $S^*_{0.l}$ accounts μ of the carrying-sediment capacity while $S^*_{1.l}$ accounts for the remaining $1-\mu$, or the carrying- sediment capacities for deposition and for scour account μ and $1-\mu$, respectively, of the carrying-sediment capacity.

The following conclusions can be made from Han's analysis and Table 11.2:

(1) The multi-valued nature of the carrying-sediment capacity was theoretically derived and quantified by Han from the formula for the carrying-sediment capacity, without directly utilizing the threshold flow velocities. Compared to

Table 11.2 Correction coefficient for multi-valued carrying- sediment capacity computed for various conditions [7]

D (mm)	$h = 2$ m, $u_* = 0.05$ m/s					$h = 10$ m, $u_* = 0.05$ m/s				
	$\varepsilon_{4.l}$	$\frac{\Omega_{1.l}}{\varepsilon_{4.l}}$	κ_l			$\varepsilon_{4.l}$	$\frac{\Omega_{1.l}}{\varepsilon_{4.l}}$	κ_l		
			$\mu = \frac{2}{3}$	$\mu = \frac{1}{2}$	$\mu = \frac{1}{3}$			$\mu = \frac{2}{3}$	$\mu = \frac{1}{2}$	$\mu = \frac{1}{3}$
0.01	0.500	0.00242	0.668	0.502	0.336	0.500	0.000577	0.667	0.501	0.334
0.025	0.497	0.0349	0.690	0.531	0.380	0.497	0.000573	0.667	0.501	0.334
0.05	0.488	0.208	0.809	0.631	0.617	0.488	0.00717	0.672	0.507	0.343
0.10	0.451	0.458	1.00	1.00	1.00	0.451	0.0342	0.692	0.537	0.383
0.25	0.272	0.197	0.908	0.862	0.816	0.272	0.0681	0.750	0.625	0.500
0.50	0.0811	0.0462	0.857	0.763	0.713	0.0811	0.0338	0.806	0.708	0.611
1.00						0.0087	0.00602	0.904	0.846	0.794
Mean	0.382		0.822	0.715	0.644	0.328		0.737	0.603	0.471

the approaches with the threshold flow velocities, this method is logically stricter with clearer derivation.

(2) Four probabilities were introduced in Han's analysis to represents the impact of the threshold flow velocities. This overcomes the disadvantages of Hou [16] analysis, in which two pairs of dual-values were formed from the three threshold flow velocities. These dual values include the threshold flow velocities for incipient movement and for ceasing movement, and the threshold velocities for suspension and for ceasing movement. It is concluded in Hou's analysis that the multi-valued nature of the carrying-sediment capacity is caused by the insistence between the incipient suspension and suspension probabilities, i.e., the multi-valued nature of the suspension probability.
(3) The mechanism of the multi-valued nature of the carrying-sediment capacity was clarified by Han after considering the cohesive force and the additional pressure of film water acting on the particles, Both the sediment influx from upstream and the particles picked up from the bed were considered. This combines Sha's analysis, which emphasized the sediment influx from upstream while ignoring the local sediment pickup from the bed, with Cao's [1] analysis that only considered the local sediment pickup from the bed while neglecting that the threshold velocity for suspending the sediment from upstream can be less than the entrainment flow velocity.

Based on Table 11.2, the coefficient for carrying-sediment capacity was computed by Han for various values of μ, which was determined by multiplying the correction coefficient κ and the coefficient for carrying-sediment capacity for equilibrium reservoir sedimentation. The results are shown in Table 11.3. Based on the field data of the Danjiangkou Reservoir and the Gezhouba Reservoir of the Yangtze River watershed, the coefficient for the carrying-sediment capacity is approximately $K = 0.025 \sim 0.030$ for equilibrium reservoir sedimentation. Let assume $K = 0.0275$. For channel deposition, $\kappa = 0.737$, and thus $K = 0.0275 \times 0.737 = 0.020$ for the deposition equilibrium condition. For channel scour, $\kappa = 0.471$, and thus $K = 0.013$ For an equilibrium channel, $\kappa = 0.603$, and $K = 0.017$. As listed in Table 11.3, the coefficient for carrying-sediment capacity computed for various equilibrium conditions are consistent with the empirical values that are used in the mathematical models for sediment transport [6]. This is also true for the Yellow River. It is noted that the value of the carrying-sediment capacity coefficient for equilibrium reservoir sedimentation is empirical However, it can be concluded that the relative difference in the coefficients for carrying-sediment capacity, which were computed based on Eqs. (4.23) and (11.5) for different equilibrium conditions, has theoretical basis and is consistent with reality.

While not directly supporting the conclusion on the multi-valued nature of carrying-sediment capacity, some studies implied the existence of this multi-valued nature. For example, Zhang's formula for carrying-sediment capacity is presented as an envelope. It is stated in Zhang et al. [22] that "the upper boundary of this envelop can be adopted for the case of deposition, while the lower boundary of the envelop can be used for the case of scour." It was further explained that "using these two boundary lines to represent the deposition and scour processes significantly simplifies the complicated process of sediment transport. The purpose is to provide inspiration for future analysis, rather than setting any fixed dogma.". It is also noted that the results on the lower and upper boundary lines are approximately 0.5 ~ 1.5 different from the results on the mean equilibrium line, which is slightly less than the variation range derived by Sha [20].

Chien and Wan [2] investigated the multi-valued nature of carrying-sediment capacity with flume experiments. The experimental data are listed in Table 11.4. Two groups of experiments were conducted. In group B experiments, firstly scour occurred in the flume until the bed reached a scour equilibrium condition, and then more sediment was added. In group C experiments, sediment was continuously added until the bed reached a deposition equilibrium condition, and then less sediment was added, resulting in a scour on the bed and various scour equilibrium conditions being achieved for different stages of the scour process.

Table 11.3 Carrying-sediment capacity coefficients for the Yellow River and the Yangtze River

Rivers	Coefficient for carrying-sediment capacity				
	Values	Equilibrium reservoir sedimentation	Deposition equilibrium	Median equilibrium	Scour equilibrium
Yellow River	Theoretical	(0.040)	0.033	0.029	0.026
	Adopted in math. model	0.040	0.030		0.026
Yangtze River	Theoretical	(0.0275)	0.020	0.017	0.014
	Adopted in mat. model	0.025 ~ 0.030	0.020	0.017	0.014

Table 11.4 Comparison of sediment transport capacity between the equilibrium deposition and equilibrium scour conditions

Test	Case	Equilibrium condition	Flow					Concentration (kg/m^3)	Bed material		
			Rate (m^3/s)	Depth (cm)	Vel (m/s)	Slope (%)	Temp (°C)		D_{90} (mm)	D_{65} (mm)	D_{35} (mm)
B[a]	1	Scour	63.2	18.7	1.09	39.5	55	12.9	0.595	0.316	0.157
	2	Deposition	63.2	20.0	1.02	41.9	58	13.5	0.540	0.245	0.132
	3	Deposition	63.0	19.0	1.03	33.3	48	16.4	0.420	0.195	0.117
	4	Deposition	63.2	19.8	1.04	35.8	52	17.2	0.440	0.195	0.112
	5	Deposition	63.2	19.9	1.03	36.6	51	18.7	0.440	0.174	0.104
C[b]	1	Deposition	63.2	19.2	1.06	30.5	61	80.0	0.470	0.225	0.104
	2	Deposition	62.5	20.2	1.00	39.3	69	118	0.490	0.208	0.097
	3	Deposition	62.5	19.5	1.04	36.8	58	149	0.475	0.190	0.086
	4	Scour	62.5	19.5	1.04	34.6	60	35.2	0.540	0.260	0.134
	5	Scour	62.5	19.3	1.05	39.4	57	12.5	0.615	0.305	0.167

[a]During the test, more sediment was added after the bed reached scour equilibrium

[b]Sediment was continuously added during the test, with transport rate being measured at various stages of the deposition equilibrium condition as bed accretion occurred; and then added less and less sediment

As listed in Table 11.4, the flow conditional was approximately the same between various tests in the group B experiments, while the bed material gradation varied. As D_{65} decreased from 0.316 to 0.174 mm, the sediment concentration (or carrying-sediment capacity) increased from 12.9 to 18.7 kg/m^3. These tests belonged to the deposition equilibrium condition. The multi-valued nature of carrying-sediment capacity was not obvious because the increase in sediment concentration was approximately consistent with the decrease in the grain size of the bed material. For group C experiments, both the hydraulic condition and the bed material gradation varied insignificantly among the tests. However, the sediment transport capacity varied from 118 to 149 kg/m^3 for the deposition equilibrium condition to 35.2 ~ 12.5 kg/m^3 for the scour equilibrium condition, obviously showing the multi-valued nature of carrying-sediment capacity. Chien and Wan [2] concluded that the carrying-sediment capacities for various sedimentation conditions should follow the same law or formulation, without the multi-valued nature, after considering the variation of the bed material gradation. However, based on our analysis, considering the variations of the water depth, flow velocity, sediment grain size (fall velocity) and roughness, Chen and Wan's [2] conclusion is not true for the group C experiments, and thus these experiments cannot be used to veto the multi-valued nature of carrying-sediment capacity.

In summary, the carrying-sediment capacity has a multi-valued nature, which is caused by the multi-valued threshold velocities and corresponding probabilities. A clear illustration of the mechanism and a detailed analysis of this multi-valued nature are presented in Han [7]. Chien and Wan [2] emphasized the variations of the flow and sediment conditions during the sedimentation process, which results in different carrying-sediment capacities. This conclusion is also worthy of attention. The multi-valued nature of carrying-sediment capacity should be clearly defined in future studies. In addition to the multi-values of the threshold flow velocities, are there any other causes for the multi-valued nature of carrying-sediment capacity? It is also of interest to investigate whether the analyses of this multi-valued nature are consistent with the real condition in natural streams. In the following section, the field data measured in the natural streams is analyzed to demonstrate the multi-valued nature of carrying-sediment capacity, and to deepen the theoretical study so that more factors and mechanism for this multi-valued nature can be illustrated.

11.2 Field Data for Multi-valued Nature of Carrying-Sediment Capacity

The field data collected in the lower Yellow River is of special interest for the investigation of the multi-valued nature of carrying-sediment capacity. This is partially because the lower Yellow River is famous for "more sediment discharge as more sediment coming", and partially because it shows an obvious trend of deposition equilibrium [8]. Based on the field data collected in the lower Yellow River between

1960 and 1996 [21], Han and Gian [14] compared the measured sediment concentration to the carrying-sediment capacity [21]. The results are listed in Table 11.5. In this table, the data are divided into three groups based on the sediment concentration. The data within each sediment concentration group are further re-grouped based on the flow discharges and then are averaged over the 36 years between 1960 and 1996. The three sediment concentration groups include $S < 20$ kg/m^3, $S = 20 \sim 80$ kg/m^3, and $S > 80$ kg/m^3, and are called low-, median-, and high- sediment concentrations, respectively, in the following for simplicity of reference. The following observations are made from Table 11.5:

(1) Comparing the low- and the high- sediment concentration conditions, the nature of "discharging more sediment as more sediment comes" can be clearly observed for the lower Yellow River. For the flow discharge of 1000 ~ 4000 m^3/s with low-sediment concentration ($S < 20$ kg/m^3), the sediment concentration averaged over the 36 years did not exceed 18 kg/m^3 at the Huayuankou station, and did not exceed 27.3 kg/m^3 at the Lijin station. For the same range of the flow discharge but with high-sediment concentration ($S > 80$ kg/m^3), the sediment concentration was approximately 71.7 ~ 93.8 kg/m^3 at the Huayuankou station, and was 60.6 ~ 79.7 kg/m^3 at the Lijin station (except for the flow discharge range of 1000 ~ 1500 m^3/s). For the same range of the flow discharge, the high sediment concentration was approximately four times higher than the low sediment concentration at the Huayuankou station, and approximately two times higher at the Lijin station. These ratios also apply to the results that are averaged over all flow discharges. This data clearly shows the typical characteristic of "discharging more sediment as more sediment comes" of the lower Yellow River.
(2) Compared to the median-sediment concentration condition ($S = 20 \sim 80$ kg/m^3), the sediment concentration for the flow with low-sediment concentration condition was more than 50% lower at the Huayuankou station, and approximately 40%-50% lower at the Lijin station for all flow discharge ranges (1000 ~ 4000 m^3/s).
(3) Compared to the median-sediment concentration condition, for the same flow discharges, the sediment concentration with high-sediment concentration was more than two times higher at the Huayuankou station, and was 1.55 to 2 times higher at the Lijin station.
(4) The similar difference in carrying-sediment capacity between various sediment concentration conditions can also be found for Gaocun and Aishan stations. These variation ranges are close to the empirical results of Sha [20], which shows the carrying-sediment capacity for scour is approximately 0.455 time of the capacity for the high saturation condition, while the carrying-sediment capacity for the high saturation condition is 2.2 times of the capacity for the mean saturation condition.

The difference between the carrying-sediment capacity and the sediment concentration not only demonstrate the characteristics of "discharging more sediment as

Table 11.5 Comparison between sediment concentration and sediment transport capacity for various sedimentation conditions

S (kg/m³)	Days	Flow discharge (m³/s)	Huayuankou			Gaocun			Aishan			Lijin			Sedimentation
			Q (m³/s)	S (kg/m³)	S* (kg/m³)	Q (m³/s)	S (kg/m³)	S* (kg/m³)	Q (m³/s)	S (kg/m³)	S* (kg/m³)	Q (m³/s)	S (kg/m³)	S* (kg/m³)	
< 20	273	1000 ~ 1500	1278	14.8	15.0	1204	16.1	15.7	1160	14.7	15.7	1013	12.9	15.4	Equilibrium
	220	1500 ~ 2000	1763	15.4	23.2	1712	19.1	22.7	1735	17.8	21.6	1621	17.0	22.9	Scour
	102	2000 ~ 2500	2420	11.9	31.9	2347	18.4	27.1	2534	20.0	40.4	2470	21.8	47.8	Scour
	76	2500 ~ 3000	2791	17.9	53.5	2819	22.8	41.0	3033	24.8	62.7	3021	27.3	54.6	Scour
	103	3000 ~ 4000	3542	15.0	41.6	3516	19.9	55.6	3589	21.1	42.5	3552	22.3	60.2	Scour
	774	All	2567	16.0	34.3	2542	21.3	32.2	2660	21.2	37.0	2570	22.5	35.0	Scour
20 ~ 80	230	1000 ~ 1500	1350	28.6	23.3	1227	27.8	23.8	1200	26.5	20.3	1123	25.4	23.7	Equilibrium
	440	1500 ~ 2000	1765	36.1	30.5	1640	31.5	28.7	1609	32.9	26.2	1473	34.5	28.7	Equilibrium
	290	2000 ~ 2500	2302	33.4	31.1	2001	32.6	31.5	2191	31.9	32.9	2116	33.9	33.0	Equilibrium
	228	2500 ~ 3000	2840	38.5	40.7	2698	36.2	47.8	2676	36.1	34.6	2509	39.3	38.8	Equilibrium
	26	3000 ~ 4000	3315	32.4	42.8	3367	34.3	44.4	3369	35.5	41.8	3208	39.2	41.5	Equilibrium
	1454	All	2490	34.4	33.5	2373	34.4	38.9	2353	34.5	32.3	2335	36.8	33.9	Equilibrium
> 80	36	1000 ~ 1500	1402	71.7	25.3	1241	48.4	23.3	1415	39.1	26.1	1276	39.3	27.8	Deposition
	138	1500 ~ 2000	1774	79.8	28.6	1577	63.5	28.9	1575	59.7	30.7	1422	60.6	26.5	Deposition
	79	2000 ~ 2500	2227	93.8	28.1	2043	73.0	39.5	2065	66.4	38.0	1944	68.7	34.3	Deposition
	29	2500 ~ 3000	2858	70.1	44.2	2430	54.8	54.0	2343	59.6	37.7	2111	65.0	30.5	Deposition
	55	3000 ~ 4000	3384	90.2	46.3	3310	76.9	66.4	3199	77.9	50.5	3064	79.7	56.6	Deposition
	333	All	2343	86.1	35.0	2171	69.8	41.5	2173	67.0	38.3	2035	69.7	35.1	Deposition

more sediment comes" for the lower Yellow River, but also illustrates the multi-valued nature of carrying-sediment capacity. This is not only because of the large variation range of the sediment concentration in the sedimentation process for the same flow discharge, but also because of the significant difference between the carrying-sediment capacity and the sediment concentration, and the wide variation range of the carrying-sediment capacity. It is noted that whether the carrying-sediment capacity is multi-valued can only be determined after considering the sediment fall velocity. For example, even for the case with constant flow condition and with significantly varying sediment concentration that is close to the corresponding carrying-sediment capacity, the multi-valued nature of carrying-sediment capacity cannot be illustrated.

As reviewed in this section, the multi-valued nature of carrying-sediment capacity is one of the important characteristics for the carrying-sediment capacity for non-uniform sediment. However, due to the incomplete analysis in the past, this multi-valued nature has not attracted enough attentions from researchers. Based on the in-depth analysis, Han and Guan [14] theoretically explained the mechanism for this multi-valued nature, and more importantly, found that the difference in the sediment saturation recovery coefficient during a sedimentation process is one of the main causes for the multi-valued nature of carrying-sediment capacity.

The difference between carrying-sediment capacity and sediment concentration has been confirmed in the law for non-equilibrium sediment transport. Is this difference caused by the process of non-equilibrium sediment transport? The distances between any two consecutive stations as listed in Table 11.5 vary between 174 and 279 km. The flow condition (such as the flow discharge) and the sediment concentration show insignificant difference along the flow path, and thus the saturation of sediment should be recovered, or the sediment concentration should approach the carrying-sediment capacity. However, this fact is not shown by the results in Table 11.5 except for the median-sediment concentration group, which indicates the computed carrying-sediment capacity is not consistent with the reality. The carrying-sediment capacity listed in Table 11.5 was computed with the formula derived and verified in Han [8, 11] for the strong equilibrium condition that reads

$$S^* = \kappa\left(\frac{V^3}{gh\omega_1}\right)^{0.92} = K\frac{Q^m}{\omega_1^{0.92}} \tag{11.11}$$

where Q is the flow discharge, V the average flow velocity, h the average water depth, ω_1 the average fall velocity computed based on the size distribution of suspended load, κ, K, and m are constants. For $\kappa/g^{0.92} = 0.029$, $K = 0.000189$ with $m = 0.805$ for the wandering reach of the Yellow River in Henan Province, and $K = 0.000113$ with $m = 0.872$ for the meandering reach in Shandong Province. The conversion between the last two parts of Eq. (11.11) involves the relationship between the river hydro-morphology and the water depth, resulting two sets of results for m and K for these two reaches. The two types of expressions for the carrying-sediment capacity, as listed in Eq. (11.11), are completely equivalent for the lower Yellow River [8, 14]. However, it is more convenient to use $S^* = f(Q, \omega_1, k, m)$ for the data analysis since it only involves the flow discharge, without the flow velocity and the water depth.

The carrying-sedimentt capacity computed with Eq. (11.11) is listed in Table 11.5. As listed in this table, the computed carrying-sediment capacity significant exceeded the sediment concentration for the data in the group of $S < 20$ kg/m^3. This was true for all stations from Gaocun to Lijin. Particularly, the sediment concentration only increased 10% from Gaocun to Lijin, which was much less than the computed carrying-sediment capacity. Similarly, for the data in the group of $S > 80$ kg/m^3, the sediment concentration showed insignificant difference for the reach downstream of the Gaocun station, but the computed carrying-sediment capacity is much less than the sediment concentration. All of this is questionable.

The multi-valued nature of carrying-sediment capacity also occurred in reservoirs, particularly during the reservoir scour process. Figure 11.6 [19] shows the scour process occurred in the Sanmenxia Reservoir during several flood events in 2003 after the operation of this reservoir was changed. Within the fifteen days between August 27 and September 10, while the flow discharge increased with time, the scour quantity decreased, and the sediment concentration of the sediment outflow from the reservoir dropped to below 20 kg/m^3. Table 11.6 [17] lists the sediment inflow and outflow during six sediment flushing events in 2005. The flushing effect of the first five events was satisfactory. The average flow discharge for the first three flushing events was less than 1000 m^3/s, the inflow sediment concentration was 96.3 kg/m^3, and the outflow sediment concentration reached 169 kg/m^3. The average flow discharge for the first five flushing events was approximately 1376 m^3/s, the inflow sediment concentration was 56.9 kg/m^3, and the outflow sediment concentration was 140 kg/m^3. However, the average flow discharge for the 6-th flushing event was 2852 m^3/s, the inflow sediment concentration was 26.4 kg/m^3, but the outflow sediment concentration was only 36.2 kg/m^3, which was unexpected. Since substantial deposition already occurred before these sediment flushing events, the bed material coarsening should not be significant. It is concluded from our analysis that consolidation of the bed material caused by scour increased the threshold flow velocity and threshold flow discharge for incipient sediment movement, resulting in a lower carrying-sediment capacity. This data also demonstrate the multi-valued nature of carrying-sediment capacity.

For the Yangtze River that has a much lower sediment concentration than the Yellow River, the channel scour in the middle and lower Yangtze River after the operation of the Three Georges Reservoir is a good evidence for the multi-value nature of carrying-sediment capacity. Table 11.7 lists the annual sediment transport rate measured at various stations after the reservoir operation in 2003, compared to the long-term averaged transport rate before the reservoir operation [3]. After the reservoir operation, a gradual channel scour occurred within the reach between Yichang and Datong, which is more than 1100 km long, and the sediment transport rate increased along the flow path. However, compared to that prior to the reservoir operation, the sediment transport rate decreased by 60 ~ 90%, and even up to 95% for a few stations. Sediment transport in the Yangtze River was approximately equilibrium before the reservoir operation. In addition to the impact of non-equilibrium sediment transport, the difference in sediment transport rate after the reservoir operation should

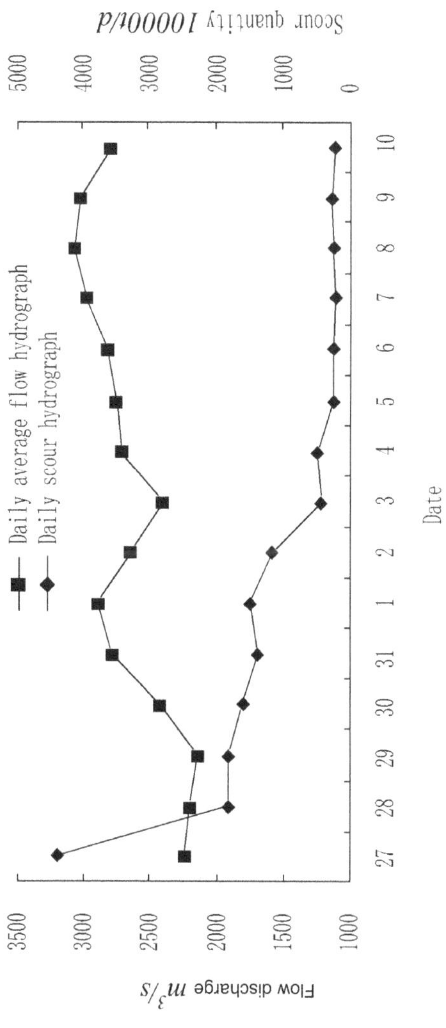

Fig. 11.6 Daily mean flow discharge and scour quantity for the Sanmenxia reservoir between August 26 and September 10, 2003 [19]

Table 11.6 Scour quantities of Sanmenxia reservoir during six sediment flushing events in 2005 [17]

Flood events	Days	Mean flow discharge (m^3/s)		Sediment quantity (10^8 tons)		Mean sediment concentration (kg/m^3)		Scour	
		In	Out	In	Out	In	Out	Quantity (10^8 tons)	Concentration (kg/m^3)
1–3	166	984	1397	0.5666	1.4268	96.3	169.0	0.8600	72.70
1–5	293	1376	1677	0.8265	2.4732	56.9	140.0	1.6465	8.31
6	214	2852	2998	0.5808	0.8354	26.4	36.2	0.2546	9.80

Table 11.7 Comparison of annual sediment transport rate in the Yangtze River downstream of the Three Georges Reservoir between the post-operation periods and the long-term averaged condition prior the reservoir operation [3]

Periods	Annual sediment transport rate (10^4 tons)							Percent for $D > 0.1$ mm at Datong
	Yi-chang	Zhi-cheng	Sha-shi	Jian-li	Luo-shan	Han-kou	Da-tong	
Before 2002	50,100	51,100	44,500	36,300	41,500	40,400	43,300	0.120
2003	9760	13,100	13,800	13,100	14,600	16,500	20,600	0.020
2004	6400	8040	9560	10,600	12,300	13,600	14,700	0.060
2005	11,000	11,700	13,200	14,000	14,700	17,400	21,600	0.110
2006	909	1200	2450	3890	5810	5760	8480	0.130
2007	5270	6800	7510	9390	9520	11,400	13,800	0.160

be mainly caused by the multi-valued nature of carrying-sediment capacity. The period of channel scour was relative short, and the hydraulic parameters did not vary significantly. This was particularly true for the reach downstream of the Luoshan station.

Bed material coarsening also had certain impact on the sediment transport rate, but should not cause as much difference as shown in the data. In fact, the most substantial bed material coarsening occurred at the Jianli station based on the size distribution of suspended load. Table 11.8 lists the size distribution of suspended load at the Jianli station for 2006 and 2007, compared to the size distribution averaged between 1986 and 2002. The bed material has been obviously coarsened since the operation of the Three Georges Reservoir. Based on the size distribution of suspended load, the average fall velocity of suspended load is estimated to be 0.00384 m/s for the period prior to the operation of this reservoir in 2003, and be 0.0167 m/s in 2006. If no coarsening occurred for suspended load, the sediment transport rate would be $\left(\frac{0.0167}{0.00384}\right)^{0.92} \times 3890 \times 10^4$ tons $= 150.41 \times 10^6$ tons in 2006, which is

Table 11.8 Size distribution of suspended load at the Jianli Station, Yangtze River, for 2006 and 2007 [3]

Periods	Percent in weight for various grain sizes				
	< 0.01 mm	0.01 mm ~ 0.025 mm	0.025 mm ~ 0.05 mm	0.05 mm ~ 0.10 mm	0.10 mm ~ 0.50 mm
1986–2002	0.500	0.140	0.140	0.090	0.130
2006	0.180	0.060	0.070	0.090	0.600
2007	0.320	0.060	2.100	0.080	0.440
$\omega_l^{0.92}$	0.000065	0.000572	0.00227	0.00814	0.0371

much less than the transport rate of 363×10^6 tons prior to the reservoir operation. Hence, the significant difference in sediment transport rate cannot be solely caused by the coarsening of suspended load. Similarly, the sediment transport rate would be $\left(\frac{0.0121}{0.00384}\right)^{0.92} \times 9390 \times 10^4$ tons $= 270 \times 10^6$ tons in 2007 if no coarsening occurred for suspended load, which is also is much less than the transport rate of 363×10^6 tons prior to the reservoir operation. Since the significant decrease in sediment transport rate after the reservoir operation cannot be explained with the single-valued carrying-sediment capacity for the Jianli station, where the most considerable sediment coarsening occurred, it is even more difficult to be explained for the downstream Luoshan, Hankou, and Datong stations. Sediment coarsening at these three stations was much weaker than the Jianli station. This is particularly true for the Datong station where negligible sediment coarsening occurred. However, the sediment transport rates at these three station after the reservoir operation were only 1/3 ~ 1/4 of the transport rates for the period prior to the reservoir operation. This is a strong evidence for the multi-valued nature of carrying-sediment capacity.

11.3 Further Verification of Multi-valued Nature of Carrying-Sediment Capacity

11.3.1 Analysis of the Adjustment Factor for Carrying-Sediment Capacity

In the analysis of carrying-sediment capacity based on the data in Table 11.5, it should be determined first whether the sediment transport is equilibrium in order to exclude the impact of non-equilibrium sediment transport. The governing equation for non-equilibrium sediment transport between two sections [5] is

$$S = S^* + \sum \left(S_0 P_{4.l.0} - S_0^* P_{4.l.0}^*\right) \exp\left(-\frac{\alpha \omega_l L}{q}\right)$$

$$+\sum \frac{q}{\alpha\omega_l L}\left[1-\exp\left(-\frac{\alpha\omega_l L}{q}\right)\right]\left(S_0^* P_{4.l.0}^* - S^* P_{4.l}^*\right) \tag{11.12}$$

where S_0 and S are the sediment concentrations at the entrance and the exit sections, respectively, S_0^* and S^* the carrying-sediment capacities at the entrance and the exit sections, respectively, $P_{4.l.0}$ the size distribution of suspended load at the entrance section, $P_{4.l.0}^*$ and $P_{4.l}^*$ the size distributions for carrying-sediment capacity at the entrance and the exit sections, respectively, ω_l the sediment fall velocity for the l-th size class, and α the saturation recovery coefficient. For the massive deposition and the equilibrium conditions, the size distribution for carrying-sediment capacity can be replaced with the size distribution of suspended load. Then Eq. (11.12) can be written as

$$\begin{aligned} S &= S^* + \sum P_{4.l.0}\left(S_0 - S_0^*\right)\exp\left(-\tfrac{\alpha\omega_l L}{q}\right) \\ &\quad + \sum \tfrac{q}{\alpha\omega_l L}\left[1-\exp\left(-\tfrac{\alpha\omega_l L}{q}\right)\right]\left(S_0^* P_{4.l.0} - S^* P_{4.l}\right) \\ &= S^* + \left(S_0 - S_0^*\right)\sum P_{4.l.0}\mu_{1.l} + \sum \left(S_0^* P_{4.l..0} - S * P_{4.l}\right)\mu_{2.l} \end{aligned} \tag{11.13}$$

where

$$\mu_{1.l} = \exp\left(-\frac{\alpha\omega_l L}{q}\right) \tag{11.14}$$

$$\mu_{2.l} = \frac{q}{\alpha\omega_l L}\left[1-\exp\left(-\frac{\alpha\omega_l L}{q}\right)\right] \tag{11.15}$$

For massive scour, $P_{4.l}^*$ can also be approximated by $P_{4.l}$. Hence, Eq. (11.13) can be used to analyze the carrying-sediment capacity for the massive deposition, equilibrium, and massive scour conditions. Based on the data in Table 11.6 and the corresponding size distribution $P_{4.l}$, the carrying-sediment capacity coefficient K, as shown in Eq. (11.11), can be solved. In order to compute the carrying-sediment capacity, with the trial-and-error method, based on the field data, Eq. (11.13) can be re-written as

$$S = S_0 \sum P_{4.l.0}\mu_{1.l} + \left[S^* - S_0^* \sum P_{4.l..0}\mu_{1.l} + \sum \left(S_0^* P_{4.l.0} - S * P_{4.l}\right)\mu_{2.l}\right] \tag{11.16}$$

Except for $S_0 \sum P_{4.l.0}\mu_{1.l}$, all terms in the square brackets of Eq. (11.16) are related to S_0^* and S^*, which are both proportional to the carrying-sediment capacity coefficient K. It can be derived from Eq. (11.16) that

$$\begin{aligned} \Delta K &= \left(S - S_0 \sum P_{4.l.0}\mu_{1.l}\right) \\ &\quad \left[S * - S_0^* \sum P_{4.l.0}\mu_{1.l} + \sum \left(S_0^* P_{4.l.0} - S * P_{4.l}\right)\mu_{2.l}\right]^{-1} \end{aligned} \tag{11.17}$$

If the computed sediment concentration at the exit section, or the right side of Eq. (11.16), equals the measured sediment concentration, or the left side of Eq. (11.16), then $\Delta K = 1$, indicating the right value is chosen for the carrying-sediment capacity coefficient. Otherwise, the coefficient needs to be adjusted. Hence, Eq. (11.11) for carrying-sediment capacity should be corrected to

$$S^* = \Delta K \kappa \left(\frac{V^3}{gh\omega_1} \right)^{0.92} = \Delta K \cdot K \frac{Q^m}{\omega_1^{0.92}} \tag{11.18}$$

where $\Delta K > 1$ indicates the carrying-sediment capacity increases, or vice versa. It is theoretically confirmed [12, 14] that $\Delta K > 1$ and the carrying-sediment capacity increases when deposition occurs, while $\Delta K < 1$ and the carrying-sediment capacity decreases when scour occurs. In the analysis of the field data with Eq. (11.17), the saturation recovery coefficient α needs to be determined while the other parameters are either measured or can be computed with the formula for non-equilibrium sediment transport. Some data were directly obtained from Han [10, 12]. The determination of α involves how to consider the impact of non-equilibrium sediment transport [10, 13]. To obtain reliable results, two values of α are considered in the analysis. One is to set $\alpha = 0.01$, which is the minimum value of α that is used in various studies for the lower Yellow River, with more impact of non-equilibrium sediment transport being considered. This results in a relatively small variation range of K or of the correction of carrying-sediment capacity. The other is to use the theoretical value of $\alpha = 0.1$, resulting in a moderate value of ΔK.

The computed sediment concentration based on $\alpha = 0.01$, compared to the field data, is listed in Table 11.9 for the lower Yellow River. Nine groups of results, representing nine combinations of various conditions. These results cover three sedimentation conditions: deposition, scour and equilibrium. Each group of results includes data sets for three reaches, which are between Huayuankou and Gaocun, between Gaocun and Aishan, and between Anshan and Lijin, respectively. In addition to the field data, this table also lists the computed sediment concentration based on $\Delta K = 1$, the value of ΔK when the computed and measured sediment concentrations are the same, and the carrying-sediment capacity and sediment concentration computed after adjusting ΔK based on the sedimentation conditions.

The following can be observed from this table:

(1) The mean value of ΔK is 1.30 if averaging over the nine data sets with channel deposition, is 1.008 if averaging over the nine data sets in equilibrium condition, and is 0.774 if averaging over the nine data sets with channel scour. This not only indicates the multi-valued nature of carrying-sediment capacity, but also shows that the carrying-sediment capacity increases when deposition occurs and decreases when channel scour occurs.
(2) Considering the errors in the data, the carrying-sediment capacity is adjusted based on $\Delta K = 1.25$ for deposition, $\Delta K = 1$for equilibrium, and $\Delta K = 0.75$ for scour. As listed in this table, the sediment concentration computed based on the adjusted carrying-sediment capacity shows better agreement with the

Table 11.9 Computed sediment concentrations

S and Q	Reach[a]	Measured		S_0^*	S^*	S^b	S^c	Post-adjustment		Sedimentation
		S_0	S					S	S	
$20 \le S \le 80$ Mean Q	R1	34.4	34.4	33.5	38.9	34.6	0.98	1.00	34.6	Equilibrium
	R2	34.4	34.5	38.9	32.3	34.9	0.95	1.00	34.9	Equilibrium
	R3	34.5	36.9	32.3	33.9	33.7	1.29	1.00	33.7	Equilibrium
$S < 20$ $2000 < Q \le 2500$	R1	11.9	18.4	31.9	27.1	21.7	0.76	0.75	19.4	Scour
	R2	18.4	20.0	27.1	40.0	27.2	0.81	0.75	19.4	Scour
	R3	20.0	21.8	44.4	47.6	25.9	0.66	0.75	22.9	Scour
$S < 20$ $3000 < Q \le 4000$	R1	15.0	19.9	41.6	55.6	22.9	0.75	0.75	19.9	Scour
	R2	19.9	21.1	55.6	42.5	27.9	0.45	0.75	24.8	Scour
	R3	21.1	22.3	42.5	60.2	27.9	0.58	0.75	24.6	Scour
$20 \le S \le 80$ $3000 < Q \le 4000$	R1	32.4	34.3	42.8	44.4	35.2	0.93	0.75	37.2	Scour
	R2	34.3	35.5	44.4	41.8	37.2	0.86	0.75	34.2	Scour
	R3	35.5	39.2	41.8	41.5	37.2	1.17	0.75	34.0	Scour
$S > 80$ $3000 < Q \le 4000$	R1	90.2	76.9	46.8	66.4	79.7	0.785	1.25	82.9	Deposition
	R2	76.9	77.9	66.4	50.5	74.0	1.295	1.25	77.3	Deposition
	R3	77.9	79.7	50.5	56.0	70.5	1.710	1.25	73.7	Deposition
$S \ge 80$ $1000 < Q \le 1500$	R1	71.7	48.4	25.3	23.3	58.2	0	1.25	60.6	Deposition
	R2	48.4	39.1	28.3	26.1	41.4	0.69	1.25	43.3	Deposition
	R3	39.1	39.3	26.1	27.8	35.1	1.52	1.25	37.1	Deposition
$S \ge 80$ Mean Q	R1	86.1	69.8	35.0	41.5	71.2	0.87	1.25	73.7	Deposition
	R2	69.8	67.0	41.5	38.3	62.4	1.43	1.25	65.0	Deposition
	R3	67.0	69.7	38.3	35.1	58.1	1.30	1.25	60.8	Deposition

(continued)

Table 11.9 (continued)

S and Q	Reach[a]	Measured		S_0^*	S^*	S^b	S^c	Post-adjustment		Sedimentation
		S_0	S					S	S	
$20 \le S \le 80$ $2000 < Q < 2500$	R1	33.4	32.6	31.1	31.5	32.3	0.93	1.00	32.3	Equilibrium
	R2	32.6	31.9	31.5	32.9	32.6	0.93	1.00	32.6	Equilibrium
	R3	31.9	33.9	32.9	33.0	32.1	1.17	1.00	32.1	Equilibrium
$20 \le S \le 80$ $2500 < Q < 3000$	R1	38.5	36.2	40.7	47.8	39.2	0.75	1.00	39.2	Equilibrium
	R2	36.2	36.1	47.8	34.6	38.7	0.77	1.00	38.7	Equilibrium
	R3	36.1	39.3	34.6	38.8	35.5	1.33	1.00	35.5	Equilibrium

[a]R1—Huayuankou to Gaocun; R2—Gaocun to Aishan; R3—Aishan to Lijin
[b]Computed sediment concentration based on $\Delta K = 1$
[c]S when computed and measured sediment concentration are the same

measured data for 13 out of the 27 data sets, while shows no improvement for nine data sets, and shows larger errors for five data sets. This inconsistence could be caused by the errors in the measured data.

(3) Based on these 27 data sets, the relative root mean squared error (RMSE) between the computed and the measured sediment concentrations is 14.2% without adjustment of the carrying-sediment capacity, and decreases to 8.8% with the adjustment. After including the adjustment of carrying-sediment capacity, the accuracy of the computed sediment concentration can be improved with a RMSE less than 10%.

11.3.2 Impact of the Variation of Saturation Recovery Coefficient on the Multi-valued Nature of Carrying-Sediment Capacity

As discussed before, the sediment saturation recovery coefficient impacts the value of ΔK computed with Eq. (11.17). When the minimum value of $\alpha = 0.01$ is used for the lower Yellow River, the impact of non-equilibrium sediment transport is highlighted while the impact of carrying-sediment capacity is underestimated. Based on the analysis [10, 13], the theoretical value of α is approximately 0.1 for the lower Yellow River. This can also be derived from Fig. 6.19 or Eq. (6.66). Averaging α corresponding to the ten values of u_* within the range of $u_* = 1 \sim 10$ cm/s yields the mean of 0.10 for α. Using the data sets of $S > 80$ kg/m^3 with $Q = 1000 \sim 1500$ m^3/s, as listed in Table 11.9 the computed ΔK based on $\alpha = 0.1$ is listed in Table 11.10.

The following can be observed from this table:

(1) For $\alpha = 0.01$, ΔK computed with Eq. (11.17) is expected to be great than 1 for this massive deposition condition. However, the computed ΔK is 0 and 0.69 for the reaches between Huayuankou and Gaocun, and between Gaocun and Aishan, respectively, and is greater than 1 only for the reach between Aishan and Lijin. This is unexpected. On the other hand, the computed ΔK is greater than 1 for all three reaches when setting $\alpha = 0.1$, which is consistent with the characteristics of sedimentation. Also, the difference in ΔK is small between different reaches.

Table 11.10 Impact of α on the carrying-sediment capacity coefficient

Reach[a]	Measured[b]		Computed[b]		$\alpha = 0.01$		$\alpha = 0.1$		Post-adjustment[b]			
	S_0	S	S_0^*	S^*	S	ΔK	S	ΔK	ΔK	S	S_0^*	S^*
R1	71.7	48.4	25.3	23.3	58.2	0	37.4	1.59	1.45	45.7	36.7	33.8
R2	48.4	39.1	23.3	26.1	41.5	0.69	34.0	1.32	1.45	41.2	33.8	37.8
R3	39.1	39.3	26.1	27.8	35.1	1.52	31.1	1.44	1.45	39.4	37.8	40.3

[a]R1—Huayuankou to Gaocun; R2—Gaocun to Aishan; R3—Aishan to Lijin
[b]Sediment concentration and sediment transport capacity in kg/m^3

(2) The average value of ΔK is 1.45 for $\alpha = 0.1$. Based on this value, the carrying-sediment capacity is adjusted and the sediment concentration is re-calculated. The results are also listed in this table. The sediment concentration at the exit section, after this adjustment, is very close to the measured data with a RMSE of approximately 5.0%. On the other hand, if using $\Delta K = 1.25$ for $\alpha = 0.01$ in the adjustment, the RMSE is approximately 15.5%.
(3) The carrying-sediment capacity, after the correction with $\Delta K = 1.45$, is still much less than the sediment concentration at Huayuankou. After adjustment along the flow path, it increases to 37.8 ~ 40.3 kg/m^3 in the reach between Gaocun nd Lijin, which is close the sediment concentration of 39.1 ~ 39.3 kg/m^3, indicating that the sediment concentration recovers the saturation condition in this reach. This is consistent with the characteristics of sediment transport in streams.
(4) Based on the rationality of ΔK, $\alpha = 0.1$ is consistent with the characteristics of channel sedimentation, while $\alpha = 0.01$ misrepresents the phenomenon of sediment transport. If a very small value is assigned to α, the multi-valued nature of carrying-sediment capacity will be hidden.

Based on more data, Guan [4] conducted more detailed analysis for $\alpha = 0.1$. The results are summarized in Tables 11.11, 11.12 and 11.13. Except for the data sets with the flow discharges of 1000 ~ 1500 m^3/s and of 1500 ~ 2000 m^3/s as listed in Table 11.12, which belongs to the mild deposition condition with $\Delta K = 1.19$ being used, the following values are assigned to the average ΔK for all other data sets:

$\Delta K = 0.67$ for channel scour
$\Delta K = 1.00$ for equilibrium condition
$\Delta K = 1.68$ for channel deposition.

These values can be used as the factors to represent the multi-valued nature of carrying-sediment capacity for the lower Yellow River during the channel evolution.

Figures 11.7, 11.8 and 11.9 show the comparison of the sediment concentration between the measured data and the computed results with and without adjusting the carrying-sediment capacity for various ranges of sediment concentration, respectively. The computed sediment concentration with the adjustment shows good agreement with data. It is concluded that setting $\alpha = 0.1$ is more reasonable for the lower Yellow River, yields more realistic results of carrying-sediment capacity and avoid the distorted relationship between carrying-sediment capacity and sediment concentration.

In order to illustrate the multi-value of carrying-sediment capacity Figs. 11.7, 11.8 and 11.9 present the comparison between the measured and computed sediment concentration before and after ΔK is adjusted based on the saturation recovery coefficient $\alpha = 0.1$ is adopted.

As shown in Figs. 11.7, 11.8 and 11.9, the computed sediment concentration without adjusting the carrying-sediment capacity show a large discrepancy from the data for both the low- and the high-sediment concentration scenarios. For the flow with $S < 20$kg/m^3, the computed sediment concentration is higher than the

Table 11.11 Computed ΔK Based on α=0.1, for $S \leq 20$ kg/m^3 [4]

Flow discharge Q (m^3/s)	Reach	Measured		Computed		Pre-adjustment		Post-adjustment				Sedimentation
		S_0 (kg/m^3)	$S_{L,real}$ (kg/m^3)	$S_0{}^*$ (kg/m^3)	$S_L{}^*$ (kg/m^3)	S_{L1} (kg/m^3)	ΔK (kg/m^3)	$\Delta\overline{K}$ (kg/m^3)	$\Delta K \times S_0{}^*$ (kg/m^3)	$\Delta K \times S_L{}^*$ (kg/m^3)	S_{L2} (kg/m^3)	
1000 ~ 1500	R1	14.79	16.06	15.00	15.70	15.14	1.07	0.67	16.10	16.86	16.10	Scour
	R2	16.06	14.70	15.70	15.70	15.53	0.93	0.67	14.64	14.64	14.68	Scour
	R3	14.70	12.85	15.70	15.40	14.71	0.85	0.67	13.34	13.08	12.77	Scour
1500 ~ 2000	R1	15.33	19.11	23.20	22.70	21.15	0.88	0.67	20.42	19.98	19.18	Scour
	R2	19.11	17.80	22.70	21.60	21.04	0.80	0.67	18.05	17.17	17.89	Scour
	R3	17.80	17.07	21.60	22.90	20.60	0.78	0.67	16.92	17.94	16.77	Scour
2000 ~ 2500	R1	11.90	18.40	31.90	27.10	24.18	0.71	0.67	22.69	19.27	19.58	Scour
	R2	18.40	20.00	27.10	40.40	28.56	0.62	0.67	16.91	25.21	16.58	Scour
	R3	20.00	21.80	40.40	47.80	34.11	0.54	0.67	22.00	26.03	18.86	Scour
2500 ~ 3000	R1	17.89	22.85	53.50	41.00	34.35	0.56	0.67	30.10	23.07	26.96	Scour
	R2	22.85	24.80	41.00	62.70	40.58	0.48	0.67	19.87	30.39	17.48	Scour
	R3	24.80	27.32	62.70	54.60	43.10	0.51	0.67	32.14	27.99	29.70	Scour
3000 ~ 4000	R1	15.04	19.97	41.60	55.60	33.84	0.50	0.67	20.85	27.87	14.43	Scour
	R2	19.97	21.13	55.60	42.50	36.32	0.43	0.67	24.08	18.40	27.26	Scour
	R3	21.13	22.30	42.50	60.20	38.49	0.46	0.67	19.70	27.91	15.16	Scour
Mean Q	R1	15.97	21.30	34.30	32.20	27.25	0.72	0.67	24.80	23.28	21.76	Scour
	R2	21.30	21.23	32.20	37.00	29.96	0.61	0.67	19.56	22.47	20.02	Scour
	R3	21.23	22.54	37.00	35.00	30.24	0.66	0.67	24.50	23.17	22.80	Scour

R1: Huayuankou to Gaocun; R2: Gaocun to Aishan; R3: Aishan to Lijin

Table 11.12 Computed ΔK based on α=0.1, for 20 kg/m^3 $\leq S \leq$ 80 kg/m^3[4]

Flow discharge Q (m^3/s)	Reach	Measured		Computed		Pre-adjustment		Post-adjustment				Sedimentation
		S_0 (kg/m^3)	$S_{L,real}$ (kg/m^3)	S_0^* (kg/m^3)	S_L^* (kg/m^3)	$S_{L,1}$ (kg/m^3)	ΔK (kg/m^3)	$\Delta\overline{K}$ (kg/m^3)	$\Delta K \times S_0^*$ (kg/m^3)	$\Delta K \times S_L^*$ (kg/m^3)	$S_{L,2}$ (kg/m^3)	
1000 ~ 1500	R1	28.62	27.76	23.30	23.80	24.76	1.17	1.19	27.21	27.79	27.81	Prone to deposition
	R2	27.76	26.50	23.80	20.30	23.32	1.19	1.19	28.39	24.22	26.11	
	R3	26.50	25.36	20.30	23.70	22.69	1.15	1.19	23.36	27.27	25.69	
1500 ~ 2000	R1	36.08	33.95	30.50	28.70	30.90	1.15	1.19	35.14	33.06	33.87	Prone to deposition
	R2	33.95	32.89	28.70	26.20	29.57	1.18	1.19	33.77	30.83	32.59	
	R3	32.89	34.48	26.20	28.70	28.55	1.29	1.19	33.93	37.16	35.03	
2000 ~ 2500	R1	33.39	32.57	31.10	31.50	31.59	1.05	1.00	32.60	33.02	32.60	Equilibrium
	R2	32.57	31.86	31.50	32.90	32.67	0.96	1.00	30.27	31.61	31.84	Equilibrium
	R3	31.86	33.95	32.90	33.00	32.53	1.06	1.00	35.01	35.11	33.96	Equilibrium
2500 ~ 3000	R1	38.51	36.21	40.70	47.80	41.69	0.80	1.00	32.39	38.04	35.13	Equilibrium
	R2	36.21	36.12	47.80	34.60	39.46	0.86	1.00	41.19	29.81	37.47	Equilibrium
	R3	36.12	39.33	34.60	38.80	36.27	1.13	1.00	38.98	43.71	39.71	Equilibrium
3000 ~ 4000	R1	32.45	34.30	42.80	44.40	38.72	0.83	1.00	35.44	36.77	34.04	Equilibrium
	R2	34.30	35.49	44.40	41.80	40.10	0.82	1.00	36.39	34.26	35.96	Equilibrium
	R3	35.49	39.17	41.80	41.50	39.22	1.00	1.00	41.72	41.42	39.17	Equilibrium
Mean Q	R1	34.73	34.40	33.40	36.40	34.98	0.97	1.00	32.54	35.46	34.35	Equilibrium
	R2	34.40	34.48	36.40	32.30	34.39	1.00	1.00	36.56	32.44	34.47	Equilibrium
	R3	34.48	36.83	32.30	33.90	33.41	1.15	1.00	37.24	39.08	37.01	Equilibrium

R1: Huayuankou to Gaocun; R2: Gaocun to Aishan; R3: Aishan to Lijin

Table 11.13 Computed ΔK based on α=0.1, for $S \geq 80$ kg/m^3 [4]

Flow discharge Q (m^3/s)	Reach	Measured		Computed		Pre-adjustment		Post-adjustment				Sedimentation
		S_0 (kg/m^3)	$S_{L,real}$ (kg/m^3)	S_0^* (kg/m^3)	S_L^* (kg/m^3)	S_{L1} (kg/m^3)	ΔK (kg/m^3)	$\Delta\overline{K}$ (kg/m^3)	$\Delta K \times S_0^*$ (kg/m^3)	$\Delta K \times S_L^*$ (kg/m^3)	S_{L2} (kg/m^3)	
1000 ~ 1500	R1	71.79	48.45	25.30	23.30	36.00	1.72	1.68	43.57	40.12	48.05	Deposition
	R2	48.45	39.32	23.30	26.10	32.13	1.40	1.68	32.71	36.64	39.89	Deposition
	R3	39.32	39.55	26.10	27.80	30.33	1.46	1.68	38.05	40.53	39.96	Deposition
1500 ~ 2000	R1	79.76	63.51	28.60	29.90	44.66	1.95	1.68	55.81	58.35	64.63	Deposition
	R2	63.51	59.69	29.90	30.70	41.56	1.88	1.68	56.23	57.74	59.90	Deposition
	R3	59.69	60.65	30.70	26.50	36.51	2.20	1.68	67.49	58.26	58.18	Deposition
2000 ~ 2500	R1	93.76	73.11	28.10	39.50	53.37	1.89	1.68	53.20	74.78	80.28	Deposition
	R2	73.11	66.52	39.50	38.00	52.55	1.58	1.68	62.59	60.21	65.71	Deposition
	R3	66.52	68.73	38.00	34.30	46.58	1.92	1.68	73.10	65.98	66.34	Deposition
2500 ~ 3000	R1	70.11	54.84	44.20	54.00	56.20	0.95	1.68	42.05	51.38	54.44	Deposition
	R2	54.84	59.63	54.00	37.70	49.85	1.38	1.68	74.59	52.08	55.14	Deposition
	R3	59.63	65.22	37.70	30.50	43.19	1.99	1.68	74.94	60.63	60.27	Deposition
3000 ~ 4000	R1	90.17	76.92	46.30	66.40	72.46	1.13	1.68	52.48	75.26	78.65	Deposition
	R2	76.92	77.83	66.40	50.50	65.80	1.39	1.68	92.38	70.26	73.72	Deposition
	R3	77.83	79.74	50.50	56.60	62.24	1.58	1.68	79.62	89.24	82.95	Deposition
Mean Q	R1	86.30	69.85	35.00	41.50	55.88	1.58	1.68	55.26	65.52	72.45	Deposition
	R2	69.85	67.02	41.50	38.30	51.89	1.63	1.68	67.53	62.32	65.56	Deposition
	R3	67.02	69.73	38.30	35.10	46.74	1.97	1.68	75.63	69.31	68.06	Deposition

R1: Huayuankou to Gaocun; R2: Gaocun to Aishan; R3: Aishan to Lijin

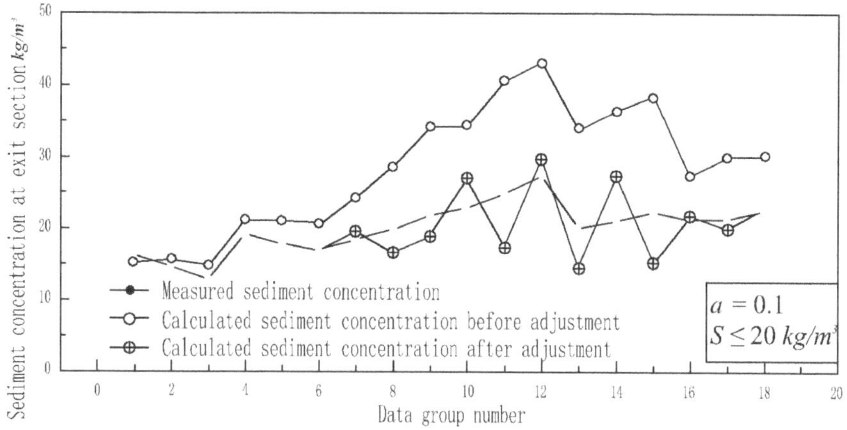

Fig. 11.7 Adjustment of multi-valued carrying-sediment capacity based on $\alpha = 0.1$, for $S \leq 20$ kg/m^3 [4]

data. For the flow with $S > 80$ kg/m^3, the computed sediment concentration is lower than the data. After adjusting the carrying-sediment capacity, the computed sediment concentration shows good agreement with the data. For the flow with the moderate sediment concentration, which is close to equilibrium sediment transport, the computed sediment concentration shows good agreement with the data no matter the carrying-sediment capacity is adjusted or not. This represents the results for the equilibrium condition. As shown in these figures, for the low-sediment concentration scenario, both the measured sediment concentration and the computed results after adjusting the carrying-sediment capacity is approximately 12 ~ 25 kg/m^3. The sediment concentration is approximately 25 ~ 40 kg/m^3 for the median-sediment concentration scenario, and is approximately 40 ~ 80 kg/m^3 for the high-sediment concentration scenario. This is a strong evidence for the characteristics of "discharging more sediment as more sediment coming" for the lower Yellow River.

In summary, by analyzing the adjustment of the carrying-sediment capacity coefficient, the multi-valued nature of carrying-sediment capacity is further verified, with the variation range being quantified. Based on $\alpha = 0.1$, the adjustment factor for the carrying-sediment capacity coefficient is $\Delta K = 0.67 \sim 1.67$. ΔK is 0.67 for scour, 1.0 for the equilibrium condition, and 1.68 for deposition.

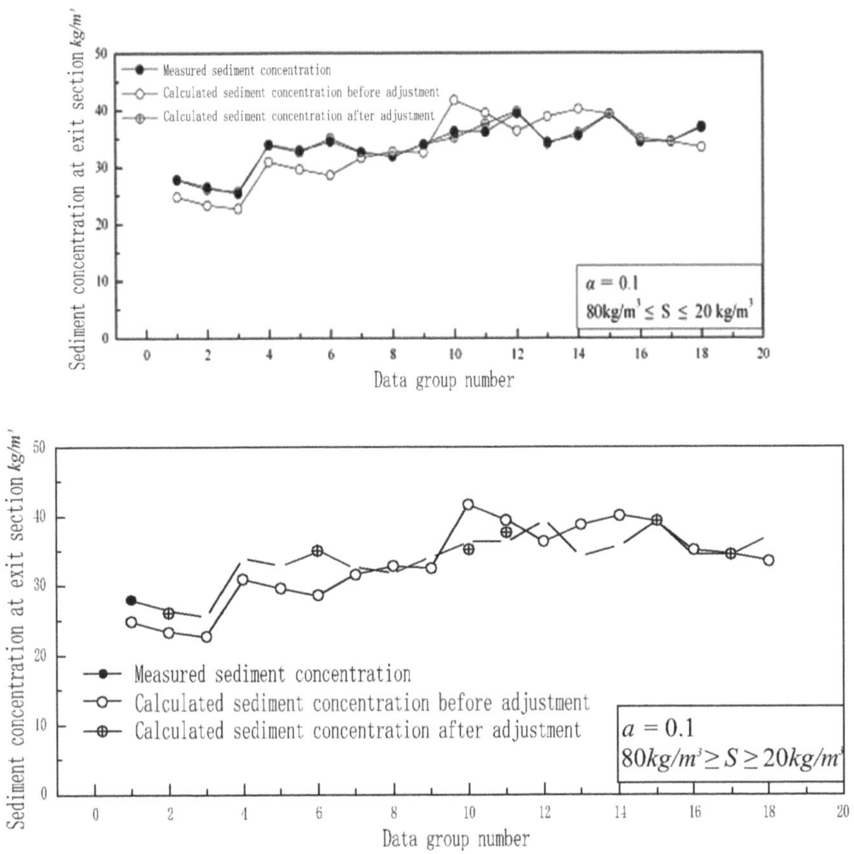

Fig. 11.8 Adjustment of multi-valued carrying-sediment capacity based on $\alpha = 0.1$, for 20 kg/m^3 $\leq S \leq$ 80 kg/m^3 [4]

11.4 Mechanism of the Multi-valued Nature of Carrying-Sediment Capacity

11.4.1 Composition and Definition of Carrying-Sediment Capacity

The carrying-sediment capacity is considered be formed due to the sediment influx from upstream and the sediment eroded from the bed [7, 9, 15]. The concentration of suspended load and the carrying-sediment capacity typically include three parts: the finer sediment particles from upstream that completely change to part of the carrying-sediment capacity, the coarser sediment particles from upstream that settle down on the bed and then are partially eroded to become part of the carrying-sediment

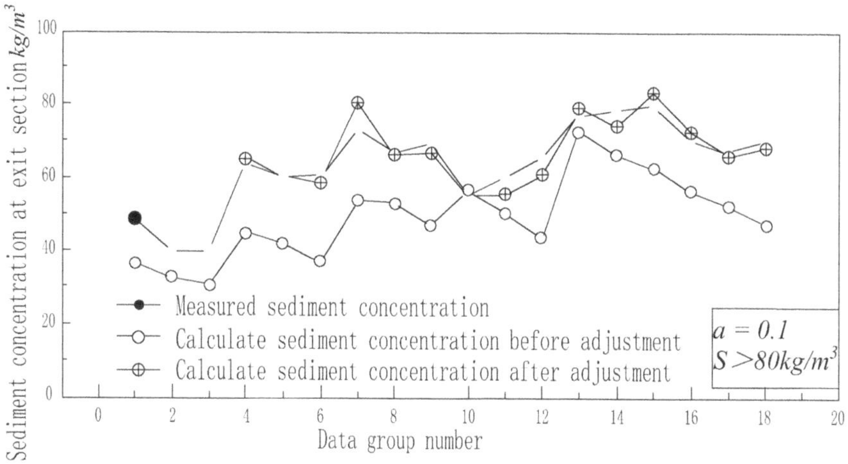

Fig. 11.9 Adjustment of multi-valued carrying-sediment capacity based on $\alpha = 0.1$, for $S > 80$ kg/m^3 [4]

capacity, and the bed material that is picked up from the bed to become part of the carrying-sediment capacity. If the sediment particles that settle down on the bed and then are partially eroded are counted as part of the sediment particles that are picked up from the bed, then the carrying-sediment capacity consists of two parts, which is described in Eq. (11.5), or

$$S_l^* = (1 - \mu_l)S_{1.l}^* + \mu_l S_{0.l}^* \tag{11.19}$$

where S_l^* is the fractional carrying-sediment capacity, $S_{0.l}^*$ part of the fractional carrying-sediment capacity that comes from the sediment influx from upstream, $S_{1.l}^*$ the other part coming from the local sediment picked up from the bed, and μ_l the proportion for $S_{0.l}^*$.

Han [5] developed the governing equation for non-equilibrium transport of non-uniform sediment

$$\frac{dS_l}{dx} = -\frac{\alpha_1 \omega_l}{q}\left(P_{4.l}S - P_{4.l}^* S^*\right) \tag{11.20}$$

and the conditions for equilibrium sediment transport. Equilibrium sediment transport can be divided into the strong equilibrium condition and the weak equilibrium condition. The strong equilibrium sediment transport is defined as

$$\frac{dS_l}{dx} = 0 \tag{11.21}$$

or

$$P_{4.l}S = P_{4.l}^* S^*$$

Integrating this equation yields

$$S = S^*$$

which leads to

$$P_{4.l} = P_{4.l}^*$$

Hence, the strong equilibrium condition can be rewritten as

$$\begin{cases} S = S^* \\ P_{4.l} = P_{4.l}^* \end{cases} \tag{11.22}$$

The weak equilibrium condition is defined as $\frac{dS}{dx} = \sum_l \frac{dS_l}{dx} = 0$. With more strict derivation, Han developed the following equation for one-dimensional non-equilibrium sediment transport

$$\frac{dS_l}{dx} = -\frac{\omega_l}{q}\left(\tilde{\alpha}_l P_{4.l} S - \alpha_l^* P_{4.l}^* S^*\right) \tag{11.23}$$

where $\tilde{\alpha}_l$ and α_l^* are the saturation recovery coefficients for sediment concentration and for carrying-sediment capacity, respectively, or the saturation recovery coefficients for sediment concentration under the non-equilibrium condition and under the equilibrium condition, respectively. $\tilde{\alpha}_l$ and α_l^* are generally not equal to each other. However, $\tilde{\alpha}_l = \alpha_l^*$ if Eq. (11.22) is satisfied. Based on Eq. (11.23), the strong equilibrium sediment transport can still be described with Eq. (11.22), while the following equation can be derived for the weak equilibrium condition:

$$\frac{dS}{dx} = -\frac{1}{q}\left[S\sum \tilde{\alpha}_l \omega_l P_{4.l} - S^* \sum \alpha_l^* \omega_l P_{4.l}^*\right] = 0 \tag{11.24}$$

It is noted from Eq. (11.24) that no deposition or scour occurs on the bed under the weak equilibrium condition. However, S does not necessarily equal S^*, $P_{4.l}$ does not necessarily equal $P_{4.l}^*$, and $\tilde{\alpha}_l$ does not necessarily equal α_l^*. Sediment transport is in the weak equilibrium condition if

$$S = \frac{\sum \alpha_l^* \omega_l P_{4.l}^*}{\sum \tilde{\alpha}_l \omega_l P_{4.l}} S^* = S^{**} \tag{11.25}$$

For a general condition, even if $S = S^*$, it is still not guaranteed for no channel deposition or erosion to occur, or it is not guaranteed for $\sum \tilde{\alpha}_l \omega_l P_{4.l}$ to be equal to $\sum \alpha_l^* \omega_l P_{4.l}^*$. The parameter to control channel sedimentation is S^{**}, and $S^{**} \neq S^*$, which is the cause for the multi-valued nature of carrying-sediment capacity.

Let further analyze the carrying-sediment capacity as described in Eq. (11.19). For the strong equilibrium condition, $S = S^*$, and $P_{4.l} = P_{4.l}^*$. This indicates that the deposited quantity of the l-th size group should equal the sediment quantity eroded from the bed for this size group in order to satisfy the equilibrium sediment exchange. For this case, the equivalent carrying-sediment capacity should be completely determined by the sediment influx from upstream, and $S_l^* = S_{0.l}^*$, or $\mu_l = 1$. In reality, the suspended sediment continuously exchanges with the bed material, with a dynamic equilibrium being maintained during sediment transport. Hence, when Eq. (11.22) is satisfied for strong equilibrium sediment transport, it also implied that $\mu_l = 1$. For non-equilibrium sediment transport, the bed material gradation and the size distribution for carrying-sediment capacity have the following relationship

$$P_{4.l}^* S^* = P_{1.l} S^*(l) \tag{11.26}$$

where $P_{1.l}$ is the effective bed material gradation. For the strong equilibrium condition, $P_{4.l} = P_{4.l}^*$, and thus

$$P_{4.l} S^* = P_{1.l} S^*(l)$$

which leads to

$$P_{4.l} = \frac{P_{1.l} S^*(l)}{S^*} = \frac{P_{1.l} S^*(l)}{\sum P_{1.l} S^*(l)} \tag{11.27}$$

or

$$P_{1.l} = \frac{P_{4.l} S}{S^*(l)} = \frac{P_{4.l}}{S^*(l)} S = \frac{P_{4.l}}{S^*(l)} \Big/ \sum \frac{P_{4.l}}{S^*(l)} \tag{11.28}$$

Hence, the strong equilibrium condition also implies that the bed material gradation $P_{1.l}$ completely corresponds to the size distribution of suspended load $P_{4.l}$. In other words, for a known $S^*(l)$, the size distribution of suspended load can be derived from the bed material gradation, or vice versa. Also, the size distribution for carrying-sediment capacity, or $P_{4.l}^*$, is no longer needed.

Except for flumes, the strong equilibrium condition and the resulting requirements for $\mu_l = 1$ and for the bed material gradation completely corresponding to the size distribution of suspended load are almost impossible to be satisfied in natural streams. However, as an important concept and tool, the strong equilibrium condition has a theoretical importance for the analysis of non-equilibrium transport of suspended load. If the carrying-sediment capacity is defined based on the strong equilibrium condition, the carrying-sediment capacity S^* is single-valued. It should be determined based on the law for carrying-sediment capacity or on the reliable formula, and at the same time satisfy Eq. (11.22). For non-equilibrium and weak equilibrium sediment transport, the carrying-sediment capacity is multi-valued, as described in Eq. (11.25). This will be further discussed in the following.

11.4.2 Mechanism and Meaning of the Multi-valued Nature of Carrying-Sediment Capacity

From the mechanism aspects, the multi-valued nature of carrying-sediment capacity is attributed to three causes. It is not only caused by the multi-valued threshold flow velocities as discussed in the previous sections, but also caused by the variation of the saturation recovery coefficient and by the multi-valued nature of the size distribution for carrying-sediment capacity. The multi-valued nature of the threshold flow velocities is eventually passed to the multi-valued nature of the composition of carrying-sediment capacity, which is shown by the inconsistence in the carrying-sediment capacities between scour from the bed material and deposition. The multi-valued natures of the saturation recovery coefficient and of the size distribution for carrying-sediment capacity are introduced based on the concept that the variable S^{**} to practically determine scour or deposition, does not equal the carrying-sediment capacity for the strong equilibrium condition. These three mechanisms for the multi-valued nature of carrying-sedimentt capacity are discussed as follows.

11.4.2.1 Multi-valued Nature of Carrying-Sediment Capacity Caused by the Structure of Carrying-Sediment Capacity

Based on Eq. (11.5)

$$S_l^* = (1-\mu_l)S_{1.l}^* + \mu_l S_{0.l}^* = \left[(1-\mu_l)\frac{\Omega_{1.l}}{\Omega_{0.l}} + \mu_l\right]S_{0.l}^*$$

From Eqs. (11.6) and (11.7), $S_{1.l}^*$ is different from $S_{0.l}^*$. Considering the difference between $\Omega_{1.l}$ and $\Omega_{0.l}$, and the variation of μ_l, S_l^* cannot be determined with the formula for the strong equilibrium condition. Then there is

$$\begin{aligned} S^{**} &= \sum S_l^* = \sum (1-\mu_l)S_{1.l}^* + \sum \mu_l S_{0.l}^* \\ &= \overline{(1-\mu_l)}\sum \frac{(1-\mu_l)}{\overline{(1-\mu_l)}} S_{1.l}^* + \overline{\mu}_l \sum \frac{\mu_l}{\overline{\mu}_l} S_{0.l}^* \\ &= \overline{(1-\mu_l)}S_{1.m}^* + \overline{\mu}_l S_{0.m}^* \end{aligned} \quad (11.29)$$

where

$$\overline{(1-\mu_l)} = \frac{1}{n}\sum_{l=1}^{n}(1-\mu_l) \quad (11.30)$$

$$\overline{\mu}_l = \frac{1}{n}\sum_{l=1}^{n}\mu_l \quad (11.31)$$

$$S^*_{1.m} = \sum \frac{(1-\mu_l)}{\overline{(1-\mu_l)}} S^*_{1.l} \tag{11.32}$$

$$S^*_{0.m} = \sum \frac{\mu_l}{\overline{\mu}_l} S^*_{0.l} \tag{11.33}$$

It is noted that S^{**}_l is different from the sediment transport capacity S^*, while $S^*_{1.l}$ and $S^*_{0.l}$ in Eqs. (11.5) and (11.29) are both computed with the formulas for carrying-sediment capacity. For the case with constant h, V, D_l, $P_{4.l}$ and n, but with varying sediment influx from upstream and bed material, S^*_l and S^* are still multi-valued that cannot be determined based on the strong equilibrium condition. If the bed material gradation does not remain constant, μ_l varies accordingly, resulting in a varying carrying-sediment capacity that cannot be simply determined with the available formula for carrying-sediment capacity. As shown in Eqs. (11.2) and (11.3), the difference between the entrainment probability $\varepsilon_{1.l}$ and the probability for not ceasing movement $\varepsilon_{0.l}$ as well as thedifference between the incipient suspension probability and the suspension probability, result in the difference between $\Omega_{1.l}$ and $\Omega_{0.l}$, with $\Omega_{1.l} < \Omega_{0.l}$. It is noted that the sediment influx from upstream as discussed in this section is limited to the case with no cohesive force or additional pressure of film water among the sediment particles.

11.4.2.2 Multi-valued Nature of Carrying-Sediment Capacity Caused by the Variation of Saturation Recovery Coefficient

Based on Eq. (11.23)

$$\frac{dS}{dx} = \sum \frac{dS_l}{dx} = -\frac{1}{q}\left(S \sum P_{4.l}\omega_l \tilde{\alpha}_l - S^* \sum P^*_{4.l}\omega_l \alpha^*_l\right) \tag{11.34}$$

Define the average fall velocity

$$\omega = \sum_{l=1}^{n} P_{4.l}\omega_l \tag{11.35}$$

$$\omega^* = \sum_{l=1}^{n} P^*_{4.l}\omega_l \tag{11.36}$$

Also rewrite

$$\sum_{l=1}^{n} P_{4.l}\omega_l \tilde{\alpha}_l = \sum_{l=1}^{n} P_{4.l}\omega_l \sum \frac{P_{4.l}\omega_l \tilde{\alpha}_l}{\sum P_{4.l}\omega_l} = \omega \tilde{\alpha}_m \tag{11.37}$$

$$\sum_{l=1}^{n} P^*_{4.l}\omega_l \alpha^* = \sum_{l=1}^{n} P^*_{4.l}\omega_l \sum \frac{P^*_{4.l}\omega_l \alpha^*_l}{\sum P^*_{4.l}\omega_l} = \omega^* \tilde{\alpha}^*_m \tag{11.38}$$

where

$$\alpha_m = \frac{\sum P_{4.l}\omega_l\tilde{\alpha}_l}{\sum P_{4.l}\omega_l} \tag{11.39}$$

$$\alpha_m^* = \frac{\sum P_{4.l}^*\omega_l\alpha^*}{\sum P_{4.l}^*\omega_l} \tag{11.40}$$

Plugging Eqs. (11.37)–(11.40) in Eq. (11.34) yields

$$\begin{aligned}\frac{dS}{dx} &= -\frac{1}{q}\left(S\omega\tilde{\alpha}_m - S^*\omega^*\alpha_m^*\right) = -\frac{\omega\tilde{\alpha}_m}{q}\left(S - \frac{\omega^*\alpha_m^*}{\omega\tilde{\alpha}_m}S^*\right)\\ &= -\frac{\omega\tilde{\alpha}_m}{q}\left(S - S_1^{**}\right)\end{aligned} \tag{11.41}$$

It is seen that sediment scour or deposition is determined by the difference between $S\omega\tilde{\alpha}_m$ and $S^*\omega^*\alpha_m^*$, not the difference between S and S^*. Defining

$$S_1^{**} = \frac{\omega^*\alpha_m^*}{\omega\tilde{\alpha}_m}S^* \tag{11.42}$$

then the difference between S_1^{**} and S controls channel sedimentation for both non-equilibrium and weak equilibrium sediment transport.

It is important to understand the exact physical meaning of S_1^{**}. It has been a long term perception of the engineers and academic researchers that no channel scour or deposition would occur if $S = S^*$. Particularly, this perception has been widely used after the research results for the non-equilibrium transport of non-uniform sediment became available. Also, during the field data collection, it is always determined as $S^* = S$ when no channel scour or deposition occurs due to the limitations on the field. As previously discussed, according to the theory for the non-equilibrium transport of non-uniform sediment, no channel scour or deposition to occur ($dS/dx = 0$) is different from $S = S^*$. Based on this concept, S_1^{**} as defined in Eq. (11.42) is called the control carrying-sediment capacity. Together with the sediment concentration, S_1^{**} controls channel sedimentation. Channel deposition occurs for $S > S_1^{**}$, channel scour occurs for $S < S_1^{**}$, and sediment transport is in weak equilibrium condition with $dS/dx = 0$ for $S = S_1^{**}$. Of course, theoretically S_1^{**} is only a virtual carrying-sediment capacity. In summary, the multi-valued nature of carrying-sediment capacity is described in Eq. (11.41) for the non-equilibrium and weak equilibrium conditions. The multi-valued carrying-sediment capacity for these two conditions cannot be determined with a simple formula.

Let consider two special cases. In the first case, assuming $P_{4.l}^* = P_{4.l}$, then $\omega^* = \omega$ based on Eqs. (11.35) and (11.36), Eq. (11.41) becomes

$$\frac{dS}{dx} = -\frac{1}{q}\left(S\sum P_{4.l}\omega_l\tilde{\alpha}_l - S^*\sum P_{4.l}\omega_l\alpha_l^*\right) = -\frac{\omega\tilde{\alpha}_m}{q_0}\left(S - \frac{\alpha_m^*}{\tilde{\alpha}_m}S^*\right) \tag{11.43}$$

where $\tilde{\alpha}_m$ and α_m^* are determined with Eqs. (11.39) and (11.40), respectively. Thus, the control carrying-sediment capacity is

$$S_2^{**} = \frac{\alpha_m^*}{\tilde{\alpha}_m} S^* \tag{11.44}$$

Case 2 is for uniform sediment. For this case, Eq. (11.41) can be rewritten as

$$\begin{aligned} \frac{dS}{dx} &= -\frac{\omega}{q}\left(S\tilde{\alpha}_l - S^*\alpha^*\right) = -\frac{\omega}{q}\left(S\tilde{\alpha} - S^*\alpha^*\right) \\ &= -\frac{\tilde{\alpha}\omega}{q}\left(S - \frac{\alpha^*}{\tilde{\alpha}}S^*\right) = -\frac{\tilde{\alpha}\omega}{q}\left(S - S_2^{**}\right) \end{aligned} \tag{11.45}$$

and thus

$$S_2^{**} = \frac{\alpha^*}{\tilde{\alpha}} S^* \tag{11.46}$$

It is noted that both Eq. (11.43) and (11.45) clearly show the multi-valued nature of carrying-sediment capacity.

Further analysis of the multi-valued nature of carrying-sediment capacity can be conducted by integrating the fractional carrying-sediment capacity as described in Eq. (11.23), which can be rewritten as

$$\frac{dS_l}{dx} = -\frac{\omega_l}{q}\left(\tilde{\alpha}_l S_l - \alpha_l^* S_l^*\right) = -\frac{\tilde{\alpha}_l\omega_l}{q}\left(S_l - \frac{\alpha_l^*}{\tilde{\alpha}_l}S_l^*\right) \tag{11.47}$$

For a short reach with the length of L, S_l^* can be assumed to vary linearly along the flow path, or

$$\frac{\alpha_l^*}{\tilde{\alpha}_l}S_l^* = \left(S_{0.l}^* + \frac{S_l^* - S_{0.l}^*}{L}x\right)\frac{\alpha_l^*}{\tilde{\alpha}_l} \tag{11.48}$$

$$\frac{\alpha_l^*}{\tilde{\alpha}_l}\frac{dS_l^*}{dx} = \left(\frac{S_l^* - S_{0.l}^*}{L}\right)\frac{\alpha_l^*}{\tilde{\alpha}_l} \tag{11.49}$$

Subtracting Eq. (11.47) from Eq. (11.49) and rearranging the terms, one obtains

$$\frac{d}{dx}\left(S_l - \frac{\alpha_l^*}{\tilde{\alpha}_l}S_l^*\right) + \frac{\tilde{\alpha}_l\omega_l}{q}\left(S_l - \frac{\alpha_l^*}{\tilde{\alpha}_l}S_l^*\right) = \frac{S_{0.l}^* - S_l^*}{L}\left(\frac{\alpha_l^*}{\tilde{\alpha}_l}\right) \tag{11.50}$$

Setting

$$y_l = S_l - \frac{\alpha_l^*}{\tilde{\alpha}_l} S_l^* \tag{11.51}$$

$$A = \frac{\alpha_l^*}{\tilde{\alpha}_l} \frac{S_{0.l}^* - S_l^*}{L} \tag{11.52}$$

then Eq. (11.50) becomes

$$\frac{dy}{dx} + \frac{\tilde{\alpha}_l \omega_l}{q} y = A \tag{11.53}$$

Integrating this equation with the boundary condition: $x = 0$, $y = y_{0.l}$ and $x = L$, $y_l = y_l$, yields

$$y_l = y_{0.l} \exp\left(-\frac{\tilde{\alpha}_l \omega_l L}{q}\right) + A\left[1 - \exp\left(-\frac{\tilde{\alpha}_l \omega_l L}{q}\right)\right] \frac{q}{\tilde{\alpha}_l \omega_l} \tag{11.54}$$

Plugging y, A in Eq. (11.54) obtaines

$$\begin{aligned} S_l - \frac{\alpha_l^*}{\tilde{\alpha}_l} S_l^* &= \left(S_{0.l} - \frac{\alpha_l^*}{\tilde{\alpha}_l} S_{0.l}^*\right) \exp\left(-\frac{\tilde{\alpha}_l \omega_l L}{q}\right) \\ &+ \frac{\alpha_l^*}{\tilde{\alpha}_l} \left(S_{0.l}^* - S_l^*\right) \frac{q}{\tilde{\alpha}_l \omega_l L} \left[1 - \exp\left(-\frac{\tilde{\alpha}_l \omega_l L}{q}\right)\right] \end{aligned} \tag{11.55}$$

Setting

$$S_l^{**} = \frac{\alpha_l^*}{\tilde{\alpha}_l} S_l^* \tag{11.56}$$

$S_l = P_{4.l} S$, and $S_l^* = P_{4.l}^* S^*$, then Eq. (11.55) becomes

$$\begin{aligned} P_{4.l} S &= P_{4.l}^* S^{**} + \left(P_{4.l.0} S_0 - P_{4.l.0}^* S^{**}\right) \exp\left(-\frac{\tilde{\alpha}_l \omega_l L}{q}\right) \\ &+ \left(P_{4.l.0}^* S_0^{**} - P_{4.l}^* S^{**}\right) \frac{q}{\tilde{\alpha}_l \omega_l L} \left[1 - \exp\left(-\frac{\tilde{\alpha}_l \omega_l L}{q}\right)\right] \end{aligned} \tag{11.57}$$

Summing up Eq. (11.57) for all size classes, one obtains

$$\begin{aligned} S &= S^{**} + \sum \left(P_{4.l.0} S - P_{4.l.0}^* S^*\right) \exp\left(-\frac{\tilde{\alpha}_l \omega_l L}{q}\right) \\ &+ \sum \left(P_{4.l.0}^* S_0^{**} - P_{4.l}^* S^{**}\right) \frac{q}{\tilde{\alpha}_l \omega_l L} \left[- \exp\left(-\frac{\tilde{\alpha}_l \omega_l L}{q}\right)\right] \end{aligned} \tag{11.58}$$

This equation is the same as Eq. (11.12) if S^* in Eq. (11.12) is replaced with S^{**}. While S^{**} is a virtual carrying-sediment capacity, it is actually the carrying-sediment capacity to decide the channel scour or deposition. It is also noted that the adjustment of the carrying-sediment capacity coefficient K, as discussed in Sect. 11.3.1, is actually to determine $\alpha_l^*/\tilde{\alpha}_l$, which has a solid theoretical basis. Based on its characteristics, $\alpha_l^*/\tilde{\alpha}_l$ is greater than 1 for deposition, and is less than 1 for scour.

11.4.2.3 Multi-valued Nature of Carrying-Sediment Capacity Caused by the Multi-valued Nature of the Size Distribution for Carrying-Sediment Capacity

Considering the difference in carrying-sediment capacity between the sediment influx from upstream and the sediment that is locally eroded from the bed, and also considering the different impact between the finer and the coarser sediment particles, a complete set of formulas were developed in Chap. 7 for the carrying-sediment capacity and corresponding size distribution. The formula for carrying-sediment capacity is

$$S^*(\omega^*) = \begin{cases} P_{4.1}S + P_{4.2}S\frac{S^*(\omega_2^*)}{S^*(\omega_{1.1}^*)} + \left[1 - \frac{P_{4.1}S}{S^*(\omega_l)} - \frac{P_{4.2}S}{S^*(\omega_{1.1}^*)}\right]P_1 S^*(\omega_{1.1}^*) \\ \qquad \left(l = 1, 2, \ldots, n;\ \frac{P_{4.1}S}{S^*(\omega_l)} + \frac{P_{4.2}S}{S^*(\omega_{1.1}^*)} < 1\right) \\ P_{4.1}S + \left[1 - \frac{P_{4.1}S}{S^*(\omega_l)}\right]S^*(\omega_2^*) \quad (l = 1, 2, \cdots, n;\quad \text{other conditions}) \end{cases}$$

As shown in this formula, the capacity for carrying non-uniform sediment, $S^*(\omega^*)$, not only depends on the sediment influx from upstream and the sediment eroded from the local reach, but also is a function of the size distribution of suspended load and the bed material gradation. The carrying-sediment capacity computed with this formula is not for the strong equilibrium condition but for a condition with scour or deposition. It cannot be determined with a simple formula.

There might be a special case for the multi-valued nature of the size distribution for carrying-sediment capacity, which is $\tilde{\alpha}_l = \alpha_l^* = \alpha$, with $P_{4.l} \neq P_{4.l}^*$. For this case, the following equation can be obtained based on Eq. (11.23)

$$\begin{aligned} \frac{dS}{dx} &= -\frac{\alpha}{q}\left(S\sum P_{4.l}\omega_l - S^*\sum P_{4.l}^* S^*\right) = -\frac{\alpha}{q}\left(S\omega - S^*\omega^*\right) \\ &= -\frac{\alpha\omega}{q}\left(S - \frac{\omega^*}{\omega}S^*\right) = -\frac{\alpha\omega}{q}\left(S - S_3^{**}\right) \end{aligned} \tag{11.59}$$

where

$$S_3^{**} = \frac{\omega^*}{\omega}S^* \tag{11.60}$$

It is seen that the channel scour or deposition is determined by S and S_3^{**}. For the weak equilibrium condition

$$S = S_3^{**} = \frac{\omega^*}{\omega} S^*$$

and the carrying-sediment capacity is also multi-valued.

11.4.3 Further Analysis of the Multi-valued Nature of Carrying-Sediment Capacity Caused by the Multi-valued Nature of the Size Distribution for Carrying-Sediment Capacity

In order to further illustrate how the multi-valued size distribution for carrying-sediment capacity cause the multi-valued nature of carrying-sediment capacity, a simplified case of scour is analyzed. This case can be described with a simplified form of Eq. (7.135) under the scour condition. In this case, the sediment influx from upstream is partially converted to the carrying-sediment capacity, while the bed material can all be suspended. For this case, the size distribution for carrying-sediment capacity is

$$\begin{aligned} P_{4.l}^* = {} & P_{4.1} P_{4.l.1} \frac{S}{S^*(\omega^*)} + P_{4.2} P_{4.l.2} \frac{S}{S^*(\omega^*)} \frac{S^*(l)}{S^*(\omega^*)} \\ & + \left[1 - \frac{P_{4.1} S}{S^*(\omega_1)} - \frac{P_{4.2} S}{S^*(\omega^*)} \right] P_{4.l.1}^* \frac{S^*(\omega_1^*)}{S^*(\omega^*)} \end{aligned}$$

It is noted that $P_{4.l}^* \neq P_{4.l}$, or the size distribution for carrying-sediment capacity $P_{4.l}^*$ is different from $P_{4.l}$ for the strong equilibrium condition. The corresponding carrying-sediment capacity is

$$S^*(\omega^*) = P_{4.1} S + P_{4.2} S \frac{S^*(\omega_2^*)}{S(\omega_1^*)} + \left[1 - \frac{P_{4.1} S}{S^*(\omega_1)} - \frac{P_{4.2} S}{S^*(\omega_1^*)} \right] S^*(\omega_1^*)$$

Thus,

$$\begin{aligned} \frac{S^*(\omega^*)}{S^*(\omega)} = {} & P_{4.1} \frac{S}{S^*(\omega)} + P_{4.l} \frac{S^*(\omega_2^*)}{S^*(\omega_1^*)} \frac{S}{S^*(\omega)} \\ & + \left[1 - \frac{P_{4.1} S}{S^*(\omega_1)} - \frac{P_{4.2} S}{S^*(\omega_1^*)} \right] \frac{S^*(\omega_1^*)}{S^*(\omega)} \end{aligned} \tag{11.61}$$

where $S^*(\omega)$ is the carrying-sediment capacity for the strong equilibrium condition, and

$$\omega = \left(\sum_{l=1}^{n} P_{4.l}\omega_l^m\right)^{\frac{1}{m}}$$

For a general condition, it is difficult to quantity the variation range of $\frac{S^*(\omega^*)}{S^*(\omega)}$ because of many parameters involved in this equation that vary within wide ranges. However, for scour, it is obviously that $\frac{S}{S^*(\omega)} < 1$, and it is proved that $\frac{S^*(\omega^*)}{S^*(\omega)} < 1$. In fact because $\frac{S^*(\omega_2^*)}{S^*(\omega_1^*)} < 1$, then

$$\frac{S^*(\omega^*)}{S^*(\omega)} < \frac{P_{4.1}S}{S^*(\omega)} + \frac{P_{4.2}S}{S^*(\omega)} + \left[1 - \frac{P_{4.1}S}{S^*(\omega)} - \frac{P_{4.2}S}{S^*(\omega_1^*)}\right]\frac{S^*(\omega_1^*)}{S^*(\omega)}$$

For the case of scour, the carrying-sediment capacity associated with the sediment picked up from the bed material $S^*(\omega_1^*)$ is less than $S^*(\omega)$ that is computed based on the sediment concentration. Thus, $\frac{S^*(\omega_1^*)}{S^*(\omega)} < 1$, and

$$\frac{S^*(\omega^*)}{S^*(\omega)} < \frac{S}{S^*(\omega)} + \left[1 - \frac{S}{S^*(\omega)}\right] = 1 \tag{11.62}$$

An example is given in the following to demonstrate the value of $\frac{S^*(\omega^*)}{S^*(\omega)}$. Assuming the sediment concentration $S = 1$ kg/m^3, the mean flow velocity $V = 2.56\, m/s$, and the mean water depth $h = 15\ m$, then $\frac{V^{3m}}{h^m} = \frac{V^{2.76}}{h^{0.92}} = 1.11$, and the carrying-sediment capacity can be expressed as

$$S^*(\omega_x) = 0.017\frac{V^{2.76}}{h^{0.92}\omega_x^{0.92}} \tag{11.63}$$

As listed in Table 11.14, the computed carrying-sediment capacity $S^*(\omega) = 2.00$ kg/m^3, $S^*(\omega_1) = 38.18$ kg/m^3, $S^*(\omega_2^*) = 0.988$ kg/m^3, $S^*(\omega_1^*) = 1.08$ kg/m^3, $P_{4.1} = 0.589$, and $P_{4.2} = 0.411$. Based on Eq. (11.60)

$$\frac{S^*(\omega^*)}{S^*(\omega)} = \frac{0.589 \times 1}{2} + \frac{0.411 \times 1 \times 0.988}{1.08 \times 2} + \left[1 - \frac{0.589 \times 1}{38.18} - \frac{0.411 \times 1}{1.08}\right]1.08 = 0.808$$

which indicates the carrying-sediment capacity for this case is approximately 80.9% of that for the strong equilibrium condition. Since $S < S^*(\omega^*) = 1.616$ kg/m^3, scour still occurs.

Table 11.14 Computed $S^*(\omega_1)$, $S^*(\omega)$, $S^*(\omega_2^*)$, and $S^*(\omega_1^*)$

Item	D (mm)								Mean fall velocity	$S^*(\omega_x)$
	< 0.005	0.005 ~ 0.01	0.01 ~ 0.025	0.025 ~ 0.05	0.05 ~ 0.10	0.10 ~ 0.25	0.25 ~ 0.50	0.50 ~ 1.00		
ω_l^m	0.000024	0.00012	0.00057	0.0023	0.0081	0.0238	0.0522	0.1011		
$P_{4.l}$	0.1842	0.1643	0.1626	0.0783	0.0542	0.3482	0.0082	0	$\omega^m =$ 0.000945	$S^*(\omega)$ = 2.00
$P_{4.l.1}$	0.312	0.279	0.276	0.133						
$P_{4.l.2}$					0.132	0.848	0.020	0	$\omega_1^m =$ 0.000494	$S^*(\omega_1)$ = 38.18
$P_{1.l.1}$	0	0	0.0002	0.0008	0.199	0.750	0.04	0.01	$\omega_2^{*m} =$ 0.0191 $\omega_1^{*m} =$ 0.0174	$S^*(\omega_2^*)$ = 0.988 $S^*(\omega_1^*)$ = 1.08

11.4.4 Impact of the Threshold Flow Velocities on Carrying-Sediment Capacity

Following the idea of Han [7], starting from the formula for carrying-sediment capacity as presented in Chap. 4, the multi-valued nature of carrying-sediment capacity caused by the multi-valued threshold flow velocities is further analyzed in this section. Based on Zhang's formula for carrying-sediment capacity combining with the analysis of the threshold flow velocities, some examples were discussed in Han [7] for the multi-valued nature of carrying-sediment capacity. In this section, Eq. (4.23), which is derived based on the statistical theory and which can represent the multi-valued nature of carrying-sediment capacity, is used. This equation is

$$S_l^* = \frac{2}{3} m_0 \gamma_s P_{1.l} \frac{D_l}{\omega_l t_{4.0.l}} \frac{L_{4.l}^*}{q} \Omega_l$$

It is noted that the composite probability Ω_l, which is defined with Eq. (11.2), is the only parameter in this equation to represent the threshold probabilities, including the incipient suspension probability β_l and the suspension probability $\varepsilon_{4.l}$. The suspension probability $\varepsilon_{4.l} = P\,[\xi_{V_{b.y}} > \omega_l]$ indicates that the sediment can be suspended as long as the vertical fluctuating velocity is greater than the sediment fall velocity. However, the finer sediment on the bed, which is acted by the cohesive force and the additional pressure of film water, needs to overcome these two forces to get loosen first before being suspended if $V_{b.y} > \omega$ is satisfied. The incipient suspension is harder to occur than the suspension, and thus $\beta_l \leq \varepsilon_{4.l}$. The equal sign is included because the cohesive force and the additional pressure of film water do not apply to the coarser particles (for example, $D > 0.25 \sim 0.5$ mm) or to the finer particles that just settle down on the bed and have not been adhesive to the bed material. Hence, β_l is used in Eq. (4.23) to represent the impact of the cohesive force and the additional pressure of film water on the bed material. For the coarser particles that can be suspended from the bed, or for the finer particles from upstream that just settle on the bed and are re-suspended before being adhesive to the bed material, the threshold condition to loosen them up can be neglected, and β_l can be substituted by $\varepsilon_{4.l}$. Hence, the composite probability has two forms. In addition to Eq. (11.2), it also has a simplified form based on $\beta_l = \varepsilon_{4.l}$, which is Eq. (11.3). The corresponding formula for carrying-sediment capacity can be written as

$$\begin{aligned} S^*(l) &= \frac{2}{3}[\mu_l \Omega_{0.l} + (1-\mu_l)\Omega_{1.l}] m_0 \gamma_s P_{1.l} \frac{D_l L_{4.l}^*}{t_{4.0.l} q} \\ &= \mu_l S_0^*(l) + (1-\mu_l) S_1^*(l) \end{aligned}$$

where $S_0^*(l)$ is part of the carrying-sediment capacity associated with the sediment influx from upstream, including the coarser particles picked up from the bed material that bear no cohesive force or additional pressure of film water. $S_1^*(l)$ is the other part of carrying-sediment capacity associated with the bed material acted by cohesive force or additional pressure of film water. $S_0^*(l)$ and $S_1^*(l)$ are determined with Eqs. (11.7), (11.6), respectively.

Another important issue is to determine μ_l. For an alluvial stream, the concentration of sediment from upstream S is typically close to the sediment transport capacity S^*. Even for this case, one cannot set $\mu_l = 1$ except for reservoirs. This is because sediment exchange still occurs between the suspended load and the bed material, with part of the bed material being eroded. For a typical alluvial stream, the proportion of the suspended sediment that exchanges with the bed material is expected not to exceed 10 ~ 30% when the channel is approximately equilibrium. Let assume the proportion is 20%. It is noted that within this 20% of carrying-sediment capacity associated with the bed material, part of it belongs to the particles with no cohesive force or additional pressure of film water. As a result, the proportion of the carrying-sediment capacity, which is associated with the particles that are picked up from the bed after overcoming the cohesive force and additional pressure of film water, is assume to be 10%, resulting $\mu_l = 90\%$. Hence, the carrying-sediment capacity for an approximately equilibrium condition can be written as

$$S_C^*(l) = 0.9S_0^*(l) + 0.1S_1^*(l) \tag{11.64}$$

For an alluvial stream with massive scour, the proportion of the suspended load that exchanges with the bed material can be assumed to be 0.5. The sediment with the cohesive force and the additional pressure of film water is assumed to account for 50%, and thus contributes 25% of the carrying-sediment capacity, or

$$S_E^*(l) = 0.75S_0^*(l) + 0.25S_1^*(l) \tag{11.65}$$

For the case of massive deposition, $\mu_l = 1$,

$$S_D^*(l) = S_0^*(l) \tag{11.66}$$

The computed results of $S_0^*(l)$ and $S_1^*(l)$ based on Eqs. (11.6) and (11.7) are listed in Tables 11.15 and 11.16 [4]. In addition to Eqs. (11.6) and (11.7), part of the 27 equations for the carrying-sediment capacity as presented in Sect. 4.3 are also needed to solve for $S_0^*(l)$ and $S_1^*(l)$. The results of $S_C^*(l)$, $S_E^*(l)$ and $S_D^*(l)$ computed with Eqs. (11.64)–(11.66), respectively, are listed in Tables 11.16, 11.17 and 11.18. In Tables 11.15 and 11.16, the results within the bold boundary lines indicate $S_{1.l}^* = S_{0.l}^*$. It is observed that the chance for $S_{1.l}^* = S_{0.l}^*$ increases as the sediment grain size increases and the flow condition $(H,\ u_*)$ become stronger.

The results of $\Delta K_{1.D.l}$ 及 $\Delta K_{1.E.l}$ computed for various sedimentation conditions are summarized in Table 11.19. These adjustment factors are defined as

$$\Delta K_{1.D.l} = \frac{S_D^*(l)}{S_C^*(l)} \tag{11.67}$$

Table 11.15 Computed $S_1^*(l)$ based on the statistical theory for carrying-sediment capacity [4]

u_*	H	Group 1	Group 2	Group 3	Group 4	Group 5	Group 6	Group 7	Group 8
0.005	1								
0.005	2								
0.005	5								
0.005	10								
0.005	20								
0.01	1								
0.01	2								
0.01	5								
0.01	10								
0.01	20								
0.02	1			0.0003	0.0037	0.0031	0.0001		
0.02	2				0.0003	0.0006			
0.02	5								
0.02	10								
0.02	20								
0.05	1	0.0116	0.2287	2.2814	18.7512	12.0662	0.9276	0.0477	0.0013
0.05	2	0.0005	0.0250	0.4668	1.5378	6.0770	0.3840	0.0230	0.0007
0.05	5		0.0002	0.0231	0.1453	0.2838	0.0983	0.0081	0.0003
0.05	10			0.0007	0.0124	0.0439	0.0278	0.0032	0.0001
0.05	20				0.0004	0.0037	0.0056	0.0011	0.0001
0.1	1	7.2735	95.3238	80.9958	69.9153	53.7994	28.2358	9.8804	2.4933
0.1	2	1.3204	6.7774	40.6767	35.0324	26.9469	14.1719	4.9852	1.2656
0.1	5	0.0469	0.5128	16.5268	14.1068	10.8216	5.6984	2.0158	0.5150
0.1	10	0.0009	0.0323	0.5139	7.1457	5.4445	2.8632	1.0159	0.2007
0.1	20		0.0006	0.0397	0.2428	2.7600	1.4421	0.5124	0.0712
0.25	1	567.2580	508.8625	440.6574	390.6322	331.2739	238.9775	144.2021	63.0472
0.25	2	284.5635	254.8192	220.4716	195.3965	165.7152	119.6387	72.3626	31.7980
0.25	5	115.1309	102.4209	88.3515	78.2400	66.3428	47.9308	29.0665	12.8440
0.25	10	6.3456	51.6883	44.3135	39.1818	33.2057	23.9953	14.5747	6.4631
0.25	20	0.5946	3.2296	22.3097	19.6526	16.6319	12.0153	7.3073	3.2502

Table 11.16 Computed $S_0^*(l)$ based on the statistical theory for carrying-sediment capacity [4]

u_*	H	Group 1	Group 2	Group 3	Group 4	Group 5	Group 6	Group 7	Group 8
0.005	1	0.7246	0.5517	0.2930	0.0959	0.0079			
0.005	2	0.3623	0.2759	0.1466	0.0482	0.0040			
0.005	5	0.1449	0.1104	0.0587	0.0194	0.0016			
0.005	10	0.0725	0.0552	0.0294	0.0097	0.0008			
0.005	20	0.0362	0.0276	0.0147	0.0049	0.0004			
0.01	1	2.8483	2.4725	1.8142	0.9215	0.1406	0.0050		
0.01	2	1.4242	1.2363	0.9074	0.4630	0.0717	0.0025		
0.01	5	0.5697	0.4945	0.3631	0.1856	0.0290	0.0010		
0.01	10	0.2848	0.2473	0.1816	0.0930	0.0146	0.0005		
0.01	20	0.1424	0.1236	0.0908	0.0465	0.0073	0.0003		
0.02	1	10.4345	9.2643	6.3725	3.7276	1.2781	0.1591	0.0074	
0.02	2	5.2173	4.6469	3.6078	2.1211	0.6924	0.0830	0.0038	
0.02	5	2.0869	1.8589	1.5364	1.0684	0.3439	0.0368	0.0016	
0.02	10	1.0435	0.9295	0.7690	0.5680	0.2122	0.0210	0.0008	
0.02	20	0.5217	0.4648	0.3845	0.2848	0.1147	0.0123	0.0004	
0.05	1	42.0100	30.1634	23.0762	18.7512	12.0662	3.8565	0.8249	0.0902
0.05	2	24.6074	16.7373	11.8937	9.4905	6.0770	1.9470	0.4187	0.0460
0.05	5	11.3385	8.7433	5.3324	3.9701	2.4831	0.7920	0.1711	0.0189
0.05	10	5.7103	5.0545	3.2754	2.1841	1.2939	0.4048	0.0874	0.0096
0.05	20	2.8552	2.5627	2.0751	1.3490	0.7168	0.2113	0.0452	0.0049
0.1	1	109.8204	95.3238	80.9958	69.9153	53.7994	28.2358	9.8804	2.4933
0.1	2	57.2101	48.3778	40.6767	35.0324	26.9469	14.1719	4.9852	1.2656
0.1	5	26.6015	20.5000	16.5268	14.1068	10.8216	5.6984	2.0158	0.5150
0.1	10	16.8424	11.5964	8.5466	7.1457	5.4445	2.8632	1.0159	0.2606
0.1	20	10.1019	7.4018	4.6719	3.6935	2.7600	1.4421	0.5124	0.1318
0.25	1	567.2580	508.8625	440.6574	390.6322	331.2739	238.9775	144.2021	63.0472
0.25	2	284.5635	254.8192	220.4716	195.3965	165.7152	119.6387	72.3626	31.7980
0.25	5	115.1309	102.4209	88.3515	78.2400	66.3428	47.9308	29.0665	12.8440
0.25	10	58.9719	51.6883	44.3135	39.1818	33.2057	23.9953	14.5747	6.4631
0.25	20	31.4465	26.4564	22.3097	19.6526	16.6319	12.0153	7.3073	3.2502

Table 11.17 Computed $S_c^*(l) = [0.90S_0^*(l) + 0.10S_1^*(l)]$ based on the statistical theory for carrying-sediment capacity [4]

u_*	H	Group 1	Group 2	Group 3	Group 4	Group 5	Group 6	Group 7	Group 8
0.005	1	0.6521	0.4965	0.2637	0.0863	0.0071			
0.005	2	0.3261	0.2483	0.1319	0.0434	0.0036			
0.005	5	0.1304	0.0994	0.0528	0.0175	0.0014			
0.005	10	0.0653	0.0497	0.0265	0.0087	0.0007			
0.005	20	0.0326	0.0248	0.0132	0.0044	0.0004			
0.01	1	2.5635	2.2253	1.6328	0.8294	0.1265	0.0045		
0.01	2	1.2818	1.1127	0.8167	0.4167	0.0645	0.0023		
0.01	5	0.5127	0.4451	0.3268	0.1670	0.0261	0.0009		
0.01	10	0.2563	0.2226	0.1634	0.0837	0.0131	0.0005		
0.01	20	0.1282	0.1112	0.0817	0.0419	0.0066	0.0003		
0.02	1	9.3911	8.3379	5.7353	3.3552	1.1506	0.1432	0.0067	
0.02	2	4.6956	4.1822	3.2470	1.9090	0.6232	0.0747	0.0034	
0.02	5	1.8782	1.6730	1.3828	0.9616	0.3095	0.0331	0.0014	
0.02	10	0.9392	0.8366	0.6921	0.5112	0.1910	0.0189	0.0007	
0.02	20	0.4695	0.4183	0.3461	0.2563	0.1032	0.0111	0.0004	
0.05	1	37.8102	27.1699	20.9967	18.7512	12.0662	3.5636	0.7472	0.0813
0.05	2	22.1467	15.0661	10.7510	8.6952	6.0770	1.7907	0.3791	0.0415
0.05	5	10.2047	7.8690	4.8015	3.5876	2.2632	0.7226	0.1548	0.0170
0.05	10	5.1393	4.5491	2.9479	1.9669	1.1689	0.3671	0.0790	0.0087
0.05	20	2.5697	2.3064	1.8676	1.2141	0.6455	0.1907	0.0408	0.0044
0.1	1	99.5657	95.3238	80.9958	69.9153	53.7994	28.2358	9.8804	2.4933
0.1	2	51.6211	44.2178	40.6767	35.0324	26.9469	14.1719	4.9852	1.2656
0.1	5	23.9460	18.5013	16.5268	14.1068	10.8216	5.6984	2.0158	0.5150
0.1	10	15.1583	10.4400	7.7433	7.1457	5.4445	2.8632	1.0159	0.2546
0.1	20	9.0917	6.6617	4.2087	3.3484	2.7600	1.4421	0.5124	0.1257
0.25	1	567.2580	508.8625	440.6574	390.6322	331.2739	238.9775	144.2021	63.0472
0.25	2	284.5635	254.8192	220.4716	195.3965	165.7152	119.6387	72.3626	31.7980
0.25	5	115.1309	102.4209	88.3515	78.2400	66.3428	47.9308	29.0665	12.8440
0.25	10	53.7093	51.6883	44.3135	39.1818	33.2057	23.9953	14.5747	6.4631
0.25	20	28.3613	24.1337	22.3097	19.6526	16.6319	12.0153	7.3073	3.2502

$$\Delta K_{1.E.l} = \frac{S_E^*(l)}{S_C^*(l)} \tag{11.68}$$

As listed in Table 11.19, these adjustment factors vary with the sediment grain size and the flow parameters. For the cases of $u^* = 2, 5, 10$ cm/s, the averaged adjustment factor $\overline{K}_{1.D.l}$ is 1.083 for $h = 2$ m, and for $h = 10$ m. Under the same conditions of

Table 11.18 Computed $S_E^*(l) = [0.75S_0^*(l) + 0.25S_1^*(l)]$ based on the statistical theory for carrying-sediment capacity [4]

u_*	H	Group 1	Group 2	Group 3	Group 4	Group 5	Group 6	Group 7	Group 8
0.005	1	0.5435	0.4138	0.2198	0.0719	0.0059			
0.005	2	0.2717	0.2069	0.1100	0.0362	0.0030			
0.005	5	0.1087	0.0828	0.0440	0.0146	0.0012			
0.005	10	0.0544	0.0414	0.0221	0.0073	0.0006			
0.005	20	0.0272	0.0207	0.0110	0.0037	0.0003			
0.01	1	2.1362	1.8544	1.3607	0.6911	0.1055	0.0038		
0.01	2	1.0682	0.9272	0.6806	0.3473	0.0538	0.0019		
0.01	5	0.4273	0.3709	0.2723	0.1392	0.0218	0.0008		
0.01	10	0.2136	0.1855	0.1362	0.0698	0.0110	0.0004		
0.01	20	0.1068	0.0927	0.0681	0.0349	0.0055	0.0002		
0.02	1	7.8259	6.9482	4.7795	2.7966	0.9594	0.1194	0.0056	
0.02	2	3.9130	3.4852	2.7059	1.5909	0.5195	0.0623	0.0029	
0.02	5	1.5652	1.3942	1.1523	0.8013	0.2579	0.0276	0.0012	
0.02	10	0.7826	0.6971	0.5768	0.4260	0.1592	0.0158	0.0006	
0.02	20	0.3913	0.3486	0.2884	0.2136	0.0860	0.0092	0.0003	
0.05	1	31.5104	22.6797	17.8775	18.7512	12.0662	3.1243	0.6306	0.0680
0.05	2	18.4557	12.5592	9.0370	7.5023	6.0770	1.5563	0.3198	0.0347
0.05	5	8.5039	6.5575	4.0051	3.0139	1.9333	0.6186	0.1304	0.0143
0.05	10	4.2827	3.7909	2.4567	1.6412	0.9814	0.3106	0.0664	0.0072
0.05	20	2.1414	1.9220	1.5563	1.0119	0.5385	0.1599	0.0342	0.0037
0.1	1	84.1837	95.3238	80.9958	69.9153	53.7994	28.2358	9.8804	2.4933
0.1	2	43.2377	37.9777	40.6767	35.0324	26.9469	14.1719	4.9852	1.2656
0.1	5	19.9629	15.5032	16.5268	14.1068	10.8216	5.6984	2.0158	0.5150
0.1	10	12.6320	8.7054	6.5384	7.1457	5.4445	2.8632	1.0159	0.2456
0.1	20	7.5764	5.5515	3.5139	2.8308	2.7600	1.4421	0.5124	0.1167
0.25	1	567.2580	508.8625	440.6574	390.6322	331.2739	238.9775	144.2021	63.0472
0.25	2	284.5635	254.8192	220.4716	195.3965	165.7152	119.6387	72.3626	31.7980
0.25	5	115.1309	102.4209	88.3515	78.2400	66.3428	47.9308	29.0665	12.8440
0.25	10	45.8153	51.6883	44.3135	39.1818	33.2057	23.9953	14.5747	6.4631
0.25	20	23.7335	20.6497	22.3097	19.6526	16.6319	12.0153	7.3073	3.2502

u_*, the averaged adjustment factor $\Delta\overline{K}_{1.E.l}$ is 0.880 for $h = 2$ m, and is 0.869 for $h =$ 10 m, which shows negligible difference. After further averaging for various water depths, it is determined that $\Delta K_{1.E} = 0.874$, and $K_{1.D} = 1.083$.

For the weak equilibrium condition, the effective bed material gradation $P_{1.l}$ can be approximated by the bed material gradation $P_{1.l.1}$. Thus.

$$S_l^* = P_{4.l}^* S^* = \mu S_{0.l}^* + (1-\mu) S_{1.l}^* = \mu P_{1.l.1} S_0^*(l) + (1-\mu) P_{1.l.1} S_1^*(l)$$

Table 11.19 Adjustment factor for carrying-sediment capacity

u_*	Water depth h (m)	D 0.005 mm	0.01 mm	0.025 mm	0.05 mm	0.10 mm	0.25 mm	0.5 mm	1.0 mm	Mean $K_{1.D}$
Adjustment factor for deposition $\Delta K_{1.D.l}$										
	2	1.11	1.11	1.11	1.11	1.11	1.11	0	1.11	1.11
0.02	10	1.11	1.11	1.11	1.11	1.11	1.11	1.11	1.11	1.11
0.05	2	1.11	1.11	1.1.1	1.09	1.00	1.09	1.10	1.11	1.11
	10	1.11	1.11	1.11	1.11	1.11	1.10	1.11	1.10	1.11
0.10	2	1.11	1.09	1.00	1.00	1.00	1.00	1.00	1.00	1.03
	10	1.11	1.11	1.10	1.00	1.00	1.00	1.00	1.02	1.03
Adjustment factor for scour $\Delta K_{1.E.l}$										Mean $K_{1.E}$
0.02	2	0.833	0.833	0.833	0.833	0.833	0.834	0.852	0	0.837
	10	0.833	0.833	0.833	0.833	0.834	0.836	0.857	0	0.837
0.05	2	0.833	0.834	0.840	0.863	1.00	0.869	0.844	0.834	0.865
	10	0.833	0.833	0.833	0.834	0.840	0.846	0.841	0.833	0.837
0.10	2	0.837	0.859	1.00	1.00	1.00	1.00	1.00	1.00	0.962
	10	0.833	0.834	0.844	1.00	1.00	1.00	1.00	0.965	0.934

$$= P_{1.l.1}\left[\mu S_0^*(l) + (1-\mu)S_1^*(l)\right] \tag{11.69}$$

which leads to

$$S^* = \mu S_0^* + (1-\mu)S_1^*$$

$$P_{4.l}^* = P_{1.l.1}\frac{\mu S_0^*(l) + (1-\mu)S_1^*(l)}{\mu S_0^* + (1-\mu)S_1^*} \tag{11.70}$$

It is seen that the actual sediment concentration S satisfies the weak equilibrium condition, $S = S^*$. The recommended values of μ are presented in Eqs. (11.64)–(11.66) for three sedimentation conditions in a general alluvial channel. These values have been thoroughly analyzed and compared, and can be used as a reference. In reality, the variation of μ is very complicated. It is hard for μ to remain constant, which results in the multi-valued nature of carrying-sediment capacity under the weak equilibrium condition. Of course, if μ is given, the carrying-sediment capacity is single-valued.

11.4.5 Multi-valued Nature of Carrying-Sediment Capacity Caused by the Multi-valued Saturation Recovery Coefficients

As presented in Eq. (11.45), for uniform sediment,

$$S^{**} = \frac{\alpha^*}{\tilde{\alpha}} S^*$$

is called the carrying-sediment capacity for scour or deposition, because it is the carrying-sediment capacity that actually controls the sedimentation condition. Here S_2^{**} in Eq. (11.45) is rewritten as S^{**}. It was stated in Sect. 11.4.2 that S^{**} does not impact the multi-valued nature of sediment transport capacity for the strong equilibrium condition. This statement will be further verified in this section. It was proved in Chap. 6 that $\alpha^*/\tilde{\alpha} > 1$ for deposition, and $\alpha^*/\tilde{\alpha} < 1$ for scour. Based on the equation for non-equilibrium sediment transport, $\tilde{\alpha} S - \alpha^* S^* > 0$ for deposition, and $\tilde{\alpha} S - \alpha^* S^* < 0$ for scour. Combining these two sets of relationships, it can be derived that for the case of deposition

$$1 < \frac{\alpha^*}{\tilde{\alpha}} < \frac{S}{S^*} \tag{11.71}$$

and for the case of scour

$$1 > \frac{\alpha^*}{\tilde{\alpha}} > \frac{S}{S^*} \tag{11.72}$$

Or, for deposition

$$S_* < \frac{\alpha^*}{\tilde{\alpha}} S^* = S^{**} < S \tag{11.73}$$

and for scour

$$S^* > \frac{\alpha^*}{\tilde{\alpha}} S^* = S^{**} > S \tag{11.74}$$

It is noted that S^{**} is always between S^* and S no matter for scour or for deposition. The characteristics of Eqs. (11.71)–(11.74) are illustrated in Table 11.20. This table lists the computed $\alpha_l^*/\tilde{\alpha}_l$ for various sediment grain size D, S/S^* and ω/u_*. The results in this table are extracted from Tables 6.7 through 6.14, and all satisfy Eqs. (11.71) and (11.72).

An important conclusion drawn from Eqs. (11.73) and (11.74) is that S^{**} does not equal S^* no matter for scour or for deposition. As a result, it does not impact the

Table 11.20 Examples for relationship between $\alpha_l^*/\tilde{\alpha}_l$ and S/S^*

$\frac{\omega}{u_*}$	D (mm)	$\alpha_l^*/\tilde{\alpha}_l$ for various S/S^*							
		0.3	0.5	0.8	1.00	1.078	1.166	2.456	5.00
0.01	0.01	0.652	0.823	0.950	1.000	1.015	1.030	1.030	
0.1	0.01	0.536	0.747	0.926	1.000	1.023	1.046	1.216	
0.5	0.01	0.424	0.635	0.876	1.000	1.042	1.086	1.473	
2.0	0.01	0.449	0.656	0.884	1.000	1.039	1.081	1.451	
0.01	0.05	0.652	0.830	0.951	1.000	1.015	1.030	1.030	
0.1	0.05	0.536	0.747	0.926	1.000	1.024	1.046	1.217	
0.5	0.05	0.423	0.635	0.876	1.000	1.042	1.086	1.473	
2.0	0.05	0.449	0.656	0.884	1.000	1.039	1.081	1.451	
0.01	0.10	0.652	0.823	0.950	1.000	1.015	1.031	1.031	
0.1	0.10	0.536	0.748	0.927	1.000	1.023	1.048	1.215	1.300
0.5	0.10	0.424	0.635	0.876	1.000	1.042	1.086	1.473	1.739
2.0	0.1	0.449	0.656	0.884	1.000	1.039	1.081	1.451	1.721
0.01	0.00	0.652	0.822	0.950	1.000	1.015	1.029	1.029	
0.1	0.00	0.535	0.746	0.926	1.000	1.022	1.045	1.211	1.310
0.5	0.00	0.424	0.634	0.876	1.000	1.041	1.086	1.472	1.743
2.0	0.00	0.449	0.656	0.884	1.000	1.039	1.081	1.451	1.721

relation of $S = S^*$, and does not cause the multi-valued nature of sediment transport capacity for the weak equilibrium condition, However, for a general condition with scour or deposition, the sediment transport capacity is multi-valued.

The following observations can be made from Table 11.20:

(1) (1) $\alpha^*/\tilde{\alpha}$ decreases as S/S^* decreases for the case of scour, and $\alpha^*/\tilde{\alpha}$ increases as S/S^* increases.

(2) While α^* and $\tilde{\alpha}$ vary with S/S^* and ω/u_*, $\alpha^*/\tilde{\alpha}$ is almost independent of the sediment grain size.

(3) For the case of scour, $\alpha^*/\tilde{\alpha}$ reaches the minimum at approximately $\omega/u_* = 0.50$. For $\omega/u_* < 0.50$, $\alpha^*/\tilde{\alpha}$ decreases as ω/u_* increases. For $\omega/u_* > 0.50$. $\alpha^*/\tilde{\alpha}$ increases as ω/u_* increases. However, $\alpha^*/\tilde{\alpha}$ does not vary significantly with ω/u_*.

(4) For the case of deposition, as listed in Table 11.20 and illustrated by further analysis, $\alpha^*/\tilde{\alpha}$ reaches the maximum for approximately $0.5 \leq \omega/u_* \leq 1$. For $\omega/u_* \leq 0.5 \sim 1.0$, $\alpha^*/\tilde{\alpha}$ increases as ω/u_* increases. For $\omega/u_* < 0.50$, $\alpha^*/\tilde{\alpha}$ decreases as ω/u_* increases.

(5) Since $\alpha^*/\tilde{\alpha}$ is independent of the sediment grain size, $\alpha^*/\tilde{\alpha}$ are computed for more values of ω/u_* and S/S^*, based on $D = 0.1$ mm. The results are listed in Table 11.21. The mean values of $\alpha^*/\tilde{\alpha}$ averaged for various ω/u_* are also listed in this table. In applications, $\alpha^*/\tilde{\alpha}$ can be looked up from Table 11.20 for a given S/S^*, and then S^{**} can be determined.

Table 11.21 Examples for relationship between $\alpha_l^*/\tilde{\alpha}_l$ and S/S^* ($D = 0.1$ mm)

$\frac{\omega}{u_*}$	$\alpha_l^*/\tilde{\alpha}_l$ for various S/S^*							
	0.3	0.5	0.8	1.00	1.076	1.166	2.456	5.000
0.001	0.665	0.830	0.953	1.000	1.003	1.003	1.003	
0.005	0.660	0.827	0.952	1.000	1.015	1.015	1.015	
0.01	0.652	0.823	0.950	1.000	1.015	1.031	1.031	
0.05	0.596	0.718	0.891	1.000	1.017	1.035	1.172	
0.100	0.536	0.748	0.927	1.000	1.023	1.048	1.215	1.300
0.500	0.424	0.635	0.876	1.000	1.042	1.086	1.473	1.739
1	0.438	0.646	0.880	1.000	1.040	1.084	1.474	1.761
2	0.449	0.656	0.884	1.000	1.039	1.081	1.451	1.721
Mean	0.553	0.735	0.914	1.000	1.024	1.048	1.230	1.630

More generally, the mean value of $\alpha^*/\tilde{\alpha}$ averaged over the results as listed in Table 11.20 is approximately 0.734 for scour, and is approximately 1.233 for deposition, These mean values can be used to estimate the averaged difference in carrying-sediment capacity for the case of scour or deposition.

The virtual carrying-sediment capacity for non-uniform sediment is

$$S_1^{**} = \frac{\omega^* \alpha_m^*}{\tilde{\alpha}_m} S^*$$

where ω^* and ω the average velocities calculated by Eqs. (11.36) and (11.37) respectively are independent of recovery saturation coefficient. The data in Table 11.20 show that $\tilde{\alpha}_m/\alpha_m^*$ independent of particle size, is only related with ω/u^* and S/S^*, thus Table 11.20 is also available for non-uniform sediment.

11.4.6 Composite Impacts of the Multi-valued Natures of Threshold Flow Velocities and of the Size Distribution on Carrying-Sediment Capacity

The impacts of the multi-valued natures of the size distribution for carrying-sediment capacity and of the threshold flow velocities on the carrying-sediment capacity are investigated in Sects. 11.4.3 and 11.4.4, respectively. The impact of the multi-valued threshold flow velocities is represented in the composition of the carrying-sediment capacity, i.e., the summation of two parts of the carrying-sediment capacity. On the other hand, the size distribution is also a representation for the composition of carrying-sediment capacity. Hence, it is reasonable to analyze the combined impacts of the multi-valued natures between the threshold flow velocities and the size distribution on the multi-valued nature of carrying-sediment capacity. By doing this, more theoretically profound results can be illustrated.

Considering the two parts of the carrying-sediment capacity, which are associated with the sediment locally eroded from the bed and with the sediment from upstream, respectively, Eq. (7.134) becomes

$$S^*(\omega^*) = \begin{cases} P_{4.1}S + P_{4.2}S\dfrac{S_0^*(\omega_2^*)}{S_0^*(\omega_{1.1}^*)} + \left[1 - \dfrac{P_{4.1}S}{S_0^*(\omega_1)} - \dfrac{P_{4.2}S}{S_0^*(\omega_{1.1}^*)}\right]P_1 S_0^*(\omega_{1.1}^*) \\ \qquad \left(\dfrac{P_{4.1}S}{S_0^*(\omega_1)} + \dfrac{P_{4.2}S}{S_0^*(\omega_{1.1}^*)} < 1\right) \\ P_{4.1}S + \left[1 - \dfrac{P_{4.1}S}{S_0^*(\omega_1)}\right]S_0^*(\omega_2^*) \quad \text{(other conditios)} \end{cases}$$

where $S_0^*(\omega_2^*)$ is the part of the carrying-sediment capacity associated with the coarser suspended particles that just settle on the bed, without being impacted by the cohesive force or the additional pressure of film water, which is denoted with the subscript "0"; $S_0^*(\omega_1)$, with the subscript "0", is the part of the carrying-sediment capacity associated with the sediment influx from upstream; $S_0^*(\omega_{1.1}^*)$ is computed based on the bed material, but is also denoted with subscript "1" because it corresponds to the coarser suspended particles that settle on the bed and subsequently are eroded. Based on Eq. (11.19), the carrying-sediment capacity associated with the bed material is

$$\begin{aligned} S^*(\omega_{1.1}) &= \sum P_{1.l.1.1}S^*(l) \\ &= \sum P_{1.l.1.1}\left[\mu S_0^*(l) + (1-\mu)S_1^*(l)\right] \\ &= P_{1.l.1.1}\left[\mu + (1-\mu)\frac{\Omega_{1.l}}{\Omega_{0.l}}\right]S_0^*(l) \end{aligned} \quad (11.75)$$

where μ is the proportion of the bed material for the coarser particles that are not impacted by the cohesive force or the additional pressure of film water, and the proportion for the finer sediment is $(1-\mu)$.

As pointed out before, Eq. (7.134) does not lead to $S^* = S$ except for a special case. Hence, it does not change the weak equilibrium condition. However, this equation clearly shows the parts of the carrying-sediment capacity and their relationships to S_0^* and S_l^* from the mechanism aspect. μ only affects the last term of this equation. Under a general condition, this term very small and μ can be determined based on the bed material gradation. For example, assuming $P_1 = 1$, the cohesive force and the additional pressure of film water can be ignored for $D = 0.1$ mm or greater, and the percent of the bed material with $D < 0.10$ mm is 80%, or $\mu = 0.8$, then Eq. (7.134) can be written as

$$
S^*(\omega^*) = \begin{cases} P_{4.1}S + P_{4.2}S\dfrac{S_0^*(\omega_2^*)}{S_0^*(\omega_1^*)} + \left[1 - \dfrac{P_{4.1}S}{S_0^*(\omega_1)} - \dfrac{P_{4.2}S}{S_0^*(\omega_1^*)}\right] \\ \quad \times\left[0.8S_0(\omega_1^*) + 0.2S_1^*(\omega_1^*)\right] \\ \qquad \left(\dfrac{P_{4.1}S}{S*(\omega_1)} + \dfrac{P_{4.2}S}{S*(\omega_{1.1}^*)} < 1\right) \\ P_{4.1}S + \left[1 - \dfrac{P_{4.1}S}{S_0^*(\omega_1)}\right]S_0^*(\omega_2^*) \quad \text{(other conditions)} \end{cases} \tag{11.76}
$$

It should be emphasized that $S_0^*(\omega_1^*) \neq S_1^*(\omega_1^*)$ can be derived not only from Eqs. (11.6) and (11.7), but also from Eq. (11.11) after considering the variation of K with scour or deposition.

Let use the data as listed in Table 11.14 as an example. Based on the last column of this table, $S_0^*(\omega) = 2.0\,\text{kg/m}^3$, $S_0^*(\omega_1) = 38.18\,\text{kg/m}^3$, $S_0^*(\omega_2^*) = 0.998\,\text{kg/m}^3$, $S_0^*(\omega_1^*) = 1.08\,\text{kg/m}^3$, Assuming $S_0^*(\omega_1^*)/S_1^*(\omega_1^*) = 2$, then $S_1^*(\omega_1^*) = 0.54\,\text{kg/m}^3$ Based on Eq. (11.76), $S^*(\omega_1^*) = 0.8 \times 1.08 + 0.2 \times 0.54 = 0.972\,\text{kg/m}^3$, and $S^*(\omega^*) = 1.606\,\text{kg/m}^3$. This is less than $S^*(\omega^*) = 2\,\text{kg/m}^3$ for the strong equilibrium condition, indicating that the carrying-sediment capacity decreases for scour.

11.4.7 Variation Range of Multi-valued Carrying-Sediment Capacity Compared with the Field Data of the Lower Yellow River

The composition and size distribution for carrying-sediment capacity only slightly affect the carrying-sediment capacity in alluvial streams, and there is barely any representative field data being measured for this size distribution. Hence, the impact of the size distribution for carrying-sediment capacity is temporally ignored, and only the multi-valued nature of carrying-sediment capacity that is caused by the multi-valued threshold flow velocities and saturation recovery coefficient are discussed in this section. While the mechanisms for the multi-valued natures are different between the threshold flow velocities and the saturation recovery coefficients, they both adjust the carrying-sediment capacity. Based on Table 11.19, the mean adjustment factors for carrying-sediment capacity caused by the mulita-valued threshold flow velocities, averaging for various water depths, are

$$
\Delta K_{1.D} = \frac{S_{D.I}^*}{S_{C.I}^*} = 1.083
$$

$$
\Delta K_{1.E} = \frac{S_{E.I}^*}{S_{C.I}^*} = 0.874
$$

Based on Table 11.21, the adjustment factors associated with the saturation recovery coefficient are

$$\Delta K_{2.D} = \frac{\alpha_l^*}{\tilde{\alpha}_l} = 1.233$$

$$\Delta K_{2.E} = \frac{\alpha_l^*}{\tilde{\alpha}_l} = 0.734$$

Considering these two causes for the multi-valued nature, the composite adjustment factors are

$$\Delta K_D = 1.083 \times 1.233 = 1.335$$

$$\Delta K_E = 0.874 \times 0.734 = 0.642$$

These theoretical values are approximately consistent with those derived based on the field data of the lower Yellow River, which are $\Delta K_D = 1.25$ and $\Delta K_E = 0.75$ for the Sanmenxia Reservoir based on Table 11.6, and are $\Delta K_D = 1.68$ and $\Delta K_E = 0.68$ for the downstream reach of the Yellow River based on Tables 11.11, 11.12 and 11.13. It is concluded that the theoretical results of the multi-valued nature of carrying-sediment capacity agrees with reality. In addition, the mulita-valued nature of carrying-sediment capacity caused by the threshold flow velocities has been successfully applied to the calculation of channel scour downstream of the Three Georges Reservoir [4].

11.4.8 Opinions on the Multi-valued Nature of Carrying-Sediment Capacity

1. **Difference and Similarity in the Impacts of the Three Multi-valued Factors on the Multi-valued Nature of Carrying-Sediment Capacity**

   (1) The multi-valued nature of carrying-sediment capacity depends on how the carrying-sediment capacity is defined. If it is defined with Eq. (11.22) for the strong equilibrium condition, or approximately by $S = S^*$, the multi-valued nature of carrying-sediment capacity that is caused by the multi-valued threshold flow velocities becomes obvious. This is particularly true when the conditions for four sediment state transitions with the four corresponding probabilities, and the variation of μ are considered. If the sediment transport capacity is defined with Eq. (11.5), together with Eqs. (11.6) and (11.7), and μ is assumed to be constants, then the carrying-sediment capacity has a form with a single-valued nature. For this case, the formula for carrying-sediment capacity is hard to use because of the new parameter that needs to

be calibrated. Rather than defining the carrying-sediment capacity single-valued, it is better to acknowledge the multi-valued nature of carrying-sediment capacity, and adjust the results computed with existing formulas to incorporate the impact of this multi-valued nature, to make sure the results are reliable in applications.

(2) Because of the characteristics as shown in Eqs. (11.73) and (11.74), the impact of the multi-valued nature of the saturation recovery coefficient on carrying-sediment capacity does not affect the carrying-sediment capacity and corresponding size distribution, and thus does not result in multi-valued carrying-sediment capacity for the strong equilibrium condition. However, it causes the multi-valued nature of carrying-sediment capacity for the case of scour or deposition and for the weak equilibrium condition.

(3) The impact of the multi-valued natures of the size distributions for carrying-sediment capacity and for suspended load on carrying-sediment capacity, which is described in Eq. (7.134) or (11.64), does not change the condition or the single-valued nature of carrying-sediment capacity for strong equilibrium sediment transport. However, it reflects the difference in carrying-sediment capacity under various sedimentation patterns and the weak equilibrium condition.

2. **Relativity for the Multi-valued Nature of Carrying-Sediment Capacity**

Due to complicated nature of the capacity for flow to carry sediment, it has been a long-time effort to illustrate its characteristics. The general procedure for the investigation includes discovering the contradictory or multi-valued nature of carrying-sediment capacity, analyzing and solving the mulita-valued nature, and then resuming to the single-valued carrying-sediment capacity. For example, a simple relationship between S^* and V, as shown in Fig. 11.1, was developed by Rosinskee and Kuzmin [18] without considering the impacts of the water depth and the flow velocity. This relationship is not single-valued, and Rosinskee and Kuzmin attributed this double-valued nature of carrying-sediment capacity to different causes. After Velikanov and Zhang proposed the relationship for $S^* \sim (V^3/gh\omega)$, the impact of the flow velocity, the water depth and the sediment fall velocity on the multi-valued nature of carrying-sediment capacity is eliminated. The carrying-sediment capacity becomes single-valued under certain conditions, as shown in Eq. (11.11) or similar formula in the form of $S^* = f[V(\tau), h, \omega, V_{b.c}(\tau)]$. Sha [20] proposed the multi-valued nature of carrying-sediment capacity caused by threshold flow velocities, and made the carrying-sediment capacity single-valued by using the empirical values of varying coefficient for carrying-sediment capacity.

As discussed in this chapter, the multi-valued nature of carrying-sediment capacity is mainly caused by the multi-valued saturation recovery coefficient, the multi-valued threshold flow velocities, and the composition variation of carrying-sediment capacity. The multi-valued natures of the first two factors have been quantified. After these results are acknowledged, and corresponding parameters are introduced as independents, the multi-valued nature of carrying-sediment capacity can further be eliminated. The multi-valued composition of carrying-sediment capacity affects the multi-valued nature of carrying-sediment through the variation of μ, which is used to combine the two parts of carrying-sediment capacity, S_1^* and S_0^*, for the weak equilibrium condition. If further analysis is conducted to reasonably determine the values of μ based on various criteria, the multi-valued nature of carrying-sediment capacity can further be illustrated, and the carrying-sediment capacity can then be theoretically changed to single-valued.

In summary, the multi-valued nature of carrying-sediment capacity is relative. The progress for the investigation of carrying-sediment capacity is in fact a procedure of discovering multi-valued nature, investigating the mechanism of this multi-valued nature and solving for its formulation, and then further changing it to single-valued. The analysis is conducted in this chapter following this procedure, with special attention paid to the definition of the multi-valued nature of carrying-sediment capacity, and to its difference from the properties of carrying-sediment capacity under various sedimentation conditions.

References

1. Cao RX, Deng XY, Qian SQ, Li Y, Chen XY (1999) Study for double-valued relation for sediment transport capacity. Res Report Xi'an Univ Technol (in Chinese)
2. Chien N, Wan ZH (1983) Mechanics of sediment transport. Science Publishing House, Beijing (in Chinese)
3. Experts Group of Sedimentation Engineering (EGSE) of Three Gorges Reservoir, Chinese Academy of Engineering (2008). Appraisal report of sedimentation engineering of Three Gorges Reservoir (in Chinese)
4. Guan JC (2009) Carrying-sediment capacity of non-uniform suspended load of non-equilibrium transport. Ph.D. Dissertation, China Institute of Water Resources and Hydropower Research (IWHR), Beijing, China (in Chinese)
5. Han QW (1979) A Study of the non-equilibrium transportation of non-uniform suspended load. Chin Sci Bull 24(17):804–808 (in Chinese)
6. Han QW (1999) Premier study on river channel scour and evolution downstream dam site of Three Gorges Reservoir (175-scheme). In: Proceedings of key technology of sedimentation and navigation of Three Gorges Reservoir (175-scheme), vol 2. Wuhan University of Industry and Technology Press, pp 75–95
7. Han QW (2003) Reservoir sedimentation. Science Press, Beijing (in Chinese)
8. Han QW (2004) Rules of sedimentation and sediment transport in the lower Yellow River. J Sediment Res 3:1–13 (in Chinese)
9. Han QW (2005) Size distribution of carrying-sediment capacity, In: Proceedings of the national symposium on basic theory of sedimentation, vol. 1. Yellow River Press, Zhengzhou, China (in Chinese)
10. Han QW (2006) Boundary condition of diffusion equation and coefficient of saturation recovery. J Changsha Univ Sci Technol 3:7–19 (in Chinese)
11. Han QW (2008) Transport capacity expression of the lower Yellow River. Yellow River (11):1–2 (in Chinese)
12. Han QW (2009) Theoretical analysis of first channel-forming discharge and transport capacity. J Yellow River 31(1):1–4 (in Chinese)
13. Han QW, Chen XJ (2008) Theory and calculation method of coefficient of saturation recovery. J Sediment Res 6:8–16 (in Chinese)

14. Han QW, Guan JC (2009) Multi-valued nature of carrying-sediment capacity and characteristic of 'more sediment discharge as more sediment coming' in lower Yellow River. Yellow River (3):1–4 (in Chinese)
15. He MM, Han QW (1990) Determination of the size distributions of carrying capacity and effective bed material. J Hydraul Eng (3):l–12 (in Chinese)
16. Hou HC (1982) Basic problems of river dynamic. Water Resources Press, Beijing (in Chinese)
17. Ministry of Water Resources of the People's Republic of China (MWR) Joint Adventure (2005) Study for control elevation at Tongguan and operation plans for Sanmenxia Reservoir. Technical Report (in Chinese)
18. Rosinskee KE, Kuzmin NA (1956) River Bed. J Sediment Res 1(1):115–151
19. Sanmenxia Bureau of Water Resources (SBWR) (2005) Application of the physical model experiments for Sanmenxia reservoir (November 2002 to October 2005) and Sedimentation Analysis. Technical Report (in Chinese)
20. Sha YQ (1965) Introduction to sediment dynamics. Industry Press, Beijing, pp 1–302 (in Chinese)
21. Yellow River Conservancy Commission (YRCC), Survey and Design Institute (1999) Characteristics of deposition and scouring in lower Yellow River. Res Report (in Chinese)
22. Zhang RJ, Xie JH, Wang MF, Huang JT (1989) Dynamics of river sedimentation. Water Resources and Electric-Power Press, Beijing (in Chinese)

Appendix A
Mechanics and Statistical Theory of Sediment Movement

The statistical theory developed by the authors is the foundation of the study on the non-equilibrium transport of non-uniform suspended load. Since most of these study results were published in Chinese, we have introduced the main contents in the appendix for the convenience of English readers.

A.1 Introduction

The movement of sediment is a highly complex process. On one hand, it follows the laws of mechanics and has a deterministic aspect. On the other hand, it has random characteristics owing to variations in the flow velocity, positions of particles on the bed surface, and non-uniform particle size. Since the middle of the last century, there has been great progress in the study of sediment movement through mechanics, probabilistic methods, and a combination of the two [1, 2, 5, 8–10, 12–17, 19, 27, 36, 38, 39, 40, 43–45, 47–50, 52–57, 60–62]. However, different methods based on various techniques have been adopted for different subjects, owing to the lack of systematic methods. Since the 1960s, the authors of this work have been developing a systematic approach that combines mechanics and probability theory, and this has led to some important results [23–26, 28].

Based on this research, particularly the theory of transition (exchange) intensities, some long standing problems have been solved theoretically.

There are parallels between this method and Einstein's. Einstein first constructed a model of the average bed load transport rate by combining mechanics with probabilistic methods [15, 16]. This pioneering work was the first to propose the concept of transition intensities from bed load to rest, and from rest to bed load. The equality of these transition intensities was used to derive an equation for the bed load transport rate. The structure of the resulting transition intensity equations, consisting of the state transition probability, step distance, and exchange period (entrainment time

Q. Han and M. He, *Non-equilibrium Transport of Non-uniform Suspended Load*, Hydroscience and Engineering, https://doi.org/10.1007/978-981-97-1123-9

should be more suitable), was reasonable. Moreover, the entrainment probability and mechanical conditions of incipient motion were introduced for the first time. From this, the entrainment probability was derived using the normal probability distribution density (PDD) of the lift force (normal PDD for flow velocity is more reasonable). Einstein's work was groundbreaking despite the fact that an intuitive probability method was adopted, rather than a stochastic process theory, and that the mechanical parameters and their expected values were mostly based on empirical assumptions, rather than theoretical derivations, and that some of these values were not reasonable. The proposed formula for the bed load transport rate was improved later, and it was used widely owing to its reasonable structure and ability to pass many tests with measured materials [35, 54].

Appendix A aims to provide a complete overview of the systematic approach to the analysis of sediment movement developed by the authors; it is divided into six Sections. Section A.2 describes the mechanisms of sediment movement, and provides a summary of the systematic approach. Then, the mechanics of step rolling, step hopping and step suspension and the expectations of the step parameters are derived along with the probabilities of transitions between states are discussed in Sects. A.3–A.5 respectively. Section A.6 confirms that the motion process is a Markov stochastic process with four random discrete states of motion and a continuous time (or distance) parameter based on the study of step motion. Thus, the transition intensities and transport rates are obtained.

A.2 Mechanisms of Single Particle Movement and Research Approaches

A.2.1 Mechanism of Sediment Motion

A precondition of this study was that particle motion can be separated into four categories: rest, rolling, hopping, and suspension. Moreover, each of these motion states can be divided into single steps and single events, as shown in Fig. A.1. Thus, dynamic equations for the different motion states can be established for each step. These equations can be used to find the mechanical parameters of single step motion, and the expectations can be derived from the PDD of the flow velocity and particle position. Under this condition, particle motion can be regarded as a Markov stochastic process with four discrete states and a continuous time parameter, so Markov process theory can be used.

Particles resting on the bed surface are called the bed material, rolling and hopping particles are called the bed load, and particles suspended in the water are called the suspended load. When the flow intensity is relatively low and the particles have low velocity, the particles generally roll after entrainment. As the flow intensity increases and the rolling particles reach a certain velocity, the particles begin to hop. Rolling

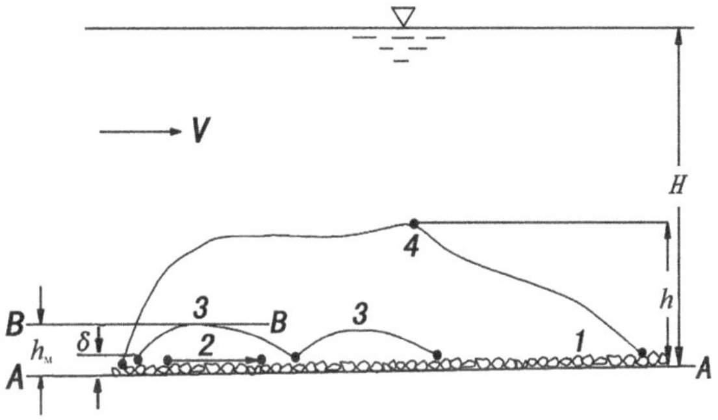

Fig. A.1 Schematic of rolling, hopping, suspension and rest particles

and hopping particles represent the total bed load. In practice, very little sliding motion is observed.

The stratified movement of bed surface particles, which occurs when the flow intensity is very high. In this case it is difficult to distinguish between hopping and rolling, so it will not be discussed here. The main motion state of the bed load is hopping, whereas the main motion state of coarse particles, such as pebbles and gravel, is rolling. When the water flow is intense, sediments can be suspended in the water and carried by turbulent vortices for a certain period of time. There are differences in the mechanical action of the bed load and suspended load. The former is mainly affected by hydrodynamic forces at the bottom of the flow (such as drag and lift forces) and the collision between particles, whereas the latter is mainly affected by the vertical velocity fraction of turbulent vortices. Therefore, the effect of the vertical velocity of turbulence is not considered when analyzing the bed load, and the bottom hydrodynamic force is not considered when analyzing the suspended load, except in a few specific cases.

In general, suspended loads, bed loads, and bed materials are all present in natural river channels. The sediment state changes constantly, and the probability of transition varies between states. Transitions between the suspension and hopping states can only occur in the bottom layer, as this is the limit of the hopping particles. For simplicity, state transitions are considered to occur only in the bed-surface layer, which is occupied by the bed material and rolling particles.

The motion of a single particle in a certain state can be divided into steps and events. For hopping and suspended particles, the threshold of a single step is the point at which the moving particle collides with the bed surface, where it may continue or change state. Rolling particles are most likely to transition to the rest state when they are at the lowest position on the bed, and they are most likely to transition to the hopping or suspension states when they are at the highest position on the bed. However, the distance between the two lowest positions and two highest positions are statistically identical. Therefore, the adjacent lowest position is used as the threshold to separate the motion steps of rolling particles. The threshold of a motion event is the contact point where the motion state actually changes. The motion of a particle from the beginning to the end of a motion state is called a motion event, and a motion event consists of several motion steps. The time period (or distance) of an event is termed as the life of sediment movement in certain state and its probability distribution can be derived from the expectation of step parameters and the transition probabilities between single step of the states. Then the transition intensities are obtained and the sediment moving process can be treated as a stochastic process.

A.2.2 Summaries and Flowchart of Research Approaches

The flowchart of research approaches of static theory of sediment movement is shown in Fig. A.2a, which can be summarized as follows:

(1) As a precondition, it can be assumed that sediment has four motion states, rest, rolling, hopping, and suspension, and each motion state can be divided into single steps and single events. Furthermore, in addition to the flow velocity and its probability distribution, the position of the particles on the bed surface and its probability distribution of uniform and non-uniform sediments are introduced as independent variables (Fig. A.2b). Consequently, no additional empirical coefficients are required to derive expressions for the sediment parameters, and all the results for uniform sediments can be extended to non-uniform sediments based on the position of non-uniform particles on the bed surface.

(2) Single motion steps can be considered as deterministic motion. Thus, mechanical methods can be used to study single step motion and establish dynamic equations that can be used to solve the motion parameters and transition conditions between different steps as functions of the flow velocity, particle position on the bed surface, and size grade of non-uniform sediment (Fig. A.2c).

(3) The steps can also be considered as a random process owing to the randomness of the flow velocity, particle position on the bed surface, and size grade of non-uniform sediment. From this, the probability distributions, expectations of step parameters, and basic transition probabilities between states can be derived theoretically (Fig. A.2c).

(4) The particle motion process can be treated as a Markov stochastic process with four discrete states and a continuous time parameter based on step motion research. The Markov stochastic process theory is used to obtain a series of important results related to sediment transport, including (Fig. A.2d) [23–25]: (a) A Markov chain with four discrete states, sixteen state transition probabilities and four limitation probabilities of states; (b) A Poisson process for the motion step and an exponential distribution for the step time period (or step distance); (c) Four life distributions (or distribution of event time period or of event distance) and twelve transition (or exchange) intensities; (d) Kolmogorov equation for the state transition probabilities in the Markov stochastic process with four discrete states and a continuous time parameter determined by the transition intensities; (e) The sediment transport rates derived from the limit of

the transition probabilities in the Kolmogorov equation as time t approaches infinity.

Fig. A.2 Diagram of research approach of statistical theory

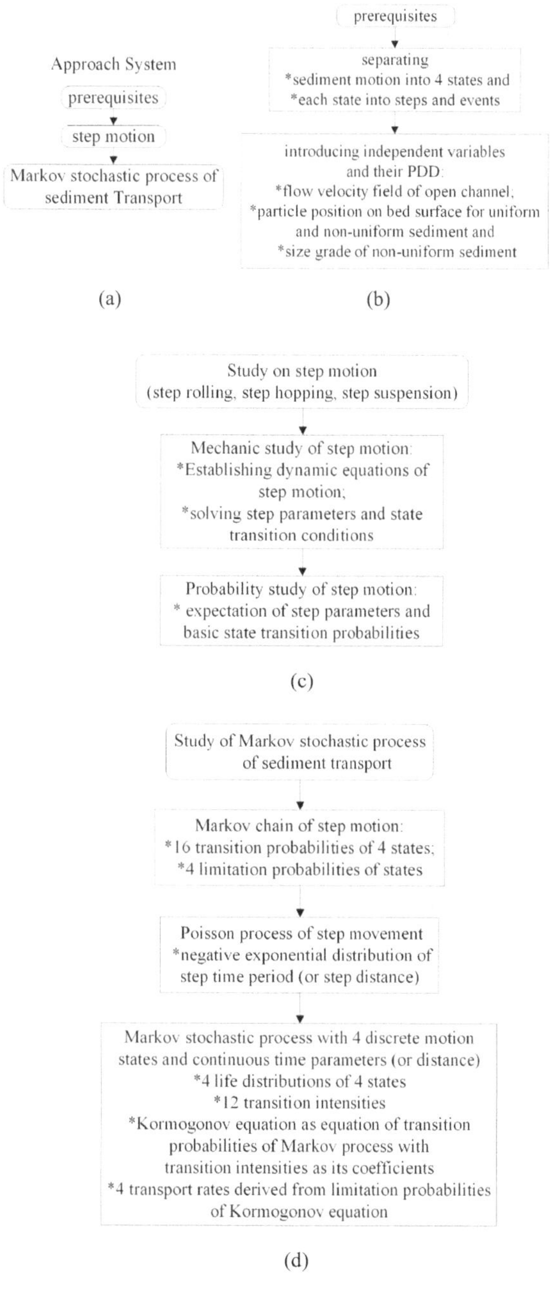

The division of motion states, each of which is divided into motion steps and events based on the sediment motion mechanism, is very important. Only in this way can the mechanical and probabilistic methods be used separately and in combination to provide all the theoretical equations for the expectations of step parameters which are used to find the transport rates and transition intensities. Moreover, the sediment transport process can be considered as a Markov stochastic process with discrete states and continuous parameters.

A.2.3 Assumptions for Simplification and Additional Assumption

1. Some assumptions are adopted for simplification: (1) The flow in the open channel is steady and uniform, and not affected by the movement of the sediment; (2) The flow and sediment movement are two dimensional quantities aligned with the longitudinal and vertical directions, and their variations over the width of the cross section are omitted; (3) The sediment particles are spherical; (4) The motion of the particles is independent except where there is contact between rolling and bed surface particles, or collisions between hopping and bed surface particles; (5) The sediment transportation is in equilibrium; (6) In general, this study is concerned with uniform sediment unless specified.
2. Additional assumption.

The first step in the proposed approach is to study single step motion mechanically. In rolling and hopping motion, the initial position of a particle is determined for each step, and it is assumed that the bottom flow velocity is constant. Thus, single step motion is deterministic, so dynamic equations can be determined for single step motion, and all the step parameters, such as the step distance, step time period, entrainment time, and conditions of state transition, can be expressed as functions of the flow velocity and initial particle position then solved.

For rolling particles, the step period and step distance are very short, so the flow velocity can be considered constant. For hopping particles, the rise and fall is determined by inertia and gravitational force once they ascend by the height of one particle. In contrast, the longitudinal movement is affected by the longitudinal flow velocity, and variations in longitudinal flow velocity over the hopping height should be considered. Thus, for hopping particles, the average step distance is dependent on the average longitudinal velocity. Moreover, although the suspension period and distance are relatively long, the bottom flow velocity only affects the incipient suspension time, which is very short. When establishing a relationship between suspended height and suspended load concentration, it is acceptable to assume that the distribution of particle positions below the suspended height is uniform with a first-order approximation. To calculate the average step distance of suspension, the expectation value for the flow velocity in the water body is used directly.

A.2.4 Flow Velocity, Particle Position and Size Distribution and Probability Distribution

A.2.4.1 Flow Velocities and PDDs

Particle motion is dependent on the flow velocity. Recent studies have shown that the skewness and kurtosis of the turbulent flow probability density function (PDF) cannot be ignored owing to the bursting phenomenon, which occurs near the boundary wall. The fourth-order Gram–Charlier PDF of the near-boundary streamwise velocity fluctuations has been expressed using universal probability distributions for turbulence in open channel flows [6, 8, 30]. However, the bursting phenomenon is highly complex and a quantitative description of its effect on particle motion has not yet been achieved [6, 7, 34]. In contrast, the expectations of the mechanical parameters of particle movement, derived from the normal distribution using empirical coefficients, are in good agreement with measured data. Therefore, the normal PDF of the flow velocity is used as an approximation in this research, and the bursting period is considered to be the only factor determining the incipient suspension period. This is discussed in Chap. 4 and in Sect. A.5. In addition, the probability distribution of the flow velocity only affects the expectation value of the mechanical parameters, and it is independent of the mechanical parameters and the structure of the formula for the transition intensities and transport rates. However, future advances in turbulence motion research will improve the study of sediment motion.

The terms ξ_x and η_x are used to express random variables in x. When ξ_{V_b} and $\xi_{V_{b.y}}$ are independent from each other, their PDDs are

$$\begin{aligned} p_{\xi_{V_b}}(V_b) &= \frac{1}{\sqrt{2\pi}\sigma_x}\exp\left[-\left(V_b-\overline{V}_b\right)^2\Big/\left(2\sigma_x^2\right)\right], \\ p_{\xi_{V_{b.y}}}\left(V_{b,y}\right) &= \frac{1}{\sqrt{2\pi}\sigma_y}\exp\left[-V_{b.y}^2/\left(2\sigma_y^2\right)\right] \end{aligned} \tag{A.2.1}$$

where $\sigma_x^2=\sigma_{V_b}^2$ and $\sigma_y^2=\sigma_{V_{b.y}}^2$ are the variance of longitudinal and vertical components of the bottom flow velocity, respectively. These values have previously been reported as $\sigma_x = 2.1u_*$, and $\sigma_y = 1.05u_*$ where u_* is the dynamic velocity [46]. The relationship between the average bottom flow velocity and the dynamic flow velocity at the top of the particles on the bed surface is $\overline{V}_b = 5.6u_*$. However, it is more appropriate to consider a point 2/3 D away from the bottom of the particles, where D is the depth of the particles, as an acting point for the bottom flow velocity, therefore

$$\overline{V}_b = 3.73u_*. \tag{A.2.2a}$$

The average longitudinal flow velocity in the vertical direction in the open channel is usually expressed as V. The relationship between average bottom flow velocity $\overline{V}_b$ and V is established through the dynamic velocity u_* [23, 26]. This is expressed as

$$\frac{V}{u_*} = \frac{V}{\sqrt{gJD}} = 6.5\left(\frac{H}{D}\right)^{\frac{1}{1+\log(H/D)}} = \psi\left(\frac{H}{D}\right) \tag{A.2.2b}$$

where H and D are the water depth and particle size, respectively. A comparison between Eqs. (A.2.2a, A.2.2b) and field data is shown in Fig. A.3. Hereafter, the longitudinal average flow velocity in the open channel V is abbreviated as the average flow velocity.

A.2.4.2 Particle Position on the Bed Surface

1. Particle position of uniform sediment

The position of a particle on the bed surface affects the force arm length of the torque exerted on rolling particles, and the collision force exerted on hopping particles. The concept of the particle position on the bed surface was first introduced by Han in 1965. Considering the velocity of a rolling particle, and a collision between a hopping particle and a particle on the bed surface, the position of particle A on the bed surface is defined as the vertical distance Δ from the lowest point on the particle a to the point of contact b between particle A and downstream particle B, as shown in Fig. A.4a. Here, Δ is a random variable with a uniform PDD given by

$$p_{\xi_{\Delta'}}(\Delta') = \begin{cases} \frac{1}{1-\Delta'_m} = 1.155, & (0.134 \le \Delta' \le 1), \\ 0, & (otherwise). \end{cases} \tag{A.2.3}$$

Here, $\Delta' = \Delta/R$ is a dimensionless variable and R is the radius of the particle. If $\Delta' > 1$ ($\Delta > R$), the particle is lower than the downstream particle and it is no longer a bed surface particle. Therefore, the range of Δ is $\Delta_m \le \Delta \le R$, Δ_m is the minimum value of Δ, when the studied particle is located on top of the two closely attached particles (see Fig. A.4b)

$$\Delta'_m = 1 - \cos\frac{\pi}{6} = 0.134 \tag{A.2.4}$$

Δ' is also the shielded value of the particle, and $1 - \Delta'$ is the prostration of the particle. Paintal [48] introduced the concept of particle exposure, which is defined as the distance from the top of the particle to the average bed surface level. Paintal considered both the shielding effects of adjacent upstream and downstream particle on the moving particle. In reality, particles on the bed surface are usually loosely packed, and the relative exposure of a downstream particle has a considerable effect on the particle motion, whereas the shielding effect of upstream particles is very small. Moreover, owing to the uncertainty in the average bed surface, the exposure can only be used with additional empirical coefficients, and the expressions for particle exposure adopted by Paintal were not reasonable. Dey et al. [8] and Lukerchenko et al. [38, 39] considered the hiding effect of downstream particles on rolling and

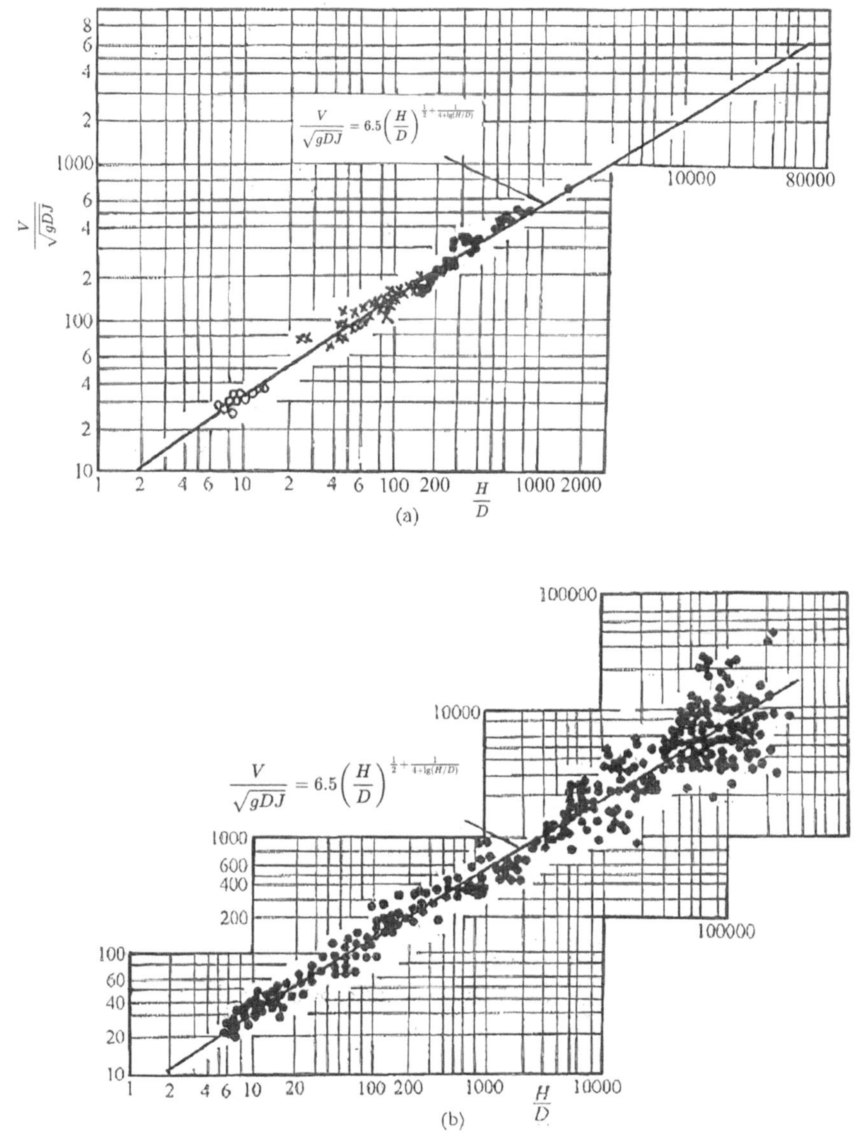

Fig. A.3 Relation between average flow velocity and dynamic velocity

jumping particles, but they still used the average bed surface to determine the particle position on the bed surface. Thus, the hiding effect is still considered in terms of the empirical hiding coefficient.

2. Particle position on the bed surface and probability distribution of non-uniform sediment

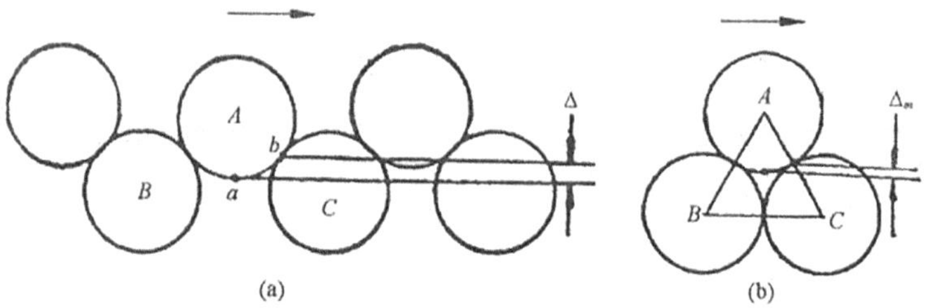

Fig. A.4 Diagram of positions of particles on bed surface

The particle position on the bed surface and the probability distribution of non-uniform sediment are very important. In the bed-surface layer, non-uniform particles usually vary by two (or even three) orders of magnitude, with a typical particle size range of 1 mm $\leq D \leq$ 100 mm. In pebble riverbeds, the bed-surface layer is controlled by the pebbles, which cannot be centralized piling, and sediments smaller than 1 mm fill the voids between larger particles. In sandy riverbeds, the bed-surface layer is controlled by sediments in the range 0.025 mm ~2 mm piling, and fine sediments smaller than 0.025 mm fill the voids.

For non-uniform sediments, the coarse particles generally do not accumulate centrally, so the relative position between coarse and fine particles increases the degree of exposure for coarse particles, and increases the degree of concealment for fine particles. Einstein [16] first introduced the concealment coefficient in his bed-load transportation rate formula, and later scholars [27, 53, 60] introduced further empirical coefficients to account for the exposure of non-uniform sediments. The authors of this paper introduced the concept of particle position on the bed surface and its probability distribution for non-uniform sediments [23, 24, 26]. This not only considers the effect of non-uniform sediment position on particle exposure, but also extends the study results for uniform sediments to non-uniform sediments.

Two types of non-uniform sediment position have been proposed by the authors. The second one [23] which is based on the first one [24] is simpler and more reasonable and will be described in this paper. The sediment can be classified into three grades with particle size ranges of $0.1 \leq \frac{\overline{D}}{D_l} \leq 0.6$, $0.6 \leq \frac{\overline{D}}{D_l} \leq 1.429$ and $1.429 \leq \frac{\overline{D}}{D_l} \leq 10$, for coarse, medium and fine size group respectively, The corresponding position ranges are $0.1 \leq \Delta_l' \leq 0.6$, $0.6 \leq \Delta_l' \leq 0.7$ and $0.7 \leq \Delta_l' \leq 1.0$ respectively. A simple expression that relates Δ_l' directly to $\overline{D}/D_l$ is introduced. For fine and medium size particles the corresponding interval of Δ_l' is very small, so it is acceptable to assume that Δ_l' is a decrease linear function with respect to $\overline{D}/D_l$. Thus, the relationship between Δ_l' and $\overline{D}/D_l$ can be expressed as

$$\Delta_l' = f_{\Delta_l'}\left(\frac{\overline{D}}{D_l}\right) = \begin{cases} \frac{\overline{D}}{D_l} & \left(\Delta_m' = 0.1 \le \frac{\overline{D}}{D_l} \le 0.6\right) \\ 0.6 + 0.121\left(\frac{\overline{D}}{D_l} - 0.6\right) & \left(0.6 \le \frac{\overline{D}}{D_l} \le 1.429\right) \\ 0.7 + 0.035\left(\frac{\overline{D}}{D_l} - 1.429\right) & \left(1.429 \le \frac{\overline{D}}{D_l} \le 10\right) \end{cases} \tag{A.2.5}$$

The PDF of particle position with particle size D_l is a one-point distribution given by:

$$F(\Delta_l') = \begin{cases} 0, & \left(\Delta_l' \le f_{\Delta_l'}(\frac{\overline{D}}{D_l})\right) \\ 1, & \left(\Delta_l' > f_{\Delta_l'}(\frac{\overline{D}}{D_l})\right) \end{cases} \tag{A.2.6}$$

This shows the great difference between the positions (and thus the incipient velocity) of particles in different size groups in the bed-surface layer, and it reflects practical observations. The particle location and the distribution of non-uniform sediment introduced by the authors differs from previous methods, such as those reported by [16, 17, 27]. These past works introduced a concealing coefficient to account for the effects of non-uniform sediments on the exposure of coarse and fine particles.

A.2.4.3 Number of Particles on the Bed Surface and Percentage Area of Non-uniform Sediments

To study the characteristics of sediments, the effects of the number of particles on the bed surface and the percentage area must be considered. According to Dou [11, 12, 13], sediments on the bed surface are loosely compacted (Fig. A.5) and, for uniform sediments, the static compactness factor in bed-surface layer is given by

$$m_0 = \frac{\pi}{4} D^2 \Big/ \left(\alpha D^2\right) = 0.4 \tag{A.2.7}$$

where α is the average distance between particles. The number of uniform sediment particles is related to the $m_0 = 0.4$ by the formula

$$n_0 = m_0 \Big/ \left(\frac{\pi}{4} D^2\right) \tag{A.2.8}$$

For non-uniform sediment, if R_l' is defined as the percentage area occupied by the l-th size group, then the particle number for the l-th size group in the unit area is given by

$$n_{0.l} = R_l' m_0 \Big/ \left(\frac{\pi}{4} D_l^2\right) \tag{A.2.9}$$

It can be demonstrated that the area percentage is equal to the volume percentage. In fact, in a volume of given thickness Δh, the number of layers occupied by particles of different sizes varies. For example, coarse particles occupy one layer, and fine particles may occupy two or more layers (Fig. A.6). Within the thickness Δh the number of layers and particles in the l-th size group are $K_l' = \frac{\Delta h}{D_l}$, and $K_l = K_l' n_{0,l}$ respectively. Substituting this into the expression for the percentage area (Eq. A.2.9) yields.

$$\begin{aligned} R_l' &= \frac{n_{0.l} D_l^2}{\sum_{l=1}^{n} n_{0.l} D_l^2} = \frac{n_{0.l} \Delta h D_l^2}{\sum_{l=1}^{n} n_{0.l} \Delta h D_l^2} \\ &= \frac{n_{0.l} K_l' D_l^3}{\sum_{l=1}^{n} n_{0.l} K_l' D_l^3} = \frac{K_l \frac{\pi}{6} D_l^3}{\sum_{l=1}^{n} K_l \frac{\pi}{6} D_l^3} \\ &= \frac{\gamma_s K_l \frac{\pi}{6} D_l^3}{\sum_{l=1}^{n} \gamma_s K_l \frac{\pi}{6} D_l^3} = P_{1.l} \end{aligned} \tag{A.2.10}$$

Therefore, the area percentage, volume percentage, weight percentage, and gradation of the bed material are all the same. Equation (A.2.9) can also be expressed as

$$n_{0,l} = m_0 R_{1,l} / \left(\frac{\pi}{4} D_l^2 \right) = n_0(l) R_{1,l} \tag{A.2.11}$$

where

$$n_0(l) = m_0 / \left(\frac{\pi}{4} D_l^2 \right) \tag{A.2.12}$$

which indicates the number of particles per unit area when all the particles have size D_l.

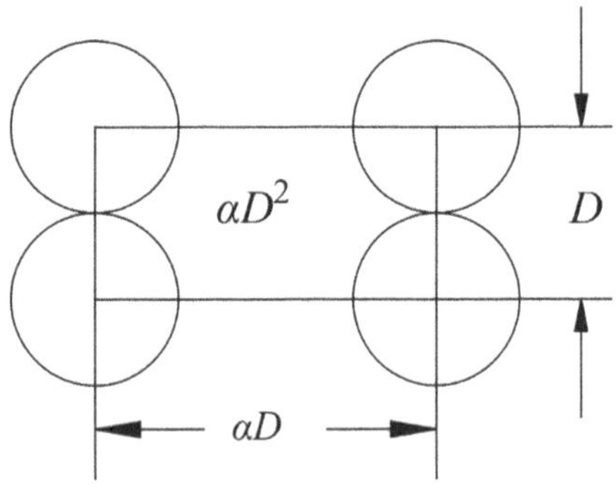

Fig. A.5 Arrangement of sediment on bed surface layer

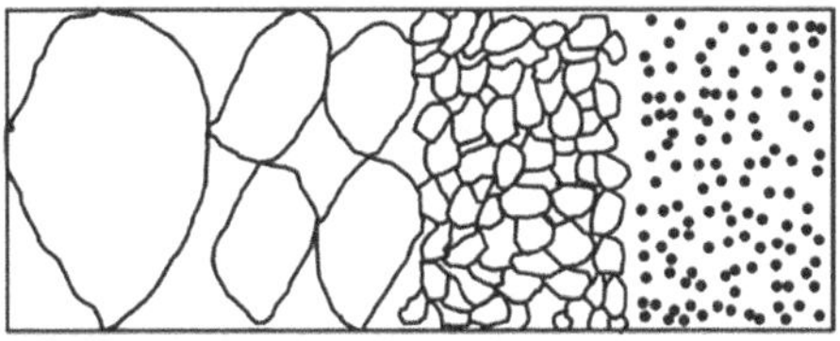

Fig. A.6 Layers of sediment with different size at bed surface layer

A.3 Step Rolling

A.3.1 Phenomenon and Mechanism of Particle Rolling

Particle rolling on the bed surface is extremely complicated and it is very difficult to study without appropriate simplifications. Progress has been made in the research of the rolling mechanism of Nakagawa at al [44], but there are some deficiencies in quantitative analysis. From the end of the last century, some research on the average rolling velocity and the rolling sediment transport rate was carried out with the help of high-speed imaging technology [43,49]. Han et al [23, 24] made an in-depth analysis of the mechanic and statistical characteristics of particle rolling and established a more complete statistical model of rolling. As shown in Fig. A.7, under the action of the instantaneous bottom flow velocity, particle *A* rolls around particle *B*, which is fixed to the bed surface. In this case, there are three possible scenarios:

(1) When the rolling velocity is low, particle *A* rolls around particle *B* and contact between the particles is maintained (Fig. A.7a).
(2) As the rolling speed increases, particle *A* may part from particle *B* and move freely under the action of centrifugal force and flow velocity. The point at which particle *A* is most likely to escape particle *B* is when it reaches the top of particle *B*, where the centrifugal force is in the vertical direction. Once particle *A* separates from particle *B*, the lifting height may be greater or less than the particle diameter. If the height exceeds the particle diameter, then the particle will no longer be in contact with the bed surface and will move freely. This is assumed to be hopping motion, which will be discussed in Part II. If the lift height is less than the particle diameter, then the particle movement is termed half-contact rolling (Fig. A.7b) to distinguish it from contact rolling.
(3) Finally, particle *A* may leave particle *B* at the beginning of a step and move freely under the action of the lifting force. If particle *A* moves at high speed, or if the lifting force is large, then particle *A* will be lifted to a height greater than the particle diameter, and it will transition to the hopping state. Otherwise, it will continue in a rolling state, and this is called non-contact rolling (Fig. A.7c).

A rolling step is defined as the roll between two adjacent lowest positions, or rolling between *A* and *A*, as shown in Fig. A.7. A particle rolling event can be divided into a number of steps based on the number of times the particle rolls over other particles on the bed. The distance covered by one rolling event is given by the summation of the distances covered by the rolling steps.

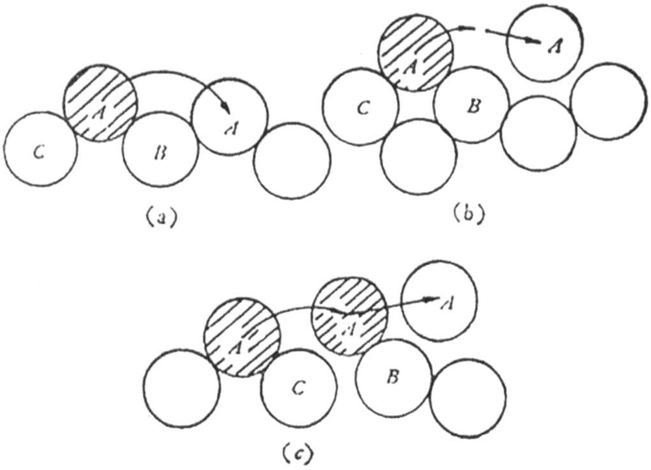

Fig. A.7 Three types of particle rolling

Based on the particle rolling phenomenon described above, a simplified model of particle rolling is developed assuming that: (1) The bottom flow velocity acting on the sediment particle remains constant for a rolling step; (2) Particle rolling is half-contact rolling; (3) The vertical displacement during rolling is much less than the longitudinal displacement and can be ignored; (4) The length of a rolling step is the rolling distance between two adjacent lowest positions; (5) The number of steps per rolling event is determined by the probability of continuous rolling.

It should be noted that the three types of rolling are simplified to the most common type (i.e., half-contact rolling), which also has medium rolling intensity. This is done to simplify the calculation of the rolling parameters, rather than to ignore the other types of rolling. As discussed later, all three types of rolling are considered when calculating the probability of a static particle transitioning to rolling motion.

A.3.2 Incipient Rolling of a Stationary Particle

A.3.2.1 Critical Condition and Flow Velocity for Incipient Rolling from a Stationary State

1. Critical condition and flow velocity for incipient rolling from a stationary state

If step rolling is considered to be a deterministic process, then the condition for incipient rolling for a particle at rest (Fig. A.8) is given by

$$P_x l_x + P_y l_y > G l_G. \tag{A.3.1}$$

Here, the forces exerted on the particle are the drag force P_x, lift force P_y, and underwater gravity G, which are and are expressed as

$$P_x = \frac{\rho C_x}{2}\pi R^2 V_b^2, \tag{A.3.2a}$$

$$P_y = \frac{\rho C_y}{2}\pi R^2 V_b^2, \tag{A.3.2b}$$

and

$$G = (\gamma_s - \gamma)\frac{4}{3}\pi R^3 = (\rho_s - \rho)g \cdot \frac{4}{3}\pi R^3, \tag{A.3.2c}$$

respectively. The terms $C_x = 0.4$ and $C_y = 0.1$ are the coefficients of the drag and lift forces, respectively; ρ and ρ_s are the densities of water and the particle, respectively; γ and γ_s are the specific gravities of water and the particle, respectively; and g is the acceleration due to gravity. In Eq. (A.3.1), l_x, l_y, and l_G are the force arms of P_x, P_y, and G about pivot point C given by

$$l_x = aR + R - \Delta, \tag{A.3.3a}$$

$$l_y = bR + \sqrt{R^2 - (R - \Delta)^2}, \tag{A.3.3b}$$

and

$$l_G = \sqrt{R^2 - (R - \Delta)^2} = \sqrt{2R\Delta - \Delta^2}, \tag{A.3.3c}$$

respectively.

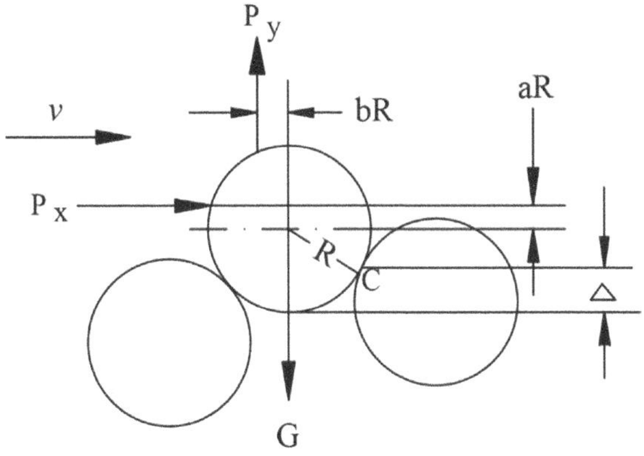

Fig. A.8 Diagrams of force for incipient particle rolling

With sufficient precision, it can be assumed that the particle velocity has a parabolic distribution for a given particle size, from which parameter $a = 1/3$ is obtained. Moreover, b can be approximated as 1/3 because the force moment of the lift force is much smaller than that of the drag force. Substituting the parameters from Eqs. (A.3.2a, A.3.2b, A.3.2c) and (A.3.3a, A.3.3b, A.3.3c) into Eq. (A.3.1) yields a critical flow velocity for incipient rolling of a particle at rest.

$$V_{b.c.0} = \omega_0 \varphi(\Delta / R) = \omega_0 \varphi(\Delta'), \tag{A.3.4}$$

where

$$\omega_0 = \sqrt{\frac{4}{3C_x} \frac{\rho_s - \rho}{\rho} g D}, \tag{A.3.5a}$$

and

$$\varphi(\Delta') = \sqrt{\frac{\sqrt{2\Delta' - \Delta'^2}}{\left(\frac{4}{3} - \Delta'\right) + \frac{1}{4}\left[\frac{1}{3} + \sqrt{2\Delta' - \Delta'^2}\right]}}. \tag{A.3.5b}$$

Here, D is the diameter of the particle and $\varphi(\Delta')$ represents the effect of the particle position on rolling. The entrainment condition becomes

$$V_b > V_{b.c.0}. \tag{A.3.6}$$

2. Critical condition and flow velocity for incipient rolling from a stationary state for fine particles

For fine particles, the cohesive force and the additional downward pressure from the water film cannot be neglected. Han proposed a formulae for these two forces and the critical flow velocity for incipient rolling that applies to both coarse and fine particles (see Appendix B) [24, 26]:

$$V_{b.c.1} = \omega_1 \varphi(\Delta'), \tag{A.3.7}$$

where

$$\begin{aligned} \omega_1 &= \sqrt{\frac{4}{3C_x} \frac{\rho_s - \rho}{\rho} g D + \frac{2\sqrt{3} q_0 \delta_0^3}{C_x \rho D}\left(3 - \frac{t}{\delta_1}\right)\left(\frac{1}{t^2} - \frac{1}{\delta_1^2}\right) + \frac{4\sqrt{3} K_2 g H}{C_x D}\left(3 - \frac{t}{\delta_1}\right)(\delta_1 - t)} \\ &= \sqrt{\frac{4}{3C_x} \frac{\rho_s - \rho}{\rho} g D + 2.98 \frac{10^{-7}}{D}(1 + 0.85H)} \\ &= \sqrt{53.9D + 2.98 \times \frac{10^{-7}}{D}(1 + 0.85H)}. \end{aligned} \tag{A.3.8}$$

Here, $\delta_0 = 3 \times 10^{-10}$ m is the diameter of a water molecule; $\delta_1 = 4 \times 10^{-7}$ m is the thickness of the water film; $q_0 = 1.52 \times 10^6$ ton/m^2 = 1.52×10^5 kg/cm^2; $K_2 = 2.67 \times 10^{-3}$; H is the depth of the water; and t is the average distance between the particles. The value of t/δ_1 depends on the dry density of the sediment γ_s', which is given by the equation (see Appendix B)

$$\gamma_s' = \left(1.85 - 0.46\frac{t}{\delta_1}\right)\left(\frac{D}{D+2t}\right)^3. \tag{A.3.9}$$

When the particles are not very fine, the cohesive force and the additional downward pressure from the water film can be neglected, thus $\omega_1 = \omega_0$. Hence, the entrainment condition for both fine and coarse particles becomes

$$V_b > V_{b.c.1}. \tag{A.3.10}$$

Owing to the randomness of the particle position on the bed surface, the critical flow velocity required for stationary particles to start rolling is a random variable. Thus, the mathematical expectation of $\eta_{\tilde{V}_{b.c.1}}$ is

$$\tilde{\overline{V}}_{b.c.1} = M\left[\eta_{\tilde{V}_{b.c.1}}\right] = \int\limits_{\Delta_m'}^{1} \varphi(\Delta')p_{\xi_{\Delta'}}(\Delta')d\Delta' = 0.916. \tag{A.3.11}$$

where $\tilde{\overline{V}}_{b.c.1} = \overline{V}_{b.c.1}/\omega_0$ is the dimensionless velocity of $\overline{V}_{b.c.1}$.

A.3.2.2 Probability of Incipient Rolling and Movement Ceasing

In Eq. (A.3.10) for the critical condition of incipient rolling, the flow velocity and particle position on the bed surface are random variables. Thus, the conditional and un-conditional probability of incipient rolling from the rest state can be expressed as

$$\begin{aligned}\varepsilon_1(\Delta') &= P\left[\xi_{V_b} > V_{b.c.0}\middle|\Delta_m' \le \xi_{\Delta'} \le 1\right] = \frac{1}{\sqrt{2\pi}\sigma_x}\int\limits_{V_{b.c.0}}^{\infty}\exp\left[-\frac{\left(V_b-\overline{V}_b\right)^2}{(2\sigma_x^2)}\right]\mathrm{d}t \\ &= \frac{1}{\sqrt{2\pi}}\int\limits_{\frac{\varphi(\Delta')\omega_1-\overline{V}_b}{\sigma_x}}^{\infty}\exp\left[-\frac{t^2}{2}\right]\mathrm{d}t = \frac{1}{0.866}\int\limits_{0.134}^{1}[\phi(t_1)]\mathrm{d}\Delta'\end{aligned} \tag{A.3.12}$$

and

$$\varepsilon_1 = P\left[\xi_{V_b} > \varphi(\xi_{\Delta'})\omega_1,\, \Delta_m' \le \xi_{\Delta'} \le 1\right]$$

$$= \frac{1}{0.866\sqrt{2\pi}} \int_{0.134}^{1} \mathrm{d}\Delta' \int_{\frac{\varphi(\Delta')\omega_1 - \overline{V}_b}{\sigma_x}}^{\infty} \exp\left[-\frac{t^2}{2}\right]\mathrm{d}t$$

$$= \frac{1}{0.866} \int_{0.134}^{1} [\phi(t_1)]\mathrm{d}\Delta' = f_{\varepsilon_1}\left(\frac{\overline{V}_b}{\omega_1}\right) \tag{A.3.13}$$

where

$$\phi(t_1) = \frac{1}{\sqrt{2\pi}} \int_{t_1}^{\infty} \exp\left[-\frac{t^2}{2}\right]\mathrm{d}t, \tag{A.3.14a}$$

and

$$t_1 = \frac{\omega_1\varphi(\Delta') - \overline{V}_b}{\sigma_x} = 2.7\left[\frac{\omega_1}{\overline{V}_b}\varphi(\Delta') - 1\right]. \tag{A.3.14b}$$

For coarse particles the probability of incipient rolling from the rest state ε_0 is similar to ε_1 only if ω_1 is substituted with ω_0 in Eqs. (A.3.12, A.3.13, A.3.14a, A.3.14b),

$$\varepsilon_0(\Delta') = P\left[\xi_{V_b} > V_{b.c.0} \middle| \Delta'_m \le \xi_{\Delta'} \le 1\right] = \frac{1}{\sqrt{2\pi}\sigma_x} \int_{V_{b.c.0}}^{\infty} \exp\left[-\frac{\left(V_b - \overline{V}_b\right)^2}{\left(2\sigma_x^2\right)}\right]\mathrm{d}t$$

$$= \frac{1}{\sqrt{2\pi}} \int_{\frac{\varphi(\Delta')\omega_0 - \overline{V}_b}{\sigma_x}}^{\infty} \exp\left[-\frac{t^2}{2}\right]\mathrm{d}t \tag{A.3.15a}$$

and

$$\varepsilon_0 = P\left[\xi_{V_b} < \varphi(\xi_{\Delta'})\omega_0, \Delta'_m \le \Delta' \le 1\right]$$

$$= \frac{1}{0.866\sqrt{2\pi}} \int_{0.134}^{1} \mathrm{d}\Delta' \int_{\frac{\varphi(\Delta')\omega_0 - \bar{V}_b}{\sigma_x}}^{\infty} \exp\left[-t^2/2\right]dt$$

$$= \frac{1}{0.866} \int_{0.134}^{1} [\phi(t_0)]d\Delta' = f_{\varepsilon_0}\left(\bar{V}_b/\omega_1\right) \tag{A.3.15b}$$

where

$$\phi(t_0) = \frac{1}{\sqrt{2\pi}} \int_{t_0}^{\infty} \exp\left[-\frac{t^2}{2}\right] \mathrm{d}t \tag{A.3.15c}$$

and

$$t_0 = \frac{\omega_0 \varphi(\Delta') - \overline{V}_b}{\sigma_x} = 2.7\left[\frac{\omega_0}{\overline{V}_b} \varphi(\Delta') - 1\right]. \tag{A.3.15d}$$

The probability that a moving particle will stop moving is $1 - \varepsilon_0$.

In contrast to Eq. (A.13), Einstein [16], Velikanov [58], and Duo [12, 13] did not consider the effect of the particle position directly in their formulae, and Einstein's assumption that the lift force has a normal distribution is not viable.

A.3.3 Dynamic Equations for a Rolling Step and Solutions

A.3.3.1 Rolling Step Distance

The distance covered by the first half of a rolling step $x_{2.1}$ is determined by the initial position of the particle on the bed Δ', and is expressed as

$$\tilde{x}_{2.1} = \frac{x_{2.1}}{D} = \cos\theta = \sqrt{2\Delta' - \Delta'^2}. \tag{A.3.16}$$

According to the static compactness factor of the bed-surface layer (Eq. A.2.7), the average dimensionless step distance and second half step distance are (Fig. A.9).

$$\tilde{x}_2 = \alpha = 1.96, \tag{A.3.17a}$$

and

$$\tilde{x}_{2.2} = \tilde{x}_2 - \tilde{x}_{2.1} = 1.96 - \sqrt{2\Delta' - \Delta'^2}. \tag{A.3.17b}$$

A.3.3.2 Equations for the First Half of a Rolling Step and Their Solutions

In the first half of a rolling step, the forces exerted on the particle (shown in Fig. A.10) include drag P_x, lift P_y, net gravity G, additional mass F_m, centrifugal $mD(\mathrm{d}\theta/\mathrm{d}t)^2$, resistance in the vertical direction $\theta_y = (\rho C/2)(\pi D^2/4)u_y^2$, normal reaction F_n, and tangential reaction F_τ forces. The last two forces both act on the contact point. The additional mass force $F_m = (\rho/2)(\pi D^3/6)(\mathrm{d}u/\mathrm{d}t) = (\rho/2)(\pi D^4/6)(\mathrm{d}^2u/\mathrm{d}t)$ is

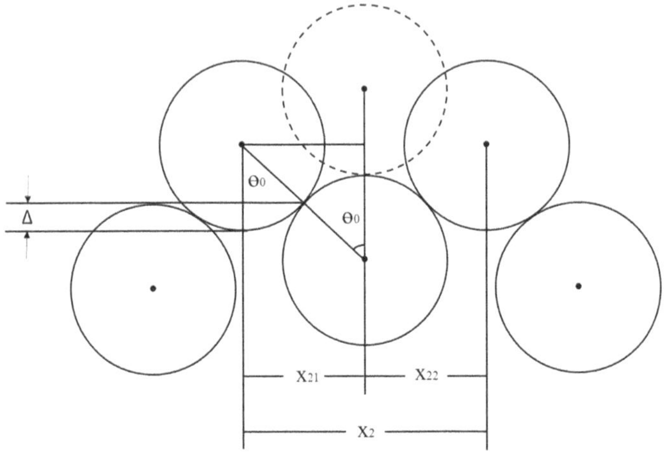

Fig. A.9 Rolling step distance

generated by the acceleration of the particle in the flow [57]. When particles roll, part of the lifting force is offset by the Magnus effect. However, on the bed surface the lifting force exerted on the particle is usually large but the rolling velocity is small, so the Magnus effect can be neglected.

Considering these forces, the dynamic equations in the tangential and normal directions with respect to the particles describe circular movement (Fig. A.10). Hence, they can be expressed in polar coordinates as

$$mD\frac{\mathrm{d}^2\theta}{\mathrm{d}t^2} = P_x \sin\theta - \left(G + \theta_y - P_y\right)\cos\theta - F_\tau - F_m, \tag{A.3.18a}$$

and

$$0 = mD\left(\frac{\mathrm{d}\theta}{\mathrm{d}t}\right)^2 + F_n - P_x\cos\theta - \left(G + \theta_y - P_y\right)\sin\theta. \tag{A.3.18b}$$

The angular momentum equation relative to the centroid is

$$(J_0 + J)\frac{\mathrm{d}^2\alpha}{\mathrm{d}t^2} = \left(P_x + P_y\right)\frac{R}{3} + F_\tau R, \tag{A.3.18c}$$

where $J = (2/5)mR^2$ is the rotation momentum of the particle inertia about the centroid; $J_0 = (1/2)(2/5)\left(\rho\pi R^2 D^3/6\right)$ is the additional rotation momentum for the inertia of the additional mass which corresponds to the additional mass force; m is the mass of the particle and $\alpha = 2\theta$ is the rotation angle of the particle around the centroid (when there is no slip, see Fig. A.10b). Equations (A.3.18a, A.3.18b, A.3.18c) give

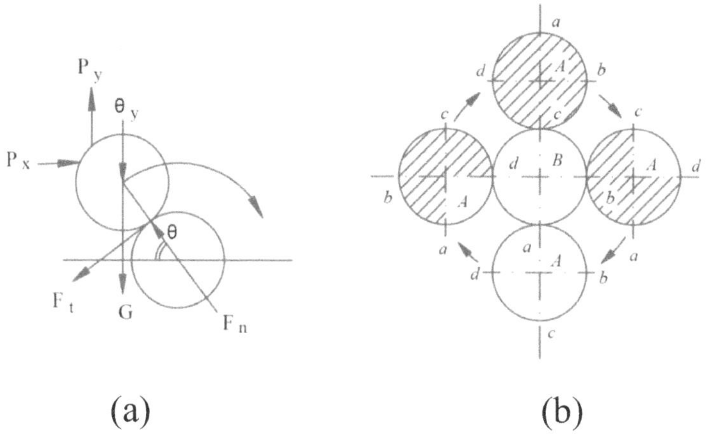

Fig. A.10 Trajectory of rolling and forces on particles in contact rolling state

$$\begin{aligned}\frac{d^2\theta}{dt^2} = \frac{5}{7D^2}\frac{\rho}{\rho_s + \frac{\rho}{2}}\Bigg\{&\frac{3C_x}{4}\left(V_b - D\frac{d\theta}{dt}\sin\theta\right)\\ &\times\left[\left(\sin\theta + \frac{1}{3}\right)\cdot\left|V_b - D\frac{d\theta}{dt}\sin\theta\right| + \frac{C_y}{C_x}\left(\cos\theta + \frac{1}{3}\right)\left(V_b - D\left(\frac{d\theta}{dt}\right)\sin\theta\right)\right]\\ &-\frac{3C}{4}D^2\left(\frac{d\theta}{dt}\right)^2\cos^3\theta - \frac{\rho_s - \rho}{\rho}gD\cos\theta\Bigg\}\end{aligned} \tag{A.3.19}$$

This is the final momentum equation in the tangential direction for a contact rolling particle.

Under the initial condition, $u_{2.x} = u_{x.0}$ and $u_{2.y} = u_{y.0}$ at $t = 0$, solving Eq. (A.3.19) using the numerical method gives the rolling speed $u_{2.x.1}$ at the end of the first half step, the half step period, $t_{2.1}$, and the average half-step rolling velocity $\tilde{U}_{2.x.1} = \tilde{x}_{2.1}/\tilde{t}_{2.1}$ as

$$f_{2.x.1} = f_{2.x.1}\left(\Delta', \tilde{V}_b, \tilde{u}_{x.0}, \tilde{u}_{y.0}\right), \tag{A.3.20}$$

where $f_{2.x.1}$ represents $\tilde{u}_{2.x.1}$, $\tilde{t}_{2.1}$, and $\tilde{U}_{2.1}$, respectively.

A.3.3.3 Simplified Equations and Their Solutions for the First Half Step of Particle Rolling

As the rolling Eq. (A.3.19) is quite complicated, a simplified equation is introduced to obtain some analytical solutions. Collisions between rolling particles and particles on the bed are relatively weak, so the velocity at the end of the previous step, $u_{x.0}$ and $u_{y.0}$, can be used to approximate the initial particle velocity. The vertical component of particle rolling is small and can be omitted, but the resistance term must still be considered. Thus, Eq. (A.3.19) can be simplified significantly, and the resulting equation is equivalent to the case of a particle rolling on a horizontal rough bed surface with a friction coefficient of 1/4. Following these assumptions, $\theta = \theta_0 = \pi/2$, $\sin\theta = \sin\theta_0 = 1$, $\cos\theta = \cos\theta_0 = 0$, $\cos\theta/\cos\theta_0 = 1$, and Eq. (A.3.19) becomes

$$\frac{\mathrm{d}u_{2.x}}{\mathrm{d}t} = a_1\left[16/17(V_b - u_{2.x})|V_b - u_{2.x}| + 1/17(V_b - u_{2.x})^2 - V_{b.c.1}^2\right], \quad \text{(A.3.21)}$$

where

$$a_1 = \frac{85}{112}\frac{\rho C_x}{(\rho_s + \rho/2)D} \approx \frac{3}{4}\frac{\rho C_x}{(\rho_s + \rho/2)D} = \frac{\rho_s - \rho}{\rho_s + \rho/2}\cdot\frac{g}{\omega_0^2} = \frac{a_2 g}{\omega_0^2} = a_0, \quad \text{(A.3.22a)}$$

and

$$a_2 = \frac{\rho_s - \rho}{\rho_s + \rho/2}. \quad \text{(A.3.22b)}$$

Equation (A.3.21) can be solved for three cases: (a) accelerated motion, where $\mathrm{d}u_{2.x}/\mathrm{d}t > 0$, i.e., $V_b - u_{2.x} > V_b - u_{2.x.1} > V_{b.c.1}$; (b) decelerated motion type I, where $\mathrm{d}u_{2.x}/\mathrm{d}t \le 0$ and $0 < \tilde{V}_b - u_{2.x} < V_{b,c.1}$; and (c) decelerated motion type II, where $\mathrm{d}u_{2.x}/\mathrm{d}t \le 0$ and $V_b - u_{x.0} < V_b - u_{2.x} < 0$. Then Eq. (A.3.21) becomes

$$\frac{1}{a_1}\frac{\mathrm{d}(V_b - u_{2.x})}{\mathrm{d}t} = V_{b.c.1}^2 - (V_b - u_{2.x})^2, (V_{b.c.1} > (V_b - u_{2.x})) \quad \text{(A.3.23a)}$$

$$\frac{1}{a_1}\frac{\mathrm{d}(V_b - u_{2.x})}{\mathrm{d}t} = V_{b.c.1}^2 - (V_b - u_{2.x})^2, (V_{b.c.1} < (V_b - u_{2.x}) \quad \text{(A.3.23b)}$$

$$\frac{\mathrm{d}(V_b - u_{2.x})}{\mathrm{d}t} = V_{b.c.1}^2 + \frac{15}{17}(V_b - u_{2.x})^2, \quad \text{(A.3.23c)}$$

for cases (a), (b), and (c), respectively. Applying the initial condition: $u_{2.x} = u_{x.0}$ for $t = 0$ yields the solution for case (a),

$$\tilde{t}_{2.1} = \frac{1}{\tilde{a}_1 \tilde{V}_{b.c.1}}\left[\operatorname{arcoth}\frac{\tilde{V}_b - \tilde{u}_{2.x.1}}{\tilde{V}_{b.c.1}} - \operatorname{arcoth}\frac{\tilde{V}_b - \tilde{u}_{x.0}}{\tilde{V}_{b.c.1}}\right], \quad \text{(A.3.24a)}$$

$$\tilde{u}_{2.x.1} = \tilde{V}_b - \tilde{V}_{b.c.1}\coth\left[\tilde{a}_1\tilde{V}_{b.c.1}\tilde{t}_{2.1} + \operatorname{arcoth}\left(\frac{\tilde{V}_b - \tilde{u}_{x.0}}{\tilde{V}_{b.c.1}}\right)\right], \quad \text{(A.3.24b)}$$

$$\tilde{x}_{2.1} = \sqrt{2\Delta' - \Delta'^2} = \frac{\tilde{V}_b}{\tilde{a}_1 V_{b.c.1}}\left(\operatorname{arcoth}\frac{\tilde{V}_b - \tilde{u}_{2.x.1}}{\tilde{V}_{b.c.1}} - \operatorname{arcoth}\frac{\tilde{V}_b - \tilde{u}_{x.0}}{\tilde{V}_{b.c.1}}\right) - \frac{1}{2\tilde{a}_1}\ln\frac{(\tilde{V}_b - \tilde{u}_{x.0})^2 - \tilde{V}_{b.c.1}^2}{(\tilde{V}_b - \tilde{u}_{2.x.1})^2 - \tilde{V}_{b.c.1}^2}, \quad \text{(A.3.24c)}$$

$$\tilde{U}_{2.x.1} = \frac{\tilde{x}_{2.x.1}}{\tilde{t}_{2.x.1}} = \tilde{V}_b - \frac{\tilde{V}_{b.c.1}}{2}\left[\frac{\ln[\frac{\tilde{V}_{b.c.1}^2-(\tilde{V}_b-\tilde{u}_{x.0})^2}{\tilde{V}_{b.c.1}^2-(\tilde{V}_b-\tilde{u}_{2.x.1})^2}]}{\operatorname{arcoth}\frac{\tilde{V}_b-\tilde{u}_{2.x.1}}{\tilde{V}_{b.c.1}} - \operatorname{arcoth}\frac{\tilde{V}_b-\tilde{u}_{x.0}}{\tilde{V}_{b.c.1}}}\right]. \quad \text{(A.3.24d)}$$

Then, the solution for case (b) is

$$\tilde{t}_{2.1} = \frac{1}{\tilde{a}_1\tilde{V}_{b.c.1}}\left(\operatorname{artanh}\frac{\tilde{V}_b - \tilde{u}_{2.x.1}}{\tilde{V}_{b.c.1}} - \operatorname{artanh}\frac{\tilde{V}_b - \tilde{u}_{x.0}}{\tilde{V}_{b.c.1}}\right], \quad \text{(A.3.25a)}$$

$$\tilde{u}_{2.x.1} = \tilde{V}_b - \tilde{V}_{b.c.1}\tanh\left[\tilde{a}_1\tilde{V}_{b.c.1}\tilde{t}_{2.1} + \operatorname{artanh}\left(\frac{\tilde{V}_b - \tilde{u}_{x.0}}{\tilde{V}_{b.c.1}}\right)\right], \quad \text{(A.3.25b)}$$

$$\begin{aligned}\tilde{x}_{2.1} = \sqrt{2\Delta' - \Delta'^2} = {} & \frac{\tilde{V}_b}{\tilde{a}_1 V_{b.c.1}}(\operatorname{artanh}\frac{\tilde{V}_b - \tilde{u}_{2.x.1}}{\tilde{V}_{b.c.1}} - \operatorname{artanh}\frac{\tilde{V}_b - \tilde{u}_{x.0}}{\tilde{V}_{b.c.1}} \\ & - \frac{1}{2\tilde{a}_1}\ln\frac{(\tilde{V}_b - \tilde{u}_{x.0})^2 - \tilde{V}_{b.c.1}^2}{(\tilde{V}_b - \tilde{u}_{2.x.1})^2 - \tilde{V}_{b.c.1}^2},\end{aligned} \quad \text{(A.3.25c)}$$

and

$$U_{2.x.1} = \tilde{V}_b - \frac{\tilde{V}_{b.c.1}}{2}\frac{\ln\frac{(\tilde{V}_b-\tilde{u}_{x.0})^2-\tilde{V}_{b.c.1}^2}{(\tilde{V}_b-\tilde{u}_{2.x.1})^2-\tilde{V}_{b.c.1}^2}}{\operatorname{artanh}\frac{\tilde{V}_b-\tilde{u}_{2.x.1}}{\tilde{V}_{b.c.1}} - \operatorname{artanh}\frac{\tilde{V}_b-\tilde{u}_{x.0}}{\tilde{V}_{b.c.1}}}. \quad \text{(A.3.25d)}$$

Finally, the solution for case (c) is

$$\tilde{t}_{2.1} = \frac{1}{\sqrt{15/17}\tilde{a}_1\tilde{V}_{b.c.1}}\left(\arctan\frac{\tilde{V}_b - \tilde{u}_{2.x.1}}{\sqrt{17/15}\tilde{V}_{b.c.1}} - \arctan\frac{\tilde{V}_b - \tilde{u}_{x.0}}{\sqrt{17/15}\tilde{V}_{b.c.1}}\right), \quad \text{(A.3.26a)}$$

$$\tilde{u}_{2.x.1} = \tilde{V}_b - \sqrt{\frac{17}{15}}\tilde{V}_{b.c.1}\operatorname{tg}\left(\sqrt{\frac{15}{17}}\tilde{a}_1\tilde{V}_{b.c.1}\tilde{t}_{2.1} + \arctan\frac{\tilde{V}_b - \tilde{u}_{2.x.0}}{\sqrt{17/15}\tilde{V}_{b.c.1}}\right), \quad \text{(A.3.26b)}$$

$$\begin{aligned}\tilde{x}_{2.1} = \sqrt{2\Delta' - \Delta'^2} = {} & \frac{\tilde{V}_b}{\sqrt{15/17}\tilde{a}_1\tilde{V}_{b.c.1}}\left[\arctan\frac{\tilde{V}_b - \tilde{u}_{2.x.1}}{\sqrt{17/15}\tilde{V}_{b.c.1}}\right. \\ & \left. - \arctan\frac{\tilde{V}_b - \tilde{u}_{x.0}}{\sqrt{17/15}\tilde{V}_{b.c.1}}\right] + \frac{1}{\tilde{a}_1 30/17}\ln\frac{(\tilde{V}_b - \tilde{u}_{2.x.1})^2 + \tilde{V}_{b.c.1}^2 17/15}{(\tilde{V}_b - \tilde{u}_{x.0})^2 + \tilde{V}_{b.c.1}^2 17/15},\end{aligned} \quad \text{(A.3.26c)}$$

$$\tilde{U}_{2.x.1} = \tilde{V}_b + \frac{\sqrt{17/15}\tilde{V}_{b.c.1}}{2}\frac{\ln[\frac{(\tilde{V}_b-\tilde{u}_{2.x.1})^2+\tilde{V}_{b.c.1}^2 17/15}{(\tilde{V}_b-\tilde{u}_{x.0})^2+\tilde{V}_{b.c.1}^2 17/15}}{\arctan\frac{\tilde{V}_b-\tilde{u}_{2.x.1}}{\sqrt{17/15}\tilde{V}_{b.c.1}} - \arctan\frac{\tilde{V}_b-\tilde{u}_{x.0}}{\sqrt{17/15}\tilde{V}_{b.c.1}}}. \quad \text{(A.3.26d)}$$

A.3.3.4 Equations for the Second Half of a Rolling Step and Their Solutions (Non-contact Rolling)

In non-contact rolling, the particles are subjected to two forces in the longitudinal direction: drag and the additional mass force. The centroid movement equation is

$$\frac{\mathrm{d}u_{2,x}}{\mathrm{d}t} = a_0\left(V_b - u_{2,x}\right)\left|V_b - u_{2,x}\right|. \tag{A.3.27}$$

There are three cases for second-half step rolling: (a) accelerated motion, where $\tilde{V}_b - \tilde{u}_{2.x} > 0$; (b) decelerated motion, where $\tilde{V}_b - \tilde{u}_{2.x} < 0$; and (c) uniform motion, where $\tilde{V}_b - \tilde{u}_{2.x} = 0$. Thus, Eq. (A.3.27) becomes

$$\frac{\mathrm{d}u_{2.x}}{\mathrm{d}t} = a_0(V_b - u_{2.x})^2, \tag{A.3.28a}$$

$$\frac{\mathrm{d}u_{2.x}}{\mathrm{d}t} = -a_0(V_b - u_{2.x})^2, \tag{A.3.28b}$$

and

$$\frac{\mathrm{d}u_{2.x}}{\mathrm{d}t} = 0, \tag{A.3.28c}$$

for cases (a), (b), and (c), respectively. Applying the initial conditions $t = 0$ and $u_{2.x} = u_{2.x.1}$ gives the solution for case (a),

$$\tilde{t}_{2,2} = \frac{1}{\tilde{a}_0} \frac{\tilde{u}_{2,x,2} - \tilde{u}_{2,x,1}}{\left(\tilde{V}_b - \tilde{u}_{2,x,2}\right)\left(\tilde{V}_b - \tilde{u}_{2,x,1}\right)}, \tag{A.3.29a}$$

$$\tilde{u}_{2.x.2} = \tilde{V}_b - \frac{\tilde{V}_b - \tilde{u}_{2.x.1}}{1 + \left(\tilde{V}_b - \tilde{u}_{2.x.1}\right)\tilde{a}_0\tilde{t}_{2.2}}, \tag{A.3.29b}$$

$$\tilde{x}_{2.2} = \frac{\tilde{V}_b}{\tilde{a}_0} \frac{\tilde{u}_{2.x.2} - \tilde{u}_{2.x.1}}{\left(\tilde{V}_b - \tilde{u}_{2.x.2}\right)\left(\tilde{V}_b - \tilde{u}_{2.x.1}\right)} - \frac{1}{\tilde{a}_0} \ln \frac{\tilde{V}_b - \tilde{u}_{2.x.1}}{\tilde{V}_b - \tilde{u}_{2.x.2}}, \tag{A.3.29c}$$

and

$$\tilde{U}_{2.x.2} = \frac{\tilde{x}_{2.2}}{\tilde{t}_{2.2}} = \tilde{V}_b - \frac{\left(\tilde{V}_b - \tilde{u}_{2.x.2}\right)\left(\tilde{V}_b - \tilde{u}_{2.x.1}\right)}{\tilde{u}_{2.x.2} - \tilde{u}_{2.x.1}} \ln \frac{\tilde{V}_b - \tilde{u}_{2.x.1}}{\tilde{V}_b - \tilde{u}_{2.x.2}}. \tag{A.3.29d}$$

For case (b) the solution is

$$\tilde{t}_{2,2} = -\frac{1}{\tilde{a}_0} \frac{\tilde{u}_{2.x.2} - \tilde{u}_{2.x.1}}{\left(\tilde{V}_b - \tilde{u}_{2,x,2}\right)\left(\tilde{V}_b - \tilde{u}_{2.x.1}\right)}, \tag{A.3.30a}$$

$$\tilde{u}_{2.x.2} = \tilde{V}_b - \frac{\left(\tilde{V}_b - \tilde{u}_{2.x.1}\right)}{1 - \left(\tilde{V}_b - \tilde{u}_{2.x.1}\right)\tilde{a}_0 \tilde{t}_{2.2}}, \tag{A.3.30b}$$

$$\tilde{x}_{2.2} = -\frac{\tilde{V}_b}{\tilde{a}_0} \frac{\tilde{u}_{2.x.2} - \tilde{u}_{2.x.1}}{\left(\tilde{V}_b - \tilde{u}_{2.x.2}\right)\left(\tilde{V}_b - \tilde{u}_{2.x.1}\right)} + \frac{1}{\tilde{a}_0} \ln \frac{\tilde{V}_b - \tilde{u}_{2.x.1}}{\tilde{V}_b - \tilde{u}_{2.x.2}}, \tag{A.3.30c}$$

and

$$\tilde{U}_{2.x.2} = \frac{\tilde{x}_{2.2}}{\tilde{t}_{2.2}} = \tilde{V}_b - \frac{\left(\tilde{V}_b - \tilde{u}_{2.x.2}\right)\left(\tilde{V}_b - \tilde{u}_{2.x.1}\right)}{\tilde{u}_{2.x.2} - \tilde{u}_{2.x.1}} \ln \frac{\tilde{V}_b - \tilde{u}_{2.x.1}}{\tilde{V}_b - \tilde{u}_{2.x.2}}. \tag{A.3.30d}$$

Finally, for case (c) the solution is

$$\tilde{u}_{2,x.1} = \tilde{u}_{2,x.2} = \tilde{V}_b - \tilde{V}_{b.c.1}. \tag{A.3.31}$$

A.3.3.5 Connection Between the First and Second Halves of Rolling Steps

The bottom flow velocity is assumed to be constant within a single step, and the parameters at the end of the first half step should be consistent with those at the beginning of the second half step, so only three types of single-step motion are possible.

Type 1: The first and second half steps are both accelerated motion. Consequently, the particle accelerates rapidly throughout the entire step. In reality, this is quite unlikely to occur.

Type 2: The first half step is type I decelerated motion, and the second half step is accelerated motion. This type of single-step motion occurs for a narrow range of rolling velocities and there is a high probability of rolling. Hence, this better represents reality.

Type 3: The first and second half steps are both type II decelerated motion. Consequently, the particle decelerates throughout the entire step.

The second type of motion is the most typical and general type of rolling. As to uniform motion, it may occur in the second half step. Calculations by [23] show that the average particle velocity for type 2 motion varies around the uniform velocity $\tilde{V}_b - \varphi(\Delta')$ with a small amplitude. Thus, the uniform velocity $\tilde{u}_{2.x} = \tilde{V}_b - \varphi(\Delta')$ can be used to approximate type 2 single step rolling.

A.3.3.6 Critical Velocities for Incipient Rolling from Stationary and from Rolling to Stationary

It can be demonstrated that particle rolling will not stop and can last for several steps as long as the particle starts rolling from a stationary state. However, if the initial step velocity of a particle is nonzero and it is decelerating, then the particle may stop rolling and become stationary in the first half step. If a rolling particle is accelerating in the first half step, then it may lose contact with the bed particle and transition to the hopping state. Related research has been published previously [23, 24].

A.3.4 Mathematical Expectation of Parameters for Rolling Step

A.3.4.1 Expectation of Distance of Rolling Step

From Eqs. (A.3.16), (A.3.17a) and (A.3.17b) the expectations for the rolling distance in the first and second half steps and a single step can be found to be

$$\tilde{\bar{x}}_{2.1} = \int_{0.134}^{1} \tilde{x}_{2.1}(\Delta')p(\Delta')\mathrm{d}\Delta' = 0.8546, \quad \text{(A.3.32a)}$$

$$\tilde{\bar{x}}_{2.2} = 1.96 - 0.855 = 1.105, \quad \text{(A.3.32b)}$$

and

$$\tilde{\bar{x}}_{2} = 1.96. \quad \text{(A.3.32c)}$$

A.3.4.2 Expectations for Particle Velocity and Time Period of Rolling Step

This section only considers the particle velocity during uniform rolling $u_{2.x} = V_b - V_{b,c,1}$. Its conditional and unconditional expectations about Δ' are

$$\begin{aligned}\tilde{U}_{2.x}\left(\tilde{\bar{V}}_b, \Delta'\right) &= \tilde{\bar{u}}_{2.x}\left(\tilde{\bar{V}}_b, \Delta'\right) = M\left[\xi_{\tilde{V}_b} - \xi_{\tilde{V}_{b.c.1}} \middle| \tilde{V}_{b.c.1} < \tilde{V}_b \le \tilde{V}_{b.c.2},\ \xi_{\Delta'} = \Delta'\right] \\ &= \frac{1}{\left(\varepsilon_0(\Delta') - \varepsilon_2(\Delta')\right)\sqrt{2\pi}\tilde{\sigma}_x} \int_{\tilde{V}_{b.c.1}}^{\tilde{V}_{b.c.2}} \left(\tilde{V}_b - \tilde{V}_{b.c.1}\right) \exp\left[-\frac{\left(\tilde{V}_b - \tilde{\bar{V}}_b\right)^2}{\left(2\tilde{\sigma}_x^2\right)}\right] \mathrm{d}\tilde{V}_b \\ &= \frac{0.148\tilde{\bar{V}}_b}{\left(\varepsilon_0(\Delta') - \varepsilon_2(\Delta')\right)} \left(\exp\left(-\frac{t_1^{*2}(\Delta')}{2}\right) - \exp\left(-\frac{t_2^{*2}(\Delta')}{2}\right)\right) - t_1^*(\Delta')\tilde{\sigma}_x\end{aligned} \quad \text{(A.3.33)}$$

and

$$\tilde{U}_{2.x}\left(\tilde{\overline{V}}_b\right) = \tilde{\overline{u}}_{2.x}\left(\tilde{\overline{V}}_b\right) = \int_{0.134}^{1} \tilde{u}_{2.x}\left(\tilde{\overline{V}}_b, \Delta'\right) p(\Delta')\mathrm{d}\Delta', \tag{A.3.34}$$

respectively. Here,

$$t_1^*(\Delta') = \frac{V_{b,c,1} - \overline{V_b}}{\sigma_x} = \frac{\varphi(\Delta')\omega_0 - \overline{V_b}}{0.373\overline{V_b}} = 2.7\left(\frac{\varphi(\Delta')}{\tilde{\overline{V}}_b} - 1\right); \tag{A.3.35a}$$

$$t_2^*(\Delta') = \frac{V_{b,c,2} - \overline{V_b}}{\sigma_x} = \frac{1.62\varphi(\Delta')\omega_0 - \overline{V_b}}{\sigma_x} = 2.7\left(\frac{1.62\varphi(\Delta')\omega_0}{\overline{V_b}} - 1\right); \tag{A.3.35b}$$

and $\varepsilon_0(\Delta')$ and $\varepsilon_1(\Delta')$ are given by Eqs. (A.3.15a) and (A.3.12). The average time period of single step is

$$\tilde{\overline{t}}_2 = \frac{\tilde{\overline{x}}_2}{\tilde{\overline{u}}_{2.x}} = \frac{1.96}{\tilde{\overline{u}}_{2.x}}. \tag{A.3.36}$$

A.3.5 *Expectations of Rolling Parameters for Non-uniform Sediment*

The results for uniform sediments can easily be extended to non-uniform sediments. In fact the parameter expressions are the same, except the particle size D and particle position Δ' in the conditional expectation for uniform sediments are replaced by the particle size D_l and particle position $\Delta_l = f\left(\overline{D}/D_l\right)$ for the l-th size group of the non-uniform sediment. Then from Eq. (A.3.31), the unconditional expectations for the particle velocity of the l-th size group can be expressed as

$$\begin{aligned}\tilde{\overline{u}}_{2.l}\left(\overline{V}_b, \frac{\overline{D}}{D_l}\right) &= \tilde{\overline{u}}_{2.l}(\overline{V}_b, \Delta')\Big|\Delta' = f_{\Delta_l'}\left(\frac{\overline{D}}{D_l}\right) = \frac{\tilde{\overline{u}}_{2.l}(\overline{V}_b, \Delta')\omega_{0.l}}{D_l}\Big|\Delta' = f_{\Delta_l'}\left(\frac{\overline{D}}{D_l}\right)\\ &= \frac{0.399}{(\varepsilon_{0.l} - \varepsilon_{2.l})} \times \int_{t_1^*(\Delta')}^{t_2^*(\Delta')} \tilde{u}_2\left(\tilde{\overline{V}}_b + \tilde{\sigma}_x t, \Delta'\right)\exp(-\frac{t^2}{2)}\mathrm{d}t\Big|\Delta' = f_{\Delta_l'}\left(\frac{\overline{D}}{D_l}\right).\end{aligned} \tag{A.3.37}$$

where

$$t_{1.l}^{*} = 2.7\left[\frac{\varphi(\Delta_l')}{\tilde{\overline{V}}_b} - 1\right]\Bigg|\Delta_l' = f_{\Delta_l'}\left(\frac{\overline{D}}{D_l}\right) = 2.7\left[\frac{\varphi\left(f_{\Delta_l'}\left(\frac{\overline{D}}{D_l}\right)\right)}{\tilde{\overline{V}}_b} - 1\right], \quad \text{(A.3.38a)}$$

$$t_{2.l}^{*} = 2.7\left[\frac{1.62\varphi(\Delta_l')}{\tilde{\overline{V}}_b} - 1\right]\Bigg|\Delta_l' = f_{\Delta_l'}\left(\frac{\overline{D}}{D_l}\right) = 2.7\left[\frac{1.62\varphi\left(f_{\Delta_l'}\left(\frac{\overline{D}}{D_l}\right)\right)}{\tilde{\overline{V}}_b} - 1\right], \quad \text{(A.3.38b)}$$

$$\begin{aligned}\varepsilon_{0.l} &= P\left[V_b > \varphi(\Delta')\omega_{0.l}\right]\Bigg|\Delta_l' = f_{\Delta_l'}\left(\frac{\overline{D}}{D_l}\right) \\ &= 1.78\left[\varphi(\Delta_l')/\tilde{\overline{V}}_b - 1\right]\Bigg|\Delta_l' = f_{\Delta_l'}\left(\frac{\overline{D}}{D_l}\right) = 1.78\left[\varphi\left(f_{\Delta_l'}\left(\frac{\overline{D}}{D_l}\right)\right)/\tilde{\overline{V}}_b - 1\right]\end{aligned} \quad \text{(A.3.39a)}$$

$$\varepsilon_{2.l} = P\left[V_b > 1.59\varphi(\Delta_l')\omega_{0.l}\right]\Bigg|\Delta' = f_{\Delta_l'}\left(\frac{\overline{D}}{D_l}\right) = 1.78\left[1.59\varphi\left(f_{\Delta_l'}\left(\frac{\overline{D}}{D_l}\right)\right)/\tilde{\overline{V}}_b - 1\right]. \quad \text{(A.3.39b)}$$

A.4 Step Hopping

Hopping is an important form of bed load movement. In particular, fine hopping particles often represent a large proportion of the bed load. Compared to rolling, hopping is relatively simple and has been studied to a greater extent. Bagnold [5], Gonchanov [19], Dou [11, 12], Yalin [61, 62], Paintal [48], Tsushiya [56], Velikanov [58] and Einstein [15, 16] have carried out a lot of research work and made progress in the hopping movement of single particle. Field data show that sediment particles can start to temporally hop on the bed, even when the lift force is less than the force of gravity. This contradicts studies by Einstein [16] and Velikanov [59]. Different explanations for this phenomena have been proposed. First, [5] proposed that collisions between moving particles and bed-surface particles affected particle hopping. However, in terms of mechanics, his description of the effects of collisions are not quite reasonable. Later, Han et al. (1984, 1987)[24,25], Han [23], Lukerchenko et al. [38, 39] and Nino et al. [47] gave an accurate mechanical description of the effects of such collisions.

The study in this Section is based on analysis of the forces acting on hopping particles and a derivation of their initial velocity after a collision. From this, complete hopping dynamic equations are established, and the parameters and conditions for transitions between motion states are obtained. The expectations for the hopping parameters and the basic transition probabilities related to the hopping state are obtained using the probability distributions of the flow velocities and particle positions on the bed surface.

A.4.1 Mechanism of Particle Hopping and a Simplified Diagram

The authors' observations of the movement of coarse particles revealed that, when the lifting force is less than the gravitational force, the particles do not immediately start hopping once they are entrained. Instead, they often roll a certain distance, obtain a certain amount of momentum, and then start to hop when they hit another particle protruding from the bed [5, 13]. Tsuchiya [56] also observed that particles travel a certain distance before they transition from the rolling state to the hopping state. The factors that cause a rolling particle to start hopping include the lifting force, collisions between rolling particles and particles on the bed, centrifugal forces when particles roll over particles on the bed, and the Magnus effect. The first two are the main factors and they are only involved in the incipient hopping stage. When a particle jumps to a height greater than its diameter, the upper and lower streamlines around the particle are approximately symmetric and the lifting force disappears. This is a gradual process, but the analysis is simplified significantly if it is assumed to be an abrupt change [13].

It is common for rolling particles to start to hop, but hopping can also occur under other conditions. For example, when the instantaneous flow intensity becomes weak, suspended particles may fall on the bed and start hopping. Alternatively, when the flow intensity remains constant, hopping particles may fall on the bed and hop continuously after bouncing back off the bed. Furthermore, if the flow intensity suddenly increases, particles resting on the bed may start to hop owing to the lift force alone. The commonality between these incipient hopping scenarios is that hopping occurs when moving particles collide with particles on the bed and get bounced back. When hopping occurs, the particle has velocity components in both the longitudinal and vertical directions prior to the collision. This is in contrast to rolling particles which primarily have velocity components in the longitudinal direction, with negligible vertical components.

From this, the process of particle hopping can be summarized as follows. Under the action of lift force and a collision with a particle on the bed surface, a moving particle can be lifted from the bed surface. The initial lift velocity is determined by the velocity after the collision. The lift force occurs when the particle is rolling, and when it is lifted from the bed surface to a height less than or equal to the diameter of the particle. When the lifting height exceeds the particle diameter, the lifting force disappears and the particle will rise to its maximum height through inertia, and then gradually fall to the bed surface owing to gravity. This entire process is considered to be one hopping step. A hopping event describes the period in which a particle is in the hopping state, which consists of multiple hopping steps. The number of continuous hopping steps is determined by the probability of continuous hopping. In addition, the threshold that differentiates between the rolling and hopping states is whether a particle is lifted to a height equivalent to its diameter.

A.4.2 Mechanics of Step Hopping

A.4.2.1 Incipient Hopping Velocity of a Moving Particle

Assume that a moving particle falls on the bed at a velocity with longitudinal component $u_{x,0}$ and vertical component $u_{y,0}$. When it collides with a particle on the bed surface, the longitudinal and vertical components become $u_{3.x,0}$ and $u_{3.y,0}$, respectively (as shown in Fig. A.11). That is,

$$\begin{aligned} u_{3.x.0} &= u_{3.\tau.0}\sin\theta - u_{3.n,0}\cos\theta = u_{x.0}f_3 + u_{y.0}f_1 \\ &= u_{x.0}[1.0, 0k_0)\left(2\Delta' - \Delta'^2\right)]+u_{y.0}[(1+k_0)(1+\Delta')\sqrt{2\Delta' - \Delta'^2}], \end{aligned} \quad \text{(A.4.1a)}$$

$$u_{3.y.0} = u_{3.\tau.0}\cos\theta + u_{3.n.0}\sin\theta = u_{x.0}f_1 + u_{y.0}f_2$$

$$\begin{aligned} u_{3.y.0} &= u_{3.\tau.0}\cos\theta + u_{3.n.0}\sin\theta = u_{x.0}f_1 + u_{y.0}f_2 \\ &= u_{x.0}(1+k_0)(10\Delta')\sqrt{2\Delta' - \Delta'^2} + u_{y.0}[1.0.0k_0)(10\Delta')^2]. \end{aligned} \quad \text{(A.4.1b)}$$

Here, Δ' is the particle position, k_0 is the recovery coefficient,

$$f_1(\Delta') = (1+k_0)(11\Delta')\sqrt{2\Delta' - \Delta'^2}, \quad \text{(A.4.2a)}$$

$$f_2(\Delta') = 14.2ak_0)(14\Delta')^2, \quad \text{(A.4.2b)}$$

and

$$f_3(\Delta') = 14.2bk_0)\left(2\Delta' - \Delta'^2\right). \quad \text{(A.4.2c)}$$

For rolling particles, the initial velocity in the vertical direction is relatively small, so its effect on the collision can be ignored. In Lukerchenko's studies [38, 39] the effects of rotation and friction were considered, but they will be ignored hereafter for simplicity.

A.4.2.2 Dynamic Equations for Step Hopping and Their Solutions

The forces acting on a hopping particle include drag P_x, (Eq. (A.3.2a)), lift P_y, (Eq. (A.3.2b)), (up to a height equivalent to one particle diameter from the bed), resistance θ_y, underwater gravity G, (Eq. (A.3.2c)), and inertia associated with the additional mass F_m. Considering that the lift force varies with height, a vertical hopping step is divided into three stages. First, the particle is lifted from the bed up to a height equivalent to the particle diameter D; second, the particle hops from height D to the

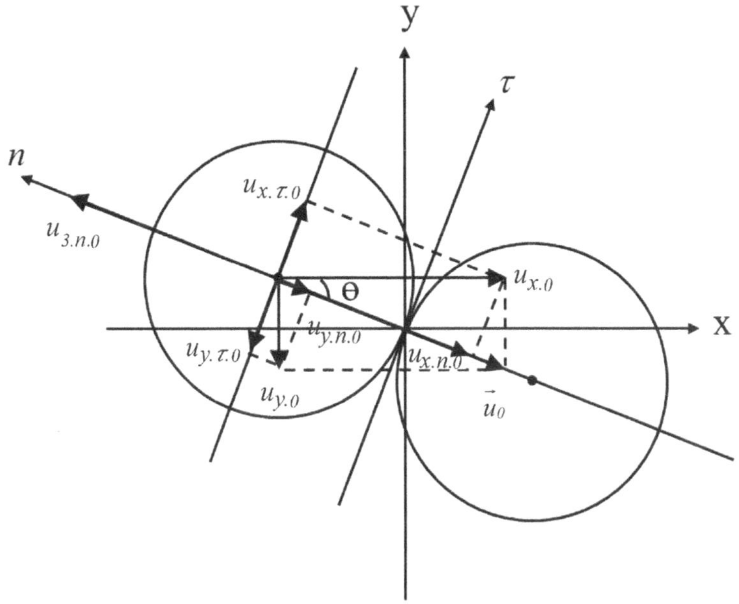

Fig. A.11 Collision between particles

maximum height y_M; third, the particle falls from the maximum height back to the bed. These stages are described in detail as follows.

(1) First stage of vertical movement ($01y \leq D$)

Considering the forces discussed above, the momentum equation for the vertical movement of a particle in the first stage is given by

$$\rho_s \frac{\pi D^3}{6} \frac{\mathrm{d}u_{3.y}}{\mathrm{d}t} = -\frac{\rho}{2} \frac{\pi D^3}{6} \frac{\mathrm{d}u_{3.y}}{\mathrm{d}t} + \frac{\rho C_y}{2} \frac{\pi D^2}{4} V_b^2 - (\rho_s - \rho) g \frac{\pi D^3}{6} - \frac{\rho C}{2} \frac{\pi D^2}{4} u_{3.y}^2, \tag{A.4.3}$$

where the terms on the right represent the inertial force associated with the additional mass, lift force, gravity force, and resistance force, respectively. In addition, $u_{3,y}$ is the vertical velocity component of particle hopping, and $C = 1.2$ is the drag coefficient. The longitudinal particle velocity $u_{3.x}$ decreases greatly following a collision with a bed particle, so V_b is used instead of $V_b - u_{3.x}$ to calculate the lift force. Thus, the momentum equation in the vertical direction is independent of the longitudinal particle movement. Eq. (A.4.3) can be rewritten as

$$\frac{\mathrm{d}u_{3.y}}{\mathrm{d}t} = -a_3 - a_4 u_{3.y}^2, \tag{A.4.4a}$$

and

$$\frac{1}{2}\frac{du_{3.y}^2}{dy} = -\left(a_3 + a_4 u_{3.y}^2\right) \tag{A.4.4b}$$

where

$$a_3 = \frac{\rho_s - \rho}{\rho_s + \rho/2}g - \frac{3C_y}{4}\frac{\rho}{\rho_s + \rho/2}\frac{V_b^2}{D} = a_2 g\left(1 - \frac{V_b^2}{V_1^2}\right), \tag{A.4.5a}$$

$$a_4 = \frac{3C}{4}\frac{\rho}{\rho_s + \rho/2}\frac{1}{D} = \frac{\gamma_s - \gamma}{\rho_s + \rho/2}\frac{1}{\omega^2} = \frac{a_2 g}{\omega^2}, \tag{A.4.5b}$$

$$a_2 = \frac{\rho_s - \rho}{\rho_s + \rho/2}, \tag{A.4.5c}$$

$$\omega = \sqrt{\frac{4}{3C}\frac{\gamma_s - \gamma}{\gamma}gD}, \tag{A.4.5d}$$

$$V_1 = \sqrt{\frac{4}{3C_y}\frac{\gamma_s - \gamma}{\gamma}gD}, \tag{A.4.5e}$$

$$\sqrt{|a_3| a_4} = \frac{a_2 g}{\omega}\sqrt{\left|1 - V_b^2/V_1^2\right|}, \tag{A.4.5f}$$

and

$$|a_3|/a_4 = \omega^2\left|1 - V_b^2/V_1^2\right|. \tag{A.4.5g}$$

To solve Eqs. (A.4.4a, A.4.4b), it is necessary to distinguish between cases where (a) $a_3 > 0$, i.e., the lift is weaker than gravity and the particle velocity decelerates; (b) $a_3 < 0$, i.e., the lift is greater than underwater gravity; and (c) $a_3 = 0$, i.e., the lift is equal to gravity and the particle motion is uniform. Integrating Eq. (A.4.4a, A.4.4b) with the conditions $u_{3.y} = u_{3.y.0}$ at $t = 0$, and $t = t_{3.0} = t_{0.3}$ and $u_{3.y} = u_{3.y.D}$ at $y = D$ for case (a) gives

$$\tilde{t}_{3.0} = \frac{1}{\sqrt{\tilde{a}_3 \tilde{a}_4}}\left(\arctan\sqrt{\frac{\tilde{a}_4}{\tilde{a}_3}}\tilde{u}_{3.y.0} - \arctan\sqrt{\frac{\tilde{a}_4}{\tilde{a}_3}}\tilde{u}_{3.y.D}\right) \tag{A.4.6}$$

and

$$\tilde{u}_{3.y.D}^2 = \tilde{u}_{3.y.0}^2 e^{-2\tilde{a}_4} - \tilde{a}_3/\tilde{a}_4\left(1 - e^{-2\tilde{a}_4}\right). \tag{A.4.7}$$

Case (b) can be further divided into three scenarios, and these give

(i) If $|a_3| - a_4 u_{3.y}^2 < 0$, the particle decelerates:

$$\tilde{t}_{3.0} = 1/\sqrt{|\tilde{a}_3|\tilde{a}_4}\left[\operatorname{arcoth}\sqrt{\tilde{a}_4/|\tilde{a}_3|}\tilde{u}_{3.y.D} - \operatorname{arcoth}\sqrt{\tilde{a}_4/|\tilde{a}_3|}\tilde{u}_{3.y.0}\right], \text{ and} \tag{A.4.8}$$

$$\tilde{u}^2_{3.y.D} = \tilde{u}^2_{y.0}e^{-2\tilde{a}_4} + \frac{|\tilde{a}_3|}{\tilde{a}_4}\left(1.e^{-2\tilde{a}_4}\right). \tag{A.4.9}$$

(ii) If $|a_3| - a_4 u^2_{3.y} > 0$, the particle accelerates:

$$\tilde{t}_{3.0} = \frac{1}{\sqrt{|\tilde{a}_3|\tilde{a}_4}}\left[\operatorname{artanh}\sqrt{\frac{\tilde{a}_4}{|\tilde{a}_3|}}\tilde{u}_{3.y.D} - \operatorname{artanh}\sqrt{\frac{\tilde{a}_4}{|\tilde{a}_3|}}\tilde{u}_{3.y.0}\right], \text{ and} \tag{A.4.10}$$

$$\tilde{u}^2_{3.y.D} = \tilde{u}^2_{y.0}e^{-2\tilde{a}_4} + \frac{|\tilde{a}_3|}{\tilde{a}_4}\left(1.e^{-2\tilde{a}_4}\right). \tag{A.4.11}$$

(iii) If $\sqrt{\frac{|a_3|}{a_4}} - u_{3.y} = 0$, the particle motion is uniform:

$$\tilde{u}_{3.y} = \tilde{u}_{3.y.D} = \tilde{u}_{3.y.0}, \text{ and} \tag{A.4.12}$$

$$\tilde{t}_{0.3} = \frac{1}{\tilde{u}_{3.y.0}}. \tag{A.4.13}$$

Finally, for case (c) the integration gives

$$\tilde{u}_{3.y.D} = \tilde{u}_{3.y.0}e^{-2\tilde{a}_4}, \tag{A.4.14}$$

and

$$\tilde{t}_{0.3} = \frac{1}{\tilde{a}_4}\left(\frac{1}{\tilde{u}_{3,y,D}} - \frac{1}{\tilde{u}_{3,y,0}}\right). \tag{A.4.15}$$

(2) Second stage of vertical movement ($D \le y \le y_M$)

Once the particle exceeds a height of $y = D$ it is detached from the bed surface layer and the lift force no longer exists. Thus, the momentum equation becomes

$$\rho_s\frac{\pi D^3}{6}\frac{du_{3,y}}{dt} = -\frac{\rho}{2}\frac{\pi D^3}{6}\frac{du_{3,y}}{dt} - (\rho_s - \rho)g\frac{\pi D^3}{6} - \frac{\rho C}{2}\frac{\pi D^2}{4}u^2_{3,y}, \tag{A.4.16}$$

where the terms on the right correspond to the inertia associated with the additional mass, gravity, and resistance, respectively. Equation (A.4.16) can be rewritten as

$$\frac{du_{3,y}}{\mathrm{d}t} = -\left(a_5 + a_4 u^2_{3,y}\right), \tag{A.4.17a}$$

and

$$\frac{1}{2}\frac{du_{3.y}^2}{dy} = -\left(a_5 + a_4 u_{3,y}^2\right), \tag{A.4.17b}$$

where

$$a_5 = \frac{\rho_s - \rho}{\rho_s + \rho/2} g = a_2 g. \tag{A.4.18}$$

Integrating Eqs. (A.4.17a, A.4.17b) under the conditions $u_{3.y.0} = u_{3.y.D}$, $t_{3.y.0} = 0$, and $u_{3.y} = 0$ at $y = y_M$ yields the hopping height and the time period for particle to reach the hopping height expressed as

$$\tilde{y}_M = 1 + \frac{1}{2\tilde{a}_4} \ln\left(1 + \frac{\tilde{a}_4}{\tilde{a}_5 \tilde{u}_{3.y.D}^2}\right), \tag{A.4.19}$$

and

$$\tilde{t}_{3.1} = \tilde{t}_{0.3} + \frac{1}{\sqrt{\tilde{a}_4 \tilde{a}_5}} \arctan \sqrt{\frac{\tilde{a}_4}{\tilde{a}_5}} \tilde{u}_{3.y.D}, \tag{A.4.20}$$

where $\tilde{a}_5 = a_5 D/\omega_0^2$.

(3) Third stage of vertical movement ($y_M \geq y \geq 0$)

When a particle reaches its highest point y_M, it will begin to fall. The forces acting on the particle include the inertia associated with the additional mass, gravity, and resistance. Thus, the momentum equation is

$$\rho_s \frac{\pi D^3}{6} \frac{du_{3,y}}{dt} = -\frac{\rho}{2} \frac{\pi D^3}{6} \frac{du_{3,y}}{dt} - (\rho_s - \rho) g \frac{\pi D^3}{6} - \frac{\rho C}{2} \frac{\pi D^2}{4} u_{3,y}^2. \tag{A.4.21}$$

This can be rewritten as

$$\frac{du_{3,y}}{dt} = -a_5 + a_4 u_{3,y}^2, \tag{A.4.22a}$$

and

$$\frac{1}{2}\frac{du_{3.y}^2}{dy} = -a_5 + a_4 u_{3,y}^2. \tag{A.4.22b}$$

Integrating Eqs. (A.4.22a, A.4.22b) with the conditions $t = t_{3.1}$ and $u_{3,y} = 0$ at $y = y_M$; and $t = t_3$ and $u_{3,y} = u_{3.y.2}$ at $y = 0$, yields

$$\tilde{t}_3 = \tilde{t}_{3,1} + \frac{1}{\sqrt{\tilde{a}_4 \tilde{a}_5}} \text{artanh} \sqrt{1 r e^{-2\tilde{a}_4 \tilde{y}_M}}, \tag{A.4.23}$$

and

$$\tilde{u}_{3,y,2} =, \sqrt{\frac{\tilde{a}_5}{\tilde{a}_4}}\sqrt{1 - e^{-2\tilde{a}_4 \tilde{y}_M}}. \tag{A.4.24}$$

(4) Longitudinal movement

Considering the longitudinal movement of a hopping particle, the particle velocity is generally different from the bottom flow velocity. The relative velocity $V_b - u_{3,x}$ between the particle and the flow causes resistance to the particle movement. In addition, there is also inertia associated with the additional mass. Hence, the momentum equation in the longitudinal direction is

$$\rho_s \frac{\pi D^3}{6}\frac{du_{3.x}}{dt} = -\frac{\rho}{2}\frac{\pi D^3}{6}\frac{du_{3.x}}{dt} + \frac{\rho C}{2}\frac{\pi D^2}{4}(V_b - u_{3.x})|V_b - u_{3.x}|, \tag{A.4.25}$$

or

$$\frac{du_{3,x}}{dt} = a_4\left|\tilde{V}_b - u_{3,x}\right|\left(\tilde{V}_b - u_{3,x}\right). \tag{A.4.26}$$

Integrating Eq. (A.4.26) once and twice with the conditions $u_{3,x} = u_{3,x,0}$ at $x = x_3$, and $u_{3,x} = v_{3,x,2}$ at $t = t_3$ for the case where $\tilde{V}_b > u_{3,x}$ gives

$$\tilde{u}_{3.x.2} = \tilde{V}_b - \frac{\tilde{V}_b - \tilde{u}_{3.x.0}}{1 + \tilde{a}_4\left(\tilde{V}_b - \tilde{u}_{3.x.0}\right)\tilde{t}_3}, \tag{A.4.27}$$

$$\tilde{x}_3 = \tilde{V}_b\tilde{t}_3 - \frac{1}{\tilde{a}_4}\ln\left[1 + \tilde{a}_4\left(\tilde{V}_b - \tilde{u}_{3,x,0}\right)\tilde{t}_3\right], \tag{A.4.28}$$

and

$$\tilde{U}_{3.x} = \frac{\tilde{x}_3}{\tilde{t}_3} = \tilde{V}_b - \frac{\left(\tilde{V}_b - \tilde{u}_{3.x.2}\right)\left(\tilde{V}_b - \tilde{u}_{3.x.0}\right)}{\tilde{u}_{3.x.2} - \tilde{u}_{3.x.0}} \times \ln\left[1 + \tilde{a}_4\left(\tilde{V}_b - \tilde{u}_{3.x.0}\right)\tilde{t}_3\right], \tag{A.4.29}$$

respectively. For the case where $\tilde{V}_b < u_{3,x}$ this is

$$\tilde{u}_{3.x.2} = \tilde{V}_b - \frac{\tilde{V}_b - \tilde{u}_{3.x.0}}{1.\tilde{a}_4\left(\tilde{V}_b - \tilde{u}_{3.x.0}\right)\tilde{t}_3}, \tag{A.4.30}$$

$$\tilde{x}_3 = \tilde{V}_b\tilde{t}_3 - \frac{1}{\tilde{a}_4}\ln\left[1n\tilde{a}_4\left(\tilde{V}_b - \tilde{u}_{3,x,0}\right)\tilde{t}_3\right], \tag{A.4.31}$$

and

$$\tilde{U}_{3.x} = \frac{\tilde{x}_3}{\tilde{t}_3} = \tilde{V}_b - \frac{(\tilde{V}_b - \tilde{u}_{3.x.2})(\tilde{V}_b - \tilde{u}_{3.x.0})}{\tilde{u}_{3.x.2} - \tilde{u}_{3.x.0}} \times \ln\left[1n\tilde{a}_4(\tilde{V}_b - \tilde{u}_{3.x.0})\tilde{t}_3\right], \tag{A.4.32}$$

respectively. Finally, for the case where $V_b = u_{3,x}$, the particle moves at a constant velocity, which gives

$$\tilde{u}_{3.x.2} = \tilde{V}_b, \tag{A.4.33}$$

$$\tilde{x}_3 = \tilde{V}_b\tilde{t}_3, \tag{A.4.34}$$

and

$$\tilde{U}_{3.x} = \tilde{V}_b, \tag{A.4.35}$$

respectively.

The hopping height may reach $6D$, so in hopping calculations the actual longitudinal flow velocity should be used instead of the bottom flow velocity. Therefore, as an approximation, the bottom flow velocity in Eqs. (A.4.27–A.4.35) can be replaced with the average flow velocity along the hopping height. According to experimental data, Nikitin [47] presented an equation for the longitudinal flow velocity in the vertical direction with a rough bed. That is,

$$\frac{V}{u_*} = f\left(\frac{y}{\delta}\right) = \begin{cases} 6.45\frac{y}{\delta} + 8.45ent\frac{\delta}{y}, & \delta \le y \le H, \\ 5.6\frac{y}{\delta}, & 0 \le y \le \delta, \end{cases} \tag{A.4.36}$$

which is more suitable for bed load transport. In Eq. (A.4.36), $\delta = D$. The average flow velocity along the hopping height V_f can be taken as the average of the flow velocity at the end points $y_b = 2/3D$ and $y = y_M - 1/3D$. That is,

$$V_f = \frac{u_*}{2}\left(2.73 + 6.45(\tilde{y}_M - 1/3) + 8.4 + \frac{2.8}{\tilde{y}_M - 1/3}\right). \tag{A.4.37}$$

There is a large difference between V_f and $\overline{V}_b$. For example if $\tilde{\overline{V}}_b = 6.7886$, then $y_M = 6.2404D$ and $V_f = 2.229\overline{V}_b$.

Besides, the longitudinal flow velocity does not affect the vertical motion of particles except the bottom flow velocity.

A.4.3 Condition and Critical Flow Velocity of Incipient Particle Hopping [23]

A.4.3.1 General condition and flow velocity for incipient particle hopping

As discussed above, hopping describes a particle that is lifted to a height exceeding its diameter. Equations (A.4.7), (A.4.9), (A.4.11), and (A.4.14) show that a particle can hop to a height greater than or equal to its diameter only if

$$\tilde{u}_{3,y,D}^2 = \tilde{u}_{3,y,0}^2 e^{-2\tilde{a}_4} - \frac{\tilde{a}_3}{\tilde{a}_4}\left(1 - e^{-2\tilde{a}_4}\right) \geq 0. \tag{A.4.38}$$

This condition is equivalent to

$$\tilde{V}_b > \tilde{V}_{b.c.2}, \tag{A.4.39}$$

where $\tilde{V}_{b.c.2}$ is the critical flow velocity for incipient hopping. This is expressed as

$$\tilde{V}_{b.c.2} = \begin{cases} \sqrt{\tilde{V}_1^2 - \mu^2 \tilde{u}_{3.y.0}^2} = 2\sqrt{12\frac{\mu^2}{4}\tilde{u}_{3.y.0}^2}, & .y. > \mu^2 \tilde{u}_{3.y.0}^2 \\ 0, & \text{(otherwise)}, \end{cases} \tag{A.4.40}$$

where

$$\mu^2 = \frac{\frac{C}{C_y}\exp\left[-\frac{3C}{2}\frac{\rho}{\rho_s+\rho/2}\right]^-}{1 - \exp\left[-\frac{3C}{2}\frac{\rho}{\rho_s+\rho/2}\right]} = 15.56, \tag{A.4.41}$$

and

$$\tilde{V}_1 = \frac{V_1}{\omega_0} = \frac{\sqrt{\frac{4}{3C_y}\frac{\rho_s-\rho}{\rho}gD}}{\sqrt{\frac{4}{3C_x}\frac{\rho_s-\rho}{\rho}gD}} = 2. \tag{A.4.42}$$

Equation (A.4.40) shows that the critical flow velocity of incipient hopping depends on the final velocity of the previous step.

A.4.3.2 Critical Flow Velocity for Incipient Hopping from Stationary State

A stationary particle will start to hop when the lift exceeds the gravitational force. In this case, from Eq. (A.4.40), the critical flow velocity for incipient hopping is

$$\tilde{V}_1 = \tilde{V}_{b.c.1.3} = \tilde{V}_{b.c.2.0} = 2, \tag{A.4.43}$$

where the subscript 1.3 in $\tilde{V}_{b.c.1.3}$ indicates the critical flow velocity required for a particle to transition to the hopping state (state 3) from the stationary state (state 1).

A.4.3.3 Critical Flow Velocity for Incipient Particle Hopping from Motion State

Transitions to the hopping state from motion states can be divided into three cases: from (1) rolling, (2) hopping, and (3) suspension. The critical flow velocity for incipient hopping is related to the final velocity of the previous step (and the initial velocity of the hopping step before a collision), but the probability distributions are difficult to determine. Therefore, a unified formula describing the flow velocities required for incipient hopping from a moving state is presented because the approximate calculation shows that the critical flow velocities are close to each other [23].

(1) The average critical flow velocity required for a particle to transition from the rolling to the hopping state can be obtained if the rolling is uniform. According to Eqs. (A.4.1a, A.4.1b), the vertical component of the hopping velocity of a rolling particle after it collides with a particle on the bed surface becomes

$$\tilde{u}_{3.y.0} = \tilde{u}_{x.0}(1 + k_0)(10\Delta')\sqrt{2\Delta' - \Delta'^2} = \tilde{u}_{x.0} f_1(\Delta'), \tag{A.4.44}$$

where $\tilde{u}_{x.0}$ is the rolling particle velocity before the collision. Assuming that the rolling is uniform, then $\tilde{u}_{x.0}$ can be expressed as

$$\tilde{u}_{x.0} = \tilde{V}_b - \tilde{V}_{b.c.1} = \tilde{V}_b - \varphi(\Delta'). \tag{A.4.45}$$

Substituting this into Eqs. (A.4.44) and (A.4.40) yields

$$\tilde{V}_{b.c.2.3} = \frac{f_1^2\mu^2\varphi(\Delta')}{1 + f_1^2\mu^2}\left[1 + \sqrt{\frac{4(1 + f_1^2\mu^2)}{\varphi^2(\Delta')f_1^4\mu^4} - \frac{1}{f_1^2\mu^2}}\right]$$

$$\tilde{V}_{b.c.2.3} = \frac{f_1^2\mu^2\varphi(\Delta')}{1 + f_1^2\mu^2}\left[1 + \sqrt{\frac{4(1 + f_1^2\mu^2)}{\varphi^2(\Delta')f_1^4\mu^4} - \frac{1}{f_1^2\mu^2}}\right]$$
$$= \frac{f_1^2\mu^2\varphi(\Delta')}{1 + f_1^2\mu^2}\left[1 + \sqrt{A(\Delta')}\right], \quad (\Delta' \neq 1). \tag{A.4.46}$$

From Eq. (A.4.46), the values of $\tilde{V}_{b.c.2.3}$ can be computed. The average is approximately $\overline{\tilde{V}}_{b.c.2.3} = 1.5035$, and the critical flow velocity for incipient particle hopping from the rolling state becomes $\overline{\tilde{V}}_{b.c.2.3} \approx (\overline{\tilde{V}}_{b.c.2.3}/\varphi(\Delta'))\varphi(\Delta') = (1.5035/0.91)\varphi(\Delta') = 1.641\varphi(\Delta')$.

(2) The critical flow velocity for the transition to incipient hopping from suspension is quite complicated. For simplicity, it is assumed that the velocity at the end of the suspension step is given by

$$u_{x.0} = V_b - \varphi(\Delta')\omega_0 \tag{A.4.47a}$$

and

$$u_{y.0} = \overline{u}_{4.y.D} = M[V_{b.y} - \omega | V_{b.y} < \omega]. \tag{A.4.47b}$$

Suspended particles usually go through a short period of rolling before they start hopping, so $u_{x.0}$ is equal the rolling velocity and $u_{y.0}$ can be approximated by the average falling velocity

$$\begin{aligned}\overline{u}_{4.y.D} &= M\left[\xi_{u_{y.0}} = \xi_{V_{b.y}} - \omega\right]\left|\xi_{V_{b.y}} - \omega \le 0\right| \\ &= \int\limits_{-\infty}^{\omega} \frac{1}{1-\varepsilon_4}(V_{b.y} - \omega)p_{V_{b.y}}(V_{b.y})\mathrm{d}V_{b.y} = -\left[\frac{u_*}{(1.\varepsilon_4)\sqrt{2\pi}}e^{-\frac{1}{2}\left(\frac{\omega}{u_*}\right)^2} + \omega\right].\end{aligned} \tag{A.4.48}$$

Here, $p_{V_{b.y}}(V_{b.y})$ is the probability distribution density (PDD) of the vertical flow velocity (Eq. A.1.1) and ε_4 is the probability of incipient suspension (Eq. A.5.10). Substituting Eqs. (A.4.47a, A.4.47b) and (A.4.48) into Eq. (A.4.1a, A.4.1b) and solving the Eq. (A.4.40) yields the flow velocity of incipient hopping from suspension $\tilde{V}_{b.c.4.3}$. The average value is $\overline{\tilde{V}}_{b.c.4.3} = 1.467$, and $\overline{\tilde{V}}_{b.c.4.3} \approx \frac{1.467}{0.916}\varphi(\Delta') = 1.602\varphi(\Delta')$ [23].

(3) The critical flow velocity for the transition to incipient hopping from the hopping state is related to the final velocity of the previous hopping step, and the final velocity distribution is difficult to derive, so a simplified flow velocity is deduced. Equation (A.4.40) shows that the incipient hopping velocity $\overline{\tilde{V}}_{b.c.4.3}$ varies over the range 2.0–0, and $\tilde{u}_{3.y.0}$ varies over the range 0–0.507. Thus, the average incipient hopping velocity is $\overline{\tilde{V}}_{b.c.3.3} = 1.597$, and the incipient hopping velocity is $\overline{\tilde{V}}_{b.c.3.3} \approx \frac{1.5035}{0.916}\varphi(\Delta') = 1.641\varphi(\Delta')$.

The incipient hopping velocity from a motion state can be presented in terms of the average velocity required to transition to incipient hopping from the rolling, hopping, and suspension states,

$$\overline{\tilde{V}}_{b.c.2} = (1.602 + 1.641 + 1.617)/3\varphi(\Delta') = 1.62\varphi(\Delta'). \tag{A.4.49}$$

A.4.4 Mathematical Expectation of Parameters for Particle Hopping Step

A.4.4.1 Expectation of Step Parameters for Hopping Process in Balance

In Sects. A.4.3.2 and A.4.3.3 a mechanical method was used to establish a dynamic system of equations which, when solved, provided all the hopping step parameters expressed as functions of the initial particle position on the bed surface, the bottom flow velocity, and the initial velocity of particle. The PDD of the initial particle position and the bottom flow velocity are given, but the PDD of the initial particle velocity is unknown. A premier study has shown that the initial particle velocity has a kind of limit probability for the stochastic process of sediment motion. Here, an approximate PDD is suggested. For given $\tilde{V}_b$ and Δ', it is assumed that the initial and the final velocities of hopping step in the longitudinal and vertical directions are the same, i.e.,

$$\tilde{u}_{x.0}^{(3)} = \tilde{u}_{3.x.2} = \tilde{u}_{x.0} \text{ and } \tilde{u}_{y.0}^{(3)} = \tilde{u}_{3.y.2} = \tilde{u}_{y.0}. \quad (A.4.50)$$

In Sects. A.4.3.2 and A.4.3.3 $\tilde{u}_{3.x.2}$ and $\tilde{u}_{3.y.2}$ were expressed as functions of Δ', $\tilde{V}_b$,$\tilde{u}_{x.0}$, and $\tilde{u}_{y.0}$, that is

$$\tilde{u}_{3.x.2} = \tilde{f}_{3.x}\left(\Delta', \tilde{V}_b, \tilde{u}_{x.0}, \tilde{u}_{y.0}\right) \text{ and } \tilde{u}_{3.y.2} = \tilde{f}_{3.y}\left(\Delta', \tilde{V}_b\tilde{u}_{x.0}, \tilde{u}_{y.0}\right). \quad (A.4.51)$$

Under the assumption given by Eq. (A.4.50), the relationship between the final velocity (initial velocity) with Δ' and $\tilde{V}_b$ becomes

$$\tilde{u}_{x.0}^{(3)} = \tilde{f}_{3.x}\left(\Delta', \tilde{V}_b\right) \text{ and } \tilde{u}_{y.0}^{(3)} = \tilde{f}_{3.y}\left(\Delta', \tilde{V}_b\right). \quad (A.4.52)$$

Then the probability distribution functions for $\tilde{u}_{x.0}^{(3)}$ and $\tilde{u}_{y.0}^{(3)}$ can be determined from the probability distribution densities of Δ' and $\tilde{V}_b$. Under this condition the step hopping is called hopping in balance.

From Eq. (A.4.52) and the results in Sect. A.4.3.3 the step time period, longitudinal distance, average velocity and time period for incipient hopping in balance can be expressed as the function of $\tilde{V}_b$ and Δ'. Thus their expectations can be expressed as

$$\begin{aligned}
\tilde{x}_3\left(\bar{\tilde{V}}_b, \Delta'\right) &= E\left[\tilde{x}_3\left(\tilde{V}_b, \Delta'\right)\middle|\tilde{V}_b > \tilde{V}_{b.c.2}, \Delta' = \Delta'\right] \\
&= \frac{1}{\varepsilon_2(\Delta')} \int\limits_{\tilde{V}_{b.c.2}(\Delta')}^{+\infty} \tilde{x}_3\left(\tilde{V}_b, \Delta'\right) p\left(\tilde{V}_b\right) d\tilde{V}_b \\
&= \frac{1}{\varepsilon_2(\Delta')} \frac{1}{\sqrt{2\pi}\tilde{\sigma}_x} \int\limits_{\tilde{V}_{b.c.2}(\Delta')}^{+\infty} \tilde{x}_3\left(\tilde{V}_b, \Delta'\right) \exp\left[-\frac{\left(\tilde{V}_b - \bar{\tilde{V}}_b\right)^2}{2\tilde{\sigma}_x^2}\right] d\tilde{V}_b
\end{aligned}$$

$$= \frac{0.399}{\varepsilon_2(\Delta')} \int_{t_2^*(\Delta')}^{+\infty} \tilde{x}_3\left(\tilde{\overline{V}}_b + \tilde{\sigma}_x t, \Delta'\right) \exp\left[-\frac{t^2}{2}\right] dt \tag{A.4.53a}$$

$$\tilde{\overline{x}}_3\left(\tilde{\overline{V}}_b\right) = \mathrm{E}\left[\tilde{x}_3\left(\tilde{\overline{V}}_b, \Delta'\right)\right] = \int_{\Delta'_m}^{1} p(\Delta')\tilde{x}_3\left(\tilde{\overline{V}}_b, \Delta'\right) d\Delta' \tag{A.4.53b}$$

$$\begin{aligned} \tilde{t}_3(\tilde{V}_b, \Delta') &= E\left[\tilde{t}_3(V_b, \Delta') \middle| V_b > V_{b.c.2,} \Delta' = \Delta'\right] \\ &= \frac{1}{\varepsilon_2(\Delta')} \int_{\tilde{V}_{b.c.2}(\Delta')}^{+\infty} \tilde{t}_3(\tilde{V}_b, \Delta') p(\tilde{V}_b) d\tilde{V}_b \\ &= \frac{0.399}{\varepsilon_2(\Delta')} \int_{t_2^*(\Delta')}^{+\infty} \tilde{t}_3\left(\tilde{\overline{V}}_b + \sigma_x t, \Delta'\right) \exp\left[-\frac{t^2}{2}\right] dt \end{aligned} \tag{A.4.54a}$$

$$\tilde{\overline{t}}_3\left(\tilde{\overline{V}}_b\right) = \int_{\Delta'_m}^{1} p(\Delta')\tilde{t}_3\left(\tilde{\overline{V}}_b, \Delta'\right) d\Delta' \tag{A.4.54b}$$

$$\tilde{\overline{t}}_{3.0}\left(\tilde{\overline{V}}_b\right) = \int_{\Delta'_m}^{1} p(\Delta')\tilde{t}_{3.0}\left(\tilde{\overline{V}}_b, \Delta'\right) d\Delta' \tag{A.4.54c}$$

$$\tilde{\overline{U}}_{3.x}\left(\tilde{\overline{V}}_b\right) = \tilde{\overline{x}}_{3.x}\left(\tilde{\overline{V}}_b\right) / \tilde{\overline{t}}_{3.x}\left(\tilde{\overline{V}}_b\right) \tag{A.4.55}$$

where the basic conditional and unconditional probability of transitioning from a motion state to the hopping state are

$$\begin{aligned} \varepsilon_2(\Delta') &= P\left[V_b > V_{b.c.2} \middle| \Delta' = \Delta'\right] = \frac{1}{\sqrt{2\pi}\tilde{\sigma}_x} \int_{\tilde{V}_{b.c.2}(\Delta')}^{+\infty} \exp\left[-\frac{\left(\tilde{V}_b - \overline{V}_b\right)^2}{2\tilde{\sigma}_x^2}\right] d\tilde{V}_b \\ \varepsilon_2 &= P\left[\tilde{V}_b > \tilde{V}_{b.c.2}\right] = \int_{\Delta'_m}^{1} \varepsilon_2(\Delta') p(\Delta') d\Delta' \end{aligned} \tag{A.4.56}$$

and

$$t_2^*(\Delta') = 2.7\left[\frac{1.62\varphi(\Delta')}{\tilde{\overline{V}}_b} - 1\right] \tag{A.4.57}$$

The expectation of time period for incipient hopping of a hopping step in balance $\tilde{\bar{t}}_{0.3} = \tilde{\bar{t}}_{3.0}$ can be obtained similarly.

A.4.4.2 Expectations for Hopping Parameters of Non-uniform Sediment

It is easy to extend the results for uniform sediment to non-uniform sediment. This is achieved by replacing particle size D and particle position Δ'_l in Eqs. (A.4.53a, A.4.53b–A.4.57) with particle size D_l and particle position $\Delta'_l = f_{\Delta'_l}\left(\frac{\overline{D}}{D_l}\right)$ for the l-th size group (see Eq. A.2.5). This yields the unconditional expectations for the particle velocity of the l-th size group. Consider $\tilde{\bar{t}}_{3.l}\left(\overline{V}_b, \overline{D}/D_l\right)$ as an example,

$$\tilde{\bar{t}}_{3.l}\left(\overline{V}_b, \frac{\overline{D}}{D_l}\right) = \tilde{\bar{t}}_3\left(\overline{V}_b, \Delta'_l\right)\Big|\Delta'_l = f_{\Delta'_l}\left(\frac{\overline{D}}{D_l}\right)$$
$$= \frac{0.399}{\varepsilon_2\left(\Delta'_l\right)} \times \int\limits_{t_2^*(\Delta'_l)}^{\infty} \tilde{t}_3\left(\tilde{\overline{V}}_b + \tilde{\sigma}_x t, \Delta'_l\right) e^{-\frac{t^2}{2}} \mathrm{d}t \Big| \Delta'_l = f_{\Delta'_l}\left(\frac{\overline{D}}{D_l}\right), \tag{A.4.58}$$

where

$$t_{2.l}^*\left(\frac{\overline{D}}{D_l}\right) = 2.7\left[\frac{1.62\varphi\left(\Delta'_l\right)}{\tilde{\overline{V}}_b} - 1\right]\Big|\Delta'_l = f_{\Delta'_l}\left(\frac{\overline{D}}{D_l}\right) = 2.7\left[\frac{1.62\varphi\left(f_{\Delta'_l}\left(\frac{\overline{D}}{D_l}\right)\right)}{\tilde{\overline{V}}_b} - 1\right], \tag{A.4.59}$$

and

$$\varepsilon_{2.l}\left(\frac{\overline{D}}{D_l}\right) = P\left[V_b > 1.62\varphi\left(\Delta'_l\right)\omega_{0.l}\right]\Big|\Delta'_l = f_{\Delta'_l}\left(\frac{\overline{D}}{D_l}\right)$$
$$= \frac{1}{\sqrt{2\pi}\tilde{\sigma}_x} \int\limits_{\tilde{V}_{b.c.2}}^{+\infty} p\left(\tilde{V}_b\right) \mathrm{d}\tilde{V} = \frac{1}{\sqrt{2\pi}} \int\limits_{t_2^*(\Delta'_l)}^{+\infty} e^{-\frac{t^2}{2}} \mathrm{d}t \Big| \Delta'_l = f_{\Delta'_l}\left(\frac{\overline{D}}{D_l}\right). \tag{A.4.60}$$

Here, the definition of particle position for non-uniform sediment was used, and the integration over the particle position was omitted.

A.4.5 Characteristics of Hopping Parameters

The numerical parameter expectations for a hopping step in balance are listed in Table A.1. Overall, the characteristics of the hopping parameters can be described as follows:

(1) The variation in the parameter expectations as $\tilde{\overline{V}}_b$ increases are shown in Table A.1.
(2) The longitudinal hopping distance $\tilde{\overline{x}}_3$, average hopping velocity $\tilde{U}_{3.x}$, and hopping height $\tilde{\overline{y}}_M$ are the three key parameters used to describe the trajectory of a hopping particle, and they are important parameters for determining the hopping transport rate. As listed in Table A.1, there are large variations in these parameters. The hopping distance varies $16D$ to $269D$, the hopping height varies from $1.75D$ to $6.35D$, and the hopping velocity varies from $1.07\omega_0$ to $13.6\omega_0$. This is quite different from the values proposed by [16] ($\tilde{\overline{y}}_m = 2D$, $\tilde{\overline{x}}_3 = 100D$ for high velocities). Some experts have modified the values for high flow velocities according to experimental results [53].
(3) The bottom flow velocities that contribute to particle hopping are acting bottom flow velocities greater than $\tilde{\overline{V}}_{b.c.2}$ and they have an average value of

$$\tilde{\overline{V}}_{b.f.3} = M\left[\xi_{\tilde{V}_b}(\tilde{V}_b)\middle|\tilde{V}_{b.c.2} \le \tilde{V}_b\right] = \int\limits_{0.134}^{1} p_{\Delta'}(\Delta')d\Delta'\frac{1}{\varepsilon_2(\Delta')}\int\limits_{\tilde{V}_{b.c.2}}^{+\infty} \tilde{V}_b p(\tilde{V}_b)d\tilde{V}_b$$

$$= \tilde{\overline{V}}_b\left\{1 + \frac{0.37\tilde{\overline{V}}_b}{\varepsilon_2\sqrt{2\pi}}\exp\left[-3.65\left(\frac{1.62\overline{\varphi}(\Delta')}{\tilde{\overline{V}}_b} - 1\right)^2\right]\right\} \quad \text{(A.4.61)}$$

which is also called the acting flow velocity. In Eq. (A.4.61), $\varphi(\Delta')$ can be substituted for its average value $\overline{\varphi}(\Delta') = 0.916$.

A.4.6 Verification of Hopping Parameters

The hopping parameters were investigated based on data measured in flume experiments, and data obtained by Hu et al. [29] was compared to the theoretical results. Nineteen sets of data published by Hu, corresponding to four dry densities, were used for the comparison. In each data set, a single-step parameter was represented by the average value of these parameters over the continuous hopping steps within a single hopping event. A comparison between Hu's data and the computed expectation values is shown in Fig. A.12 for the hopping distance, hopping height, and hopping speed. It should be noted that the acting flow velocity $\tilde{\overline{V}}_f$ and acting bottom flow velocity $\tilde{\overline{V}}_{b.f.3}$ were used in the authors' calculations. The measured data and theoretical results (relation lines) generally agree with each other. Among them, the

Table A.1 Expectations of parameters for hopping particle in balance

$\tilde{\bar{V}}_b$	$\tilde{\bar{V}}_{b.f.3}$	$\tilde{\bar{u}}_{3.y.0}$	$\tilde{\bar{u}}_{3.x.0}$	$\tilde{\bar{u}}_{3.x.2}$	$\tilde{\bar{u}}_{3.y.D}$	$\tilde{\bar{y}}_m$	$\tilde{\bar{u}}_{3.y.2}$	$\tilde{\bar{t}}_3$	$\tilde{\bar{t}}_{0.3}$	$\tilde{\bar{t}}_{3.1}$	$\tilde{\bar{t}}_{3.2}$	$\tilde{\bar{x}}_3$	$\tilde{U}_{3.x.} = \tilde{\bar{x}}_3/\tilde{\bar{t}}_3$
0.2238	NaN	NaN	NaN	NaN	NaN	NaN	NaN	NaN	NaN	NaN	NaN	NaN	
0.2984	NaN	NaN	NaN	NaN	NaN	NaN	NaN	NaN	NaN	NaN	NaN	NaN	
0.4476	0.65317	0.60558	− 0.9655	1.32353	0.37646	1.75619	− 0.4474	15.7012	6.07309	3.11664	6.51147	16.8205	1.07129
0.5222	0.76781	0.62318	− 0.9782	1.35634	0.39962	1.81419	− 0.4502	16.0031	6.12661	3.24929	6.62725	17.6993	1.10599
0.6714	1.00638	0.66784	− 1.0096	1.43719	0.45796	1.95261	− 0.4537	16.0856	5.65587	3.55152	6.87825	18.9365	1.17723
0.8206	1.25968	0.72204	− 1.046	1.53284	0.5269	2.10449	− 0.4521	15.7267	4.76956	3.84665	7.11052	20.0771	1.27662
0.9698	1.52894	0.7875	− 1.0843	1.64138	0.60443	2.26573	− 0.4462	15.2721	3.83524	4.11094	7.32595	21.7356	1.42322
1.119	1.81359	0.867	− 1.1246	1.76409	0.69037	2.43737	− 0.4385	14.9071	3.01921	4.34242	7.54544	24.1605	1.62074
1.2682	2.11193	0.96109	− 1.1696	1.90315	0.78416	2.61833	− 0.4308	14.698	2.37186	4.54313	7.78299	27.3886	1.86343
1.492	2.58049	1.12637	− 1.2513	2.14445	0.93773	2.90047	− 0.4216	14.668	1.69309	4.79573	8.17922	33.5976	2.29053
1.6412	2.90395	1.24953	− 1.3163	2.32578	1.04739	3.09062	− 0.4172	14.7898	1.38741	4.93877	8.46365	38.5091	2.60376
2.0142	3.73995	1.5891	− 1.5102	2.8351	1.3408	3.55358	− 0.4112	15.3535	0.92299	5.23211	9.1984	52.8555	3.44257
2.3872	4.60208	1.95785	− 1.7357	3.39864	1.65322	3.98143	− 0.409	16.0461	0.67861	5.45757	9.90989	69.3797	4.32378
3.4316	7.08819	3.06341	− 2.4452	5.11204	2.57854	4.9678	− 0.4089	17.8462	0.38271	5.86503	11.5984	122.807	6.8814
4.0284	8.53589	3.71703	− 2.8737	6.13274	3.12279	5.41634	− 0.4094	18.689	0.30525	6.00932	12.3745	156.446	8.37102
4.9982	10.9138	4.72812	− 3.5574	7.75558	3.96586	5.99199	− 0.4094	19.7535	0.22985	6.1567	13.367	211.951	10.7298
5.8188	12.9442	5.33486	− 3.9916	8.79199	4.47539	6.25864	− 0.4048	20.147	0.19221	6.16164	13.7931	248.803	12.34494
6.7886	15.3606	5.59569	− 4.1919	9.29211	4.69751	6.24039	− 0.3891	19.7786	0.16394	5.96884	13.6459	268.765	13.588888886

test data for $\gamma_s = 2.65$, 2.20, and 1.043 g/cm^3, and for the average velocity of step hopping show the best agreement with the theoretical results. The velocity is a key parameter in determining the bed load transport rate, so its reliability is important.

The larger errors in some points may be related to the assumption that hopping is a transport process in balance, and further analysis of the adopted rebound coefficient is required. However, these errors are acceptable for sediment studies.

A.5 Step Suspension

A.5.1 Mechanism of Sediment Suspension and a Simplified Diagram

Particle suspension is also very complicated. Thus, the concepts of critical condition and critical flow velocity for incipient suspension of fine particles from the rest state, the incipient suspension time, and the suspension height are introduced for the first time. Subsequently, expressions for these concepts are derived. The critical condition for the transition from rest to the suspended state is based on a dynamic equation, and the forces acting on the particle include the cohesive force and the additional downward press of water. The incipient suspension time is derived from the bursting period. The suspension height is related to the concentration of the suspended load, from which the step suspension distance in longitudinal direction is derived.

The suspension of sediment is caused by the vertical component of flow turbulence. However, turbulent vortices have different scales, and they are constantly generated and broken. Consequently, the vertical component at a given position varies with time, and there is no correlation between different positions unless they are a sufficiently small distance apart. It is conceivable that the movement of suspended sediment is extremely complicated: sometimes rising and sometimes falling. The sediment can move upward until it reaches the surface of the water, and it can fall to the bed surface and remain on the bottom for a period of time. However, the vertical distribution of the suspended sediment concentration is stable for equilibrium sediment transport in steady uniform flow. This indicates that the suspended sediment movement has certain statistical characteristics.

Measured data shows that the vertical component of the flow velocity (such as its variance or fluctuation intensity) only varies slightly. As discussed before, the vertical component of the flow velocity follows a normal distribution with a mean value of zero. These facts indicate that the non-uniform vertical distribution of the sediment concentration is not caused by variations in the vertical velocity component with depth. Instead, it is caused by diffusion from the large sediment source on the bed combined with the vortices. The vortices break as they move upward, hence the sediment concentration decreases as the vertical distance from the bed increases. The fact that the vertical distribution of suspended sediment is relatively uniform in channels with pebble beds reflects the role of the diffusion source. An individual

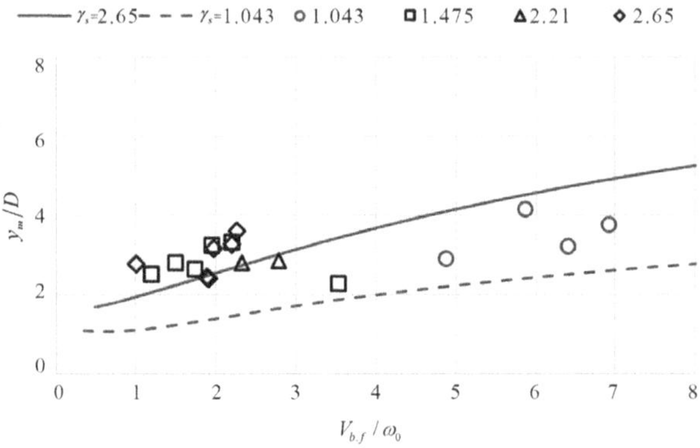

(a) hopping height

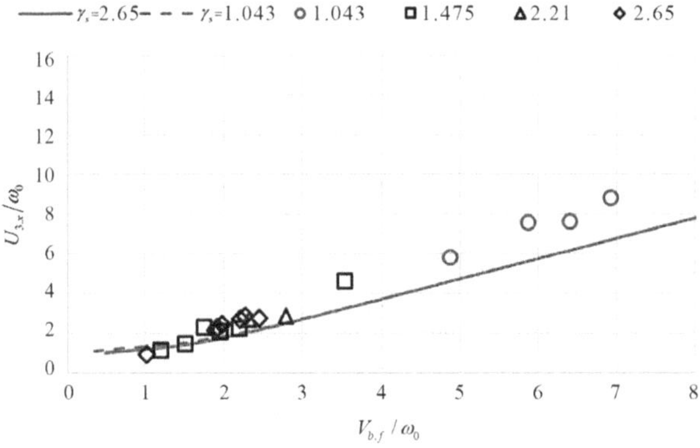

(b) average hopping velocity

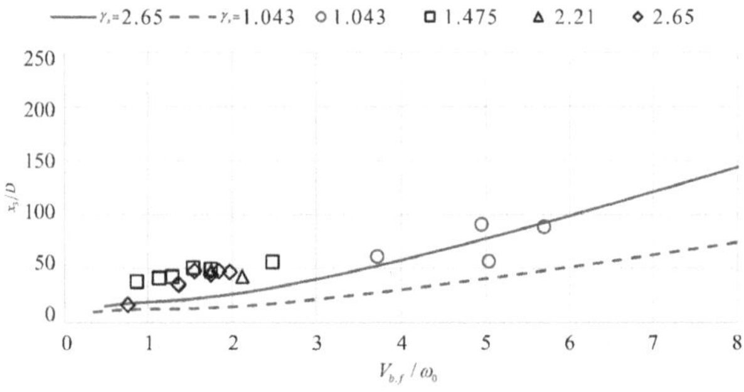

(c) hopping distance

Fig. A.12 Comparison of average hopping parameters between experiments and theory

sediment particle has different probabilities of being suspended at different vertical positions. Sediment transitions from the bed material or bed load to the suspended load in the bed surface layer, and vortices break as they rise, so the probability of a particle being suspended decreases as the height from the bed increases. Moreover, the maximum suspension height of a sediment particle is random, but higher the suspension height, the smaller the probability of suspension.

The variance in the vertical fluctuating velocity $\sigma^2_{V_{b,y}}$ does not change significantly in the vertical direction, and there are no reliable data on the vertical distribution of $\sigma^2_{V_{b,y}}$. As a first approximation, it is assumed that $\sigma^2_{V_{b,y}}$ is constant over the depth of the water, with a mathematic expectation of zero in the vertical direction.

Sedimentary particles have greater density than water, so there is always relative movement between the sediment and the surrounding fluid with a certain acceleration. For simplicity, this relative movement is assumed to be uniform. That is, the relative movement has uniform velocity in the vertical direction, and the particles move with the fluid in the longitudinal direction. This can be expressed as

$$\left.\begin{array}{l} |u_{4.x}(y) = V(y) \\ |u_{4.y} = V_y - \omega \end{array}\right\}, \tag{A.5.1}$$

where variations in the longitudinal flow velocity component $V(y)$ in the vertical direction is not ignored.

The simplified diagram of sediment suspension can be summarized as follows. The fluctuating vertical velocity component of turbulent flow causes sedimentary particles on the bottom to transition from one of four states (resting, rolling, hopping, and suspension) to the suspension state. The suspended particles move upward, reach a maximum suspension height h, then descend to the bed surface where they transition to the hopping, rolling, or resting state, or continue to be suspended with a certain probability. The suspension height for a given particle is random. A suspension step is defined as the period in which a particle is first suspended from the bed, reaches the suspension height, then falls back to the bed surface. A suspension event is defined as the period from when a particle enters the suspension state until it leaves. One suspension event is composed of multiple suspension steps. The number of suspension steps in a given suspension event is random, and it is determined by the probability of continuous suspension. In the vertical direction, there is relative motion between the particle and the surrounding fluid as the particle rises and falls. In contrast, the motion of the particle in the longitudinal direction is consistent with the fluid.

A.5.2 Critical Condition and Basic Probabilities of Incipient Suspension

The additional downward pressure and cohesive force experienced by moving particles or coarse particles suspended from static state can be omitted, and the suspension condition can be expressed as

$$\tilde{V}_{b.y} > \tilde{\omega} = \tilde{V}_{b.y.c.4}. \tag{A.5.2}$$

However, for fine particles suspended from the stationary state, it is necessary to overcome the adhesion force between particles and the additional downward pressure of the water film, so the particles can be suspended only after loosening. When a fine particle is at rest, it is subjected to underwater gravity force G (Eq. A.3.2c), lift force P_y (see Eq. A.3.2b), the hydrodynamic force of water flow $\theta_y. = (\rho C/2)(\pi D^2/4)V_{b.y}^2$, adhesion force P_μ, Eq. (B.3.12) and the additional downward pressure of the water film ΔG(Eq. B.3.15). Therefore, the loosening condition is

$$P_\mu + \Delta G + G < P_y + \theta_y. \tag{A.5.3}$$

Suppose that the longitudinal and vertical components of the bottom flow velocity are linearly related with a correlation coefficient of − 1; then,

$$\tilde{V}_b = -\frac{\sigma_x}{\sigma_y}\tilde{V}_{b.y} + \tilde{\tilde{V}}_b = -2\tilde{V}_{b.y} + \tilde{\tilde{V}}_b. \tag{A.5.4}$$

Substituting this and the expressions of each force into Eq. (A.5.3) gives

$$\left(\frac{4C_y}{C} + 1\right)\tilde{V}_{b.y}^2 - \frac{4C_y}{C}\tilde{\tilde{V}}_b\tilde{V}_{b.y} + \left(\frac{C_y}{C}\tilde{\tilde{V}}_b^2 - \frac{C_x}{C}\tilde{\omega}_1^2\right) > 0, \tag{A.5.5}$$

where $\tilde{\omega}_1 = \omega_1/\omega_0$ and ω_1 is the term after considering the adhesion and additional downward pressure of the water film (see Eq. A.3.8). The roots of the quadratic equation in Eq. (A.5.5) are

$$\tilde{V}_{b.y.c.3}^{(1),(2)} = \frac{\tilde{\tilde{V}}_b}{8} \pm \sqrt{\frac{1}{4}\tilde{\omega}_1^2 - \frac{3}{64}\tilde{\tilde{V}}_b^2}. \tag{A.5.6}$$

From this, the event of a stationary particle entering the suspension state can be expressed as

$$B = \begin{cases} \left\{\xi_{\tilde{V}_{b.y}} > \omega\right\}, & \text{if } \left(\tilde{\overline{V}}_b \geq \frac{4\tilde{\omega}_1}{\sqrt{3}}, \text{ or } \overline{V}_b \leq \frac{4\omega_1}{\sqrt{3}} \text{ and } V_{b.y.c.3}^{(1)} \leq \omega\right), \\ \left\{\xi_{\tilde{V}_{b.y}} > \tilde{V}_{b.y.c.3}^{(1)}\right\}, & \text{if } \left(\tilde{\overline{V}}_b < \frac{4\tilde{\omega}_1}{\sqrt{3}} \text{ and } \tilde{V}_{b.y.c.3}^{(2)} \leq \tilde{\omega} < \tilde{V}_{b.y.c.3}^{(1)}\right), \\ \left\{\xi_{\tilde{V}_{b.y}} > \omega\right\} \cap \left\{\xi_{\tilde{V}_{b.y}} < \tilde{V}_{b.y.c.3}^{(2)}\right\} \cup \left[V_{b.y} > V_{b.y.c.3}^{(1)}\right], & \text{if} \left(\tilde{\overline{V}}_b < \frac{4\tilde{\omega}_1}{\sqrt{3}} \text{ and } \tilde{\omega} < V_{b.y.c.3}^{(2)}\right). \end{cases} \tag{A.5.7}$$

Thus, the basic probability of incipient suspension from the stationary state for a fine particle is $\beta = P\left\{V_{b.y} \in B\right\}$. That is,

$$\beta = \begin{cases} \frac{1}{\sqrt{2\pi}} \int_{\frac{\omega}{u_*}}^{\infty} e^{-\frac{t^2}{2}} dt = \varepsilon_4 & \left(\overline{V}_b > \frac{4\omega_1}{\sqrt{3}} \text{ or } \overline{V}_b \leq \frac{4\omega_1}{\sqrt{3}} \text{ and } V_{b.y.c.3}^{(1)} \leq \omega\right) \\ \frac{1}{\sqrt{2\pi}} \int_{\frac{V_{b.y.c.3}^{(1)}}{u_*}}^{\infty} e^{-\frac{t^2}{2}} dt = \varepsilon_{5.1} & \left(\overline{V}_b \leq \frac{4\omega_1}{\sqrt{3}} \text{ and } V_{b.y.c.3}^{(2)} \leq \omega < V_{b.y.c.3}^{(1)}\right) \\ \frac{1}{\sqrt{2\pi}} \int_{V_{b.y.c.3}^{(1)}}^{\infty} e^{-\frac{t^2}{2}} dt + \frac{1}{\sqrt{2\pi}} \int_{\frac{\omega}{u_*}}^{\frac{V_{b.y.c.3}^{(2)}}{u_*}} e^{-\frac{t^2}{2}} dt = \varepsilon_{5.1} + \varepsilon_4 - \varepsilon_{5.2}, & \\ .2 & \left(\overline{V}_b \leq \frac{4\omega_1}{\sqrt{3}} \text{ and } \omega < V_{b.y.c.3}^{(2)}\right) \end{cases} \tag{A.5.8}$$

where

$$\varepsilon_{5.2} = \frac{1}{\sqrt{2\pi}} \int_{\frac{V_{b.y.c.3}^{(2)}}{u_*}}^{\infty} e^{-\frac{t^2}{2}} dt. \tag{A.5.9}$$

The basic probabilities of incipient suspension from rolling, hopping and suspension are

$$\varepsilon_4 = \frac{1}{\sqrt{2\pi}} \int_{\frac{\omega}{u_*}}^{\infty} e^{-\frac{t^2}{2}} dt. \tag{A.5.10}$$

A.5.3 Incipient Suspension Period

An important aspect of suspended load movement is its exchange with the material on the bed surface. The transfer between the bed material and the suspended load can be characterized by the intensity of the incipient suspension, which is defined as the number of particles that are picked up from a unit area per unit time. However, the incipient suspension intensity of a particle is affected by the incipient suspension time, which is the time for a particle to be lifted from the bed surface to a height equivalent to the particle diameter. A longer incipient suspension time means that the particles below the suspended particles are protected against being picked up for longer, so the incipient suspension intensity is lower.

To calculate the incipient suspension time $\tilde{t}_{0.4}$, a dynamic equation was established by considering the forces acting on a particle [24]. However, test data and a thorough investigation of the theory and concepts revealed that the computed time for particles to be suspended from the bed $t_{0.4}$ was too short for fine particles, and that busting is a major mechanism in sediment suspension. The effect of turbulence bursting on incipient suspension time has been discussed in Sect. 4.2 of Chap. 4.

A.5.4 Suspension Height and Step Distance of Suspension

The single step suspension distance is an important parameter for the transition intensity and transport rate. To calculate it, the suspension height must first be introduced.

A.5.4.1 Suspension Height [24]

According to the diagram of sediment suspension, within a single suspension step a particle is lifted from the bed surface, reaches suspension height h, and then falls back down to the bed surface. As a particle moves vertically above the bed surface, it moves up with speed $u_{y.u} = V_{b.y} - \omega, \quad \left(V_{b.y} > \omega\right)$ and falls with speed $u_{y.d} = V_{b.y} + \omega\left(V_{b.y} < \omega\right)$. As $V_{b.y}$ is a random variable, $u_{y.u}$ and $u_{y.d}$ are both random variables.

Assuming that the conditional probability distribution of particle position y below a certain elevation h on the vertical line is uniform, it can be approximated as

$$F_{\eta_y|\eta_h}(y) = P\left[\eta_y < y|\eta_h = h\right] = \begin{cases} 1, \ h < y \le H, \\ \frac{y}{h}, \ 0 < y \le h, \\ 0, \ y = 0, \end{cases} \tag{A.5.11}$$

where H is the water depth, h is the suspension height, and y is the particle position. Then the distribution function of the particle position on the vertical line can be expressed as

$$\begin{aligned} F_{\eta_y}(y) &= P\left[\eta_y < y\right] = \int_0^H F_{\eta_y|\eta_h}(y)\mathrm{d}F_{\eta_h}(h) \\ &= \int_0^y F_{\eta_y|\eta_h}(y)\mathrm{d}F_{\eta_h}(h) + \int_y^H F_{\eta_y|\eta_h}(y)\mathrm{d}F_{\eta_h}(h) = F_{\eta_h}(y) + \int_y^H \frac{y}{h}p_{\eta_h}(h)\mathrm{d}h, \end{aligned} \tag{A.5.12}$$

which demonstrates the relationship between $F_{\eta_y}(y)$, the probability of particle position being below the water depth y and the probability distribution function of the suspension height.

Now the relationship between $F_{\eta_y}(y)$ and the sediment concentration will be established. Consider a unit water column with unit area and height equal to the water depth. In the column, the particle number is

$$n = \frac{\int_0^H S(y)\mathrm{d}y}{\pi\,\gamma_s D^3/6}. \tag{A.5.13}$$

The particle number in a unit water column with height y is

$$n_y = \frac{\int_0^y S(y)\mathrm{d}y}{\pi\,\gamma_s D^3/6}. \tag{A.5.14}$$

Based on Borel's theorem of large numbers

$$P\left[\frac{n_y}{n} \xrightarrow[n\to\infty]{} F_{\eta_y}(y)\right] = 1. \tag{A.5.15}$$

Thus,

$$\frac{\int_0^y S(y)\mathrm{d}y}{\int_0^H S(y)\mathrm{d}y} = F_{\eta_h}(y) + \int\limits_y^H \frac{y}{h} p_{\eta_h}(h)\mathrm{d}h. \tag{A.5.16}$$

Differentiating twice with respect to y and using the expression for the average concentration in the y direction $\overline{S} = (1/H)\int_0^H S(y)\mathrm{d}y$, gives the distribution density

$$p_{\eta_h}(h) = -\frac{S'(h)}{\overline{S}}\frac{h}{H}, \tag{A.5.17}$$

the probability function of the suspension height

$$F_{\eta_h}(h) = \int\limits_0^h p_{\eta_h}(h)\mathrm{d}h = -\int\limits_0^h \frac{hS'(h)}{H\overline{S}}\mathrm{d}h = \frac{1}{H\overline{S}}\int\limits_0^h S(h)\mathrm{d}h - \frac{hS(h)}{H\overline{S}}. \tag{A.5.18}$$

and the mathematical expectation of the suspension height is

$$\overline{h} = M[\eta_h] = \int\limits_0^H hp_{\eta_h}(h)\mathrm{d}h = -\int\limits_0^H \frac{S'(h)h^2}{H\overline{S}}\mathrm{d}h = \int\limits_0^H \frac{2h}{H\overline{S}}S(h)\mathrm{d}h. \tag{A.5.19}$$

Because $F_{\eta_h}(H) = 1cS(H)/\overline{S}$ satisfies $F_{\eta_h}(H) = 1$, a boundary condition for the water surface $S(H) = 0$ is required, which indicates that the sediment concentration at the surface is zero. Equations (A.5.17)–(A.5.19) relate the probability distribution of suspension height to the vertical distribution of sediment concentration and demonstrate the statistical significance of the vertical distribution of sediment concentration.

Assuming the suspension height has an exponential probability distribution (Eq. 6.36 in Sect. 6.3 of Chap. 6)

$$S(y) = S_b \exp\left(-\frac{6\omega C y}{\kappa u_* h}\right) \tag{A.5.20}$$

yields the mathematical expectation of the suspension height (Eq. 6.39)

$$\eta_{\overline{h}} = \frac{\overline{h}}{H} = 2\left\{\frac{\kappa u_*}{6\omega C} - \exp\left(-\frac{6\omega C}{\kappa u_*}\right)\left[1 - \exp\left(-\frac{6\omega C}{\kappa u_*}\right)\right]^{-1}\right\} \tag{A.5.21}$$

A.5.4.2 Expectation of Step Distance and Step Velocity for Particle Suspension

The step distance is related to the suspension height. The horizontal distances over which a particle travels as it moves from the bed surface to the suspension height $\overline{x}_{4.u}$, and from the suspension height to the bed surface $\overline{x}_{4.d}$ can be expressed as

$$\overline{x}_{4.u} = \frac{1}{\overline{u}_{4.y.u}} \int_0^{\overline{h}} V(y)\mathrm{d}y = \frac{q(\overline{h})}{\overline{u}_{4.y.u}}, \tag{A.5.22a}$$

and

$$\overline{x}_{4.d} = \frac{1}{\overline{u}_{4.y.d}} \int_0^{\overline{h}} V(y)\mathrm{d}y = \frac{q(\overline{h})}{\overline{u}_{4.y.d}}, \tag{A.5.22b}$$

where $\overline{u}_{4.y.u}$ and $\overline{u}_{4.y.d}$ are the average velocity in the upward and downward vertical direction respectively (see Eq. A.5.1) and $q(\overline{h})$ is the flow discharge per unit distance between the bed and the suspension height $\overline{h}$. The step distance for a sediment particle $\overline{x}_4$, is the sum of $\overline{x}_{4.u}$ and $\overline{x}_{4.d}$, or

$$\overline{x}_4 = q(\overline{h})\left(1/\overline{u}_{4.y.u} + 1/\overline{u}_{4.y.\mathrm{d}}\right) \tag{A.5.23}$$

Now, $q\left(\overline{h}\right)$ will be derived. Assuming that the flow velocity distribution in the longitudinal direction is parabolic yields

$$V(\eta) = V_b + \frac{3u_*}{\kappa}[1((1\left(\eta\right)^2] = \overline{V} + \frac{u_*}{\kappa} - \frac{3u_*}{\kappa}(1(\eta)^2, \tag{A.5.24}$$

where $\eta = y/H$ and $\overline{V} = \int_0^1 V(\eta)\mathrm{d}\eta = V_b + (2u_*/\kappa)$ are the relative height and average flow velocity with depth, respectively. The relative flow discharge cumulatively from the bed to suspension height $\overline{h}$ is

$$\begin{aligned} \frac{q\left(\overline{h}\right)}{q} &= \frac{1}{H\overline{V}} \int\limits_0^{\overline{h}} V\mathrm{d}y = \frac{\overline{h}}{H\overline{V}} \int\limits_0^{\eta_h} V\mathrm{d}\eta = \frac{\overline{h}}{H\overline{V}} \left\{ \left(\overline{V} + \frac{u_*}{\kappa} \right) \eta_{\overline{h}} + \frac{u_*}{\kappa} \left[(1(\eta_{\overline{h}})^3 - 1 \right] \right\} \\ &= \left(1 \frac{2u_*}{\kappa \overline{V}} \right) \eta_{\overline{h}} + \frac{3u_*}{\kappa \overline{V}} \eta_{\overline{h}}^2 - \frac{u_*}{\kappa \overline{V}} \eta_{\overline{h}}^3 = q_1\left(\eta_{\overline{h}}\right), \end{aligned} \tag{A.5.25}$$

where $q = H\overline{V}$ is the flow discharge of the cross section per unit width and $\eta_{\overline{h}} = \overline{h}/H$ the relative suspension height. In order to determine the relation between $\overline{V}$ and u_* the Eqs. (6.28) or (4.36) can be used.

The flow velocity is random, so from Eq. (A.5.1) the particle velocities in vertical direction $u_{4.y.u}$ and $u_{4.y.d}$ are random variables, and their mathematical expectations are

$$\frac{\overline{u}_{4.y.u}}{\omega} = M\left[\frac{V_{b.y}}{\omega} - 1 \middle| \frac{V_{b.y}}{\omega} - 1 \geq 0\right] = \frac{1}{\sqrt{2\pi}\,\varepsilon_4} \frac{u_*}{\omega} \exp\left[-\frac{1}{2}\left(\frac{\omega}{u_*}\right)^2\right] - 1, \tag{A.5.26a}$$

and

$$\frac{\overline{u}_{4.y.d}}{\omega} = \frac{1}{\sqrt{2\pi}\,(1 - \varepsilon_4)} \frac{u_*}{\omega} \exp\left[-\frac{1}{2}\left(\frac{\omega}{u_*}\right)^2\right] + 1, \tag{A.5.26b}$$

respectively. Substituting Eqs. (A.5.25) and (A.5.26a, A.5.26b) into Eq. (A.5.23) gives the single-step distance for a suspended particle as

$$\begin{aligned} \overline{x}_4 = \frac{q_1(\overline{h}/H)}{\omega} \Bigg\{ & \left\{ \frac{1}{\sqrt{2\pi}} \frac{u_*}{\omega\varepsilon_4} \exp\left[-\frac{1}{2}\left(\frac{\omega}{u_*}\right)^2\right] - 1 \right\}^{-1} \\ & + \left\{ \frac{1}{\sqrt{2\pi}} \frac{u_*}{\omega(1 - \varepsilon_4)} \exp\left[-\frac{1}{2}\left(\frac{\omega}{u_*}\right)^2\right] + 1 \right\}^{-1} \Bigg\} \end{aligned} \tag{A.5.27}$$

A.6 Stochastic Process of Sediment Transport

A.6.1 Introduction

In recent decades studies have shown that sediment transport may be described as a Markov process with discrete states and a continuous time (or distance) parameter (Sun [52–54], Tsai [55], Ancey [1–4], Wu [60],Lisle [36])]. However, either some parameters and their distributions resulted in a half-empirical method, or appropriate assumptions were not made.

By analyzing step motion using a combination of mechanical and probabilistic methods, the authors have developed a complete Markov stochastic model of sediment transport, as shown in Sect. A.2.2 and Fig. A.1 [23, 24]. In this Section the stochastic model of sediment transport will be discussed in detail.

The sediment transport rate is one of the important topics in sediment movement. As discussed in Sect. A.1, Einstein's model of the average transport rate of the bed load pioneered the use of probability theory in the study of sediment movement. Compared with equations for the transport rate derived using mechanical methods the work by Einstein and his successors represented great progress (Dou [13], Kalinski [31], Rosinski and Lubomirova [50], Velikanov [58], Yalin [62]. Strictly speaking,the average transport rate model is not a random model. There are also some problems in the structure and parameters of established average models. In general, the formulae obtained using mechanical methods are only applicable to scenarios with a high transport rate, because incipient motion probability is not considered. Moreover, such methods have a zero point and are unreasonable. The model of the Markov stochastic process is more complete in theory than the average transport rate model of bed load. Based on stochastic theory, it not only includes formulae for the transport rate under weak and strong equilibrium conditions, but also their probability distributions [23, 24]. Furthermore, the parameters involved in the formulae for the transport rare can all be derived through the study of step motion using a combination of mechanical and probabilistic methods. This is a complete theoretical model of transport rates.

The transition intensity is a more important concept than the transport rate. All the expressions of transport rate can be derived from transition intensities and the application of transition intensity has solved a series of long-standing problems. such as the boundary conditions for the diffusion equations for suspended loads, non-equilibrium transport of non-uniform suspended loads, long-distance scouring of the reaches downstream from dam sites, similarities and differences between wash loads and bed loads, and the multi-value of carrying sediment capacity [23]. This Appendix will list some of the study results verified using experimental and field data.

A.6.2 *Markov Chain of Particle Movement with Probabilities of State Transitions and Stationary Probabilities of State*

A.6.2.1 Markov Chain of Particle Movement

Before studying a stochastic process with a continuous parameter, a simple model is introduced: the Markov chain with four states. Here, $\{Z_n = i; i = 1, 2, 3, 4; n \in I\}$, where $i = 1, 2, 3, 4$ represents the rest, rolling, hopping, and suspension states, respectively. The sixteen transition probabilities are given by

$$\beta_{i.j} = P\{Z_{n+1} = j|Z_n = i\} \geq i|i, j = 1, 2, 3, 4), \tag{A.6.1}$$

which is not dependent on n. There exists

$$\sum_{j \in I} \beta_{i.j} = 1, \quad (i = 1, 2, 3, 4), \tag{A.6.2}$$

and usually $\beta_{i.j} \neq 0$. Moreover, the movement of the $(n + 1)$-th step depends on that of the n-th step, but not on the steps before n. Therefore the process is a homogeneous, aperiodic, and positive recurrence Markov chain, or in summary, an ergodicity Markov chain. Thus, the transition probabilities through n steps have the limits

$$\lim_{\text{nim}} \beta_{i.j}^{(n)} = \lim_{n \to \infty} P[Z_n = j/Z_0 = i] = R_j. \tag{A.6.3}$$

The limitation probabilities R_j are stationary probabilities of the states (or state probabilities) and satisfy the Kolmogorov equation

$$\begin{pmatrix} R_1 \\ R_2 \\ R_3 \\ R_4 \end{pmatrix} = (R_1 R_2 R_3 R_4) \begin{pmatrix} \beta_{1.1}\beta_{1.2}\beta_{1.3}\beta_{1.4} \\ \beta_{2.1}\beta_{2.2}\beta_{2.3}\beta_{2.4} \\ \beta_{3.1}\beta_{3.2}\beta_{3.3}\beta_{3.4} \\ \beta_{4.1}\beta_{4.2}\beta_{4.3}\beta_{4.4} \end{pmatrix}, \tag{A.6.4}$$

where the stationary probabilities satisfy

$$\sum_{j \in I} R_j = 1. \tag{A.6.5}$$

These equations can be solved using the state transition matrix $\{\beta_{i.j}\}$, which will be described later. The transition chains for sediment motion state in the bed layer is shown in Fig. A.13.

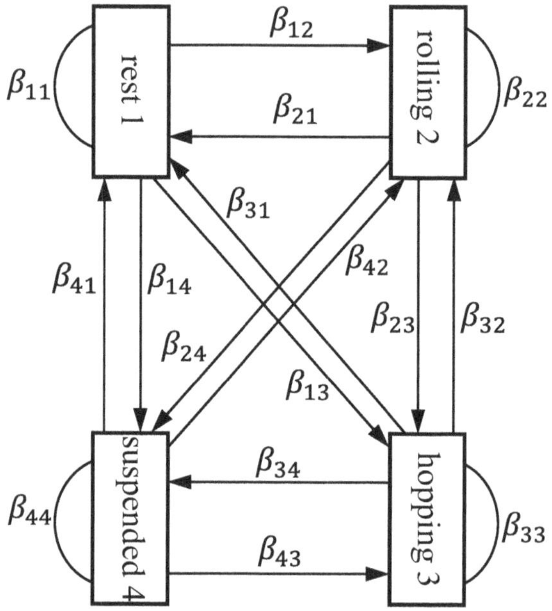

Fig. A.13 Probability of state transition on bed surface layer

A.6.2.2 Transition Probabilities of Step State

The sixteen transition probabilities of step state $\beta_{i.j}$ can be derived from the six basic state transition probabilities, as shown in Sects. A.3–A.5. These probabilities are the probability of transitioning to incipient rolling from rest ε_1, ceasing movement probability $1-\varepsilon_0$, to incipient hopping from a motion state ε_2, to incipient hopping from rest $\varepsilon_{0.2}$, to suspension from a motion state ε_4, and to incipient suspension from rest, β (see Sects. A.3–A.4).

The probabilities of a static particle remaining at rest on the bed, or transitioning to the rolling, hopping, or suspension states are

$$\beta_{1.1} = P\left[\xi_{V_b} \le V_{b.c.1} \cap \xi_{V_y} \le \omega\right] = (1-\varepsilon_1)(1-\beta), \tag{A.6.6a}$$

$$\beta_{1.2} = P[(V_{b.c.0} < \xi_{V_b} < V_{b.c.2.0}) \cap \xi_{V_y} \le \omega_5] = (\varepsilon_1 - \varepsilon_{2.0})(1-\beta), \tag{A.6.6b}$$

$$\beta_{1.3} = P[(\xi_{V_b} > V_{b.c.2.0}) \cap \xi_{V_y} \le \omega] = \varepsilon_{2.0}(10\varepsilon_4), \tag{A.6.6c}$$

and

$$\beta_{1.4} = P\left\{V_{b.y} \in B\right\} = \beta, \tag{A.6.6d}$$

respectively.

The probabilities of a particle in a motion state i ($i \neq 1$) transitioning to the rest, rolling, hopping, or suspension states are

$$\beta_{i.1} = P\big[\big(\xi_{V_b} \leq V_{b.c.1} \cap \xi_{V_y} \leq \omega\big] = (1 - \varepsilon_0)(1]\varepsilon_4\big), \quad (i = 2, 3, 4), \qquad \text{(A.6.7a)}$$

$$\beta_{i.2} = P\big[\big(V_{b.c.1} < \xi_{V_b} \leq V_{b.c.2}\big) \cap \xi_{V_{b.y}} \leq \omega\big] = (\varepsilon_0 - \varepsilon_2)(1 - \varepsilon_4), \quad (i = 2, 3, 4), \qquad \text{(A.6.7b)}$$

$$\beta_{i.3} = P\big[\big(\xi_{V_b} > V_{b.c.2}\big) \cap \xi_{V_{b.y}} \leq \omega\big] = \varepsilon_2(1.\varepsilon_4), \quad (i = 2, 3, 4), \qquad \text{(A.6.7c)}$$

and

$$\beta_{i.4} = P[\xi_{V_b} > \omega] = \varepsilon_4 \quad (i = 2, 3, 4), \qquad \text{(A.6.7d)}$$

respectively. Hence, the sixteen state transition probabilities have been derived, and their matrix can be written as

$$\{\beta_{i.j}\} = \begin{pmatrix} (1-\varepsilon_1)(1-\beta) & (\varepsilon_1 - \varepsilon_{2.0})(1)\beta) & \varepsilon_{2.0}(10\beta) & \beta \\ (1-\varepsilon_1)(1-\varepsilon_4) & (\varepsilon_0 - \varepsilon_2)(1-\varepsilon_4) & \varepsilon_2(11\varepsilon_4) & \varepsilon_4 \\ (1-\varepsilon_0)(1-\varepsilon_4) & (\varepsilon_0 - \varepsilon_2)(1-\varepsilon_4) & \varepsilon_2(11\varepsilon_4) & \varepsilon_4 \\ (1-\varepsilon_0)(1-\varepsilon_4) & (\varepsilon_0 - \varepsilon_2)(1-\varepsilon_4) & \varepsilon_2(11\varepsilon_4) & \varepsilon_4 \end{pmatrix}. \qquad \text{(A.6.8)}$$

Obviously there are constraints on the transition probabilities $\beta_{i,j}$, which are given by

$$\sum_{j=1}^{4} \beta_{i,j} = 1, (i = 1, 2, 3, 4). \qquad \text{(A.6.9)}$$

A.6.2.3 Stationary Probabilities of Step State

From Eqs. (A.6.4) and (A.6.5) the stationary probabilities for the rest, rolling, hopping, and suspension can be solved:

$$R_1 = (1m\varepsilon_1)(1m\beta)R_1 + (1m\varepsilon_0)(1m\varepsilon_{4.})(1mR_1) = \frac{A_1}{A}, \qquad \text{(A.6.10a)}$$

$$R_2 = (\varepsilon_1 - \varepsilon_{2.0})(16\beta)R_1 + (\varepsilon_0 - \varepsilon_2)(16\varepsilon_4)(16R_1) = \frac{A_1}{A}\beta_{1.2} + \frac{A_2}{A}\beta_{2.2}, \qquad \text{(A.6.10b)}$$

$$R_3 = \varepsilon_{2.0}(1.\beta)R_1 + \varepsilon_2(1 - \varepsilon_4)(1 - R_1) = \frac{A_1}{A}\beta_{1.3} + \frac{A_2}{A}\beta_{3.3}, \qquad \text{(A.6.10c)}$$

and

$$R_4 = \beta R_1 + \varepsilon_4(1 - R_1) = \frac{A_2}{A}\varepsilon_4 + \frac{A_1}{A}\beta, \tag{A.6.10d}$$

respectively, where

$$A = 1 + (1 - \varepsilon_0)(11\varepsilon_4) + (1 + \varepsilon_1)(11\beta), \tag{A.6.11a}$$

$$A_1 = (16\varepsilon_0)(16\varepsilon_4), \tag{A.6.11b}$$

and

$$A_2 = 1A6.1\varepsilon_1)(16\beta) = A - A_1. \tag{A.6.11c}$$

A.6.3 *Transition (Exchange) Intensities*

A.6.3.1 Markov Process of Particle Movement

Particles continuously transition between states. These transitions occur at the end of a single event, and they depend on the transition probabilities. If a particle is not at the end of an event, then the time required to reach the end of the event should be considered. Therefore a Markov chain, which only considered the probability of state transition, cannot describe the motion processes of a particle completely. Instead, a homogeneous Markov process with a continuous time (or distance) parameter and discrete motion states as random variables should be used. This can be expressed as

$$\{X(t), t \geq 0;\ X(t) \in I = \{1, 2, 3, 4\};\ 02t < +\infty\}. \tag{A.6.12}$$

The probability of transitioning from state i to state j is

$$p_{i.j}(t) = P[X(t) = j | X(0) = i], \tag{A.6.13}$$

which satisfies $p_{i.j}(t) \geq 0$, $\sum_{j=1}^{4} p_{i.j}(t) = 1$, and the Kolmogorov equation

$$p'_{i.j}(t) = \sum_{r \neq j} p_{i.r}(t)\lambda_{r.j} - p_{i.j}(t)\lambda_{j.j} \tag{A.6.14}$$

with initial conditions

$$p_{i.i}(0) = 1; \quad p_{i.j}(0) = 0, i \neq j. \tag{A.6.15}$$

Here,

$$\Lambda = \begin{Bmatrix} -\lambda_{11} & \lambda_{12} & \lambda_{13} & \lambda_{14} \\ \lambda_{21} & -\lambda_{2.2} & \lambda_{23} & \lambda_{24} \\ \lambda_{31} & \lambda_{32} & \lambda_{3.3} & \lambda_{34} \\ \lambda_{41} & \lambda_{42} & \lambda_{43} & \lambda_{4.4} \end{Bmatrix} \tag{A.6.16}$$

is the matrix of transition intensities and it satisfies the conditions

$$\sum_{j \neq i}^{4} \lambda_{i.j} - \lambda_{i.i} = 0, \tag{A.6.17a}$$

$$\lambda_{i.i} = \lim_{\Delta t \to 0} \frac{1 - p_{i.i}(\Delta t)}{\Delta t}, \tag{A.6.17b}$$

and

$$\lambda_{i.j} = \lim_{\Delta t \to 0} \frac{p_{i.j}(\Delta t)}{\Delta t}, (i \neq j). \tag{A.6.17c}$$

Equations (A.6.14)–(A.6.17a, A.6.17b, A.6.17c) show that the matrix of transition intensities determines the Markov process. Of particular interest are the stationary probabilities, which form the limits for the transition probability of the Markov process. Owing to the ergodicity of the Markov process of sediment transport, the moving states $p_{ij}(t)$ have the limitation probabilities

$$\lim_{t \to \infty} p_{ij}(t) = \pi_i, (i = 1, 2, 3, 4), \tag{A.6.18}$$

where π_i denotes the stationary probabilities which satisfy the system of Kolmogorov equations

$$\begin{pmatrix} \pi_1 & \pi_2 & \pi_3 & \pi_4 \end{pmatrix} \begin{Bmatrix} -\lambda_{11} & \lambda_{12} & \lambda_{13} & \lambda_{14} \\ \lambda_{21} & -\lambda_{2.2} & \lambda_{23} & \lambda_{24} \\ \lambda_{31} & \lambda_{32} & -\lambda_{3.3} & \lambda_{34} \\ \lambda_{41} & \lambda_{42} & \lambda_{43} & -\lambda_{4.4} \end{Bmatrix} = 0, \text{ and } \pi_1 + \pi_2 + \pi_3 + \pi_4 = 1 \tag{A.6.19}$$

Following this discussion, a stochastic process of particle motion can be determined from a matrix of transition intensities, which is related to the life distribution and will be investigated later in this Appendix.

A.6.3.2 Probability Distribution for the Movement Duration of a Single Step

If a particle moves in the same state (such as the rolling state) for several steps, then the moving process can be described as a stochastic process $\{N(t), t \geq 0\}$, $N(t) \in I = \{0, 1, 2, i\}$, where $N(t)$ is the step number at instant t. Obviously, $N(t)$ satisfies the following requirements: (1) $N(0) = 0$; (2) $N(t)$ is a stationary Markov process with independent increments; (3) The probability of one rolling step within an infinitesimal time interval Δt is proportional to Δt, i.e., $P\{N(\Delta t) = 1\} = \lambda \Delta t + o(\Delta t)$, and the probability of a particle rolling more than one step is a higher order term of Δt or $P\{N(\Delta t) \geq 2\} = o(\Delta t)$. Hence, $\{N(t), t \geq 0\}$ has a Poisson distribution and is expressed as

$$P\{N(t) = k\} = \frac{[\lambda t]^k}{k!} \mathrm{e}^{-\lambda t}, \ (t > 0; k = 0, 1, 2, \ldots). \tag{A.6.20}$$

Let the time interval of the n-th step be denoted by $t^{(n)} (n \geq 1)$, a random variable with the same distribution for all n that is independent of n. Then, $t^{(n)}$ has the same distribution as $t^{(1)}$. The first step duration $t^{(1)}$ implies that there is no rolling in interval $t < t^{(1)}$. Then, from Eq. (A.6.20) the probability density function (PDF) of the duration of a single step is

$$F_{t^{(1)}}(t) = P\left\{t^{(1)} < t\right\} = 1 - P\left\{t^{(1)} \geq t\right\} = 1 - P\{N(t) = 0\} = 1(\mathrm{e}^{-\lambda t}, \tag{A.6.21}$$

which is a negative exponential distribution with mean period $\bar{t}^{(1)} = \lambda^{-1}$. The durations associated with single steps in the hopping and suspension states are similarly expressed as

$$F_{t_i}(t) = P\{t_i < t\} = 1 - \mathrm{e}^{-\lambda_i t} = 1 - \mathrm{e}^{-\frac{t}{\bar{t}_i}}, \quad (i = 2, 3, 4), \tag{A.6.22}$$

where $i = = 2, 3, 4$ denotes the rolling, hopping, and suspension states, respectively. In general $\bar{t}_i = \mu_i U_i$ is the step duration expectation of the i-th moving state, where $\mu_i = 1/\bar{x}_i$ is the reciprocal of the expectation of single step distance, and U_i is the average velocity of single step. When t is replaced by distance x, the PDF of the distance of a single step is

$$F_{x_i}(x) = P\{x_i < x\} = 1 - \mathrm{e}^{-\mu_i x}. \tag{A.6.23}$$

However, there appears to be a contradiction here. In Sects. A.3–A.5, the probability distribution of a single-step period was obtained from the distribution of the flow velocity and particle position on the bed surface, but here the distribution of a single-step period is obtained as an exponential distribution. In fact, these two distributions form a composite distribution. For simplicity, the expected value of the single step period is obtained from the single step motion, and the exponential distribution of the single step period is obtained from the stochastic process.

A.6.3.3 Life Distributions of Moving Particles

The life distribution is actually the PDF of a single event distance (or duration), and it can be derived from the summation of the distribution of the single-step distance. The random process of single-step particle motion is given by $\{Z(x) = i, (i = 2, 3, 4), 0 \le x < +\infty\}$, where x is the distance the particle travels and $Z(x) = i, (i = 2, 3, 4)$ is the state i (rolling, hopping, or suspension) of the particle at x. The distance distribution $P_i^{(k)}$ of a particle after k steps in state i when it leaves state i at step $k + 1$ is expressed as

$$\begin{aligned} P_i^{(k)} &= P\left[X_i^{(k)} < x, Z^{(2)} = i, \ldots Z^{(k)} = i,\ Z^{(k+1)} \ne i \middle| Z^{(1)} = i\right] \\ &= P\left[X_i^{(k)} < x \middle| Z^{(1)} = i\ , Z^{(2)} = i, \ldots Z^{(k)} = i\right] \\ &\quad \times P\left[Z^{(k+1)} \ne i|Z^{(k)} = i\right] P\left[Z^{(k)} = i|Z^{(k-1)} = i\right] \cdots P\left[Z^{(2)} = i|Z^{(1)} = i\right] \\ &= (1|\beta_{ii})\beta_{ii}^{k-1} P\left[X_i^{(k)} \langle x|Z^{(1)} = i,\ Z^{(2)} = i, \ldots Z^{(k)} = i\right]. \end{aligned} \quad \text{(A.6.24)}$$

Here, $X_i^{(k)} = \sum_{l=1}^{k} x_i^{(l)}$ is the distance summation for k steps in state i, $x_i^{(l)}$ is the distance of the l-th step of a particle in state i with probability given by Eq. (A.6.22), and $Z^{(l)}$ is the moving state i at l-th step; i.e., $Z^{(l)} = Z(x), \left(\sum_{l'=1}^{l} x_i^{(l')} \le x < \sum_{l'=1}^{l+1} x_i^{(l')}\right)$. The probability of reaching distance $X_i^{(k)}$ is

$$\begin{aligned} & P\left[X_i^{(k)} < x \middle| Z^{(1)} = i, Z^{(2)} = i, \ldots Z^{(k)} = i\right] \\ &= \int\limits_{\sum_{l=1}^{k} x_i^{(i)} < x} \mu_i^k \mathrm{e}^{-\mu_i\left(x_i^{(1)} + \ldots + x_i^{(k)}\right)} \mathrm{d}x_i^{(1)} \ldots \mathrm{d}x_i^{(k)} \\ &= 1 - \sum_{l=0}^{k-1} \exp(x\mu_i x) \frac{(\mu_i x) l}{l!} = \mathrm{e}^{-\mu_i x} \sum_{l=k}^{\infty} \frac{(\mu_i x) l}{l!}. \end{aligned} \quad \text{(A.6.25)}$$

Thus, the PDF of the distance of a single event is given by

$$\begin{aligned} F_{X_i}(x) &= \sum_{k=1}^{\infty} P_i^{(k)} = \sum_{k=1}^{\infty} e^{-\mu_i x} \sum_{l=k}^{\infty} \frac{(\mu_i x) l}{l!} \beta_{i,i}^{k-1} (1 - \beta_{i.i})_i \\ &= \sum_{l=1}^{\infty} e^{-\mu_i x} \frac{(\mu_i x) l}{l!} (1 - \beta_{i.i})_i \sum_{k=1}^{l} \beta_{i.i}^{k-1} = \sum_{l=1}^{\infty} e^{-\mu_i x} \frac{(\mu_i x) l}{l!} (1 - \beta_{i.i}) \frac{1 - \beta_{i.i}^l}{1 - \beta_{i,i}} \alpha_i \end{aligned}$$

$$= e^{-\mu_i x}\left(e^{\mu_i x} - 1\right) - e^{-\mu_i x}\left(e^{-\mu_i \beta_{i,i} x} - 1\right) = 1 - e^{-(1-\beta_{i,i})\mu_i x} \quad \text{(A.6.26)}$$

This is also a negative exponential distribution. The corresponding PDD of the distance of a single event is

$$p_{X_i}(x) = (1 - \beta_{i.i})\mu_i \mathrm{e}^{-(1-\beta_{i.i})\mu_i x}, \quad (i = 2, 3, 4). \quad \text{(A.6.27)}$$

The expectation for the distance of single event is

$$\overline{X_i} = \int_0^\infty x p_{X_i}(x)\mathrm{d}x = \frac{1}{\mu_i(1 - \beta_{i.i})}, \quad (i = 2, 3, 4), \quad \text{(A.6.28)}$$

which is $1/\left(1/\beta_{i,i}\right)$ times the single step distance. Because $x = U_i t$, the parameter x can be replaced by t, then Eqs. (A.6.26–A.6.28) becomes

$$F_{T_i}(t) = 1 - e^{-\left(1-\beta_{i,i}\right)\mu_i U_i t} \quad \text{(A.6.29)}$$

$$p_{T_i}(t) = (1 - \beta_{i.i})\mu_i e^{-(1-\beta_{i.i})\mu_i U_i t}, \quad (i = 2, 3, 4) \quad \text{(A.6.30)}$$

$$\overline{T_i} = \int_0^\infty t p_{T_i}(t)dt = \frac{1}{\mu_i U_i(1 - \beta_{i.i})} = \frac{\bar{t}_i}{(1 - \beta_{i.i})}, \quad (i = 2, 3, 4) \quad \text{(A.6.31)}$$

where $\overline{T}_i$, the average duration of single event, reflects the life (time interval) of a particle in state *i*. This is called the life of particle in state *i*, and its PDF (Eq. A.6.29) is called the life distribution.

The reciprocal of $\overline{T}_i$ can be understood as the probability of a particle changing state over the duration of a single motion event. When a particle remains in state *i* over $[0, t)$, the probability that it will leave state *i* over $[t, t + \Delta t)$ is then

$$\begin{aligned} P[t + \Delta t > T_i | T_i \geq t] &= \frac{P[t + \Delta t > T_i, T_i \geq t]}{P[T_i \geq t]} = \frac{P[T_i > t] - P[T_i > t + \Delta t]}{P[T_i \geq t]} \\ &= \frac{e^{-\left(1-\beta_{i,i}\right)\mu_i U_i t} - e^{-\left(1-\beta_{i,i}\right)\mu_i U_i (t+\Delta t)}}{e^{-\left(1-\beta_{i,i}\right)\mu_i U_i t}} \\ &= \left(1 - \beta_{i,i}\right)\mu_i U_i \Delta t + 0(\Delta t) = P[\Delta t > T_i]. \end{aligned} \quad \text{(A.6.32)}$$

Hence, when neglecting higher order terms of Δt, the probability of a particle in state *i* changing state over time period Δt is approximately $\left(1 i \beta_{i,i}\right)\mu_i U \Delta t$, and per unit time period it is

$$\lambda_i' = (1 - \beta_{i.i})\mu_i U_i = \lambda_{i.i}'. \quad \text{(A.6.33)}$$

This can be interpreted as the probability of one particle in state i changing state per unit time. Therefore, it is an index of the transition probability of an individual particle, also known as the transition intensity of a single particle. Furthermore,

$$\lambda'_{i.j} = \beta_{i,j}\mu_i U_i \tag{A.6.34}$$

is the probability of one particle in state i transitioning to state j per unit time.

For K_i particles moving simultaneously in state i, the transition intensities from state i are

$$\lambda_i = (1 - \beta_{i.i})\mu_i U_i K_i = \sum_{\substack{|j=1 \\ |j \neq i}}^{4} \beta_{i.j}\mu_i U_i K_i = \lambda_{i.i}, \quad (i = 2,\ 3,\ 4) \tag{A.6.35}$$

This is also the number of particles that transition from state i per unit time. The transition intensities from state i to j are

$$\lambda_{i.j} = \beta_{i.j}\mu_i U_i K_i (i = 2, 3, 4; i \neq j). \tag{A.6.36}$$

A.6.3.4 Life Distributions of Rest Particles

Particles at rest also have a life distribution, and it is related to the incipient motion process for rest particles in the bed surface layer. The process by which a rest particle starts moving can be expressed by the stochastic process $\{N(t), t \geq 0\}$, $N(t) \in I = \{0, 1, 2, \ldots\}$, where time t is continuous parameter, and $N(t)$ is the probability of a certain number of particles transitioning from rest to a motion state in period (0, t). Similar to Sect. A.6.3.2, $\{N(t), t \geq 0\}$ is a Poisson process that can be expressed as

$$P\{N(t) = k\} = \frac{[\lambda_1 t]^k}{k!} \mathrm{e}^{-\lambda_1 t}, (t \geq 0, k = 0, 1, 2, \ldots). \tag{A.6.37}$$

Consider a time interval in which no particle starts moving $T_n (n \geq 1)$. This is a random variable independent of n, and can be written as T_1. The presence of a rest particle in interval T_1 implies that no movement happens in interval $t < T_1$. Thus, from Eq. (A.6.37) the probability $P\{T_1 > t\} = P\{N(t) = 0\} = e^{-\lambda t}$, and the PDF of the time interval T_1 is

$$F_{T_1}(t) = P\{T_1 \leq t\} = P\{N(t) \geq 1\} = 1(\mathrm{e}^{-\lambda_1 t}, \tag{A.6.38}$$

where $\overline{T}_1 = 1/\lambda_1$. Similar to the derivation of the transition intensity of moving particles, the transition intensity of static particles is λ_1. Evidently, λ_1 is proportional to $R_1 n_0$, the number of rest particles per unit area in bed surface layer, where R_1 is the stationary probability of a particle in the rest state. Moreover, λ_1 is proportional to $\beta_{1.i}$, the probability that a particle will transition from rest to motion state i, and inversely proportional to $t_{0.i}$, the average period required for a rest particle to leave the bed surface. Here, $t_{0.2} = \bar{t}_2$ (Eq. A.3.36) is the step rolling time, $t_{0.3} = \bar{t}_{3.0}$ is the time required for hopping a particle to leave the bed surface and achieve a height equivalent to the particle diameter (Eq. A.4.54c), and $t_{0.4} = T_{0.4}$ (Eq. 4.34). In fact, if a particle leaves the bed surface layer, the particle underneath is able to start moving. Thus, the incipient intensity or incipient number of rest particles per unit area of the bed surface layer per unit time can be expressed as

$$\lambda_1 = \sum_{j=2}^{4} \lambda_{1,j} = \left(\frac{\beta_{1,2}}{t_{0.2}} + \frac{\beta_{1,3}}{t_{0.3}} + \frac{\beta_{1,4}}{t_{0.4}} \right) R_1 n_0 = \lambda_{1.1}, \tag{A.6.39}$$

where

$$\lambda_{1,j} = \frac{\beta_{1.j}}{t_{0.j}} R_1 n_0 \quad (j = 2, 3, 4), \tag{A.6.40}$$

are the intensities for incipient rolling, hopping, and suspension, respectively. The transition intensities for one particle are

$$\lambda_1' = \sum_{j=2}^{4} \lambda_{1,j}' = \left(\frac{\beta_{1,2}}{t_{0.2}} + \frac{\beta_{1,3}}{t_{0.3}} + \frac{\beta_{1,4}}{t_{0.4}} \right) R_1 = \lambda_{1.1}', \tag{A.6.41a}$$

and

$$\lambda_{1,j}' = \frac{\beta_{1.j} R_1}{t_{0.j}}, \quad (j = 2, 3, 4). \tag{A.6.41b}$$

The life distribution for one rest particle per unit area in the bed surface layer is

$$F_{T_1'}(t) = P\left[T_1' < t\right] = 1 - \exp\left[-\left(\frac{\beta_{1,2}}{t_{0.2}} + \frac{\beta_{1,3}}{t_{0.3}} + \frac{\beta_{1,4}}{t_{0.4}} \right) R_1 t \right], \tag{A.6.42}$$

where

$$T_1' = \frac{1}{\left(\frac{\beta_{1,2}}{t_{0.2}} + \frac{\beta_{1,3}}{t_{0.3}} + \frac{\beta_{1,4}}{T_{0.4}} \right) R_1} = 1/\lambda_{1.1}' \tag{A.6.43}$$

is the duration of incipient motion for an individual particle. Hereafter, $t_{0.4} = T_{0.4}$ is adopted. Under weak hydraulic conditions the incipient intensity is low and the bottom sediment is in the so-called incipient motion state. It is reasonable to assume that $\beta_{1.3} = 0$ and $\beta_{1.4} = 0$, thus the incipient intensity becomes

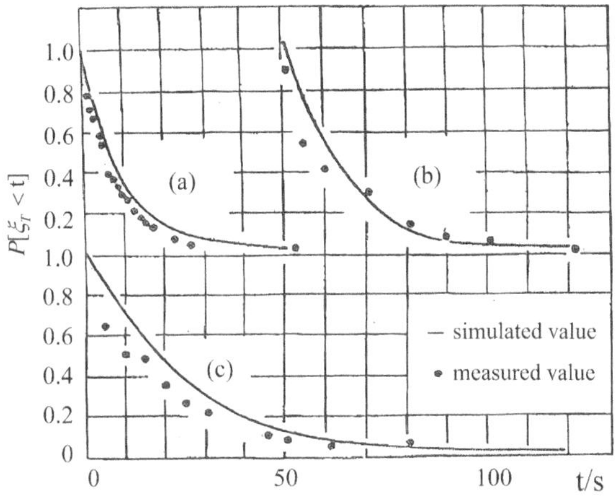

Fig. A.14 Verification of probability of non-incipient motion under a low flow velocity for a rest particle

$$\lambda_1 = \frac{\beta_{1.2}}{t_{0.2}} R_! n_0. \tag{A.6.44}$$

The average number of incipient particles in time t is

$$\overline{K} = \lambda_1 t = \frac{\beta_{1.2}}{t_{2.0}} R_! n_0 t. \tag{A.6.45}$$

Thus,

$$\beta_{1.2} = \varepsilon_1 = \frac{t_{2.0}}{t} \frac{\overline{K}}{R_! n_0}. \tag{A.6.46}$$

It can be seen that the incipient probability is proportional to the ratio between the number of incipient particles and the number of particles in the bed surface layer. Because $t_{0.2}/t << 1$, the incipient probability is much less than $\overline{K}/R_! n_0$. Thus , from Eq. (A.6.46), the incipient probability can be measured experimentally.

Figure A.14 shows the verification of the probability of non-incipient motion for a rolling particle at x, which is expressed as $P\{X_2 \geq x\} = P\{N(x) = 0\} = e^{-\mu_2 x}$ (see Eq. (A.6.38) if x is substituted for t under a low flow velocity).

A.6.3.5 Transition Intensities

In Sects. 3.3 and 3.4 the transition intensities were derived. Substituting the probability of state transition $\beta_{i.j}$, expressed in Eq. (A.6.8). into Eqs. (A.6.38) and (A.6.40) yields

$$\begin{cases} \lambda_{1.2} = (\varepsilon_1 - \varepsilon_{2.0})(1-\beta)\frac{n_0 R_1}{t_{0.2}}, \quad \lambda_{1.3} = \varepsilon_{2.0}(1-\beta)\frac{n_0 R_1}{t_{0.3}}, \\ \lambda_{1.4} = \beta\frac{n_0 R_1}{T_{0.4}}, \\ \lambda_{2.1} = (1-\varepsilon_0)(1-\varepsilon_4)\overline{K}_2\mu_2 U_2,\ \lambda_{2.3} = \varepsilon_2(1-\varepsilon_4)\overline{K}_2\mu_2 U_2, \\ \lambda_{2.4} = \varepsilon_4\overline{K}_2\mu_2 U_2, \\ \lambda_{3.1} = (1-\varepsilon_0)(1-\varepsilon_4)\overline{K}_3\mu_3 U_3,\ \lambda_{3.2} = (\varepsilon_0-\varepsilon_2)(1-\varepsilon_4)\overline{K}_3\mu_3 U_3, \\ \lambda_{3.4} = \varepsilon_4\overline{K}_3\mu_3 U_3, \\ \lambda_{4.1} = (1-\varepsilon_0)(1-\varepsilon_4)\overline{K}_4\mu_4 U_4,\ \lambda_{4.2} = (\varepsilon_0-\varepsilon_2)(1-\varepsilon_4)\overline{K}_4\mu_4 U_4, \\ \lambda_{4.3} = \varepsilon_2(1-\varepsilon_4)\overline{K}_4\mu_4 U_4. \end{cases} \tag{A.6.47}$$

These twelve transition intensities are derived for $j \neq i$. For $j = i$, the intensities can be expressed as

$$\lambda_{1.1} = \frac{(1-\beta_{1.1})n_0 R_0}{T_1} = \left[\frac{(\varepsilon_1-\varepsilon_{2.0})(1-\beta)}{t_{0.2}} + \frac{\varepsilon_{2.0}(1-\beta)}{t_{0.3}} + \frac{\beta}{T_{0.4}}\right] n_0 R_0, \tag{A.6.48a}$$

$$\lambda_{2.2} = [16(\varepsilon_0-\varepsilon_2)(1-\varepsilon_4)]\overline{K}_2\mu_2 U_2, \tag{A.6.48b}$$

$$\lambda_{3.3} = [13\varepsilon_2(11\varepsilon_4)]\overline{K}_3\mu_3 U_3, \tag{A.6.48c}$$

and

$$\lambda_{4.4} = (1-\varepsilon_4)\overline{K}_4\mu_4 U_4, \tag{A.6.48d}$$

where

$$\frac{1}{T_1} = \frac{1}{1-\beta_{1.1}}\sum_{j=2}^{4}\frac{\beta_{1.j}}{t_{0.j}}. \tag{A.6.49}$$

These transition intensities were derived from the life distribution. This derivation was detailed and mathematically strict, but it was also complicated and not intuitive. To understand the concept more easily, the random state-transition process is considered as a deterministic process, and the transition intensity from state i to state j is derived using a more intuitive approach.

Assume that there are $\overline{K}_i$ particles in state i in a water column, and that the average distance moved in a single event is $1/[(1i\beta_{i.i})\mu_i]$, then all $\overline{K}_i$ particles will change their states within the average time interval $1/[(1h\beta_{i.i})\mu_i U_i]$. Therefore, the number of particles that change state per unit time is

$$\lambda_i = \frac{\overline{K}_i}{\frac{1}{[(1-\beta_{i.i})_i\mu_i U_i]}} = (1-\beta_{i.i})\overline{K}_i\mu_i U_i, \qquad (i = 2, 3, 4). \tag{A.6.50}$$

These particles will change from state i to j according to the ratio of the transition probabilities

$$\frac{\beta_{i.j}}{(1.\beta_{i.i})} = \frac{\beta_{i.j}}{\sum_{j\neq i}\beta_{i.j}}, \quad (i = 2, 3, 4;\ j = 1, 2, 3, 4;\ j \neq i). \tag{A.6.51}$$

Then,

$$\lambda_{i.j} = \frac{\beta_{i.j}}{(1.\beta_{i.i})}\lambda_i = \beta_{i.j}\overline{K}_i\mu_i U_i, \quad (i = 2, 3, 4;\ j = 1, 2, 3, 4;\ j \neq i), \tag{A.6.52}$$

which is the same as Eq. (A.6.47). As for the sediment at rest, $t_{0.j}$, $(j = 2, 3, 4)$ gives the time required for a particle at rest to enter the rolling, hopping, and suspension states, respectively. A particle can be incipient on $t/t_{0.j}$ occasions over time t, and the probability of a particle transitioning from rest to state j is $\beta_{1.j}$. Assuming that the number of particles at rest in a unit area of the bed surface layer is $K_1 = R_1 n_0$, then among K_1 particles, the number of particles transitioning to state j per unit time is

$$\lambda_{1.j} = \beta_{1.j}\frac{K_1}{t_{j.0.}}, \quad (j = 2, 3, 4) \tag{A.6.53}$$

which are the same as Eq. (A.6.47). Moreover, the total number of particles that achieve incipient motion from the rest state per unit time is

$$\lambda_1 = \sum_{j\neq 1}\frac{\beta_{1.j}}{t_{0.j}}. \tag{A.6.54}$$

A.6.3.6 Transition Intensities for Non-uniform Sediment

Based on the characteristics of non-uniform sediment (Sect. 2.4.2) the transition intensities of non-uniform sediment can be derived. Dividing the non-uniform sediment into n size classes, the basic state transition probabilities are $\varepsilon_{1.l}$, $\varepsilon_{0.l}$, $\varepsilon_{2.0.l}$ $\varepsilon_{2.l}$, $\varepsilon_{4.l}$, and β_l for each class, the corresponding state transition probabilities are $\beta_{i.j.l}$, and the number of particles in l-th size class at rest on the bed surface per unit area is

$$\overline{K}_{1.l} = R_{1.l}n_{0.l} = \frac{P_{1.l}R_{1.l}m_0}{\left(\frac{\pi}{4}D_l^2\right)}. \tag{A.6.55}$$

Here, the subscript l represent the l-th size class with the particle size D_l, $R_{1.l}$ is the stationary probability for rest particles in the bed surface layer, $P_{1.l}$ is the gradation of the bed material, and $n_{0.l} = P_{1.l}m_0/[(\pi/4)D_l^2]$ is the number of particles per unit area in the bed surface layer. The number of particles from the l-th size class in motion state i is

$$\overline{K}_{i.l} = P_{i.l}\overline{K}_i, \quad (i = 2, 3, 4), \tag{A.6.56}$$

where $P_{i.l}$ is the gradation for rolling, hopping, and suspended sediment, respectively. The transition intensities are

$$\lambda_{i.j.l} = \begin{cases} \beta_{i.j.l}\overline{K}_{i.l}\mu_{i.l}U_{i.l}, & (i = 2, 3, 4;\ j \neq i;\ l = 1, 2, \cdots n), \\ \beta_{1.j.l}\overline{K}_{1.l}\frac{1}{t_{0.j.l}}\beta_{1.j.l}\frac{R_{1.l}P_{1.l}m_0}{\left(t_{0.j.l}\frac{\pi}{4}D_l^2\right)}, & (j = 2, 3, 4;\ l = 1, 2, \cdots n). \end{cases} \tag{A.6.57}$$

Substituting in the expressions for state transition probabilities $\beta_{i.j.l}$, the transition intensity for non-uniform sediment becomes

$$\begin{cases} \lambda_{1.2.l} = (\varepsilon_{1.l} - \varepsilon_{2.0.l})(1.\beta_l)\frac{m_0 R_{1.l} P_{1.l}}{\frac{\pi}{4}D_l^2 t_{0.2.l}}, \quad \lambda_{1.3.l} = \varepsilon_{2.0.l}(1.\beta_l)\frac{m_0 R_{1.l} P_{1.l}}{\frac{\pi}{4}D_l^2 t_{0.3.l}}, \\ \lambda_{1.4.l} = \beta_l \frac{m_0 R_{1.l} P_{1.l}}{\frac{\pi}{4}D_l^2 T_{0.4.l}}, \\ \lambda_{2.1.l} = (1.\varepsilon_{0.l})(1.\varepsilon_{4.l})\overline{K}_{2.l}\mu_{2.l}U_{2.l}, \quad \lambda_{2.3.l} = \varepsilon_{2.l}(1.\varepsilon_{4.l}) = \overline{K}_{2.l}\mu_{2.l}U_{2.l}, \\ \lambda_{2.4.l} = \varepsilon_{4.l}\overline{K}_{2.l}\mu_{2.l}U_{2.l}, \\ \lambda_{3.1.l} = (1.\varepsilon_{0.l})(1.\varepsilon_{4.l})\overline{K}_{3.l}\mu_{3.l}U_{3.l}, \quad \lambda_{3.2.l} = (\varepsilon_{0.l} - \varepsilon_{2.l})(1.\varepsilon_{4.l})\overline{K}_{3.l}\mu_{3.l}U_{3.l}, \\ \lambda_{3.4.l} = \varepsilon_4\overline{K}_{3.l}\mu_{3.l}U_{3.l}, \\ \lambda_{4.1.l} = (1.\varepsilon_{0.l})(1.\varepsilon_{4.l})\overline{K}_{4.l}\mu_{4.l}U_{4.l}, \quad \lambda_{4.2.l} = (\varepsilon_{0.l} - \varepsilon_{2.l})(1.\varepsilon_{4.l})\overline{K}_{4.l}\mu_{4.l}U_{4.l}, \\ \lambda_{4.3.l} = \varepsilon_{2.l}(1.\varepsilon_{4.l})\overline{K}_{4.l}\mu_{4.l}U_{4.l}. \end{cases} \tag{A.6.58}$$

Since for non-uniform sediment the particle position on bed surface depends on the particle size, the transition intensity is also a function of $D_l/\overline{D}$ in addition to $\overline{V}_b/\omega_0$.

A.6.4 Transport Rate of Sediment

A.6.4.1 Transport Rate Under Equilibrium Conditions

The movement of an individual particle can be described as an ergodicity Markov process. Here, the transition intensities for one particle $\lambda'_{i.j}$ (Eqs. A.6.33, A.6.34, and A.6.41a, A.6.41b) are used as the coefficients in the Kolmogorov equation (Eq. A.6.19), which then becomes

$$-(\beta_{2,1} + \beta_{2,3} + \beta_{2,4})\mu_2 U_2\pi_2 + \beta_{3,2}\mu_3 U_3\pi_3 + \beta_{4,2}\mu_4 U_4\pi_4 = -\frac{\beta_{1,2}}{t_{0.2}}\pi_1, \tag{A.6.59a}$$

$$\beta_{2,3}\mu_2U_2\pi_2 - (\beta_{3,1} + \beta_{3,2} + \beta_{3,4})\mu_3U_3\pi_3 + \beta_{4,3}\mu_4U_4\pi_4 = -\frac{\beta_{1,3}}{t_{0.3}}\pi_1, \quad \text{(A.6.59b)}$$

and

$$\beta_{2,4}\mu_2U_2\pi_2 + \beta_{3,4}\mu_3U_3\pi_3 - (\beta_{4,1} + \beta_{4,2} + \beta_{4,3})\mu_4U_4\pi_4 = -\frac{\beta_{1,4}}{T_{0.4}}\pi_1. \quad \text{(A.6.59c)}$$

The corresponding solutions are

$$\pi_2 = \frac{\Delta_2}{\Delta_0}\frac{\pi_1}{\mu_2U_2}, \quad \text{(A.6.60a)}$$

$$\pi_3 = \frac{\Delta_3}{\Delta_0}\frac{\pi_1}{\mu_3U_3}, \quad \text{(A.6.60b)}$$

and

$$\pi_4 = \frac{\Delta_4}{\Delta_0}\frac{\pi_1}{\mu_4U_4}, \quad \text{(A.6.60c)}$$

or

$$\pi_2 = \frac{\pi_1}{\mu_2U_2}\left[\frac{(\varepsilon_1 - \varepsilon_{2.0})(1.\beta)(1.\varepsilon_2)}{(1-\varepsilon_0)}\frac{1}{t_{0.2}} + \frac{(\varepsilon_0 - \varepsilon_2)(1-\beta)\varepsilon_{2.0}}{(1.\varepsilon_0)}\frac{1}{t_{0.3}} + \frac{(\varepsilon_0 - \varepsilon_2)\beta}{(1-\varepsilon_0)}\frac{1}{T_{0.4}}\right], \quad \text{(A.6.61a)}$$

$$\pi_3 = \frac{\pi_1}{\mu_3U_3}\left[\frac{\varepsilon_2(\varepsilon_1 - \varepsilon_{2.0})(1.\beta)}{(1.\varepsilon_0)}\frac{1}{t_{0.2}} + \frac{\varepsilon_{2.0}(1.\beta)(1.\varepsilon_0 + \varepsilon_2)}{(1-\varepsilon_0)}\frac{1}{t_{0.3}} + \frac{\varepsilon_2\beta}{(1-\varepsilon_0)}\frac{1}{T_{0.4}}\right], \quad \text{(A.6.61b)}$$

and

$$\pi_4 = \frac{\pi_1}{\mu_4U_4}\left[\frac{\varepsilon_4(\varepsilon_1 - \varepsilon_{2.0})(1.\beta)}{(1.\varepsilon_4)(1-\varepsilon_0)}\frac{1}{t_{0.2}} + \frac{\varepsilon_4\varepsilon_{2.0}(1.\beta)}{(1.\varepsilon_4)(1-\varepsilon_0)}\frac{1}{t_{0.3}} + \frac{\beta(1-\varepsilon_0+\varepsilon_0\varepsilon_4)}{(1-\varepsilon_4)(1-\varepsilon_0)}\frac{1}{T_{0.4}}\right], \quad \text{(A.6.61c)}$$

where Δ_0 and $\Delta_i, i = (2, 3, 4)$ are the determinants of the coefficients in Eqs. (A.6.60a, A.6.60b, A.6.60c).

The physical meaning of π_i is the limitation probability of a particle in state i, or the probability of a particle being in state i when it is in equilibrium during motion. According to the law of large numbers, similar results are available for multiple particles. If the total number of particles in a unit water column is $\overline{K}$, of which $\overline{K}_i$ particles are in state i, then

$$\overline{K}_i = \pi_i\overline{K}, \quad (i = 2, 3, 4), \quad \text{(A.6.62a)}$$

and

$$\overline{K}_1 = \pi_1 \overline{K} = R_1 n_0. \tag{A.6.62b}$$

In terms of the particle number, the sediment transport rate is

$$\overline{q}_{n.i} = \overline{K}_i U_i \quad (i = 2, 3, 4). \tag{A.6.63}$$

Substituting Eqs. (A.6.62a, A.6.62b) and (A.6.63) into Eq. (A.61) yields an expression for the transport rate in terms of the particle number,

$$\begin{aligned}\overline{q}_{n.2} &= \frac{R_1 n_0}{\mu_2}\left[\frac{(\varepsilon_1 - \varepsilon_{2.0})(1.\beta)(1.\varepsilon_2)}{(1-\varepsilon_0)}\frac{1}{t_{0.2}} + \frac{(\varepsilon_0 - \varepsilon_2)(1-\beta)\varepsilon_{2.0}}{(1.\varepsilon_0)}\frac{1}{t_{0.3}} + \frac{(\varepsilon_0 - \varepsilon_2)\beta}{(1-\varepsilon_0)}\frac{1}{T_{0.4}}\right] \\ &= \frac{\overline{K}_1}{\mu_2} f_1.\end{aligned} \tag{A.6.64a}$$

$$\begin{aligned}\overline{q}_{n.3} &= \frac{R_1 n_0}{\mu_3}\left[\frac{\varepsilon_2(\varepsilon_1 - \varepsilon_{2.0})(1.\beta)}{(1.\varepsilon_0)}\frac{1}{t_{0.2}} + \frac{\varepsilon_{2.0}(1.\beta)(1.\varepsilon_0 + \varepsilon_2)}{(1-\varepsilon_0)}\frac{1}{t_{0.3}} + \frac{\varepsilon_2\beta}{(1-\varepsilon_0)}\frac{1}{T_{0.4}}\right] \\ &= \frac{\overline{K}_1}{\mu_3} f_2,\end{aligned} \tag{A.6.64b}$$

$$\begin{aligned}\overline{q}_{n.4} &= \frac{R_1 n_0}{\mu_4}\left[\frac{\varepsilon_4(\varepsilon_1 - \varepsilon_{2.0})(1.\beta)}{(1.\varepsilon_4)(1-\varepsilon_0)}\frac{1}{t_{0.2}} + \frac{\varepsilon_4\varepsilon_{2.0}(1.\beta)}{(1.\varepsilon_4)(1-\varepsilon_0)}\frac{1}{t_{0.3}} + \frac{\beta(1-\varepsilon_0+\varepsilon_0\varepsilon_4)}{(1-\varepsilon_4)(1-\varepsilon_0)}\frac{1}{T_{0.4}}\right] \\ &= \frac{\overline{K}_1}{\mu_4} f_3.\end{aligned} \tag{A.6.64c}$$

Then, replacing $\overline{K}_1$ with $n_0 = m_0/[(\pi/4)D^2]$ (see Eqs. A.6.64a, A.6.64b, A.6.64c), yields the transport rate in terms of the weight $\overline{q}_i = \frac{\gamma_s \pi D^3}{6}\overline{q}_{n.i}$, $(i = 2, 3, 4)$ as

$$\overline{q}_2 = \frac{2}{3} m_0 \gamma_s D \frac{R_1}{\mu_2} f_1, \tag{A.6.65a}$$

$$\overline{q}_3 = \frac{2}{3} m_0 \gamma_s D \frac{R_1}{\mu_3} f_2, \tag{A.6.65b}$$

and

$$\overline{q}_4 = \frac{2}{3} m_0 \gamma_s D \frac{R_1}{\mu_4} f_3. \tag{A.6.65c}$$

By using a more intuitive method, the transport rate can be derived directly from the transition intensities. If the particles in a unit water column are in a weak equilibrium state, then in a unit time period the number of particles transitioning from state i should equal the number transitioning to state i. The definition of transition intensity gives

$$\left.\begin{aligned}\lambda_{1,2}+\lambda_{3,2}+\lambda_{4,2}=\lambda_{2,1}+\lambda_{2,3}+\lambda_{2,4}\\ \lambda_{1,3}+\lambda_{2,3}+\lambda_{4,3}=\lambda_{3,1}+\lambda_{3,2}+\lambda_{3,4}\\ \lambda_{1,4}+\lambda_{2,4}+\lambda_{3,4}=\lambda_{4,1}+\lambda_{4,2}+\lambda_{4,3}\end{aligned}\right\}, \tag{A.6.66}$$

which is the same as Eqs. (A.6.59a, A.6.59b, A.6.59c).

As for the transport rate under strong equilibrium conditions, the number of particles per unit time transitioning from motion state i to rest on the bed surface is equal to the number transitioning from rest to motion state i. From the transition intensities

$$\lambda_{1,i}=\overline{\lambda}_{i.1},\ (i=2,3,4). \tag{A.6.67}$$

This leads directly to the number of particles in motion within a unit water column per unit time,

$$\overline{K}_2=\frac{\beta_{1,2}}{\beta_{2,1}}\frac{R_1 n_0}{\mu_2 t_{0.2} U_2}, \tag{A.6.68a}$$

$$\overline{K}_3=\frac{\beta_{1,3}}{\beta_{3,1}}\frac{R_1 n_0}{\mu_3 t_{0.3} U_3}, \tag{A.6.68b}$$

and

$$\overline{K}_4=\frac{\beta_{1,4}}{\beta_{4,1}}\frac{R_1 n_0}{\mu_4 T_{0.4} U_4}. \tag{A.6.68c}$$

Further, the transport rates in terms of the weight are

$$\overline{q}_2=\frac{2}{3}m_0\gamma_s D\frac{\psi_2}{\mu_2 t_{0.2.}}, \tag{A.6.69a}$$

$$\overline{q}_3=\frac{2}{3}m_0\gamma_s D\frac{\psi_3}{\mu_3 t_{0.3.}}, \tag{A.6.69b}$$

and

$$\overline{q}_4=\frac{2}{3}m_0\gamma_s D\frac{\psi_4}{\mu_3 T_{0.4.}}, \tag{A.6.69c}$$

where the composite probabilities are expressed as

$$\left.\begin{aligned}\psi_2\frac{(\varepsilon_1-\varepsilon_{2.0})(1-\beta)}{1+(1+\varepsilon_0)(1-\varepsilon_4)-(1-\varepsilon_1)(1-\beta)}\\ \psi_3\frac{\varepsilon_{2.0}(1-\beta)}{1+(1+\varepsilon_0)(1-\varepsilon_4)-(1-\varepsilon_1)(1-\beta)}\\ \psi_4\frac{\beta}{1+(1+\varepsilon_0)(1-\varepsilon_4)-(1-\varepsilon_1)(1-\beta)}\end{aligned}\right\}. \tag{A.6.70}$$

The transport rate derived above is under equilibrium conditions, and it is also called the transport capacity.

A.6.4.2 Transport Rate of Bed Load Under Equilibrium Condition

Previous study usually considered three states of sediment movement including stationary, bed load and suspension. Moreover the bed load is taken as hopping particles only. In the authors' research beside the stationary state three moving states, rolling, hopping and suspension are included, and the bed load refers to both rolling and hopping particles. The transport rate of bed load is the superposition of transport rate of rolling and hopping particles. There are three equalities of bed load: particle number in unit water column, transition intensity and transport rate:

$$\begin{aligned} &\overline{K}_b = \overline{K}_2 + \overline{K}_3, \ \lambda_b = \overline{K}_b \mu_b U_b = \overline{K}_2 \mu_2 U_2 + \overline{K}_3 \mu_3 U_3. \\ &\overline{q}_b = K_b U_b = K_2 U_2 + K_3 U_{3.} \end{aligned} \tag{A.6.71}$$

Then μ_b and U_b are derived as

$$U_b = \frac{K_2 U_2 + K_3 U_{3.}}{K_2 + K_3}, \tag{A.6.72a}$$

and

$$\mu_b = \frac{K_2 U_2 \mu_2 + K_3 U_3 \mu_3}{K_2 U_2 + K_3 U_{3.}}. \tag{A.6.72b}$$

The parameters in expression of bed load $q_b^{\frac{2}{3}0s\frac{\psi_b}{\mu_b t_{b.0}}}$ under strong equilibrium condition can be derived as

$$q_b = q_2 + q_3, \tag{A.6.73a}$$

$$U_b = \frac{q_2 + q_3}{\frac{q_2}{U_2} + \frac{q_3}{U_3}}, \tag{A.6.73b}$$

$$\mu_b = \frac{q_2 \mu_2 + q_3 \mu_3}{q_2 + q_3}, \tag{A.6.73c}$$

$$\psi_b = \psi_2 + \psi_3 \tag{A.6.74a}$$

$$\frac{1}{t_{0.b}} = \frac{\mu_b}{\psi_b}\left[\frac{\psi_2}{\mu_2 t_{0.2}} + \frac{\psi_3}{\mu_3 t_{0.3}}\right] \tag{A.6.74b}$$

where q_b, U_b, μ_b, ψ_b, and $t_{0.b}$ are transport rate, average step velocity, reciprocal of step distance, composite probability and incipient time period of bed load respectively.

A.6.4.3 Transport Rate for Non-uniform Sediment

The transport rate for uniform sediment can easily be extended to non-uniform sediment. The expressions for the transport rate under weak equilibrium are

$$\begin{aligned} q_{2.l} &= \frac{2}{3} m_0 \gamma_s D_l \frac{R_{1.l} P_{1.l}}{\mu_{2.l}} \left[\frac{(\varepsilon_{1.l} - \varepsilon_{2.0.l})(1.\beta_l)(1 - \varepsilon_{2.l})}{(1.\varepsilon_{0.l}) t_{0.2.l}} + \frac{(\varepsilon_{0.l} - \varepsilon_{2.l})(1.\beta_l)\varepsilon_{2.0.l}}{(1.\varepsilon_{0.l}) t_{0.3.l}} + \frac{(\varepsilon_{0.l} - \varepsilon_{2.l})\beta_l}{(1 - \varepsilon_{0.l}) T_{0.4.l}} \right] \\ &= P_{1.l} q_2(l), \end{aligned} \tag{A.6.75a}$$

$$\begin{aligned} q_{3.l} &= \frac{2}{3} m_0 \gamma_s D_l \frac{R_{1.l} P_{1.l}}{\mu_{3.l}} \left[\frac{\varepsilon_{2.l}(\varepsilon_{1.l} - \varepsilon_{2.0.l})(1.\beta_l)}{(1 - \varepsilon_{0.l}) t_{0.2.l}} + \frac{\varepsilon_{2.0.l}(1.\varepsilon_{0.l} + \varepsilon_{2.l})(1.\beta_l)}{(1 - \varepsilon_{0.l}) t_{0.3.l}} + \frac{\varepsilon_{2.l}\beta_l}{(1 - \varepsilon_{0.l}) T_{0.4.l}} \right] \\ &= P_{1.l} q_3(l), \end{aligned} \tag{A.6.75b}$$

and

$$\begin{aligned} q_{4.l} &= \frac{2}{3} m_0 \gamma_s D_l \frac{R_{1.l} P_{1.l}}{\mu_{4.l}} \left[\frac{\varepsilon_{4.l}(\varepsilon_{1.l} - \varepsilon_{2.0.l})(1.\beta_l)}{(1 - \varepsilon_{0.l})(1.\varepsilon_{4.l}) t_{0.2.l}} + \frac{\varepsilon_{4.l}\varepsilon_{2.0.l}(1.\beta_l)}{(1 - \varepsilon_{0.l})(1.\varepsilon_{4.l}) t_{0.3.l}} + \frac{(1.\varepsilon_{0.l} + \varepsilon_{0.l}\varepsilon_{4.l})\beta_l}{(1 - \varepsilon_{0.l})(1.\varepsilon_{4.l}) T_{0.4.l}} \right] \\ &= P_{1.l} q_4(l). \end{aligned} \tag{A.6.75c}$$

The transport rates under strong equilibrium are

$$q_{i.l} = \frac{2}{3} m_0 \gamma_s D_l P_{1.l} \frac{\psi_{i.l}}{\mu_{i.l} t_{0.i.l}} = P_{1.l} q_i(l), \quad (i = 2, 3, 4, b), \tag{A.6.76}$$

where $P_{1.l}$ is the size distribution of the bed material, $q_i(l)$ is the transport capacity of uniform sediment with particle size $D = D_i$ under equilibrium conditions, and the composite probabilities are

$$\left. \begin{aligned} &\psi_{2.l} \frac{(\varepsilon_{1.l} - \varepsilon_{2.0.l})(1.\beta_l)}{1 + (1 + \varepsilon_{0.l})(1.\varepsilon_{4.l}) - (1.\varepsilon_{1.l})(1.\beta_l)} \\ &\psi_{3.l} \frac{\varepsilon_{2.0.l}(1.\beta_l)}{1 + (1 + \varepsilon_{0.l})(1.\varepsilon_{4.l}) - (1.\varepsilon_{1.l})(1.\beta_l)} \\ &\psi_{4.l} \frac{\beta_l}{1 + (1 + \varepsilon_{0.l})(1.\varepsilon_{4.l}) - (1.\varepsilon_{1.l})(1.\beta_l)} \end{aligned} \right\}, \tag{A.6.77}$$

$$\psi_{b.l} = \psi_{2.l} + \psi_{3.l} = \frac{\varepsilon_{1.l}(1.\beta_l)}{1 + (1 + \varepsilon_{0.l})(1.\varepsilon_{4.l}) - (1.\varepsilon_{1.l})(1.\beta_l)}. \tag{A.6.78}$$

All the parameters for l-th size class presented above are functions of $\tilde{\overline{V}}_b$, D_l and $\frac{\overline{D}}{D_l}$.

So far, a series of equations for the transport rate have been derived from the Markov process or from the transition intensities directly. Thus, it has been demonstrated that the transition intensity plays an important role in the theoretical analysis of sediment transport.

Compared to previous results, which were based on assumptions only concerned with the transport capacities of the bed load and suspended load, this study generalizes the cases of rolling, hopping, suspension, and bed load, and it has a profound theoretical basis. Although limited to equilibrium conditions, the results for the transition intensities can be extended to non-equilibrium transport of non-uniform sediment [23, 73] (Han 2013). Furthermore, the probability distribution of the transport rate has been derived directly [23, 24], from which the same equations for the average transport rate were obtained.

A.6.5 Verification of Study Results

A.6.5.1 Verification for Transport Capacities of Uniform Bed Load

Three sets of data are quoted to verify the expressions for the transport capacity of a uniform bed load. Herein, the dimensionless transport rates

$$\lambda_{q_b^*} = \frac{q_b^*}{\gamma_s \omega_0 D} \tag{A.6.79}$$

are adopted. Furthermore, the terms q_b^* and $q_b^*(l)$ are used to denote the transport capacity of q_b and $q_b(l)$, or the transport rate under equilibrium conditions.

Figure A.15 shows experimental data for pebble motion collected from the left canal of Dujianyan Baitiao River in Guan County by IYRWR (Institute of Yangtze River of Water Resources Research) [22], USWES [57] and Liu [37]. The experimental and computed results are in accordance. This data mainly verifies the rolling bed load under low velocity. The dimensionless value of low transport rate is approximate to $\lambda_{q_2^*} = q_2^*/(\gamma_s D\omega_0) = 2/3m_0\tilde{U}_2\varepsilon_1 = f_{\lambda_{q_2^*}}(V_b/\omega_0)$. The data from flume experiment with coarse sediment and small gravel conducted by [18] are in good agreement with the calculated results. The comparison is shown in Fig. A.16. Finally, Fig. A.17 compares the theoretical results with the bed load measurements collected at Yichang Hydrologic Station, where the water is more than 10 m deep.

The three sets of data cover a variety of situations, including different sediment sizes (fine sand, medium sand, pebbles), water depths (a few meters up to twenty meters), relative bottom velocities (from 0.2 to 7), and relative transport capacities (from 10^{-10} to 300). The data broadly shows that the theoretical equations for the transport rate of the bed load are reliable. Further verification data for uniform and non-uniform bed loads have been reported by [23].

A.6.5.2 Verification of Transport Capacities of Non-uniform Bed Loads

The dimensionless transport rate for the total bed load and each size class can be expressed as

$$\lambda_{q_b^*(l)} = \frac{q_b^*(l)}{\gamma_s \omega_{0.l} D_l} = \frac{q_{b.l}^*}{P_{1.l}\gamma_s \omega_{0.l} D_l} = f_l\left(\frac{\overline{V}_b}{\omega_{0.l}}, \frac{\overline{D}}{D_l}\right), \tag{A.6.80a}$$

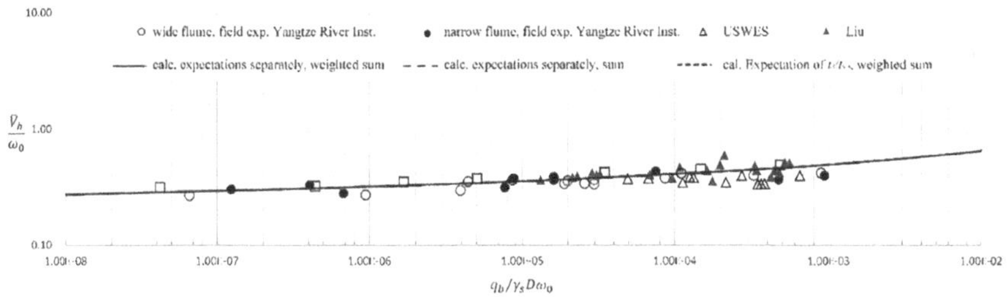

Fig. A.15 Comparison of low transport rate for uniform-pebble bed load between experiment and calculated data

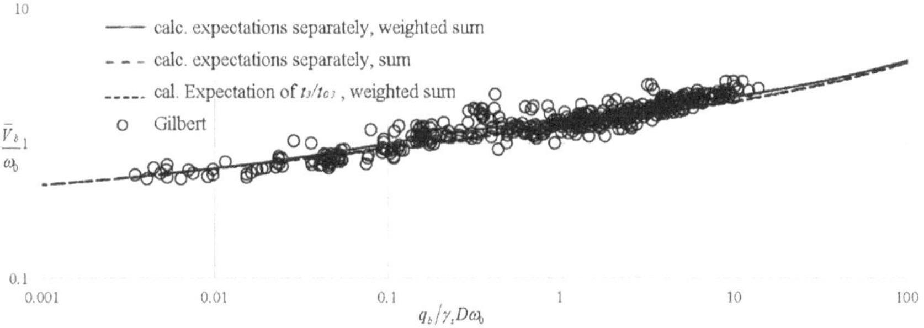

Fig. A.16 Comparison of low transport rate for uniform bed load of coarse sand and small gravel between experiment and calculated data (Gibert [18])

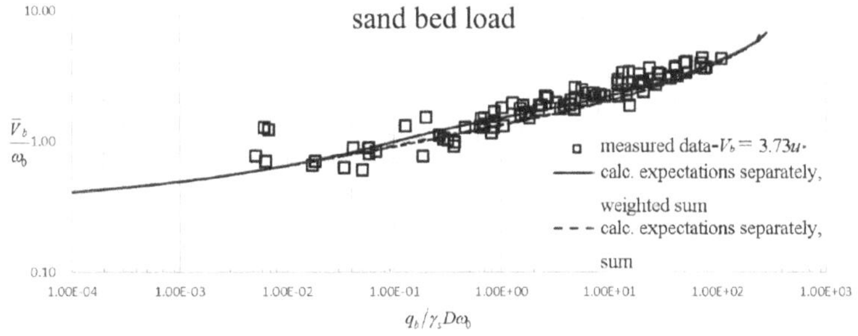

Fig. A.17 Comparison of sand bed load between calculated and measured data at Yichang Hydrologic Station

and

$$\lambda_{q_b^*} = \sum P_{1.l} \lambda_{q_b^*(l)} = \sum P_{1.l} f_l \left(\frac{\overline{V}_b}{\omega_{0.l}}, \frac{\overline{D}}{D_l} \right), \tag{A.6.80b}$$

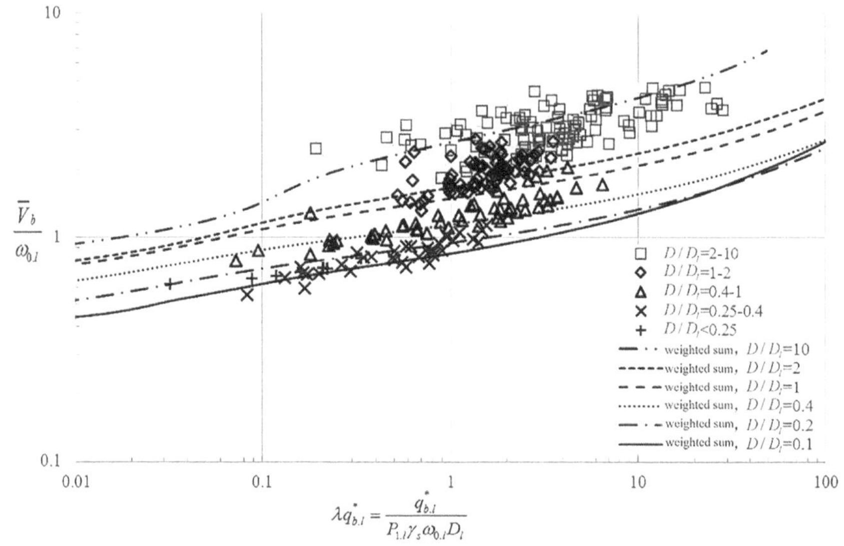

Fig. A.18 Comparison of transport capacities of non-uniform bed load [51]

respectively. These equations were verified using experimental data reported by Samaga et al. (1985). The experiment were conducted with $\overline{D}/D_l$ ranges of 0.5–0.4, 0.4–1.0, 1–2, and 2–10. The experimental data are in good agreement with the theoretical calculations, and a comparison is shown in Fig. A.18.

The carrying-sediment capacity for a non-uniform suspended load and verification are presented in Chap. 4.

References

1. Ancey C (2010) Stochastic moldering in sediment dynamics: Exner equation for planar bed incipient bed load transport conditions. J Geophys Res 115(F2):F00A11
2. Ancey C (2019) Bed load transport: a walk between randomness and determinism. Part 1. The state of the art. J Hyd Res 58(1):1–17. https://doi.org/10.1080/00221686.2019.1702594
3. Ancey C, Bohm T, Jodeau M, Frey P (2006) Statistical description of sediment transport experiment. Phys Rev 74(1):011302
4. Ancey C, Davison AC, Bohm T, Jodeau M, Frey P (2008) Entrament and motion of coarse particles in a shallow water stream down a steep slope. J Fluid Mech 595:83–114
5. Bagnold RA (1973) The nature of saltation and bed load transport in water. Proc R Soc Ser A 332:473–502
6. Bose SK, Dey S (2010) Universal probability distributions of turbulence in open channel flows. J Hyd Res 48(3):388–394. https://doi.org/10.1080/00221686.2010.481832

7. Cheng NS, Law A (2003) Fluctuations of turbulent bed shear stress. J Eng Mech 129:126–130. https://doi.org/10.1061/(ASCE)0733-9399
8. Dey S, Ali SJ (2017) Stochastic mechanics of loose boundary particle transport in turbulent flow. Phys Fluids 29:055103. https://doi.org/10.1063/1.4984042
9. Dey S, Ali SJ (2018) Review article: advances in modeling of bed particle entrainment sheared by turbulent flow. Phys Fluids 30:061301. https://doi.org/10.1063/1.5030458
10. Dey S, Ali SJ, Padhi E (2020) Hydrodynamic lift on sediment particles at entrainment: present status and its prospect. J Hyd Eng ASCE 146(6). https://doi.org/10.1061/(ASCE)HY.1943-7900.0001751
11. Dou GR (1963) Theory of sediment movement. Science report, Nanjing Hydraulic Research Institute (NHRI), Nanjing, China
12. Dou GR (1964) Movement regularity of bed load. Science report, Nanjing Hydraulic Research Institute (NHRI), Nanjing, China
13. Dou GR (1965) Movement regularity of bed load. In: Proceeding of river and harbour. Nanjing Hydraulic Research Institute (NHRI), Nanjing, China
14. Dou, GR (1999) Further study on incipient flow velocity of sediment. J Sed Res 6:1–9
15. Einstein HA (1942) Formulas for the transportation of bed load. Trans ASCE 107:561–597
16. Einstein HA (1950) The bed-load function for sediment transportation in open channel flows. Tech Bull 1026, US Dept. of Agriculture
17. Egiazaroff IV (1965) Calculation of non-uniform sediment concentrations. J Hyd Div ASCE 1965(91):225– 247
18. Gilbert GK (1954) The transportation of debris by running water. U.S. Geologic Survey. Prof. Pap. 263
19. Gonchanov VN (1954) Basis of river dynamics. National Power Press
20. Guan JC (2009) Carrying sediment capacity of non-equilibrium transport of non-uniform suspended load. Ph.D. Thesis, Dept. of Sedimentation Res. IWHR
21. Han QW (1965) Field flume experiment of Manxi River in Wutong Bridge, Sichuan Province. Res. Report, Yanze River Ins. of Water Resources Res. (in Chinese)
22. Han QW (1982) Regularity of incipient sediment motion and incipient veloity. J Sediment Res 1982(2):11–26 (in Chinese)
23. Han QW (2021) Statistical theory of bed load transport. Science Press, Beijing, China (in Chinese)
24. Han QW, He MM (1984) Statistical theory of sediment motion. Science Press, Beijing, China (in Chinese)
25. Han QW, He MM (1987) Exchange and transport rate of bed load. Encyclopedia of fluid mechanics, vol 6. Chapter 18, 761–791. Gulf Pub.com Book Div
26. Han QW, He MM (1999) Regularity of incipient sediment motion and incipient veloity. Science Press, Beijing, China (in Chinese)
27. Hayashi T, Qsaki S, Ichibashi T (1980) Study on bed load transport of sediment mixture. In: Proceedings of 24th Japanese conference on hydraulic

28. He MM, Han QW (1982) Model of incipient motion. J Hyd Div ASCE 108(2):1–11
29. Hu CH, Hui YJ (1995) Mechanic and statistical regularities of carrying sediment flow in open channel. Science Press, Beijing, China
30. Jiménez J (1998) Turbulent velocity fluctuations need not be Gaussian. J Fluid Mech 376:139–147
31. Kalinske AA (1947) Movement of sediment as bed load in rivers. Trans Am Geophys Union 28(4):615–620
32. Kline SJ, Reynolds WC, Schraub FA (1967) The structure of turbulent boundary layers. J Fluid Mech 50:233–255
33. Lavelle JW, Mofjeld HO (1987) Do critical stresses for incipient motion and erosion really exist? J Hyd Res 113(3):370–385
34. Lelouvetel J, Bigillon F, Doppler D, Vinkovic L, Champagne JY (2009) Experimental investigation of ejections and sweeps involved in particle suspension. Water Res Res 45:W02416. https://doi.org/10.1029/2007WR006520
35. Li JD, Sun J, Lin B (2018) Bed-load transport rate based on the entrainment probabilities of sediment grains by rolling and lifting. J Sediment Res 33(2):126–136
36. Lisle IG, Rose CW, Hogarth WL, Hairsine PB, Sander G, Parlange JY (1998) Stochastic sediment transport in soil erosion. J Hyd Eng ASCE 204(1–4):217–230. https://doi.org/10.1016/S0022-1694(97)00123-6
37. Liu XI, Cao SY, Fang D (1997) Incipient flow velocity of non-uniform sediment in refining and coarsening processes. Geological Press Beijing, China, 18–24
38. Lukerchenko N, Chara Z, Vlasak P (2002) Mathematical model of spherical particle saltation over rough bed in a two-dimensional open channel flow. In: Proceedings of the 15th international scientific conference on MMET-15, Tambov, Russia, pp 36–39
39. Lukerchenko N, Chara Z, Vlasak P (2006) 2D numerical model of particle–bed collision in fluid particle flows over bed. J Hyd Eng ASCE. https://doi.org/10.1080/00221686.2006.9521662
40. Mao Y, Zhang ZJ, Xu XR (1999) Review of the study on coherent structure of turbulence in open channel flow and its interaction with sediment movement. J Hohai Univ 27(4):24–29
41. Mao Y, Yang H, Yuan XM (2002a) Characteristics of turbulent bursting phenomenon in open channel with a flat and smooth bottom. J Hydrodyn 17:413–421
42. Mao Y, Yang H, Yuan XM (2002b) Bed surface roughness and bursting phenomena in open channel. J Hydraul Eng 6:53–69
43. Meier CI (1995) Transport velocities of single bed load grains in hydraulically smooth open channel flow. M. S. thesis, Dept. of Civil Engineering, Colorado State Univ., Fort Collins, CO
44. Nakagawa N, Tsujinmoto T (1976) On probabilistic characteristics of motion of individual sediment particles on stream beds. In: Proceedings of 2nd international symposium on stochastic hydraulics, IAHR, 1976, Lund Institute of Technology, Univ. of Lund, 219

45. Nakagawa N, Tsujinmoto T (1980) A stochastic model for bed load transport and its application to alluvial phenomena. In: Shen HW, Kikkawa H (eds) Application of stochastic process in sediment transport. Water Resources Publications, Littleton, Colo, 1–54
46. Nikitin IK (1963) Turbulence process near river bed layer. Press of Falkland Academy of Sciences, Kiev
47. Nino Y, Garcia M (1994) Gravel saltation 2, modeling. Water Resour Res 30(6):1915–1924
48. Paintal AS (1971) A stochastic model of bed load transport. J Hyd Res 9(4):91–113
49. Pierre YJ, Bounthanh B (2013) Velocity of rolling bed load particles. J Hyd Eng 139(2):177–286
50. Rosinski KI, Lubomirova KS (1976) Mechanism of river sediment movement. In: Proceedings of hydrology conference on 4th series, vol 10, 122–131
51. Samaga BR, Raju KG, Garde RJ (1986) Bed load transport of sediment mixures. J Hyd Eng ASCE 112(11):1003–1018
52. Sun Z (1989) A stochastic model of sediment interchanges. J Sediment Res 1989(3):1–9
53. Sun Z, Donahue J (2000) Stochastically derived bed load formula for any fraction of non-uniform sediment. J Hyd Eng 126(2):105–111
54. Sun Z, Zhu Y (1991) A study on Einstein bed-load transport formula. J Sed Res 1991(2):20–26
55. Tsai C, Yang FN (2013) Modeling bed-load transport by a three-state continuous-time. Markov chain model J Hyd Eng 139(12):1265–1276
56. Tsuchiya Y (1969) On the mechanics of solution of a spherical sand particle in a turbulent stream. IAHR 13th Cong 191
57. USWES (1935) Studies of river bed materials and their movement with special reference to the lower Mississippi River. Pap. 17. Vicksburg, Miss, 161
58. Velikanov MA (1955) Dynamics of river. Sci. and Tech. Lib. Moscow, Chap. II
59. Wang DC (2000) On the spatial structure of low-speed streaks and particle motion in the near wall region of turbulent open channel flow. Thesis of Tsinghua University, Department of Hydraulic Engineering, Beijing
60. Wu FC, Yang KH (2004) Entrainment and probabilities of mixed-size sediment: incorporating near bed coherent flow structures. J Hyd Eng 130(12):1187–1197
61. Yalin MS (1963) An expression for bed-load transportation. J Hyd Div Proc ASCE 89(Hy3), Part 2:221–247
62. Yalin MS (1972) Mechanics of sediment transport. Oxford Pergamon Press

Appendix B
Rules of Sediment Entrainment and Incipient Velocity

B.1 Introduction

Herein, the entrainment is referred to as incipient rolling under low flow velocity, which is also called incipient rolling. Sediment entrainment is one of the most important topics in sediment research, and it is widely used in engineering technology. For example, erosion and deposition at variable backwater region of reservoir, channel erosion downstream dam site, river bed deformation, bank protection work, channel stability, physical model experiment and mathematical model calculation, etc. There has been great deal of research on the flow velocity required for incipient motion, and entrainment threshold criterion have been developed along with many other achievements [2, 15, 18, 25]. Nevertheless, some questions remained unanswered, such as, what is an appropriate value of the critical flow velocity for incipient motion, and is there a unified entrainment threshold criterion. However, using statistical theory these problems have been well solved and is discussed in Sect. B.2 [5, 15, 18]. The relevant research results for the influence of adhesion force and additional film water pressure on the entrainment of fine sediment must be considered and are given in Sect. B.3. Moreover these forces are closely related to the dry density of bed material, the study results of which is described in Sect. B.4. Finally the effect of pebble particle shape on the entrainment motion is discussed in Sect. B.5. Most of the study results in Appendix B were presented by the authors for the first time.

B.2 Rules of Sediment Entrainment and Incipient Velocity

For a long time, the exact meaning of sediment entrainment standard and the determination of entrainment velocity have been problems to be solved. Based on the statistical theory of sediment movement, These problems are well solved.

Q. Han and M. He, *Non-equilibrium Transport of Non-uniform Suspended Load*, Hydroscience and Engineering, https://doi.org/10.1007/978-981-97-1123-9

The study results of statistical theory of sediment movement show that the average flow velocity for incipient motion is different from the transient incipient flow velocity

(1) Sediment moving in water flow is a random process. The critical flow velocity for incipient rolling $\tilde{V}_{b.c.1} = V_{b.c.1}/\omega_0$ introduced in Eq. (A.3.7) describes a certain flow velocity acting on a certain particle occupying a certain position on the bed surface. If the flow velocity exceeds this threshold, the particle will roll. This critical flow velocity is instantaneous and applies to a certain particle. It is different from the critical flow velocity in terms of the longitudinal flow velocity V averaged along vertical direction (see Sect. A.2.3.1 and Eq. A.2.2a, A.2.2b). The average flow velocity V, the average bottom flow velocity V_b and some other average parameters including incipient probability, ε_1, transition intensity, λ_1 and dimensionless transport rate, λ_{q_2}, can be used as indicator of incipient motion. The transient flow velocity is also a random variable and cannot be an indicator to express the incipient intensity, but is a necessary parameter to establish the relationship between the average incipient flow velocity V (or V_b) and the other indicators.
(2) In an open channel with steady and uniform flow, the turbulent flow velocity field corresponds to an average flow velocity V, and the dimensionless average velocity $\tilde{V} = V/\omega_1$ is an indicator for the sediment transport intensity. If $\tilde{V}$ is small, the particles have only two states, static and rolling, and the basic transition probability between hopping and suspension states can be ignored compared to the transition probability from rest to rolling state (Eq. A.3.13).

$$\varepsilon_1 = f\varepsilon_1\left(\overline{V}_b/\omega_1\right) \tag{B.2.1}$$

The transition intensity (Eq. A.6.47) and dimensionless transport rate in terms of sediment weight per unit with (Eqs. A.6.69a and A.6.70) can be approximated by

$$\lambda_1 = \frac{\varepsilon_1}{t_{0.2}} n_0 = f_{\lambda_1}\left(\frac{\overline{V}_b}{\omega_1}\right), \tag{B.2.2}$$

and

$$\lambda_{q_2} = \frac{q_2}{\gamma_s D \omega_1} = \frac{2}{3} m_0 U_2 \varepsilon_1 = f_{\lambda_{q_2}}\left(\frac{\overline{V}_b}{\omega_1}\right), \tag{B.2.3a}$$

or

$$\lambda_{q_2} = \frac{q_2}{\gamma_s D \omega_0} = \frac{2}{3} m_0 U_2 \varepsilon_0 = f_{\lambda_{q_2}}\left(\frac{\overline{V}_b}{\omega_0}\right), \tag{B.2.3b}$$

From Eqs. (A.2.2a, A.2.2b) the average bottom flow velocity V_b can be converted into average flow velocity V. Eqs. (B.2.1)–(B.2.3a, B.2.3b) show that $\tilde{V}, V_b$, ε_1, λ_2, and λ_{q2} are related to each other, and any one of those parameters can be taken as the indicator for incipient intensity.

(3) Most studies of incipient motion have finally settled on the incipient flow velocity. Up to now, more than one hundred formulas of incipient velocity of non-viscous particles have been proposed. There are more than ten representative or widely circulated ones [5, 6, 8, 9, 15, 18, 30, 31, 35, 37] and the supplemented and modified Shields curves [1, 8, 28].

Consider the question of which V should be chosen as V_c, the critical flow velocity of incipient rolling. Different researchers have proposed different equations for the critical velocity of incipient rolling based on a range of flume experiments and field data. As there is no unified criterion for the critical average flow velocity of incipient motion, the criteria for the entrainment threshold seems to be a convention of the human mind or depends on the engineering requirement.

(4) Dou [4], Tang [31], Hua [22], and WHIWH [33] have collected a large amount of data from flume experiments, which they compared with their proposed equations. All the relevant data are very concentrated. Owing to space limitations, only the Tang's data [31] is presented here (Fig. B.1). However, if there is no unified criterion, why are the incipient velocities so concentrated? It is because small variations in the average flow velocity correspond to the large variations in the rolling transport rate under low flow velocities. As shown in Table B.1, the transport rate increases by a factor of 3.31, from 0.383×10^{-4} to 0.123×10^{-2}, when the average flow velocity increases by just 38%, from 0.375 to 0.518, in the range of $-$ 16.1 to $+$ 16.1%. Thus, visual observations of incipient rolling particles vary greatly, but observations of the average flow velocity do not. The rapid increase in the transport rate is mainly caused by the incipient probability. The critical flow velocity for incipient rolling is usually determined from the number of particles that transition from rest to rolling per unit area per unit time (i.e., the transition intensity) recorded via visual observation in flume experiments, or from the transport rates measured for rivers. The transport rate may vary greatly, whereas the average flow velocity does not. Fortunately, the average flow velocity, rather than the transport rate, has been used as the entrainment threshold criteria. The low transport rate (Eq. A.6.69a) has been verified experimentally (Fig. A.15).

(5) In addition to the average flow velocity, other parameters have been used to determine the threshold for incipient rolling, including the probability of incipient rolling [5, 6], dimensionless transport rate (Taylor 1971), and sediment transport intensity (Yalin 1972). The transport rate can be observed or measured directly, so the dimensionless transport rate is a more appropriate parameter for the threshold of incipient motion and has been proposed by the authors [12, 15, 18]. The entrainment threshold criteria for different cases are as follows:

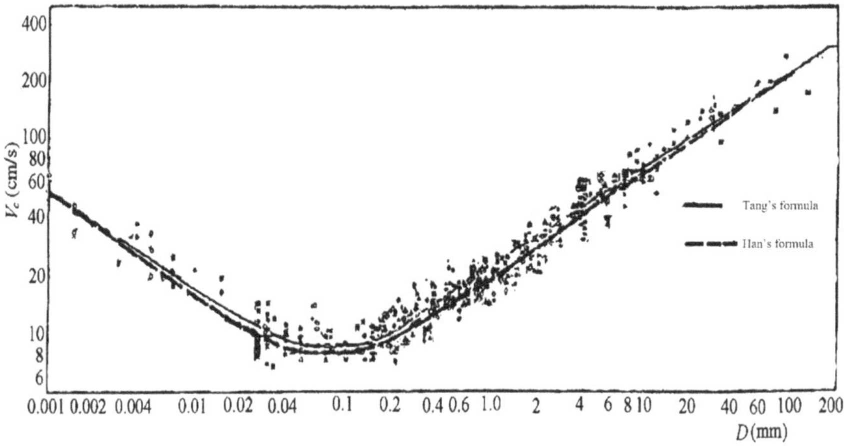

Fig. B.1 Diagram of incipient velocity (Tang 1965)

Table B.1 Parameters for rolling particles

V_b/ω_1	ε_1	λ_{q2}	U_2/ω_1
0.174	0.2340E − 12	0.4410E − 15	0.7066E − 02
0.205	0.1687E − 08	0.4583E − 11	0.1018E − 01
0.231	0.1798E − 06	0.6314E − 09	0.1316E − 01
0.260	0.6343E − 05	0.2838E − 07	0.1677E − 01
0.288	0.6726E − 04	0.3702E − 06	0.2064E − 01
0.319	0.4367E − 03	0.2964E − 05	0.2545E − 01
0.346	0.1469E − 02	0.1182E − 04	0.3015E − 01
0.375	0.4017E − 02	0.3834E − 04	0.3578E − 01
0.404	0.8851E − 02	0.9938E − 04	0.4210E − 01
0.433	0.4017E − 01	0.2186E − 03	0.4912E − 01
0.462	0.2803E − 01	0.4256E − 03	0.5693E − 01
0.519	06134E − 01	0.1227E − 02	0.7503E − 01
0.550	0.8545E − 01	0.1966E − 02	0.8628E − 01
0.577	0.1095	0.2828E − 02	0.9680E − 01
0.600	0.1321	0.3745E − 02	0.1062
0.700	0.2457	0.1002E − 01	0.1529

(a) Entrainment threshold criterion for flume experiments. By fitting a large amount of experimental data from flume experiments, the critical dimensionless transport rate for incipient rolling is found to be

$$\lambda_{q_{2.c}} = 0.219 \times 10^{-3}. \tag{B.2.4a}$$

The corresponding critical average bottom flow velocity and critical average flow velocities are

$$F_b^{-1}(\lambda_{q_{2.c}}) = \frac{\overline{V}_{b.c}}{\omega_1} = 0.433, \tag{B.2.4b}$$

$$V_c = 0.268 \times 0.433 \times \psi\omega_1 = 0.116\psi\omega_1, \tag{B.2.4c}$$

respectively, where water depths of $H = 0.15$ m or $H = 0.10$ m are adopted, and the coefficient of sediment compactness is $t/\delta = 0.375$. This criterion agrees well with the data compiled by Duo et al. and their formulae (see Fig. B.1: [15, 18]).

(b) Entrainment threshold criterion for natural river channels. A natural river channel with medium and fine sediment can be approximated as having uniform sediment. Owing to the depth of the water, the river bed cannot be seen clearly, so the criterion for incipient motion often relies on data with low sediment transport rates. In the past, it was general practice to use the zero point of the sediment transport rate; that is, to plot the relationship between the transport rate and the average flow velocity, and then extrapolate the relationship to a sediment transport rate of zero. However, it is difficult to determine the zero point, and it does not exist in reality. From data on the low sediment transport rate of the Yangtze River, the entrainment threshold criterion is found to be

$$\lambda_{q_{2.c}} = 0.00197, \tag{B.2.5}$$

which corresponds to a critical flow velocity of

$$F_{b.c}^{-1}(\lambda_{q_{2.c}}) = \frac{\overline{V}_{b.c}}{\omega_1} = 0.55, \tag{B.2.6a}$$

$$V_c = 0.268 \times 0.55 \times \psi\omega_1 = 0.147\psi\omega_1. \tag{B.2.6b}$$

Equations (B.2.5) and (B.2.6a, B.2.6b) were verified using data for the Yangtze River measured at different depths and dry density conditions [18, 19].

(c) For river beds consisting of pebbles or pebbles mixed with sand, the particle size range typically extends over 2–3 orders of magnitude. Therefore, it is not appropriate to use a uniform sediment approximation with a representative particle size. Instead, it is necessary to calculate the incipient flow velocity for each size group, to consider the effects of particle size on the particle location at the bed surface and the shape of the particles [15, 18].

At this time, due to the different exposure, the incipient flow velocity of coarse particles of sediment mixture on the bed surface is smaller than that of the uniform sediment, and the fine particles are larger. Under certain hydraulic conditions, the effects of coarse and fine particle size and of their large and small exposure on bed surface can partially offset each other, resulting in the difference in incipient flow velocity between coarse and fine particles being greatly reduced, or even similar.

According to the field data in the Yangtze River the incipient criteria should be greatly reduced for the case of non-uniform pebble sediment [14, 17]. Its incipient criteria can be taken as

$$\lambda_{q_{b.c.l}} = F_b\left(\frac{V_{b.c.l}}{\omega_{0.l}}, \frac{D_l}{\overline{D}}\right) = 0.3 \times 10^{-6} \tag{B.2.7}$$

The available incipient flow velocity is

$$V_{c.l} = 0.268 F_b^{-1}\left(0.3 \times 10^{-6}, \frac{D_l}{\overline{D}}\right)\psi\left(\frac{H}{D_l}\right)\omega_{0.l} \tag{B.2.8}$$

For the size group with average particle size $\overline{D}$, i.e. $D_l = \overline{D}$ there is

$$V_{c.m} = 0.268 F_b^{-1}(0.3 \times 10^{-6}, 1)\psi\left(\frac{H}{\overline{D}}\right)\omega_{0.m} \tag{B.2.9}$$

The ratio of the two is

$$\begin{aligned}\frac{V_{c.l}}{V_{c.m}} &= \frac{F_b^{-1}\left(0.3 \times 10^{-6}, \frac{D_l}{\overline{D}}\right)}{F_b^{-1}(0.3 \times 10^{-6}, 1)} \frac{\psi\left(\frac{H}{D_l}\right)}{\psi\left(\frac{H}{\overline{D}}\right)} \sqrt{\frac{3.33\frac{(\gamma_s-\gamma)}{\gamma} g D_l}{3.33\frac{(\gamma_s-\gamma)}{\gamma} g \overline{D}}} \\ &= K\left(\frac{\overline{D}}{D_l}\right)^{0.143}\left(\frac{D_l}{\overline{D}}\right)^{0.5} = K\left(\frac{D_l}{\overline{D}}\right)^{0.357}\end{aligned} \tag{B.2.10}$$

where

$$K = \frac{F_b^{-1}(0.3 \times 10^{-6}, \frac{D_l}{\overline{D}})}{F_b^{-1}(0.3 \times 10^{-1}, 1)} \tag{B.2.11}$$

and

$$\frac{\psi\left(\frac{H}{D_l}\right)}{\psi\left(\frac{H}{\overline{D}}\right)} = \frac{6.5\left(\frac{H}{D_l}\right)^{\frac{1}{4+\lg\frac{H}{D_l}}}}{6.5\left(\frac{H}{\overline{D}}\right)^{\frac{1}{4+\lg\frac{H}{\overline{D}}}}} \approx \left(\frac{\overline{D}}{D_l}\right)^{\frac{1}{7}} \tag{B.2.12}$$

This is because the natural sandy riverbed is mostly 1/8, and the pebble riverbed is mostly 1/6, on average, it is 1/7.

For uniform sediment with $D = D_l$, the incipient flow velocity is

Table B.2 Ratio of incipient flow velocity between non-uniform and uniform sediment with the same particle size

$\frac{D_l}{D}$	$V_{c.l}/V_c$				
	Qin	Zhang	Egiazaroff	Hayashi	Han
0.25	1.16	1.18	1.89	1.00	1.08
0.50	0.978	1.05	1.31	1.00	1.04
1.00	0.858	1.00	1.00	1.00	1.00
3.00	0.782	0.987	0.728	0.654	0.929
10.00	0.824	0.992	0.561	0.475	0.787

Table B.3 Ratio of incipient flow velocity between different size group

$\frac{D_l}{D}$	$V_{c.l}/V_{c.m}$				
	Qin	Zhang	Egiazaroff	Hayashi	Han
0.25	0.851	0.742	0.946	1.00	0.658
0.50	0.904	0.831	0.926	1.00	0.811
1.00	1.00	1.00	1.00	1.00	1.00
3.00	1.31	1.42	1.26	1.13	1.38
10.00	2.07	2.14	1.78	1.50	1.79

$$V_{c.u} = 0.268 F_b^{-1}(0.3 \times 10^{-6}, 1)\psi\left(\frac{H}{D_l}\right)\omega_{0.l} \tag{B.2.13}$$

Thus, the ratio of incipient flow velocity between non-uniform and uniform sediment with the same particle size is

$$\frac{V_{c.l}}{V_{c.u}} = \frac{F_b^{-1}\left(0.3 \times 10^{-6}, \frac{D_l}{D}\right)}{F_b^{-1}(0.3 \times 10^{-6}, 1)} = K \tag{B.2.14}$$

The ratios $V_{c.l}/V_c$ and $V_{c.l}/V_{c.m}$ calculated according to the above equations are listed in Tables B.2 and B.3 and compared with the others' results [7, 23, 29, 36]. Obviously Han's results is more reasonable.

B.3 Adhesion Force and Additional Film Water Pressure

B.3.1 Mechanism of Adhesion Force and the Presence of Additional Film Water Pressure

Several concepts and results from physiochemistry and electrochemistry of the bound water around the soil particles can be used as references for our study of incipient sediment motion. Some of these concepts are basic, as per the following description (1) The bound water on the outside of the soil particles has a thickness of generally not more than 0.25–0.5 μm, and is divided into two layers. The inner layer is adsorptive water (firmly bound water) with a thickness of about 0.01–0.1 μm, having solid-state properties and not complying with Pascal's theorem of hydrostatic pressure transfer. Its attractive force is about 10,000 atm and density is about 2 g/cm^3. Such tightly packed water molecules are impossible to split with solid particles; therefore, it is called firmly bound water. (2) The outer layer is film water (loosely bound water). Film water has the property of transition and does not fully possess the property of unidirectional pressure transmission. Hence, while considering a single soil particle molecule, one cannot ignore its unidirectional pressure characteristics which do not entirely comply with Pascal's theorem. Moreover, the entire contacted area of the film water also cannot be considered in the calculation of the additional downward pressure caused by the unidirectional pressure. It is only a part of the fully contacted area that is subject to the additional downward pressure. (3) The adhesion force is produced when the film water between particles comes in contact with each other. From a micro perspective, the adhesion force is mainly a macroscopic manifestation of Van der Waals gravitational force. (4) As for the existence of additional downward pressure of film water, some experiment results have shown that the force that influences underwater particle stability does exist and is related to the water depth, given by the progress of relevant research study [3, 30]. It is only possible that this force is the additional downward pressure of the film water. Here, we analyze the role of water depth on the incipient flow velocity based on the measured data of incipient flow velocity at different water depths in the river channel to prove that this role is not owing to the influence of flow velocity distribution [18, 19].

The relationships between measured incipient flow velocity and water depth at Yichang and other hydrological stations of the Yangtze River are shown in Figs. B.2, B.3 and B.4. These data are obtained from the sampling results of sand bed load, and the transport rate of bed load $q_s = 1$ g/(m s) is taken as the criterion of incipient motion. The range of water depth is wide, approximately from 2.5 to 20 mm. It is clear from these three figures that the incipient flow velocity is roughly proportional to the 0.35th–0.64th power of the water depth. But if the water depth is only to affect the flow velocity distribution, then from its exponential distribution:

$$V(y) = V_0 \eta^{\frac{1}{m}} = V_0 \left(\frac{y}{H}\right)^{\frac{1}{m}} \tag{B.3.1}$$

Fig. B.2 Measured $V_c \sim H$ at Yichang station (1975–1976)

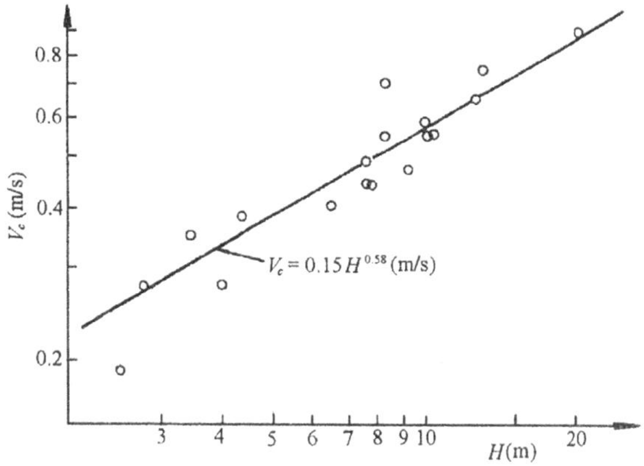

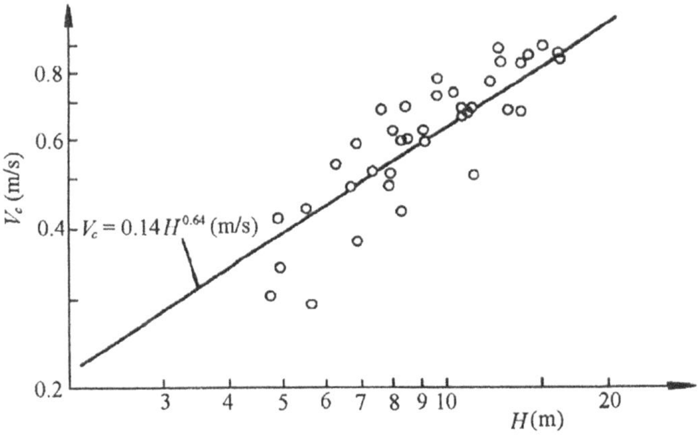

Fig. B.3 Measured $V_c \sim H$ at Yichang station (1989, 1990)

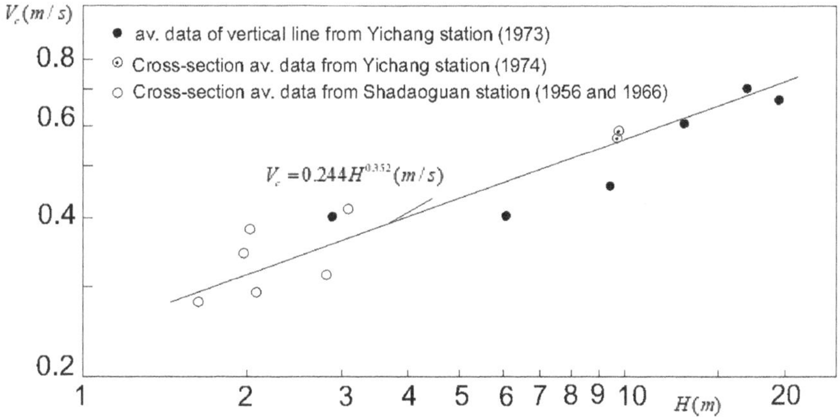

Fig. B.4 Measured $V_c \sim H$ at Shadaoguan and Yichang station

the average flow velocity is:

$$\overline{V} = V_0 0 \int^1 \eta^{\frac{1}{m}} d\eta = V_0 \frac{m}{m+1}$$

where V_0 is the flow velocity at water surface. The bottom flow velocity can be expressed as:

$$V_b = \frac{m+1}{m} \overline{V} \left(\frac{\alpha D}{H} \right)^{\frac{1}{m}}$$

The incipient flow velocity can be expressed as:

$$V_c = \frac{m}{m+1} V_{b,c} \left(\frac{H}{\alpha D} \right)^{\frac{1}{m}} = \frac{m}{m+1} \frac{V_{b,c}}{(\alpha D)^{\frac{1}{m}}} H^{\frac{1}{m}} \tag{B.3.2}$$

Here H is the water depth, y is the height from the bottom of the river, y/H is the relative height, m is the index, $V(y)$ and V_o are the flow velocity at the height y and the water surface respectively, V is the average flow velocity of $V(y)$, V_b is the average flow velocity at the bottom of the river, and V_c and $V_{b.c}$, are the incipient flow velocity expressed by the average flow velocity and the bottom flow velocity, respectively and $\alpha = 1/2 \sim 1$ is the coefficient of the flow acting point. Since the effect of the force associated with the water depth is not taken into account, $V_{b.c}/D^{1/m}$ is mainly determined by the sediment particle size, and is independent of the water depth. Eq. (B.3.2) shows that the incipient flow velocity is proportional to 1/m power of H. For natural river 1/m = 1/6–1/8 = 0.167–0.125. It is obviously different from the results indicated by the previous measured data, where the incipient flow velocity is proportional to the 0.35th–0.64th power of the water depth. Thus, it indicates that if only the flow velocity distribution is considered but the force related to the water depth is ignored, the most significant part of the effect of water depth cannot be reflected.

B.3.2 Derivation of Adhesion Force

The adhesion force can be intuitively derived from the fact that the adhesion force is inversely proportional to the higher power (> 2) of the distance between the particles. It is assumed that the adhesion force per unit area between the corresponding two points on two particles A and A′ is given as follows (Fig. B.5).

$$q = \frac{K}{(2h)^3} \tag{B.3.3}$$

Then the adhesive force can be obtained as.

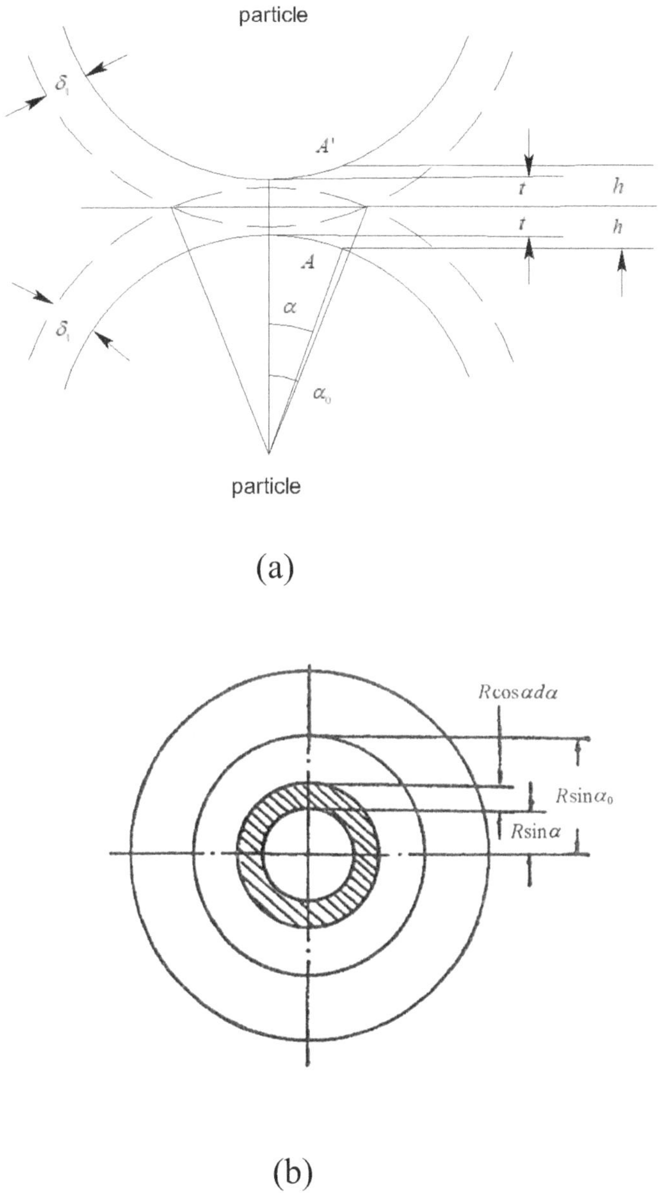

Fig. B.5 Schematic representation of adhesive force

$$P'_\mu = \int\limits_\Omega \frac{Kd\Omega}{(2h)^3} = \int\limits_\Omega \frac{q_0\delta_0^3}{h^3} d\Omega \tag{B.3.4}$$

Here $\delta_0 = 3 \times 10^{-10}$ m is the thickness of a water molecule, q_0 is the adhesion force per unit area when $h = \delta_0$, Ω is the projection of the film water contact area between particles. From Fig. B.5b

$$d\Omega = 2\pi R^2 \sin\alpha \cos\alpha d\alpha \tag{B.3.5}$$

where R is the radius of a particle. From Fig. B.5a, we have

$$h = t + R(1 - \cos\alpha) \tag{B.3.6}$$

$$dh = R \sin\alpha d\alpha \tag{B.3.7}$$

Substituting Eqs. (B.3.5)–(B.3.7) into Eq. (B.3.4) gives:

$$\begin{aligned} P'_\mu &= q_0\delta_0^3 \int\limits_0^{\alpha_0} \frac{2\pi R^2 \sin\alpha \cos\alpha}{h^3} d\alpha = 2\pi q_0\delta_0^3 \int\limits_t^{h_0} \frac{R-(h-t)}{h^3} dh \\ &= 2\pi q_0\delta_0^3 \left[-\frac{(R+t)}{2h^2}\bigg|_t^{h_0} + \frac{1}{h}\bigg|_t^{h_0} \right] \\ &= 2\pi q_0\delta_0^3 \left[\left(\frac{1}{t^2} - \frac{1}{h_0^2}\right)\frac{R+t}{2} - \left(\frac{1}{t} - \frac{1}{h_0}\right) \right] \end{aligned} \tag{B.3.8}$$

in which,

$$h_0 = t + R(1 - \cos\alpha_0) \tag{B.3.9}$$

Note that $t \ll R$, therefore after omitting the infinitely small quantities of higher order Eq. (B.3.8) reduces to

$$P'_\mu = \pi q_0\delta_0^3 R\left(\frac{1}{t^2} - \frac{1}{h_0^2}\right) \tag{B.3.10}$$

On the other hand, note that $\delta_1 \ll R$; hence, from Fig. B.5a:

$$h_0 = \delta_1 \cos\alpha_0 = \delta_1 \frac{R+t}{R+\delta_1} \approx \delta_1$$

Therefore, Eq. (B.3.10) is reduces to:

$$P'_{\mu} = \pi q_0 \delta_0^3 R\left(\frac{1}{t^2} - \frac{1}{\delta_1^2}\right) \tag{B.3.11}$$

Here $\delta_1 = 4 \times 10^{-7}$ m is the thickness of the film water, and the constant $q_0 = 1.52 \times 10^6$ ton/m^2 $= 1.52 \times 10^5$ kg/cm^2.

Along with Van der Waals gravitational forces between two particles, bilayer repulsive force, Brown's motion force, and electric field force are also present. Generally, only the sediment particles with D $\geq$ 0.001 mm or above are studied, where Brown's motion and the electric field force can be neglected. The bilayer repulsive force has a significant effect only when the gap between the particles $2t = S - 2R <$ 10^{-6} mm in Fig. B.5 [27]. But for bed sediments that are not very dense, generally, the inter-particle gap is unlikely to be smaller than this value, so the repulsive force can also be neglected. Thus, it is reasonable to consider only Van der Waals gravitational force.

Equation (B.3.11) is in the case of one particle in positive contact with another particle (Fig. B.6b). When the bed particle arrangement is dense enough, generally the entrainment particle are in oblique contact with three particles on the bed (Fig. B.6a), and then the total adhesion force will be

$$P_{\mu} = 3P'_{\mu}\cos 30^{\circ} = \frac{3\sqrt{3}}{2}P'_{\mu}$$

When the particles are loosely arranged, $t = \delta_1$, one entrainment particle is mostly in contact with two particles, and then the total adhesion force will be $P_{\mu} = 2P'_{\mu}\cos 30° = \sqrt{3}P'_{\mu}$. Thus, as t increases, it can be assumed that the number of particles in contact with the entrainment particle linearly decreases from three to two, and thus the adhesion force is

$$\begin{aligned} P_{\mu} &= \frac{\sqrt{3}}{2}\left(3 - \frac{t}{\delta_1}\right)P'_{\mu} \\ &= \frac{\sqrt{3}}{2}\pi q_0 \delta_0^3 R \times \left(3 - \frac{t}{\delta_1}\right)\left(\frac{1}{t^2} - \frac{1}{\delta_1^2}\right) \quad (t \leq \delta_1) \end{aligned} \tag{B.3.12}$$

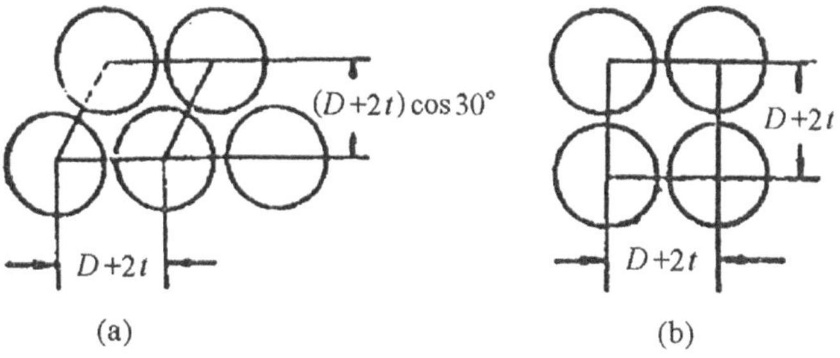

Fig. B.6 Diagram of particle arrangement

Using the dry density equation in the next section relates the adhesion force with the dry density and eliminates t from the above equation. For the verification of Eq. (B.3.12), please refer to Tan [31] and Han and He [18].

B.3.3 Additional Downward Pressure of Film Water

According to Fig. B.6, when the considered particle is in positive contact with one particle on the bed surface, the additional downward pressure as per Han [13] is:

$$\Delta G' = \int_{\Omega} K_2 \gamma H d\Omega \tag{B.3.13}$$

Here γ is the specific gravity of the water; H is the water depth, and K_2 is the percentage of area occupied by unidirectional pressure transfer in the contact area of the water film. Substituting Eq. (B.3.5) yields

$$\begin{aligned}\Delta G' &= \int_0^{\alpha_0} K_2 \gamma H 2\pi R^2 \cos\alpha \sin\alpha d\alpha = -2\pi K_2 \gamma H R^2 \int_0^{\alpha_0} \cos\alpha d(\cos\alpha) \\ &= \pi K_2 \gamma H R^2 [1 - \cos^2 a_0] = \pi K_2 H R^2 \left[1 - \left(\frac{R+t}{R+\delta_1}\right)^2\right] \\ &= \pi K_2 \gamma H R^2 \left[\frac{2R(\delta_1 - t)}{(R+\delta_1)^2} + \frac{(\delta_1 - t)(\delta_1 + t)}{(R+\delta_1)^2}\right]\end{aligned}$$

Omitting the infinitive small value of higher-order obtains the following

$$\Delta G' = 2\pi K_2 \gamma H R(\delta_1 - t) \tag{B.3.14}$$

Similar to the derivation of adhesion force when the entrainment particle is in oblique contact with more than one particle on the bed surface, the additional downward pressure of water film can be expressed as:

$$\begin{aligned}\Delta G &= \frac{\sqrt{3}}{2}\left(3 - \frac{t}{\delta_1}\right)\Delta G' \\ &= \sqrt{3}\pi K_2 \gamma H R\left(3 - \frac{t}{\delta_1}\right)(\delta_1 - t)\end{aligned} \tag{B.3.15}$$

where $K_2 = 2.285 \times 10^{-3}$. There is no direct data to test the above equation. However, the incipient flow velocity obtained from this equation is in good agreement with the measured data and reflects the influence of water depth on the incipient flow velocity (Figs. B.2, B.3 and B.4).

B.4 Dry Density of Bed Material and Deposited Sediment

The dry density of bed material and deposited sediment have a significant influence on the sediment entrainment, especially on the adhesion force and additional downward pressure of water film. Here, we will mainly focus on the mechanism of initial dry density and the final compactness of deposited sediment [12, 19]. The variation process of dry density is referred to [13, 14, 17].

B.4.1 Dry Density of Uniform Sediment

The dry density of uniform sediment, γ_s', is completely determined by the mean gap t between particles and the particles arrangement patterns (Fig. B.6). There are three cases for particle arrangement:

(1) If the bed particles in three directions are not stagger arranged (Fig. B.6b), the particles arrangement is loose, the dry density is

$$\gamma_s' = \frac{\frac{\pi}{6} D^3 \gamma_s}{(D+2t)^3} = \frac{\pi}{6}\gamma_s\left(\frac{D}{D+2t}\right)^3 \tag{B.4.1}$$

where γ_s is the specific weight of particle.

(2) If the bed particles in two directions are stagger arranged (Fig. B.6a), the bed particle arrangement is dense enough,

$$\begin{aligned}\gamma_s' &= \frac{\frac{\pi}{6} D^3 \gamma_s}{(D+2t)[(D+2t)\cos 30^\circ]^2}\\ &= \frac{\pi}{6}\frac{3}{4}\gamma_s\left(\frac{D}{D+2t}\right)^2 = \frac{2}{9}\pi\gamma_s\left(\frac{D}{D+2t}\right)^2\end{aligned} \tag{B.4.2}$$

(3) If the bed particles in one direction are stagger arranged (Fig. B.6a), then

$$\begin{aligned}\gamma_s' &= \frac{\frac{\pi}{6} D^2 \gamma_s}{(D+2t)^2(D+2t)\cos 30^\circ}\\ &= \frac{\pi}{6}\frac{2}{\sqrt{3}}\gamma_s\left(\frac{D}{D+2t}\right)^2 = \frac{\pi}{3\sqrt{3}}\pi\gamma_s\left(\frac{D}{D+2t}\right)^3\end{aligned} \tag{B.4.3}$$

Considering the above three cases, the dry density can be expressed as

$$\gamma_s' - k\pi\gamma_s\left(\frac{D}{D+2t}\right)^2 \tag{B.4.4}$$

where k is the arrangement coefficient, dependent on the compactness of particle arrangement and linearly increased as t decreases. Then there is

$$\gamma_s' = \frac{2}{9}\pi\gamma_s\left(1 - \frac{t}{4\delta_1}\right)\left(\frac{D}{D + 2t}\right)^2 \quad (t \le \delta_1) \tag{B.4.5}$$

B.4.2 Initial Dry Density of Fine Particles

Equation (B.4.5) only applies when the arrangement of particles is not very loose. If the fine sediment has just silted down and started to fixed its structure it may be honeycombed with the large inter-particle gap (Fig. B7) and its initial dry density can be expressed as

$$\gamma_s' = \frac{\pi}{6}\gamma_s\left(\frac{D}{D + 2t}\right)^3 = 1.41\left(\frac{D}{D + 4\delta_1}\right)^3 \tag{B.4.6}$$

which is derived from Eq. (B.4.1) with $t = 2\delta_1$ and $\delta_1 = 4 \times 10^{-7}$ m, the thickness of the film water.

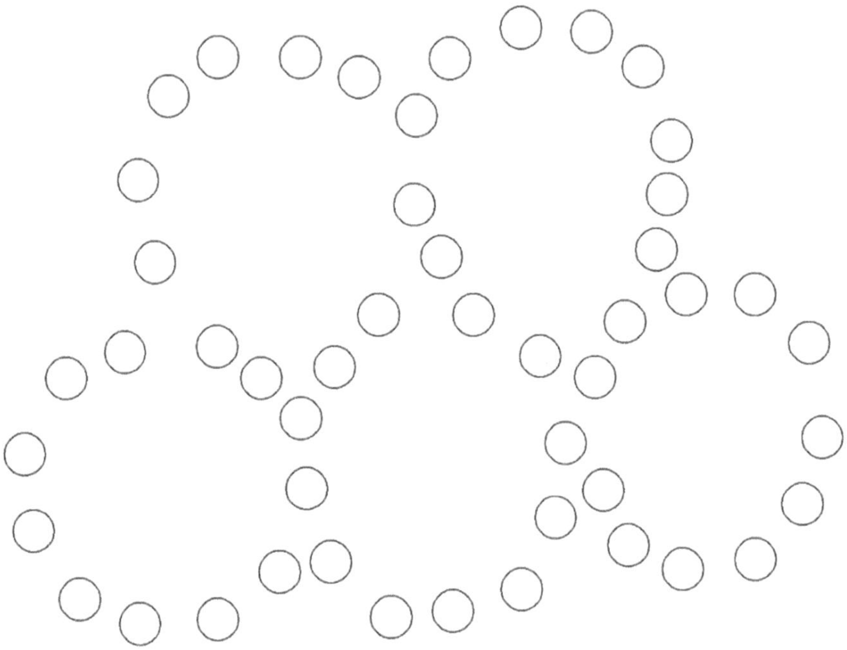

Fig. B7 Honeycomb structure

B.4.3 Initial Dry Density of Coarse Particles

When $D \geq 1\text{mm}$, $\left(\frac{D}{D+4\delta_1}\right)^3 \approx 1$, and the effect of film water on the dry density can be ignored. The initial dry density is the final dry density and there is no compactness problem. On the other hand, the particles almost all transport as bed load. The coarser the particles, the more likely they are to roll. Due to the friction between the rolling particles, the particles tend to have rounder shape and smoother surface, which increases the possibility of the particles stagger arrangement and finally increases the compact and the dry density of the sediment. Therefore the dry density is related with the particle size and it can be assumed that the increment of the difference of dry densities is proportional to the difference of dry densities and the increment of particle size, i.e.

$$\Delta(\gamma_2' - \gamma') = -\alpha(\gamma_2' - \gamma')\Delta D \tag{B.4.7}$$

integrating which under the condition of $\gamma' = \gamma_1'$ if $D = D_1 = 1$ mm obtains the dry density of pebble particles

$$\gamma' = \gamma_2' - (\gamma_2' - \gamma_1')\exp[-\alpha(D - D_1)] \quad (D \geq 1\,\text{mm}) \tag{B.4.8}$$

where $\gamma_2' = 2\pi/9\gamma_s = 0.698\gamma_s$ is the dry density for the case of compact arrangement of particles, $\gamma_1' = \gamma_s\pi/6$ is the dry density for the case of loosely compact particles while $D = D_1 = 1$ mm, and $\alpha = 0.095/D_1$ is an empirical coefficient. Thus, when$\gamma_s = 2.7\,\text{t/m}^3$ the formula above becomes:

$$\begin{aligned}\gamma_0' &= \left\{0.698 - 0.175\exp\left[-0.095\frac{D-D_1}{D_1}\right]\right\}\gamma_s \\ &= 1.88 - 0.472\exp\left[-0.095\frac{D-D_1}{D_1}\right] = \gamma_c'\end{aligned} \quad (D \geq 1\,\text{mm}) \tag{B.4.9}$$

which shows that when $D \geq 1$ mm, there is no compactness problem of bed sediment, the initial dry density γ_0' is the same as stabilized dry density γ_c' and the maximum dry density of the uniform pebbles is $1.88\,\text{t/m}^3$.

The initial dry density obtained by combining Eqs. (B.4.6) and (B.4.9) is

$$\gamma_0' = \begin{cases} 1.41\left(\frac{D}{D+4\delta_1}\right)^3 & (D \leq D_1) \\ 1.88 - 0.472\exp\left[-0.095\frac{D-D_1}{D_1}\right] & (D \geq D_1) \end{cases} \tag{B.4.10}$$

The comparison between Eq. (B.4.11) and the results of the initial dry density indoor test is shown in Fig. B8, which illustrates that the theoretical equations are consistent with the measured data [18, 20].

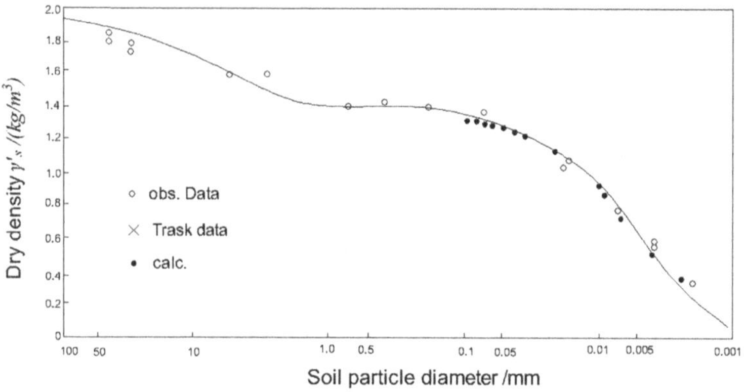

Fig. B8 Verification of dry density

B.4.4 Dry Density of Non-uniform Sediment

In the literature, Han et al. [20] and Han [13] have studied the dry density of non-uniform sediment. According to the difference of size distribution there are two cases for dry densities. One is the case for no mutual filling for different size groups, the other is the case for the voids of coarser groups random filled with finer particles.

B.4.4.1 Determination of Dry Density for Non-uniform Sediment with no Mutual Filling

If the non-uniform sediment particles are sized not widely different, for example, no more than one order of magnitude, then the sediment of each size group is piled together without mutual filling of sediments with different sizes as shown in Fig. B9. Let V_l, G_l, $\gamma'_{s.l}$ and P_l denote the volume, the weight, the dry density and the weight percentage of *lth* size group, respectively. Thus, there are the following equations of the total volume V and the volume V_l of *lth* size group:

$$V_i = \frac{G_l}{\gamma'_{s,l}} \tag{B.4.11}$$

and

$$V = \sum_l V_l = \sum_l \frac{G_l}{\gamma'_{s,l}} \tag{B.4.12}$$

Considering γ'_s, as the dry density of the mixed sediment yields

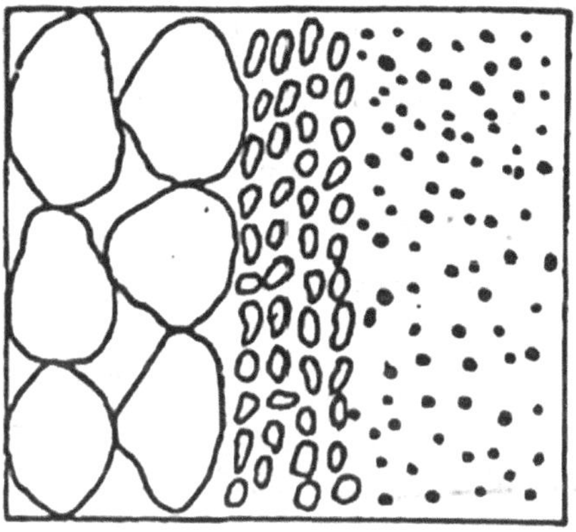

Fig. B9 Non-uniform sediment with no mutual filling

$$V = \frac{G}{\gamma_s'} \tag{B.4.13}$$

Equating Eqs. (B.4.12) and (B.4.13) obtains

$$V = \frac{G}{\gamma_s'} = \sum_l \frac{G_l}{\gamma_{s.l}'}$$

or

$$\frac{1}{\gamma_s'} = \sum_l \frac{G_l}{G} \frac{1}{\gamma_{s.l}'} = \sum_l \frac{P_l}{\gamma_{s.l}'} \tag{B.4.14}$$

where the gradation (or weight percentage) $P_l = G_l/G$. When the sediment of different size groups is not mutually filled with each other, the reciprocal of the dry density of the mixed sediment equals to the weighted sum of the reciprocals of the dry density of each size group with the gradation as weighted factor. The above formula had been verified by the field data from Danjiangkou Reservoir [20].

B.4.4.2 Determination of Dry Density with Randomly Filled Sediment

If the size range of the non-uniform sediment is large the case of randomly filling should be considered. Here, we only study the random filling problem of two size groups: a fine and a coarse size group. For the filling of multiple size groups, the filling of the smallest two size groups is considered first, and then filling which into the coarser size group is considered.

First we introduce the contact probability. Let the two size groups of particles, coarse and fine, have particle sizes D_1 and D_2, and weight percentages P_1 and P_2, respectively. Then, their contact probabilities with a certain particle are expected to be proportional to their corresponding surface areas. In fact in the mixed sand with a total weight of G, the number of coarse particles is $GP_1/(\frac{\pi}{6}D_1^3)$, their total surface area is $\frac{\pi}{4}D_1^2 \times \left[GP_1/(\frac{\pi}{6}D_1^3) = \frac{3}{2}\frac{GP_1}{D_1}\right]$. Similarly, the total surface area of fine particles is $\frac{3}{2}\frac{GP_2}{D_2}$. Therefore, the probabilities that the certain particle contacting with the coarse and fine particles, are given by Q_1 and $Q_{2,}$ respectively as:

$$Q_1 = \frac{\frac{3}{2}\frac{GP_1}{D_1}}{\frac{3}{2}\frac{GP_1}{D_1} + \frac{3}{2}\frac{GP_2}{D_2}} = \frac{\frac{P_1}{D_1}}{\frac{P_1}{D_1} + \frac{P_2}{D_2}} \tag{B.4.15}$$

$$Q_2 = \frac{\frac{P_2}{D_2}}{\frac{P_1}{D_1} + \frac{P_2}{D_2}} \tag{B.4.16}$$

Additionally, if all the coarse particles are concentrated, their void volume is:

$$V_t = GP_1\left(\frac{1}{\gamma'_{s.1}} - \frac{1}{\gamma_s}\right) \tag{B.4.17}$$

However, the coarse particles cannot be entirely concentrated. Thus, considering the contact probability that has just been calculated, the concentrated part of coarse group is:

$$V_1 = V_t Q_1 = GP_1 Q_1\left(\frac{1}{\gamma'_{s.1}} - \frac{1}{\gamma_s}\right) \tag{B.4.18}$$

That is, some of the total void volumes of coarse particles are not filled with fine particles.

Now, we consider the case where the coarse particle voids are filled with fine particles, which can be further divided into two situations. First, the voids are fully filled with fine particles as shown in Fig. B10 and the second, when voids are partially filled. Assume that the maximum sand thickness at which fine particles fill the voids of coarse particles is

$\Delta h_M = 2\Delta h$ (Fig. B11), therefore

$$\Delta h = \frac{\frac{D_1}{2} + \frac{D_2}{2}}{\cos 30^\circ} - \frac{D_1}{2} = \frac{D_2}{2} + \frac{1}{2}(D_1 + D_2)\left(\frac{2}{\sqrt{3}} - 1\right)$$
$$= \frac{D_2}{2} + 0.078(D_1 + D_2) \tag{B.4.19}$$

Thus, the number of layers n when the void space is partially filled is:

$$n = \frac{\Delta h}{D_2} = \frac{1}{2} + \frac{0.078(D_1 + D_2)}{D_2} \tag{B.4.20}$$

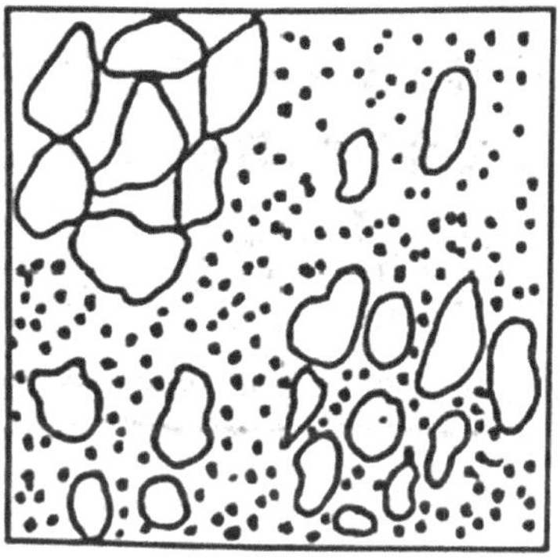

Fig. B.10 Coarse groups filled with fine particles

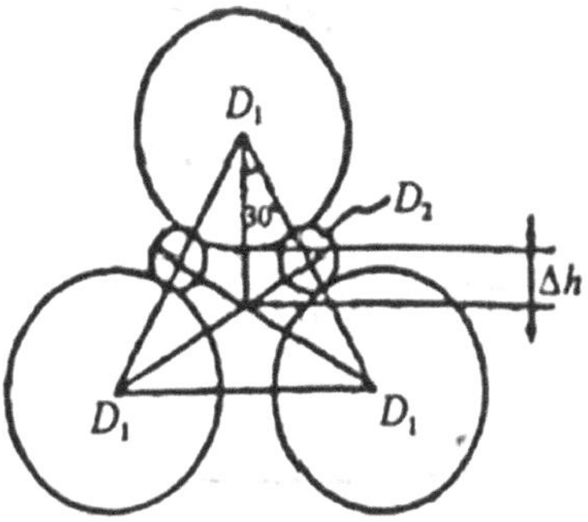

Fig. B11 coarse particles filled with fine particle layers

The number of layers N when the void space is completely filled is:

$$N = \frac{\Delta h_M}{D_2} = 2n \tag{B.4.21}$$

In order to simplify the results, an outright assumption is made: if the number of filled layers is $\leq n$, then the voids are all unfilled; if the number of filled layers is $> n$, then the voids are all filled. Thus, the probability that the voids are completely filled with fine particles is $P[\xi_n > n]$, and the probability that the voids are completely unfilled with fine particles is $P[\xi_n \leq n]$. Exploiting the contact probability obtains that the probability of exactly one layer being filled with fine particles is $Q_2[1 - Q_2]$, and the probability of exactly two layers being filled with fine particles is $Q_2^2(1 - Q_2)$. Therefore,

$$\begin{aligned} P[\xi \leq n] &= P[0 < \xi_n \leq 1] + P[1 < \xi_n \leq 2] + \cdots \\ &\quad + P[n-1 < \xi_n \leq n] \\ &= \sum_{k=1}^{n} Q_2^k (1 - Q_2) \end{aligned}$$

$$= Q_2(1 - Q_2^n) \quad \text{(B.4.22)}$$

and

$$P[\xi_n > n] = \sum_{k=n+1}^{+\infty} Q_2^k(1 - Q_2) = Q_2^{n+1} \quad \text{(B.4.23)}$$

Thus, the unfilled portion of the void volume, in which the coarse particles in contact with the fine particles is

$$V_{2.1} = V_t Q_2(1 - Q_2^n) \quad \text{(B.2.24)}$$

The filled portion $V_{2,2}$ of the void volume is

$$V_{2.2} = V_t Q_2^{n+1} \quad \text{(B.4.25)}$$

Thus, the volume V_n that is not filled by fine particles in the total void volume V_t of coarse particles is

$$\begin{aligned} V_n &= V_1 + V_{2.1} = \left[Q_1 + Q_2(1 - Q_2^n)\right]V_t \\ &= (1 - Q_2^{n+1})\left(\frac{GP_1}{\gamma'_{s.1}} - \frac{GP_1}{\gamma_s}\right) = QGP_1\left(\frac{1}{\gamma'_{s.1}} - \frac{1}{\gamma_s}\right) \end{aligned} \quad \text{(B.4.26)}$$

where the probability of being unfilled with the fine particles is $Q = (1 - Q_2^{n+1})$, and the volume filled with fine particles is $Q_2^{n+1} V_t$. The sum of the two is exactly V_t, which is as expected.

According to the above discussion, under the random filling of fine particles, the non-uniform sediment volume can be expressed as

$$\frac{G}{\gamma'_s} = GP_1\left(\frac{1}{\gamma'_{s,1}} - \frac{1}{\gamma,}\right)Q + \frac{GP_1}{\gamma_s} + \frac{GP_z}{\gamma'_{s,2}}$$

or

$$\frac{1}{\gamma'_s} = P_1\left(\frac{Q}{\gamma'_{s,1}} + \frac{1 - Q}{\gamma_s}\right) + \frac{P_2}{\gamma'_{s,2}} \quad \text{(B.4.27)}$$

Set

$$\frac{1}{\tilde{\gamma}'_{s,1}} = \frac{Q}{\gamma'_{s.1}} + \frac{1 - Q}{\gamma_s} \quad \text{(B.4.28)}$$

Then, Eq. (B.4.27) turns to

$$\frac{1}{\gamma_s'} = \frac{P_1}{\tilde{\gamma}'_{s.1}} + \frac{P_2}{\gamma'_{s.2}} \tag{B.4.29}$$

where, $\tilde{\gamma}'_{s.1}$ is the dry density of coarse particles with fillings of finer sediment, and is between the individual particle density γ_s and the dry density $\gamma'_{s,1}$ of coarse group without filling.

If $Q = 0$, all the voids of coarse particles are filled, then

$$\frac{1}{\gamma_s'} = \frac{P_1}{\gamma_s} + \frac{P_2}{\gamma'_{s,2}} \tag{B.4.30}$$

which is the case when the proportion of coarse particles is small. Otherwise, if Q = 1, all the voids of coarse particles are unfilled, then Eq. (B.4.27) is transformed into Eq. (B.4.14).

The question now is under what conditions can filling not be considered. The condition that the void of coarse particles is filled with fine particles is that at least one layer of fine particles can be filled in the void of coarse particles. Then from Eq. (B.4.20) exits the inequality

$$\frac{1}{2} + 0.078\frac{D_1 + D_2}{D_2} \geq 1 \tag{B.4.31}$$

Solving Eq. (B.4.31) yields

$$\mathrm{D}_1 \geq 5.55\mathrm{D}_2 \tag{B.4.32}$$

which shows that the filling should be disregarded if the particle size range of the sediment is within one order of magnitude.

Equation (B.4.29) gives the dry density of non-uniform sediment mixed with two size groups of particles. When three or more size groups of non-uniform sediment is considered, the average dry density of the two smallest size groups should be derived first, and then the third group of the coarser sediment should be filled with it to set the average dry density of non-uniform sediment mixed with three size groups, and so on.

The dry density data of non-uniform sediment with three size groups: 80 mm > $D_1 \geq 10$ mm, 10 mm > $D_2 \geq 1.0$ mm, 1.0 mm > $D_3 \geq 0.1$ mm, have been observed in the field and are in good agreement with the calculation results of Eq. (B.4.29) (Fig. B.12) [20], which also show that the maximum dry density of the non-uniform sediment appears at a certain gradation of the mixed size groups of particles, and is always greater than the dry density of the uniform sediment that is involved in filling.

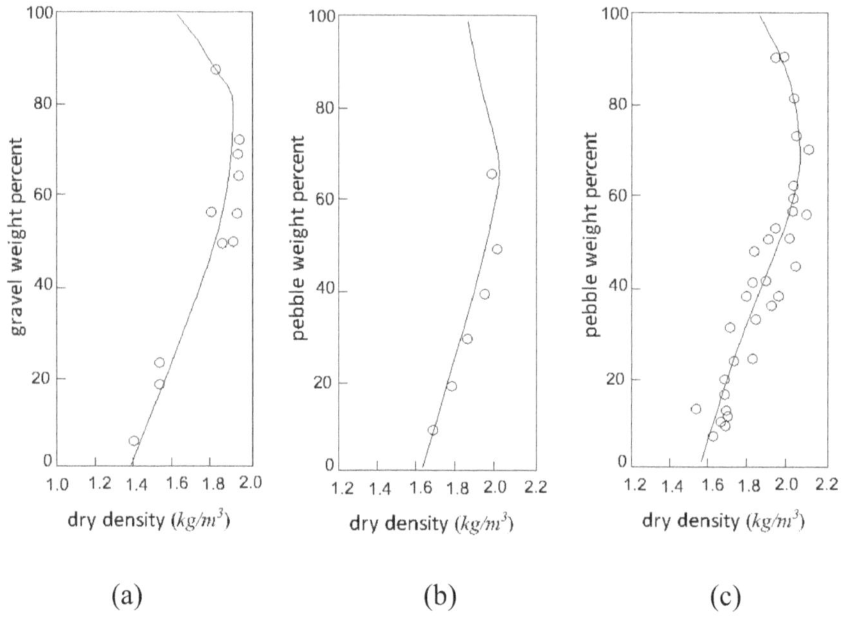

(a) gravel filled with sand;

(b) pebble filled with gravel;

(c) pebble filled with sand and gravel;

Fig. B12 Verification of dry density (Eq. B.4.29)

B.5 Influence of Pebble Shape on Incipient Motion

For the pebble bed, the incipient particle number or incipient transport rate in terms of the particle number is very small, so the influence of the shape of a single pebble particle on its entrainment is difficult to ignore. The shape of a pebble particle can generally be approximated by flatness,

$$\lambda = \frac{\sqrt{ab}}{c} \tag{B.5.1}$$

where a, b, and c are the long, middle, and short axes of the pebble particle. This formula was first introduced by [10]. Based on the measurement results of 1946 pebble particles in the Chuanjiang River and Jinsha River from 1960 to 1961, the following empirical relationship is obtained

$$\frac{a}{c} = \lambda^{1.23} \tag{B.5.2}$$

$$\frac{b}{c} = \lambda^{0.77} \tag{B.5.3}$$

$$\frac{D}{c} = \frac{\lambda^{1.23} + \lambda^{0.77} + 1}{3} \tag{B.5.4}$$

where $D = (a + b + c)/3$ is the particle size.. Assume that the short axis of the pebble particle is perpendicular to the bed, and the long axis and the middle axis of the particle each have a half chance of being perpendicular to the water flow, then the torque balance equation during incipient rolling is obtained as Eq. (A.3.1):

$$P_x l_x + P_y l_y > G l_G \tag{B.5.5}$$

or

$$\begin{aligned} &\frac{\rho C_x}{2} V_{b.c}^2 \frac{\pi}{4} \sqrt{abc}\alpha c + \frac{\rho C_y}{2} V_{b.c}^2 \frac{\pi}{4} ab\beta\sqrt{ab} \\ &= \frac{\pi}{6}(\gamma_s - \gamma) D^3 \frac{\sqrt{ab}}{2} \end{aligned} \tag{B.5.6}$$

where $\pi/4\sqrt{abc}$ and $\pi/4ab$ are the action area of drag force and lift force, respectively, αc and $\beta\sqrt{ab}$ are the force arm of $P_x and P_y$ respectively, $\sqrt{ab}/2$ is the force arm of gravity in water, α and β are the undetermined coefficients. From Eq. (B.5.6) the instantaneously incipient flow velocity $V_{b,c}$ is obtained as

$$\begin{aligned} V_{b,c}^2 &= \left[\frac{4}{3C_x} \frac{\gamma_s - \gamma}{\gamma} gD\right) \left[\frac{\lambda}{2\alpha\lambda + \frac{1}{2}\beta\lambda^3}\right] \frac{D^2}{c^2} \\ &= \left(\frac{4}{3C_x} \frac{\gamma_s - \gamma}{\gamma} gD\right) \left(\frac{1}{2\alpha + \frac{1}{2}\beta\lambda^2}\right) \left(\frac{\lambda^{1.23} + \lambda^{0.77} + 1}{3}\right)^2 \end{aligned}$$

or

$$V_{b.c} = \sqrt{\frac{1}{2\alpha + \frac{\beta}{2}\lambda^2}} \left(\frac{\lambda^{1.23} + \lambda^{0.77} + 1}{3}\right) \omega_0 = f_1(\lambda)\omega_0 \tag{B.5.7}$$

where

$$f_1(\lambda) = \sqrt{\frac{1}{2\alpha + \frac{\beta}{2}\lambda^2}} \left(\frac{\lambda^{0.23} + \lambda^{0.77} + 1}{3}\right) \tag{B.5.8}$$

In order not to make the formula too complicated in the derivation, the influence of particle position Δ' at bed surface and the particle flatness λon instantaneously incipient flow velocity are considered separately. The former Δ' is not taken into account in the derivation of Eq. (B.5.6), and the latter λ is not taken into account in the derivation of Eq. (A.3.4). If both are considered simultaneously then

$$V_{b.c} = \varphi(\Delta') f_1(\lambda)\omega_0 \tag{B.5.9}$$

When $\lambda = 1$ the particle is a sphere and

$$f_1(\lambda) = \sqrt{\frac{1}{2\alpha + \beta/2\lambda^2}} = 1 \tag{B.5.10}$$

In general $\alpha = \beta$ can be assumed, then from Eq. (B.5.10) $\alpha = \beta = 0.4$ and Eq. (B.5.8) turns to

$$f_1(\lambda) = \sqrt{\frac{1}{0.8 + 0.2\lambda^2}} \left(\frac{\lambda^{1.23} + \lambda^{0.77} + 1}{3} \right) \tag{B.5.11}$$

The shape of pebble particle also influences the conversion between average bottom flow velocity and average longitudinal flow velocity. If the particle shape is considered the conversion Eqs. (A.2.2a, A.2.2b) turns to

$$\overline{V}'_b = 3.73u_s = \frac{3.73}{\psi\left(\frac{H}{c}\right)} V \tag{B.5.12}$$

where

$$\begin{aligned}
\psi\left(\frac{H}{c}\right) &= 6.5\left(\frac{H}{c}\right)^{\frac{1}{4+k\frac{H}{c}}} = 6.5\left(\frac{H}{Da}\right)^{\frac{1}{4+\lg\left(\frac{H}{D}\frac{D}{c}\right)}}\left(\frac{D}{c}\right)^{\frac{1}{4+\lg\frac{H}{c}}} \\
&\approx 6.5\left[\left(\frac{H}{D}\right)^{\frac{1}{4+k\frac{H}{D}}}\right]\left(\frac{\lambda^{1.23} + \lambda^{0.77} + 1}{3}\right)^{\frac{1}{4+\lg\frac{H}{D}}} \\
&= \psi\left(\frac{H}{D}\right)\left(\frac{\lambda^{1.23} + \lambda^{a.n} + 1}{3}\right)^{\frac{1}{4+\lg\frac{H}{D}}}
\end{aligned} \tag{B.5.13}$$

since $4 + \lg\frac{H}{c} = 4 + \lg\frac{H}{D} + \lg\frac{D}{c} and \lg D/c \ll (4 + \lg H/D)$, the approximate term exist. If $H = 5$–30 m and $D = 0.03$–0.15 m the exponent $1/(4 + \lg H/D)$ is between 1/5.5 and 1/7.0, which can be approximated by 1/6, then Eqs. (B.5.12) and (B.5.13) turn to

$$\psi\left(\frac{H}{c}\right) = \psi\left(\frac{H}{D}\right)\left(\frac{\lambda^{1.23} + \lambda^{0.77} + 1}{3}\right)^{\frac{1}{6}} \tag{B.5.14}$$

and

$$\overline{V_b'} = \frac{3.73V}{\varphi\left(\frac{H}{D}\right)}\left(\frac{\lambda^{1.23} + \lambda^{0.77} + 1}{3}\right)^{-\frac{1}{6}} = \overline{V}_b\left(\frac{\lambda^{1.23} + \lambda^{0.77} + 1}{3}\right)^{-\frac{1}{6}} \tag{B.5.15}$$

respectively. The above equation shows that for the same size particle the larger the particle flatness, the smaller the bottom velocity. This is obvious because the larger the flatness the smaller the short axis of the particle and the closer the flow action point to the bed and therefore the smaller the bottom flow velocity acted on the particle.

Now turns to the study of the influence of flatness λ on the formula of low transport rate. For a spherical particle with a given particle size D under the action of the average longitudinal flow velocity V its dimensionless transport rate in terms of the weight is (see Eq. B.2.3b):

$$\lambda_{q_2} = \frac{2}{3}m_0\varepsilon_0\left(\frac{\overline{V}_b}{\omega_0}\right)U_2\left(\frac{\overline{V}_b}{\omega_0}\right) = F\left(\frac{\overline{V}_b}{\omega_0}\right) \tag{B.5.16}$$

If under the same condition but the particle has a flatness λ Eq. (B.5.16) is extended to

$$\lambda'_{q_2'} = \frac{2}{3}m_0\varepsilon_0\left(\frac{\overline{V}'_b}{\omega'_0}\right)U_2\left(\frac{\overline{V}'_b}{\omega'_0}\right) = F\left(\frac{\overline{V}'_b}{\omega'_0}\right) \tag{B.5.17}$$

where (see Eqs. 5.7 and 5.8 and B.5.15)

$$\frac{\overline{V}'_b}{\omega'_0} = \frac{1}{f(\lambda)}\frac{\overline{V}_b}{\omega_0} = \frac{\left(\frac{\lambda^{1.23}+\lambda^{0.77}+1}{3}\right)^{-\frac{1}{6}}}{f_1(\lambda)}\frac{\overline{V}_b}{\omega_0} \tag{B.5.18}$$

and

$$f(\lambda) = f_1(\lambda)\left(\frac{\lambda^{1.23} + \lambda^{0.77} + 1}{3}\right)^{\frac{1}{6}} = \sqrt{\frac{1}{0.8 + 0.2\lambda^2}}\left(\frac{\lambda^{1.23} + \lambda^{0.77} + 1}{3}\right)^{\frac{7}{6}} \tag{B.5.19}$$

The above results Eqs. (B.5.17)–(B.5.19) show that the dimensionless transport rate of particle with flatness is less than that of particle without flatness. The decrease of transport rate corresponds to the increase of particle weight by $f_1(\lambda)$ times and the decrease of bottom flow velocity by $\left[\left(\lambda^{1.23} + \lambda^{0.77} + 1\right)/3\right]^{-\frac{1}{6}}$ times.

The calculation results shows that the approximate formula

$$f(\lambda) = \lambda^{0.45} \tag{B.5.20}$$

is valid.

The influence of pebble particle flatness on incipient transport rate and incipient flow velocity has been verified by field flume experiment [9, 17].

References

1. Akers P, White WR (1973) Sediment transport: new approach and analysis. J Hyd Viv ASCE 99(HY11):2041–2060
2. Dey S, Ali SJ (2018) Review article: advances in modeling of bed particle entrainment sheared by turbulent flow. Phys Fluids 30:061301. https://doi.org/10.1063/1.5030458
3. Dou GR (1960) Sediment incipient velocity. J Water Resour 1960(4):44–59
4. Dou GR (1963) Theory of sediment movement. Science Report, Nanjing Hydraulic Research Institute (NHRI), Nanjing, China
5. Dou GR (1964) Movement regularity of bed load. Science Report, Nanjing Hydraulic Research Institute (NHRI), Nanjing, China
6. Dou GR (1999) Further study on incipient flow velocity of sediment. J Sed Res 1999(6):1–9
7. Egiazaroff IV (1965) Calculation of non-uniform sediment concentration. J Hyd Div ASCE 91(Hy4):225–247
8. Gessler J (1971) Critical shear stress for sediment mixture. Proc 14th Cong Inter Assoc Hyd Res 3:1–8
9. Gonchanov VI (1962) Fundamentals of River Channel dynamics. Hydrometeorological Publishing house, pp 214–216
10. Han (1965) Pebble experiment in field flume. Science Report, Institute of Yangzi River Research, 100–200
11. Han (1980) Investigation of pebble morphology and characteristics of pebbles pilling up. In: Proceedings of bed load testing technology. National Key Sediment Research Collaboration
12. Han QW (1982) Rules of sediment entrainment and incipient velocity. J Sed Res 1982(2):11–22
13. Han QW (1997) Dry density distribution and its application. J Sed Res 1997(2):10–16
14. Han QW (2003) Reservoir sedimentation. Science Press, Beijing, China
15. Han QW (2021) Statistical theory of bed load transport. Science Press, Beijing, China (in Chinese)
16. Han QW, He MM (1984) Statistical theory of sediment motion. Science Press, Beijing, China (in Chinese)
17. Han QW, He MM (1997) Sedimentation compactness and variation of dry density. In: Proceedings of application study on the three gorge reservoir project engineering. Geology Press, Beijing, China, 44–52
18. Han QW, He MM (1999) Regularity of incipient sediment motion and incipient veloity. Science Press, Beijing, China (in Chinese)
19. Han QW, He MM, Wang CH (1995) Study on incipient flow velocity. Res. Report. China Institute of Water resources and Hydro-Electric Power Research
20. Han QW, Wang YC, Xing XL (1981) Initial dry density of sedimentation. J Sed Res 1981(1):1–12

21. He MM, Han QW (1982) Model of incipient motion. J Hyd Div ASCE 108(2):1–11
22. Hua GX (1965) Incipient flow velocity of sediment. Bull Chengdu Ind Univ 1965(1):1–12
23. Hayashi T, Qsaki S, Ichibashi T (1980) Study on bed load transport of sediment mixture. In: Proceedings of 24th Japanese conference on hydraulic
24. Koelzer VA, Lara JM (1958) Densities and compaction on rates of deposited sediment. J Hyd Div ASCE 84(Hy2):1958
25. Lavelle JW, Mofjeld HO (1987) Do critical stresses for incipient motion and erosion really exist? J Hyd Res 113(3):370–385
26. Levi II (1962) Dynamics of river channel flow. Hydrometeorological Publishing House, Moscow, pp 214–216
27. Mantz PA (1977) Packing and angle of response of naturally sedimented fine silica solids immersed in natural aqueous electrolytes. Sedimentology 24(6):810–832
28. Meyer-Peter E, Muller R (1948) Formula for bed load transport. In: Proceedings of international association, hydraulic structuring 2nd meeting, vol 3. Stockholm
29. Qin RY, Wang CH (1980) Rules of incipient motion of non-uniform sediment. J Sediment Res 1965(1):1–12
30. Sha YQ (1956) Basic rules of sediment movement: incipient velocity. J Sed Res 1956(1, 2)
31. Tang CB (1963) Entrainment laws of sediment motion. J Water Resour 1963(2):1–12
32. Taylor BD (1971) Temperature effects in alluvial streams. Report No KH-R-27, W. M. Keck Laboratory of Hydraulics and Water Resources, California Institute of Technology
33. WHIWH (Wuhan Inst. of water Resources and Hydra-Electric Power) (1960) River dynamics. China Industrial Press
34. Yalin MS (1972) Mechanics of sediment transport. Oxford Pergaman Press
35. Yang CT (1973) Incipient motion and sediment transport. J Hyd Div ASCE 99(Hy10):1674–1704
36. Zhang QW (1992) Incipient flow velocity of non-uniform sediment. In: Proceedings of 1st national conference on basic theory of sediment research. Sediment Professional Committee of Water Resources Society
37. Zhang RJ, WHIWH (Wuhan Inst. of water Resources and Hydra-Electric Power) (1960) River dynamics. China Industrial Press

SPRINGER NATURE

GPSR Compliance

The European Union's (EU) General Product Safety Regulation (GPSR) is a set of rules that requires consumer products to be safe and our obligations to ensure this.

If you have any concerns about our products, you can contact us on ProductSafety@springernature.com

In case Publisher is established outside the EU, the EU authorized representative is:

Springer Nature Customer Service Center GmbH
Europaplatz 3
69115 Heidelberg, Germany

The manufacturer's authorised representative in the EU is Springer Nature Customer Service Centre GmbH, Europaplatz 3, 69115 Heidelberg, Germany. If you have any concerns regarding our products, please contact ProductSafety@springernature.com

Printed and bound by CPI Group (UK) Ltd, Croydon, CR0 4YY
01/10/2026
02225849-0019